Lecture Notes in Computer Science 13164

More information about this subseries at https://link.springer.com/bookseries/7409

Giuseppe Nicosia · Varun Ojha ·
Emanuele La Malfa · Gabriele La Malfa ·
Giorgio Jansen · Panos M. Pardalos ·
Giovanni Giuffrida · Renato Umeton (Eds.)

Machine Learning, Optimization, and Data Science

7th International Conference, LOD 2021
Grasmere, UK, October 4–8, 2021
Revised Selected Papers, Part II

Editors
Giuseppe Nicosia
University of Catania
Catania, Italy

Emanuele La Malfa
Department of Computer Science
University of Oxford
Oxford, UK

Giorgio Jansen
Department of Biochemistry
University of Cambridge
Cambridge, UK

Giovanni Giuffrida
University of Catania
Catania, Italy

Varun Ojha
Department of Computer Science
University of Reading
Reading, UK

Gabriele La Malfa
Cambridge Judge Business School
University of Cambridge
Cambridge, UK

Panos M. Pardalos
Department of Industrial and Systems Engineering
University of Florida
Gainesville, FL, USA

Renato Umeton
Department of Informatics
Dana-Farber Cancer Institute
Boston, MA, USA

ISSN 0302-9743 ISSN 1611-3349 (electronic)
Lecture Notes in Computer Science
ISBN 978-3-030-95469-7 ISBN 978-3-030-95470-3 (eBook)
https://doi.org/10.1007/978-3-030-95470-3

LNCS Sublibrary: SL3 – Information Systems and Applications, incl. Internet/Web, and HCI

This Springer imprint is published by the registered company Springer Nature Switzerland AG
The registered company address is: Gewerbestrasse 11, 6330 Cham, Switzerland

Preface

LOD is an international conference embracing the fields of machine learning, optimization, and data science. The seventh edition, LOD 2021, took place during October 4–8, 2021, in Grasmere (Lake District), UK. LOD 2021 was held successfully online and onsite to meet challenges posed by the worldwide outbreak of COVID-19. This year the scientific program of the conference was even richer than usual; LOD 2021 hosted the first edition of the Advanced Course and Symposium on Artificial Intelligence & Neuroscience – ACAIN 2021. In fact, this year, in the LOD proceedings we decided to also include the papers of the first edition of the Symposium on Artificial Intelligence and Neuroscience (ACAIN 2021). The symposium was scheduled for 2020, but due to the COVID-19 pandemic until we were forced to postpone it to 2021.

The review process for the papers submitted to ACAIN 2021 was double blind, performed rigorously by an international Program Committee consisting of leading experts in the field. The following three articles in this volume comprise the articles accepted to ACAIN 2021:

- Effect of Geometric Complexity on Intuitive Model Selection by Eugenio Piasini, Vijay Balasubramanian, and Joshua Gold.
- Training Convolutional Neural Networks with Competitive Hebbian Learning Approaches by Gabriele Lagani, Giuseppe Amato, Fabrizio Falchi, and Claudio Gennaro.
- Towards Understanding Neuroscience of Realisation of Information Need in Light of Relevance and Satisfaction Judgement by Sakrapee Paisalnan, Frank Pollick, and Yashar Moshfeghi.

Since 2015, the LOD conference has brought academics, researchers, and industrial researchers together in a unique multidisciplinary community to discuss the state of the art and the latest advances in the integration of machine learning, optimization, and data science to provide and support the scientific and technological foundations for interpretable, explainable, and trustworthy AI. In 2017, LOD adopted the Asilomar AI Principles.

The annual conference on machine Learning, Optimization, and Data science (LOD) is an international conference on machine learning, computational optimization, and big data that includes invited talks, tutorial talks, special sessions, industrial tracks, demonstrations, and oral and poster presentations of refereed papers.

LOD has established itself as a premier multidisciplinary conference in machine learning, computational optimization, and data science. It provides an international forum for presentation of original multidisciplinary research results, as well as exchange and dissemination of innovative and practical development experiences.

The manifesto of the LOD conference is as follows:

> "The problem of understanding intelligence is said to be the greatest problem in science today and "the" problem for this century – as deciphering the genetic code

was for the second half of the last one. Arguably, the problem of learning represents a gateway to understanding intelligence in brains and machines, to discovering how the human brain works, and to making intelligent machines that learn from experience and improve their competences as children do. In engineering, learning techniques would make it possible to develop software that can be quickly customized to deal with the increasing amount of information and the flood of data around us."

The Mathematics of Learning: Dealing with Data

Tomaso Poggio (MOD 2015 and LOD 2020 Keynote Speaker) and Steve Smale

"Artificial Intelligence has already provided beneficial tools that are used every day by people around the world. Its continued development, guided by the Asilomar principles of AI, will offer amazing opportunities to help and empower people in the decades and centuries ahead."

The Asilomar AI Principles

The Asilomar AI Principles were adopted by the LOD conference following their inception (January 3–5, 2017). Since then these principles have been an integral part of the manifesto of the LOD conferences.

LOD 2021 attracted leading experts from industry and the academic world with the aim of strengthening the connection between these institutions. The 2021 edition of LOD represented a great opportunity for professors, scientists, industry experts, and research students to learn about recent developments in their own research areas and to learn about research in contiguous research areas, with the aim of creating an environment to share ideas and trigger new collaborations.

As chairs, it was an honour to organize a premier conference in these areas and to have received a large variety of innovative and original scientific contributions.

During LOD 2021, 12 plenary talks were presented by leading experts:

LOD 2021 Keynote Speakers:

- Ioannis Antonoglou, DeepMind, UK
- Roberto Cipolla, University of Cambridge, UK
- Panos Pardalos, University of Florida, USA
- Verena Rieser, Heriot Watt University, UK

ACAIN 2021 Keynote Lecturers:

- Timothy Behrens, University of Oxford, UK
- Matthew Botvinick, DeepMind, UK
- Claudia Clopath, Imperial College London, UK
- Ila Fiete, MIT, USA
- Karl Friston, University College London, UK
- Rosalyn Moran, King's College London, UK
- Maneesh Sahani, University College London, UK
- Jane Wang, DeepMind, UK

LOD 2021 received 215 submissions from authors in 68 countries in five continents, and each manuscript was independently reviewed by a committee formed by at least five members. These proceedings contain 86 research articles written by leading scientists in the fields of machine learning, artificial intelligence, reinforcement learning, computational optimization, neuroscience, and data science presenting a substantial array of ideas, technologies, algorithms, methods, and applications.

At LOD 2021, Springer LNCS generously sponsored the LOD Best Paper Award. This year, the paper by Zhijian Li, Bao Wang, and Jack Xin, titled "An Integrated Approach to Produce Robust Deep Neural Network Models with High Efficiency", received the LOD 2021 Best Paper Award.

This conference could not have been organized without the contributions of exceptional researchers and visionary industry experts, so we thank them all for participating. A sincere thank you goes also to the 41 subreviewers and to the Program Committee, comprising more than 250 scientists from academia and industry, for their valuable and essential work of selecting the scientific contributions.

Finally, we would like to express our appreciation to the keynote speakers who accepted our invitation, and to all the authors who submitted their research papers to LOD 2021.

October 2021

Giuseppe Nicosia
Varun Ojha
Emanuele La Malfa
Gabriele La Malfa
Giorgio Jansen
Panos Pardalos
Giovanni Giuffrida
Renato Umeton

Organization

General Chairs

Giorgio Jansen	University of Cambridge, UK
Emanuele La Malfa	University of Oxford, UK
Renato Umeton	Dana-Farber Cancer Institute, MIT, Harvard T.H. Chan School of Public Health and Weill Cornell Medicine, USA

Conference and Technical Program Committee Co-chairs

Giovanni Giuffrida	University of Catania and NeoData Group, Italy
Varun Ojha	University of Reading, UK
Panos Pardalos	University of Florida, USA

Special Sessions Chair

Gabriele La Malfa	University of Cambridge, UK

ACAIN 2021 Chairs

Giuseppe Nicosia	University of Cambridge, UK, and University of Catania, Italy
Varun Ojha	University of Reading, UK
Panos Pardalos	University of Florida, USA

Steering Committee

Giuseppe Nicosia	University of Cambridge, UK, and University of Catania, Italy
Panos Pardalos	University of Florida, USA

Program Committee

Adair, Jason	University of Stirling, UK
Adesina, Opeyemi	University of the Fraser Valley, Canada
Agra, Agostinho	University of Aveiro, Portugal
Allmendinger, Richard	University of Manchester, UK
Alves, Maria Joao	University of Coimbra, Portugal
Amaral, Paula	University Nova de Lisboa, Portugal
Arany, Adam	University of Leuven, Belgium
Archetti, Alberto	Politecnico di Milano, Italy

Aringhieri, Roberto	University of Turin, Italy
Baggio, Rodolfo	Bocconi University, Italy
Baklanov, Artem	International Institute for Applied Systems Analysis, Austria
Bar-Hen, Avner	CNAM, France
Bentley, Peter	University College London, UK
Bernardino, Heder	Universidade Federal de Juiz de Fora, Brazil
Berrar, Daniel	Tokyo Institute of Technology, Japan
Berzins, Martin	University of Utah, USA
Beyer, Hans-Georg	Vorarlberg University of Applied Sciences, Austria
Blekas, Konstantinos	University of Ioannina, Greece
Boldt, Martin	Blekinge Institute of Technology, Sweden
Bonassi, Fabio	Politecnico di Milano, Italy
Borg, Anton	Blekinge Institute of Technology, Sweden
Borrotti, Matteo	University of Milano-Bicocca, Italy
Boscaini, Davide	Fondazione Bruno Kessler, Italy
Braccini, Michele	University of Bologna, Italy
Browne, Will	Queensland University of Technology, Australia
Cagliero, Luca	Politecnico di Torino, Italy
Cagnoni, Stefano	University of Parma, Italy
Campbell, Russell	University of the Fraser Valley, Canada
Canim, Mustafa	IBM, USA
Carletti, Timoteo	University of Namur, Belgium
Carrasquinha, Eunice	Universidade de Lisboa, Portugal
Carta, Salvatore	University of Cagliari, Italy
Castellini, Alberto	Verona University, Italy
Cavicchioli, Roberto	Universita' di Modena e Reggio Emilia, Italy
Ceci, Michelangelo	Universita degli Studi di Bari, Italy
Cerveira, Adelaide	INESC-TEC, Portugal
Chakraborty, Uday	University of Missouri - St. Louis, USA
Chen, Keke	Marquette University, USA
Chen, Ying-Ping	National Yang Ming Chiao Tung University, Taiwan
Chinneck, John	Carleton University, Canada
Chlebik, Miroslav	University of Sussex, UK
Cho, Sung-Bae	Yonsei University, South Korea
Chretien, Stephane	National Physical Laboratory, France
Cire, Andre Augusto	University of Toronto, Canada
Codognet, Philippe	University of Tokyo, Japan
Consoli, Sergio	European Commission, Joint Research Centre, Italy
Costa, Juan Jose	Universitat Politecnica de Catalunya, Spain
Damiani, Chiara	University of Milano-Biocca, Italy
Dandekar, Thomas	University of Wuerzburg, Germany
Daraio, Elena	Politecnico di Torino, Italy
De Brouwer, Edward	Katholieke Universiteit Leuven, Belgium
De Leone, Renato	University of Camerino, Italy
Del Buono, Nicoletta	University of Bari Aldo Moro, Italy

Dell'Amico, Mauro	University of Modena and Reggio Emilia, Italy
Dhaenens, Clarisse	Université Lille 1, France
Di Fatta, Giuseppe	University of Reading, UK
Di Gaspero, Luca	University of Udine, Italy
Dias, Joana	University of Coimbra and INESCC, Portugal
Dionisio, Joao	University of Porto, Portugal
Dobre, Ciprian	University Politehnica of Bucharest, Romania
Doerfel, Stephan	Kiel University of Applied Sciences, Germany
Donaghy, John	University of New Hampshire, USA
Drezewski, Rafal	AGH University of Science and Technology, Poland
Durillo, Juan	Leibniz Supercomputing Centre, Germany
Ebecken, Nelson	COPPE/UFRJ, Brazil
Eftimov, Tome	Jozef Stefan Institute, Slovenia
Engelbrecht, Andries	University of Stellenbosch, South Africa
Esposito, Flavia	University of Bari Aldo Moro, Italy
Esposito, Roberto	University of Torino, Italy
Farinelli, Alessandro	Verona University, Italy
Fasano, Giovanni	University Ca'Foscari of Venice, Italy
Ferrari, Carlo	University of Padova, Italy
Filisetti, Alessandro	Explora Biotech Srl, Italy
Fillatre, Lionel	University of Nice Sophia Antipolis, France
Finck, Steffen	Vorarlberg University of Applied Sciences, Austria
Fliege, Joerg	University of Southampton, UK
Formenti, Enrico	Universite Cote d'Azur, France
Franchini, Giorgia	University of Modena and Reggio Emilia, Italy
Franco, Giuditta	University of Verona, Italy
Frandi, Emanuele	Cogent Labs Inc., Japan
Fraternali, Piero	Politecnico di Milano, Italy
Freschi, Valerio	University of Urbino, Italy
Frohner, Nikolaus	TU Wien, Austria
Gajek, Carola	Augsburg University, Germany
Gallicchio, Claudio	University of Pisa, Italy
Garza, Paolo	Politecnico di Torino, Italy
Gauthier, Bertrand	Cardiff University, UK
Gendreau, Michel	Ecole Polytechnique de Montreal, Canada
Giannakoglou, Kyriakos	National Technical University of Athens, Greece
Gnecco, Giorgio	IMT School for Advanced Studies Lucca, Italy
Goncalves, Teresa	University of Evora, Portugal
Granitzer, Michael	University of Passau, Germany
Grishagin, Vladimir	Nizhni Novgorod State University, Russia
Guidolin, Massimo	Bocconi University, Italy
Gurbani, Vijay	Illinois Institute of Technology, USA
Hanse, Gideon	Leiden University, The Netherlands
Hao, Jin-Kao	University of Angers, France
Heidrich-Meisner, Verena	Kiel University, Germany
Henggeler, Antunes Carlos	University of Coimbra, Portugal

Hernandez-Diaz, Alfredo	Pablo de Olvide University, Spain
Herrmann, Michael	University of Edinburgh, UK
Hoogendoorn, Mark	Vrije Universiteit Amsterdam, The Netherlands
Ianovski, Egor	Higher School of Economics, Russia
Jakaite, Livija	University of Bedfordshire, UK
Johnson, Colin	University of Nottingham, UK
Jourdan, Laetitia	Inria, France
Kalinichenko, Vera	UCLA, USA
Kalyagin, Valeriy	Higher School of Economics, Russia
Karakostas, George	McMaster University, Canada
Kavsek, Branko	University of Primorska, Slovenia
Khachay, Michael	Krasovsky Institute of Mathematics and Mechanics, Russia
Kiani, Shahvandi Mostafa	ETH Zurich, Switzerland
Kiziltan, Zeynep	University of Bologna, Italy
Kochetov, Yury	Sobolev Institute of Mathematics, Russia
Kouchak, Shokoufeh	Arizona State University, USA
Kruger, Hennie	North-West University, South Africa
Kumar, Chauhan Vinod	University of Cambridge, UK
Kvasov, Dmitri	University of Calabria, Italy
La Malfa, Gabriele	University of Cambridge, UK
Landa-Silva, Dario	University of Nottingham, UK
Lanz, Oswald	Fondazione Bruno Kessler, Italy
Le Thi, Hoai An	Universite de Lorraine, France
Lera, Daniela	University of Cagliari, Italy
Lombardo, Gianfranco	University of Parma, Italy
Lu, Yun	Kutztown University of Pennsylvania, USA
Lu, Paul	University of Alberta, Canada
Luukka, Pasi	Lappeenranta University of Technology, Finland
Maalouf, Eliane	University of Neuchatel, Switzerland
Manzoni, Luca	University of Trieste, Italy
Maratea, Marco	University of Genova, Italy
Marinaki, Magdalene	Technical University of Crete, Greece
Marinakis, Yannis	Technical University of Crete, Greece
Martins de Moraes, Rafael	New York University, USA
Matsatsinis, Nikolaos	Technical University of Crete, Greece
Matsuura, Shun	Keio University, Japan
Meyer, Angela	Bern University of Applied Sciences, Switzerland
Meyer-Nieberg, Silja	Universitaet der Bundeswehr Muenchen, Germany
Milani, Federico	Politecnico di Milano, Italy
Milne, Holden	University of the Fraser Valley, Canada
Mongiovi, Misael	Consiglio Nazionale delle Ricerche, Italy
Montanez, George	Harvey Mudd College, USA
Nadif, Mohamed	Université de Paris, France
Nanni, Mirco	ISTI-CNR, Italy
Nicosia, Giuseppe	University of Catania, Italy

Nowaczyk, Slawomir	Halmstad University, Sweden
Nunez-Gonzalez, David	University of the Basque Country, Spain
Ojha, Varun	University of Reading, UK
Orchel, Marcin	AGH University of Science and Technology, Poland
Otero, Beatriz	Polytechnic University of Catalonia, Spain
Pacher, Mathias	Goethe-Universität Frankfurt am Main, Germany
Palar, Pramudita Satria	Bandung Institute of Technology, Indonesia
Papastefanatos, George	ATHENA Research Center, Greece
Paquet, Eric	National Research Council, Canada
Paquete, Luis	University of Coimbra, Portugal
Pardalos, Panos	University of Florida, USA
Parsopoulos, Konstantinos	University of Ioannina, Greece
Patane, Andrea	University of Oxford, UK
Pazhayidam, George Clint	Indian Institute of Technology Goa, India
Pedio, Manuela	University of Bristol, UK
Pedroso, Joao Pedro	University of Porto, Portugal
Peitz, Sebastian	Universität Paderborn, Germany
Pelta, David	University of Granada, Spain
Pereira, Ivo	University Fernando Pessoa, Portugal
Perrin, Dimitri	Queensland University of Technology, Australia
Petkovic, Milena	Zuse Institute Berlin, Germany
Ploskas, Nikolaos	University of Western Macedonia, Greece
Podda, Alessandro Sebastian	University of Cagliari, Italy
Poggioni, Valentina	University of Perugia, Italy
Polyzou, Agoritsa	Florida International University, USA
Pravin, Chandresh	University of Reading, UK
Prestwich, Steve	Insight Centre for Data Analytics, Ireland
Qian, Buyue	Xi'an Jiaotong University, China
Qiao, Ting	University of Auckland, New Zealand
Quadrini, Michela	University of Camerino, Italy
Radzik, Tomasz	King's College London, UK
Raidl, Gunther	Vienna University of Technology, Austria
Rauch, Jan	Prague University of Economics and Business, Czech Republic
Rebennack, Steffen	Karlsruhe Institute of Technology, Germany
Regis, Rommel	Saint Joseph's University, USA
Reif, Wolfgang	University of Augsburg, Germany
Requejo, Cristina	University of Aveiro, Portugal
Rinaldi, Francesco	University of Padua, Italy
Ripamonti, Laura Anna	University of Milan, Italy
Rocha, Humberto	University of Coimbra, Portugal
Rodrigues, Maia Jose Gilvan	Universidade Federal do Ceara, Brazil
Roy, Arnab	Fujitsu Laboratories of America, USA
Ruan, Hang	University of Edinburgh, UK

Santana, Roberto	University of the Basque Country, Spain
Sartor, Giorgio	SINTEF, Norway
Sartori, Claudio	University of Bologna, Italy
Saubion, Frederic	University of Angers, France
Schaefer, Robert	AGH University of Science and Technology, Poland
Schaerf, Andrea	University of Udine, Italy
Schetinin, Vitaly	University of Bedfordshire, UK
Schuetze, Oliver	CINVESTAV-IPN, Mexico
Scotney, Bryan	University of Ulster, UK
Selicato, Laura	University of Bari Aldo Moro, Italy
Serani, Andrea	CNR-INM, Italy
Sevaux, Marc	Université Bretagne-Sud, France
Shenmaier, Vladimir	Sobolev Institute of Mathematics, Russia
Shui, Zeren	University of Minnesota, USA
Shukla, Saurabh	National University of Ireland Galway, Ireland
Siarry, Patrick	Université Paris-Est Créteil Val de Marne, France
Silvestrin, Luis Pedro	Vrije Universiteit Amsterdam, The Netherlands
Simm, Jaak	Katholieke Universiteit Leuven, Belgium
Smith, Cole	New York University, USA
So, Anthony Man-Cho	Chinese University of Hong Kong, Hong Kong
Soja, Benedikt	ETH Zurich, Switzerland
Stanciu, Maria	Universita degli Studi di Cagliari, Italy
Stukachev, Alexey	Sobolev Institute of Mathematics, Russia
Suykens, Johan	Katholieke Universiteit Leuven, Belgium
Tangherloni, Andrea	University of Bergamo, Italy
Tchemisova, Tatiana	University of Aveiro, Portugal
Thimmisetty, Charanraj	Palo Alto Networks, USA
Tolomei, Gabriele	Sapienza University of Rome, Italy
Tonda, Alberto	Inria, France
Torres, Rocio	Politecnico di Milano, Italy
Tosetti, Elisa	Ca' Foscari University of Venice, Italy
Tsan, Brian	University of California, Merced, USA
Tuci, Elio	Universite de Namur, Belgium
Ulm, Gregor	Fraunhofer-Chalmers Research Centre for Industrial Mathematics, Sweden
Vadillo, Jon	University of the Basque Country, Spain
Van Geit, Werner	EPFL, Switzerland
Van Stein, Bas	Leiden University, The Netherlands
Varela, Carlos	Rensselaer Polytecnic Institute, USA
Viktor, Herna	University of Ottawa, Canada
Villani, Marco	University of Modena and Reggio Emilia, Italy
Viroli, Mirko	University of Bologna, Italy
Vito, Domenico	Politecnico di Milano, Italy
Werner, Ralf	Augsburg University, Germany
Wu, Ouyang	HAW Hamburg, Germany
Xu, Dachuan	Beijing University of Technology, China

Yuen, Shiu Yin	City University of Hong Kong, Hong Kong
Zabinsky, Zelda	University of Washington, USA
Zaidi, Moayid Ali	Ostfold University College, Norway
Zese, Riccardo	University of Ferrara, Italy
Zhang, Yongfeng	Rutgers University, USA
Zhigljavsky, Anatoly	Cardiff University, UK

Best Paper Awards

LOD 2021 Best Paper Award

"An Integrated Approach to Produce Robust Deep Neural Network Models with High Efficiency"
Zhijian Li[1], Bao Wang[2], and Jack Xin[1]
[1] University of California, Irvine, USA
[2] University of Utah, USA
Springer sponsored the LOD 2021 Best Paper Award with a cash prize.

Special Mention

"Statistical Estimation of Quantization for Probability Distributions: Best Equivariant Estimator of Principal Points"
Shun Matsuura[1] and Hiroshi Kurata[2]
[1] Keio University, Japan
[2] University of Tokyo, Japan

"Neural Weighted A*: Learning Graph Costs and Heuristics with Differentiable Anytime A*"
Alberto Archetti, Marco Cannici, and Matteo Matteucci
Politecnico di Milano, Italy

LOD 2021 Best Talk

"Go to Youtube and Call me in the Morning: Use of Social Media for Chronic Conditions"
Rema Padman[1], Xiao Liu[2], Anjana Susarla[3], and Bin Zhang[4]
[1] Carnegie Mellon University, USA
[2] Arizona State University, USA
[3] Michigan State University, USA
[4] University of Arizona, USA

Contents – Part II

Contents – Part I

Boosted Embeddings for Time-Series Forecasting

Sankeerth Rao Karingula(✉), Nandini Ramanan, Rasool Tahmasbi, Mehrnaz Amjadi, Deokwoo Jung, Ricky Si, Charanraj Thimmisetty, Luisa F. Polania, Marjorie Sayer, Jake Taylor, and Claudionor Nunes Coelho Jr

Advanced Applied AI Research, Palo Alto Networks, Santa Clara, USA
https://www.paloaltonetworks.com/

Abstract. Time-series forecasting is a fundamental task emerging from diverse data-driven applications. Many advanced autoregressive methods such as ARIMA were used to develop forecasting models. Recently, deep learning based methods such as DeepAR, NeuralProphet, and Seq2Seq have been explored for the time-series forecasting problem. In this paper, we propose a novel time-series forecast model, `DeepGB`. We formulate and implement a variant of gradient boosting wherein the weak learners are deep neural networks whose weights are incrementally found in a greedy manner over iterations. In particular, we develop a new embedding architecture that improves the performance of many deep learning models on time-series data using a gradient boosting variant. We demonstrate that our model outperforms existing comparable state-of-the-art methods using real-world sensor data and public data sets.

Keywords: Time-series · Forecasting · Deep learning · Gradient boosting · Embedding

1 Introduction

Time-series forecasting plays a key role in many business decision-making scenarios and is one of the central problems in engineering disciplines. In particular, many prediction problems arising in financial data [36], weather data [20], econometrics [11] and medical data [21] can be modeled as time-series forecasting problems.

Time-series forecasting models can be developed using various autoregressive (AR) methods. Classical linear models such as autoregressive integrated moving average (ARIMA) [6] are used to explain the past behavior of a given time-series data and then used to make predictions of the time-series. ARIMA is one of the most widely used forecasting methods for univariate time-series data forecasting. To account for seasonality in time-series data, ARIMA models can be further extended to seasonal autoregressive integrated moving average (SARIMA) [6]. In turn, SARIMA models can be extended with covariates or other regression variables to Seasonal AutoRegressive Integrated Moving Averages with eXogenous regressors, referred to as SARIMAX model, where the X

G. Nicosia et al. (Eds.): LOD 2021, LNCS 13164, pp. 1–14, 2022.
https://doi.org/10.1007/978-3-030-95470-3_1

stands for "exogenous". The exogenous variable can be a time-varying measurement like the inflation rate, or a categorical variable separating the different days of the week, or a Boolean representing special festive periods. Some limitations of AR models are that they become impractically slow when attempting to model long-range dependencies and do not scale well for large volumes of training data due to the strong assumptions they impose on the time-series [37].

Recurrent neural networks (RNNs), and in particular Long Short Term Memory (LSTM) networks, have achieved success in time-series forecasting due to their ability to capture long-range dependencies and to model nonlinear functions. Hewamalage et al. [17] ran an extensive empirical study of the existing RNN models for forecasting. They concluded that RNNs are capable of modeling seasonality directly if the series contains homogeneous seasonal patterns; otherwise, they recommended a deseasonalization step. They demonstrated that RNN models generally outperform ARIMA models. However, RNN models require more training data than ARIMA models as they make fewer assumptions about the structure of the time series and they lack interpretability.

A popular RNN-based architecture is the Sequence to Sequence (Seq2Seq) architecture [18], which consists of the encoder and the decoder, where both act as two RNN networks on their own. The encoder uses the encoder state vectors as an initial state, which is how the decoder gets the information to generate the output. The decoder learns how to generate target $y[t+1,\ldots]$ by matching the given target $y[\ldots,t]$ to the input sequence. The DeepAR model for probabilistic forecasting, recently proposed by Salinas et al. [30], uses a Seq2Seq architecture for prediction. DeepAR can learn seasonal behavior and dependencies on given covariates and makes probabilistic forecasts in the form of Monte Carlo samples with little or no history at all. Also, it does not assume Gaussian noise and the noise distribution can be selected by users.

In an effort to combine the best of traditional statistical models and neural networks, the AR-Net was proposed [37]. It is a network that is as interpretable as Classic-AR but also scales to long-range dependencies. It also eliminates the need to know the true order of the AR process since it automatically selects the important coefficients of the AR process. In terms of computational complexity with respect to the order of the AR process, it is only linear for AR-Net, as compared to quadratic for Classic-AR.

Facebook Prophet [34] is another forecasting method, which uses a decomposable time-series model with three main model components: trend, seasonality, and holidays. Using time as a regressor, Prophet attempts to fit several linear and non-linear functions of time as components. Prophet frames the forecasting problem as a curve-fitting exercise rather than explicitly looking at the time-based dependence of each observation within a time series. NeuralProphet [1], which is inspired by Facebook Prophet [34] and AR-Net [37], is a neural network-based time-series model. It uses Fourier analysis to identify the seasons in a particular time series and can model trends (autocorrelation modeling through AR-Net), seasonality (yearly, weekly, and daily), and special events.

Inspired by the success of learning using *Gradient Boosting* (GB) [9] and *Deep Neural Networks* [12,39], we present a novel technique for time-series forecasting called Deep Neural Networks with Gradient Boosting or DeepGB. Neural networks can represent non-linear regions very well by approximating them with a combination of linear segments. It is proven by the universal approximation theorem [7] that neural networks can approximate any non-linear region but it could potentially need an exponential number of nodes or variables in order to do this approximation. On the other hand, methods such as gradient boosting that build decision trees are very good at representing non-linear regions but they cannot handle complex arithmetic relations, such as multiplication of elements. The main idea of the proposed approach is to combine the strengths of gradient boosting and neural networks.

The proposed approach consists of building a series of regressors or classifiers and solving the problem for the residual at each time. This approach enables the generation of a number of small parallel models for a single task, instead of creating a large deep learning model that attempts to learn very complex boundary regions. Because each subsequent model attempts to solve only the gradient of the loss function, the task becomes simpler than attempting to perform a full regression or classification on the output range. Eventually, the error becomes small enough and indistinguishable from the data noise.

Our main contributions are as follows.

- We propose DeepGB, an algorithm for learning temporal and non-linear patterns for time-series forecasting by efficiently combining neural network embedding and gradient boosting.
- We propose *boosted embedding*, a computationally efficient embedding method that learns residuals of time-series data by incrementally freezing embedding weights over categorical variables.
- In our empirical evaluation, we demonstrate how the proposed approach DeepGB scales well when applied to standard domains and outperforms state-of-the-art methods including Seq2Seq and SARIMA in terms of both efficiency and effectiveness.

The rest of the paper is organized as follows: First, the background on gradient boosting is introduced in Sect. 2. Section 3 outlines the need for neural network embedding for time-series modelling, followed by the detailed description of the proposed approach, DeepGB. In Sect. 4, we present the experimental evaluation of the proposed method. Finally, conclusions and directions for future work are presented in Sect. 5.

2 Gradient Boosting

In Gradient Boosting, the solution is comprised of simpler parallel models that are trained sequentially but added together at the end. One of the main ideas of gradient boosting is that each subsequent model (which may be as small as a small tree or as large as a full deep neural network) needs to only solve for the residue between the output and the previous regressors built, thus making it a much easier problem to solve. Gradient Boosting [15] is motivated by the intuition that finding multiple weak hypotheses to estimate local probabilistic predictions can be easier than finding a highly accurate model. Friedman [15] in the seminal work, proposed a gradient boosting framework to train decision trees in sequence, such that each tree is modeled by fitting to the gradient/error from the tree in the previous iteration. Consider for $Data = \{\langle \mathbf{x}, y \rangle\}$, in the gradient boosting process, a series of approximate functions F_m are learned to minimize the expected loss $Loss := E_x[L(y, F_m(x))]$ in a greedy manner:

$$F_m := F_{m-1} + \rho_m * \psi_m \tag{1}$$

where ρ_m is the step size and ψ_m is a tree selected from a series of $\mathbf{\Psi}$ functions to minimize L in the m-th iteration:

$$\psi_m = \arg\min_{\psi \in \mathbf{\Psi}} E[L(y, F_{m-1}(x)) + \psi(x)] \tag{2}$$

where the loss function is usually least-squares in most works and a negative gradient step is used to solve for the minimization function.

In recent years three highly efficient and more successful gradient-based ensemble methods became popular, namely, XGBoost, LightGBM, and CatBoost [23,25]. However, both XGBoost and LightGBM suffer from overfitting due to biased point-wise gradient estimates. That is, gradients at each iteration are estimated using the same instances that were used by the current model for learning, leading to a bias. Unbiased Boosting with Categorical Features, known as CatBoost, is a machine learning algorithm that uses gradient boosting on decision trees [8]. CatBoost gains significant efficiency in parameter tuning due to the use of trees that are balanced to predict labels. The algorithm replaces the gradient step of the traditional gradient boosting with ordered boosting, leading to reduced bias in the gradient estimation step. It additionally transforms the categorical features as numerical characteristics by quantization, i.e., by computing statistics on random permutations of the data set and clustering the labels into new classes with lower cardinality. In this paper, we employ CatBoost, which is known to demonstrate effectiveness from literature, as the gradient boosting algorithm of choice.

In the field of Neural Networks (NNs), He et al. [16] introduced a deep Residual Network (ResNet) learning architecture where trained ResNet layers are fit to the residuals. Although ResNet and gradient boosting are methods designed for different spaces of problems, there has been significant research that has gone into formalizing the gradient boosting perspective of ResNet [27,38]. Inspired by the complementary success of deep models and gradient boosting, we in this work propose gradient boosting of NNs, `DeepGB`, where we fit a sequence of NNs using gradient boosting. NNs have previously been used as weak estimators although the methods mostly focused on majority voting for classification tasks [13] and uniform or weighted averaging for regression tasks [28,29]. However, we are the first to investigate gradient boosting of NNs in the context of time-series forecast in this work.

Algorithm 1 describes a generic algorithm of gradient boosting. It shows how a combination of simple models (not just trees) can be composed to create a

Algorithm 1. Generic Gradient Boosting Algorithm

```
function GradientBoosting(Data, Models)
    where Data = {⟨x_i, y_i⟩}_{i=1}^N
    models = []
    F_0 := y
    for 1 ≤ m ≤ |Models| do                          ▷ Iterate over the Models
        model.fit(x, F_{m-1})                         ▷ Fit the generic weak learner
        F_m = F_{m-1} − ρ_m · model.predict(x)        ▷ Compute the residual
        if abs(F_m − F_{m-1}) < ε then                ▷ Check termination condition
            break
        end if
        models.append(model)
    end for
    return models
end function
```

better model by just adding them together, effectively assuming each subsequent model can solve the difference. Interestingly enough, nonlinear autoregressive moving average model with exogenous inputs - NARMAX [31] has proposed this approach by clustering increasingly more complex models together. Note that Algorithm 1 does not make any distinction between how to select the order of the models. In NARMAX [31], for example, it is suggested that we should seek simpler models first, and start using more complex models only when required.

3 DeepGB Algorithm

In this section, we present our main algorithm `DeepGB` that is a combination of gradient boosting with embeddings.

It is often not straightforward to select the best model $F_1, \ldots, F_m$ in Algorithm 1 as the models need to account for complex temporal dependency and non-linear patterns in the time-series. Our approach is to use a neural network with an embedding layer to learn the underlying residual features from data with time categorical variables extracted from the covariate's time stamps. For example, low dimensional embeddings are fit on time-related variables like month, week of the year, day of the week, day of the month, holidays, etc. to extract this meaningful temporal dependence.

The first modification we will perform in the general Gradient Boosting algorithm using general models is to consider:

$$Models = \mathbf{list}(\mathbf{Embedding_1}, \ldots, \mathbf{Embedding_M}, \mathbf{Residual}),$$

where $Embedding_i$ is an embedding model that models categorical data, such as dates, holidays, or user-defined categories, and *Residual* is a traditional machine learning model, such as deep neural networks, Gradient Boosting or Support Vector Machines.

An example of an embedding model can be seen in Fig. 1 using Keras. In this example, `embedding_size` can be computed experimentally as $(\texttt{number\_of\_categories} + 1)^{0.25}$ [35] or as $\min(50, (\texttt{number\_of\_categories} + 1)//2)$ [3], where 1 was added to represent one more feature that is required by the Keras Embedding layer.

```
categorical_input_shape = (window_size, 1)
embedding_size = min(50, (number_of_categories+1)/ 2)
x_inp = Input(categorical_input_shape)
x_emb = Embedding(
    number_of_categories + 1,
    embedding_size,
    input_length=window_size)(x_inp)
x_out = Dense(1)(x_emb)
```

Fig. 1. Keras model for single output dense layer

A standard gradient boosting approach to train these simple embedding models, $Embedding_1, \ldots, Embedding_M$, is to select one of those models at each iteration, train it and then compute its residual error. In the next iteration, the next selected embedding model is trained on the previous residual error.

The key insight of this paper is that we can greatly reduce computational complexity by freezing embedding weights as we iterate through embedding training. The motivation behind this iterative freezing is that we preserve the gradient boosting notion of training weak learners on residuals. Each time an embedding is learned, we remove the effect of that embedding from the target data and train the next iteration on the residual.

This idea is shown in Fig. 2 where we show the sequential process in which the different embedding layers are trained. At any step i, we freeze the weights of the embedding layers $E_1, E_2, \ldots, E_{i-1}$ and then train the full model that concatenates all of the i embedding layers. Finally, a series of dense layers, D, are applied on the concatenated output.

DeepGB Algorithm Description. Pseudocode for the `DeepGB` algorithm is presented in Algorithm 2. The steps of the algorithm are described as follows

- Line 6: We first initiate a sequence of embedding models, $E_1, \ldots, E_M$.
- Line 7: We specify the input shape of the categorical variable that goes into the current embedding model.
- Line 8: We concatenate the current embedding with the previously concatenated embedding models.
- Line 10 - 13: We add dense layers that take the concatenated output of the embedding models as input.
- Line 15: In the current iteration, the model learns the weights of the current embedding layer while the previous embedding models are frozen.

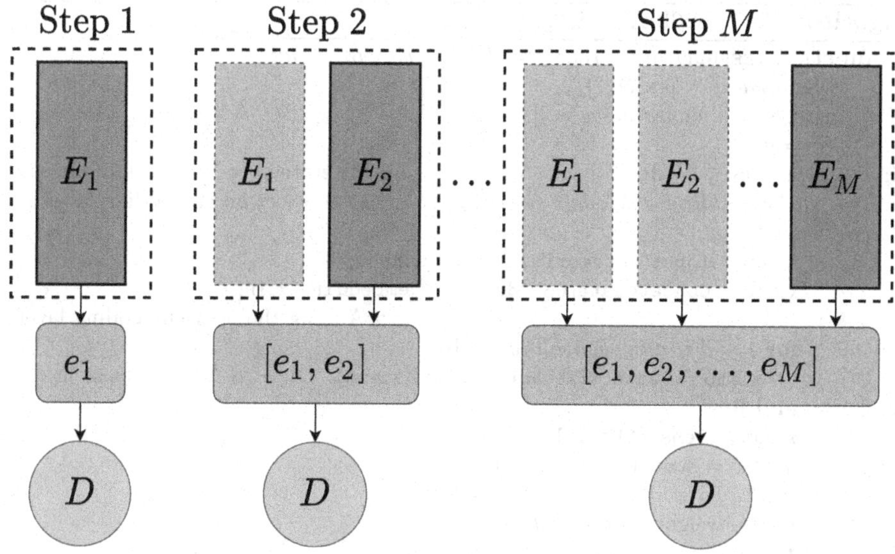

Fig. 2. Training sequence by freezing of the embedding layers

- Line 18: We don't need to learn for all M iterations, stop once the error converges below a certain threshold ϵ.
- Line 21: The current frozen weights of the embedding model are appended to the embedding list.
- Line 24: This is where the external model gets trained. In this paper, CatBoost is used as the external model.

Figure 3 corresponds to plots of the outputs of the algorithm. In particular, Fig. 3 shows embedding layers' prediction outputs (*model.predict*(x) in Algorithm 2) for two different data sets of server CPU utilization. In the figures, embedding layers are incrementally trained with data input from two time categorical variables, namely $dayofweek$ (days of a week) and *hour* (hours of a day). In the first iteration, the embedding layer learns weekly patterns using $dayofweek$ (i.e., F_1) as a categorical variable. Then, it uses the next categorical variable *hour* (i.e., F_2) to further improve the model by learning hourly patterns. The bottom plots shows the final residual (or irregular) time-series feature after embedding F_m in Algorithm 2.

Figure 3(a) shows that CPU utilization follows a highly correlated weekly pattern and that the embedding model accurately captures the weekly and hourly regular temporal pattern through 1st and 2nd iterations. Note that the final residual features capture irregular daily pattern in 2020–09–08 (Monday). In the bottom plot of Fig. 3(b), three irregular hourly patterns in 2020–09–10, 09–11, and 09–23 are accurately captured in the residual feature and shown as spikes.

Algorithm 2. DeepGB Algorithm

```
function DEEPGB(Data, Models = [E_1, ..., E_M, Residual])
    where Data = {⟨x_i, y_i⟩}_{i=1}^N
    models = [], embeddings = []
    F_0 = y
    for 1 ≤ m ≤ M do                                          ▷ Iterate over N embedding layers
        c, embedding = Models.select(E_1, ..., E_M)           ▷ Select an embedding layer at position c
        x_inp = Input(categorical_input_shape)
        x_emb = Concatenate(embeddings + [embedding(x_inp)])
                                                              ▷ Adding the next embedding layer
        for 1 ≤ d ≤ num_dense_layers do
            x_emb = Dense(size, activation)(x_emb)            ▷ Adding dense layers
        end for
        x_out = Dense(1)(x_emb)
        model = Model(x_inp, x_out)
        model.fit(x, y)
        Freeze weights of embedding
        F_m = F_{m-1} - model.predict(x)
        if abs(F_m - F_{m-1}) < ε then                        ▷ Check termination condition
            break
        end if
        embeddings.append(embedding)
    end for
    models = [model]
    Residual.fit(x, F_m)                        ▷ Fit Residual (Catboost) model on the error
    models.append(Residual)
    return models
end function
```

3.1 Gradient Boosting, Forward Stagewise Additive Models, and Structural Time Series Analysis

There is a fundamental connection between Gradient Boosting and Additive Regression Models fit in a greedy forward stepwise manner [10,15]. In terms of the presentation of Gradient Boosting in Eqs. (1) and (2) from Sect. 2, we can express our final Gradient Boosted model as an additive model:

$$F_M(x_t) = \sum_{m=1}^{M} \rho_m F_m(x_t, \psi_m) + w_t; w_t \sim \mathcal{D}(0, \sigma^2)$$

Where the updates to $\{\rho_m, \psi_m\}, m = 1, ..., M$ are computed based upon the previous $\{\rho_k, \psi_k\}_{k=1}^{M-1}$ trained weights and the functional form of F_m is set before training begins. In the special case of DeepGB derived in Sect. 3, $F_m(\cdot)$ has the same architecture as discussed in Sect. 2, ψ_m are the weights of the learned Neural Network, and m is an index of a categorical variable which corresponds to a specific set of time covariates, x_t. Further, the ordering of categories is no longer

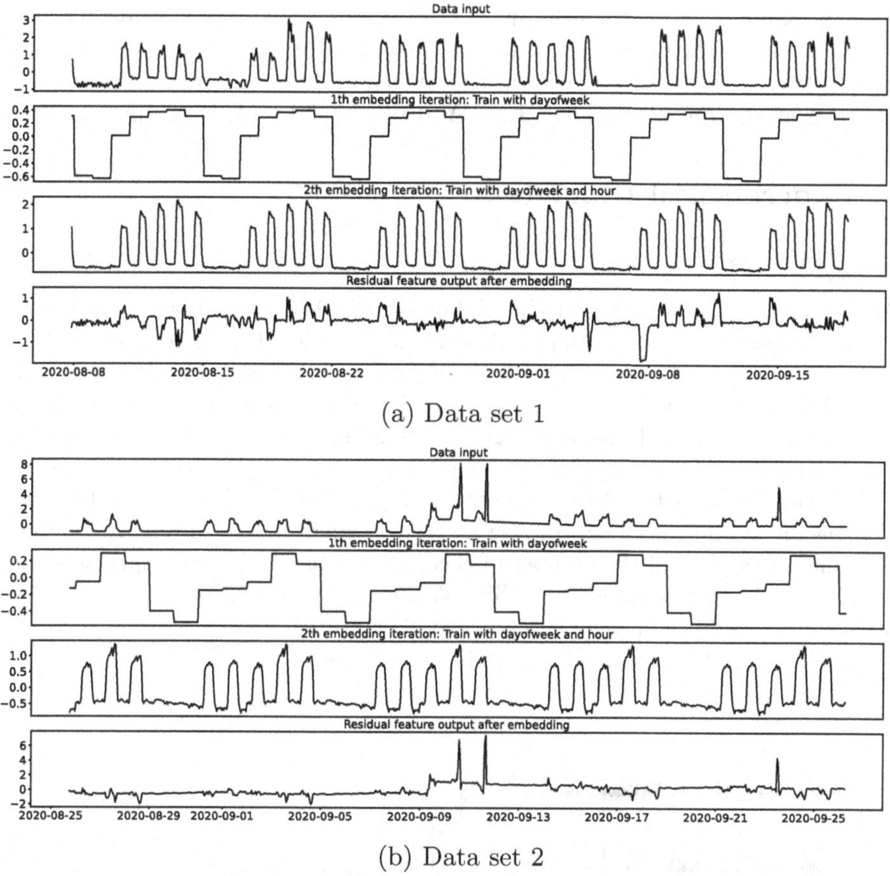

(a) Data set 1

(b) Data set 2

Fig. 3. Embedding comparison over data sets

arbitrarily selected from a committee of weak learners, but related to a specific ordering of the categorical time features (e.g. first hour, then day of the week, month, year, holidays, and so on). Finally, adding a noise model w_t, the `DeepGB` additive model can be seen as a greedy temporal basis expansion in functional space. This formulation can also be seen as a Structural Time Series (STS) model commonly found in the time series analysis literature [14,19,33] where the (possibly log-transformed) observed time series is related to an additive decomposition of several components (trend, seasonality, holidays, etc.) plus an additive error:

$$F_{\text{STS}}(x_t) = F_{\text{trend}}(x_t) + F_{\text{seasonality}}(x_t) + \ldots + F_{\text{holidays}}(x_t) + w_t; w_t \sim \mathcal{D}(0, \sigma_t^2)$$

In `DeepGB`, these additive features are substituted with learned embeddings of categorical time features with the goal of modeling the same temporal dependence. However, unlike a typical STS model, the additive components in `DeepGB` are not required to be smooth (in fact, they are often highly nonlinear *and* discontinuous as shown in Fig. 3). In this light, `DeepGB` can be seen as a discontinuous alternative to Facebook's Prophet which is also a STS model that can be viewed from an equivalent Generalized Additive Model (GAM) perspective [34].

4 Experimental Evaluation

4.1 Datasets

We conducted experiments on the Wikipedia [2] data available to the public via the Kaggle platform and internal networking device data. The internal data measures *connections per second* to a device, a rate-based variable which in general is hard to forecast. Figure 4(a) depicts the time-series for *connections per second* over a month, from Dec 2020 to Jan 2021. For the sake of simplicity, we use $I1, \ldots, I9$ for representing internal data sets.

The time series in the Wikipedia dataset represents the number of daily views of a specific Wikipedia article during a certain timeframe. Figure 4(b) depicts a few time-series plots of the number of accesses of a particular wiki page starting from July 2015 to September 2019. The Wikipedia dataset also represents a rate variable, the number of page accesses per day. Similar to the internal dataset, we use $P1, \ldots, P9$ for representing public data sets.

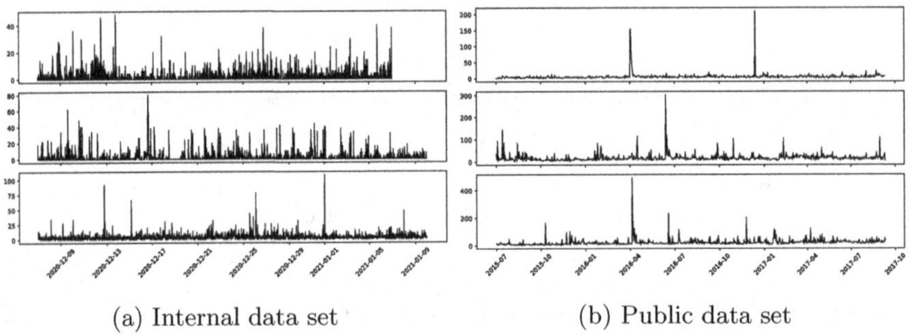

(a) Internal data set (b) Public data set

Fig. 4. A few sample time-series plots from the internal and public data sets

4.2 Model Setup

For benchmarking, `DeepGB` is compared with `SARIMA` and `Seq2Seq`. To keep comparisons as fair as possible, the following protocol is used: while employing `SARIMA`, the implementation in `statsmodels` [32] is used. For `SARIMA`, the best parameters are selected via grid search and `SARIMA(p=3, d=1, q=0)(P=1, D=0, Q=1)` is the final model employed for our analysis. For `Seq2Seq`, we started with the architecture in Hwang et al. [18] and further tuned to the setting that gives the best empirical performance. The implementation of the `Seq2Seq` model includes a symmetric encoder-decoder architecture with an LSTM layer of dimension 128, a RepeatVector layer that acts as a bridge between the encoder and decoder modules, a global soft attention layer [5] for many-to-one sequence tasks, and a self-attention layer [24]. Also, Time2Vec [22] and categorical embeddings [39], that enable the use of categorical features based on the properties of the time-series signature, were added to `Seq2Seq`. These enhancements improved the performance of `Seq2Seq`. The proposed `DeepGB` model consists of two parts: layers in the first half of the model implement embeddings and the later half uses a Boosting model. In particular, `CatBoost` with 800 trees of depth 3 was used as the gradient boosting model for the experiments. The proposed `DeepGB` model consists of 4 embedding layers followed by a concatenation layer and 4 dense layers with ReLu activation function after each layer. A dropout layer is also added to prevent overfitting. Time2Vec [22] and categorical embedding [39] were also used in `DeepGB`. Additionally, Root Mean Square Propagation with step size=0.0002 was used as the optimization algorithm for the experiments. For CatBoost, the implementation on `scikit-learn 0.24.1` with default parameter setting was used.

For all the methods, 30 d and 3 d of data were used for training and testing, respectively. Only `SARIMA` used 14 d of data for training, as its time for convergence increases significantly with the length of data. The widely used metric for time-series error analysis, Symmetric mean absolute percentage error (SMAPE), is employed to evaluate forecast accuracy in the experiments [4]. SMAPE is the main metric used in the M3 forecasting competition [26], as it computes the relative error measure that enables comparison between different time series, making it suitable for our setting.

4.3 Results

The results of our experiments for the task of time-series forecasting are summarized in Tables 1 and 2 for the Wikipedia and internal data sets, respectively. Table 1 shows SMAPE results for `DeepGB`, Seq2Seq and `SARIMA`. Bold font indicates the lowest SMAPE for the given dataset. It can be observed that `DeepGB` is significantly better than `SARIMA` and Seq2Seq in the majority of the cases. These results help in stating that `DeepGB` is on par or significantly better than state-of-the-art approaches, in real domains, at scale. The SMAPE scores presented in Table 2 indicate that `DeepGB` outperforms the other two methods in 6 out of 8 data sets. A closer inspection of the results suggests that the efficiency of `DeepGB` is significantly higher (about 3 times faster) for dataset $P1$.

In general, it can be observed that deep models take more time than statistical methods. However, `DeepGB` is significantly faster than the other deep model, Seq2Seq, which is reported to have the second-best performance after `DeepGB`. The results indicate that the proposed `DeepGB` approach does not sacrifice efficiency (time) for effectiveness (error percentage).

Table 1. SMAPE (lower ⇒ better) and training time (lower ⇒ better) for DeepGB, Seq2Seq and SARIMA methods on the internal dataset.

Dataset	Error (SMAPE)			Training time (secs)		
	DeepGB	Seq2seq	SARIMA	DeepGB	Seq2Seq	SARIMA
I1	**1.04**	6.12	1.44	10.38	85.80	3.19
I2	1.75	2.00	**1.10**	9.40	86.48	6.32
I3	**5.44**	21.95	20.77	8.66	86.63	5.99
I4	**2.71**	6.92	29.59	8.28	87.63	2.49
I5	6.33	**6.11**	6.58	8.74	81.76	5.64
I6	**4.59**	9.39	11.01	8.16	82.28	3.10
I7	**6.98**	19.03	17.61	11.72	193.54	5.00
I8	6.81	**4.45**	17.61	13.35	195.19	4.95
I9	61.08	62.97	**59.14**	12.95	196.26	3.37

Table 2. SMAPE (lower ⇒ better) and training time (lower ⇒ better) for DeepGB, Seq2Seq and SARIMA methods on the public dataset (wikipedia.org_all-access_spider).

Dataset	Original series	Error (SMAPE)			Training time (secs)		
		DeepGB	Seq2Seq	SARIMA	DeepGB	Seq2Seq	SARIMA
P1	2NE1_zh	7.98	**7.93**	16.12	5.86	17.04	0.22
P2	3C_zh	16.85	**6.11**	15.43	6.40	20.71	0.36
P3	4minute_zh	**2.34**	4.38	5.90	6.22	20.36	0.36
P4	5566_zh	**4.39**	8.01	12.44	6.05	20.73	0.30
P5	AND_zh	**5.36**	13.16	25.86	8.29	24.51	0.29
P6	AKB48_zh	**2.27**	5.81	6.78	6.30	19.48	0.34
P7	ASCII_zh	**2.64**	8.90	7.93	7.46	20.53	0.23
P8	Ahq_e-Sports_Club_zh	**3.44**	5.02	12.54	6.26	21.07	0.27

5 Conclusion

Time-series forecasting is a central problem in machine learning that can be applied to several real-world scenarios, such as financial forecasts and weather forecasts. To the best of our knowledge, this is the first work employing gradient boosting of deep models, infused with embeddings, in the context of time-series forecasting. To validate the performance of the proposed approach, we evaluated DeepGB on a public Wikipedia data set and an internal networking device data set. The experimental results showed that DeepGB outperforms SARIMA and Seq2Seq using SMAPE as performance metric. It was also shown in the empirical evaluations that the proposed method scales well when applied to standard domains as it offers a faster convergence rate when compared to other deep learning techniques.

Finally, this paper opens up several new directions for further research. Extending the proposed approach to multivariate time-series forecasting and deriving theoretical bounds and convergence properties for `DeepGB`, remain open problems and are interesting to us from a practical application standpoint.

References

1. Neural prophet. https://github.com/ourownstory/neural_prophet
2. Wikipedia web traffic time series forecasting. https://www.kaggle.com/c/web-traffic-time-series-forecasting/
3. Arat, M.M.: How to use embedding layer and other feature columns together in a network using keras? (2019). https://mmuratarat.github.io/2019-06-12/embeddings-with-numeric-variables-Keras
4. Armstrong, J.S.: Long-range Forecasting. Wiley, Hoboken (1985)
5. Bahdanau, D., Chorowski, J., Serdyuk, D., Brakel, P., Bengio, Y.: End-to-end attention-based large vocabulary speech recognition. In: ICASSP, pp. 4945–4949 (2016)
6. Box, G., Jenkins, G.M.: Time Series Analysis: Forecasting and Control. Holden-Day, San Francisco (1976)
7. Csáji, B.C.: Approximation with artificial neural networks. Fac. Sci. Eötvös Loránd Univ. Hungary **24**, 7 (2001)
8. Dorogush, A.V., Ershov, V., Gulin, A.: Catboost: gradient boosting with categorical features support. arXiv:1810.11363 (2018)
9. Friedman, J.: Stochastic gradient boosting. Comput. Stat. Data Anal. **38**, 367–378 (2002)
10. Friedman, J., Hastie, T., Tibshirani, R.: Additive logistic regression: a statistical view of boosting (With discussion and a rejoinder by the authors). Ann. Stat. **28**(2), 337–407 (2000)
11. Fuleky, P. (ed.): Macroeconomic Forecasting in the Era of Big Data. ASTAE, vol. 52. Springer, Cham (2020). https://doi.org/10.1007/978-3-030-31150-6
12. Goodfellow, I., Bengio, Y., Courville, A.: Deep Learning. The MIT Press, Cambridge (2016)
13. Hansen, L.K., Salamon, P.: Neural network ensembles. IEEE Trans. Pattern Anal. Mach. Intell. **12**, 993–1001 (1990)
14. Harvey, A., Peters, S.: Estimation procedures for structural time series models. J. Forecasting. **9**, 89–108 (1990)
15. Hastie, T., Tibshirani, R., Friedman, J.: Boosting and additive trees. In: The Elements of Statistical Learning, pp. 337–387. Springer, New York (2009). https://doi.org/10.1007/978-0-387-21606-5_10
16. He, K., Zhang, X., Ren, S., Sun, J.: Deep residual learning for image recognition. In: CVPR (2016)
17. Hewamalage, H., Bergmeir, C., Bandara, K.: Recurrent neural networks for time series forecasting: Current status and future directions. Int. J. Forecast. **37**(1), 388–427 (2021)
18. Hwang, S., Jeon, G., Jeong, J., Lee, J.: A novel time series based seq2seq model for temperature prediction in firing furnace process. Procedia Comput. Sci. **155**, 19–26 (2019)
19. Hyndman, R., Athanasopoulos, G.: Forecasting: Principles and Practice, 3rd edn. OTexts, Australia (2021)

20. Karevan, Z., Suykens, J.A.: Transductive LSTM for time-series prediction: an application to weather forecasting. Neural Netw. **125**, 1–9 (2020)
21. Kaushik, S., et al.: AI in healthcare: time-series forecasting using statistical, neural, and ensemble architectures. Front. Big Data **3**, 4 (2020)
22. Kazemi, S.M., et al.: Time2vec: learning a vector representation of time (2019)
23. Ke, G., et al.: LIGHTGBM: a highly efficient gradient boosting decision tree. In: Advances in Neural Information Processing Systems 30: Annual Conference on Neural Information Processing Systems 2017, 4–9, December 2017, Long Beach, CA, USA, pp. 3146–3154 (2017)
24. Lin, Z., et al.: A structured self-attentive sentence embedding. arXiv:1703.03130 (2017)
25. Makridakis, S., Spiliotis, E., Assimakopoulos, V.: The m5 accuracy competition: results, findings and conclusions. Int. J. Forecast. (2020)
26. Makridakis, S., Hibon, M.: The m3-competition: results, conclusions and implications. Int. J. Forecast. **16**(4), 451–476 (2000)
27. Nitanda, A., Suzuki, T.: Functional gradient boosting based on residual network perception. In: International Conference on Machine Learning, pp. 3819–3828. PMLR (2018)
28. Opitz, D.W., Shavlik, J.W.: Actively searching for an effective neural network ensemble. Connection Sci. **8**, 337–354 (1996)
29. Perrone, M.P., Cooper, L.N.: When networks disagree: ensemble methods for hybrid neural networks. Brown University Institute for Brain and Neural Systems, Tech. rep. (1992)
30. Salinas, D., Flunkert, V., Gasthaus, J., Januschowski, T.: DeepAR: probabilistic forecasting with autoregressive recurrent networks. Int. J. Forecast. **36**(3), 1181–1191 (2020)
31. Chen, S., Billings, S.A.: Representations of non-linear systems: the NARMAX model. Int. J. Control **49**(3), 1013–1032 (1989)
32. Seabold, S., Perktold, J.: Statsmodels: econometric and statistical modeling with Python. In: 9th Python in Science Conference (2010)
33. Shumway, R., Stoffer, D.: Time series analysis and its applications with R examples, vol. 9, January 2011. https://doi.org/10.1007/978-1-4419-7865-3
34. Taylor, S.J., Letham, B.: Forecasting at scale. Am. Stat. **72**(1), 37–45 (2018)
35. Team, T.: Introducing tensorflow feature columns (2017). https://developers.googleblog.com/2017/11/introducing-tensorflow-feature-columns.html
36. Timmermann, A.: Forecasting methods in finance. Ann. Rev. Financ. Econ. **10**, 449–479 (2018)
37. Triebe, O., Laptev, N., Rajagopal, R.: AR-Net: a simple auto-regressive neural network for time-series. arXiv:1911.12436 (2019)
38. Veit, A., Wilber, M.J., Belongie, S.: Residual networks behave like ensembles of relatively shallow networks. NIPS **29**, 550–558 (2016)
39. Wen, Y., Wang, J., Chen, T., Zhang, W.: Cat2vec: learning distributed representation of multi-field categorical data (2016). http://openreview.net/pdf?id=HyNxRZ9xg

Deep Reinforcement Learning for Optimal Energy Management of Multi-energy Smart Grids

Dhekra Bousnina(✉) and Gilles Guerassimoff

MINES ParisTech, PSL Research University, CMA - Centre de Mathématiques Appliquées, Sophia Antipolis, France
dhekra.bousnina@mines-paristech.fr

Abstract. This paper proposes a Deep Reinforcement Learning approach for optimally managing multi-energy systems in smart grids. The optimal control problem of the production and storage units within the smart grid is formulated as a Partially Observable Markov Decision Process (POMDP), and is solved using an actor-critic Deep Reinforcement Learning algorithm. The framework is tested on a novel multi-energy residential microgrid model that encompasses electrical, heating and cooling storage as well as thermal production systems and renewable energy generation. One of the main challenges faced when dealing with real-time optimal control of such multi-energy systems is the need to take multiple continuous actions simultaneously. The proposed Deep Deterministic Policy Gradient (DDPG) agent has shown to handle well the continuous state and action spaces and learned to simultaneously take multiple actions on the production and storage systems that allow to jointly optimize the electrical, heating and cooling usages within the smart grid. This allows the approach to be applied for the real-time optimal energy management of larger scale multi-energy Smart Grids like eco-distrits and smart cities where multiple continuous actions need to be taken simultaneously.

Keywords: Deep Reinforcement Learning · Actor-critic · Energy management · Smart grids · Multi-energy

Nomenclature

P_{Grid}	Grid power consumption
P_{gen}	Distributed power generation
C_{Grid}	Cost of power purchase from the grid
C_{gen}	Cost of distributed power generation
P_{Load}	Load power

Supported by the program Investissement d'Avenir, operated by l'Agence de l'Environnement et de la Maitrise de l'Energie ADEME, France.

G. Nicosia et al. (Eds.): LOD 2021, LNCS 13164, pp. 15–30, 2022.
https://doi.org/10.1007/978-3-030-95470-3_2

P_{pv}	PV power generation
P_{Bat}	Battery power
P_{H2}	Hydrogen storage power
P_{TRHP}	Electric power consumed by TRHP
$Q_{\text{TRHP},t}^{H-prod}$	Heat produced by TRHP
$Q_{\text{TRHP},t}^{C-prod}$	Cold produced by TRHP
COP_{TRHP}	Coefficient of performance of TRHP
Q_{H-load}	Heating load
Q_{C-load}	Cooling load
t	Time step
$P_{\text{(i)}}$	Power of a storage system i
$P_{\text{Ch}}^{(i)}$	Charging power of a storage system i
$P_{\text{Disch}}^{(i)}$	Discharging power of a storage system i
$P_{\text{min}}^{(i)}$	Minimum power of storage system i
$P_{\text{max}}^{(i)}$	Maximum power of storage system i
$\eta_{\text{Ch}}^{(i)}$	Charging efficiency of a storage system i
$\eta_{\text{Disch}}^{(i)}$	Discharging efficiency of a storage system i
$k_{\text{sd}}^{(i)}$	Self-discharge rate of a storage system i
$E_{init}^{(i)}$	Energy initially stored in storage system i
$E^{(i)}$	Energy stored in storage system i

Acronyms

PV	Photo-voltaic
SoC	State of Charge
MG	Microgrid
SG	Smart Grid
TRHP	Thermo-Refrigerating Heat Pump
SDHS	Smart District Heating System
MPC	Model Predictive Control
MDP	Markov Decision Process
ML	Machine Learning
DL	Deep Learning
RL	Reinforcement Learning
DRL	Deep Reinforcement Learning
DQN	Deep Q-Networks
DQL	Deep Q-Learning
DPG	Deep Policy Gradient
DDPG	Deep Deterministic Policy Gradient

1 Introduction

1.1 Context of the Problem

Within the radical changes that the energy landscape is currently undergoing, Smart Grids are playing a major role in the modernization of the electric grid [5]. These smart electricity networks have the great advantage of integrating in a cost-effective way the behavior and actions of all the users connected to it, including consumers, producers and prosumers, to ensure a cost-efficient and sustainable operation of the power system while guaranteeing quality and security of supply [36]. Besides electrical networks, district heating and cooling systems also play a paramount role in the implementation of the new smart energy systems [39]. In fact, the concept of smart thermal grids also comes up with numerous advantages including flexibility potentials and ability to adapt to the changes that affect the thermal demand and supply in short, medium and long terms. Thus, Smart Thermal Grids, as well, are expected to be an integrated part of the future energy system [4,33]. However, research works on the optimal control and energy management within the smart grid context traditionally focus solely on the electrical usages. Though, jointly optimizing the electrical networks together with other energy vectors interacting with them like heating and cooling networks has a great potential to increase the overall economic and environmental efficiency and flexibility of the energy systems. This idea brings about a generalization of the Smart Grid concept to Smart Multi Energy Grids [22] that lies on the interaction between electricity and other energy sectors (like heating, cooling, gas and hydrogen) as well as other sectors that electricity might interact with like the transport sector. Considering all these interactions in the optimal management of energy systems allows to unlock considerable efficiency and flexibility potentials and represents one of the main advantages of Smart Multi Energy Grids.

Optimal control of smart (multi-energy) grids is essential to guarantee a reliable operation for the smart grid components and ensure an optimal management of controllable loads, production units and storage systems while minimizing energy and operational costs [21]. One of the most popular and widely used optimal control techniques is Model Predictive Control (MPC), also referred to as Receding Horizon Control [10,25]. MPC is a feedback control method where the optimal control problem is solved at each time step to determine a sequence of control actions over a fixed time horizon. Only the first control actions of this sequence are then applied on the system and the resulting system state is measured. At the next time step, the time horizon is moved one step forward and a new optimization problem is then solved, taking into account the new system state and updated forecasts of future quantities. This receding time horizon and periodic adjustment of the control actions make the MPC robust against the uncertainties inherent to the model and forecasts [11]. MPC has been used in many successful applications in the field of Microgrid/ Smart Grid energy management including [1,26,27,38]. Nevertheless, MPC and model-based approaches in general, rely on the development of accurate models and predictors and on the usage of appropriate solvers. This does not only require domain expertise but also needs to re-design these components each time that a change occurs on the architecture or scale of the Smart Grid [12]. Furthermore, classical optimization approaches based on Mixed Integer Linear Programming (MILP), Dynamic Programming (DP) or heuristic methods like Particle Swarm Optimization (PSO) generally suffer from time-consuming procedures. In fact, they have to compute all or part of possible solutions in order to choose the optimal one, and have to re-run a generally time-consuming optimization procedure each time that an optimal decision needs be taken. Therefore, such methods, despite their ability to provide quite accurate results, generally fail to consider on-line solutions for large-scale real data-bases [32].

Learning-based techniques, on the other hand, do not need accurate system models and uncertainty predictors and can, thus, be an alternative to model-based approaches. Reinforcement Learning (RL) [34] has been gaining popularity over the past few years when it comes to dealing with challenging sequential decision making tasks [6]. Nevertheless, RL-based approaches fail to handle large state and actions spaces owing to the curse of dimensionality [41]. This major limitation of RL can be overcome by Deep Reinforcement Learning (DRL) which is a state-of-the art Machine Learning (ML) technique evolving through the interface between RL and Deep Learning (DL) [23]. In other words, it combines the strong nonlinear perceptual capability of deep neural networks (DNNs) with the robust decision making ability of RL [7]. Unlike RL, it therefore exhibits strong generalization capabilities in problems with complex state spaces. One of the main advantages of DRL compared to other classical optimization approaches is that, once it learned an optimal strategy, it can take optimal decisions in a few milliseconds without having to re-compute any costly optimization procedure. This makes DRL algorithms less time-consuming than classical optimization approaches and makes them, as a consequence more suitable for real-time optimization problems. DRL has, this way, shown successful applications in various real-life problems with large state spaces like Atari and Go games [30], robotics [2,37], autonomous driving [16,29] and other complex control tasks [23]. More recently, [3] proposed a novel assembling methodology of Q-learning agents trained several times with the same training data for stock market forecasting. The use of DQN aimed at avoiding problems that may occur when using supervised learning-based classifiers like over-fitting. Other recent successful applications of DRL include intrusion detection systems as presented in [20]. Furthermore, [19] proposed a new ensemble DRL model for predicting wind speed and the comparison of the proposed model with nineteen alternative mainstream forecasting models showed that the DRL-based approach provided the best accuracy. Moreover, Google has announced in 2018 that it gave control over the cooling of several of its data centers to a DRL algorithm [13].

1.2 Deep Reinforcement Learning in Smart Grids: Related Work

When it comes to the energy field, there have recently been several successful applications of DRL for instance in the context of microgrids, smart homes and

Smart Grids, mainly for the development of cost optimization and energy management strategies. For example, [8] considers an electricity microgrid featuring PV generation, a Battery Energy Storage System (BESS) and a hydrogen storage, and adresses the problem of optimally operating these storage systems using a Deep Q-Learning (DQL) architecture. The developed Deep Q-Network (DQN) agent was tested on the case of a residential customer microgrid located in Belgium and showed to successfully extract knowledge from the past PV production and electricity consumption time series. However, it only takes discrete actions for the hydrogen storage (whether to charge at maximum rate, discharge at maximum rate or stay idle). The operation of the BESS, on the other hand, is not a direct action of the DRL agent but is rather dynamically adapted based on the balance equation of the microgrid. Similarly, [12] proposed a DQN approach to develop real-time generation schedules for a microgrid while optimizing its daily operational costs. DQL algorithms have also been applied in [28] for the coordinated operation of wind farms and energy storage and in [18] for the on-line optimization of a microgrid featuring PV and wind generation, diesel generators, fuel cells, electric load and a BESS. Among the various DRL algorithms, the conventional DQL remains the most widely used approach and algorithms such as Policy Gradient (PG) and Actor-Critic (AC) are rarely investigated. This is primarily due to the simplicity of the DQL and to the fact that it handles well discrete action spaces. Meanwhile, DQL can not be directly applied to problems with continuous action spaces since they need to discretize the action space which leads to an explosion of the number of actions and, as a consequence, to a decreased performance [9,17]. Indeed, considering only discrete actions for the planning and control of the Smart Grid components significantly restrains their flexibility potentials and prevents from obtaining the best optimal scheduling and control strategies. Unlike DQL, Deep Policy Gradient (DPG) algorithms are capable of dealing with environments with continuous actions spaces. In this respect, [24] proposed the use of DQL and DPG for online building energy optimization through the scheduling of electricity consuming devices. The results showed that DPG algorithms are more suitable than DQN to perform online energy resources scheduling. Even though this work pioneered the use of DRL for online building energy optimization, the actions it considers are restricted to the on/off status of flexible load devices in a smart building. Besides, the DPG algorithms are also often criticized for their low sampling efficiency as well as the fact that their gradient estimator may have a large variance, which is likely to lead to slow convergence [14]. In order to overcome this limitation, Actor-Critic (AC) algorithms were proposed to combine the strong points of DPG and DQL approaches by estimating both the policy and the Q-value function during the training. In this respect, two DRL algorithms were designed for Smart Grid optimization in [32]: on the one hand, DQL was applied for the discrete action control tasks like charging/discharging the BESS or switching the buy/sell modes of the grid. On the other hand, an AC algorithm named H-DDPG (Hybrid-Deep Deterministic Policy Gradient) was developed to deal with continuous state and action spaces. Yet, only the results of the DQN approach were presented in the paper and benchmarked with the results of a Mixed Integer Linear Programming (MILP) optimization Matlab tool. Even though DDPG algorithms were proposed for some applications in the energy systems context namely for dealing with cost optimization problems in Smart Home energy systems in [40], for flow rate control in Smart District Heating Systems (SDHS) in [42], and for solving the Nash Equilibrium in energy sharing mechanisms in [15], most of these applications consider mono-action and/or mono-fluid use-cases. In other words, they consider solely electrical or thermal Smart grids and do not consider jointly optimizing the uses of several energy vectors within a multi-energy Smart Grid. Besides, most of the previous works consider applications on the Smart Home or building level and do not consider testing these approaches on a larger smart district-level. Finally, thourough comparisons of the performance of DDPG-based approaches with other widely used techniques like MPC for dealing with energy management systems in Smart Grids have rarely been reported in the literature.

In the present work, we propose a DDPG-based approach to deal with the real-time energy management of multi-energy Smart Grids. More specifically, we formulated the optimal control problem as a POMDP and developed a DDPG agent to perform real-time scheduling of the multi-energy systems within a Smart Grid. The main contributions of the present work are the following:

- Unlike most of previous works where mono-fluid (electrical or thermal) Smart Grids are considered, we focus on multi-energy (electrical, heating, cooling, hydrogen) smart grids that interact with the main utility grid. A variable electricity price signal is considered and a DRL-based energy management system is developed to take price-responsive control actions.
- The DDPG algorithm is proposed instead of the mostly used DQN to deal with the continuous action and state spaces inherent to the multi-energy smart grid model. At each time step of the control horizon, multiple continuous actions are simultaneously taken by the DDPG agent to optimally schedule the various storage systems as well as the thermal production units.
- The proposed approach is tested on a residential multi-energy smart grid model and will be applied on a real-life district-level multi-energy smart grid which is being currently under construction in France. More specifically, the developed DDPG agent is aimed at operating real-time energy management of the various energy systems within an eco-district: BESS, heating and cooling storage systems, controllable loads of the buildings, heated water storage tanks, as well as District Heating and Cooling production units, Electric Vehicle Charging Stations and the public lighting of the district.
- The proposed approach is benchmarked with an MPC-based approach.

The remainder of this paper is organized as follows: in Sect. 2, the considered multi-energy smart grid model is described together with the optimal energy management problem addressed in this work. In Sect. 3, the problem is formulated as an MDP and a DRL-based approach is proposed to solve it. Section 4 presents the simulations and results and finally conclusions and future work are asserted in Sect. 5.

2 The Multi-energy Smart Grid Model and Optimal Control Mechanism

The Smart multi-energy grid model considered in this paper is shown in Fig. 1. It is composed of residential electric, heating and cooling loads, distributed energy generators (PV panels), heating and cooling production units consisting of geothermal Thermo-Refrigerating Heat Pumps (TRHPs), a BESS, a heat storage system (by phase-change materials) and a cold storage system (by ice storage tanks). The Smart Grid components are related to the main utility grid. In fact, besides the residential electrical usages, the TRHPs also consume electric power to produce heat and cold for the thermal needs of the buildings. At each time step, the electric loads of the buildings are met by the local PV generation, by discharging the BESS or by withdrawing electricity from the public utility grid. Thermal needs in terms of heating, on the other hand, are met whether by directly producing heat via thermo-refrigerating heat pumps or by discharging the heat storage system. Similarly, cooling loads are ensured by directly producing cold via TRHPs or by discharging the cooling storage system.

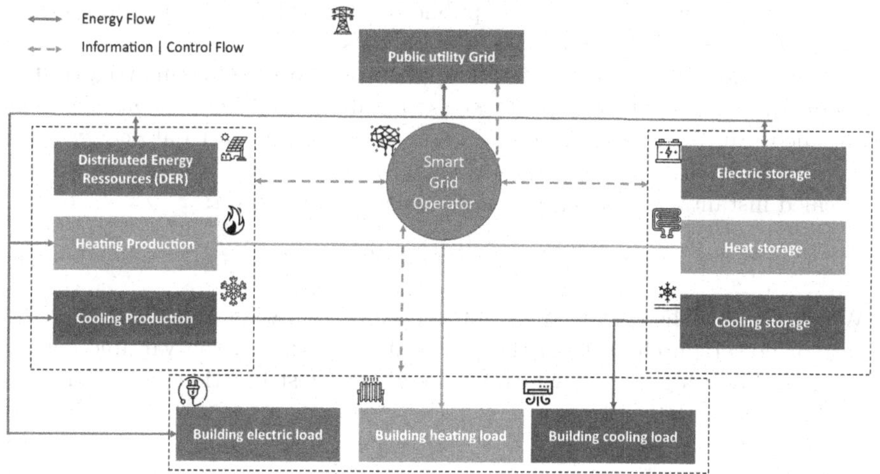

Fig. 1. Architecture of the multi-energy Smart Grid

In order to jointly optimize the operation of the multi-energy systems of the Smart Grid, an energy management system is needed to schedule the different controllable units while minimizing the daily operational costs. To solve this sequential decision making problem, we formulate it as a Markov Decision Process (MDP). In fact, the energy level of each energy storage system, at each time step, depends only on the current energy level, together with the current charge/discharge power, and as a consequence, the scheduling of the different energy storage systems and production units can be formulated as an MDP

$M = (S, A, T, R, \gamma)$ where its key components, the state space S, the action space A, the reward R and the transition function T are designed as follows:

- State: the environment state at each time step $t \in H$ is denoted by s_t and is composed of six types of information: $s_t = (s_t^{Storage}, s_t^{Load}, s_t^{DER}, s_t^{Grid}, s_t^{Prod}, s_t^{Temp})$ where $s_t^{Storage} \in S^{Storage}$ denotes the storage operation of the Smart Grid and describes the amount of energy stored in each of the battery, hydrogen, heating and cooling storage systems $s_t^{Storage} = (s_t^{Bat}, s_t^{H2}, s_t^{HS}, s_t^{CS})$, $s_t^{Load} \in S^{Load}$ contains the h past realizations of the electric, heating and cooling loads, where h, the history length is set as 24, so that the history length covers one day of past realizations with time steps $\Delta t = 1$ hour. Similarly, $s_t^{DER} \in S^{DER}$ contains the h past realizations of PV generation, $s_t^{Grid} \in S^{Grid}$ contains the h past realizations of the electricity prices as well as the amount of power withdrawn from the main utility grid at time step t, $s_t^{Prod} \in S^{Prod}$ contains the quantities of heat and cold produced by the TRHPs at time step t. Finally, $s_t^{Temp} \in S^{Temp}$ contains both the indoor and outdoor temperatures.
- Action: the aim of the energy management system is to decide the charging/discharging power of each energy storage system P^{SS}, the amount of energy to be purchased from the public utility grid P^{Grid} and the thermal energy (heat or cold) produced by the TRHPs Q^{TRHP}.
- Reward: when an action $a_t \in A_t$ is applied on the system, this triggers the environment to move from state s_{t-1} to state s_t and hence a reward r_t is obtained. Since the aim of the agent is to minimize the total energy costs within the Smart Grid, the reward signal r_t corresponds to the negative of rescaled instantaneous operational revenues at time step t:

$$r_t = -\alpha.[C_{gen}.P_{gen}(t) + C_{grid}(t).P_{grid}(t)] \tag{1}$$

  Where C_{gen} is the cost of distributed power generation and $C_{grid}(t)$ is the cost of power purchase from the public utility grid i.e. the variable energy price, and α is a factor by which we rescale the cost function, such that

$$0 < \alpha \leq 1 \tag{2}$$

3 The Proposed Deep Reinforcement Learning-Based Approach

RL is an Artificial Intelligence (AI) paradigm where the AI agent interacts with its environment by taking actions over a sequence of time steps in order to maximize a cumulative reward signal [34]. At each time step t, the agent performs a control action a_t based on the measure of the current state of the environment s_t and receives, in return, a reward r_t and information on the new state of the environment s_{t+1} for the next time step $t+1$. This way, the RL agent learns an optimal control policy through the interaction with the environment as shown in Fig. 2.a.

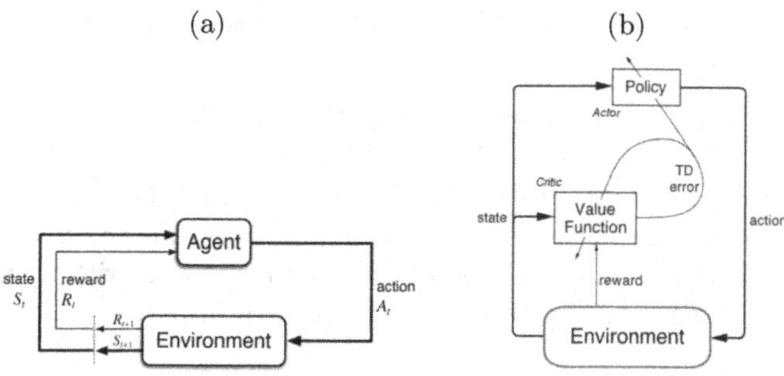

Fig. 2. (a): The agent-environment interaction in reinforcement learning; (b): The actor-critic architecture, from [34].

DRL [23] is a family of methods which evolve through the interface between RL and DL. Such a combination of RL and DL has recently shown its ability to learn complex tasks directly from high-dimensional inputs. DRL methods are divided into two main types, namely value-based and policy-based methods. In value-based methods, the neural network learns the optimal Q-function $Q^*(s,a)$ of each action a given a state s, which is the maximum sum of rewards r_t discounted by a factor γ at each time step t achievable by a policy $\pi = P(a_t|s_t)$ after taking and action a_t given a state s_t:

$$Q^*(s,a) = \max_{\pi} E[r_t + \gamma.r_{t+1} + \gamma^2.r_{t+2} + ...|s_t = s, a_t = a, \pi] \tag{3}$$

Meanwhile, for policy-gradient methods, the artificial neural network learns a probability distribution of the action a at a given state s instead of computing the Q-function. Value-based methods are known to be suitable for discrete action spaces whereas policy-based algorithms handle well continuous actions spaces.

This work proposes an application of the DDPG (Deep Deterministic Policy Gradient) algorithm which is a policy-based algorithm belonging to the actor-critic (AC) family [31]. AC methods rely on the idea of combining DPG and DQN: the policy function $\mu(s,\theta^\mu)$ is referred to as the actor where θ^μ represent the weights of the actor network. It specifies the current policy by deterministically mapping states to a specific action. The value-function $Q(s,a)$ is known as the critic and produces an error signal given the state, the output of the actor and the resultant reward signal as shown in Fig. 2.b [17].

When the agent takes an action a_t, under a given state s_t, according to a policy $\mu(s,\theta^\mu)$, the value of reward is given by the Bellman equation [35]:

$$Q^\mu(s_t,a_t) = E^\mu[r_t + \gamma.Q^\mu(s_{t+1},\mu(s_{t+1},\theta^\mu))] \tag{4}$$

The Q-network loss function is then given by:

$$L(\theta^Q) = E^\mu[(Q(s_t,a_t|\theta^Q) - y_t)^2] \tag{5}$$

where

$$y_t = r_t + \gamma.Q(s_{t+1},\mu(s_{t+1}|\theta^Q)) \tag{6}$$

The performance objective which measures the performance of the policy μ is given by:

$$J_\beta(\mu) = \int_S \rho^\beta(s) Q^\mu(s, \mu(s)) ds \tag{7}$$

Where ρ^β is the probability-distribution function of s_t. The aim of the training process is to maximize performance objective $J_\beta(\mu)$ while minimizing the loss function $L(\theta^Q)$. The training process of the used DDPG algorithm implemented in this work is given by Algorithm 1, also described in [15].

Algorithm 1: DDPG algorithm

Initialize the actor network μ and the critic network Q with random weights θ^μ and θ^Q ;
Initialize target network μ' and Q' with the weights $\theta^{\mu'} \leftarrow \theta^\mu$ and $\theta^{Q'} \leftarrow \theta^Q$;
Initialize the experience replay Buffer B ;
for *episode* $\leftarrow 0$ **to** $N_{episodes}$ **do**
  Initialize a random process R for action exploration;
  Get initial observation of state S_1 at time step $t = 1$;
  for $T \leftarrow 1$ **to** N_{steps} **do**
    Select action $a_t = \mu(s_t|\theta^\mu) + R_t$ according to the current policy and exploration noise ;
    Execute action a_t in the environment and observe the resulting reward r_t and the new state s_{t+1} ;
    Store the transition (s_t, a_t, r_t, s_{t+1}) in experience replay buffer B;
    Sample a random mini-batch of N transitions (s_i, a_i, r_i, s_{i+1}) from B ;
    Set $y_i(r_i, s_{i+1}) = r_i + \gamma.Q'(s_{i+1}, \mu'(s_{i+1}|\theta^{\mu'})|\theta^{Q'})$;
    Update the critic by minimising the loss $L = 1/N \sum_i Q(s_i, a_i|\theta^Q) - y_i)^2$;
    Update the actor policy using the policy gradient $\nabla_{\theta^\mu} 1/N \sum_{s \in B} Q(s, \mu(s|\theta^\mu)|\theta^Q)$;
    Update the target networks: $\theta^{Q'} \leftarrow (1-\rho).\theta^Q + \rho.\theta^{Q'}$ and $\theta^{\mu'} \leftarrow (1-\rho).\theta^\mu + \rho.\theta^{\mu'}$
  end
end

This algorithm was integrated in a specifically-designed multi-energy Smart Grid energy management framework where the DDPG agent interacts with the Smart Grid environment to generate an optimal schedule of its various energy systems. The Smart Grid environment describes the dynamics of the energy systems within the Smart Grid and is modeled as follows:

$$\min \quad \sum_{t=1}^{H} C_{gen}.P_{gen}(t) + C_{grid}(t).P_{grid}(t) \tag{8a}$$

$$\text{s.t. } P_{\text{Grid},t} = P_{\text{Load},t} + P_{\text{Bat},t} + P_{\text{H2},t} + P_{\text{pv},t} + P_{\text{TRHP},t} \qquad \forall t \tag{8b}$$

$$Q^{H-prod}_{\text{TRHP},t} + Q^{C-prod}_{\text{TRHP},t} = COP_{TRHP}.P_{\text{TRHP},t}\forall t \tag{8c}$$

$$Q^{H-prod}_{\text{TRHP},t} = Q_{H-load,t} + Q_{\text{HS},t}\forall t \tag{8d}$$

$$Q^{C-prod}_{\text{TRHP},t} = Q_{C-load,t} + Q_{\text{CS},t} \tag{8e}$$

$$P^{(i)}_{t} = P^{(i)}_{\text{ch,t}} + P^{(i)}_{\text{disch, t}} \qquad \forall i \in SS, \forall t \tag{8f}$$

$$E^{(i)}_{1} = E^{(i)}_{\text{init}}.(1 - k^{(i)}_{\text{sd}}) + \Delta_t \left(P^{(i)}_{\text{Ch},0}\eta_{\text{Ch}} - P^{(i)}_{\text{Disch},0}\frac{1}{\eta_{\text{Disch}}} \right) \forall i \in SS \tag{8g}$$

$$E^{(i)}_{\text{t+1}} = E^{(i)}_{\text{t}}.(1 - k^{(i)}_{\text{sd}}) + \Delta_t(P^{(i)}_{\text{ch,t}}.\eta_{\text{ch}} - \frac{1}{\eta_{\text{disch}}}.P^{(i)}_{\text{disch,t}} \qquad \forall i \in SS, \forall t \tag{8h}$$

$$E^{(i)}_{\text{min}} \leq E^{(i)}_{t} \leq E^{(i)}_{\text{max}} \qquad \forall i \in SS, \forall t \tag{8i}$$

$$P^{(i)}_{\text{min}} \leq P^{(i)}_{t} \leq P^{(i)}_{\text{max}} \qquad \forall i \in SS, \forall t \tag{8j}$$

Where Δt is the time slot (set to 1hour) and H is the optimization time horizon. Eq. (8a) represents the cost function to be minimized, Eqs. (8b) to (8e) express the electrical and thermal power balance within the Smart Grid, Eqs. (8f) to (8h) express the dynamics of the multi-energy storage systems within the Smart Grid and Eqs. (8i) to (8j) express the limitations on energy and charge and discharge power of each storage system, while SS represents the set of energy storage systems within the Smart-Grid.

4 Implementation Details, Simulations and Results

A framework was developed based on the previously described DDPG algorithm and tested on the designed environment of a residential consumer multi-energy smart grid which parameters are given in Table 1. During the training process, the DDPG agent was provided with three years of actual past realizations of PV generation, electric loads and electricity prices, as well as simulated data of heating and cooling loads and indoor and outdoor temperatures, for a residential consumer located in France. The historical data of a typical day in winter and in summer can be visualized in Figs. 3.a and 3.b. As in [8], we split the time series into a training set and a validation set that correspond to a different one year of historical data each. The Deep Neural Network (DNN) obtained at the end of the training process is then used in a test environment to provide an independent estimation of the final policy. Finally, to evaluate the performance of the proposed DRL approach, we use a benchmark solution that we refer to as "theoretical MPC". In this solution, we use an MPC controller that is supposed to have, at each day, a "perfect knowledge" of the stochastic system variables for the next 24 h. Unlike the DDPG, the MPC was given the actual future realizations of the unknown quantities in the predictor. The MPC with a time step $t = 1$ h and a time horizon $H = 24$ h was run for a one-year simulation, with the objective of minimizing the total operational costs. As shown

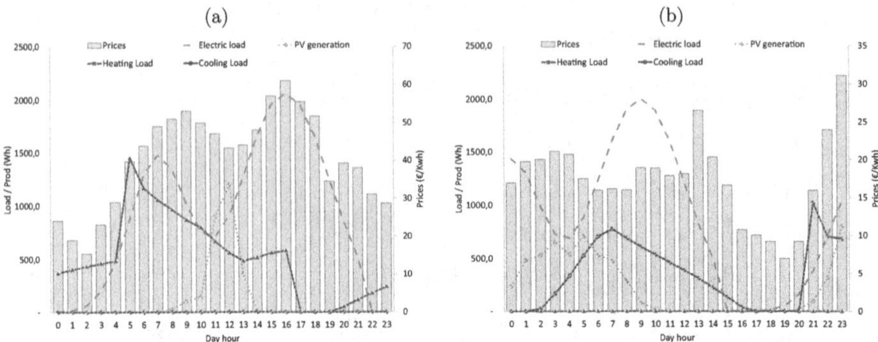

Fig. 3. Historical data used for (a): a typical winter day; (b): a typical summer day.

Table 1. Implementation details.

Parameters of the smart grid	Optimization parameters
Size of the battery ξ_{bat}: 15 kWh	DDPG number of training episodes: 5000
Battery charge/discharge efficiency η_{bat}: 90%	DDPG number of training steps 438.10^5
Size of the hydrogen ξ_{H2}: 1,1 kWh	DDPG learning rate of the actor:
Hydrogen charge/discharge efficiency η_{H2}: 65%	DDPG learning rate of the critic: 0,0001
Size of the heat storage ξ_{HS}: 1,2 kWh	DDPG learning rate of the critic: 0.0002
HS charge/discharge efficiency η_{HS}: 75%	DDPG discount factor γ: 0, 99
Size of the cooling storage ξ_{CS}: 0,8 kWh	DDPG and MPC time step: 1 h
CS charge/discharge efficiency η_{CS}: 75%	DDPG reward rescale factor α : 0, 001
Average electric consumption/day: 18 kWh/day	MPC time-horizon: $H = 24$ h
Peak power generation of PV: 15 kWp	Solver used in MPC optimization: GLPK
Maximal heat/cold generated by TRHP:50 kWh	

in Fig. 4, the performance of the proposed DDPG-based approach is close to "theoretical MPC" optimum. These results demonstrate the effectiveness of the proposed DRL approach for multi-energy management of Smart Grids under uncertainty. The DDPG was able to take multiple scheduling continuous actions simultaneously and succeeded to extract knowledge from the past realizations of the stochastic variables. The DRL agent learnt a strategy similar to the optimal strategy given by the MPC-based approach. We notice for instance that the DRL agent successfully learnt to purchase electricity from the main utility grid at low price periods and to rather discharge the storage systems during peak price periods. It also successfully learnt to maintain the power balance within the multi-energy Smart Grid.

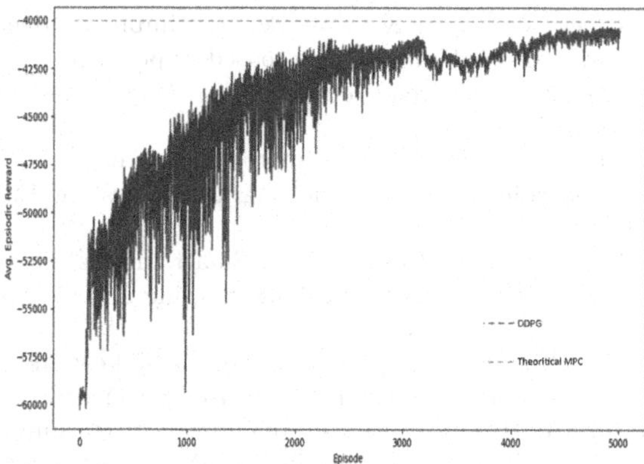

Fig. 4. Learning curve of the DDPG approach for a 5000-episode training process

5 Conclusion

This paper presented a DRL-based approach to deal with optimal energy management of multi-energy Smart Grids. The considered sequential decision making problem was formulated as an MDP and is addressed using a Deep Deterministic Policy Gradient (DDPG) algorithm. The developed framework was tested on the model of a multi-energy smart grid where the DDPG agent was designed to optimally schedule the various energy storage and thermal production units. The simulations showed that the agent handles well the continuous state and action spaces and learns to take multiple control actions simultaneously. Benchmark tests were conducted using a "theoretical MPC" solution to evaluate the performance of the proposed approach. Results showed that the total rewards obtained by the DDPG algorithm were close to the theoretical optimum and thus showed the effectiveness of the proposed DRL-based approach for dealing with optimal energy management of multi-enenrgy Smart Grid. More detailed results regarding the behavior of the policy will be given at the conference and will be the subject of upcoming papers. Future works also include the extension of the smart grid model to a district level smart grid where further devices are to be controlled including Heated water Storage tanks and other buildings controllable loads, Electric Vehicle Charging Stations and public lighting of the district. The proposed framework will also be applied on a real-life project of a multi-energy smart grid currently under construction in France.

References

1. Bousnina, D., de Oliveira, W., Pflaum, P.: A stochastic optimization model for frequency control and energy management in a microgrid. In: Nicosia, G., et al. (eds.) LOD 2020. LNCS, vol. 12565, pp. 177–189. Springer, Cham (2020). https://doi.org/10.1007/978-3-030-64583-0_17

2. de Bruin, T., Kober, J., Tuyls, K., Babuška, R.: Improved deep reinforcement learning for robotics through distribution-based experience retention. In: 2016 IEEE/RSJ International Conference on Intelligent Robots and Systems (IROS), pp. 3947–3952. IEEE (2016)
3. Carta, S., Ferreira, A., Podda, A.S., Recupero, D.R., Sanna, A.: Multi-DQN: An ensemble of deep q-learning agents for stock market forecasting. Expert Syst. Appl. **164**, 113820 (2021)
4. van den Ende, M., Lukszo, Z., Herder, P.M.: Smart thermal grid. In: 2015 IEEE 12th International Conference on Networking, Sensing and Control, pp. 432–437. IEEE (2015)
5. Fang, X., Misra, S., Xue, G., Yang, D.: Smart grid-the new and improved power grid: a survey. IEEE Commun. Surv. Tutorials **14**(4), 944–980 (2011)
6. François-Lavet, V.: Contributions to deep reinforcement learning and its applications in smartgrids. Ph.D. thesis, Université de Liège, Liège, Belgique (2017)
7. François-Lavet, V., Henderson, P., Islam, R., Bellemare, M.G., Pineau, J.: An introduction to deep reinforcement learning. arXiv preprint arXiv:1811.12560 (2018)
8. François-Lavet, V., Taralla, D., Ernst, D., Fonteneau, R.: Deep reinforcement learning solutions for energy microgrids management. In: European Workshop on Reinforcement Learning (EWRL 2016) (2016)
9. Gao, G., Li, J., Wen, Y.: Energy-efficient thermal comfort control in smart buildings via deep reinforcement learning. arXiv preprint arXiv:1901.04693 (2019)
10. Garcia, C.E., Prett, D.M., Morari, M.: Model predictive control: theory and practice-a survey. Automatica **25**(3), 335–348 (1989)
11. Gelleschus, R., Böttiger, M., Stange, P., Bocklisch, T.: Comparison of optimization solvers in the model predictive control of a PV-battery-heat pump system. Energ. Procedia **155**, 524–535 (2018)
12. Ji, Y., Wang, J., Xu, J., Fang, X., Zhang, H.: Real-time energy management of a microgrid using deep reinforcement learning. Energies **12**(12), 2291 (2019)
13. Knight, W.: Google just gave control over data center cooling to an AI (2018)
14. Konda, V.R., Tsitsiklis, J.N.: Actor-critic algorithms. In: Advances in Neural Information Processing Systems, pp. 1008–1014 (2000)
15. Kuang, Y., Wang, X., Zhao, H., Huang, Y., Chen, X., Wang, X.: Agent-based energy sharing mechanism using deep deterministic policy gradient algorithm. Energies **13**(19), 5027 (2020)
16. Liaw, R., Krishnan, S., Garg, A., Crankshaw, D., Gonzalez, J.E., Goldberg, K.: Composing meta-policies for autonomous driving using hierarchical deep reinforcement learning. arXiv preprint arXiv:1711.01503 (2017)
17. Lillicrap, T.P., et al.: Continuous control with deep reinforcement learning. arXiv preprint arXiv:1509.02971 (2015)
18. Lin, S., Yu, H., Chen, H.: On-line optimization of microgrid operating cost based on deep reinforcement learning. In: IOP Conference Series: Earth and Environmental Science, vol. 701, p. 012084. IOP Publishing (2021)
19. Liu, H., Yu, C., Wu, H., Duan, Z., Yan, G.: A new hybrid ensemble deep reinforcement learning model for wind speed short term forecasting. Energy **202**, 117794 (2020)
20. Lopez-Martin, M., Carro, B., Sanchez-Esguevillas, A.: Application of deep reinforcement learning to intrusion detection for supervised problems. Expert Syst. Appl. **141**, 112963 (2020)
21. Ma, T., Wu, J., Hao, L., Lee, W.J., Yan, H., Li, D.: The optimal structure planning and energy management strategies of smart multi energy systems. Energy **160**, 122–141 (2018)

22. Mancarella, P.: Smart multi-energy grids: concepts, benefits and challenges. In: 2012 IEEE Power and Energy Society General Meeting, pp. 1–2. IEEE (2012)
23. Mnih, V., et al.: Human-level control through deep reinforcement learning. Nature **518**(7540), 529–533 (2015)
24. Mocanu, E., et al.: On-line building energy optimization using deep reinforcement learning. IEEE Trans. Smart Grid **10**(4), 3698–3708 (2018)
25. Morari, M., Lee, J.H.: Model predictive control: past, present and future. Comput. Chem. Eng. **23**(4–5), 667–682 (1999)
26. Parisio, A., Rikos, E., Glielmo, L.: A model predictive control approach to microgrid operation optimization. IEEE Trans. Control Syst. Technol. **22**(5), 1813–1827 (2014)
27. Pflaum, P., Alamir, M., Lamoudi, M.Y.: Comparison of a primal and a dual decomposition for distributed MPC in smart districts. In: 2014 IEEE International Conference on Smart Grid Communications (SmartGridComm), pp. 55–60. IEEE (2014)
28. Qin, J., Han, X., Liu, G., Wang, S., Li, W., Jiang, Z.: Wind and storage cooperative scheduling strategy based on deep reinforcement learning algorithm. In: Journal of Physics: Conference Series, vol. 1213, p. 032002. IOP Publishing (2019)
29. Sallab, A.E., Abdou, M., Perot, E., Yogamani, S.: Deep reinforcement learning framework for autonomous driving. Electron. Imaging **2017**(19), 70–76 (2017)
30. Silver, D., et al.: Mastering the game of go with deep neural networks and tree search. Nature **529**(7587), 484–489 (2016)
31. Silver, D., Lever, G., Heess, N., Degris, T., Wierstra, D., Riedmiller, M.: Deterministic policy gradient algorithms. In: International Conference on Machine Learning, pp. 387–395. PMLR (2014)
32. Sogabe, T., et al.: Smart grid optimization by deep reinforcement learning over discrete and continuous action space. In: 2018 IEEE 7th World Conference on Photovoltaic Energy Conversion (WCPEC)(A Joint Conference of 45th IEEE PVSC, 28th PVSEC & 34th EU PVSEC), pp. 3794–3796. IEEE (2018)
33. Stănişteanu, C.: Smart thermal grids-a review. The Scientific Bulletin of Electrical Engineering Faculty 1(ahead-of-print) (2017)
34. Sutton, R.S., Barto, A.G.: Reinforcement Learning: An Introduction. MIT Press, Cambridge (2018)
35. Sutton, R.S., Barto, A.G., et al.: Introduction to Reinforcement Learning, vol. 135. MIT Press, Cambridge (1998)
36. Tuballa, M.L., Abundo, M.L.: A review of the development of smart grid technologies. Renew. Sustain. Energ. Rev. **59**, 710–725 (2016)
37. Vecerik, M., et al.: Leveraging demonstrations for deep reinforcement learning on robotics problems with sparse rewards. arXiv preprint arXiv:1707.08817 (2017)
38. Wang, T., Kamath, H., Willard, S.: Control and optimization of grid-tied photovoltaic storage systems using model predictive control. IEEE Trans. Smart Grid **5**(2), 1010–1017 (2014)
39. Yang, L., Entchev, E., Rosato, A., Sibilio, S.: Smart thermal grid with integration of distributed and centralized solar energy systems. Energy **122**, 471–481 (2017)
40. Yu, L., et al.: Deep reinforcement learning for smart home energy management. IEEE Internet Things J. **7**(4), 2751–2762 (2019)

41. Zhang, B., Hu, W., Cao, D., Huang, Q., Chen, Z., Blaabjerg, F.: Deep reinforcement learning-based approach for optimizing energy conversion in integrated electrical and heating system with renewable energy. Energ. Convers. Manage. **202**, 112199 (2019)
42. Zhang, T., Luo, J., Chen, P., Liu, J.: Flow rate control in smart district heating systems using deep reinforcement learning. arXiv preprint arXiv:1912.05313 (2019)

A k-mer Based Sequence Similarity for Pangenomic Analyses

Vincenzo Bonnici(✉), Andrea Cracco, and Giuditta Franco

Department of Computer Science, University of Verona, 37134 Verona, Italy
{vincenzo.bonnici,giuditta.franco}@univr.it

Abstract. In this work we propose an approach to improve the performance of a current methodology, computing k-mer based sequence similarity via Jaccard index, for pangenomic analyses. Recent studies have shown a good performance of such a measure for retrieving homology among genetic sequences belonging to a group of genomes.

Our improvement is obtained by exploiting a suitable k-mer representation, which enables a fast and memory-cheap computation of sequence similarity. Experimental results on genomes of living organisms of different species give an evidence that a state of the art methodology is here improved, in terms of running time and memory requirements.

Keywords: AF sequence similarity · Genomic dictionary · Jaccard index · k-mer content · Pangenome

1 Introduction

The importance of pangenomic analysis has been shown in several fields, including clinical applications, where it is employed to identify drug targets in vaccines and antibacterials [15,18], to investigate pathogens in epidemic diseases [11], to detect strain-specific virulence factors [9]. More specifically, gene-based pangenomic analysis aims at identifying homologous genes, and their biological relevance, by means of their presence or absence within a group of genomes [20]. Genetic inheritance, internal gene duplication, loss of genetic material as well as horizontal transmission of genetic information make the identification of gene families a difficult task [17,19]. Namely, it has been shown to be an NP-hard problem [16], mainly due to the fact that all-against-all comparisons between gene sets are required to solve the task.

Given a living organism (also called an *isolate*), the genetic material contained in its genome can be transmitted in two different ways. In vertical transmission, genes are transmitted from one organism to its descendant during the reproduction process. In horizontal transmission, frequent in simple organisms such as viruses and bacteria, the exchange of genetic material occurs between two contemporary living organisms. The transmission copies the genetic sequence from the donor to the receiver. Alterations may appear in the copied sequence in form of deletions, insertions, and substitutions. Moreover, one or more copies

G. Nicosia et al. (Eds.): LOD 2021, LNCS 13164, pp. 31–44, 2022.
https://doi.org/10.1007/978-3-030-95470-3_3

of the genes can be produced. The percentage of alteration may vary from one gene to another, thus some genes may be more altered than others. A copy of a gene g that has been transmitted from one genome to another is said to be orthologous to g. Different copies of a gene within the same genome are said to be paralogous. Genes transmitted via horizontal gene transfer are a particular type of orthologs called xenologs.

Gene-oriented pangenomic analyses aim at reconstructing the homology relationships (paralogy, orthology and xenology) among the genes contained in given genomes. The result is the clustering of genes into groups, also called gene families, such that the genes of a family are linked each other by an homology relation. In this context, homology is intended as sequence similarity among genetic sequences, that does not necessary reflect the functional homology of genes. Several measures can be applied for computing sequence similarity. The traditional and most known one is the BLAST score, computed by aligning two sequences, while other more recent measures are based on alignment free approaches.

In a recent study [5] several methodologies for discovery of pangenomic content have been compared by investigating their ability to capture pangenomic information. Genetic sequences of genomes were built by simulating an evolutionary process and by varying simulation parameters, in such a way that the obtained sequences showed similar properties of real ones. The sensibility of several methods was estimated over benchmarks with these synthetic genes, by varying the number of involved genomes, and by dealing with gene loss/acquisition and sequence alteration probability. The analysis in [5] showed that the two methods GET_HOMOLOGUES [8] and PanDelos [3] outperform the other approaches. The first one uses an alignment-based sequence similarity, computed via BLAST scores. PanDelos instead computes an alignment-free based sequence similarity, by means of a Jaccard index, which measures the k-mer content of genetic sequences. In general, very efficient solutions for the computation of BLAST score exist in the literature, while few efforts have been made for optimizing most recent k-mer based approaches. Both the algorithms are made of several steps in which similarity is computed and then exploited for retrieving the homology relations. PanDelos has shown a better performance in those experiments in which synthesized genomes have their compositional properties most similar to real organisms. However, in some situations it requires a relatively high amount of computational resources.

In this paper we focus our attention to radically reduce the computational requirements of PanDelos methodology, by replacing algorithms and data structures for computing the sequence similarity with more efficient ones. Besides, the proposed approach for k-mer content representation and processing may be implemented to other k-mer based similarity computation. Fast solutions for computing Jaccard index have been proposed in the literature [13], also for computing such measures in set of genomes [1,14]. However, such solutions take into account the problem of computing the measures among sets rather than among multisets, namely only k-mer presence/absence is considered and k-mer multiplicity is discarded. On the other hand, the approach here developed, based on Jaccard index to compute sequence similarity among multisets, shows an efficient performance in terms of time and space complexity on benchmarks of real genomes. In Sect. 2 a background of terminology and essential notions on PanDelos approach are given, while main results on our new methodology are detailed in Sect. 3. Experimental results on real pangenomes are reported in Sect. 4, whereas Sect. 5 concludes the paper with some relevant comments, also on our future work.

2 Background: Notations and PanDelos

A genome may be represented as the whole set of genetic material that is contained in it, that is, a set of genetic sequences, which are represented as strings over a given alphabet. Genetic sequences are *stored* in the genomic sequence of an organism as genes, namely contiguous linear sub-regions of its DNA (or RNA) material. The alphabet, here denoted by Γ, may be composed by four or twentyone symbols, as in the case of (DNA or RNA) nucleotides or aminoacids, respectively. In the next section we will report our experiments executed on both nucleotidic and amonoacidic sequences. The well known genetic code assignes to specific triplets of nucleotides a corresponding aminoacid.

We assume to be given a pangenome, which is a set of genomes (isolates from one of more species for example) with all their respective genes collected in $G = \{g_1, g_2, \ldots, g_n\}$. PanDelos is a computational tool for retrieving gene families from a given set of genomes. It uses an alignment-free measure of sequence similarity that is based on the k-mer content of genetic sequences, that is the set of all its k-factors, also called genetic k-dictionary. For a specific value of k, the complete set of k-mers (substrings long k) within a string g is denoted by $D_k(g)$.

Given a k-mer α, the set of positions of the occurrences of α in a gene g is given by $p(\alpha, g) = \{i : g[i, i+k] = \alpha\}$. The multiplicity of α in g (i.e., the number of its occurrences) is defined as

$$m(\alpha, g) = |p(\alpha, g)|$$

Depending on the chosen value of k and on the (nucleotidic or amino acidic) level of analysis, the number of k-mers that can be extracted in a pangenomic analysis can be very high. Let us recall that for a sequence of length m, at most $m-k+1$ k-mers may be extracted, which corresponds to the peculiar case that all the k-mers of the dictionary are unique (also called *hapaxes*), for each position of the sequence. Factors with multiplicity greater than one are called *repeats* and are very informative for pangenomic studies. Indeed, in a pangeomic analysis, the aim is to capture the similarity among sequences, thus the chosen value of k must ensure that a certain level of repetitiveness among the sequences can be assumed [10]. Thus, in our computations, where multiplicity of each k-mer is recorded, the number of k-mers $|D_k(g)|$ of a gene $g \in G$ usually is much less than its maximum value $(m-k+1)$.

A critical aspect in the entire analysis is therefore the choice of the value k, which is in PanDelos is set to be $log_{|\Gamma|} \sum_{i=1}^{n} |g_i|$ [3]. This choice is fruit of information theoretic based investigations and experiments on several real genomes developed in some previous works (see for example [4,7]). It is the logarithm of the sum of the lengths of all the genes in G present in the given pangenome, having as a base the cardinality of the alphabet of the sequences. For example, in real applications involving bacterial amino acid sequences, the value of k usually ranges between 5 and 8, depending on the length of the genetic sequences and on the number of genomes examined. However, due to possible frameshift mutations, it is helpful to analyse nucleotidic sequence too, which means that k can be 3 times larger. In conclusion, the number of k-mers involved in pangenomic analyses is relatively high and requires specialized data structures for retrieving them.

Representation of *k*-mer Dictionaries

PanDelos uses an enhanced suffix array data structure, that is described in [2], by which it is possible to enumerate the complete set of k-mers of a string, and their multiplicity, in linear time w.r.t. the length of the string. The data structure enhances a classical suffix array with an LCP (longest common prefix) array for efficient k-mer enumeration. Moreover, the data structure is equipped with a further document listing array for determining the set of sequences in which a k-mer occurs, and with an additional array for discarding k-mers that are not in the given alphabet. In fact, real genomic sequences can contains ambiguous loci in which the correct nucleotide has not been identified, thus an extra symbol is inserted (usually an N). However, pangenomes presented in the experiments of next section contain all perfectly sequenced strings (that is over a regular alphabet of four or twentyone symbols).

Because a genome can be thousands or millions nucleotide longs, and because thousands of genes can be contained in a genomes, these arrays are implemented as arrays of integers, which means that 4×4 bytes of memory for genome position are required. Searching a k-mer in such a data structure involves a recursive binary search that has $k \cdot logm$ complexity, where m is the length of the indexed string.

Sequence Similarity Computation

Sequence similarity between two genes, g_i and g_j with $i, j = 1, \ldots n$, is computed as the generalized Jaccard similarity, given by

$$J(g_i, g_j) = \frac{\sum_{\alpha \in D_k(g_i) \cup D_k(g_j)} min(m(\alpha, g_i), m(\alpha, g_j))}{\sum_{\alpha \in D_k(g_i) \cup D_k(g_j)} max(m(\alpha, g_i), m(\alpha, g_j))}$$

It equals 1 if the two sequences have the same k-mer content, and 0 otherwise.

For each pair of distinct genes involved in the analysis, the similarity score must be computed. Then, a threshold is applied to the score for retrieving an initial set of homology relationships. The threshold takes into account the percentage of k-mers that is in common between two sequences (by computing such a percentage via dictionary intersection weighted by k-mer multiplicities). Finally, PanDelos builds a network based on such relations, and applies a community detection algorithm for retrieving the final set of gene families. This last computational phase is due to the case that a common ancestor may exist, for a given set of living organisms, even for organisms evolved in different environments. This fact results into a difference of the alteration levels that are own by a given genome. Thus, the level of similarity among genes belonging to the same family may not be homogeneous. For this reason, pangenomic discovery methodologies often involve the application of clustering techniques or community detection algorithms, or approaches for detecting the most suitable value for thresholds that are applied in order to define two gene as homologues.

PanDelos computes gene similarity by a two-by-two genome comparison, in which two sets of genetic sequences $\{g_1, \ldots, g_n\}$ and $\{h_1, \ldots, h_p\}$ belonging to two genomes (G and H respectively) are compared. An index structure is built for the concatenation of the genetic sequences contained in both genomes. Only bidirectional best hits (BBH) are extracted, that is an homology relation between two genes, $g \in G$ and $h \in H$, such that g is the most similar gene in G for h, and viceversa. For this purpose, a similarity matrix M is built for storing gene-by-gene similarities and for determining BBHs. The matrix has $|G| + |H|$ rows and $|G| + |H|$ columns, because orthologs of both genomes are also investigated.

The reason behind the choice of a two-by-two instead of an all-vs-all comparison is mainly due to the memory requirement of the enhanced array. The enhanced suffix array used by PanDelos is an efficient way for enumerating the k-mers of a sequence independently of the value of k, without re-indexing the sequence on changing such a value. However, for pangenomic analyses where the value of k is initially fixed the complexity of the data structure may result excessive. In next section a neat improvement of the computational efficiency in terms of time and space complexity is introduced.

3 A Computationally Efficient Approach

The main goals of the present study is to propose an alternative methodology for computing sequence similarity that allows to run pangenomic analyses on larger inputs by reducing the computational requirements (time and memory). To reach the goals, the suffix-array structure (namely used by PanDelos) is substituted by a data structure that is less expensive in memory and that allows to list the k-mers that are present in a given set of sequences for a specific value of k. Moreover, an inverted list allows to determine the list of sequences in which a given k-mer is present. Differently than in previous approaches, the similarity matrix M is computed row-by-row rather than entirely.

Representation of k-mer Dictionaries

The proposed methodology extracts the k-dictionary, containing a portion of repeats, from a all given genetic sequences. A compact representation of the dictionary turns out useful, since a huge number of k-mers has to be extracted, specially in nucleotidic analyses.

We identify each k-mer by its position in a given order induced over Γ^k (which comprehends the number of all different k-mers). The first k-mer in such an order will be identified by the number 0, the second one by the number 1, and so forth, by means of the function $ord(\alpha) : \Gamma^k \rightarrow \mathbb{N}$ which assigns a number in the interval $[0 \dots |\Gamma|^k - 1]$ to each k-mer. For $k = 1$ the function in manually defined, for example, for the nucleotidic alphabet it is set as $A \mapsto 0, C \mapsto 1, G \mapsto 2, T \mapsto 3$. For $k > 1$, given a k-mer $\alpha \in \Gamma^k$, the function is defined as

$$ord(\alpha) = \sum_{i=0}^{k-1} ord(\alpha[i]) |\Gamma|^i$$

where $\alpha[i]$ is the $(i + 1)$-th character of α. In the following, $ord(\alpha)$ is simply referred to as the *order* of the k-mer α.

Modern computer architectures use a 64-bits number representation. It means that given an alphabet Γ, the maximum k for which k-mers can be identified with at most 64 bits equals $\lfloor log_{|\Gamma|} 2^{64} \rfloor$. It is 32 for the nucleotidic alphabet, and 14 for 21 amino acids. PanDelos is manly applied to bacterial and viral pangenomes. Bacteria have millions long genomes, with an high percentage of genetic coverage. Since k is chosen as the logarithm (with base equal to the cardinality of the alphabet) of the total genetic material analysed, it is reasonable to estimate that a pangenomic analysis with hundreds or thousands of bacteria can exceed the limit given by the 64-bits representation. For this reason, we implemented an **alternative hash-based order representation**. It exploits the rolling hash function that has been originally proposed for the Rabin-Karp string search algorithm [12]. Given the $(i + 1)$-th k-mer α_i in the lexicographic order induced over Γ^k, the hashing H of α_i is defined as:

$$H(\alpha_i) = R(\alpha_i) \bmod m$$

where $R(\alpha_i)$ is defined as

$$R(\alpha_i) = (R(\alpha_{i-1}) - \alpha_{i-1}[1] |\Gamma|^{k-1}) |\Gamma| + \alpha_i[k - 1].$$

We set the values of m as 18446744073709551557, that is the largest prime number within 2^{64} (see https://primes.utm.edu/lists/2small/0bit.html). Thus, for a chosen resolution k such that $|\Gamma|^k > 2^{64}$, we set $ord(\alpha_i) = H(\alpha_i)$.

Hashing introduces collisions among hashed objects. In our case, it means that two different k-mers are identified by the same hash code, thus they are recognized as the same k-mer. An upper-bound to collision probability can be estimated as the *birthday problem* [6], which originally aims at calculating the probability that two persons, in a finite group of people, are born in the same day. Since the choice of k is made such that a certain level of repetitiveness is produced, it is reasonable to estimate that, for large values of k, the size of the dictionary is much smaller than $|\Gamma|^k$. Namely, a relatively small portion of the possible k-mers is present in the input pangenome, and because of their nature they may not be uniformly distributed in $[0 \ldots |\Gamma|^k - 1]$. Therefore, theoretical collision calculation results to be not suitable in practice.

Sequence Similarity Computation

The proposed similarity computes **one gene-vs-group similarity at time**, instead of two genomes at time (thus storing the big matrix M). This means that the similarity between a specific gene and a group of genetic sequences, that can be a part of or the whole complete input, is computed in a single task.

Because BBHs must be extracted, for each gene we need to take trace of the most similar genes to it in each input genome. Then, when the similarity is reciprocally conformed to be the highest one for the compared genes, than a BBH record is extracted. It implies that we need to store the list of most similar genetic sequences for each gene. Let us consider two observations at this point. For each gene, very few genes in another genome have the highest similarity with it. In fact, we can assume that an ortholog plus a very few copies of it are present in the genome. Moreover, a threshold on the Jaccard similarity between two genes is applied by PanDelos, consequently the number of homology relations to be investigated for BBHs is further reduced.

Given a subgroup $\hat{G}$ of the input genetic sequences G, a list $L(\hat{G})$ is built such that it contains triplets in the form $(ord(\alpha), id(g), m(\alpha, g))$, where $\alpha \in D_k(g)$, with $g \in \tilde{\mathbb{G}}$, and $id(g)$ is a numerical identifier for each input gene g of the pangenome. The list contains the complete set of k-mers that are present in the input group. If a k-mer is repeated in more than one gene, then multiple triplets report its multiplicity in each gene in which it is present. Subsequently, the list is sorted by using the radix sort algorithm such that the order between two triplets if defined by the lexicographic order among their contained k-mers. At this point, an inverted list $\overleftarrow{L}(\hat{G}))$ is built. The aim of the inverted list is to report for each k-mer, identified by its order, the limits of the range corresponding to it in the list L. In fact, given a k-mer α, since L is sorted by $ord(\alpha)$, all the occurrences of α are stored consecutively. L is implemented as an array of triplets, instead, $\overleftarrow{L}$ is more similar to an hash table. Since multiple k-mers may fall into the same position, a list of k-mers (by means of their orders) and their corresponding range limits are stored for that position.

Once $L(\hat{G})$ and $\overleftarrow{L}(\hat{G})$ are built, for each gene $g \in G$ its similarity with each gene in $\hat{G}$ is computed. A vector of size $|\hat{G}|$ is initialized to be 0 in each position. Then, for each k-mer $\alpha \in D_k(g)$, it is searched in $\overleftarrow{L}(\hat{G})$ such that its interval in L is retrieved. Subsequently, for each triplet in the interval, the corresponding counter in the similarity vector is updated. Then, the threshold is applied.

If $\hat{G}$ equals G, simply the most similar genes are extracted. Otherwise, the pangenome is partitioned into some disjoint subgroups of genes. The most similar genes in one first subgroup, together with their similarity measures, are temporarily stored. Then, such similarities are iteratively refined by scanning on the successive subgroup. If one or more genes with higher similarity are found in a successive group, then the list of most similar genes to a given g is entirely redefined by those new genes.

Since $\overleftarrow{L}(\hat{G})$ is very similar to an hash table, it can be accessed in amortized constant time. Moreover, the retrieving of k-dictionaries, and the calculation of the multiplicity of the k-mers in any $D_k(g)$, is performed in a way similar to the construction of L. A list of k-mers orders, with duplicates, is built and sorted, then successive runs of the same order identify the k-mers and their multiplicity.

Discussion on the Asymptotic Complexity

For a given input of n genomic positions, the original PanDelos methodology needs to compute 4 indexing arrays, each one of size n. Each array can be computed in a time that is linear to n. The total resulting memory cost is four time the cost of an integer for each position. If 64 bits are used for representing integers, 8×4 bytes per position are required. However, the current Java implementation uses 32-bits integers, thus the requirement is 4×4 bytes per position. The similarity computation is performed by a two-by-two genome comparison. For each comparison an indexing structure is built, then it is iterated in linear time, and temporary similarities are stored in a matrix which size depends on the number of genes that are present in the two compared genomes.

Relevantly, the complexity of the proposed approach does not linearly depends on n but it depends on the number of distinct k-mers that are extracted in each sequence. In fact, the array L stores, for each distinct k-mer, the order of the k-mers, the id of the gene where it is present and its multiplicity within the gene. In addition, the hash table $\overleftarrow{L}$ stores for each k-mer the interval limits. Thus, for each k-mer 64 bits (for the order) plus 5×4 bytes for the other information are required, plus an additional cost for structuring the hash table. If all the extracted k-mers are hapaxes, an overhead of 3 bytes per position/k-mer is introduced by the new approach. However, the choice of k ensures a certain level of repetitiveness of the extracted k-mers [7], thus it is expected that the number of distinct k-mers is relatively low w.r.t. the sequence length. Thus, the overhead should be amortized. The new approach does not require the storing of the complete matrix M because each gene is evaluated separately, but a few additional memory cost is required for storing temporary homologies. For what concerns the running time, the new approach uses radix sort for sorting L. A linear time complexity is required for extracting the k-mers and for searching them in $\overleftarrow{L}$. Thus, the asymptotic complexity is still linear. The number of distinct k-mers expected to be present in a genetic sequence can be difficulty predicted, which makes the definition of the theoretical bounds of the approach a difficult task. However, previous studies suggest that k-mers repetitiveness should be relatively high [3]. In any case, the original methodology computes two-by-two genomes comparisons, which introduces a quadratic factor in the analysis, whereas the new approach is not affected by such a quadratic factor.

Implementation

The proposed approach is implemented in C++ and its is interfaced with the original Java source of PanDelos via the Java Native Interface (JNI). Because no external libraries are required, the portability of the software is maintained, since nowadays a version of the g++ compiler exists for almost every modern operating system.

4 Experimental Results

We compared the proposed approach with the original PanDelos methodology on real pangenomes. This first comparison shows the performance difference over the case studies that were evaluated in the original work [3]. In the following experiments, 4 different pangenomes were investigated. Each pangenome regards a different bacterial species or genus and involves a different number of isolates and thus genetic sequences. Their composition is reported in Table 1.

Table 1. Statistics regarding the 4 investigated real pangenomes.

Pangenome	Genomes	Sequences	Nucleotides
Escherichia	10	46,587	43,764,342
Mycoplasma	64	46,760	49,109,138
Salmonella	7	30,074	27,368,957
Xanthomonas	14	55,189	56,770,477

Table 2 reports main statistics regarding the comparison of the two approaches (PanDelos and the new methodology explained in previous section) on the pangenome in Table 1 in terms of temporal and storage performance in searching homologies among amino acidic sequences. Results show that there is a speed-up of the new approach w.r.t. the original one (see the Time speed-up column of Table 2) from 43 to 77 the running times. Moreover, the proposed approach saves from 2 to 8 times the memory that is used by the original methodology (see the last RAM gain column of Table 2).

Table 3 reports statistics obtained by searching for homologies among nucleotidic sequences of the same pangenome. Results show that there is a speed-up of the new approach w.r.t. the original one from 19 to 53 the running times (see the Time speed-up column of Table 3). Moreover, the proposed approach is able to save up to 7.31 times the memory that is used by the original methodology. An exception regards the *Mycoplasma* benchmark where the new approach used a slight larger memory portion. Most likely this phenomenon is due to the high variability of several strains in *Mycopaslma* pangenome, that produces a low percentage of shared k-mers.

Table 2. Running time and memory requirement (in MB) of the proposed approach (*New time* and *New RAM*) and the original PanDelos methodology (*Old time* and *old RAM*). Time is measured as user elapsed time in seconds and memory requirements are measured as max peak of RAM occupancy. Tests were run by comparing amino acid sequences. Time speed-ups (*Time speed-up*) were calculated as *Old time*/*New time*, and memory gains (*RAM gain*) were calculated as *Old RAM*/*New RAM*.

Pangenome	New time	New RAM	Old time	Old RAM	Time speed-up	RAM gain
Escherichia	5.34	1035	355.48	7736	66.56	7.47
Mycoplasma	6.40	1267	487.54	2980	76.17	2.35
Salmonella	3.26	697	143.00	5802	43.86	8.32
Xanthomonas	7.64	1193	507.77	7349	66.46	6.16

Table 3. Running time and memory requirement (in MB) of the proposed approach (*New time* and *New RAM*) and the original PanDelos methodology (*Old time* and *old RAM*). Time is measured as user elapsed time in seconds and memory requirements are measured as max peak of RAM occupancy. Tests were run by comparing nucleotidic genetic sequences. Time speed-ups were calculated as *Old time*/*New time*, and memory gains were calculated as *Old RAM*/*New RAM*.

Pangenome	New time	New RAM	Old time	Old RAM	Time speed-up	RAM gain
Escherichia	16.85	2,177,056	458.35	10,339,360	27.20	4.75
Mycoplasma	19.48	2,491,312	1047.95	1,852,404	53.80	0.74
Salmonella	9.67	1,316,500	192.18	9,618,568	19.87	7.31
Xanthomonas	22.64	2,818,496	688.42	6,835,976	30.41	2.43

In the following we show the results obtained by evaluating the performance of the proposed method while varying the input pangenome size.

Figure 1 reports the running time of the two compared approaches. The time of the original PanDelos versions is proportional to the number of genomes that are involved in the pangenomic analysis. In fact, it compares two genomes at time inducing a quadratic time complexity for the analysis. On the other hand, the running time of the proposed approach is not affected by the quadratic complexity, and its running time depends only on the input size.

The quadratic complexity does not affect the memory requirements of both approaches, that are shown in Fig. 2. The original PanDelos implementation is based on an enhanced suffix array data structure which size is correlated with the number of indexed nucleotides. However, since a part of both implementations is developed in Java, the exact memory requirement cannot be measured because it is affected by the Java garbage collector policy. This fact may be the

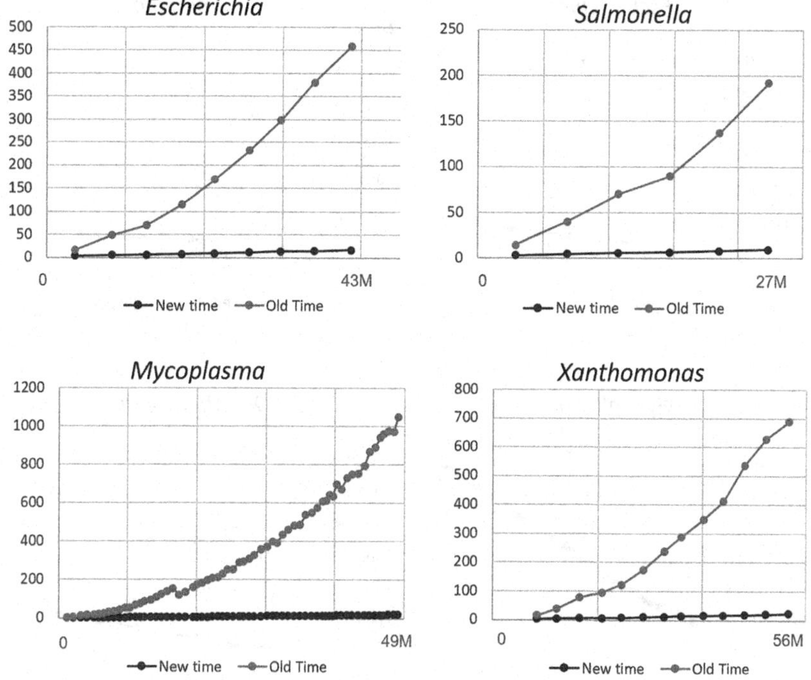

Fig. 1. Running time (Y axis, seconds) of the proposed approach (*New time*) and the original PanDelos version (*Old time*). The occupancy is measured by varying the number of input genomes, which correspond to a variation on the total number of nucleotides (X axis) involved in the analysis.

cause of a slight increase in memory occupancy of the original version. The new approach has a clear dependency of the RAM occupancy w.r.t. the input size. However, its performance slightly increases w.r.t. that of the original version, with only one exception given by the *Mycoplasma* collection. Here we notice that the *Mycoplasma* case requires less than 3 Gigabytes, while the other three pangenomes require a RAM occupancy closed to 14 Gigabytes.

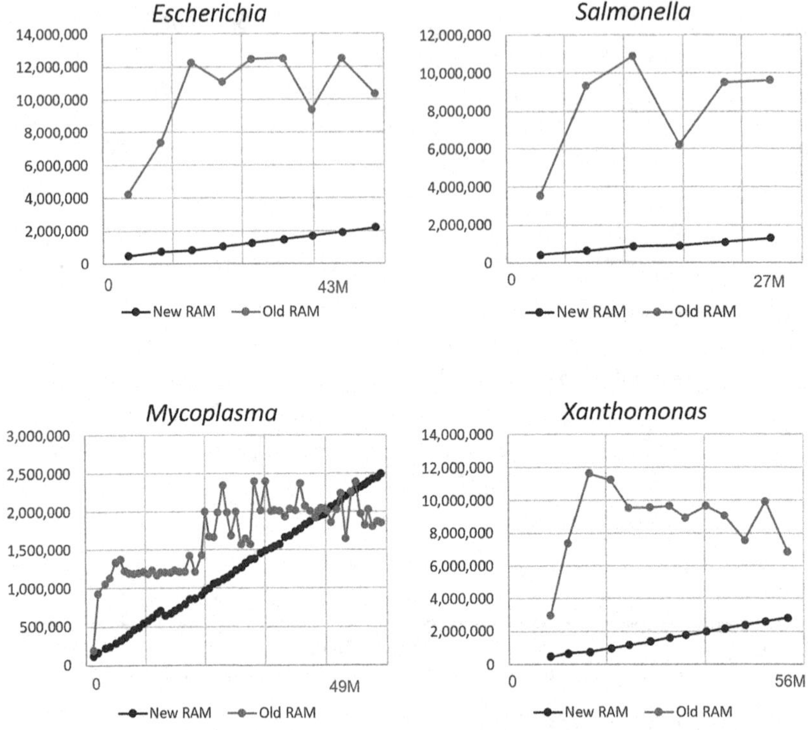

Fig. 2. RAM occupancy (Y axis, peak max in bytes) of the proposed approach (*New RAM*) and the original PanDelos version (*Old RAM*). The occupancy is measured on varying the number of input genomes which correspond to a variation on the total number of nucleotides (X axis) involved in the analysis.

5 Conclusions

The search for sequence homology is a current challenge in computational analyses of pangenomes, concerning both computational performance and ability to capture pangenomic information. k-mer-based approaches have recently shown good performance regarding the detection of genetic homology. However, such innovative solutions often lack in reducing computational requirements. Here, we proposed an approach for improving the performance of an existing methodology. A suitable representation of the k-mer content of a pangenome allows the development of a procedure for computing genetic sequence similarity that results fast and with low memory requirements. Temporal speed-ups from 19x to 76x are shown on real pangenomes, as well as a gain in RAM memory occupancy (up to 7.31 times) for almost every benchmark.

Finally, we point out that the proposed approach regards a efficiency improvement only for the computation of sequence similarity, while the complete PanDelos methodology is composed of two steps: sequence similarity computation and community detection phase. We are interested to further improve the whole

PanDelos performance in terms of parallelization. An initial trivial parallelization of the similarity computation procedure has been here implemented by using a Java thread pool, where each Thread processes a genome-vs-pangenome at time. Namely, each thread compares the genes of a given genome with the genetic sequences of the entire pangenomes. Indeed, the inverted list allows to determine the list of sequences in which a given k-mer is present. Differently than in previous approaches, the similarity matrix M in this paper is computed row-by-row rather than entirely, and this dynamical behaviour of the algorithm allowed us to also develop a simple procedure for parallel computation of sequence similarity, which will be further developed in future work.

References

1. Besta, M., et al.: Communication-efficient jaccard similarity for high-performance distributed genome comparisons. In: 2020 IEEE International Parallel and Distributed Processing Symposium (IPDPS), pp. 1122–1132. IEEE (2020)
2. Bonnici, V., Manca, V.: Infogenomics tools: a computational suite for informational analysis of genomes. J. Bioinforma Proteomics Rev. **1**, 8–14 (2015)
3. Bonnici, V., Giugno, R., Manca, V.: PanDelos: a dictionary-based method for pangenome content discovery. BMC Bioinformatics **19**(15), 437 (2018)
4. Bonnici, V., Manca, V.: Informational laws of genome structures. Sci. Rep. **6**, 28840 (2016). http://www.nature.com/articles/srep28840
5. Bonnici, V., Maresi, E., Giugno, R.: Challenges in gene-oriented approaches for pangenome content discovery. Brief. Bioinformatics **22**(3), bbaa198 (2020)
6. Borja, M.C., Haigh, J.: The birthday problem. Significance **4**(3), 124–127 (2007)
7. Castellini, A., Franco, G., Milanese, A.: A genome analysis based on repeat sharing gene networks. Nat. Comput. **14**(3), 403–420 (2014). https://doi.org/10.1007/s11047-014-9437-6
8. Contreras-Moreira, B., Vinuesa, P.: GET_HOMOLOGUES, a versatile software package for scalable and robust microbial pangenome analysis. Appl. Environ. Microbiol. **79**(24), 7696–7701 (2013)
9. D'Auria, G., Jiménez-Hernández, N., Peris-Bondia, F., Moya, A., Latorre, A.: Legionella pneumophila pangenome reveals strain-specific virulence factors. BMC Genom. **11**(1), 181 (2010)
10. Franco, G., Milanese, A.: An investigation on genomic repeats. In: Bonizzoni, P., Brattka, V., Löwe, B. (eds.) CiE 2013. LNCS, vol. 7921, pp. 149–160. Springer, Heidelberg (2013). https://doi.org/10.1007/978-3-642-39053-1_18
11. Holt, K.E., et al.: High-throughput sequencing provides insights into genome variation and evolution in salmonella typhi. Nat. Genet. **40**(8), 987–993 (2008)
12. Karp, R.M., Rabin, M.O.: Efficient randomized pattern-matching algorithms. IBM J. Res. Dev. **31**(2), 249–260 (1987)
13. Kobayakawa, M., Kinjo, S., Hoshi, M., Ohmori, T., Yamamoto, A.: Fast computation of similarity based on jaccard coefficient for composition-based image retrieval. In: Muneesawang, P., Wu, F., Kumazawa, I., Roeksabutr, A., Liao, M., Tang, X. (eds.) PCM 2009. LNCS, vol. 5879, pp. 949–955. Springer, Heidelberg (2009). https://doi.org/10.1007/978-3-642-10467-1_87
14. Lees, J.A., et al.: Fast and flexible bacterial genomic epidemiology with poppunk. Genome Res. **29**(2), 304–316 (2019)

15. Muzzi, A., Masignani, V., Rappuoli, R.: The pan-genome: towards a knowledge-based discovery of novel targets for vaccines and antibacterials. Drug Discov. Today **12**(11), 429–439 (2007)
16. Nguyen, N., et al.: Building a pan-genome reference for a population. J. Comput. Biol. **22**(5), 387–401 (2015)
17. Puigbò, P., Lobkovsky, A.E., Kristensen, D.M., Wolf, Y.I., Koonin, E.V.: Genomes in turmoil: quantification of genome dynamics in prokaryote supergenomes. BMC Biol. **12**(1), 66 (2014)
18. Serruto, D., Serino, L., Masignani, V., Pizza, M.: Genome-based approaches to develop vaccines against bacterial pathogens. Vaccine **27**(25), 3245–3250 (2009)
19. Soucy, S.M., Huang, J., Gogarten, J.P.: Horizontal gene transfer: building the web of life. Nat. Rev. Genet. **16**(8), 472–482 (2015)
20. Tettelin, H., Medini, D.: The Pangenome: Diversity, Dynamics and Evolution of Genomes. Springer, Cham (2020). https://doi.org/10.1007/978-3-030-38281-0

A Machine Learning Approach to Daily Capacity Planning in E-Commerce Logistics

Barış Bayram[1(✉)], Büşra Ülkü[1], Gözde Aydın[1], Raha Akhavan-Tabatabaei[2], and Burcin Bozkaya[2,3]

[1] HepsiJET, Istanbul, Turkey
{baris.bayram,busra.ulku,gozde.aydin}@hepsijet.com

[2] Sabanci Business School, Sabanci University, Istanbul, Turkey
{akhavan,bbozkaya}@sabanciuniv.edu

[3] New College of Florida, Sarasota, USA
bbozkaya@ncf.edu

Abstract. Due to the accelerated activity in e-commerce especially since the COVID-19 outbreak, the congestion in the transportation systems is continually increasing, which affects on-time delivery of regular parcels and groceries. An important constraint is the fact that a given number of delivery drivers have a limited amount of time and daily capacity, leading to the need for effective capacity planning. In this paper, we employ a Gaussian Process Regression (GPR) approach to predict the daily delivery capacity of a fleet starting their routes from a cross-dock depot and for a specific time slot. Each prediction specifies how many deliveries in total the drivers in a given cross-dock can make for a certain time-slot of the day. Our results show that the GPR model outperforms other state-of-the-art regression methods. We also improve our model by updating it daily using shipments delivered within the day, in response to unexpected events during the day, as well as accounting for special occasions like Black Friday or Christmas.

Keywords: Transportation · E-commerce logistics · Capacity planning · Gaussian process regression · Continual learning

1 Introduction

Recently, the increasingly widespread use of e-commerce sites in the presence of the COVID-19 pandemic has brought about a huge increase in online shopping [1]. The package delivery industry has been significantly affected as a result. The number of deliveries in the transportation sector had already been on the rise and this increase accelerated further with the COVID-19 pandemic. HepsiJET is a fast and practical, technology-driven e-commerce logistics system that aims to provide better quality service to its customers. It was launched in May 2017 and the size of its operations (shipments delivered) rose by 91% from 2018 to 2019 and by 171% from 2019 to 2020.

G. Nicosia et al. (Eds.): LOD 2021, LNCS 13164, pp. 45–50, 2022.
https://doi.org/10.1007/978-3-030-95470-3_4

Machine learning techniques can be applied in a variety of domains in transportation planning and optimization, including capacity planning to cope with the congestion in the logistics systems. To reduce slack capacity in vehicles, a deep reinforcement learning based approach has been developed based on the Deep Q network algorithm to assign shipments to the relevant couriers for efficient delivery [2]. In the logistics area, an optimization model was developed for capacity planning based on the distance of delivery areas, earnings per delivery of carriers, and delivery time to improve short-term multiple food deliveries [3]. In machine learning based studies that deal with this food delivery problem, the aim is to estimate the delivery time and reduce costs [4]. Also, regression models [5] are used to estimate the delivery demands of online orders. The Gaussian Process Regression (GPR) is a non-parametric Bayesian approach [6] that has been employed for real-world capacity prediction problems of electricity load [7] and pile bearing [8], among others. Various studies of real-world applications have proven the advantages of GPR such as capturing all the uncertainties in the data and interpretability between the predictions and observations [6].

In our study, we employ GPR to predict the daily total number of deliveries for each HepsiJET cross-dock in each delivery time slot. Our goal is to reduce the number of delivery delays, improve capacity planning, optimize the necessary delivery actions in advance of possible congestion, and also provide input to route optimization. The GPR model is further updated daily to train on the most recent customer ordering behavior and the unexpected number of deliveries. The predicted value from the model is utilized as the combined capacity of the delivery drivers of the cross-dock to obtain maximum efficiency on the deliveries. For each day, the shipments loaded in the morning on a vehicle are certain, but new grocery and food delivery requests may be received in real-time. Therefore, using the predicted capacity, one can better balance the assignment of deliveries among the drivers operating from the same cross-dock.

The contributions of this study can be listed as follows; i) conducting the first national work in the transportation industry to estimate daily delivery capacity for cross-docks using machine learning, ii) developing one of the first ML-based methods for capacity planning to adapt the abnormal conditions in special situations, iii) the realization of a regression method that continues to learn from new data on shipments delivered within the day, iv) the use of the adaptive method by retraining daily for the capacity planning in the field of transportation.

2 Proposed Approach

For the cross-dock and time slot based prediction of total daily delivery capacity, the first step is data preparation that includes outlier removal, feature vector construction, daily aggregation, and extraction of additional features. In step 2, other features are extracted that include deliveries made in one day, three days and one week prior. Next, a training set is constructed, which is composed of all the deliveries of all the cross-docks for each slot in a training time period. The initial regression model is trained using this training set.

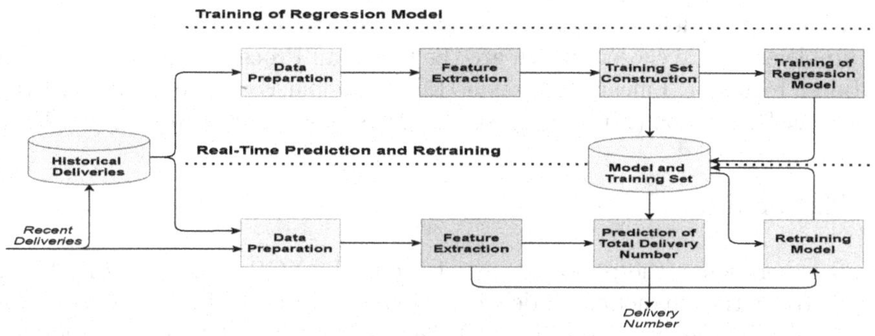

Fig. 1. The proposed approach for predicting total daily delivery capacity.

For data preparation, individual deliveries are aggregated by cross-dock, day and time-slot, and the records with null values are removed. The raw features processed at this stage are the cross-dock, time slot, delivery id, courier id, date, district, and delivery address. These are used to extract additional features such as the time-based attributes (e.g. day of week, month, or year) as well as several historical aggregations of deliveries. All together, 32 features are generated and further reduced to 14 using a p-value based feature selection process for each cross-dock and delivery time slot (see Table 1). The target variable is the total deliveries ("capacity") for a given day ("today"), time-slot, and cross-dock.

Table 1. Selected input features and their descriptions.

Input feature	Description
xdock_id	id of the cross-dock
slot	the time slot in the day
dow	day of week for today, the target prediction day
total_del_num_yest	total deliveries made yesterday
mean_district_num_yest	mean number of districts delivered yesterday
mean_attempt_num_yest	mean delivery attempt made by all couriers yesterday
mean_capacity_yest	mean deliveries of all couriers yesterday
max_capacity_yest	maximum delivery made a courier yesterday
mean_total_del_num_3d	mean total delivery made in the last 3 d
mean_courier_num_3d	mean courier count in the last 3 d
mean_attempt_num_3d	mean number of delivery attempts in the last 3 d
mean_address_num_3d	mean number of unique addresses in the last 3 d
mean_capacity_3d	mean number of deliveries by all couriers in last 3 d
mean_total_del_num_1week	mean total delivery made in the last week
mean_attempt_num_1week	mean number of delivery attempts in the last week
mean_capacity_1week	mean deliveries of all couriers in the last week

For predicting delivery capacity, several regression models employed in different problems [9] are chosen, which are the Gaussian Process Regression (GPR), XGBoost regressor, Linear Regression (LR), Random Forest Regressor (RFR), Multilayer Perceptron (MLP) Regression, and Support Vector Regressor (SVR).

3 Experiments

In our experiments, the performances of the GPR, XGBoost, LR, RFR, MLP, and SVR for the prediction of delivery capacity are evaluated in terms of R^2 and Root Mean-Square Error (RMSE). By testing different kernel functions, a non-stationary kernel is selected for GPR. After parameter tuning for XGBoost, the hyper-parameters, *learning_rate*, *max_depth* and *gamma* are estimated as 0.03, 4 and 0.3, respectively, and for RFR, *max_depth* is selected as 2. Also, the parameters of Random Forest are estimated similarly. For MLP, LR, and linear SVR, the best performances are observed with the default parameters. After the initial experiments, the most appropriate regression model to be utilized in the next step is chosen, for continual learning that uses additional daily training.

4 Results

The experiments for predicting delivery capacity were performed on nine different training and test set pairs specified by the number of months in 2020 included in the training period, to account for unexpected market shifts due to the special occasions and days, COVID-19 restrictions, etc. For the test month m, the deliveries in months January 2020 to $m-1$ are used in training. The average R^2 and RMSEs over all training and test sets are given in Table 2 in which we find that the GPR model has the best performance. We further illustrate these models' performances in Fig. 2(a) for delivery predictions in November and December 2020 further broken down by day of year. The GPR model provided the best R^2 on most of the days in Nov. and Dec. 2020 with noticeable fluctuations.

We used the GPR model with additional recent daily delivery records in 2021 included in a "learning" training set to further refine the performance of our proposed model. In this experiment, improvements on R^2 shown in the Fig. 2(b) were observed in most of the days. The overall R^2 values of the GPR and the learning GPR models were 0.776 and 0.819, respectively, an increase of 5.5%.

Table 2. The average of overall R^2 values and RMSE of the regression algorithms.

Algorithms	Avg. R^2 on training set	Avg. RMSE on training set	Avg. R^2 on test set	Avg. RMSE on test set
GPR	0.863	167.8	**0.833**	**198.2**
XGBoost	0.920	96.1	0.817	228.4
LR	0.884	153.9	0.795	376.6
RFR	0.911	98.0	0.801	251.9
MLP	0.874	157.8	0.807	238.2
SVR	0.852	199.4	0.741	649.2

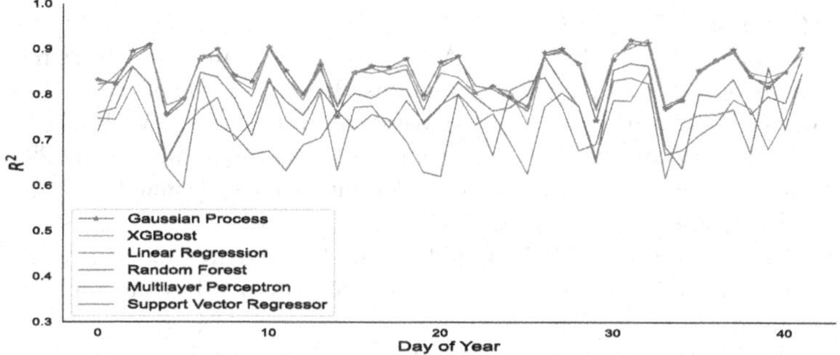

(a) The R^2 values of the regression models.

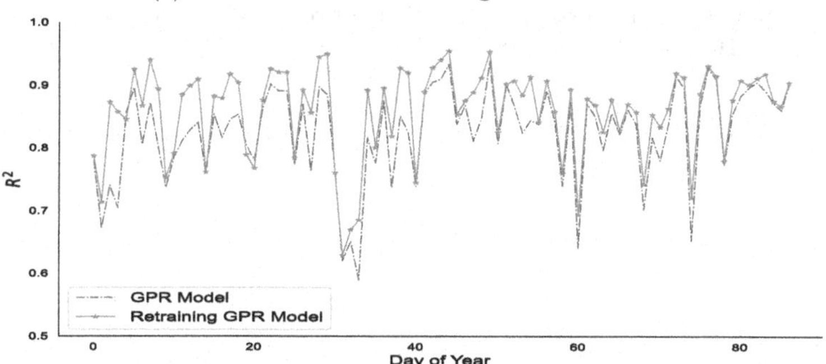

(b) The R^2 values of the GPR and retraining GPR models.

Fig. 2. Prediction performance in weekdays of (a) Nov-Dec 2020 and (b) 2021.

5 Conclusion

In this work, we presented a Gaussian Process Regression (GPR) model for daily delivery capacity planning for a transportation company, HepsiJET. We showed that the performance of the GPR outperforms several state-of-the-art regression models. In future work, we plan to use the predictions as capacities of couriers to make decisions on the optimized delivery of new on-demand food shipments.

References

1. Viu-Roig, M., Alvarez-Palau, E.J.: The impact of E-commerce-related last-mile logistics on cities: a systematic literature review. Sustainability **12**(16), 6492 (2020)
2. Ahamed, T., et al.: Deep Reinforcement Learning for Crowdsourced Urban Delivery: System States Characterization, Heuristics-guided Action Choice, and Rule-Interposing Integration. arXiv preprint arxiv:2011.14430 (2020)
3. Yildiz, B., Savelsbergh, M.: Service and capacity planning in crowd-sourced delivery. Transp. Res. Part C Emerg. Technol. **100**, 177–199 (2019)
4. Moghe, R.P., Rathee, S., Nayak, B., Adusumilli, K.M.: Machine learning based batching prediction system for food delivery. In: 8th ACM IKDD CODS and 26th COMAD, pp. 316–322 (2021)
5. Fabusuyi, T., et al.: Estimating small area demand for online package delivery. J. Transp. Geogr. **88**, 102864 (2020)
6. Schulz, E., Speekenbrink, M., Krause, A.: A tutorial on Gaussian process regression: modelling, exploring, and exploiting functions. J. Math. Psychol. **85**, 1–16 (2018)
7. Karunaratne, P., Moshtaghi, M., Karunasekera, S., Harwood, A.: PSF+-Fast and improved electricity consumption prediction in campus environments. In: 2017 IEEE International Conference on Smart Grid Communications (SmartGridComm), pp. 241–246. IEEE (2017)
8. Momeni, E., Dowlatshahi, M.B., Omidinasab, F., Maizir, H., Armaghani, D.J.: Gaussian process regression technique to estimate the pile bearing capacity. Arab. J. Sci. Eng. **45**(10), 8255–8267 (2020)
9. Sharifzadeh, M., Sikinioti-Lock, A., Shah, N.: Machine-learning methods for integrated renewable power generation: A comparative study of artificial neural networks, support vector regression, and Gaussian Process Regression. Renew. Sustain. Energy Rev. **108**, 513–538 (2019)

Explainable AI for Financial Forecasting

Salvatore Carta, Alessandro Sebastian Podda(✉), Diego Reforgiato Recupero, and Maria Madalina Stanciu

Department of Mathematics and Computer Science, University of Cagliari, Cagliari, Italy
{salvatore,sebastianpodda,diego.reforgiato,madalina.stanciu}@unica.it

Abstract. One of the most important steps when employing machine learning approaches is the feature engineering process. It plays a key role in the identification of features that can effectively help modeling the given classification or regression task. This process is usually not trivial and it might lead to the development of handcrafted features. Within the financial domain, this step is even more complex given the general low correlation between features extracted from financial data and their associated labels. This represents indeed a challenging task that it is possible to explore today through the explainable artificial intelligence approaches that have recently appeared in the literature. This paper examines the potential of machine learning automatic feature selection process to support decisions in financial forecasting. Using explainable artificial intelligence methods, we develop different feature selection strategies in an applied financial setting where we want to predict the next-day returns for a set of input stocks. We propose to identify the relevant features for each stock individually; in this way, we take into account the heterogeneous stocks' behavior. We demonstrate that our approach can separate important features from unimportant ones and bring prediction performance improvements as shown by our performed comparisons between our proposed strategies and several state-of-the-art baselines on real-world financial time series.

Keywords: XAI · Machine learning · Financial forecasting · Time-series

1 Introduction

Given the increasing amount of labeled data generated for different tasks, Machine learning (ML) methods were employed and proved to be enormously successful for making predictions, as opposed to solutions based on canonical regressive models [6]. As such, we have witnessed their increased adoption in several domains, among which the financial one. Despite these advances, relying only on sophisticated ML models trained on massive annotated datasets introduces the risk of creating and using decision systems that work as black boxes that nobody can truly understand. This fact has a direct impact on

G. Nicosia et al. (Eds.): LOD 2021, LNCS 13164, pp. 51–69, 2022.
https://doi.org/10.1007/978-3-030-95470-3_5

ethics, accountability, safety, and industrial liability [16,18]. For relatively simple decision-making applications, such as an online retail recommender system, an algorithm employing classical ML techniques can be considered an acceptable strategy. However, the use of ML in life-changing situations such as risk decisions in the finance sector, diagnostic decisions in healthcare, and safety-critical systems in autonomous vehicles are sensitive and of great interest for businesses and society, requiring to unveil how the black boxes of ML systems work [1]. Therefore, it is essential to have frameworks that understand how ML methods produce their findings and suggestions. As a consequence, the field of eXplainable Artificial Intelligence (XAI) has emerged, that is, techniques developed to "X-ray the black-box" and provide insights about which input features are more important and how they affect the predictions [22]. Explainability might appear straightforward to be applied within the financial domain, given that ML approaches for finance tasks have been extensively used in both research [13] and industry [24]. Nevertheless, explainability is far from easy as XAI research demonstrates [2], and particularly challenging for financial forecasting, where the low signal-to-noise ratio is typical for the finance data. Nevertheless, complementing the ML models with explainability methods in the financial domain leads to the understanding of the key signals the ML models use for their prediction, as well as to the interpretation of the output.

Several scientific communities studied the problem of explaining ML decision models. As such, different perspectives and motivations related to the *explanationability* have been established: explanations to justify, explanations to control, explanations to discover, and finally, explanations to improve classification or regression tasks [1]. In this work, we focus on the latter. In a few words, given that high-dimensional financial data are becoming more available, the need to efficiently select which subset of data represents more valuable information for a forecasting problem is of great interest. Evaluation and verification of such a process are challenging tasks, as financial time-series data are hardly interpretable even by domain experts. Also, when performing the forecasting, it is well known that using too many features can lead to overfitting, which can significantly hinder the performance of predictive models [24]. Understanding which input features are most relevant for the outcome of an ML approach is an essential first step towards its interpretability.

In this paper, we investigate the ability to select relevant features in an applied setting. Specifically, for a large set of stocks, we train Random Forest (RF) models, with lagged returns as features, and aim at forecasting the next-day return. We formulate our problem by building a model for each of the stocks as, in many cases, stocks behavior is highly heterogeneous, and the relevant features may differ between stocks, across market regimes, or geographies [24]. Naturally, by performing feature selection at a stock level, we can select the features that are the most relevant to each stock. Globally selecting the features may perform well, on average, across all stocks, but we prove that choosing the input features at an individualized level outperforms the former.

Therefore, the contributions we bring in this work are:

- We apply an XAI technique based on permuting the values of each feature in the out-of-bag sample to derive the feature importance. On top of it, we propose three strategies to select the best subset of features to remove *for each stock* so to improve the predictive performance both at an individual level (stock) and globally (across the stock set);
- For each proposed strategy, we aim at determining the feature importance *threshold* such that we can divide the features into two subsets: important features that meaningfully contribute to the forecasting task and non-important ones that do not. Thus, each strategy learns across the stock set which features can be removed for which stock, and moreover, whose removal improves the overall prediction performance;
- We determine the efficacy of our strategies by evaluating the change in predictive performance by cross-referencing the generated model with removed features against the base model without any input feature removed. We further evaluate our strategies by comparing their prediction performance against two other feature importance inference baselines: (i) the well-known Mean Decrease Impurity (MDI) of an RF [3], and the (ii) Local Interpretable Model agnostic Explanations (LIME) [25];
- In contrast with some of the most recent works in finance [10,15], we are interested in predicting rather than finding the causes of stock return behavior. Thus, we do not employ any techniques to establish causality claims. We restrict our exercise to a prediction effort and its explainability, in alignment with the most recent ML applications in the industry.

The remainder of this paper is organized as follows. Section 2 briefly describes relevant related work in the literature, and Sect. 3 outlines state-of-the-art methods for feature importance estimation. Section 4 introduces the methodology of our proposal, whereas Sect. 5 presents the dataset that we have used to validate our approach and details of our experimental setup. Section 6 presents the evaluation results and highlights our findings. Finally, Sect. 7 ends the paper with conclusions and future directions where we are heading.

2 Related Work

Most previous works that seek to explain model predictions assess the relative contribution of features. Examples include techniques based on Sensitivity Analysis [9], linear explanations of the vicinity of points of interest [25], or based on the game theory [20,28]. Another direction of work within the explainability domain is represented by explicitly designed explainers such as L2X [7] and INVASE [31]. These methods select a subset of relevant features approximating the black-box model using mutual information and Kullback-Leibler-divergence, respectively. Most of these methods have been designed for static data like images and need to be carefully considered for the time series contexts.

XAI for time series explains models by evaluating the importance of each feature on the output using attention models. The parameters in these models, called attention weights, are used to explain model behavior in time [8,26]. In [30] the authors propose FIT, a framework that determines the importance of observations over time, based on their contribution to the temporal shift of the model output distribution. Another family of XAI methods is known as perturbation-based methods. They assign importance to a feature based on changes to the model output by perturbing the input features, often replacing the feature value with its mean [12] or random uniform noise [29]. The work in [11] considered averaging over all possible permutations and derived concentration bounds based on an unbiased class of statistics, i.e., U-statistics [14]. When it comes to explainable methods applied to the financial domain, we could mention the work performed by authors in [15]. The author uses four predefined feature sets to predict the next day's stock returns and infers the feature importance by assessing the performance of an algorithmic trading strategy. The most salient finding in this work is that increasing the number of features does not translate into better performance. Moreover, authors in [17] use the feature importance of Random Forest and Gradient Boosted Trees to understand the prominent features of a long-short strategy. Moreover, in [10] the authors use the coefficients of a Carhart regression applied to the returns and first and second-order time-series statistics to explain the dependence between the market regime, stocks behavior, and long-short trading strategy obtained results. Furthermore, authors in [4] use the decision tree feature importance to infer links between news and stock behavior. Finally, in [21] the authors propose an "instability index" strategy of selecting features based on the features' ranks variance. Our proposed approach takes inspiration from the last-mentioned work, in the sense that the authors derive the feature importance using the same XAI technique of permuting the feature values in the out-of-bag sample. However, there are two differences between our work and theirs. The authors remove input features from the feature set regardless of their category, i.e., informative or uninformative. In other words, they do not propose a strategy to differentiate the features. Secondly, their investigation focuses on identifying a convergence point for feature importance stability. To this end, for the entire stock set, the authors train a model and, by using different configurations of the underlying XAI technique, compute the feature importance for each of these configurations, then derive the *instability* index. We argue that such an approach is computationally expensive (with a $O(n \log n)$ complexity) and intractable for a high number of stocks such as ours.

3 Standard XAI Methods

Let $X \in \mathbb{R}^{D \times T}$ be a training set consisting of observations of a multivariate time-series, where D is the number of features, and T is the number of observations over time. Further, $\mathbf{x_i} \in \mathbb{R}^D$ denotes the ith observations at time $i \in \{1, .., T\}$. For a single feature, j, an observation at time i is indicated by $x_{i,j}$. Our goal is to predict a target variable Y based on the multivariate predictive variable X. We use a learning algorithm to output a model function $f(\cdot)$ representing the estimate of Y based on X. In this context, in the remainder of this section, we present three methods related to our work for determining the feature importance. Feature importance is defined as an array of numbers where each number corresponds to a feature, indicating how much the feature contributed to the model prediction [28].

Permutation importance (PI), originally introduced by [3], provides a score for each feature based on how much the replacement of the feature with noise impacts the predictive power of the estimator. Given the matrix of feature values X and the corresponding response variable Y, let $X^{\pi,j}$ be a matrix achieved by randomly permuting the jth column of X and $L(Y, f(X))$ be the loss for predicting the target variable Y from the data X using the model $f(\cdot)$. Then, the feature importance can be mathematically expressed as follows:

$$VI_j^{\pi} = \frac{1}{N}\sum_{\pi} L(Y, f(X^{\pi,j}) - L(Y, f(X)), \tag{1}$$

where N represents the number of permutations applied to each feature. Specifically, the importance of the feature j is given by the increase in loss due to replacing $X_{:,j}$ with values randomly chosen from the distribution of feature j.

Mean decrease impurity (MDI) is a tree-specific feature importance method computed as the average of feature importance across all decision trees in the forest. The structure of the trees has a direct impact on determining the feature importance. Briefly, the RF algorithm constructs a decision tree set where each decision tree is a set of internal nodes and leaves. An internal node is constructed by splitting the data of a specific feature into two separate partitions, with similar target variables. The algorithm splits the data by taking an input feature (from a randomly picked subset of features) and determining which cut-point minimizes the variance of the target variable. Variance minimization occurs when the data points in the nodes have very similar values to the target variable. After determining the optimal cut-point per feature, the algorithm selects the best input feature for splitting, i.e. the feature that would result in the best partition with the lowest variance. Finally, it adds this split to the tree. The feature importance is computed by iterating all the splits generated by a feature and measuring how much that feature reduced the variance w.r.t. the parent node. The more the variance decreases, the more significant the input feature is.

Local Interpretable Model agnostic Explanation (LIME) [25] locally approximates a black-box model $f()$ around the instance of interest $\mathbf{x_i}$ with a linear model $g()$, also known as a local model. Formally, it can be expressed as $g(\mathbf{x_i}) = w_0 + \sum_{j=1}^{D} w_j {x^*}_j$, where w_j represents the regression coefficients of a feature j of the perturbed sample $\mathbf{x}^*$. In a nutshell, to obtain observation by observation explanations, the algorithm follows a suite of steps. Given the observation of interest (for which an explanation of its black box prediction is sought), it perturbs the instance and gets the black box predictions for these new points, $\mathbf{x}^*$. It then trains a weighted, interpretable model $g()$ on the dataset with the variations. Finally, it explains the prediction $\mathbf{x_i}$ by interpreting the local model.

4 The Proposed Strategies

The goal of this work is to provide means of identifying uninformative features for a prediction task and propose those features for removal, so to increase the models' predictive performance. To classify the input features into the two categories, i.e., informative and uninformative, we proceed with the following steps: (i) we compute the feature importance using the PI method, introduced in Sect. 3. (ii) we identify a feature importance threshold below which the features are considered uninformative. The motivation in choosing the PI approach is threefold: (i) it is model agnostic, i.e., applicable to any model derived from any learning method; (ii) it computes the feature importance on the test set, i.e., out-of-bag samples (OOS), which makes it possible to highlight which feature contributes the most to the generalization power of the inspected model [27]; (iii) it has no tuning parameters, and it relies only on averages which makes it statistically very stable.

For computing the feature importance, we modified the Eq. 1 by scaling the changes in loss due to permuting feature j values with the loss of the base regressor as follows (in bold):

$$VI_j = \frac{1}{N} \sum_{\pi} \frac{L(Y, f(X^{\pi,j})) - L(Y, f(X))}{\mathbf{L(Y, f(X))}} \tag{2}$$

The feature importance given by Eq. 2 can be interpreted as the average change in prediction loss relative to the loss of the base regressor. We chose to work with the relative error change instead of the absolute error as it is scale-free and leads to a fair comparison between feature importance for different models.

The feature importance will have positive values for an informative feature as replacing the corresponding feature values with others at random increases the loss compared to the original regressor. On the other hand, unimportant input features will have feature importance values relatively closer or below 0, as their replacement with noise-like information, either will not produce significant changes, or will increase the prediction performance. Keeping the unimportant features in the feature set leads to overfitting. Having this in mind, we propose three strategies to remove the features:

1. *PI best* - identify the features which have the feature importance lower than 0 and propose for removal the one which has the *highest* feature importance of them, that is features that are close to 0. Then, we remove that feature if its importance is lower than a certain threshold.
2. *PI worst* - identify the features which have the feature importance lower than 0 and propose for removal the one which has the *lowest* feature importance of them. Similarly to *PI best*, if that feature has its importance below a certain threshold, we remove it.
3. *PI running* - identify the features which have the feature importance lower than a variable threshold and propose for removal the one which has the *highest* feature importance of them.

Stock	Feature	Feature importance
FP.PA	Returns_2	-0.1077
FP.PA	Returns_63	-0.3906
FP.PA	Returns_5	-0.5298
FP.PA	Returns_3	-1.1571
0001.HK	Returns_1	0.1060
0003.HK	Returns_5	-0.2911
0003.HK	Returns_252	-0.3211
0003.HK	Returns_3	-0.3420
0003.HK	Returns_63	-0.7658
0003.HK	Returns_126	-1.1547

(a) Feature importance

	PI best		*PI worst*		*PI running*	
Stock	Feature	Feature importance	Feature	Feature importance	Feature	Feature importance
FP.PA	N/A		Returns_3	-1.1571	Returns_63	-0.3906
0001.HK	N/A		N/A		N/A	
0003.HK	Returns_5	-0.2911	Returns_126	-1.1547	Returns_5	-0.2911

(b) Selected features

Fig. 1. Illustration of the proposed feature selection method. Panel (a) shows the features and their importance for each stock. Note that in this example, we show only a limited number of features, whereas, in a real-world scenario, panel (a) would include all the features associated with a stock. Panel (b) shows the selected features for each method for a threshold of −0.15.

The main difference between *PI running* and the others is that *PI running* uses the threshold to identify the best feature to remove. *PI best* and *PI worst* use the threshold to identify whether the best feature (the feature with the highest importance below 0) or the worst feature (the feature with the lowest importance below 0) will be removed. To clearly explain how the proposed methods work, in Fig. 1 we show a simple example applied to three stocks and one walk. The left-hand side panel, (a) Feature importance, for stocks *FP.PA* and *0003.HK* displays all the features that have feature importance lower than 0. For stock *0001.HK* we show the least important feature as the stock has no features whose importance is lower than 0. The features are sorted in descending order by their feature importance. The right-hand side panel (b) Selected features shows the features selected for each method provided that we used a threshold $= -0.15$. In this setup, for stock *FP.PA* the feature with the highest negative feature importance is *Returns_2*. However, since it does not have the feature importance below the threshold −0.15, *PI best* for stock *FP.PA* will not remove any features. *PI worst* will remove *Returns_3* which has the lowest value. *PI running* will remove *Returns_63* which is the first feature whose importance value is below the threshold. For the stock *0001.HK* the feature *Returns_1* does not have the feature importance lower than 0, and as a consequence, none of the methods remove features. Finally, for stock *0003.HK* all the methods propose one feature for removal using a similar process described for the other stocks. As a final remark, *PI best* and *PI running* focus on removing the input features with feature importance close to 0, whereas *PI worse* removes the features whose feature importance is below 0. Also, given that the feature importance score is an estimation and PI may underestimate the feature importance, with the introduction of *PI best* and *PI running*, we aim to find the feature importance value below which features are indeed uninformative and whose removal does not affect the overall prediction performance.

Optimal Threshold Selection

For each of the proposed approaches, we select the optimal threshold according to the following steps:

1. for threshold values within a predefined range determine the features to remove as given by each of the methods *PI best*, *PI worst*, and *PI running*;
2. for each model/stock remove the indicated feature and retrain a new model;
3. for each threshold value, compute the mean loss difference between the base regressor (without any feature removed) and the corresponding newly trained model.
4. compute the optimal threshold as the one that maximizes the mean loss difference computed above.

5 Experimental Setup

We apply our feature selection approach to a financial dataset with the goal is to improve the prediction performance by discarding features for each stock according to the methodology explained in Sect. 4 and use the remaining input features to predict the next day's returns.

5.1 Dataset

We carried out our experiments using a set of 300 stocks given by the constituents of the S&P100 Index[1], CAC40[2], FTSE 100[3], S&P Asia 50 Index[4], and Dow Jones Global Titans 50. We justify our choice by the fact that we aim at constructing a highly heterogeneous stock set spread out among different continents, i.e., Europe, the U.S., and Asia-Pacific. For each stock, we collect daily raw financial data[5] such as opening (*open*) and closing (*close*) prices. Based on these data we construct the following information:

Features - Lagged returns (computed with respect to day i), each expressed as

$$Return\_j = \frac{close_{i-1} - open_{i-j}}{open_{i-j}} \text{ for } j \in \{1, 2, 3, 4, 5, 21, 63, 126, 252\},$$

[1] https://en.wikipedia.org/wiki/S%26P_100.
[2] https://en.wikipedia.org/wiki/CAC_40.
[3] https://en.wikipedia.org/wiki/FTSE_100_Index.
[4] https://en.wikipedia.org/wiki/S%26P_Asia_50.
[5] We use publicly available data from Yahoo Finance.

where j and denotes the length of the time-window for which the return was computed[6]. *Return_j* generates 9 features.

Target - Intra-day return

$$y_i = \frac{close_i - open_i}{open_i}.$$

5.2 Forecasting and Feature Selection

As prediction models, we use Random Forest (RF) for two compelling reasons. First, it is a machine learning model that delivers competitive results within financial prediction tasks [17,24]. Second, it is an explainable model [22]. Hence, For each stock, we train a model with the following hyperparameters: *n_estimators = 500* - we set the number of decision trees, *min_samples_leaf = 5* - we limit the depth of the trees, and finally *max_features = 1* - we limit the maximum number of features in each split.

Finally, for determining the feature importance, we have permuted the feature values $N = 100$ times (in Eq. 2). To determine the optimal feature importance threshold, for each feature removal strategy, we have performed a threshold swipe in the range of 0 to -0.025 with a step of 0.0001, as presented in Sect. 4.

5.3 Backtesting

To validate our assumption that the influential features may differ across different market regimes, i.e., in time, we backtested our strategies. We choose a study period of January 2007 to January 2018. We use the *walk-forward* validation, a common approach for backtesting in finance [5,17]. It consists of splitting the study period into overlapping training periods and non-overlapping test periods, i.e., walks. For each walk, we considered four years of training and a year of testing. Under this setup, we form seven walks.

5.4 Baselines

We compared the proposed strategies with three baselines:

1. **Base regressor** - for each stock, we train a model including all the features in the feature set (we do not perform any removal);
2. **MDI** - for each stock, we compute the MDI as presented in Sect. 3 using scikit-learn implementation[7]. Specifically, MDI is the average feature importance of all decision trees (in our case 500) composing the forest. Moreover, using the

[6] For example, $j = 1$ denotes the return in the previous day for each observation, whereas $j = 252$ denotes the return in the past year, considering that there are 252 trading days in a year.

[7] https://scikit-learn.org/stable/modules/generated/sklearn.ensemble.RandomForestRegressor.html#sklearn.ensemble.RandomForestRegressor.feature_importances_.

RF hyperparameter *max_features = 1* we overcome RF's inability to correctly estimate feature importance when dealing with features that are correlated with each other, as it often happens for financial time series. Then, given the obtained features' importance, we discard the feature with the lowest importance.

3. **LIME** - for each stock, given a trained base regressor, we explain each sample in the test period with a linear model, as detailed in Sect. 3. We proceed as such, as LIME is not designed to assign feature importance globally. Then, each observation's feature importance is equal to the absolute value of the regression coefficient of that feature. For each sample, we order the input features by their importance and rank them (where rank 1 denotes the highest feature importance). To obtain the global feature importance, i.e., the feature importance for the whole test period, we average the ranks across all test observations and remove the least important feature, that is, the feature with the lowest average rank. For the LIME baseline, we use the python LIME package[8].

Note that for the proposed strategies and MDI and LIME baselines, for a fair comparison, we perform the same procedure: we compute the feature importance for each model (i.e., one per stock) and each input feature; then, we remove - from the feature set - the features selected by the strategies, or the least important one as indicated by MDI or LIME. Finally, we train a new model and evaluate the performance.

Implementation Details

The strategies proposed in this paper have been developed in *Python*, by using the scikit-learn library [23]. The code of our solution has been made publicly available at https://github.com/Artificial-Intelligence-Big-Data-Lab/feature-selection. The experiments have been executed on a desktop system with the following specifications: an Intel(R) Xeon(R) Gold 6136 CPU @ 3.00 GHz, 32 GBytes of RAM, and 64-bit Operating System (Linux Ubuntu). In terms of *Big O* notation, our algorithm entails an $O(|S| \times m^{-})$ complexity proportional to the number of stocks $|S|$ and the number of features to remove for each stock, m^{-}.

5.5 Evaluation Metrics

In this work, three performance metrics are considered. Point forecasting performance is determined by using the mean squared error (MSE), similar to [21]. Evaluating the performance of feature selection methods on real data is difficult

[8] https://github.com/marcotcr/lime/tree/master/lime.

since ground truth relevance is not known. Therefore, we focus on the prediction performance. Specifically, we measure the effectiveness of removing one feature by assessing the number of improvements over the base regressor in terms of MSE and the ratio given by the number of improvements over the base regressor out of the number of models (stocks) which had a feature removed. Explicit expressions for the metrics are provided as follows:

- $MSE = \frac{1}{T}\sum_i(y_i - \hat{y}_i)^2$, where $\hat{y}_i$ represents the ith predicted value, y_i the ith target value, and T the number of observations in the test period;
- $no\ improvements = \sum_{i=1}^{|S|} I(MSE_{base\ regressor_i} - MSE_{M_i^-} > 0)$, where $i \in \{1, ..., |S|\}$, S is the set of stocks, $MSE_{base\ regressor_i}$ is the error of the base regressor of stock i (i.e., no features have been removed), $MSE_{M_i^-}$ is the error of the regressor for which a feature has been removed, and finally, I is the indicator function. In other words, *no improvements* measures how many models have an MSE improvement (lower than the MSE of the base regressor) when removing a feature;
- $ratio = \frac{no\ improvements}{\mathcal{M}^-}$, where $\mathcal{M}^-$ is the number of models that had a feature removed, or, differently put, the number of models that have one feature whose importance is lower than a given threshold.

6 Results

In this section, we examine the proposed strategies for each of the walks as presented in Sect. 5 by performing two different experiments: (A) finding the optimal feature importance threshold in order to select features to be discarded, and (B) qualitatively evaluating the improvement over the baselines.

For experiment A, in Fig. 2 we show the performance of the proposed strategies under the three evaluation metrics presented in Sect. 5.5 for different thresholds and against the base regressor. Figure 2a shows the average MSE difference between the base regressor and the model with features removed (the feature whose importance is lower than the threshold). Figure 2b, analogously, shows the number of improved models by comparison to the corresponding base regressor. Finally, Fig. 2c presents the ratio of improved models (i.e., number of times the MSE decreases) out of the total number of models that had a feature removed. Each figure shows metrics values for thresholds ranging from 0 to -0.025 with a step of 0.0001. Furthermore, for a subset of threshold values $\{0, -0.005, -0.01, -0.015, -0.02, -0.025\}$ we also show in the graphs the corresponding metric values, i.e. *average MSE difference*, *noimprovements*, and *ratio* obtained by each of the proposed methods.

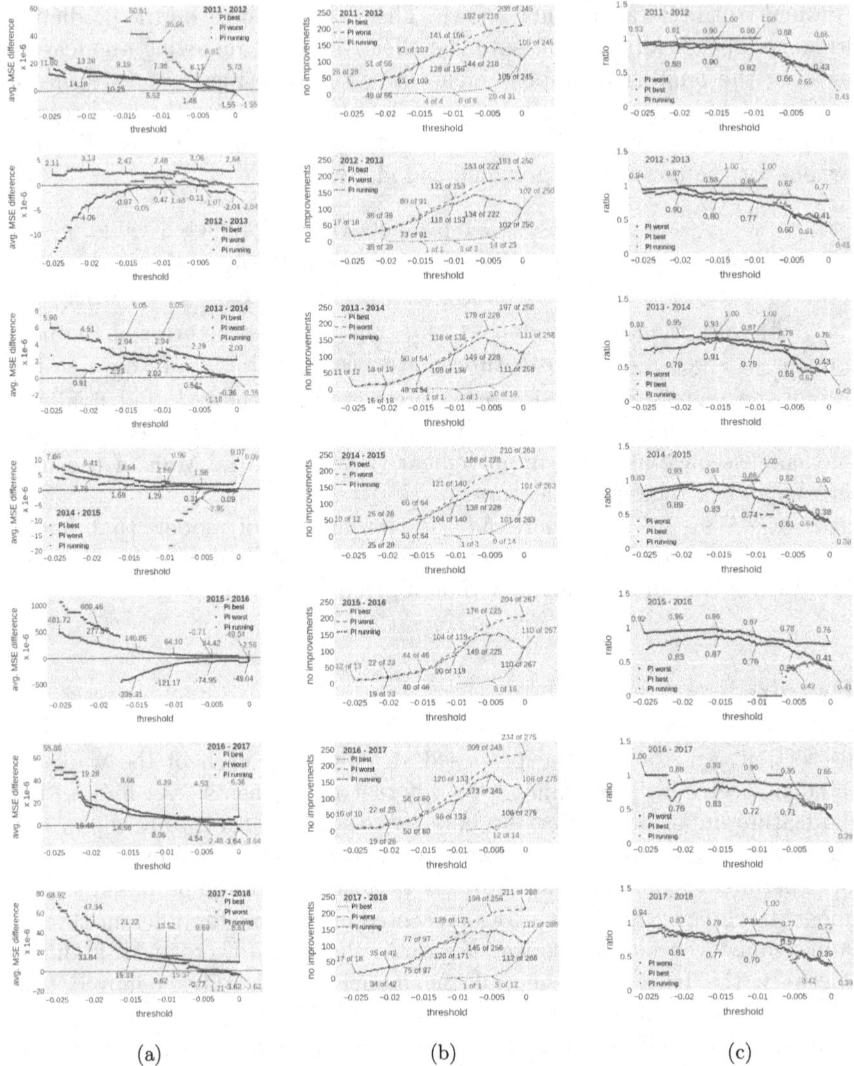

(a) (b) (c)

Fig. 2. (a) Average MSE difference between the base regressor and the model when discarding unimportant features according to each method for different threshold values. To note that positive values indicate a better performance of our strategies. (b) The number of models having an MSE improvement when discarding unimportant features according to each method for different threshold values. (c) Percentage of models having an MSE improvement when discarding unimportant features according to each method for different threshold values.

Table 1. Threshold across walks and the corresponding average MSE difference between the base regressor (all features are used) and the models when a feature was removed according to the proposed methods.

Walk	Thresh.	*PI best*			*PI worst*				*PI running*			
		Average MSE difference	No improvements	Ratio	Thresh.	Average MSE difference	No improvements	Ratio	Thresh.	Average MSE difference	No improvements	Ratio
2011–2012	−0.0142	5.05e−05	4.0	1.0000	−0.0198	1.3e−05	101.0	0.9018	−0.0249	2.39e−05	26.0	0.8965
2012–2013	−0.0076	2.65e−06	6.0	0.8571	−0.0079	3.0e−06	158.0	0.8316	−0.0079	0.7e−06	135.0	0.7219
2013–2014	−0.0084	5.04e−06	1.0	1.0000	−0.0188	0.5e−05	144.0	0.8521	−0.0249	2.70e−06	9.0	0.7500
2014–2015	−0.0095	9.58e−07	1.0	1.0000	−0.0001	0.9e−05	129.0	0.8600	−0.0003	7.04e−06	110.0	0.4247
2015–2016	−0.0027	−1.76e−07	30.0	0.5000	−0.0198	27.8e−05	197.0	0.7787	−0.0246	1.06e−03	9.0	0.6923
2016–2017	−0.0083	1.02e−05	2.0	1.0000	−0.0199	1.90e−05	151.0	0.8629	−0.0249	4.89e−05	7.0	0.7000
2017–2018	−0.0079	1.53e−05	1.0	1.0000	−0.0195	4.7e−05	169.0	0.8009	−0.0214	4.17e−05	28.0	0.8235

Table 1 presents the optimal threshold per walk for each of the proposed methods and reports the metrics presented in Sect. 5.5. With regards to finding the optimal threshold, we identify it as the one that maximizes the average MSE difference between the base regressor and the corresponding models trained with a feature removed, provided that the feature has the feature importance lower than the threshold. Average MSE differences with values higher than 0 are those showing a clear improvement of the proposed strategies.

For experiment B, Fig. 3 presents the average MSE across each walk for the models when removing the features according to the proposed strategies. If for *PI best* and *PI worst* no feature was removed, we consider the MSE of the base regressor. We compare the obtained results against the three presented baselines in Sect. 5.4.

6.1 Discussion

Experiment A. In Fig. 2a we have inspected the variation of the average MSE difference between the base regressor and the model trained without the feature identified as irrelevant by the proposed strategies. In this context, naturally, for the *average MSE difference*, values higher than 0 are sought, as they clearly indicate a better prediction performance of our strategies by comparison to the base regressor. Values close to 0 indicate that removing features does not bring any improvement to the prediction performance. However, having the same performance with a lower number of features is better for the computational time and simplicity of the feature engineering process. Conversely, values lower than 0 indicate a prediction performance degradation and that most of the features removed were, in fact, informative. In Fig. 2a the "0" value is marked with a black, continuous line. In Fig. 2a, the reader can notice that the *average MSE difference* shows different values across the presented walks. This fact may be due to data-shift that makes the identification of non-representative features difficult. It negatively stands out the walk between 2015–2016 when *PI best* has the lowest performance with values asymptotically close to 0 or below. The results show that *PI best* systematically misclassifies important features as unimportant. Similarly, for high threshold values, *PI running* suggests removing features that

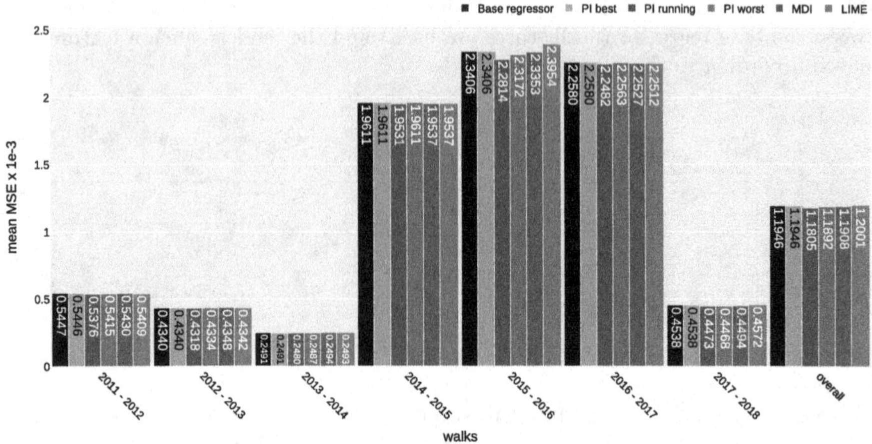

Fig. 3. Average MSE across different walks for the proposed methods when, for each stock we remove the feature whose feature importance is lower than the optimal threshold. Comparison to the three baselines: all feature are used (base regressor), remove the feature with lowest feature importance as given by MDI (MDI), and remove the feature with lowest feature importance as given by LIME (LIME).

are important and, as such, worsens the prediction performance of the regressors. For the other walks, all the methods demonstrate their capability in identifying unimportant features, thus providing a prediction performance improvement. As a final remark, *PI worst* is the most stable strategy, and it outperforms the base regressor in all cases.

Analyzing Fig. 2b, where we show the number of models that have an increase in predictive power (i.e., the feature removal was effective), the reader can notice that the best performing method is *PI worst*, followed by *PI running* and *PI best*. The latter produces satisfactory results only for high threshold values. One can expect such behavior since *PI best* removes the uninformative features (with importance lower than 0), but the input features with the highest feature importance among them.

Figure 2c shows the ratio of the number of models whose predictive performance has increased (with the feature removal) by the total number of models that had a feature removed. There we can see a similar behavior of *PI worst* and *PI running* for all the walks, but *PI worst* shows a higher ratio. Desirable values are close to 1. At the opposite pole, we find the *PI best* method, which displays discordant values: (i) below 0.5, meaning that less than half of the models have a significant improvement, (ii) asymptotically close to 1 which means that all feature removals resulted in a loss decrease. Although this might seem an optimal result, in Fig. 2b we notice a low number of feature removal actions. Therefore, the *ratio*, in this case, does not provide informative insights.

Table 1 shows the optimal thresholds for each of the strategies together with the corresponding evaluation metrics. Judging by the *average MSE difference*,

PI running obtains the best results overall. Furthermore, the *no improvements* indicates that *PI worst* is the best performing method. Thus, when removing the feature with the lowest feature importance we obtain the highest number of improvements. *PI best* is the worst performing of the three strategies, accounting for a negative *average MSE difference* for the walk 2015–2016. Moreover, for the same walk, the *ratio* of improvements over the total number of feature removals performed is 0.5. Such a value means that half of the feature removals resulting in lower performance than the base regressor are worst than the other half, which has better performance. The different values of the optimal threshold across different walks confirm the prediction models' time-dependency.

Experiment B. Results presented in Fig. 3 highlight that the proposed strategies outperform the three baselines. Specifically, *PI best* achieves similar average MSE values of the base regressor on most of the walks and overall. Such a result is encouraging as, although the prediction performance does not increase much, employing a smaller number of features leads to a lower training and prediction time.

Overall, *PI worst* has the second-best behavior, while *PI running* achieves the best results out of the three proposed methods. Such a behavior meets our expectations, as the MDI is known to be biased, especially when it comes to time-series data-sets [19]. Consequently, MDI may over- or under-estimate feature importance, which we hypothesize is the explanation for the close average MSE to the base regressors' (using all the features). As for LIME, the reader can notice that it has the worst performance amongst all methods. Therefore, we can conclude that LIME consistently underestimates feature importance and mistakenly assigns lower feature importance to otherwise important input features. A possible explanation of this behavior lies in the fact that LIME relies on perturbing the input observation around its neighborhood and observes the model's behaviors (predictions). Originally, LIME was designed with specific implementations for text and images but without focus on time-series data. Moreover, financial time series have a low signal-to-noise ratio and exhibit a random-walk behavior with a low degree of predictability, and opening the black-box models by "peaking" into the randomly generated neighborhood yields under-par results.

Also, to be noted that in the case of financial time-series forecasting, in the particular case of predicting the daily returns, both the input features and the target variable values are small with an order of magnitude of 1e-4. Consequently, the MSE has small values too, which constitutes a challenge for model training or when it comes to distinguishing a performing model from an under-performing one.

Statistical Significance. To test the statistical relevance of our results, we exploit the Wilcoxon signed-rank test, a non-parametric test that determines whether two samples were selected from populations having the same distribution. Such a test is an alternative to the paired Student's t-test, when the distribution of the difference between two samples' means cannot be assumed to be normally distributed, as in our case. We evaluate the null hypothesis $\mathcal{H}_0$ asserting that the MSE of each feature selection method, i.e. *PI best*, *PI worst*, and *PI running* has the same distribution to the MSE of the base regressors. We obtain the p-values: 0.0678, $7.3761e-21$, and 0.0179, respectively. In all tests, we set the confidence level to 95%, that is, if p-value is below 0.05, then $\mathcal{H}_0$ can be rejected. Under this assumption, we can reject $\mathcal{H}_0$ for *PI worst* and *PI running*, but not for *PI best*. However, the latter outcome is expected and due to low number of feature removals, as shown in Table 1.

7 Conclusions

Algorithmic decision-making systems are becoming very popular in the financial domain, prompting us to rely more and more on their decisions, with potentially negative consequences for the affected users. Automatic feature selection might be a step towards a more reliable and robust explainable artificial intelligence. In this paper, we have made a step closer towards this goal by introducing three strategies to efficiently remove features and, in doing so, increase the forecasting power of the machine learning models. We employ the proposed approaches over a set of stocks, and we determine for stock individually the important and unimportant features. We achieve this goal by identifying an optimal threshold for which: (i) we discard the features whose *negative* importance is the *highest*, *only* if it is also below the optimal threshold, (ii) we discard the features whose *negative* importance is the *lowest*, *only* if it is also below the optimal threshold, or (iii) we *always* discard the feature whose importance is below the optimal threshold. The predictive performance of the models trained with feature selection is improved compared to the models with no feature selection. Specifically, we find that the predictive performance increases for the models trained without the features indicated by the second and third strategies. Furthermore, we compare the proposed strategies to state-of-the-art approaches such as the Random Forest feature importance and local interpretable models and show that our permutation-based strategies are superior at discovering unimportant features for each time-series data.

A future line of work is to find the optimal subsets of features to discard and not focusing on just one as in the proposed paper. Also, a series of experiments can be performed by enriching the study with other machine learning models and examining their behavior under the proposed strategies. Last but not least, we would like to apply the proposed approaches to trading strategies such as the statistical arbitrage or include them as a prediction step in the MultiCharts trading platform, and in doing so, yet proving their effectiveness.

Acknowledgements. The research performed in this paper has been supported by the "Bando "Aiuti per progetti di Ricerca e Sviluppo" - POR FESR 2014-2020-Asse 1, Azione 1.1.3, Strategy 2- Program 3, Project AlmostAnOracle - AI and Big Data Algorithms for Financial Time Series Forecasting".

References

1. Adadi, A., Berrada, M.: Peeking inside the black-box: a survey on explainable artificial intelligence (XAI). IEEE Access **6**, 52138–52160 (2018). https://doi.org/10.1109/ACCESS.2018.2870052
2. Arrieta, A., et al.: Explainable artificial intelligence (XAI): concepts, taxonomies, opportunities and challenges toward responsible AI. Inf. Fusion **58**, 82–115 (2020)
3. Breiman, L.: Random forests. Mach. Learn. **45**(1), 5–32 (2001). https://doi.org/10.1023/A:1010933404324
4. Carta, S.M., Consoli, S., Piras, L., Podda, A.S., Recupero, D.R.: Explainable machine learning exploiting news and domain-specific lexicon for stock market forecasting. IEEE Access **9**, 30193–30205 (2021). https://doi.org/10.1109/ACCESS.2021.3059960
5. Carta, S.M., Consoli, S., Podda, A.S., Recupero, D.R., Stanciu, M.M.: Ensembling and dynamic asset selection for risk-controlled statistical arbitrage. IEEE Access **9**, 29942–29959 (2021). https://doi.org/10.1109/ACCESS.2021.3059187
6. Carta, S., Medda, A., Pili, A., Reforgiato Recupero, D., Saia, R.: Forecasting e-commerce products prices by combining an autoregressive integrated moving average (arima) model and google trends data. Future Internet **11**, 5 (2019)
7. Chen, J., Song, L., Wainwright, M.J., Jordan, M.I.: Learning to explain: an information-theoretic perspective on model interpretation. CoRR abs/1802.07814 (2018). http://arxiv.org/abs/1802.07814
8. Choi, E., Bahadori, M., Kulas, J., Schuetz, A., Stewart, W., Sun, J.: Retain: an interpretable predictive model for healthcare using reverse time attention mechanism. In: Advances in Neural Information Processing Systems, 30th Annual Conference on Neural Information Processing Systems, NIPS 2016, 05 December 2016 Through 10 December 2016, pp. 3512–3520, January 2016
9. Cortez, P., Embrechts, M.J.: Opening black box data mining models using sensitivity analysis. In: 2011 IEEE Symposium on Computational Intelligence and Data Mining (CIDM), pp. 341–348 (2011). https://doi.org/10.1109/CIDM.2011.5949423
10. Fischer, T., Krauss, C.: Deep learning with long short-term memory networks for financial market predictions. Eur. J. Oper. Res. **270**(2), 654–669 (2018)
11. Fisher, A.J., Rudin, C., Dominici, F.: Model class reliance: variable importance measures for any machine learning model class, from the "rashomon" perspective (2018)
12. Fong, R.C., Vedaldi, A.: Interpretable explanations of black boxes by meaningful perturbation. In: 2017 IEEE International Conference on Computer Vision (ICCV), pp. 3449–3457 (2017). https://doi.org/10.1109/ICCV.2017.371
13. Henrique, B.M., Sobreiro, V.A., Kimura, H.: Literature review: machine learning techniques applied to financial market prediction. Expert Syst. Appl. **124**, 226–251 (2019)
14. Hoeffding, W.: A class of statistics with asymptotically normal distribution. Ann. Math. Stat. **19**(3), 293–325 (1948). https://doi.org/10.1214/aoms/1177730196
15. Huck, N.: Large data sets and machine learning: applications to statistical arbitrage. Eur. J. Oper. Res. **278**(1), 330–342 (2019). https://doi.org/10.1016/J.EJOR.2019.04.013

16. Kingston, J.K.C.: Artificial intelligence and legal liability. In: Bramer, M., Petridis, M. (eds.) Research and Development in Intelligent Systems XXXIII, pp. 269–279. Springer, Cham (2016). https://doi.org/10.1007/978-3-319-47175-4_20
17. Krauss, C., Do, X.A., Huck, N.: Deep neural networks, gradient-boosted trees, random forests: statistical arbitrage on the S&P 500. Eur. J. Oper. Res. **259**(2), 689–702 (2017)
18. Kroll, J., et al.: Accountable algorithms. Univ. Pennsylvania Law Rev. **165**, 633–705 (2017)
19. Louppe, G., Wehenkel, L., Sutera, A., Geurts, P.: Understanding variable importances in forests of randomized trees. In: Burges, C.J.C., Bottou, L., Welling, M., Ghahramani, Z., Weinberger, K.Q. (eds.) Advances in Neural Information Processing Systems, vol. 26, pp. 431–439. Curran Associates, Inc. (2013). https://proceedings.neurips.cc/paper/2013/file/e3796ae838835da0b6f6ea37bcf8bcb7-Paper.pdf
20. Lundberg, S.M., et al.: Explainable machine-learning predictions for the prevention of hypoxaemia during surgery. Nat. Biomed. Eng. **2**(10), 749–760 (2018). https://doi.org/10.1038/s41551-018-0304-0
21. Man, X., Chan, E.P.: The best way to select features? Comparing MDA, LIME, and SHAP. J. Financ. Data Sci. **3**(1), 127–139 (2020). https://doi.org/10.3905/jfds.2020.1.047. https://jfds.pm-research.com/content/early/2020/12/04/jfds.2020.1.047
22. Molnar, C., Casalicchio, G., Bischl, B.: Interpretable machine learning-a brief history, state-of-the-art and challenges (2020)
23. Pedregosa, F., et al.: Scikit-learn: machine learning in Python. J. Mach. Learn. Res. **12**, 2825–2830 (2011)
24. de Prado, M.L.: Advances in Financial Machine Learning, 1st edn. Wiley Publishing, Hoboken (2018)
25. Ribeiro, M.T., Singh, S., Guestrin, C.: "Why should i trust you?": explaining the predictions of any classifier. In: Proceedings of the 22nd ACM SIGKDD International Conference on Knowledge Discovery and Data Mining, KDD 2016, pp. 1135–1144. Association for Computing Machinery, New York (2016). https://doi.org/10.1145/2939672.2939778
26. Song, H., Rajan, D., Thiagarajan, J., Spanias, A.: Attend and diagnose: clinical time series analysis using attention models. In: 32nd AAAI Conference on Artificial Intelligence, AAAI 2018, 32nd AAAI Conference on Artificial Intelligence, AAAI 2018, pp. 4091–4098. AAAI Press (2018)
27. Strobl, C., Boulesteix, A.L., Kneib, T., Augustin, T., Zeileis, A.: Conditional variable importance for random forests. BMC Bioinform. **9**(1), 307 (2008). https://doi.org/10.1186/1471-2105-9-307
28. Strumbelj, E., Kononenko, I.: An efficient explanation of individual classifications using game theory. J. Mach. Learn. Res. **11**, 1–18 (2010)
29. Suresh, H., Hunt, N., Johnson, A., Celi, L.A., Szolovits, P., Ghassemi, M.: Clinical intervention prediction and understanding with deep neural networks. In: Doshi-Velez, F., Fackler, J., Kale, D., Ranganath, R., Wallace, B., Wiens, J. (eds.) Proceedings of the 2nd Machine Learning for Healthcare Conference. Proceedings of Machine Learning Research, Boston, Massachusetts, 18–19 August 2017, vol. 68, pp. 322–337. PMLR (2017). http://proceedings.mlr.press/v68/suresh17a.html

30. Tonekaboni, S., Joshi, S., Campbell, K., Duvenaud, D.K., Goldenberg, A.: What went wrong and when? Instance-wise feature importance for time-series black-box models. In: Larochelle, H., Ranzato, M., Hadsell, R., Balcan, M.F., Lin, H. (eds.) Advances in Neural Information Processing Systems, vol. 33, pp. 799–809. Curran Associates, Inc. (2020)
31. Yoon, J., Jordon, J., van der Schaar, M.: INVASE: instance-wise variable selection using neural networks. In: International Conference on Learning Representations (2019). https://openreview.net/forum?id=BJg_roAcK7

Online Semi-supervised Learning from Evolving Data Streams with Meta-features and Deep Reinforcement Learning

Parsa Vafaie[1], Herna Viktor[1(✉)], Eric Paquet[1,2], and Wojtek Michalowski[3]

[1] School of Electrical Engineering and Computer Science, University of Ottawa, Ottawa, Canada
{pvafaie,hviktor}@uottawa.ca

[2] Digital Technologies, National Research Council, Ottawa, Canada
eric.paquet@nrc-cnrc.gc.ca

[3] Telfer School of Management, University of Ottawa, Ottawa, Canada
wojtek@telfer.uottawa.ca

Abstract. Online semi-supervised learning (SSL) from data streams is an emerging area of research with many applications due to the fact that it is often expensive, time-consuming, and sometimes even unfeasible to collect labelled data from streaming domains. State-of-the-art online SSL algorithms use clustering techniques to maintain micro-clusters, or, alternatively, employ wrapper methods that utilize pseudo-labeling based on confidence scores. Current approaches may introduce false behaviour or make limited use of labelled instances, thus potentially leading to important information being overlooked. In this paper, we introduce the novel Online Reinforce SSL algorithm that uses various K Nearest Neighbour (KNN) classifiers to learn meta-features across diverse domains. Our Online Reinforce SSL algorithm features a meta-reinforcement learning agent trained on multiple-source streams obtained by extracting meta-features and subsequently transferring this meta-knowledge to our target domain. That is, the predictions of the KNN learners are used to select pseudo-labels for the target domain as instances arrive via an incremental learning paradigm. Extensive experiments on benchmark datasets demonstrate the value of our approach and confirm that Online Reinforce SSL outperforms both the state-of-the-art and a self-training baseline.

1 Introduction

Data streams produced by sensor networks, customer click streams and scientific data are ubiquitous in our society, and extracting knowledge from these fast-evolving repositories poses a significant research challenge. To this end, online supervised learning from streaming data is an active area of research with the objective of building near-real-time models capable of continuously adapting to the changes taking place in these fast-evolving repositories. In many real-world

G. Nicosia et al. (Eds.): LOD 2021, LNCS 13164, pp. 70–85, 2022.
https://doi.org/10.1007/978-3-030-95470-3_6

situations, a major challenge is dealing with a scarcity of class labels since such labels are often difficult to obtain during training. Semi-supervised learning is a sub-field of machine learning aimed at creating accurate models when not all labels are immediately available [1]. That is, semi-supervised learning algorithms feature improved performances due to training based on a small set of labelled instances, combined with a large number of unlabelled instances [2].

Online semi-supervised learning (SSL) from data streams is an emerging area of research, with existing approaches mainly employing cluster analysis methods. For instance, online reliable semi-supervised learning (OReSSL) [3] is an online SSL algorithm that maintains micro-clusters of labelled and unlabelled instances and updates them as instances arrive. Semi-supervised pool and accuracy-based stream classification (SPASC) [4] is another cluster-based method used in online SSL; it maintains an ensemble of cluster-based classifiers covering all the concepts in the data. Some online SSL algorithms use model predictions for pseudo-labelling unlabelled instances. For instance, streaming co-forest (SCo-Forest) [5] utilizes the most confident prediction of classifiers for training each classifier in the forest. The authors in [6] introduce the Improved Online Ensemble (IOE) algorithm and a distance-based approach for dealing with missing labels which employs the one nearest neighbour (1NN) classifier to predict the labels of unlabelled instances. Evolving streams via self-training windowing ensembles (LESS-TWE) [7] is another online SSL method that uses the predictions of ensembles as the labels for unlabelled instances. A drawback of existing approaches is that they make limited use of the labelled data. In addition, these approaches have the potential to introduce false behaviour by selecting incorrect instances, potentially leading to sub-optimal results [1,2].

In our work, we introduce the Online Reinforce SSL algorithm that utilizes meta-features to train a meta-reinforcement learning agent. The meta-features are extracted from multiple-source datasets via the predictions of various KNN classifiers. That is, our approach uses extracted meta-features to train a meta-reinforcement (m-RL) agent to select the most appropriate predictions from the KNN classifiers. We demonstrate that this algorithm accurately learns to distinguish between correct and incorrect predictions and show that our results outperform the current state-of-the-art.

This paper is organised as follows. Section 2 introduces related work, while Sect. 3 details our Online Reinforce algorithm. An experimental evaluation follows in Sect. 4. We conclude our paper in Sect. 5.

2 Background and Related Work

This section describes the background and related work for meta-learning and online semi-supervised learning.

2.1 Meta-learning

Meta-learning, or *learning how to learn*, is defined as training a model on source tasks to achieve generalization for new tasks. In other words, meta-learning focuses on improving the performance of a model on a new task by training the model on previous tasks. In meta-learning, it is assumed that the source and target tasks come from similar distributions of tasks $p(\tau)$ [8]. More formally, given meta-training set $\mathcal{D}_{meta-training} = \{\mathcal{D}_1, ..., \mathcal{D}_n\}$, where $\mathcal{D}_i$ is defined as the training data for task the τ_i, meta-learning leverages the $\mathcal{D}_{meta-training}$ for learning parameters θ: $P(\theta|\mathcal{D})$ that can be generalized to new tasks [9].

Meta-learning has many applications and has been successfully used in numerous domains. For instance, Fin et al. introduced a method for meta-learning called model agnostic meta-learning (MAML) [9], in which the model parameters are directly trained by gradient optimizations and adapted to new tasks via optimizations or high-tuning based on the gradient of the target task. Meta-learning has also been employed in reinforcement learning (RL) [8]. For instance, the authors in [10] evaluate their Meta-RL model by training it on small mazes and evaluating it on previously-unseen, larger mazes. The authors in [11] introduced an algorithm called Meta-AAD for dealing with anomaly detection using a meta-deep RL model for detecting anomalies in data streams. Their approach incorporates active learning in the labeling process [12]. Through the extraction of meta-features from data, the trained model can be transferred and applied to new datasets. Meta-RL selects possible anomalies, then the selected instances are sent to an oracle, whereas Meta-AAD outperforms state-of-the-art methods in terms of dealing with anomaly detection while reducing the false positives rates.

2.2 Online Semi-supervised Learning

Online SSL is an active and emerging area of research [3]. A number of approaches are based on the so-called wrapper-based methods, such as self-training, an approach that selects the most confident predictions of a base classifier to predict labels. For instance, [7] employed self-training for pseudo-labelling in data streams by adding the predictions of an ensemble as the labels for unlabelled instances. A major limitation of this approach is that it could emphasize false behaviour [1,2]. The authors in [6] introduced a window-based SSL approach for data streams employing the 1NN algorithm for labelling unlabelled instances. In this method, labels are only added if the distance between neighbours is closer than a threshold ϵ. A drawback of this technique is that it is highly sensitive to the value of ϵ and that 'missed close-second' neighbors may lead to incorrect results. In [13], the authors use a mixture of active learning and self training. When the confidence scores are lower than the threshold for one instance, an oracle is asked to label the unlabelled instance. However, in a streaming setting, asking an oracle for labels might not be feasible and could end up being very expensive. Temporal label propagation (TLP) is an online graph-based SSL algorithm introduced in [14]. In TLP, τ unlabelled instances are propagated in a graph H by computing the harmonic solution. Although the TLP algorithm achieves good performances with a minimal number of labelled instances, traversing the graph may be prohibitive in a streaming environment.

Other recent research on online SSL focuses on using clustering techniques for dealing with scarce labelled data [4,15]. Semi-supervised pool and accuracy-based stream classification (SPASC) uses an ensemble of classifiers introduced in [4] for dealing with online SSL. The base classifiers in SPASC are cluster-based and updated with each labelled and unlabelled batch of data. Each classifier is used to describe a single concept. When a new batch arrives, a new classifier is added to the pool if the batch contains a new concept. Otherwise, a pre-existing classifier related to the concept represented by the new batch will be updated based on this batch. There are many hyper-parameters involved in SPASC, namely the batch-size, number of clusters, and method for detecting the similarity between a concept and cluster. A limitation of this approach is that clusters with similar characteristics are not merged to create larger clusters. Thus, clusters may arise that cannot perfectly distinguish between classes.

Reliable SSL (ReSSL) [15] is another cluster-based algorithm. In this method, micro-clusters are generated from labelled and unlabelled data. However, in order to deal with changes in data streams, as each instance arrives, the weights of the older micro-clusters are reduced. Online Re SSL (OReSSL) [3] extends ReSSL in that each micro-cluster's importance is based on its reliability in predicting instances. For classification, an ensemble of KNN classifiers is used to classify each incoming instance. SPASC and OReSSL both employ unsupervised learning during the labeling process, which may result in reduced performances for some datasets, as shown in our experimental Sect. 4. Next, we introduce our Online Reinforce SSL algorithm.

3 Online Reinforce Algorithm

Our Online Reinforce framework takes advantage of both meta-learning and reinforcement learning, as follows. Given a labelled data stream from a source domain, transferable meta-features are learned by multiple KNN classifiers, with k ranging from 1 to K. Thereafter, meta-reinforcement learning (m-RL) is employed in order to predict the labels of unlabelled data streams pertaining to different target domains. The newly labelled instances are buffered in a sliding window of size w, allowing a base classifier to be trained incrementally. The m-RL agent may be assimilated with an active-learning oracle that can attribute labels to unlabelled instances. Our meta-features are described in the next section.

3.1 Meta-features

In order to be transferable, meta-features should share a common feature space and be, as much as possible, domain-independent. The meta-features employed in this work are based on the (i) Euclidean distance in the feature space between unlabelled instances and their nearest labelled neighbors, (ii) confidence scores of the KNN classifiers, and (iii) level of agreement among classifiers.

1 – Meta-distance

Given the feature vector of an unlabelled instance and its K nearest labelled neighbors, the meta-distance is defined as

$$\chi_1\left(\mathbf{x}_u\right) = \{\|\mathbf{x}_u - k\text{-}NN\left(\mathbf{x}_u\right)\|_2\}_{k=1}^{K} \tag{1}$$

Therefore, the K nearest labelled neighbors are inferred for each unlabelled instance. Next, the Euclidean distances in between these neighbors and each unlabelled instance are evaluated. Note that these distances are not necessarily transferable from one domain to another, as the distance ranges may vary considerably. Therefore, the meta-distance defined in Eq. 1 is standardized by applying a z-score:

$$\chi_1 \leftarrow \left\{\frac{\chi_{1,i} - \mu}{\sigma}\right\}_{i=1}^{|\chi_1|}, \tag{2}$$

where μ and σ are the mean and the standard deviation of the meta-distance features during the warm-up period. Standardization ensures that feature vectors can be transferred between domains.

2 – Confidence score

The next meta-feature consists of all the confidence scores (probabilities) attributed to the unlabelled data by the K classifiers:

$$\chi_2\left(\mathbf{x}_u\right) = \{\Pr_{k\text{-}NN}\left(\mathbf{x}_u\right)\}_{k=1}^{K} \tag{3}$$

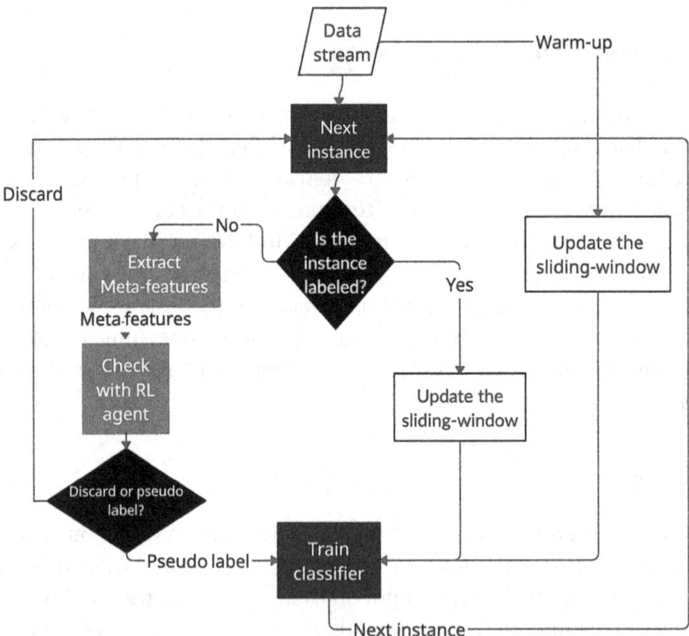

Fig. 1. The workflow of our Online Reinforce SSL algorithm.

3 – Pairwise agreements between KNN classifiers

The various classifiers do not necessarily attribute the same label to an unlabelled instance (disagreement among classifiers). Therefore, a third meta-feature is introduced in order to consider pairwise agreements between classifiers. This feature is defined as:

$$\chi_3 = \{\chi_{3,i}\}_{i=1}^{\binom{k}{2}}, \quad \begin{matrix} k_1 \in [1, K-1] \\ k_2 \in [k_1+1, K] \end{matrix} : \left\{ i = 1, 2, \ldots, \binom{k}{2} : \begin{matrix} k_1\text{-}NN = k_2\text{-}NN \Rightarrow \chi_{3,i} = 1 \\ k_1\text{-}NN \neq k_2\text{-}NN \Rightarrow \chi_{3,i} = 0 \end{matrix} \right. \tag{4}$$

It follows that the resulting meta-feature is obtained by concatenating the features above, that is:

$$\chi = \{\chi_1 \| \chi_2 \| \chi_3\} \tag{5}$$

3.2 Pseudo-labelling with Meta-reinforcement Learning

Meta-reinforcement learning is employed for pseudo-labelling. Initially, a sliding window of size w is filled with the first w labelled instances in the source data stream. Once the sliding window is initialized, the data stream is processed instance by instance. If a new instance is labelled, then it is added to the sliding window, and the oldest instance is discarded. Otherwise, if the instance is unlabelled, the KNN classifiers predict its first K nearest labelled neighbors via the current sliding window. These nearest neighbors are employed to calculate the meta-features defined in Sect. 3.1. According to the meta-features, the m-RL selects the best predictions from the KNN classifiers and assigns a label to the unlabelled instance. Subsequently, the base classifier is trained with the newly-labelled instance. If the prediction confidence level is too low, then the unlabelled instance is discarded. This approach is outlined in Fig. 1.

3.3 Training of the Meta-reinforcement Learning Model

The training of the m-RL model is performed with data streams from multiple-source domains. Once trained, the m-RL model can be employed for pseudo-labelling in various target domains. In order to improve its robustness and generalization capability, the m-RL model is trained against several data streams in a sequential manner. Initially, meta-feature vectors are extracted from the source data streams, as described in Sect. 3.1. These data streams contain both labelled and unlabelled instances. If an instance is unlabelled, a label is assigned by the KNN classifier following the procedure described in Sect. 3.2. The m-RL model is used to determine which classifiers are accurate when predicting labels. As there are many classifiers, the model must be able to determine which subsets of classifiers (many classifier may converge to the same prediction) correctly predict the labels and which do not. In order to train the m-RL model, a training data stream is created. To avoid any bias toward a particular source data stream, an equal number of instances are randomly sampled from each data stream.

Algorithm 1. Online Reinforce SSL Framework

```
Initialize sliding-window
Initialize base-classifier
for (x,y) in warm-up do
    base-classifier.fit(x,y)
    sliding-window.append(x,y)
end for
for (x_i, y_i) in all_instances do
    if y_i is not null then
        base-classifier.fit(x_i, y_i)
        sliding-window.add(x_i, y_i)
    else
        all-KNNs.fit(sliding-window)                          ▷ Train all KNN classifiers
        Meta-features = Obtain-Meta-features(x_i,sliding-window)
        action = m-RL(Meta-features)
        if action == discard-instance then
            Continue
        else
            ŷ = get-KNN-prediction(action)               ▷ Prediction of the KNN_action
            base-classifier.fit(x_i,ŷ)
        end if
    end if
end for
```

The feature space for the training set consists of the meta-feature vectors associated with these instances. The corresponding training label space consists of binary vectors indicating which subsets of classifiers accurately predict a label: for each classifier, a value of one indicates a good prediction, while a value of zero indicates an erroneous one:

$$\psi = \{\psi_k\}_{k=1}^{K}, \quad \begin{cases} \psi_k = 1 \Rightarrow \hat{y}_{i,k} = y_i \\ \psi_k = 0 \Rightarrow \hat{y}_{i,k} \neq y_i, \end{cases} \tag{6}$$

where y_i is the real label associated with instance i, $\hat{y}_{i,k}$ is the label predicted by the KNN classifier, and ψ is the so-called m-RL label; that is, the label employed to train the m-RL model. These m-RL labels should be distinguished from the original source labels. Indeed, their role is to indicate which KNN classifiers correctly label the instances, while the original labels refer to the classes associated with the meta-feature vectors.

Reinforcement learning involves four essential elements: the state S_t, the policy π_θ, the action A_t, and the reward R_t. The state consists of all extracted meta-features (Sect. 3.1). The policy, which is a learnable neural network, determines which action the agent should take to increase the expected reward. Given an unlabelled instance, the agent selects a classifier or a subset of classifiers for label prediction (recall that the m-RL label ψ allows a subset of classifiers to agree over the same label). Alternatively, it may discard an unlabelled instance if the probability associated with the action is too small. The reward is equal to one if the selected classifiers properly predict the label (positive reward), zero if the instance is discarded (neutral reward), and $-\rho$ when the prediction is incorrect (negative reward); the lower the value of $-\rho$, the greater the number of instances discarded.

The problem may be formulated in terms of a Markov decision process (MDP). This process aims to maximize the expected cumulative reward given either the current state (value function) or the current state and action (value-action function):

$$V_\theta(s_t) = \mathbb{E}_t\left[\sum_{k=0}^{\infty} \gamma^k R_{t+k+1} \,|s_t\right] \tag{7}$$

$$Q_\theta(s_t, a_t) = \mathbb{E}_t\left[\sum_{k=0}^{\infty} \gamma^k R_{t+k+1} \,|s_t, a_t\right], \tag{8}$$

where θ represents the parameters of the neural network associated with the policy, $\gamma \in [0,1]$ is a discount factor (weighting between the short-term and long-term rewards), $V_\theta(S)$ is the value function, $Q_\theta(S, A)$ is the value-action function, and $\mathbb{E}_t$ is the mathematical expectation. Reinforcement learning solves the MDP optimization problem by learning a value or a value-action function approximating the expected cumulative reward. In the present work, proximal policy optimization (PPO) is employed [16]. This policy is an actor-critic model, in which the critic estimates the value function, while the policy is inferred by the actor. The parameters are learned by minimizing a loss function consisting of three components: the actor loss function, the critic loss function, and an entropy term fostering exploration. The loss function associated with the critic is given by:

$$\mathcal{L}_t^C = \mathbb{E}_t\left[\left(V_\theta(s_t) - \hat{V}_\tau\right)^2\right], \tag{9}$$

where $\hat{V}_\tau$ is an estimate of the value function based on the data. The loss function associated with the actor is the clipped surrogate objective:

$$\mathcal{L}_t^A = \mathbb{E}_t\left[\min\left(r_t(\theta)\hat{A}_t,\ \Xi\left(r_t(\theta);1-\varepsilon,1+\varepsilon\right)\hat{A}_t\right)\right], \tag{10}$$

where

$$r_t(\theta) \triangleq \frac{\pi_\theta(a_t\,|s_t)}{\pi_{\theta^-}(a_t\,|s_t)}, \tag{11}$$

in which π_{θ^-} is the policy prior to the update, $\Xi(r_t(\theta);1-\varepsilon,1+\varepsilon)$ is a function that clips $r_t(\theta)$ in the interval $[1-\varepsilon, 1+\varepsilon]$, ε is the clipping hyperparameter, and $\hat{A}_t$ is the generalized advantage estimator [17], which is defined as:

$$\hat{A}_t \triangleq \delta_t + \sum_{t'}^{T} (\gamma\lambda)^{t'} \delta_{t+t'}, \tag{12}$$

where $\delta_t = r_t + \gamma V_\theta(s_{t+1}) - V_\theta(s_t)$ and λ is a hyperparameter. The advantage estimator determines the efficacy of the various actions while the Ξ function ensure that the policy remains stable by impeaching large updates [11,16]. The total loss function is presented as follows:

$$\mathcal{L}_t = \mathbb{E}_t \left[\mathcal{L}_t^A(\theta) + \mu_1 \mathcal{L}_t^C(\theta) + \mu_2 S\left(\pi_\theta\left(\cdot \,|s_t\right)\right) \right], \tag{13}$$

where S is the entropy, while μ_1 and μ_2 are weighting hyperparameters. The mathematical expectation is approximated via data sampling [11]. Once the training is completed, the PPO model can be employed for pseudo-labelling in a target data stream. Algorithm 1 outlines the generation of the pseudo-labels with the trained m-RL model in a streaming environment. Our experimental results are presented in the next section.

4 Experimental Evaluation

In this section, we describe our experimental setup. We will first detail the datasets used for the evaluation. Consequently, we will describe the state-of-the-art algorithms and the hyper-parameters used for comparing the methods.

4.1 Datasets

Table 1 summarizes the data streams used in our evaluation. We employed the following synthetic datasets:
Random radial basis function (RBF) generator is a synthetic dataset employing random centroids for generating new instances: 50,000 instances partitioned into 50 clusters were created [18].
Random RBF generator drift is a variation of RBF featuring concept drifts. In this variation, we can choose any number of centroids to contain drifts with a change speed [18]. As in RBF, we created 50,000 instances by setting the number of centroids to 50 and setting the number of drifting centroids to 25. Furthermore, the changing speed for the drifting centroids was set to 0.89 [18].
Waveform is another synthetic dataset used in our evaluation. It uses waveform formulas for creating synthetic instances, and we set the number of instances to 30,000 [18].

In addition, our evaluation also included the following real datasets:
Gas sensor array drift (GSD) is a multi-class dataset used to characterize the behavior of six gases at various pressures [19].
Electrical is a binary dataset that relates to whether the prices of electricity will go up or down [18].
Shuttle is an imbalanced NASA dataset containing space shuttle re-entry parameters: 80% of the instances belong to the rad flow class, while the remaining instances pertain to the remaining five classes [20].
Weather is a binary dataset containing 18,159 instances predicting whether it will rain or not.
Room occupancy aims predicting whether an office is occupied or not based on lighting, temperature, humidity, and carbon dioxide level [20].

PAMAP tracks six subjects performing 18 different activities [21].
Nursery concerns the ranking of applications for access to nurseries. It has 12,960 instances, eight features, and five classes [20].

Table 1. Characteristics of the data streams.

Data stream	Number of features	Number of instances	Number of classes
GSD	130	13616	6
Electrical	9	45311	2
RBF	11	50000	5
RBF drift	11	50000	5
Waveform	22	30000	3
Shuttle	10	58000	7
Weather	9	18159	2
PAMAP	54	59425	6
Room occupancy	6	20560	2
Nursery	9	12959	5

4.2 SSL Algorithms

Online Reinforce: For our method, we used the IOE algorithm [6] with Hoeffding trees (HTs) as base learners [18]. The parameters for the base models were set according to the recommended values in [6]. For the implementation of PPO, we used the stable baseline's [22] implementation with the recommended hyper-parameters[1]. Two fully-connected neural networks, each consisting of two hidden layers of 64 neurons with tangent hyperbolic (*tan-h*) activation functions, were used as the network architecture in both the actor and critic models. A linear activation function was used for the output layer of the critic model, while a *softmax* activation function was used for the output layer of the actor model. The learning rate for training the neural networks was set to 0.0003, while the sliding window size w was set to 1000. Further, the value for ρ was set to 15. For each dataset, we train ORSSL based on the other datasets and set the episode length to 50000. The KNN classifiers used were 1NN, 3NN, 5NN, and 7NN. All of the above-mentioned values were set by inspection.
OReSSL [3], introduced earlier, is a state-of-the-art method for dealing with SSL in data streams. Hyper-parameters were selected based on the recommended values in the OReSSL paper.
SPASC [4] is another online ensemble algorithm using cluster-based algorithms as base estimators. The batch size and the number of clusters are tuned for each individual dataset based on the values suggested in the original paper. SPASC

[1] Our repository is available at https://github.com/pvafaie/Online-Reinforce-SSL.

has two modes, namely Heuristic and Bayesian. In the results section, we show the results for the mode with the highest accuracy.

Self-training is a wrapper method widely used in the literature due to its low complexity and high performance [1]. Thus, we used it as our baseline. The confidence scores of the unlabelled instances, along with their predictions, were calculated. When the confidence was higher than a threshold, the prediction was used as the label for the unlabelled instance. In our experiments, we paired self-training with the IOE algorithm and set the confidence threshold to 91%.

4.3 Experimental Results

This section will compare the results of our algorithm with those of the algorithms described in Sect. 4.2. For each dataset, we use four different labelled rates, namely 30%, 10%, 3%, and 1%. For all the evaluations, we used prequential test-then-train [18] to obtain the accuracy of the classifiers. Table 2 demonstrates the accuracies of the SSL algorithms on all the datasets.

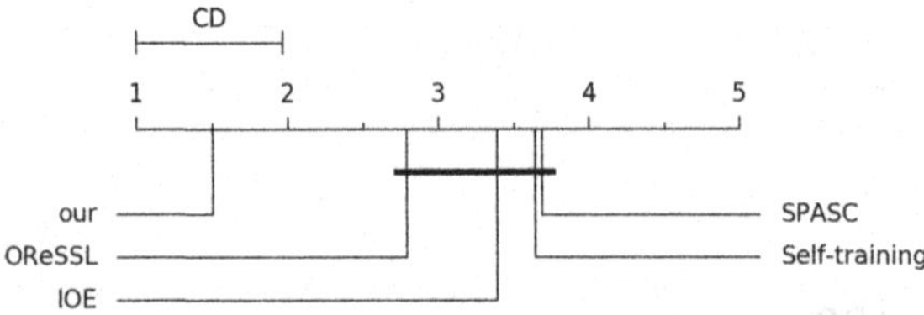

Fig. 2. The Nemenyi test results with α set to 0.05.

Based on the obtained results, we conclude that our Online Reinforce SSL algorithm often outperforms the state-of-the-art in terms of model accuracy, with OReSSL coming in second. This is the case for all levels of unlabelled data across the target datasets. Our results also confirm that combining the IOE algorithm with Online Reinforce SSL benefits learning. For instance, in the PAMAP dataset with 1% labelled data, using our algorithm with IOE resulted in 76% accuracy, while the standalone IOE had a 42.84% accuracy; that is 34% improvement by incorporating IOE into Online Reinforce.

4.4 Discussion

Our Online Reinforce SSL algorithm produces the highest accuracy values against most datasets, which had varying percentages of labelled instances. However, OReSSL yielded the highest values for the RBF and Gas (GSD) streams, and SPASC produced slightly higher values than Online Reinforce SSL for the Nursery stream, implying that, for these datasets, clustering benefits learning. The lower accuracies obtained when employing self-training indicates that pseudo-labelling adds too many incorrect labels, which leads to a reduction in the performance of the IOE algorithm [1].

Table 2. Accuracies of SSL algorithms for different label percentages. OR is shorthand for Online Reinforce, while ST stands for self-training.

Dataset	Label%	Acc				
		OR + IOE	OReSSL	SPASC	ST + IOE	IOE
GSD	1%	57.90	**67.17**	51.20	53.41	55.65
	3%	67.01	**81.77**	63.74	56.00	55.34
	10%	77.03	**91.19**	79.10	61.40	58.99
	30%	82.98	**96.12**	87.90	76.42	70.31
Electrical	1%	72.50	66.02	67.73	64.62	**73.24**
	3%	**75.06**	69.97	69.18	72.78	74.96
	10%	**79.83**	74.32	72.63	77.40	79.32
	30%	**83.28**	80.71	78.51	82.46	82.25
RBF	1%	86.43	**91.49**	63.71	50.91	55.67
	3%	85.60	**91.48**	66.12	51.37	61.12
	10%	87.03	**91.71**	69.77	69.06	76.77
	30%	86.65	**91.86**	71.39	82.26	81.63
RBF drift	1%	**56.83**	52.42	44.88	35.53	37.84
	3%	**58.62**	56.18	47.44	47.44	40.59
	10%	**61.06**	59.10	50.05	50.32	49.88
	30%	59.65	**60.74**	53.7	56.91	54.85
Waveform	1%	**84.40**	80.12	78.46	80.92	83.45
	3%	**83.53**	83.05	80.56	79.82	82.00
	10%	**83.02**	82.47	79.82	80.56	82.37
	30%	83.08	**83.47**	78.46	82.45	81.58
Shuttle	1%	**99.19**	98.00	96.88	96.07	94.07
	3%	**99.12**	98.18	96.20	94.70	94.76
	10%	**99.18**	98.87	96.65	97.38	97.91
	30%	**99.53**	99.20	97.42	99.40	99.20
Weather	1%	**73.85**	67.35	71.80	67.49	67.53
	3%	**72.63**	66.75	70.54	57.97	67.24
	10%	71.47	**72.71**	70.58	67.65	68.66
	30%	71.21	**76.42**	71.78	70.14	70.41
PAMAP	1%	**76.76**	56.86	42.27	64.29	42.83
	3%	**83.77**	61.26	43.30	75.79	60.08
	10%	**93.94**	66.39	46.58	83.59	91.77
	30%	**97.10**	66.66	49.30	94.97	96.58
Room occupancy	1%	**98.89**	97.39	98.05	98.67	98.73
	3%	**98.90**	98.36	98.06	98.65	98.39
	10%	**98.87**	98.52	97.84	98.82	98.83
	30%	**98.89**	98.68	98.42	98.87	98.84
Nursery	1%	73.78	60.80	**81.71**	71.50	70.36
	3%	80.51	67.59	**84.59**	75.74	78.49
	10%	83.33	73.77	**86.30**	79.85	81.38
	30%	87.35	78.28	**87.65**	84.84	86.71
AVG rank		**1.5**	2.7875	3.6875	3.6375	3.3875

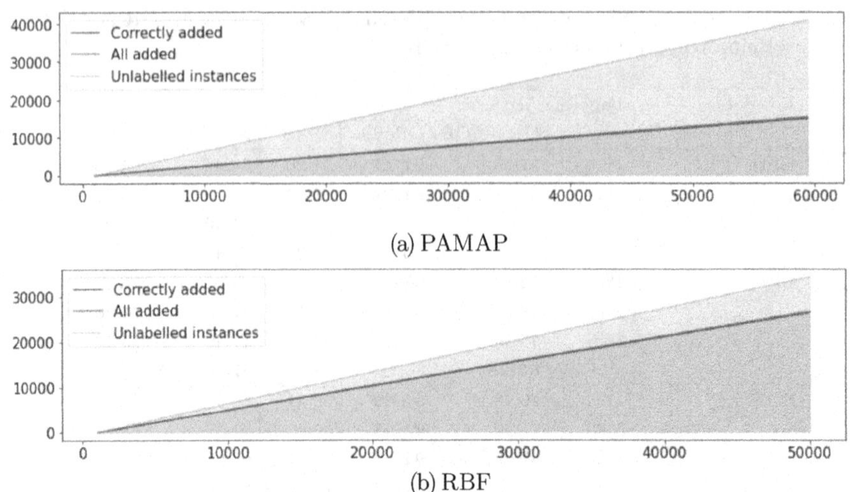

(a) PAMAP

(b) RBF

Fig. 3. Number of unlabelled instances correctly labelled by the Online Reinforce SSL algorithm. Green show all unlabelled instances, blue depicts correctly labelled instances, while red corresponds to instances labelled incorrectly. (Color figure online)

Next, we present the results of the Nemenyi posthoc test shown in Fig. 2. This test was used to highlight the contrasts in the algorithms against all the datasets. The figure shows that there is a critical difference between our Online Reinforce SSL algorithm and OReSSL with α set to 0.05. This difference can also be seen in the average ranks, which were 1.5 for Online Reinforce SSL, 2.78 for OReSSL, and 3.68 for SPASC.

Recall that the Online Reinforce SSL algorithm is a wrapper method that is meant to be paired with numerous base classifiers. When compared to other wrapper methods, such as self-training, our algorithm performs significantly better, indicating that the use of meta-features increases the percentage of correct instances added to the training set.

For instance, Fig. 3 depicts the number of Online Reinforce SSL labels against the PAMAP and RBF datasets, each with 70% missing labels. The results indicate that the resultant model is clearly able to distinguish between data that can and cannot be labelled, thus leading to a reduction in the number of false positives. That is, in both figures, the red section is insignificant compared to the green and blue sections, which confirms the value of our meta-reinforcement learning approach.

Figure 4 illustrates the percentage of each action selected by the Online Reinforce SSL algorithm for the PAMAP and RBF datasets with 70% unlabelled data. Note that for the PAMAP dataset, the agent mostly chose 1NN, while for RBF, the most chosen action was 3NN. Interestingly, the percentages for the actions are different for each dataset, meaning - as may be expected - that selecting actions is dependent on the characteristics of the data.

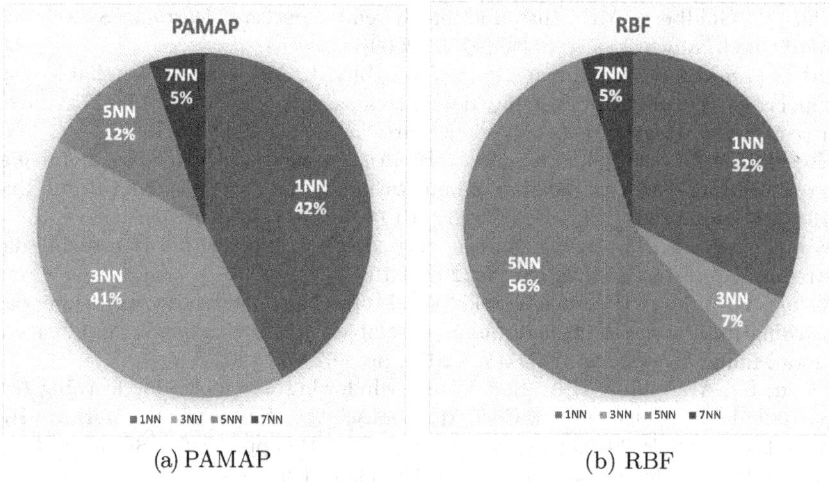

(a) PAMAP (b) RBF

Fig. 4. Percentages of the actions (KNN learners) chosen by Online Reinforce SSL for the PAMAP and RBF datasets with 30% labelled data.

5 Conclusion and Future Work

This paper introduced the Online Reinforce methodology for dealing with missing labels in data streams. Our algorithm employs KNN classifiers to construct meta-features that subsequently act as input to an m-RL agent, thus enabling our online learning algorithm to select the best pseudo-label for an unlabelled instance as it arrives in the stream. A crucial component of our approach is the fact that learning of the meta-RL agent is occurring across multiple domains while using transferable meta-features. We compared our Online Reinforce method with state-of-the-art Online SSL approaches and showed that our algorithm consistently yielded promising results for numerous benchmark data streams.

In the future, we plan to investigate the use of the Online Reinforce SSL algorithm in highly imbalanced domains, where the number of instances in one or more classes is far less than in the others. It follows that using online classifiers, other than KNN, should be investigated in our future work. Moreover, we plan to conduct ablation studies to explore the influence of the various meta-features on the outcome of the meta-RL agent. In addition, handling stream susceptible to evolving data distributions and emerging concepts is another area of future research. Finally, the use of co-training in a streaming environment requires further investigation [5].

References

1. van Engelen, J.E., Hoos, H.H.: A survey on semi-supervised learning. Mach. Learn. **109**(2), 373–440 (2019). https://doi.org/10.1007/s10994-019-05855-6

2. Zhu, X., Goldberg, A.B.: Introduction to semi-supervised learning. Synth. Lect. Artif. Intell. Mach. Learn. **3**(1), 1–130 (2009)
3. Ud Din, S., Shao, J., Kumar, J., Ali, W., Liu, J., Ye, Y.: Online reliable semi-supervised learning on evolving data streams. Inf. Sci. **525**, 153–171 (2020). https://www.sciencedirect.com/science/article/pii/S0020025520302322
4. Hosseini, M.J., Gholipour, A., Beigy, H.: An ensemble of cluster-based classifiers for semi-supervised classification of non-stationary data streams. Knowl. Inf. Syst. **46**(3), 567–597 (2015). https://doi.org/10.1007/s10115-015-0837-4
5. Wang, Y., Li, T.: Improving semi-supervised co-forest algorithm in evolving data streams. Appl. Intell. **4**(10), 3248–3262 (2018)
6. Vafaie, P., Viktor, H., Michalowski, W.: Multi-class imbalanced semi-supervised learning from streams through online ensembles. In: International Conference on Data Mining Workshops (ICDMW) 2020, pp. 867–874 (2020)
7. Floyd, S.L.A., Viktor, H.L.: Soft voting windowing ensembles for learning from partially labelled streams. In: Ceci, M., Loglisci, C., Manco, G., Masciari, E., Ras, Z. (eds.) NFMCP 2019. LNCS (LNAI), vol. 11948, pp. 85–99. Springer, Cham (2020). https://doi.org/10.1007/978-3-030-48861-1_6
8. Hospedales, T., Antoniou, A., Micaelli, P., Storkey, A.: Meta-learning in neural networks: a survey, arXiv preprint arXiv:2004.05439 (2020)
9. Finn, C., Abbeel, P., Levine, S.: Model-agnostic meta-learning for fast adaptation of deep networks. In: International Conference on Machine Learning. PMLR, pp. 1126–1135 (2017)
10. Mishra, N., Rohaninejad, M., Chen, X., Abbeel, P.: A simple neural attentive meta-learner, arXiv preprint arXiv:1707.03141 (2017)
11. Zha, D., Lai, K.-H., Wan, M., Hu, X.: Meta-AAD: active anomaly detection with deep reinforcement learning, arXiv preprint arXiv:2009.07415 (2020)
12. Settles, B.: Active learning. Synth. Lect. Artif. Intell. Mach. Learn. **6**(1), 1–114 (2012)
13. Haque, A., Khan, L., Baron, M.: Sand: semi-supervised adaptive novel class detection and classification over data stream. In: Thirtieth AAAI Conference on Artificial Intelligence (2016)
14. Wagner, T., Guha, S., Kasiviswanathan, S., Mishra, N.: Semi-supervised learning on data streams via temporal label propagation. In: International Conference on Machine Learning. PMLR, pp. 5095–5104 (2018)
15. Shao, J., Huang, C., Yang, Q., Luo, G.: Reliable semi-supervised learning. In: 2016 IEEE 16th International Conference on Data Mining (ICDM), pp. 1197–1202. IEEE (2016)
16. Schulman, J., Wolski, F., Dhariwal, P., Radford, A., Klimov, O.: Proximal policy optimization algorithms, arXiv preprint arXiv:1707.06347 (2017)
17. Schulman, J., Moritz, P., Levine, S., Jordan, M., Abbeel, P.: High-dimensional continuous control using generalized advantage estimation, arXiv preprint arXiv:1506.02438 (2015)
18. Bifet, A., Gavaldà, R., Holmes, G., Pfahringer, B.: Machine Learning for Data Streams with Practical Examples in MOA. MIT Press (2018). https://moa.cms.waikato.ac.nz/book/
19. Vergara, A., Vembu, S., Ayhan, T., Ryan, M.A., Homer, M.L., Huerta, R.: Chemical gas sensor drift compensation using classifier ensembles. Sens. Actuators B: Chem. **166–167**, 320–329 (2012). http://www.sciencedirect.com/science/article/pii/S0925400512002018
20. Dua, D., Graff, C.: UCI machine learning repository (2017). http://archive.ics.uci.edu/ml

21. Reiss, A., Stricker, D.: Introducing a new benchmarked dataset for activity monitoring. In: 2012 16th International Symposium on Wearable Computers, pp. 108–109 (2012)
22. Raffin, A., Hill, A., Ernestus, M., Gleave, A., Kanervisto, A., Dormann, N.: Stable baselines3 (2019). https://github.com/DLR-RM/stable-baselines3

Dissecting FLOPs Along Input Dimensions for GreenAI Cost Estimations

Andrea Asperti[1(✉)], Davide Evangelista[2], and Moreno Marzolla[1]

[1] Department of Informatics: Science and Engineering (DISI), University of Bologna, Bologna, Italy
asperti@cs.unibo.it

[2] Department of Mathematics, University of Bologna, Bologna, Italy

Abstract. The term GreenAI refers to a novel approach to Deep Learning, that is more aware of the ecological impact and the computational efficiency of its methods. The promoters of GreenAI suggested the use of Floating Point Operations (FLOPs) as a measure of the computational cost of Neural Networks; however, that measure does not correlate well with the energy consumption of hardware equipped with massively parallel processing units like GPUs or TPUs. In this article, we propose a simple refinement of the formula used to compute floating point operations for convolutional layers, called α-FLOPs, explaining and correcting the traditional discrepancy with respect to different layers, and closer to reality. The notion of α-FLOPs relies on the crucial insight that, in case of inputs with multiple dimensions, there is no reason to believe that the speedup offered by parallelism will be uniform along all different axes.

1 Introduction

Artificial Intelligence, especially in its modern incarnation of Deep Learning, has achieved remarkable results in recent years, matching – and frequently trespassing – human capabilities in a number of different tasks. These techniques usually require the deployment of massive computational resources, with huge implications in terms of energy consumption. To make a couple of examples the hyper-realistic Generative Adversarial Network for face generation in [19] required training on 8 Tesla V100 GPUs for 4 days; the training of BERT [12], a well known generative model for NLP, takes about 96 h on 64 TPU2 chips. Researchers at the University of Massachusetts [26] have recently performed a life cycle assessment relative to the training of large state-of-the-art AI models, discovering that the process can emit a quantity of carbon dioxide roughly equivalent to the lifetime emissions of five medium cars. Other authors reached similar conclusions [20].

Until a few years ago, the ecological impact of artificial intelligence was entirely neglected by researchers and industry, who were mostly focused on improving performance at any cost. However, this has changed in recent years, with a growing awareness that this trend of research is not sustainable any more [28], and an increased attention towards energetic efficiency [27].

G. Nicosia et al. (Eds.): LOD 2021, LNCS 13164, pp. 86–100, 2022.
https://doi.org/10.1007/978-3-030-95470-3_7

The GreenAI paper [25] summarizes well the goal and objectives of the new philosophy: it promotes a new practice in Deep Learning, that is more focused on the social costs of training and running models [2,7,15], encouraging the investigation of increasingly efficient models [5,21].

To this aim, it is essential to identify widely acceptable and reliable metrics to assess and compare the cost and efficiency of different models. Several metrics are investigated and discussed in [25]; in conclusion, the number of Floating Point Operations (FLOPs) is advocated and promoted, since it is easily computed for Neural Networks while offering a hardware independent, schematic but meaningful indication of the actual computation cost of the model [20].

Unfortunately, the mere computation of FLOPs does not cope well with the massively parallel architectures (GPU and TPU) typically used in Deep Learning [17]. Efficient implementation of neural networks on these architectures depends both on complex algorithms for General Matrix Multiplication (GEMM) [18] and sophisticated load balancing techniques [13] splitting the workload on the different execution units. As we shall see, these algorithms usually perform better for specific layers and, especially, *along specific axes* of the input dimension of these layers.

Our claim is that it is possible to study the performance of neural layers (especially, convolutions) as "black boxes", measuring the execution time for a number of different configurations, and separately investigating the execution time for increasing dimensions along different axis.

As a result, we propose a simple correction to the formula used to compute FLOPs for convolutional layers, that provides better estimations of their actual cost, and helps to understand the discrepancy with respect to the cost of different layers.

Organization of the Article. This paper has the following structure. In Sect. 2 we briefly discuss some possible metrics for measuring the efficiency of models; we particularly focus on FLOPs, discussing their computation for some basic operations relevant for Neural Networks. In Sect. 3 we introduce the GEMM (GEneral Matrix Multiply) operation, that helps to understand the canonical computation of FLOPs for the Convolution layers. In Sect. 4 we present some experiments which show that, if Convolutions are executed on GPU, FLOPs are not a good measure for efficiency. That is the motivation for introducing a correction, that we call α-FLOPs, defined and discussed in Sect. 5. Section 6 offers more experimental results, validating the formula with respect to growing input dimensions along specific axes.

2 Measures of Efficiency

In this section we review some of the metrics that can be used to measure the efficiency of an AI algorithm, following the discussion of [25].

Carbon Emission. As already remarked in the introduction, the present work is motivated by the need to reduce the energy consumption of training large state-of-the-art AI models. Unless a significant fraction of such energy comes from renewable sources, reducing the power required for AI training means that less carbon dioxide is released into the atmosphere. Unfortunately, precise quantification of carbon emission associated with computational tasks is impractical, since it depends both on the hardware hosting the computation, and also on the local energy production and distribution infrastructure.

Number of Parameters. The number of parameters of a Deep Learning model is an interesting and hardware-independent measure of the complexity of models. Unfortunately, the number of parameters alone is poorly correlated with the total training time, since parameters may refer to different operations. For example, Convolutional Layers have relatively few parameters, relative to the kernel of the convolution; this does not take into account the actual cost of convolving the kernel over the input.

Execution Time. The total running time is a natural measure of efficiency: faster algorithms are better. Execution time depends on the number of instructions executed and hence is strictly correlated with the total energy consumption [24]; therefore, it is a good proxy of power usage when direct energy measurement is impractical. There are a couple of important considerations that must be made when considering execution time as a metric: (*i*) it requires an implementation of the algorithm being measured, which may take time and effort to be developed; (*ii*) execution time is hardware- and language-dependent, since it depends on both the underlying hardware and on the efficiency of the compiler/interpreter.

FLOPs. The number of FLoating Point OPerations (FLOPs) is a metric that is widely used in the context of numerical computations [14,22,23,29]. It is defined as the total count of elementary machine operations (floating point additions and multiplications) executed by a program. Floating point operations have a latency of several CPU cycles on most current processor architectures [3,9,10], although the use of pipelining, multiple-issue and SIMD instructions significantly increase the throughput. In general, floating point operations have higher latency than most of the other CPU instructions (apart from load/stores from/to main memory, where memory access is the bottleneck); therefore, they tend to dominate the execution time of numerical algorithms. For this reason, the number of floating point operations is used as a proxy for the execution time of a program.

As an example, suppose that v and w are n-dimensional arrays. Then, the inner product between v and w

$$\langle v; w \rangle = \sum_{i=1}^{n} v_i w_i \tag{1}$$

requires n multiplications and $n-1$ additions, for a total of $2n-1$ FLOPs. Similarly, the matrix-vector product between an $m \times n$ matrix A and an n-dimensional vector v requires m inner product, for a total of $2mn - m$ FLOPs.

Since operations similar to (1), where a sequence of multiplications are added together, are very common, modern CPUs supports FMA (Fused Multiply-Add) instructions, where a multiplication followed by an addition are executed as a single operation and require less time than two separate instructions. For this reason, the definition of FLOPs is usually modified to be the total number of FMA operations needed for a full iteration of an algorithm. With this definition (that it is usually followed by some authors), the inner product of two n-dimensional arrays requires n FLOPs, while the product between an $m \times n$ matrix with an n-dimensional vector requires nm FLOPs. Nonetheless, since we are interested in measuring the performance under massively parallel architectures, through this paper we will follow the classical definition of FLOPs.

3 Computation of FLOPs for Basic Layers

The basic operation that dominates training of Neural Network models is the dense matrix-matrix product. This operation is often referred in the technical literature as GEMM (for *GEneral Matrix Multiply*), owing its name to the x`GEMM` family of functions provided by the Basic Linear Algebra Subprograms (BLAS) library [6]. BLAS is a widely used collection of subroutines implementing basic operations involving vectors and matrices, such as vector addition, dot product, vector-matrix multiplication and so on; these functions act as building blocks on which more complex linear algebra computations can be programmed. Being at the core of many applications, the performance of BLAS primitives are critical, so most hardware vendors provide their own optimized implementations, e.g., `cuBLAS` for nVidia GPUs [11], and `clBLAS` for OpenCL devices [8], including various brands of GPUs and multicore processors.

A GEMM operation takes the general form:

$$\mathbf{C} \leftarrow \alpha\mathbf{AB} + \beta\mathbf{C} \tag{2}$$

where $\mathbf{A}, \mathbf{B}, \mathbf{C}$ are matrices of compatible size, and α, β are scalars. The matrix-matrix product $\mathbf{C} \leftarrow \mathbf{AB}$ is a special case of (2) where $\alpha = 1$, $\beta = 0$.

Assuming that the size of $\mathbf{A}$ is $m \times k$ and the size of $\mathbf{B}$ is $k \times n$, then the size of $\mathbf{C}$ must be $m \times n$ and the direct computation of (2) using vector dot products requires:

- $2mkn + mn$ FLOPs for the matrix product $\alpha\mathbf{AB}$, assuming that dot products are implemented with an inner loop involving a multiply-accumulate operation like $s \leftarrow s + x_i y_i$
- mn FLOPs for the computation of $\beta\mathbf{C}$
- mn additional FLOPs for the computation of the matrix sum $\alpha\mathbf{AB} + \beta\mathbf{C}$

from which we get that a total count of $2mkn + mn + mn + mn = mn(2k + 3)$ FLOPs are required for the general GEMM. Neglecting lower-order terms we can approximate the operation count with $2mkn$.

We can apply this result for the layers of a Neural Network. Consider a Dense layer, with input and output dimensions D_{in} and D_{out}, respectively. We need to compute the product between the weight matrix of size $D_{out} \times D_{in}$ and the input, plus a bias term B of dimension D_{out}; therefore, the number of FLOPs is

$$2D_{in}D_{out} - D_{in} + D_{out}$$

As above, we omit the lower order terms as they are asymptotically negligible. As a consequence, we will consider a Dense layer to have a number of FLOPs equal to

$$2D_{in}D_{out} \tag{3}$$

The case of a convolutional layer is slightly more complex. Let us consider the case of a 2D convolution. Let (W_{in}, H_{in}, C_{in}) the dimension of the input (written with the notation (*Width, Height, Channels*)), $(W_{out}, H_{out}, C_{out})$ the dimension of the output (depending on the stride and number of kernels), and let K_1, K_2 be the dimensions of the kernel. Then, the number of FLOPs is given by

$$2 \cdot \underbrace{K_1 \cdot K_2 \cdot C_{in}}_{\text{kernel dim}} \cdot \underbrace{W_{out} \cdot H_{out}}_{\text{input dim}} \cdot \underbrace{C_{out}}_{\text{output dim}} \tag{4}$$

In the following, we shall frequently consider the case of convolutions with stride 1 in "same" padding modality. In this case, $W_{in} = W_{out}$ and $H_{in} = H_{out}$, so we shall drop the subscripts, and just write W and H. Moreover, in the frequent case kernels are squared, we drop the subscripts in K_1, K_2 and just write K.

4 The Problem of Convolutions

A dense layer of dimension $D_{in} \times D_{out}$ is the same as a unary convolution ($K = 1$) with $C_{in} = D_{in}$, $C_{out} = D_{out}$ and $H = W = 1$; it is easy to experimentally check that both require the same time to be evaluated. However, as soon as we distribute the total number of FLOPs of Eq. (4) across different dimensions, we observe a significant speedup, that has no justification in terms of FLOPs. This raises concerns about the use of FLOPs for estimating running time (and hence energy consumption). In this section we provide empirical evidence of this phenomenon.

In Fig. 1, we compare the time required to evaluate a dense layer with several different convolutional layers with a same amount of FLOPs computed according to (4); the execution time has been measured on an NVIDIA Quadro T2000 graphics card and a Intel Core i7-9850H CPU. Times are averaged over 2000 experiments for each scenario.

327.68 M FLOPs	
Dense layer	
(D_1, D_2)	time (ms)
$(12800, 12800)$	6.401
Convolutional layers	
$(W, H, C_{in}, C_{out}, K_1, K_2)$	time (ms)
$(1, 1, 12800, 12800, 1, 1)$	6.392
$(1, 2, , 6400, 12800, 1, 1)$	3.224
$(2, 2, 6400, 6400, 1, 1)$	1.626
$(4, 4, 3200, 3200, 1, 1)$	0.454

(a)

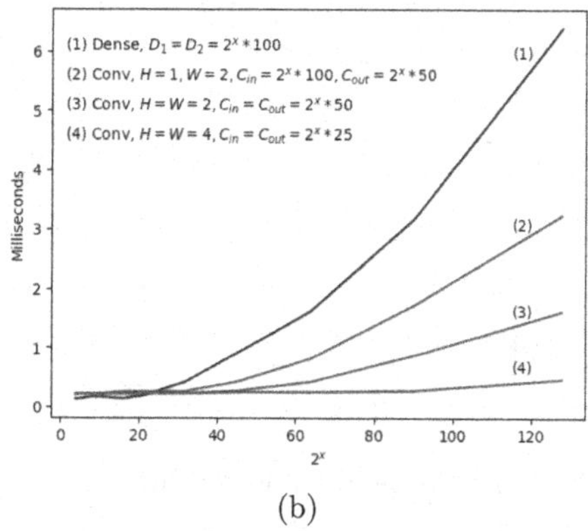

(b)

Fig. 1. Comparison of execution times for Dense and Convolutional layers with the *same* amount of FLOPs. In Table (a) we provide numerical values for layers with 327.68 Million FLOPs; in the right we show the execution time of similar configurations for increasing dimensions. All layers for a given value of 2^x (i.e. along any vertical section) have the same amount of FLOPs.

In particular, in Table 1a we evaluate a scenario of maximum size compatible with our hardware, corresponding to a Dense layer of size 12800×12800 $(163, 852, 800$ parameters), and compare it with several different convolutional layers with the same total amount of FLOPs. The dense layer takes about 6.4 milliseconds (ms), while a unary convolution with $C_{in} = C_{out} = 3200$ on an input of spatial dimension 4×4 just takes 0.46 ms, approximately 16 times faster.

In Fig. 1b, we repeat the same experiment, varying the total amount of flops with powers of 2. For the dense layer we go from dimension 100×100 to dimension $(100 \times 2^7) \times (100 \times 2^7)$.

In the following experiments, we keep the number of FLOPs constant while we increase some dimensions and proportionally decrease others. If (4) had a good correlation with time, we should observe straight horizontal lines.

In all experiments, we consider four different amounts of FLOPs identified by different colors: 2025×10^6 (red line in Fig. 2), 900×10^6 (green line), 490×10^6 (orange line) and 225×10^6 (blue line). We progressively increase K from 1 to 30. In the first experiment, we compensate it by enlarging the input and output dimension of channels (C_{in} and C_{out}), keeping a constant (small) spatial dimension 10×10.

In the second test we compensate the growing convolutions by reducing the spatial dimensions, starting from an initial dimension of 300×300. Channels are constant, in this case. Result are reported in Fig. 2.

In the case of the first experiment (Fig. 2a), apart from the exceptional performance of 1×1 convolutions already discussed in [17], we observe the expected constant behavior. However, we have a completely different result in the case of the second experiment (Fig. 2b). Here the execution time increases with the

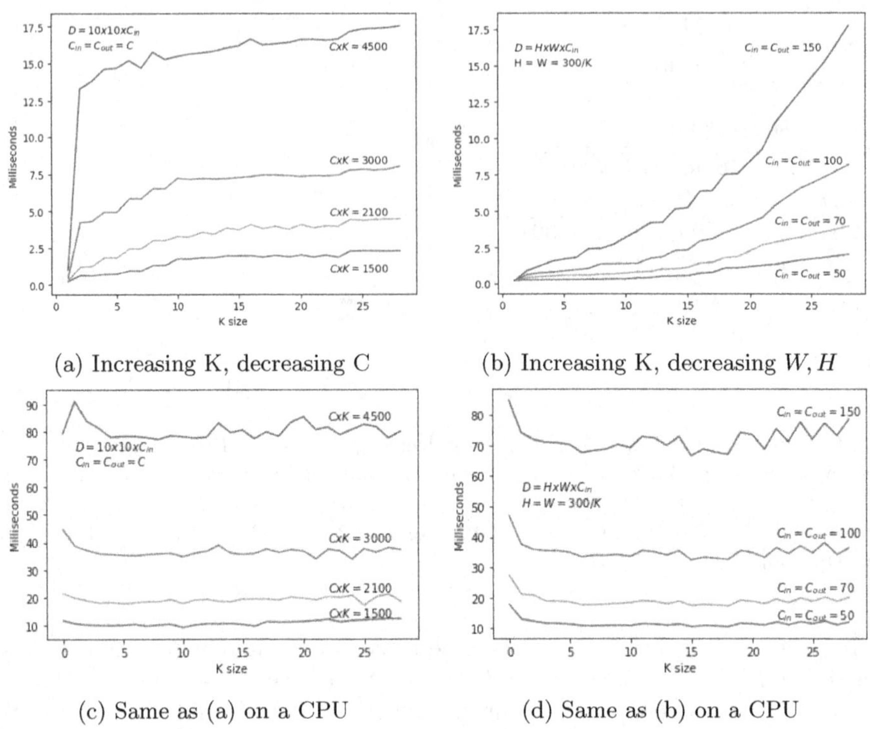

(a) Increasing K, decreasing C

(b) Increasing K, decreasing W, H

(c) Same as (a) on a CPU

(d) Same as (b) on a CPU

Fig. 2. Execution time vs different input dimensions, keeping the number of FLOPs constant. In plot (a) we increase K and proportionally decrease C_{in} and C_{out}. In plot (b) we increase K and proportionally decrease W and H. We would expect *constant lines*, but this is not the case. In plots (c) and (d) we repeat the experiment on a (single core) CPU, instead of a GPU. (Color figure online)

kernel dimension, possibly at a quadratic rate; this growth should have been compensated by the simultaneous decrease along both spatial dimensions, but clearly this is not the case.

By comparing the results of the two experiments, we can draw another conclusion. Remember that the number of FLOPs along lines of the same color is the same; therefore, the nonlinear behaviour in Fig. 2b is not due to an *overhead* but, on the contrary, there is an important *speed up* of the computation of increasing relevance for small kernels. In other words, the formula computing FLOPs is *overestimating* the total number of operations, presumably because it does not take into consideration the fact that convolutions can be *easily parallelized* along spatial dimensions (but not quite so along kernel dimensions).

The goal of the work is to derive a simple correction to the formula for computing FLOPs explaining the observed behaviours. The correction might depend on the specific hardware, but it should be possible to evaluate the relevant parameters in a simple way.

5 α-FLOPs

In this section we introduce our correction to the formula for computing FLOPs, that we call α-FLOPs. Instead of FLOPs, that count the total number of floating point operations, α-FLOPs provide an estimation of the "perceived" FLOPs, that are less than FLOPS due to parallelism. The crucial idea is that when we run in parallel an algorithm with a multidimensional input there is no reason to suppose that the total number of operations have similar latency along different dimensional axes. Our proposal is to adjust the formula for computing FLOPs by multiplying it by the following scaling factor:

$$\alpha_K(S) = \left(\frac{S_K + \beta_K(S - S_K)}{S}\right)^{\gamma_K} \tag{5}$$

where $S = W \times H$, and $0 < \beta_K \ll 1, 0 < \gamma_K \leq 1$, and $1 \leq S_K \leq S$ ($S_1 = 1$) are parameters (moderately) depending from K. We call α-FLOPs the correction to the usual formula for FLOPs by the previous factor.

The parameters β_K and γ_K can be easily evaluated by regression on a given GPU/TPU. Although they are hardware dependent, some preliminary investigations seem to suggest that fluctuations are smaller than expected.

For the purposes of this article, using an Invida Quadro T2000 GPU we obtained good predictions just distinguishing two cases: $K = 1$ and $K > 1$. For $K = 1$, $\beta_K = 0.02$ and $\gamma_K = .99$; for $K > 1$, $\beta_K = 0.001$ and $\gamma_K = .56$.

Before discussing the main properties of $\alpha_K(S)$, let us have a look at the prediction of the execution time (dashed line) for the problematic experiments shown above. More examples will be presented in Sect. 6.

The experiment in Fig. 1 is replicated, with the time predicted by means of α-FLOPs, in Fig. 3b. In the Table on the left, we give the computed and predicted times for the convolutional configurations (1–4) with 327.68M FLOPs.

Similarly, in Fig. 4 we show the predicted execution time for the experiments of Fig. 2.

5.1 Main Properties of the α-correction

Before discussing our intuition behind (5), let us point out some of its distinctive properties. First of all the equation can be rewritten in the following, possibly more readable form:

$$\alpha_K(S) = \left((1 - \beta_K) \times \frac{S_K}{S} + \beta_K\right)^{\gamma_K} \tag{6}$$

From that, the following properties can be observed:

1. $\alpha_K(S) < 1$ for any K and S. This is evident, given the constraint $S_k < S$.

327.68 M FLOPs		
Convolutional layers		
config.	time	predicted
(1)	6.392	6.154
(2)	3.224	3.351
(3)	1.626	1.847
(4)	0.454	0.611

(a)

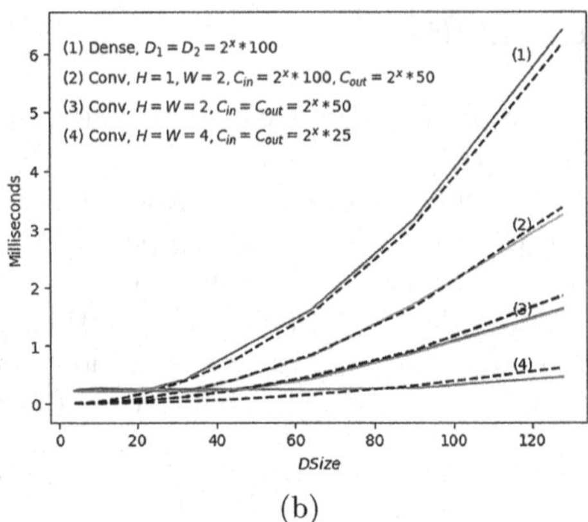

(b)

Fig. 3. Predicted execution time by means of α-FLOPs for the same convolutional configurations of Fig. 1; in (b) predictions are depicted as dashed lines.

2. If $\beta_K = 1$, then $\alpha_K(S) = 1$, independently from γ_K and S. In this case, we recover the original expression for FLOPs, that is hence a subcase with no additional speedup.
3. $\alpha_1(1) = 1$ independently from β_K and γ_K. This is due to the fact that $S_1 = 1 = S$. The case $S = 1, K = 1$ is important since, as discussed at the beginning of Sect. 3, it gives the relation between convolutional and dense layers, and for a dense layer we want no correction. Moreover, the fact that the fundamental equation $\alpha_1(1) = 1$ holds independently from β and γ improves the stability of the property.
4. The formula with $\gamma = 1$ already gives reasonable approximations. However, it tends to underestimate the execution time for large S, in a more sensible way for increasing values of K. By raising the correction to a power smaller than 1 we mitigate this phenomenon without giving up the appealing properties provided by β.
5. The parameter S_K increases slowly with K. The point is to take into account a plausible overhead for growing dimensions of the kernel, especially when passing from $K = 1$ to $K > 1$. This constant can be possibly understood as a minimum expected spatial dimension for kernels larger than 1. It does not make much sense to apply a kernel of dimension 3×3, on an input of dimension 1×1, and it is hard to believe that reducing the dimension of the input below the dimension of the kernel may result in any speedup. However, fixing $S_K = K$ does not seem to be the right solution.

5.2 Rationale

We now provide a possible explanation for the α-FLOP formula (6). Let us consider a computational task requiring a given amount of work W. Let $\beta \in [0, 1]$ be the fraction of that work that can be executed in parallel; therefore, the

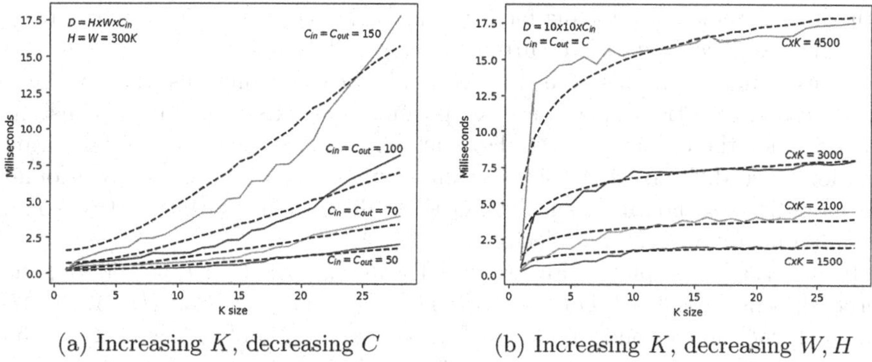

(a) Increasing K, decreasing C

(b) Increasing K, decreasing W, H

Fig. 4. Predicted execution time by means of α-FLOPS, depicted as dashed lines, for the same convolutional configurations of Fig. 2

sequential portion of the task is $(1-\beta)W$. Let us scale the problem by a factor $N > 1$; in a purely sequential framework, the amount of work would become NW. However, Gustafson's law [16] suggests that when we scale the size of a parallel task, *the sequential part tend to remain the same.* This means that the amount of work *actually* done by the parallel program is $(1-\beta)W + \beta NW$. The ratio between the *actual* amount of work done by the parallel version versus the *expected* amount of work done by the serial version is:

$$\frac{(1-\beta)W + \beta NW}{NW} = \frac{1-\beta}{N} + \beta \tag{7}$$

where we readily recognize the backbone of Eq. (6). We already discussed the small adjustments we had to do to this formula to fit it to the empirical observations.

Gustafson's law describes the theoretical speedup of a parallel task in terms of growing resources, on the reasonable assumption that programmers tend to set the problem size to fully exploit the available computing power. Gustafson's law was meant to address a shortcoming of a similar law formulated by Amdahl [1], that on the other hand assumes a fixed workload that does not depend on the number of execution units used.

In our case, computational resources are fixed and we focus on different input dimensions. Our assumption is that suitable programs and load balancing techniques will optimize the use of resources, eventually resulting in different speedups along different spatial dimensions.

6 Additional Experimental Results

We conducted several experiments to assess the rate of grow of the execution time along different input dimensions. Data providing the base for this paper are available on Github (https://github.com/asperti/alpha_flops_dataset), together with analysis tools and plotting facilities, including the predictions by means of α-FLOPs. Additional data are currently being collected.

All experiments discussed in this Section involve convolutions where we progressively increase the dimension of a specific input axis x, keeping a constant dimension for the others axes X_c. For each experiment, we draw the execution time for three different dimensions of an auxiliary axis x_{aux} in X_c. We found this more understandable than plotting three dimensional surfaces, that would be quite difficult to draw and decipher.

In the case of the plot in Fig. 5a, x is W, and x_{aux} is C_{in}; $H = 100$ and the Kernel dimension is 3×3. For Fig. 5b, x is C_{out}, and x_{aux} is C_{in}; $H = W = 100$ and the kernel dimension is 3×3. For Fig. 6a, x is C_{out}, and x_{aux} is H; $C_{in} = 50$ and the kernel dimension is 1×1. Finally, for Fig. 6b, x is C_{in}, and x_{aux} is K; $H = W = 10$ and $C_{out} = 1000$.

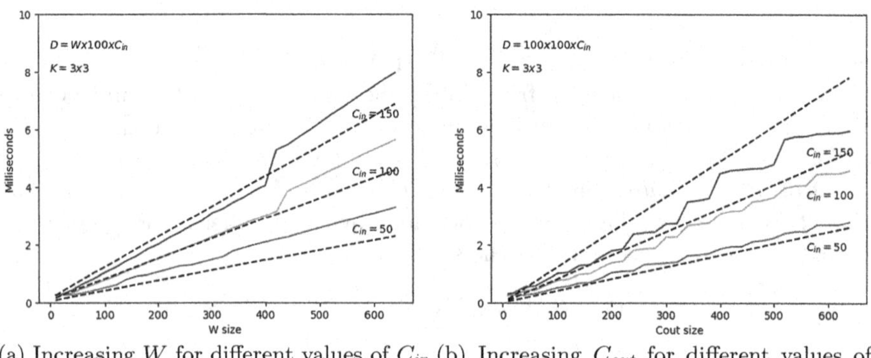

(a) Increasing W for different values of C_{in} and a kernel of dimension 3x3

(b) Increasing C_{out} for different values of C_{in} *and a kernel of dimension* 3x3

Fig. 5. Execution time and predictions by means of α-FLOPs (dashed lines)

6.1 Dense Layers vs Batchsize

We already observed that a dense layer can be assimilated to a convolutional layer with kernel 1×1 and spatial dimension 1. In this perspective, it is plausible to conjecture that the batchsize can be assimilated to a spatial dimension. Indeed, the general wisdom across Deep Learning researchers and practitioners is that, for making predictions over a large set of data – e.g., over the full training or test set – it is convenient to work with a batchsize as large as possible, compatibly with the resource constraints of the underlying hardware, e.g., memory. This has no justification in terms of FLOPs, since the total number of operations is always the same; however, using a large batchsize is much more efficient.

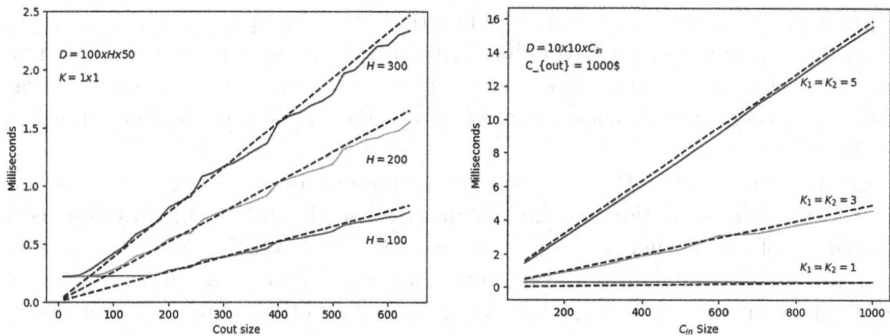

(a) Increasing C_{out} for different values of H, with a kernel of dimension 1x1

(b) Increasing C_{in} for different kernels 1x1, 3x3 and 5x5

Fig. 6. Execution time and predictions by means of α-FLOPS (dashed lines)

To test this behaviour, we take large dense layers (at the limit of our hardware capacity), and then apply them to inputs with increasing batch size.

The results are summarized in Fig. 7. Under the FLOPs assumption, the lines should be straight lines *departing from the origin*. In terms of α-FLOPs we start with the cost relative to batchsize 1, and then slowly grow along the batchsize dimension, reflecting the experimental behaviour.

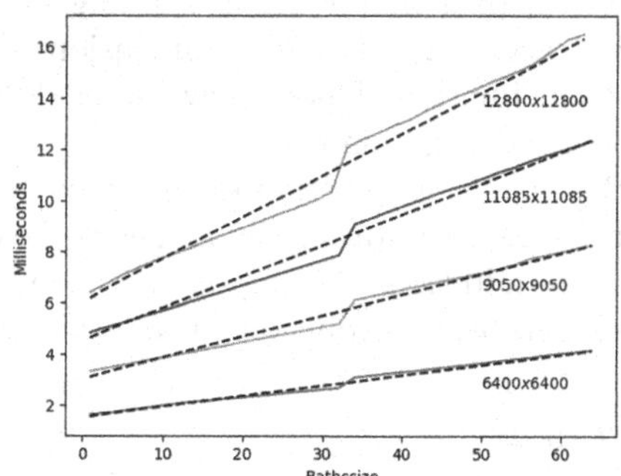

Fig. 7. Computational time for dense layer increasing the batchsize. The jump between 32 and 33 is probably due to some discretization in the software.

7 Conclusions

In this paper we introduced the notion of α-FLOPs that is meant to provide a simple numerical correction to the mismatch between FLOPs and execution time in case of hardware equipped with massively parallel processing units like GPUs or TPUs. Since this kind of hardware is the norm for AI applications based on Deep Neural Networks, α-FLOPS may become an important tool to compare the efficiency of different networks on a given hardware.

The definition of α-FLOPs is based on the crucial observation that, in case of an input with multiple dimensions, the computational speedup offered by parallelism is typically far from uniform along the different axes. In particular, we provided extensive empirical evidence that growing spatial (and batchsize) dimensions in convolutional layers has less impact than growing different dimensions. The idea of dissecting the cost along the different input dimensions was inspired by recent investigations of the first author on computational complexity over finite types [4].

The notion of α-FLOPs lays between the number of parameters of the layer, and the traditional notion of FLOPs; in a sense, it can be understood as a revaluation of the former as a measure of cost: if it is true that, in the case of convolutions, the number of parameters does not take into account the actual cost of the convolution, the traditional notion of FLOPs seems to largely overestimate it.

Much work is still ahead. On the experimental side, we are currently collecting more data, on architectures with different computing capabilities. On the theoretical side, it would be interesting to provide a low-level algorithmic justification of α-FLOPs. The formula itself, that was derived empirically, can be eventually fine-tuned and possibly improved, both in view of additional observations, and of a better understanding of the phenomenon. In particular, we mostly focused on the spatial dimension, since it is the axis most affected by parallelism, but the dependency along different axes does eventually deserve additional investigation.

In this article, we mostly focused on convolutional and dense layers, since they are the most computationally intensive layers in Neural Networks. Extending the work to additional layers, or more sophisticated forms on convolutions, like Depth-Separable Convolutions, is another major research direction.

References

1. Amdahl, G.M.: Validity of the single processor approach to achieving large-scale computing capabilities. In: AFIPS Conference Proceedings, vol. 30, pp. 483–485 (1967)
2. Anthony, L.F.W., Kanding, B., Selvan, R.: Carbontracker: tracking and predicting the carbon footprint of training deep learning models. CoRR, abs/2007.03051 (2020)
3. Arm cortex-r8 mpcore processor (2018). https://developer.arm.com/documentation/100400/0002/floating-point-unit-programmers-model/instruction-throughput-and-latency?lang=en. Accessed 26 Apr 2021
4. Asperti, A.: Computational complexity via finite types. ACM Trans. Comput. Log. **16**(3), 26:1–26:25 (2015)
5. Asperti, A., Evangelista, D., Piccolomini, E.L.: A survey on variational autoencoders from a green AI perspective. SN Comput. Sci. **2**(4), 301 (2021)
6. Blackford, L.S., et al.: An updated set of basic linear algebra subprograms (BLAS). ACM Trans. Math. Softw. **28**(2), 135–151 (2002)
7. Cao, Q., Balasubramanian, A., Balasubramanian, N.: Towards accurate and reliable energy measurement of NLP models. CoRR, abs/2010.05248 (2020)
8. clBLAS. http://clmathlibraries.github.io/clBLAS/. Accessed 26 Apr 2021
9. AMD Corporation. Software optimization guide for AMD family 19h processors (pub), November 2020. https://www.amd.com/system/files/TechDocs/56665.zip. Accessed 25 Apr 2021

10. Intel Corporation. Intel® Xeon scalable processor® instruction throughput and latency, August 2017. https://software.intel.com/content/dam/develop/public/us/en/documents/intel-xeon-scalable-processor-throughput-latency.pdf. Accessed 25 Apr 2021
11. cuBLAS. https://docs.nvidia.com/cuda/cublas/index.html. Accessed 26 Apr 2021
12. Devlin, J., Chang, M.W., Lee, K., Toutanova, K.: BERT: pre-training of deep bidirectional transformers for language understanding. In: Burstein, J., Doran, C., Solorio, T. (eds.) Proceedings of the 2019 Conference of the North American Chapter of the Association for Computational Linguistics: Human Language Technologies, NAACL-HLT 2019, Minneapolis, MN, USA, 2–7 June 2019, vol. 1 (Long and Short Papers), pp. 4171–4186. Association for Computational Linguistics (2019)
13. Gadou, M., Banerjee, T., Arunachalam, M., Ranka, S.: Multiobjective evaluation and optimization of CMT-bone on multiple CPU/GPU systems. Sustain. Comput.: Inform. Syst. **22**, 259–271 (2019)
14. Gordon, A., et al.: Morphnet: fast & simple resource-constrained structure learning of deep networks. In: 2018 IEEE Conference on Computer Vision and Pattern Recognition, CVPR 2018, Salt Lake City, UT, USA, 18–22 June 2018, pp. 1586–1595. IEEE Computer Society (2018)
15. Gupta, U., et al.: Chasing carbon: the elusive environmental footprint of computing. CoRR, abs/2011.02839 (2020)
16. Gustafson, J.L.: Reevaluating Amdahl's law. Commun. ACM **31**(5), 532–533 (1988)
17. Jeon, Y., Kim, J.: Constructing fast network through deconstruction of convolution. In: Bengio, S., Wallach, H.M., Larochelle, H., Grauman, K., Cesa-Bianchi, N., Garnett, R. (eds.) Advances in Neural Information Processing Systems 31: Annual Conference on Neural Information Processing Systems 2018, NeurIPS 2018, 3–8 December 2018, Montréal, Canada, pp. 5955–5965 (2018)
18. Jhurani, C., Mullowney, P.: A GEMM interface and implementation on NVIDIA GPUs for multiple small matrices. J. Parallel Distrib. Comput. **75**, 133–140 (2015)
19. Karras, T., Aila, T., Laine, S., Lehtinen, J.: Progressive growing of GANs for improved quality, stability, and variation. In: 6th International Conference on Learning Representations, ICLR 2018, Vancouver, BC, Canada, 30 April–3 May 2018, Conference Track Proceedings. OpenReview.net (2018)
20. Lacoste, A., Luccioni, A., Schmidt, V., Dandres, T.: Quantifying the carbon emissions of machine learning. CoRR, abs/1910.09700 (2019)
21. MacAvaney, S., Nardini, F.M., Perego, R., Tonellotto, N., Goharian, N., Frieder, O.: Efficient document re-ranking for transformers by precomputing term representations. In: Proceedings of the 43rd International ACM SIGIR Conference on Research and Development in Information Retrieval, SIGIR 2020, Virtual Event, China, 25–30 July 2020, pp. 49–58. ACM (2020)
22. Molchanov, P., Tyree, S., Karras, T., Aila, T., Kautz, J.: Pruning convolutional neural networks for resource efficient inference. In: 5th International Conference on Learning Representations, ICLR 2017, Toulon, France, 24–26 April 2017, Conference Track Proceedings. OpenReview.net (2017)
23. Patterson, D.A., Hennessy, J.L.: Computer Architecture: A Quantitative Approach. Morgan Kaufmann, Burlington (2017)
24. Rodrigues, R., Annamalai, A., Koren, I., Kundu, S.: A study on the use of performance counters to estimate power in microprocessors. IEEE Trans. Circuits Syst. II Express Briefs **60**(12), 882–886 (2013)

25. Schwartz, R., Dodge, J., Smith, N.A., Etzioni, O.: Green AI. Commun. ACM **63**(12), 54–63 (2020)
26. Strubell, E., Ganesh, A., McCallum, A.: Energy and policy considerations for modern deep learning research. In: The Thirty-Fourth AAAI Conference on Artificial Intelligence, AAAI 2020, The Thirty-Second Innovative Applications of Artificial Intelligence Conference, IAAI 2020, The Tenth AAAI Symposium on Educational Advances in Artificial Intelligence, EAAI 2020, New York, NY, USA, 7–12 February 2020, pp. 13693–13696. AAAI Press (2020)
27. Tan, M., Le, Q.V.: Efficientnet: rethinking model scaling for convolutional neural networks. In: Chaudhuri, K., Salakhutdinov, R. (eds.) Proceedings of the 36th International Conference on Machine Learning, ICML 2019, 9–15 June 2019, Long Beach, California, USA, Proceedings of Machine Learning Research, vol. 97, pp. 6105–6114. PMLR (2019)
28. van Wynsberghe, A.: Sustainable AI: AI for sustainability and the sustainability of AI. AI Ethics **1**, 213–218 (2021). https://doi.org/10.1007/s43681-021-00043-6
29. Veniat, T., Denoyer, L.: Learning time/memory-efficient deep architectures with budgeted super networks. In: 2018 IEEE Conference on Computer Vision and Pattern Recognition, CVPR 2018, Salt Lake City, UT, USA, 18–22 June 2018, pp. 3492–3500. IEEE Computer Society (2018)

Development of a Hybrid Modeling Methodology for Oscillating Systems with Friction

Meike Wohlleben[1(✉)], Amelie Bender[1], Sebastian Peitz[2], and Walter Sextro[1]

[1] Faculty of Mechanical Engineering, Dynamics and Mechatronics, Paderborn University, Warburger Str. 100, 33098 Paderborn, Germany
meike.wohlleben@uni-paderborn.de

[2] Department of Computer Science, Data Science for Engineering, Paderborn University, Warburger Str. 100, 33098 Paderborn, Germany

Abstract. Modeling of dynamical systems is essential in many areas of engineering, such as product development and condition monitoring. Currently, the two main approaches in modeling of dynamical systems are the physical and the data-driven one. Both approaches are sufficient for a wide range of applications but suffer from various disadvantages, e.g., a reduced accuracy due to the limitations of the physical model or due to missing data. In this work, a methodology for modeling dynamical systems is introduced, which expands the area of application by combining the advantages of both approaches while weakening the respective disadvantages. The objective is to obtain increased accuracy with reduced complexity. Two models are used, a physical model predicts the system behavior in a simplified manner, while the data-driven model accounts for the discrepancy between reality and the simplified model. This hybrid approach is validated experimentally on a double pendulum.

Keywords: Hybrid modeling · Gray-box modeling · Double pendulum · Modeling of dynamical systems with friction

1 Introduction

For a very long time, the standard approach to modeling dynamical systems in physics and engineering has been via first principles such as Newton's law or the conservation of energy, mass and so on. These equations usually result in differential equations which – depending on the accuracy of the underlying assumptions – allow to predict the system behavior more or less accurately. In addition, the use of parameters allows describing entire classes of systems in different operating conditions. One disadvantage of this approach is that complex phenomena such as friction can only be considered with great computational effort or simplifying assumptions, which will lead to decreasing accuracy [1].

With the increases in computational power seen in recent years, another modeling approach has been getting ever more popular: data-driven modeling. This class of models is based exclusively on measurements and can therefore only be

G. Nicosia et al. (Eds.): LOD 2021, LNCS 13164, pp. 101–115, 2022.
https://doi.org/10.1007/978-3-030-95470-3_8

used for already existing systems. The measurements are used to generate the data-driven model, e.g., using machine learning algorithms [2–5]. The parameters of the model are numerical values, often without any empirical or physical interpretation (also referred to as hyperparameters), hence it is also referred to as black-box model [6,7].

The measurements correspond to an explicit system for explicit operating conditions. Therefore, data-driven models are often only suitable under operating conditions represented in the data. A variety of modeling algorithms are available that are independent of the specific system under consideration, i.e., that are of black-box nature. Taking the relationship between input and output of the model into account, different algorithms can be used so that the accuracy and complexity of the model can be adapted to the respective needs. Once the model is generated, modeling runs quickly [8]. However, this approach has the disadvantage that a large amount of data is often required, which can be computationally and financially expensive [8,9].

Since both approaches have their advantages and disadvantages, the idea is to introduce another category between black- and white-box modeling, in which the disadvantages are minimized while the respective advantages are retained. Various modeling approaches fall into the category of gray-box models, such as differential equations with parameter estimators [10] or black-box models with constraints based on physical laws [11,12]. Another example is hybrid modeling [6,12], which will be discussed in this paper. While hybrid and gray-box modeling are used as synonyms in other works [13], these two terms are clearly distinguished here. The term "hybrid" describes a bundling or also a crossing. In engineering, this is understood as a combination of different technologies. Thus, it can be a coupling of a physical submodel with a data-driven submodel, regardless of the type of coupling, but the model boundaries of the submodels have to be drawn clearly.

The objective of this work is to combine a white- with a black-box model to obtain hybrid models of higher accuracy for complex systems. Reasons for this may either be that not all phenomena can be described physically, or that there is an insufficient amount of data to build an accurate data-driven model. Another goal is to obtain increased accuracy and decreased complexity, and thus, an increase in efficiency concerning the prediction of new data.

2 Hybrid Modeling: The Framework

Hybrid modeling is defined in several ways, depending on the discipline – in material technology, hybrid modeling is understood differently than in process engineering. But even within the same discipline, the definition of hybrid modeling is not consistent. In this work, hybrid modeling is understood as the coupling of a parametric model with a non-parametric model [6,7,14].

2.1 Parametric Modeling

In a parametric model, the structure is defined a priori through knowledge of the underlying system [15]. The model parameters have a physical or empirical interpretation, in contrast to the hyperparameters in black-box models mentioned above. The physical models considered in this paper represent a subgroup of these parametric models, which are obtained via conservation laws, thermodynamics, kinetics, or transport laws. Such a model can be expressed mathematically by

$$Y_f = f(X_f, w_f), \tag{1}$$

where Y_f is the output, X_f is the input, and w_f is a set of physical parameters [14].

2.2 Non-parametric Modeling

The structure of a non-parametric model is defined by data [15]. While these models do not have parameters with physical or phenomenological interpretation, they may still depend on hyperparameters such as the number of layers in a neural network or the degree of a polynomial. Non-parametric models can for example be generated with splines [16], wavelets [17] or also with machine learning algorithms [18]. Expressed in mathematical form, a non-parametric model can be written as

$$Y_g = g(X_g, w_g). \tag{2}$$

Here Y_g is the output, X_g is the input and w_g is a set of hyperparameters [14]. The data-driven model considered in this paper can be regarded as such a non-parametric model.

2.3 Hybrid Modeling

As addressed in Sect. 1, a hybrid model (sometimes called hybrid semi-parametric model) is the combination of a parametric and a non-parametric model [6,15]. These two single models can be combined in two different ways: the two models can be connected in parallel so that the following applies

$$Y = f(X_f, w_f) * g(X_g, w_g), \tag{3}$$

where the symbol $*$ can be an arbitrary operation like addition or multiplication.

A serial concatenation is also possible so that either

$$Y = f \circ g = f(g(X_g, w_g), w_f) \tag{4}$$

or

$$Y = g \circ f = g(f(X_f, w_f), w_g) \tag{5}$$

is valid [10,19]. Mixed forms are also conceivable [14,19].

So far, hybrid modeling has mainly been used in the context of chemical processes, bioprocesses or process engineering [6,7,15]. The first sources that address an application to mechanical systems date back to the 90s [19]. However, hybrid modeling has seldomly been applied to mechanical systems in practice so far. This opens up an area of modelling that has been little explored so far, but where the multitude of approaches from the fields of process engineering can be used for orientation. The proposed method in this paper is closely related to [20], where analytical equations for a discrepancy model are obtained via the popular SINDy algorithm [21]. Thus, it is no longer a hybrid model in the strict sense like the method presented in this paper, as two equation-based models are used simultaneously.

3 Proposed Hybrid Methodology for Oscillating Systems

A dynamical system with friction is considered for which measurements of the form $(x(t_k), t_k)$ for $k \in \{0, 1, \ldots, N\}$ are available.

3.1 Approach

For the considered system, the physical model is given in the form of an ordinary differential equation (ODE)

$$\dot{x}(t) = F(t, x(t)). \tag{6}$$

This ODE can be used to generate the physical model f from Eq. (1) that maps the state $x(t)$ at time t to the subsequent state $x(t + \Delta t)$. Thus, f is defined via $x(t + \Delta t) = f(x(t), \Delta t)$, where Δt is a model parameter.

However, it is assumed that the present physical model f cannot exactly represent the real system, be it due to a lack of modeling approaches for certain phenomena or due to an infeasible complexity. Hence, the notation for this imprecise trajectory is $(\tilde{x}(t_k), t_k)$ for $k \in \{0, 1, \ldots, N\}$ and $\tilde{x}(t_k) = f(x(t_{k-1}), \Delta t)$. It should be emphasized that the calculation for $\tilde{x}(t_{k+1})$ is based on the measured value $x(t_k)$ and not on the value of the physical model $\tilde{x}(t_k)$.

This leads to a discrepancy between the physical model and the measurement of

$$\Delta x(t_k) = x(t_k) - \tilde{x}(t_k) \tag{7}$$

for a time increment $\Delta t = t_k - t_{k-1}$, which is shown in Fig. 1.

The idea behind the proposed hybrid methodology is that it should be possible to reconstruct the measured value $x(t_k)$ by jointly using a physical model f (see Eq. (1)) and a data-driven model g (see Eq. (2)) that predicts $\Delta x(t_k)$. By rearranging, equation (7) can be written as

$$x(t_k) = \tilde{x}(t_k) + \Delta x(t_k) \tag{8}$$

$$\Rightarrow x(t_k) = f(x(t_{k-1}), \Delta t) + g(X_g, w_g). \tag{9}$$

This leads to a hybrid (or discrepancy) model. The parametric model f is based on the differential equation in Eq. (6) and provides $\tilde{x}(t_k)$ based on the previous

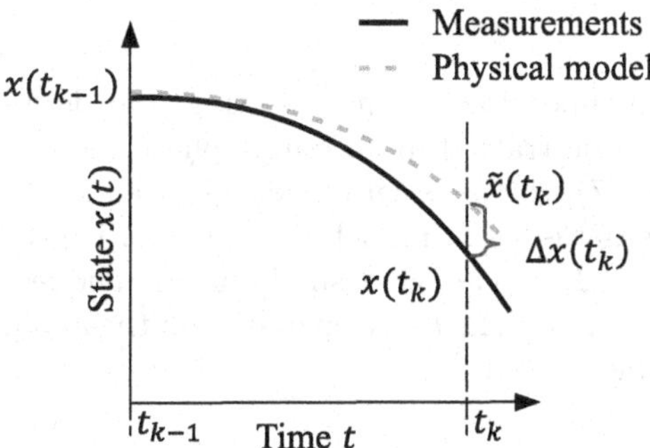

Fig. 1. Discrepancy between measurements and an inaccurate physical model.

time step $x(t_{k-1})$. The non-parametric model g provides $\Delta x(t_k)$ and is trained with real data.

The proposed methodology is shown in Fig. 2. The process is divided into two parts: The training of the data-driven model g and the prediction of a new trajectory.

The structure is neither purely parallel, nor purely serial. The output of the physical model f forms the input of the data-driven model g, implying a serial structure, but the two models are also subsequently summed, corresponding to a parallel structure

$$Y = f(X_f, w_f) + g(f(X_f, w_f), w_g) = f + (g \circ f). \tag{10}$$

Here, the arbitrary operation $*$ from Eq. (3) is replaced by the addition $+$.

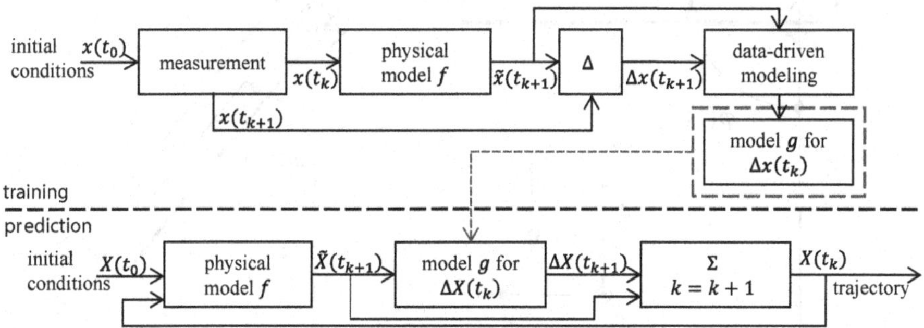

Fig. 2. Schematic representation of the proposed approach for hybrid modeling of dynamical systems.

3.2 Training

As is common in the supervised learning setting, the data-driven part of the hybrid model needs to be trained in an offline phase. To calculate the discrepancy $\Delta x(t_k)$ from Eq. (7), the physical model f is used to obtain the trajectory $(\tilde{x}(t_k), t_k)$. Then, the data-driven modeling algorithm is given the values $\tilde{x}(t_k)$ as input and the values $\Delta x(t_k)$ as the desired output. The result is a data-driven model g which maps $\tilde{x}(t_k)$ on $\Delta x(t_k)$. Depending on the application, other input parameters are possible as well.

3.3 Prediction

For the prediction, only the initial conditions $X(t_0)$ of the system under consideration are known. The remainder of the trajectory $(X(t_k), t_k)$ is calculated iteratively. In the first step, the physical model f is used to calculate $\tilde{X}(t_{k+1})$ based on the previous state $X(t_k)$. This is then the input for the data-driven model g, which provides $\Delta X(t_{k+1})$. Then, $X(t_{k+1})$ is obtained according to Eq. (8). This leads to the next iteration loop, where $\tilde{X}(t_{k+2})$ is computed with the help of $X(t_{k+1})$, and the process is repeated until the oscillation has completely decayed or its amplitude falls below a certain threshold.

3.4 Challenges in Real Experiments

The hybrid procedure presents several challenges, which can best be illustrated by using the physical pendulum as a simple example, shown in Fig. 3.

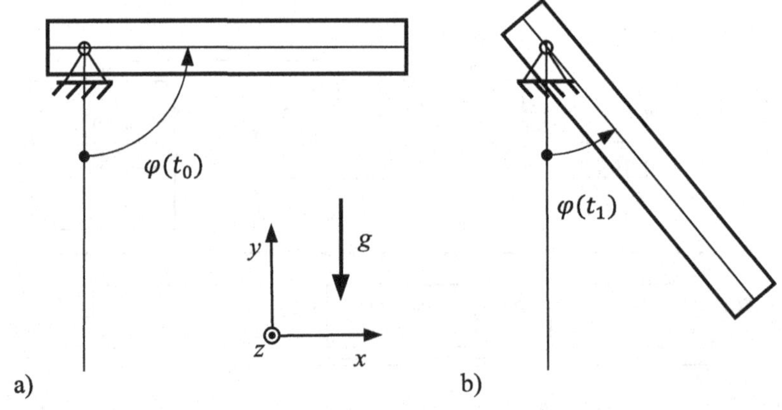

Fig. 3. The physical pendulum a) at time t_0 in its starting position with angular velocity $\dot{\varphi}(t_0) = 0\,\text{rad/s}$ and b) at time t_1 with angular velocity $\dot{\varphi}(t_1) > 0\,\text{rad/s}$.

The pendulum is initially deflected and then released from rest. In the data collection phase, the position (x, y, z) of the pendulum is measured at discrete points in time t_k. The physical model f is defined by an ODE in polar coordinates, i.e., for the pendulum angle $\ddot{\varphi}(t_k) = F(\dot{\varphi}(t_k), \varphi(t_k), t_k)$, where the angle between the pendulum arm and the vertical axis is denoted by φ.

Based on euclidean measurements, the first challenge is to determine the desired physical quantity, i.e., the angle φ. In the example, it is rather simple to obtain the angle $\varphi(t_k)$ from $(x, y, z)(t_k)$. However, there are cases in which this is not as simple, for example if the component under consideration is part of a larger system and can therefore only be measured indirectly. For the data-driven discovery of suitable coordinates, see also [22]. Since only the position $(x, y, z)(t_k)$ is measured, time derivatives need to be calculated via numerical differentiation. For the given ODE, the first derivative $\dot{\varphi} = d\varphi/dt$ is required. Measurement noise is increased by numerical differentiation, and the measurements may have to be smoothed in advance, cf. [21] for an extensive discussion on numerical differentiation of noisy data.

By introducing $\Phi(t_k) = (\varphi, \dot{\varphi})(t_k)$, the physical model f allows predicting $\Phi(t_{k+1})$. As the dimension of Φ is greater than one, there are several options for the discrepancy model:

a) $\Delta\varphi$ and $\Delta\dot{\varphi}$ are both determined using a data-driven model,
b) use a single data-driven model for $\Delta\varphi$, then calculate $\dot{\varphi}$ via numerical differentiation, or
c) use a single data-driven model for $\Delta\dot{\varphi}$, then calculate φ via numerical integration.

For a), a data-driven model with multiple outputs (in this case, two) has to be trained, which can be computationally expensive. Note the special case where both outputs are fully decoupled and thus independent. Due to the independent discrepancy, $\dot{\varphi}$ might no longer correspond to the actual derivative of φ, which would pose another problem.

For b), only a single data-driven model with the target $\Delta\varphi$ has to be trained. Thus, φ can be calculated and $\dot{\varphi}$ is obtained using numerical differentiation. Although this guarantees compatibility, the accuracy of numerical differentiation is problematic. The accuracy of the rather simple central difference quotient is $\mathcal{O}(\Delta t^2)$ [23], and the accuracy for the backward difference quotient, which would have to be used in the prediction, is $\mathcal{O}(\Delta t)$ [23]. As with the generation of the derivates, the errors due to the uncertainties of the data-driven model also increase due to the numerical differentiation.

As before, for c), only a single data-driven model with the target $\Delta\dot{\varphi}$ has to be learned. Thus, $\dot{\varphi}$ can be calculated and φ is obtained using numerical integration. This guarantees compatibility. The accuracy of the numerical integration is higher than with the numerical differentiation. Using the closed Newton-Cotes quadrature rules, the error for $n = 1$ is $\mathcal{O}(\Delta t^3)$, where n indicates the order and also the number of reference points $(n + 1)$. The error for $n = 4$ is only $\mathcal{O}(\Delta t^6)$, and may be further decreased by further increasing n [16].

After the target for the data-driven model has been selected, the input and the algorithm need to be considered. This process can be represented as the composition of two functions

$$g(h(\tilde{x}), w_g) = g(h(\tilde{\varphi}, \dot{\tilde{\varphi}}), w_g). \tag{11}$$

As previously defined, g is the non-parametric function representing the black-box model and – if necessary – h can be a function for preprocessing or selecting quantities. This choice depends on the specific application and cannot be answered in a general fashion.

4 Validation: Double Pendulum

A double pendulum is used for the validation of the methodology, as illustrated in Fig. 4. The double pendulum consists of two pendulum arms which are connected with a pivot bearing, all of which have non-zero masses. The first pendulum arm is also connected to the surrounding by a grounded pivot bearing with non-zero mass. The distance between the two pivot bearings is denoted by l_1, the distance from the second pivot bearing to the end of the second pendulum arm by l_2. The centers of mass of the two pendulum arms S_1 and S_2 are defined by s_1 and s_2. The two degrees of freedom are φ_1 and φ_2, which are defined as the angles between the pendulum arms under the vertical axis, respectively [24]. Both angles are measured in radians.

The double pendulum dynamics are non-linear and it is well known that they exhibit chaotic behavior for large deflections [25]. This means that the system behavior is highly dependent on the initial conditions, and small uncertainties have a significant effect on the course of the trajectory [26]. This considerably complicates the modeling of such a system. The equations of motion of the undamped double pendulum, and thus the basis of the physical model f, are as follows [24]

$$\left[\left(\frac{s_1}{l_1}\right)^2 + \frac{J_{S1}}{m_1 + l_1^2} + \frac{m_2}{m_1}\right]\ddot{\varphi}_1 + \left[\frac{m_2}{m_1}\frac{s_2}{l_1}\cos(\varphi_1 - \varphi_2)\right]\ddot{\varphi}_2$$
$$= -\frac{m_2}{m_1}\frac{s_2}{l_1}\dot{\varphi}_2^2\sin(\varphi_1 - \varphi_2) - \left(\frac{s_1}{l_1} + \frac{m_2}{m_1}\right)\frac{g}{l_1}\sin(\varphi_1), \tag{12}$$

$$\left[\frac{m_2}{m_1}\frac{s_2}{l_1}\cos(\varphi_1 - \varphi_2)\right]\ddot{\varphi}_1 + \left[\frac{m_2}{m_1}\left(\frac{s_2}{l_1}\right)^2 + \frac{J_{S2}}{m_1 l_1^2}\right]\ddot{\varphi}_2$$
$$= \frac{m_2}{m_1}\frac{s_2}{l_1}\dot{\varphi}_1^2\sin(\varphi_1 - \varphi_2) - \frac{m_2}{m_1}\frac{s_2}{l_1}\frac{g}{l_1}\sin(\varphi_2), \tag{13}$$

where $g \approx 9.81\,\mathrm{m/s^2}$ is the gravitational constant. The mass moments of inertia related to the respective centers of gravity are denoted by J_{S1} and J_{S2}. The masses of the two pivot bearings together with the mass of the first pendulum arm is denoted by m_1, the mass of the second pendulum arm is denoted by m_2.

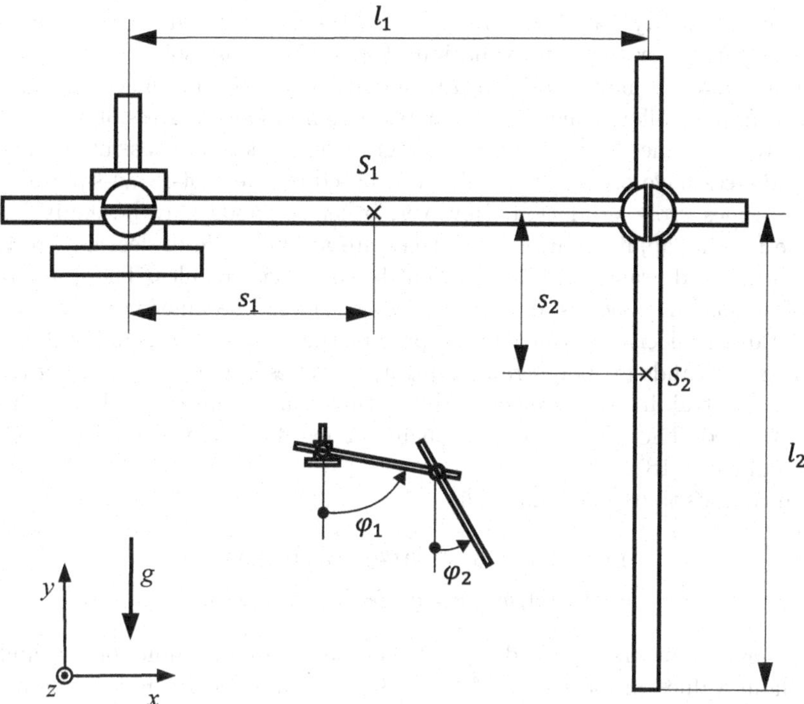

Fig. 4. Sketch of the double pendulum under consideration.

For the experimental validation, friction is introduced into the system via the pivot bearings. Furthermore, the initial conditions chosen in Sects. 4.1 and 4.2 are such that the dynamic behavior of the double pendulum is largely reproducible except for small deviations.

4.1 Synthetic Data

The first validation step is the proof of concept based on synthetic measurements. These synthetic measurements are simulative results obtained by exact equations of motion, including damping. There is no noise in the data and the damping can be adjusted as required. Viscous damping is used for the two pivot bearings.

The double pendulum has two degrees of freedom and is described by a second order ODE, which means that $\Phi = (\varphi_1, \dot{\varphi}_1, \varphi_2, \dot{\varphi}_2)$. After reviewing the data of the double pendulum and considering the advantages and disadvantages in Sect. 3.4, it was decided to define two decoupled data-driven models g_1 and g_2, where g_1 models $\Delta\dot{\varphi}_1$ and g_2 models $\Delta\dot{\varphi}_2$. The angles φ_1 and φ_2 are then determined via numerical integration using Newton-Cotes quadrature. Since the damping is neglected in the physical model and therefore has to be integrated into the hybrid model using the data-driven one, it is straightforward to choose the angular velocity $\dot{\varphi}_1$, respectively $\dot{\varphi}_2$ as the input parameters for g_1 and g_2, respectively. For the approximation of g_1 and g_2, any suitable data-driven modeling approach may be chosen. In comparison to 19 other machine learning methods from 5 different fields – linear regression models, regression trees, support vector machines, Gaussian process regression models and ensembles of trees – bagged trees have proven to be effective, based on the root-mean-square error and five-times cross validation. However, other methods may be equivalently suitable for this application. Bagged trees are an ensemble of decision trees. A tree is a directed graph used to represent binary decision rules. The graph consists of a root, inner nodes and end nodes or leaves. The inner nodes represent observations or decisions, the leaves represent the resulting targets [18,27]. The term bagged stands for bootstrap aggregated. By resampling the existing training sets, new training sets are generated so that a large number of decision trees can be trained. The overall model is then composed as the weighted sum of the individual trees [18,27,28].

The compositions are defined as

$$g_1(h_1(\tilde{\Phi}), w_{g,1}) = g_1(\dot{\tilde{\varphi}}_1, w_{g,1}) = \Delta\dot{\varphi}_1, \tag{14}$$

$$g_2(h_2(\tilde{\Phi}), w_{g,2}) = g_2(\dot{\tilde{\varphi}}_2, w_{g,2}) = \Delta\dot{\varphi}_2. \tag{15}$$

The hyper-parameters $w_{g,1}$ and $w_{g,2}$ of the non-parametric functions g_1 and g_2 are default values proposed by MATLAB and contain information about the number of trees that are bagged, which is 30, the associated weight of each tree, which is one, and the minimal number of leaves of each tree, which is eight.

The results are shown in Fig. 5. The trajectory which was used for the training is shown in blue, the initial conditions were $\Phi(t_0) = (2\pi/3\,\text{rad}, 0\,\text{rad}, 0\,\text{rad/s}, 0\,\text{rad/s})$. For prediction, the initial conditions $\Phi(t_0) = (\pi/3\,\text{rad}, 0\,\text{rad}, 0\,\text{rad/s}, 0\,\text{rad/s})$ are chosen, the corresponding synthetic measurement is shown in black. The time increment is $\Delta t = 0.1\,\text{s}$. The trajectory predicted with the introduced method is shown in red dotted lines. It can be seen that the discrepancies $\Delta\dot{\varphi}_1$ and $\Delta\dot{\varphi}_2$ are almost identical for the target trajectory and the prediction. Due to the large time increment, the discrepancies are also large and ensure that the hybrid model decays within about 5 s. This example under ideal conditions shows that the hybrid methodology used does indeed deliver very good results under the right conditions, e.g., when the initial conditions of the target trajectory are comparable to the initial conditions of the training.

4.2 Measurements

After the proof of concept of the proposed methodology has been carried out, the procedure is applied to real measurements of a double pendulum.

The experimental setup can be seen in Fig. 6. A camera-based measurement system is used to record the positions of the three points p_0, p_1 and p_2. These points are chosen in such a way that the angles φ_1 and φ_2 can be calculated. The angular velocities $\dot{\varphi}_1$ and $\dot{\varphi}_2$ are determined by means of numerical differentiation. A time increment of $\Delta t = 0.01\,\text{s}$ is used.

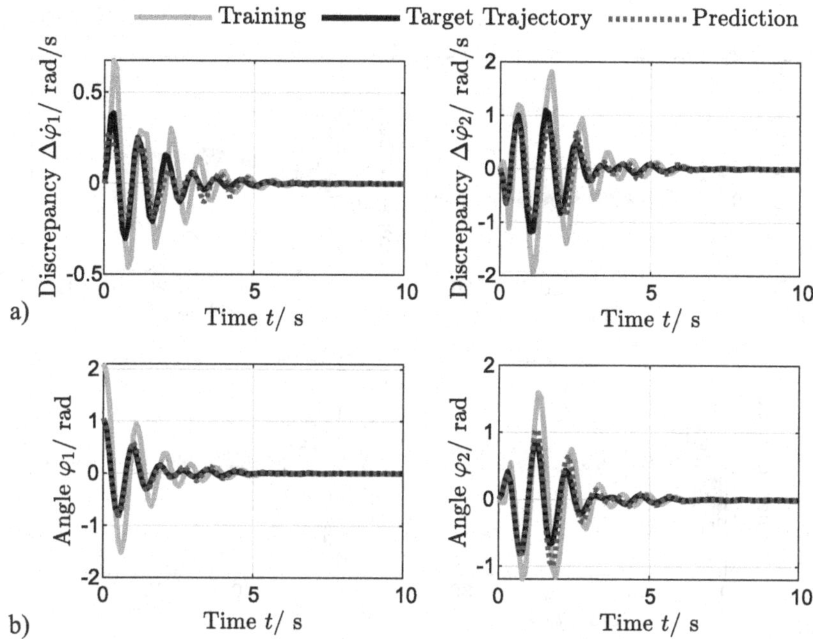

Fig. 5. a) The discrepancies $\Delta\dot{\varphi}_1$ and $\Delta\dot{\varphi}_2$ of the training data (blue), the target trajectory (black) and those calculated with the data-driven models g_1 and g_2 (dotted red). b) The angles φ_1 and φ_2 of the training data (blue) and the prediction (dotted red) as a result of the hybrid modeling with time increment $\Delta t = 0.1$ s. (Color figure online)

The measurements are carried out with two different initial conditions: In series A (shown in Fig. 6), the first pendulum arm is deflected at an angle of $\varphi_1^A(t_0) = \pi/2$ rad, while the second pendulum arm has an angle of $\varphi_2^A(t_0) = 0$ rad to the vertical axis. In series B, the first pendulum arm is deflected at an angle of $\varphi_1^B(t_0) = \pi/2$ rad, and the second pendulum arm is deflected at an angle of $\varphi_2^B(t_0) = \pi/2$ rad.

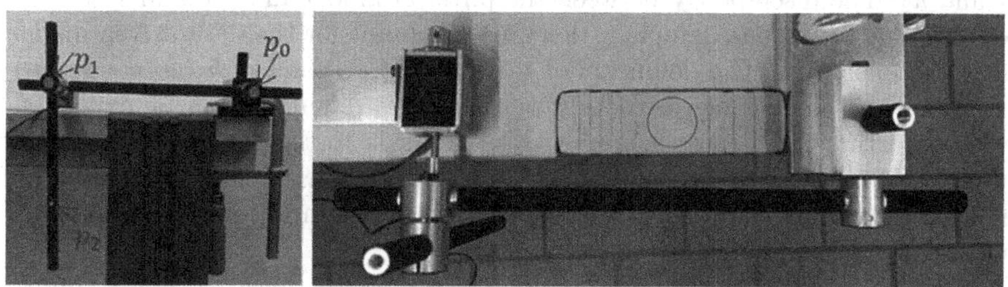

Fig. 6. Front view (left) and top view (right) of the experimental setup of the double pendulum. The initial conditions correspond to series A.

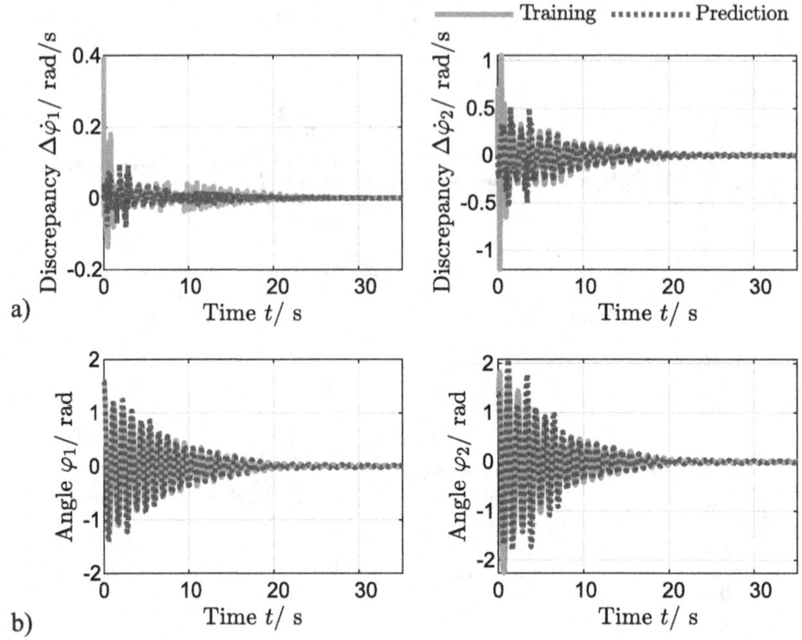

Fig. 7. a) The discrepancies $\Delta\dot{\varphi}_1$ and $\Delta\dot{\varphi}_2$ of the training data (series B, blue) and those calculated with the data-driven models g_1 and g_2 (dotted red). b) The angles φ_1 and φ_2 of the training data (series B, blue) and the prediction (dotted red) as a result of the hybrid modeling with time increment $\Delta t = 0.01\,\text{s}$. (Color figure online)

In both series the double pendulum is released from rest so that the initial angular velocities $\dot{\varphi}_1(t_0)$ and $\dot{\varphi}_2(t_0)$ are 0 rad/s for both pendulum arms.

For the physical model, the same equations of motion are used as for the synthetic data, see Eqs. (12) and (13). The model parameters are determined computationally based on the geometry (measured with a caliper gauge) and are subject to a large uncertainty. The main difference between the application and synthetic data lies in the choice of the functions for preprocessing and selecting h_1 and h_2. The discrepancy between the physical model and the real system is not only due to viscous damping, therefore the input for the data-driven model has to be expanded. The definition of $h_1(\tilde{\Phi}) = (\tilde{\varphi}_1, \dot{\tilde{\varphi}}_1)$ and of $h_2(\tilde{\Phi}) = (\tilde{\varphi}_2, \dot{\tilde{\varphi}}_2)$ has proven to be suitable in the application. This difference can be explained by the fact that in these measurements, in addition to the obvious discrepancy due to the lack of friction in the two pivot bearings (which cannot necessarily be modeled by viscous damping), there are uncertainties in the model parameters and aerodynamic drag is also neglected.

Figure 7 shows the case where one time series of series B was used for training. The initial conditions of series B were used for the prediction. Ideally, the prediction should accurately reconstruct the training data. Figure 7 shows that the prediction is very close to the trajectory that was used for training. Occasionally there are small differences, but the frequency and decay match very well.

Figure 8 shows the case in which different initial conditions are used for training and in prediction. One time series of series B was used to train the hybrid model. The target trajectory - and thus the initial conditions for the prediction - is from series A. Even though another trajectory was used in training, the predicted trajectory is very close to the target trajectory at almost all points in time; Small differences in amplitude can only be seen in isolated cases.

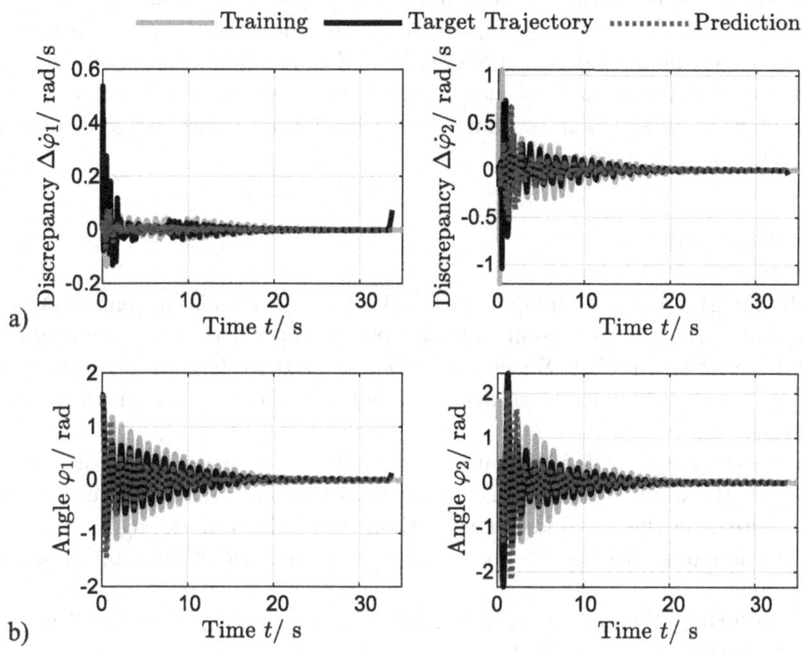

Fig. 8. a) The discrepancies $\Delta\dot{\varphi}_1$ and $\Delta\dot{\varphi}_2$ of the training data (series B, blue), the target trajectory (series A, black) and those calculated with the data-driven models g_1 and g_2 (dotted red). b) The angles φ_1 and φ_2 of the training data (series B, blue), the target trajectory (series A, black) and the prediction (dotted red) as a result of the hybrid modeling with time increment $\Delta t = 0.01$ s. (Color figure online)

The advantage of the second combination is that the amplitude of the training trajectory is higher than the amplitude of the target trajectory. Hence, the states of the target trajectory are well covered. When looking at the opposite scenario, series A is used for training and the prediction is applied to series B, then the results are significantly less accurate. This is in the nature of data-driven models: depending on the method used, extrapolation can be a difficult matter. Thus, the goal should be to cover the space of input variables for the data-driven model as complete as possible.

5 Discussion

The results generated with the presented hybrid modeling methodology underline the potential for accurate long-term prediction. For a nonlinear system like a double pendulum, good results are obtained despite large uncertainties in the system's parameters. This shows that in addition to obvious errors in the physical model, unknown errors can also be handled by the hybrid approach.

An advantage over other hybrid modeling methods, such as the method in [20], is that there are no restrictions on either the choice of the physical model f or the choice of the data-driven model g, as well as regarding their compatibility. For the physical model, an ODE can be used, but it is also possible to work with a finite element model. For the data-driven model, an entire range of machine learning methods is available.

6 Summary

A methodology for hybrid modeling of oscillating systems is presented. A data-driven model is used to represent the discrepancy that exists between a simplified physical model and reality. Special attention is paid to data preprocessing, the choice of the output for the data-driven model, and the choice of the data-driven model itself.

The methodology is validated using synthetic and measured data of a nonlinear system, the double pendulum. The focus is less on reducing the complexity, but on increasing the accuracy, since without the hybrid modeling at this point it is very challenging to make accurate long-term predictions using physics-based models only.

In the future, the reduction of complexity is a main goal. In addition, the study of chaotic systems will also be of great interest.

References

1. Sextro, W.: Dynamic Contact Problems with Friction, 2nd edn. Springer, Heidelberg (2007)
2. Klus, S., Nüske, F., Peitz, S., Nieman, J.-H., Clementi, C., Schütte, C.: Data-driven approximation of the Koopman generator: model reduction, system identification, and control. Physica D **406**, 132416 (2020)
3. Vlachas, P.R., Byeon, W., Wan, Z.Y., Sapsis, T.P., Koumoutakos, P.: Data-driven forecasting of high-dimensional chaotic systems with long short-term memory network. Proc. R. Soc. A Math. Phys. Eng. Sci. **474**, 20170844 (2018)
4. Lukoševičius, M., Jaeger, H.: Reservoir computing approaches to recurrent neural network training. Comput. Sci. Rev. **3**(3), 127–149 (2009)
5. Kaiser, E., et al.: Cluster-based reduced-order modelling of a mixing layer. J. Fluid Mech. **754**, 365–414 (2014)
6. von Stosch, M., Oliveira, R., Peres, J., Feyo de Azevedo, S.: Hybrid semi-parametric modeling in process systems engineering: past, present and future. Comput. Chem. Eng. **60**, 86–101 (2014)

7. von Stosch, M., Glassey, J.: Benefits and challenges of hybrid modeling in the process industries: an introduction. In: Hybrid Modeling in Process Industries, pp. 1–12. CRC Press, Boca Raton (2018)
8. Isermann, R., Münchof, M.: Identification of Dynamic Systems: An Introduction with Applications, 1st edn. Springer, Heidelberg (2011). https://doi.org/10.1007/978-3-540-78879-9
9. Bieker, K., Peitz, S., Brunton, S.L., Kutz, J.N., Dellnitz, M.: Deep model predictive flow control with limited sensor data and online learning. Theoret. Comput. Fluid Dyn. **34**(4), 577–591 (2020). https://doi.org/10.1007/s00162-020-00520-4
10. Halfmann, C., Holzmann, H.: Adaptive Modelle für die Kraftfahrzeugdynamik, 1st edn. Springer, Heidelberg (2003). https://doi.org/10.1007/978-3-642-18215-0
11. Kramer, M.A., Thompson, M.L., Bhagat, P.M.: Embedding theoretical models in neural networks. In: American Control Conference, pp. 475–479 (1992)
12. Sohlberg, B., Jacobsen, E.W.: Grey box modelling - branches and experiences. In: Proceeding of the 17th IFAC World Congress, pp. 11415–11420 (2008)
13. Matei, I., de Kleer, H., Feldman, A., Rai, R., Chowdhury, S.: Hybrid modeling: applications in real-time diagnosis. arXiv:2003.02671 (2020)
14. von Stosch, M., Portela, R.M.C., Oliveira, R.: Hybrid model structures for knowledge integration. In: Hybrid Modeling in Process Industries, pp. 13–35. CRC Press, Boca Raton (2018)
15. Thompson, M.L., Kramer, M.A.: Modeling chemical processes using prior knowledge and neural networks. AIChE J. **40**(8), 1328–1340 (1994)
16. Deuflhard, P., Hohmann, A.: Numerische Mathematik 1, 4th edn. De Gruyter, Berlin (2008)
17. Daubechies, I.: Ten Lectures on Wavelets, 3rd edn. Society for Industrial and Applied Mathematics, Philadelphia (1994)
18. Alpaydin, E.: Maschinelles Lernen, 2nd edn. De Gruyter Oldenbourg, Berlin, Boston (2019)
19. Agarwal, M.: Combining neural and conventional paradigms for modelling, prediction and control. Int. J. Syst. Sci. **28**(1), 65–81 (1997)
20. Kaheman, K., Kaiser, E., Strom, B., Kutz, N., Brunton, S. L.: Learning discrepancy models from experimental data. arXiv:1909.08574 (2019)
21. Brunton, S.L., Proctor, J.L., Kutz, J.N.: Discovering governing equations from data by sparse identification of nonlinear dynamical systems. Proc. Natl. Acad. Sci. **113**(15), 3932–3937 (2016)
22. Champion, K., Lusch, B., Kutz, N., Brunton, S.L.: Data-driven discovery of coordinates and governing equations. Proc. Natl. Acad. Sci. **116**(45), 22445–22451 (2019)
23. Forster, O.: Analysis 1: Differential- und Integralrechnung einer Veränderlichen, 12th edn. Springer Spektrum, Wiesbaden (2016)
24. Dankert, H., Dankert, J.: Technische Mechanik: computerunterstützt, 2nd edn. B. G. Teubner, Stuttgart (1995)
25. Shinbrot, T., Grebogi, C., Wisdom, J., Yorke, J.: Chaos in a double pendulum. Am. J. Phys. Am. Assoc. Phys. Teachers **60**, 491–499 (1992)
26. Magnus, K., Popp, K., Sextro, W.: Schwingungen, 10th edn. Springer Vieweg, Wiesbaden (2016). https://doi.org/10.1007/978-3-658-13821-9
27. Breiman, K., Friedman, J.H., Olshen, R.A., Stone, C.J.: Classification and Regression Trees, 1st edn. CRC Press, Boca Raton (1984)
28. Breiman, L.: Bagging predictors. Mach. Learn. **24**, 123–140 (1996)

Numerical Issues in Maximum Likelihood Parameter Estimation for Gaussian Process Interpolation

Subhasish Basak[1], Sébastien Petit[1,2], Julien Bect[1], and Emmanuel Vazquez[1(✉)]

[1] Laboratoire des Signaux et Systèmes, CentraleSupélec, CNRS, Univ. Paris-Saclay, Gif-sur-Yvette, France
{subhasish.basak,sebastien.petit,julien.bect, emmanuel.vazquez}@centralesupelec.fr

[2] Safran Aircraft Engines, Moissy-Cramayel, France

Abstract. This article investigates the origin of numerical issues in maximum likelihood parameter estimation for Gaussian process (GP) interpolation and investigates simple but effective strategies for improving commonly used open-source software implementations. This work targets a basic problem but a host of studies, particularly in the literature of Bayesian optimization, rely on off-the-shelf GP implementations. For the conclusions of these studies to be reliable and reproducible, robust GP implementations are critical.

Keywords: Gaussian processes · Maximum likelihood estimation · Optimization

1 Introduction

Gaussian process (GP) regression and interpolation (see, e.g., Rasmussen and Williams 2006), also known as kriging (see, e.g., Stein 1999), has gained significant popularity in statistics and machine learning as a non-parametric Bayesian approach for the prediction of unknown functions. The need for function prediction arises not only in supervised learning tasks, but also for building fast surrogates of time-consuming computations, e.g., in the assessment of the performance of a learning algorithm as a function of tuning parameters or, more generally, in the design and analysis computer experiments (Santner et al. 2003). The interest for GPs has also risen considerably due to the development of Bayesian optimization (Mockus 1975; Jones et al. 1998; Emmerich et al. 2006; Srinivas et al. 2010...).

This context has fostered the development of a fairly large number of open-source packages to facilitate the use of GPs. Some of the popular choices are the Python modules scikit-learn (Pedregosa et al. 2011), GPy (Sheffield machine learning group 2012–2020), GPflow (Matthews et al. 2017), GPyTorch (Gardner et al. 2018), OpenTURNS (Baudin et al. 2017); the R package DiceKriging

G. Nicosia et al. (Eds.): LOD 2021, LNCS 13164, pp. 116–131, 2022.
https://doi.org/10.1007/978-3-030-95470-3_9

Table 1. Inconsistencies in the results across different Python packages. The results were obtained by fitting a GP model, with constant mean and a Matérn kernel ($\nu = 5/2$), to the Branin function, using the default settings for each package. We used 50 training points and 500 test points sampled from a uniform distribution on $[-5, 10] \times [0, 15]$. The table reports the estimated values for the variance and length scale parameters of the kernel, the empirical root mean squared prediction error (ERMSPE) and the minimized negative log likelihood (NLL). The last row shows the improvement using the recommendations in this study.

LIBRARY	Version	Variance	Lengthscales	ERMSPE	NLL
SCIKIT-LEARN	0.24.2	$9.9 \cdot 10^4$	(13, 43)	1.482	132.4
GPY	1.9.9	$8.1 \cdot 10^8$	(88, 484)	0.259	113.7
GPYTORCH	1.4.1	$1.1 \cdot 10^1$	(4, 1)	12.867	200839.7
GPFLOW	1.5.1	$5.2 \cdot 10^8$	(80, 433)	0.274	114.0
OPENTURNS	1.16	$1.3 \cdot 10^4$	(8, 19)	3.301	163.1
GPY "IMPROVED"	1.9.9	$9.4 \cdot 10^{10}$	(220, 1500)	0.175	112.0

(Roustant et al. 2012); and the Matlab/GNU Octave toolboxes GPML (Rasmussen and Nickisch 2010), STK (Bect et al. 2011–2021) and GPstuff (Vanhatalo et al. 2012).

In practice, all implementations require the user to specify the mean and covariance functions of a Gaussian process prior under a parameterized form. Out of the various methods available to estimate the model parameters, we can safely say that the most popular approach is the *maximum likelihood estimation* (MLE) method. However, a simple numerical experiment consisting in interpolating a function (see Table 1), as is usually done in Bayesian optimization, shows that different MLE implementations from different Python packages produce very dispersed numerical results when the default settings of each implementation are used. These significant differences were also noticed by Erickson et al. (2018) but the causes and possible mitigation were not investigated. Note that each package uses its own default algorithm for the optimization of the likelihood: GPyTorch uses ADAM (Kingma and Ba 2015), OpenTURNS uses a truncated Newton method (Nash 1984) and the others generally use L-BFGS-B (Byrd et al. 1995). It turns out that none of the default results in Table 1 are really satisfactory compared to the result obtained using the recommendations in this study[1].

Focusing on the case of GP interpolation (with Bayesian optimization as the main motivation), the first contribution of this article is to understand the origin of the inconsistencies across available implementations. The second contribution is to investigate simple but effective strategies for improving these implementations, using the well-established GPy package as a case study. We shall propose recommendations concerning several optimization settings: initialization and restart strategies, parameterization of the covariance, etc. By anticipation of our numerical results, the reader is invited to refer to Fig. 1 and Table 2, which show that significant improvement in terms of estimated parameter values and prediction errors can be obtained over default settings using better optimization schemes.

[1] Code available at https://github.com/saferGPMLE.

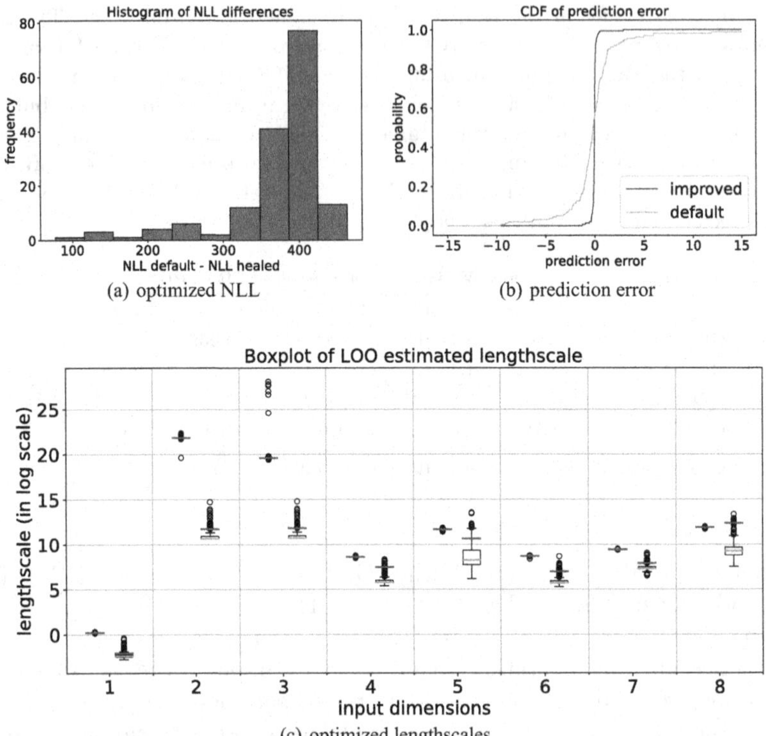

(a) optimized NLL

(b) prediction error

(c) optimized lengthscales

Fig. 1. *Improved* (cf. Sect. 6) vs *default* setups in GPy on the Borehole function with $n = 20d = 160$ random training points. We remove one point at a time to obtain (a) the distribution of the differences of negative log-likelihood (NLL) values between the two setups; (b) the empirical CDFs of the prediction error at the removed points; (c) pairs of box-plots for the estimated range parameters (for each dimension, indexed from 1 to 8 on the x-axis, the box-plot for *improved* setup is on the left and the box-plot for *default* setup is on the right; horizontal red lines correspond to the estimated values using the whole data set without leave-one-out). Notice that the parameter distributions of the *default* setup are more spread out.

Even though this work targets a seemingly prosaic issue, and advocates somehow simple solutions, we feel that the contribution is nonetheless of significant value considering the widespread use of GP modeling. Indeed, a host of studies, particularly in the literature of Bayesian optimization, rely on off-the-shelf GP implementations: for their conclusions to be reliable and reproducible, robust implementations are critical.

The article is organized as follows. Section 2 provides a brief review of GP modeling and MLE. Section 3 describes some numerical aspects of the evaluation and optimization of the likelihood function, with a focus on GPy's implementation. Section 4 provides an analysis of factors influencing the accuracy of numerical MLE procedures. Finally, Sect. 5 assesses the effectiveness of our solutions through numerical experiments and Sect. 6 concludes the article.

Table 2. *Improved* (cf. Sect. 6) vs *default* setups in GPy for the interpolation of the Borehole function (input space dimension is $d = 8$) with $n \in \{3d, 5d\}$ random data points (see Sect. 5.3 for details). The experiment is repeated 50 times. The columns report the leave-one-out mean squared error (LOO-MSE) values (empirical mean over the repetitions, together with the standard deviation and the average proportion of the LOO-MSE to the total standard deviation of the data in parentheses).

Method	$n = 3d$	$n = 5d$
Default	17.559 (4.512, 0.387)	10.749 (2.862, 0.229)
Improved	3.949 (1.447, 0.087)	1.577 (0.611, 0.034)

2 Background

2.1 Gaussian Processes

Let $Z \sim \mathrm{GP}(m, k)$ be a Gaussian process indexed by $\mathbb{R}^d$, $d \geq 1$, specified by a mean function $m : \mathbb{R}^d \to \mathbb{R}$ and a covariance function $k : \mathbb{R}^d \times \mathbb{R}^d \to \mathbb{R}$.

The objective is to predict $Z(x)$ at a given location $x \in \mathbb{R}^d$, given a data set $D = \{(x_i, z_i) \in \mathbb{R}^d \times \mathbb{R}, 1 \leq i \leq n\}$, where the observations z_is are assumed to be the outcome of an additive-noise model: $Z_i = Z(x_i) + \varepsilon_i$, $1 \leq i \leq n$. In most applications, it is assumed that the ε_is are zero-mean Gaussian i.i.d. random variables with variance $\sigma_\varepsilon^2 \geq 0$, independent of Z. (In rarer cases, heteroscedasticity is assumed.)

Knowing m and k, recall (see, e.g. Rasmussen and Williams 2006) that the posterior distribution of Z is such that $Z \mid Z_1, \ldots, Z_n, m, k \sim GP(\hat{Z}_n, k_n)$, where $\hat{Z}_n$ and k_n stand respectively for the posterior mean and covariance functions:

$$\begin{aligned} \hat{Z}_n(x) &= m(x) + \textstyle\sum_{i=1}^n w_i(x; \underline{\mathrm{x}}_n)\,(z_i - m(x_i))\,, \\ k_n(x, y) &= k(x, y) - \mathrm{w}(y; \underline{\mathrm{x}}_n)^{\mathsf{T}} \mathrm{K}(\underline{\mathrm{x}}_n, x)\,, \end{aligned}$$

where $\underline{\mathrm{x}}_n$ denotes observation points $(x_1, \ldots, x_n)$ and the weights $w_i(x; \underline{\mathrm{x}}_n)$ are solutions of the linear system:

$$(\mathrm{K}(\underline{\mathrm{x}}_n, \underline{\mathrm{x}}_n) + \sigma_\varepsilon^2 \mathrm{I}_n)\, \mathrm{w}(x; \underline{\mathrm{x}}_n) = \mathrm{K}(\underline{\mathrm{x}}_n, x)\,, \tag{1}$$

with $\mathrm{K}(\underline{\mathrm{x}}_n, \underline{\mathrm{x}}_n)$ the $n \times n$ covariance matrix with entries $k(x_i, x_j)$, I_n the identity matrix of size n, and $\mathrm{w}(x; \underline{\mathrm{x}}_n)$ (resp. $\mathrm{K}(\underline{\mathrm{x}}_n, x)$) the column vector with entries $w_i(x; \underline{\mathrm{x}}_n)$ (resp. $k(x_i, x)$), $1 \leq i \leq n$.

It is common practice to assume a zero mean function $m = 0$—a reasonable choice if the user has taken care to center data—but most GP implementations also provide an option for setting a constant mean function $m(\,\cdot\,) = \mu \in \mathbb{R}$. In this article, we will include such a constant in our models, and treat it as an additional parameter to be estimated by MLE along with the others. (Alternatively, μ could be endowed with a Gaussian or improper-uniform prior, and then integrated out; see, e.g., O'Hagan (1978).)

The covariance function, aka covariance kernel, models similarity between data points and reflects the user's prior belief about the function to be learned.

Table 3. Some kernel functions available in GPy. The Matérn kernel is recommended by Stein [1999]. Γ denotes the gamma function, $\mathscr{K}_\nu$ is the modified Bessel function of the second kind.

KERNEL	$r(h),\ h \in [0, +\infty)$
SQUARED EXPONENTIAL	$\exp(-\frac{1}{2}r^2)$
RATIONAL QUADRATIC	$(1+r^2)^{-\nu}$
MATÉRN WITH PARAM. $\nu > 0$	$\frac{2^{1-\nu}}{\Gamma(\nu)}\left(\sqrt{2\nu}r\right)^{\nu}\mathscr{K}_\nu\left(\sqrt{2\nu}r\right)$

Most GP implementations provide a couple of stationary covariance functions taken from the literature (e.g., Wendland 2004; Rasmussen and Williams 2006). The *squared exponential*, the *rational quadratic* or the *Matérn* covariance functions are popular choices (see Table 3). These covariance functions include a number of parameters: a variance parameter $\sigma^2 > 0$ corresponding to the variance of Z, and a set of range (or length scale) parameters $\rho_1, \ldots, \rho_d$, such that

$$k(x, y) = \sigma^2 r(h), \tag{2}$$

with $h^2 = \sum_{i=1}^{d}(x_{[i]} - y_{[i]})^2/\rho_i^2$, where $x_{[i]}$ and $y_{[i]}$ denote the elements of x and y. The function $r : \mathbb{R} \to \mathbb{R}$ in (2) is the stationary correlation function of Z. From now on, the vector of model parameters will be denoted by $\theta = (\sigma^2, \rho_1, \ldots, \rho_d, \ldots, \sigma_\varepsilon^2)^\mathsf{T} \in \Theta \subset \mathbb{R}^p$, and the corresponding covariance matrix $\mathrm{K}(\underline{\mathrm{x}}_n, \underline{\mathrm{x}}_n) + \sigma_\varepsilon^2 \mathrm{I}_n$ by K_θ.

2.2 Maximum Likelihood Estimation

In this article, we focus on GP implementations where the parameters $(\theta, \mu) \in \Theta \times \mathbb{R}$ of the process Z are estimated by maximizing the likelihood $\mathcal{L}(\underline{Z}_n|\theta, \mu)$ of $\underline{Z}_n = (Z_1, \ldots, Z_n)^\mathsf{T}$, or equivalently, by minimizing the negative log-likelihood (NLL)

$$-\log(\mathcal{L}(\underline{Z}_n|\theta, \mu)) = \frac{1}{2}(\underline{Z}_n - \mu\mathbb{1}_n)^\mathsf{T}\mathrm{K}_\theta^{-1}(\underline{Z}_n - \mu\mathbb{1}_n) + \frac{1}{2}\log|\mathrm{K}_\theta| + \text{constant}. \tag{3}$$

This optimization is typically performed by gradient-based methods, although local maxima can be of significant concern as the likelihood is often non-convex. Computing the likelihood and its gradient with respect to (θ, μ) has a $O(n^3 + dn^2)$ computational cost (Rasmussen and Williams 2006; Petit et al. 2020).

3 Numerical Noise

The evaluation of the NLL as well as its gradient is subject to numerical noise, which can prevent proper convergence of the optimization algorithms. Figure 2 shows a typical situation where the gradient-based optimization algorithm stops before converging to an actual minimum. In this section, we provide an analysis on the numerical noise on the NLL using the concept of local condition numbers. We also show that the popular solution of adding *jitter* cannot be considered as a fully satisfactory answer to the problem of numerical noise.

Numerical noise stems from both terms of the NLL, namely $\frac{1}{2}\underline{Z}_n^\top \mathrm{K}_\theta^{-1}\underline{Z}_n$ and $\frac{1}{2}\log|\mathrm{K}_\theta|$. (For simplification, we assume $\mu = 0$ in this section.)

First, recall that the condition number $\kappa(\mathrm{K}_\theta)$ of K_θ, defined as the ratio $|\lambda_{\max}/\lambda_{\min}|$ of the largest eigenvalue to the smallest eigenvalue (Press et al. 1992), is the key element for analyzing the numerical noise on $\mathrm{K}_\theta^{-1}\underline{Z}_n$. In double-precision floating-point approximations of numbers, $\underline{Z}_n$ is corrupted by an error ϵ whose magnitude is such that $\|\epsilon\|/\|\underline{Z}_n\| \simeq 10^{-16}$. Worst-case alignment of $\underline{Z}_n$ and ϵ with the eigenvectors of K_θ gives

$$\frac{\|\mathrm{K}_\theta^{-1}\epsilon\|}{\|\mathrm{K}_\theta^{-1}\underline{Z}_n\|} \simeq \kappa(\mathrm{K}_\theta) \times 10^{-16}, \tag{4}$$

which shows how the numerical noise is amplified when K_θ becomes ill-conditioned.

The term $\log|\mathrm{K}_\theta|$ is nonlinear in K_θ, but observe, using the identity $\mathrm{d}\log|\mathrm{K}_\theta|/\mathrm{d}\mathrm{K}_\theta = \mathrm{K}_\theta^{-1}$, that the differential of $\log|\cdot|$ at K_θ is given by $H \mapsto \mathrm{Trace}(\mathrm{K}_\theta^{-1}H)$. Thus, the induced operator norm with respect to the Frobenius norm $\|\cdot\|_F$ is $\|\mathrm{K}_\theta^{-1}\|_F$. We can then apply results from Trefethen and Bau (1997) to get a local condition number of the mapping $\mathrm{A} \mapsto \log|\mathrm{A}|$ at K_θ:

$$\kappa(\log|\cdot|, \mathrm{K}_\theta) \triangleq \lim_{\epsilon\to 0} \sup_{\|\delta_\mathrm{A}\|_F \le \epsilon} \frac{\big|\log|\mathrm{K}_\theta + \delta_\mathrm{A}| - \log|\mathrm{K}_\theta|\big|}{\big|\log|\mathrm{K}_\theta|\big|} \frac{\|\mathrm{K}_\theta\|_F}{\|\delta_\mathrm{A}\|_F} = \frac{\sqrt{\sum_{i=1}^n \frac{1}{\lambda_i^2}}\sqrt{\sum_{i=1}^n \lambda_i^2}}{|\sum_{i=1}^n \log(\lambda_i)|} \tag{5}$$

where $\lambda_1, \cdots, \lambda_n$ are the (positive) eigenvalues of K_θ. Then, we have

$$\frac{\kappa(\mathrm{K}_\theta)}{|\sum_{i=1}^n \log(\lambda_i)|} \le \kappa(\log|\cdot|, \mathrm{K}_\theta) \le \frac{n\kappa(\mathrm{K}_\theta)}{|\sum_{i=1}^n \log(\lambda_i)|}, \tag{6}$$

which shows that numerical noise on $\log|\mathrm{K}_\theta|$ is linked to the condition number of K_θ.

The local condition number of the quadratic form $\frac{1}{2}\underline{Z}_n^\top \mathrm{K}_\theta^{-1}\underline{Z}_n$ as a function of $\underline{Z}_n$ can also be computed analytically. Some straightforward calculations show that it is bounded by $\kappa(\mathrm{K}_\theta)$.

(When the optimization algorithm stops in the example of Fig. 2, we have $\kappa(\mathrm{K}_\theta) \simeq 10^{11}$ and $\kappa(\log|\cdot|, \mathrm{K}_\theta) \simeq 10^{9.5}$. The empirical numerical fluctuations are measured as the residuals of a local second-order polynomial best fit, giving noise levels 10^{-7}, 10^{-8} and $10^{-7.5}$ for $\mathrm{K}_\theta^{-1}\underline{Z}_n$, $\frac{1}{2}\underline{Z}_n^\top \mathrm{K}_\theta^{-1}\underline{Z}_n$ and $\log|\mathrm{K}_\theta|$ respectively. These values are consistent with the above first-order analysis.)

Thus, when $\kappa(\mathrm{K}_\theta)$ becomes large in the course of the optimization procedure, numerical noise on the likelihood and its gradient may trigger an early stopping of the optimization algorithm (supposedly when the algorithm is unable to find a proper direction of improvement). It is well-known that $\kappa(\mathrm{K}_\theta)$ becomes large when $\sigma_\varepsilon^2 = 0$ and one of the following conditions occurs: 1) data points are close, 2) the covariance is very smooth (as for instance when considering the

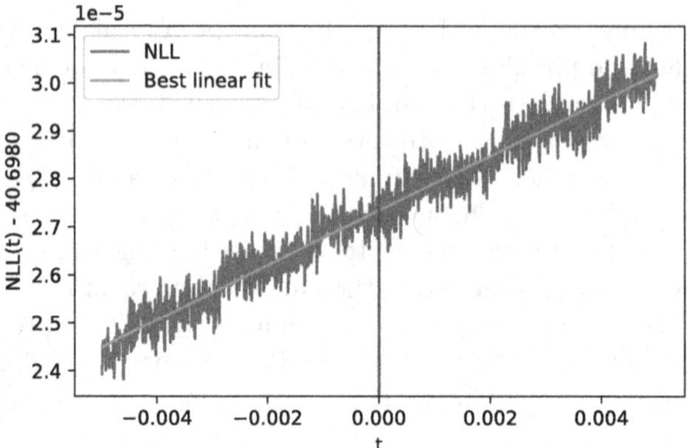

Fig. 2. Noisy NLL profile along a particular direction in the parameter space, with a best linear fit (orange line). This example was obtained with GPy while estimating the parameters of a Matérn 5/2 covariance, using 20 data points sampled from a *Branin function*, and setting $\sigma_\varepsilon^2 = 0$. The red vertical line indicates the location where the optimization of the likelihood stalled. (Color figure online)

squared exponential covariance), 3) when the range parameters ρ_i are large. These conditions arise more often than not. Therefore, the problem of numerical noise in the evaluation of the likelihood and its gradient is a problem that should not be neglected in GP implementations.

The most classical approach to deal with ill-conditioned covariance matrices is to add a small positive number on the diagonal of the covariance matrix, called *jitter*, which is equivalent to assuming a small observation noise with variance $\sigma_\varepsilon^2 > 0$. In GPy for instance, the strategy consists in always setting a minimal jitter of 10^{-8}, which is automatically increased by an amount ranging from $10^{-6}\sigma^2$ to $10^{-1}\sigma^2$ whenever the Cholesky factorization of the covariance matrix fails (due to numerical non-positiveness). The smallest jitter making K_θ numerically invertible is kept and an error is thrown if no jitter allows for successful factorization. However, note that large values for the jitter may yield smooth, non-interpolating approximations, with possible unintuitive and undesirable effects (see Andrianakis and Challenor 2012), and causing possible convergence problems in Bayesian optimization.

Table 4 illustrates the behaviour of GP interpolation when σ_ε^2 is increased. It appears that finding a satisfying trade-off between good interpolation properties and low numerical noise level can be difficult. Table 4 also supports the connection in (4) and (6) between noise levels and $\kappa(\mathrm{K}_\theta)$. In view of the results of Fig. 1 based on the default settings of GPy and Table 4, we believe that adaptive jitter cannot be considered as a do-it-all solution.

4 Strategies for Improving Likelihood Maximization

In this section we investigate simple but hopefully efficient levers/strategies to improve available implementations of MLE for GP interpolation, beyond the

Table 4. Influence of the jitter on the GP model (same setting as in Fig. 2). The table reports the condition numbers $\kappa(\mathrm{K}_\theta)$ and $\kappa(\log|\cdot|, \mathrm{K}_\theta)$, and the impact on the relative empirical standard deviations δ_{quad} and δ_{logdet} of the numerical noise on $\underline{Z}_n^T \mathrm{K}_\theta^{-1} \underline{Z}_n$ and $\log|\mathrm{K}_\theta|$ respectively (measured using second-order polynomial regressions). As σ_ε increases, δ_{quad} and δ_{logdet} decrease but the interpolation error $\sqrt{\mathrm{SSR}/\mathrm{SST}} = \sqrt{\frac{1}{n}\sum_{j=1}^n (Z_j - \hat{Z}_n(x_j))^2}/\mathrm{std}(Z_1, ..., Z_n)$ and the NLL increase. Reducing numerical noise while keeping good interpolation properties requires careful attention in practice.

$\sigma_\varepsilon^2 \ / \ \sigma^2$	0.0	10^{-8}	10^{-6}	10^{-4}	10^{-2}
$\kappa(\mathrm{K}_\theta)$	10^{11}	10^{9}	$10^{7.5}$	$10^{5.5}$	$10^{3.5}$
$\kappa(\log\|\cdot\|, \mathrm{K}_\theta)$	$10^{9.5}$	$10^{8.5}$	$10^{6.5}$	$10^{4.5}$	$10^{2.5}$
δ_{quad}	10^{-8} $(= 10^{11-19})$	$10^{-9.5}$ $(= 10^{9-18.5})$	$10^{-10.5}$ $(= 10^{7.5-18})$	10^{-12} $(= 10^{5.5-17.5})$	10^{-14} $(= 10^{3.5-17.5})$
δ_{logdet}	$10^{-7.5}$ $(= 10^{9.5-17})$	10^{-9} $(= 10^{8.5-17.5})$	10^{-11} $(= 10^{6.5-17.5})$	$10^{-13.5}$ $(= 10^{4.5-18})$	$10^{-15.5}$ $(= 10^{2.5-18})$
$-\log(\mathcal{L}(\underline{Z}_n\|\theta))$	40.69	45.13	62.32	88.81	124.76
$\sqrt{\mathrm{SSR}/\mathrm{SST}}$	$3.3 \cdot 10^{-10}$	$1.2 \cdot 10^{-3}$	0.028	0.29	0.75

control of the numerical noise on the likelihood using jitter. We mainly focus on 1) initialization methods for the optimization procedure, 2) stopping criteria, 3) the effect of "restart" strategies and 4) the effect of the parameterization of the covariance.

4.1 Initialization Strategies

Most GP implementations use a gradient-based local optimization algorithm to maximize the likelihood that requires the specification of starting/initial values for the parameters. In the following, we consider different initialization strategies.

Moment-Based Initialization. A first strategy consists in setting the parameters using empirical moments of the data. More precisely, assuming a constant mean $m = \mu$, and a stationary covariance k with variance σ^2 and range parameters $\rho_1, \ldots, \rho_d$, set

$$\mu_{\text{init}} = \mathrm{mean}\,(Z_1, \ldots, Z_n), \tag{7}$$
$$\sigma^2_{\text{init}} = \mathrm{var}\,(Z_1, \ldots, Z_n), \tag{8}$$
$$\rho_{k,\,\text{init}} = \mathrm{std}\,(x_{1,\,[k]}, \ldots, x_{n,\,[k]}), \quad k = 1, \ldots, d, \tag{9}$$

where mean, var and std stand for the empirical mean, variance and standard deviation, and $x_{i,\,[k]}$ denotes the kth coordinate of $x_i \in \mathbb{R}^d$. The rationale behind (9) (following, e.g., Rasmussen and Williams 2006) is that the range parameters can be thought of as the distance one has to move in the input space for the function value to change significantly and we assume, a priori, that this distance is linked to the dispersion of data points.

In GPy for instance, the default initialization consists in setting $\mu = 0$, $\sigma^2 = 1$ and $\rho_k = 1$ for all k. This is equivalent to the *moment-based* initialization scheme when the data (both inputs and outputs) are centered and standardized. The practice of standardizing the input domain into a unit length hypercube has been proposed (see, e.g., Snoek et al. 2012) to deal with numerical issues that arise due to large length scale values.

Profiled Initialization. Assume the range parameters $\rho_1, \ldots, \rho_d$ (and more generally, all parameters different from σ^2, σ_ε^2 and μ) are fixed, and set $\sigma_\varepsilon^2 = \alpha\sigma^2$, with a prescribed multiplicative factor $\alpha \geq 0$. In this case, the NLL can be optimized analytically w.r.t. μ and σ^2. Optimal values turn out to be the generalized least squares solutions

$$\mu_{\text{GLS}} = (\mathbb{1}_n^\mathsf{T} \mathrm{K}_{\tilde{\theta}}^{-1} \mathbb{1}_n)^{-1} \mathbb{1}_n^\mathsf{T} \mathrm{K}_{\tilde{\theta}}^{-1} \underline{Z}_n\,, \tag{10}$$

$$\sigma_{\text{GLS}}^2 = \frac{1}{n} (\underline{Z}_n - \mu_{\text{GLS}}\, \mathbb{1}_n)^\mathsf{T} \mathrm{K}_{\tilde{\theta}}^{-1} (\underline{Z}_n - \mu_{\text{GLS}}\, \mathbb{1}_n)\,, \tag{11}$$

where $\tilde{\theta} = (\sigma^2, \rho_1, \ldots, \rho_d, \ldots, \sigma_\varepsilon^2)^\mathsf{T} \in \Theta$, with $\sigma^2 = 1$ and $\sigma_\varepsilon^2 = \alpha$. Under the *profiled* initialization scheme, $\rho_1, \ldots, \rho_d$ are set using (9), α is prescribed according to user's preference, and μ and σ^2 are initialized using (10) and (11).

Grid-Search Initialization. Grid-search initialization is a *profiled* initialization with the addition of a grid-search optimization for the range parameters.

Define a nominal range vector ρ_0 such that

$$\rho_{0,[k]} = \sqrt{d} \left(\max_{1 \leq i \leq n} x_{i,[k]} - \min_{1 \leq i \leq n} x_{i,[k]} \right), \quad 1 \leq k \leq d.$$

Then, define a one-dimensional grid of size L (e.g., $L = 5$) by taking range vectors proportional to ρ_0: $\{\alpha_1 \rho_0, \ldots, \alpha_L \rho_0\}$, where the α_is range, in logarithmic scale, from a "small" value (e.g., $\alpha_1 = 1/50$) to a "large" value (e.g., $\alpha_L = 2$). For each point of the grid, the likelihood is optimized with respect to μ and σ^2 using (10) and (11). The range vector with the best likelihood value is selected. (Note that this initialization procedure is the default initialization procedure in the Matlab/GNU Octave toolbox STK.)

4.2 Stopping Condition

Most GP implementations rely on well-tested gradient-based optimization algorithms. For instance, a popular choice in Python implementations is to use the limited-memory BFGS algorithm with box constraints (L-BFGS-B; see Byrd et al. 1995) of the SciPy ecosystem. (Other popular optimization algorithms include the ordinary BFGS, truncated Newton constrained, SQP, etc.; see, e.g., Nocedal and Wright (2006).) The L-BFGS-B algorithm, which belongs to the class of quasi-Newton algorithms, uses limited-memory Hessian approximations and shows good performance on non-smooth functions (Curtis and Que 2015).

Regardless of which optimization algorithm is chosen, the user usually has the possibility to tune the behavior of the optimizer, and in particular to set the stopping condition. Generally, the stopping condition is met when a maximum number of iterations is reached or when a norm on the steps and/or the gradient become smaller than a threshold.

By increasing the strictness of the stopping condition during the optimization of the likelihood, one would expect better parameter estimations, provided the numerical noise on the likelihood does not interfere too much.

4.3 Restart and Multi-start Strategies

Due to numerical noise and possible non-convexity of the likelihood with respect to the parameters, gradient-based optimization algorithms may stall far from the global optimum. A common approach to circumvent the issue is to carry out several optimization runs with different initialization points. Two simple strategies can be compared.

Table 5. Two popular reparameterization mappings τ, as implemented, for example, in GPy and STK respectively. For *invsoftplus*, notice parameter $s > 0$, which is introduced when input standardization is considered (see Sect. 5).

Reparam. method	$\tau : \mathbb{R}_+^\star \to \mathbb{R}$	$\tau^{-1} : \mathbb{R} \to \mathbb{R}_+^\star$
invsoftplus(s)	$\log(\exp(\theta/s) - 1)$	$s \log(\exp(\theta') + 1)$
log	$\log(\theta)$	$\exp(\theta')$

Restart. In view of Fig. 2, a first simple strategy is to restart the optimization algorithm to clear its memory (Hessian approximation, step sizes...), hopefully allowing it to escape a possibly problematic location using the last best parameters as initial values for the next optimization run. The optimization can be restarted a number of times, until a budget N_{opt} of restarts is spent or the best value for the likelihood does not improve.

Multi-start. Given an initialization point $(\theta_{\text{init}}, \mu_{\text{init}}) \in \Theta \times \mathbb{R}$, a multi-start strategy consists in running $N_{\text{opt}} > 1$ optimizations with different initialization points corresponding to perturbations of the initial point $(\theta_{\text{init}}, \mu_{\text{init}})$. In practice, we suggest the following rule for building the perturbations: first, move the range parameters around $(\rho_{1,\,\text{init}}, \ldots, \rho_{d,\,\text{init}})^T$ (refer to Sect. 5 for an implementation); then, propagate the perturbations on μ and σ^2 using (10) and (11). The parameter with the best likelihood value over all optimization runs is selected.

4.4 Parameterization of the Covariance Function

The parameters of the covariance functions are generally positive real numbers $(\sigma^2, \rho_1, \rho_2 \ldots)$ and are related to scaling effects that act "multiplicatively" on the predictive distributions. Most GP implementations introduce a reparameterization using a monotonic one-to-one mapping $\tau : \mathbb{R}_+^\star \to \mathbb{R}$, acting component-wise on the positive parameters of θ, resulting in a mapping $\tau : \Theta \to \Theta'$. Thus, for carrying out MLE, the actual criterion J that is optimized in most implementations may then be written as

$$J : \theta' \in \Theta' \mapsto -\log(\mathcal{L}(\underline{Z}_n | \tau^{-1}(\theta'), c)). \tag{12}$$

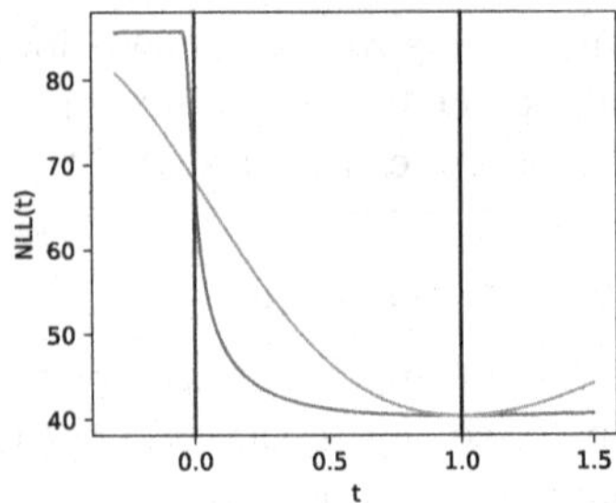

Fig. 3. Profiles of the NLL along a linear path t through the *profiled* initialization point (at zero, blue vertical line) and the optimum (at one, black vertical line). Orange (resp. blue) line corresponds to the *log* (resp. *invsoftplus*) reparameterization. (Color figure online)

Table 5 lists two popular reparameterization mappings τ.

The effect of reparameterization is to "reshape" the likelihood. Typical likelihood profiles using the *log* and the so-called *invsoftplus* reparameterizations are shown on Fig. 3. Notice that the NLL may be almost flat in some regions depending on the reparameterization. Changing the shape of the optimization criterion, combined with numerical noise, may or may not facilitate the convergence of the optimization.

5 Numerical Study

5.1 Methodology

The main metric used in this numerical study is based on empirical cumulative distributions (ECDFs) of differences on NLL values.

More precisely, consider $N+1$ optimization schemes $S_0, S_1, \ldots, S_N$, where S_0 stands for a "brute-force" optimization scheme based on a very large number of multi-starts, which is assumed to provide a robust MLE, and $S_1, \ldots, S_N$ are optimization schemes to be compared. Each optimization scheme is run on M data sets D_j, $1 \le j \le M$, and we denote by $e_{i,j}$ the difference

$$e_{i,j} = \mathrm{NLL}_{i,j} - \mathrm{NLL}_{0,j}, \quad 1 \le i \le N, \quad 1 \le j \le M,$$

where $\mathrm{NLL}_{i,j}$ the NLL value obtained by optimization scheme S_i on data set D_j.

A good scheme S_i should concentrate the empirical distribution of the sample $E_i = \{e_{i,j}, j = 1, \ldots, M\}$ around zero—in other words, the ECDF is close to the ideal CDF $e \mapsto \mathbb{1}_{[0,\infty[}(e)$. Using ECDF also provides a convenient way to compare performances: a strategy with a "steeper" ECDF, or larger area under the ECDF, is better.

5.2 Optimization Schemes

All experiments are performed using GPy version 1.9.9, with the default L-BFGS-B algorithm. We use a common setup and vary the configurations of the optimization levers as detailed below.

Common Setup. All experiments use an estimated constant mean-function, an anisotropic Matérn covariance function with regularity $\nu = 5/2$, and we assume no observation noise (the adaptive jitter of GPy ranging from $10^{-6}\sigma^2$ to $10^2\sigma^2$ is used, however).

Initialization Schemes. Three initialization procedures from Sect. 4.1 are considered.

Stopping Criteria. We consider two settings for the stopping condition of the L-BFGS-B algorithm, called *soft* (the default setting: `maxiter`$= 1000$, `factr`$=10^7$, `pgtol`10^{-5}) and *strict* (`maxiter`$= 1000$, `factr`$=10$, `pgtol`$= 10^{-20}$).

Restart and Multi-start. The two strategies of Sect. 4.3 are implemented using a *log* reparameterization and initialization points $(\theta_{\text{init}}, \mu_{\text{init}})$ determined using a *grid-search* strategy. For the *multi-start* strategy the initial range parameters are perturbed according to the rule $\rho \leftarrow \rho_{\text{init}} \cdot 10^{\eta}$ where η is drawn from a $\mathcal{N}(0, \sigma_\eta^2)$ distribution. We take $\sigma_\eta = \log_{10}(5)/1.96$ (≈ 0.35), to ensure that about 0.95 of the distribution of ρ is in the interval $[1/5 \cdot \rho_{\text{init}},\ 5 \cdot \rho_{\text{init}}]$.

Reparameterization. We study the *log* reparameterization and two variants of the *invsoftplus*. The first version called *no-input-standardization* simply corresponds to taking $s = 1$ for each range parameter. The second version called *input-standardization* consists in scaling the inputs to a unit standard deviation on each dimension (by taking the corresponding value for s).

5.3 Data Sets

The data sets are generated from six well-known test functions in the literature of Bayesian optimization: the Branin function ($d = 2$; see, e.g. Surjanovic and Bingham 2013), the Borehole function ($d = 8$; see, e.g. Worley 1987), the Welded Beam Design function ($d = 4$; see Chafekar et al. 2003), the g10 function ($d = 8$; see Ahmed 2004, p. 128), along with two modified versions, g10mod and g10modmod (see Feliot 2017).

Each function is evaluated on Latin hypercube samples with a multi-dimensional uniformity criterion (LHS-MDU; Deutsch and Deutsch 2012), with varying sample size $n \in \{3d, 5d, 10d, 20d\}$, resulting in a total of $6 \times 4 = 24$ data sets.

5.4 Results and Findings

Figure 4 shows the effect of reparameterization and the initialization method. Observe that the *log* reparameterization performs significantly better than the

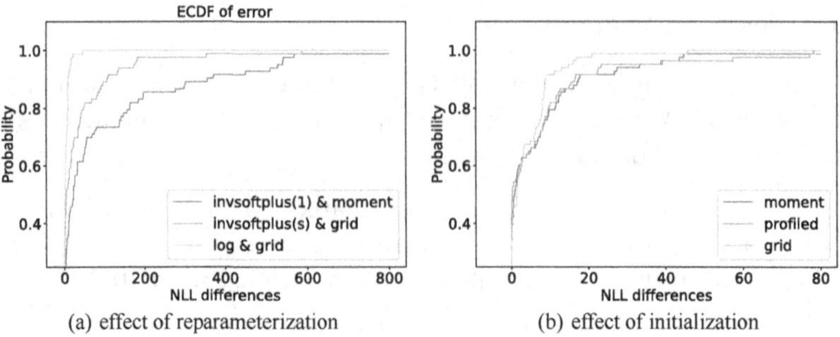

(a) effect of reparameterization (b) effect of initialization

Fig. 4. Initialization and reparameterization methods. (a) ECDFs corresponding to the best initialization method for each of the three reparameterizations—red line: *log* reparam. with *grid-search* init.; green line: *invsoftplus* with *input-standardization* reparam. and *grid-search* init; blue line: *invsoftplus* with *no-input-standardization* reparam. and *moment-based* init. (b) ECDFs for different initialization methods for the log reparameterization. (Color figure online)

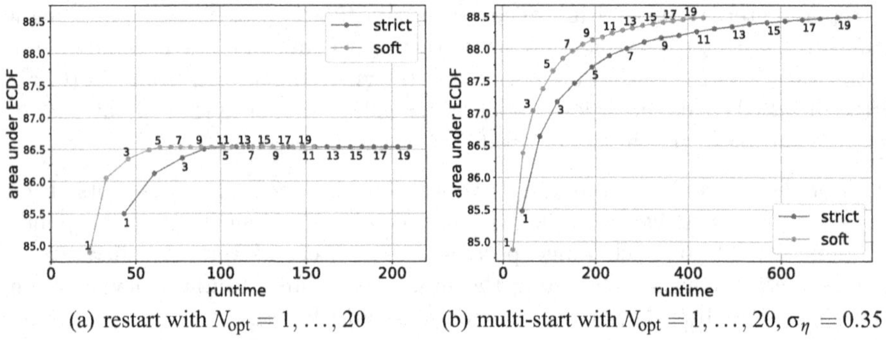

(a) restart with $N_{\text{opt}} = 1, \ldots, 20$ (b) multi-start with $N_{\text{opt}} = 1, \ldots, 20$, $\sigma_\eta = 0.35$

Fig. 5. Area under the ECDF against run time: (a) *restart* strategy; (b) *multi-start* strategy. The maximum areas obtained are respectively 86.538 and 88.504.

invsoftplus reparameterizations. For the *log* reparameterization, observe that the *grid-search* strategy brings a moderate but not negligible gain with respect to the two other initialization strategies, which behave similarly.

Next, we study the effect of the different restart strategies and the stopping conditions, on the case of the *log* reparameterization and *grid-search* initialization. The metric used for the comparison is the area under the ECDFs of the differences of NLLs, computed by integrating the ECDF between 0 and $\text{NLL}_{\text{max}} = 100$. Thus, a perfect optimization strategy would achieve an area under the ECDF equal to 100. Since the *multi-start* strategy is stochastic, results are averaged over 50 repetitions of the optimization procedures (for each N_{opt} value, the optimization strategy is repeated 50 times). The areas are plotted against the computational run time. Run times are averaged over the repetitions in the case of the *multi-start* strategy.

Figure 5 shows that the *soft* stopping condition seems uniformly better. The *restart* strategy yields small improvements using moderate computational overhead. The *multi-start* strategy is able to achieve the best results at the price of higher computational costs.

6 Conclusions and Recommendations

Our numerical study has shown that the parameterization of the covariance function has the most significant impact on the accuracy of MLE in GPy. Using *restart*/*multi-start* strategies is also very beneficial to mitigate the effect of the numerical noise on the likelihood. The two other levers have second-order but nonetheless measurable influence.

These observations make it possible to devise a recommended combination of improvement levers—for GPy at least, but hopefully transferable to other software packages as well. When computation time matters, an improved optimization procedure for MLE consists in choosing the combination of a *log* reparameterization, with a *grid-search* initialization, the *soft* (GPy's default) stopping condition, and a small number, say $N_{\text{opt}} = 5$, of restarts.

Figure 1 and Table 2 are based on the above optimization procedure, which results in significantly better likelihood values and smaller prediction errors. The *multi-start* strategy can be used when accurate results are sought.

As a conclusion, our recommendations are not intended to be universal, but will hopefully encourage researchers and users to develop and use more reliable and more robust GP implementations, in Bayesian optimization or elsewhere.

References

Ahmed, A.R.H.A.: Studies on metaheuristics for continuous global optimization problems. Ph.D. thesis, Kyoto University (2004)

Andrianakis, I., Challenor, P.G.: The effect of the nugget on Gaussian process emulators of computer models. Comput. Stat. Data Anal. **56**(12), 4215–4228 (2012)

Baudin, M., Dutfoy, A., Iooss, B., Popelin, A.L.: OpenTURNS: an industrial software for uncertainty quantification in simulation. In: Ghanem, R., Higdon, D., Owhadi, H. (eds.) Handbook of Uncertainty Quantification, pp. 2001–2038. Springer, Switzerland (2017)

Bect, J., Vazquez, E., et al.: STK: a small (Matlab/Octave) toolbox for Kriging. Release 2.6 (2011–2021). http://kriging.sourceforge.net

Byrd, R., Lu, P., Nocedal, J., Zhu, C.: A limited memory algorithm for bound constrained optimization. SIAM J. Sci. Comput. **16**(5), 1190–1208 (1995)

Chafekar, D., Xuan, J., Rasheed, K.: Constrained multi-objective optimization using steady state genetic algorithms. In: Cantú-Paz, E., et al. (eds.) GECCO 2003. LNCS, vol. 2723, pp. 813–824. Springer, Heidelberg (2003). https://doi.org/10.1007/3-540-45105-6_95

Curtis, F.E., Que, X.: A Quasi-Newton algorithm for nonconvex, nonsmooth optimization with global convergence guarantees. Math. Program. Comput. **7**(4), 399–428 (2015). https://doi.org/10.1007/s12532-015-0086-2

Deutsch, J.L., Deutsch, C.V.: Latin hypercube sampling with multidimensional uniformity. J. Stat. Plann. Inference **142**(3), 763–772 (2012)

Emmerich, M.T.M., Giannakoglou, K.C., Naujoks, B.: Single- and multiobjective evolutionary optimization assisted by Gaussian random field metamodels. IEEE Trans. Evol. Comput. **10**(4), 421–439 (2006)

Erickson, C.B., Ankenman, B.E., Sanchez, S.M.: Comparison of Gaussian process modeling software. Eur. J. Oper. Res. **266**(1), 179–192 (2018)

Feliot, P.: Une approche bayésienne pour l'optimisation multi-objectif sous contrainte. Ph.D. thesis, University of Paris-Saclay (2017)

Gardner, J.R., Pleiss, G., Bindel, D., Weinberger, K.Q., Wilson, A.G.: GPyTorch: blackbox matrix-matrix Gaussian process inference with GPU acceleration. In: Advances in Neural Information Processing Systems, vol. 31. Curran Associates (2018)

Jones, D.R., Schonlau, M., Welch, W.J.: Efficient global optimization of expensive black-box functions. J. Global Optim. **13**(4), 455–492 (1998)

Kingma, D.P., Ba, J.: Adam: a method for stochastic optimization. In: Bengio, Y., LeCun, Y. (eds.) 3rd International Conference on Learning Representations, ICLR 2015, San Diego, USA (2015)

de G. Matthews, A.G., et al.: GPflow: a Gaussian process library using TensorFlow. J. Mach. Learn. Res., **18**(40), 1–6 (2017)

Mockus, J.: On Bayesian methods for seeking the extremum. In: Marchuk, G.I. (ed.) Optimization Techniques IFIP Technical Conference Novosibirsk. July 1–7, 1974, pp. 400–404. Springer, Heidelberg (1975)

Nash, S.G.: Newton-type minimization via the Lanczos method. SIAM J. Numer. Anal. **21**(4), 770–788 (1984)

Nocedal, J., Wright, S.J.: Numerical Optimization. Springer, New York (2006). https://doi.org/10.1007/978-0-387-40065-5

O'Hagan, A.: Curve fitting and optimal design for prediction. J. R. Stat. Soc. B **40**, 1–24 (1978)

Pedregosa, F., et al.: Scikit-learn: machine learning in Python. J. Mach. Learn. Res. **12**, 2825–2830 (2011)

Petit, S., Bect, J., Da Veiga, S., Feliot, P., Vazquez, E.: Towards new cross-validation-based estimators for Gaussian process regression: efficient adjoint computation of gradients. arXiv:2002.11543 (2020)

Press, W.H., Teukolsky, S.A., Vetterling, W.T., Flannery, B.P.: Numerical Recipes in C. The Art of Scientific Computing. Cambridge University Press, New York (1992)

Rasmussen, C.E., Nickisch, H.: Gaussian processes for machine learning (GPML) toolbox. J. Mach. Learn. Res. **11**, 3011–3015 (2010)

Rasmussen, C.E., Williams, C.K.I.: Gaussian Processes for Machine Learning. MIT Press, Cambridge (2006)

Roustant, O., Ginsbourger, D., Deville, Y.: DiceKriging, DiceOptim: two R packages for the analysis of computer experiments by Kriging-based metamodeling and optimization. J. Stat. Software **51**(1), 1–55 (2012)

Santner, T.J., Williams, B.J., Notz, W.I.: The Design and Analysis of Computer Experiments. SSS, Springer, New York (2003). https://doi.org/10.1007/978-1-4757-3799-8

Sheffield machine learning group. GPy: a Gaussian process framework in Python, version 1.9.9 (2012–2020). http://github.com/SheffieldML/GPy

Snoek, J., Larochelle, H., Adams, R.P.: Practical Bayesian optimization of machine learning algorithms. In: 25th International Conference on Neural Information Processing Systems, vol. 2, pp. 2951–2959. Curran Associates Inc. (2012)

Srinivas, N., Krause, A., Kakade, S., Seeger, M.: Gaussian process optimization in the bandit setting: no regret and experimental design. In: 27th International Conference on Machine Learning (ICML), pp. 1015–1022 (2010)

Stein, M.L.: Interpolation of Spatial Data: Some Theory for Kriging. Springer, New York (1999). https://doi.org/10.1007/978-1-4612-1494-6

Surjanovic, S., Bingham, D.: Virtual library of simulation experiments: test functions and datasets (2013). http://www.sfu.ca/~ssurjano/branin.html. Accessed 13 Oct 2020

Trefethen, L.N., Bau, D.: Numerical Linear Algebra. SIAM (1997)

Vanhatalo, J., Riihimäki, J., Hartikainen, J., Jylänki, P., Tolvanen, V., Vehtari, A.: Bayesian modeling with Gaussian processes using the MATLAB toolbox GPstuff (v3.3). CoRR, abs/1206.5754 (2012)

Wendland, H.: Scattered Data Approximation. Cambridge Monographs on Applied and Computational Mathematics, Cambridge University press, Cambridge (2004)

Worley, B.A.: Deterministic uncertainty analysis. Technical report, ORNL-6428, Oak Ridge National Laboratory, TN, USA (1987)

KAFE: Knowledge and Frequency Adapted Embeddings

Awais Ashfaq[1,2(✉)], Markus Lingman[2,3], and Slawomir Nowaczyk[1]

[1] Center for Applied Intelligent Systems Research, Halmstad University, Halmstad, Sweden
awais.ashfaq@hh.se

[2] Halland Hospital, Region Halland, Sweden

[3] Department of Molecular and Clinical Medicine/Cardiology, Institute of Medicine, Sahlgrenska Academy, University of Gothenburg, Gothenburg, Sweden

Abstract. Word embeddings are widely used in several Natural Language Processing (NLP) applications. The training process typically involves iterative gradient updates of each word vector. This makes word frequency a major factor in the quality of embedding, and in general the embedding of words with few training occurrences end up being of poor quality. This is problematic since rare and frequent words, albeit semantically similar, might end up far from each other in the embedding space.

In this study, we develop KAFE (Knowledge And Frequency adapted Embeddings) which combines adversarial principles and knowledge graph to efficiently represent both frequent and rare words. The goal of adversarial training in KAFE is to minimize the spatial distinguishability (separability) of frequent and rare words in the embedding space. The knowledge graph encourages the embedding to follow the structure of the domain-specific hierarchy, providing an informative prior that is particularly important for words with low amount of training data. We demonstrate the performance of KAFE in representing clinical diagnoses using real-world Electronic Health Records (EHR) data coupled with a knowledge graph. EHRs are notorious for including ever-increasing numbers of rare concepts that are important to consider when defining the state of the patient for various downstream applications. Our experiments demonstrate better intelligibility through visualisation, as well as higher prediction and stability scores of KAFE over state-of-the-art.

Keywords: Word embeddings · Knowledge graphs · Adversarial learning

1 Introduction

Distributed representation of words (also referred to as word embeddings or dense/continuous representations) have become de facto standard as high-quality inputs to neural-network (NN) based models built for natural language processing (NLP) applications. Examples include text classification and summarization, machine translation, sentiment analysis and more [1]. Word embeddings

G. Nicosia et al. (Eds.): LOD 2021, LNCS 13164, pp. 132–146, 2022.
https://doi.org/10.1007/978-3-030-95470-3_10

are typically learned from big unlabelled corpora using co-occurrence statistics. In general terms, the learning principle hinges on the assumption that words with similar context (surrounding words) are semantically similar and should be placed close in an arbitrary embedding space. Contrary to the traditional one-hot format, word embeddings are low-dimensional and semantically similar [2]. In recent years, the popularity of word embedding has expanded beyond text corpora into structured (non-text) data sources such as the Electronic Health Records (EHRs) to represent medical concepts and patients [3].

EHRs are real-time digital patient-centred records that allow responsible access to authorized care-providers when required. Patient data populates EHRs in two ways - structured and unstructured. Structured data means documented using a controlled vocabulary rather than free text (unstructured information). In general, structured data curbs ambiguity about what data means and facilitates data interoperability among different clinical systems. For that matter, several classification schema and ontologies exist to record clinical information. For instance, the International Classification of Diseases (ICD-10) (Table 1) defines alpha-numeric codes corresponding to different diagnoses [4]. Concurrently a patient's disease profile is coded as time-ordered lists of ICD-10 codes registered along his/her visit to the care centres. Given that, the analogy between free text and structured clinical profiles is simple: a visit profile is considered as a sentence or context and clinical codes as the words in it. Ordered set of visit profiles gives the patient profile - analogous to documents in text. Following this analogy, several researchers have applied NLP techniques to represent medical concepts and patients as low-dimensional dense vectors for downstream predictive and descriptive tasks [3,5]. The most notable among the techniques include skip-gram and continuous-bag-of-words (publicized as word2vec [2]) by Mikolov (more details later).

However, the training of word2vec models is typically based on the idea of iterative gradient updates for each word vector. Put differently, it assumes that each word appears a sufficient amount of times in the training data which is often not the case in real world. Word frequency is a major factor contributing to the quality of the embedding. For words with fewer training samples (occurrences), the quality of the learned representations will be poor [6]. Rare and frequent words often appear in different sub-regions in the embedding space with rare words having mostly rare and semantically unrelated neighbours and frequent words having frequent and semantically related neighbours [7].

In the context of NLP-inspired EHR representation, there are two common approaches to this problem. First, ignore the rare words by limiting the vocabulary size to the N most frequent words. Second, leverage the hierarchical clinical schema to group medical codes to high-level concepts since high-level concepts are likely to appear more times in the data than low-level specific concepts [8]. The former approach is problematic since it conflates all the meanings of rare concepts into a single representation, thus losing individual characteristics of clinical concepts. The problem is exacerbated when a rare clinical concept for instance, is fatal (B20: Human immune virus HIV, A41: Sepsis) fuelling imperfections in downstream applications such as treatment assignment prediction. The latter approach ignores important differences among concepts. For instance, E10.22 (Table 1) and E10.32 (Diabetes Mellitus with mild non-proliferative diabetic retinopathy), albeit in the same disease category, describe different characteristics and thus require different set of therapies.

Table 1. ICD-10 hierarchy - example

Class	I *Diseases of circulatory system*	E *Endocrine/metabolic diseases*
Category	I50 *Heart failure (HF)*	E10 *Diabetes mellitus I*
Characteristics (etiology, severity, anatomic site, other vital details)	I50.2 *Systolic HF*	E10.2 *with kidney complications*
	150.21 *Acute systolic HF*	E10.22 *with chronic kidney complications*

Moreover, it is no surprise that the medical knowledge is expanding rapidly [9]. The ICD-10 schema specified over 68,000 (detailed and more specific) diseases which was 5 times the size of ICD-9 and it is much likely that the list will grow as we await ICD-11[1]. A greater granularity in coding schema means that the frequency of rare concepts gets even smaller than general concepts and the number of unique rare concepts gets much larger than popular concepts. In the context of learning medical concept vectors, it means even fewer occurrences of rare concepts in the training data. Thus, in order to continue reaping the benefits of clinical embeddings for downstream applications, there exists a clear need to develop new embedding techniques - techniques that are capable of efficiently representing both high and low-frequency concepts.

To supplement the need, we develop KAFE (Knowledge And Frequency adapted Embeddings) which couples adversarial principles and clinical knowledge to efficiently represent concepts in EHRs. The goal of the adversarial training is to minimize the spatial distinguishability (separability) of frequent and rare words in the embedding space. The clinical knowledge graph facilitates the representation learning process by driving the concepts close in the hierarchy to also being close in the embedding space. Our work emphasizes that joint learning from both knowledge graph and adversarial training leads to better quality embeddings - particularly for rare words.

There are two primary contributions of our work. First, we propose a novel technique that leverages a domain knowledge graph to limit the influence of word frequency when learning embedding of medical concepts. Second, we evaluate the effectiveness of the proposed model on a comprehensive real-world dataset over

[1] https://www.who.int/standards/classifications/classification-of-diseases.

three different qualitative and quantitative tasks. The results demonstrate higher performance of KAFE over state-of-the-art.

2 Background

Distributed representation models are known to struggle with small data. Several approaches have been proposed to deal with the rare word problem in NLP and in general be grouped into decomposition, ontext-based and knowledge-based approaches.

Decomposition approaches rely on breaking down a rare word into sub-word level linguistic units such as morphemes and learning their embedding [10,11]. A word embedding is typically an accumulation of its morpheme embeddings. Whilst these works have shown promising results in different NLP tasks, they primarily depend on a well-established morphology which is not readily available in the context of clinical concepts. An even finer decomposition approach on rare words (syllables and characters) has also been studied [12–14]. These approaches, albeit being light on learnable parameters, fail to distinguish between semantically unrelated words such as *cat* and *can*. Reiterating our earlier example, E10.22 and E10.32 have a different disease characteristics, yet a higher character-level similarity.

Knowledge based approaches leverage available lexical sources such as WordNet to guide the embedding learning process [15,16]. Given the availability of extensive classification schema in medicine, this approach has gained much recognition from researchers to learn better quality medical concept embeddings and simultaneously achieve better alignment with the domain knowledge. Specifically, schema that encode the relationships among the medical concepts in well-structured formats (e.g. a knowledge graph (KG) or a hierarchical tree (HT)) can be utilized to guide the representation learning process. The approach is primarily built upon the principles of attention-mechanism [17] to evaluate what parts in the KG or HT weigh more when learning latent representation of specific medical codes. Among the pioneers in this area were [18] who proposed a graph-based attention model (GRAM) that learns representations of medical concepts as a convex combination of the code embedding and its ancestors in the knowledge graph. Later, [19] extended the concept and proposed an end-to-end knowledge-based attention model (KAME) that (in addition to learning embedding of medical concepts) leverages them to influence prediction performance. [20] contributed to the domain by learning multiple representations of a concept in the KG that were expected to correspond to a particular *sense* and carry distinct semantic meanings (e.g. in terms of therapy and etiology). Whilst these approaches have demonstrated success in various downstream applications, an inherent assumption (in the context of rare word embedding) when using KG or HT is that the higher the concept in the hierarchy, the more times it is likely to appear during training. Put differently, higher-level concepts are never rare and thus their embeddings are reliable. Moreover, since the final code embedding is an accumulation of itself and its ancestors in the hierarchy, a greater attention (weight) is given to the ancestors of rare concepts when learning their embedding. Whilst the assumption may be true for ICD-9 (as demonstrated in [18–20]), it does not quite hold true for its chronological successor which is far more specific. For instance, Table 2 shows that over 40% of ICD-10 categorical codes appear less than 10 times in a real-world EHR from Sweden that includes nearly 25 million healthcare visits [21].

Table 2. Frequency distribution of ICD-10 category codes

No. of occurrences	$<10^1$	$[10^1,10^2)$	$[10^2,10^3)$	$[10^3,10^4)$	$[10^4,10^5)$
ICD-10 category codes	832 (40.4%)	519 (25.2%)	491 (23.9%)	196 (9.5%)	19 (0.9%)

Additionally, there has also been some work on context-based approaches that focus on accumulating the embeddings of surrounding (context) words of the rare word to generate the embedding of it [22,23]. Attentive Mimicking [24,25] is a notable approach in this direction that uses self-attention to filter a subset of reliable and informative context, instead of using the entire context of the word. This approach has been mostly used in connection to transformer-based contextualized word embedding models. Contrary to traditional word embeddings that learn a global (or fixed) representation of each word, contextualized word embeddings are aimed to learn different representations of words depending on context-level semantics. Thus, they are particularly useful for polysemous words such as *bank*. However, when representing well-defined clinical concepts such as *I50.21*, polysemy is trivial. In this paper, we focus on the traditional global word embedding models.

Recently, [6] proposed an adversarial approach to learn frequency agnostic embeddings of words yielding a good mix of rare and frequent words in the embedding space. They introduce a binary discriminator in the neural language model that is optimized to classify words into rare/frequent categories. The final word embeddings are optimized towards minimizing a task-specific loss while maximizing the discriminator loss. While the approach is shown to lead improvements in machine translation and text classification task, it is unclear if (and why) maximizing indistinguishability among rare-frequent embeddings would result in semantically similar rare words coming closer together. We address this conundrum by integrating the adversarial framework with a knowledge base to ensure semantic similarity among rare concepts as they are intermingled with frequent concepts in the embedding space by the discriminator during training.

3 Method

KAFE constitutes two components: knowledge injection and frequency expulsion. Figure 1 shows the overall framework.

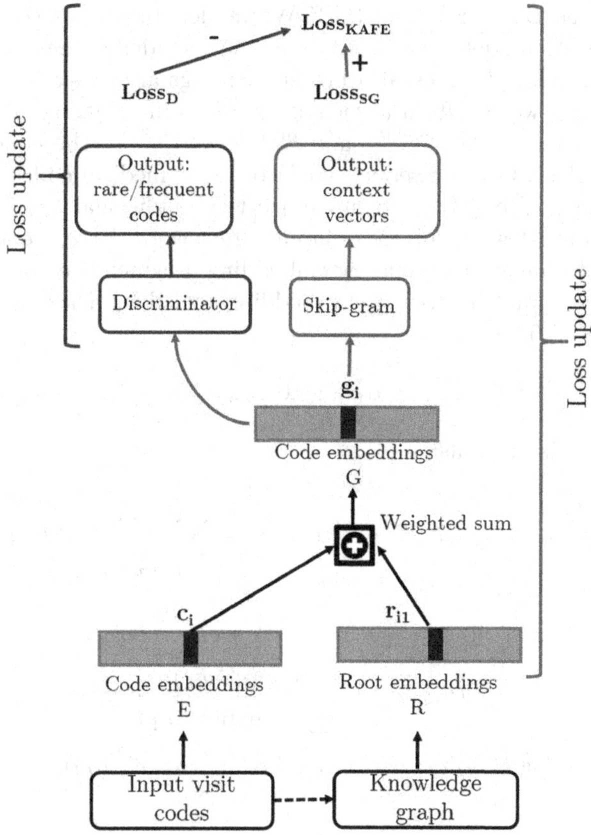

Fig. 1. Proposed learning framework KAFE.

3.1 Notations

Let V and C denote a set of all visits and medical codes where $|V|$ and $|C|$ are the total number of visits and medical codes respectively. Each visit $v \in V$ is represented as a subset of heterogeneous medical codes $v \subseteq C$. Let G denote a directed acyclic graph (DAG), corresponding to the ICD-10 hierarchy (Table 1). Since the ICD-10 codes are generally reported between 3 and 5 alpha-numeric characters, there exists a mapping $c \underset{G}{\rightarrow} r$ of code c_i to its root or ancestors r_{c_i}. Let R denote a set of all ancestors where $|R|$ is the total number of ancestors.

3.2 Knowledge Injection into word2vec

We begin with the original word2vec architecture proposed by Mikolov [2]. Given a word in a free text, the goal of word2vec is to predict the neighbouring words - often known as the context - or the other way around: given the context words, predict the centre word. The former is known as Skip Gram (SG) and the latter is the Continuous Bag Of Words (CBOW) model. In this work, we arbitrarily selected the SG model, since there is no clear evidence that either of these outperforms the other. In general, word embeddings are real-valued vector representations $\mathbf{c}$ of a word c from a vocabulary C in an arbitrary D dimensional space. Let's denote $\theta^{code} \in \mathbb{R}^{D\text{x}|C|}$ and $\theta^{root} \in \mathbb{R}^{D\text{x}|R|}$ as the codes and roots embedding matrices. Given a sequence of words (or medical codes in a visit v), $c_1, c_2, ..., c_T$, the goal of SG is to maximize the conditional log probability of observing the context words given an input target word. Here, we adapt the SG architecture to formulate a secondary embedding g_i which is a convex combination of the code c_i and its root r_{c_i} embedding scaled by a learnable attention value α_i and $\alpha_i \in (0,1)$.

$$g_i = \alpha_i c_i + (1 - \alpha_i) r_{c_i} \tag{1}$$

The SG loss L_{SG} is given as:

$$L_{SG}(g_1, g_2, ..., g_{|C|}; \theta^{root}, \theta^{code}) = -\frac{1}{|C|} \sum_{i=1}^{|C|} \sum_{j \in v, j \neq i} \log P(g_j | g_i) \tag{2}$$

where

$$P(c_O | c_I) = \frac{\exp(\mathbf{c'}_O^T \mathbf{c_I})}{\sum_{c \in C} \exp(\mathbf{c'}_c^T \mathbf{c_I})} \quad \text{and,}$$
$$\log P(c_O | c_I) = \mathbf{c'}_O^T \mathbf{c_I} - \log(\sum_{c \in C} \exp(\mathbf{c'}_c^T \mathbf{c_I}))$$

where $\mathbf{c}$ and $\mathbf{c'}$ are the input and output vector representation of word c.

Computationally, the summation in Eq. 2 is expensive since at every iteration, the algorithm goes through the entire vocabulary C whose number can be in the order of millions. [26] proposed a negative sampling objective to approximate $P(c_O|c_I)$ by generating k negative samples (non-context words) from a unigram distribution of word frequency P_n. The embeddings are computed by approximating $\log P(c_O|c_I)$ with

$$\log \sigma\, (\mathbf{c'}_O^T \mathbf{c_I}) + \sum_{k=1}^{K} \mathbb{E}_{c_I \sim P_n(c)} [\log \sigma\, (-\mathbf{c'}_k^T \mathbf{c_I})$$

where $\sigma(x)$ is the sigmoid function.

3.3 Frequency Expulsion

Let θ^D denote the discriminator's weights. Given all the secondary embeddings g_i's as input, the output of the discriminator $\hat{y}_i \in (0,1)^{|C|}$ is a score indicating the likeliness of g_i being a rare code. The discriminator loss L_D is given as the binary cross-entropy:

$$L_D(g_1, g_2, ..., g_{|C|}; \theta^D) = -(y\log(\hat{y}) + (1-y)\log(1-\hat{y})) \tag{3}$$

where $y_i \in 0, 1^{|C|}$ are the code labels generated based on their occurrence frequency in V. Here, we labelled top 25% of codes as frequent and the rest as rare.

3.4 KAFE

Following the idea of adversarial training, KAFE aims to minimize the skip-gram loss and simultaneously mislead the discriminator by adjusting the learned embeddings to be indistinguishable in terms of frequency.

$$L_{\text{KAFE}}(g_1, g_2, ..., g_{|C|}; \theta^{root}, \theta^{code}) = L_{SG} - \lambda L_D \tag{4}$$

where λ determines the trade-off between the two losses. Of note, when performing a gradient descent on L_{KAFE}, the negative term L_D would specify the direction of ascent (or increase) for L_D.

Algorithm 1: KAFE optimization

Input: Dataset V of vocabulary size C; Ontology as a DAG G
Output: Code embeddings θ^{code}, Root embeddings θ^{root}, Attention α
Initialization: Randomly initialize code and root embedding matrices θ^{code} and θ^{root}, attention parameter α, and discriminator's weights θ^D.

while *iterations or until convergence* **do**
  Sample a minibatch V' from V
  for *visit v in V'* **do**
    for *c_i in v* **do**
      Get the root of c_i from G
      Compute g_i using Eq. 1
  Calculate L_{SG} using Eq. 2
  Calculate L_D using Eq. 3
  Calculate L_{KAFE} using Eq. 4
  Update θ^D wrt. L_D //gradient descent wrt. L_D
  Update $\theta^{root}, \theta^{code}$ wrt. L_{KAFE} //gradient descent wrt. L_{SG} and ascent wrt. L_D

---

4 Experiments and Results

We performed qualitative and quantitative evaluation of KAFE. We compared its performance with three state-of-the-art embedding models: GRAM [18], FRAGE [6] and SG [2].

4.1 Data

We used the Regional Healthcare Information Platform (RHIP) which includes clinical information on healthcare visits in Halland, Sweden [21]. All healthcare visits between 2013 and 2017 were included. We populated each visit with time ordered ICD-10 diagnoses within ±4 weeks period centred around the date of visit. This is an intuitive way of capturing a nearly complete medical concept in each visit by adjusting for delays due to referrals to or from other care facilities. Only the visits with at least two diagnoses were included.

4.2 Reproducibility

The source codes (with default hyperparameters) and dummy data is available at https://github.com/caisr-hh/KAFE. Request for real data collaboration can be directed to FoU [21].

4.3 Quantitative Tasks

Disease Onset Prediction. We used two year (2015–2016) diagnostic information of patients between age 40 and 90 in RHIP to predict onset of 10 critical diseases (as highlighted by [27]) in 2017. Only patients with at least one recorded diagnosis and alive during 2015–2017 were included. The final dataset comprised of 106,207 patients of which $2/3^{\text{rd}}$ were used for training and the remaining for testing. We used three-layer feed-forward neural network (NN) with time-aware attention [28] to adjust for the chronological order of diagnostic codes and corresponding delays in between to represent patients. The inputs (concept embeddings) of the NN were pre-trained via the different embedding models which were then fine-tuned together with other NN weights for the specific tasks. For each prediction task and embedding model, we repeated the experiment 10 times and report the mean area under the receiver operating characteristic curve (i.e., AUC-ROC) on test patients for each diagnosis. The ROC curve is a plot of true positive rate versus false positive rate found over the set of predictions. AUC is computed by integrating the ROC curve and it is bounded between 1 (perfect predictions) and 0.5 (random predictions).

Table 3 shows the disease onset prediction scores by NNs with similar architectures and training parameters, but different embedding initialization. We found KAFE to outperform onset prediction of 7 out of 10 chronic diseases followed by GRAM evidencing an effective initialization technique for the task. In general, knowledge-guided embeddings prove to be a better initialization for NN-based prediction model. Nevertheless, initialization via pre-trained embeddings

Table 3. 1 year disease onset prediction-ROC AUC's (SD) reported

Disease	Random	Skip-gram	FRAGE	GRAM	KAFE
Dementia	0.603 (0.0224)	0.677 (0.012)	0.682 (0.014)	0.718 (0.019)	**0.737 (0.014)**
Congestive heart failure	0.640 (0.011)	0.689 (0.009)	0.698 (0.010)	0.721 (0.011)	**0.725 (0.011)**
Peripheral vascular disease	0.631 (0.012)	0.672 (0.009)	0.670 (0.010)	0.670 (0.017)	**0.697 (0.009)**
Myocardial infarction	0.592 (0.019)	0.640 (0.008)	0.632 (0.011)	0.645 (0.015)	0.645 (0.009)
Cerebrovascular disease	0.560 (0.009)	0.614 (0.023)	0.603 (0.020)	**0.632 (0.008)**	0.627 (0.011)
Any malignancy/tumors	0.556 (0.012)	0.594 (0.004)	0.616 (0.047)	0.619 (0.005)	**0.635 (0.004)**
Diabetes	0.525 (0.010)	0.586 (0.020)	0.588 (0.017)	0.610 (0.014)	**0.617 (0.011)**
Hemiplegia or paraplegia	0.502 (0.013)	0.568 (0.020)	0.570 (0.037)	**0.617 (0.023)**	0.612 (0.017)
Liver disease	0.492 (0.039)	0.548 (0.013)	0.557 (0.047)	0.577 (0.017)	**0.605 (0.024)**
Chronic pulmonary disease	0.515 (0.009)	0.543 (0.015)	0.560 (0.012)	0.561 (0.016)	**0.575 (0.009)**

(learned from a much richer data) clearly help lift the prediction performance compared to random initialization.

It is worth emphasizing that there does not exist a well-established suite for quantifying the quality of embeddings [29]. Thus, coupling the embeddings with a well-defined prediction task is often the norm, where the output of the task serves as a proxy for embedding quality. To avoid any additional bias in the experiments, we only used the diagnostic information for predicting future disease onset which also explains the low AUCs reported relative to other studies [30,31]. Integrating other relevant patient information from EHRs would help improve the prediction scores [32], however it is beyond the scope of this methodological study.

Stability Analysis. Due to the stochastic nature of the training process, subsequently applying the same technique to the same data twice, can produce entirely different embeddings. The order in which the training data is fed to the embedding model has also been shown to be a major contributor to embedding instability in word2vec based architectures [33]. While the raw values themselves are of no interest, we expect that the relative positions of the concepts to other similar or dissimilar concepts is preserved.

Given two different embeddings of code c learned from two randomly initialized embedding spaces, let θ_1, θ_2 represent the n (5 in this study) nearest neighbours of c in each embedding space extracted via cosine similarity. Then the stability of c is given by the overlap of neighbours in the two spaces. Stability is bounded between 0 (no overlap) and 1 (complete overlap).

$$stability(c) = \frac{|\theta_1 \cup \theta_2|}{n} \quad (5)$$

Table 4 shows the mean stability of all the frequent and rare codes calculated via three different embeddings generated from each model. Each time, the embedding model was initialized with random weights and the training corpus

Table 4. Embedding stability. Reporting mean (SD)

Stability	Skip-gram	FRAGE	GRAM	KAFE
Frequent codes	0.582(0.002)	0.567 (0.008)	**0.754 (0.003)**	0.752 (0.005)
Rare codes	0.317 (0.001)	0.332 (0.005)	0.709 (0.003)	**0.723 (0.004)**

(visit order) was shuffled. The results generalize that knowledge guided embeddings tend to have higher stability for both frequent and rare words with performance being more significant for the latter. FRAGE improve the stability of rare concepts, compromising slightly for frequent concepts. KAFE significantly outperforms the state-of-art for rare words stability and competes almost equally with GRAM for frequent codes.

4.4 Qualitative Tasks

We randomly selected 1000 (each) rare and frequent clinical codes and visualized (Figs. 2, 3, 4 and 5) their relative positions in embedding space after dimensionality reduction via UMAP [34]. The colours (left) correspond to the ICD-10 classes and (right) frequency labels. The fact that not all diagnoses within an ICD-10 class follow the same characteristics is reiterated. Figure 2 confirms the separation of frequent and rare codes in the embedding space which is consistent with the findings in [6]. This effect is countered in Fig. 3 via a discriminator that filters out the frequency component from the embeddings.

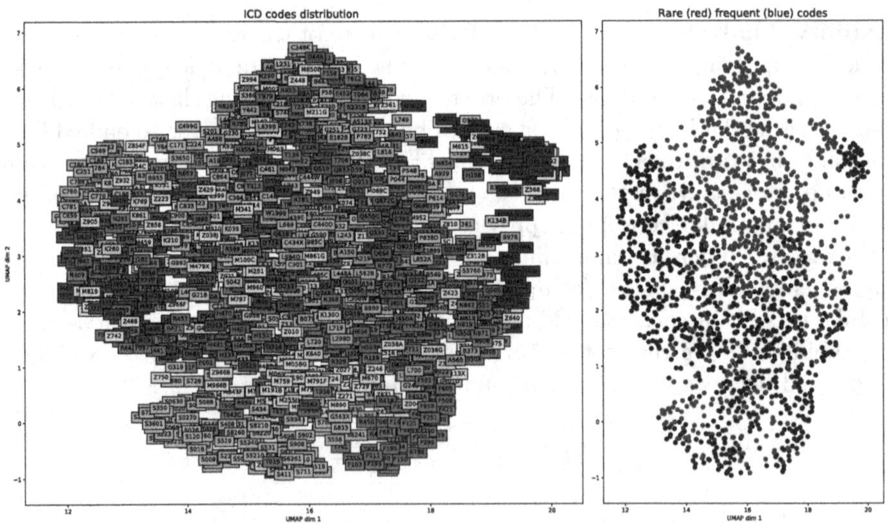

Fig. 2. SG embeddings

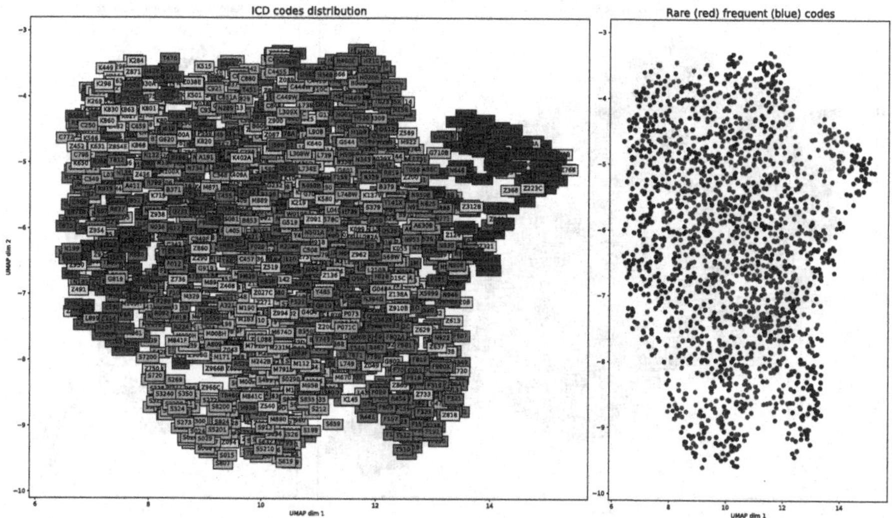

Fig. 3. FRAGE embeddings

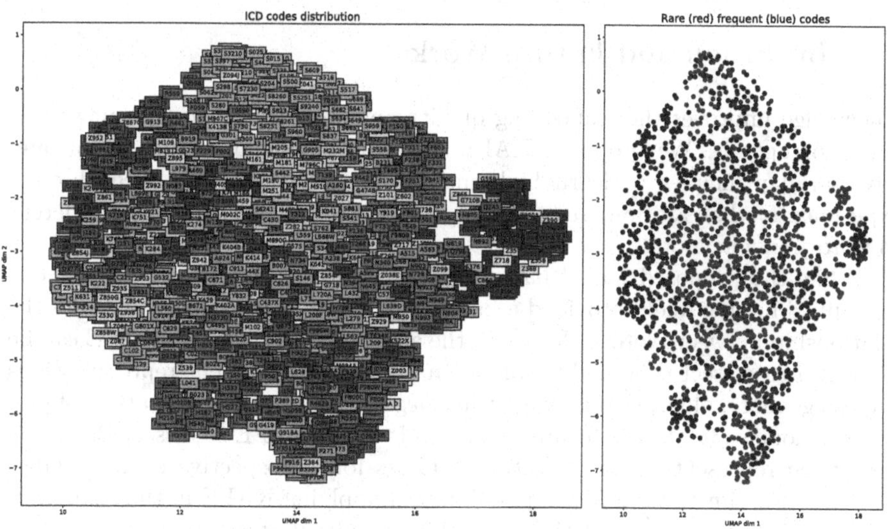

Fig. 4. GRAM embeddings

While the most obvious clusters (0: Pregnancy, childbirth; S: Injury due to external causes etc.) are clearly visible in all embeddings, GRAM (Fig. 4) and KAFE (Fig. 5) present a more consistent view with the ICD-10 knowledge schema which is not possible using only co-occurrence statistics. KAFE further supplements the embedding quality by ensuring a good mix of rare and frequent concepts in the embedding space.

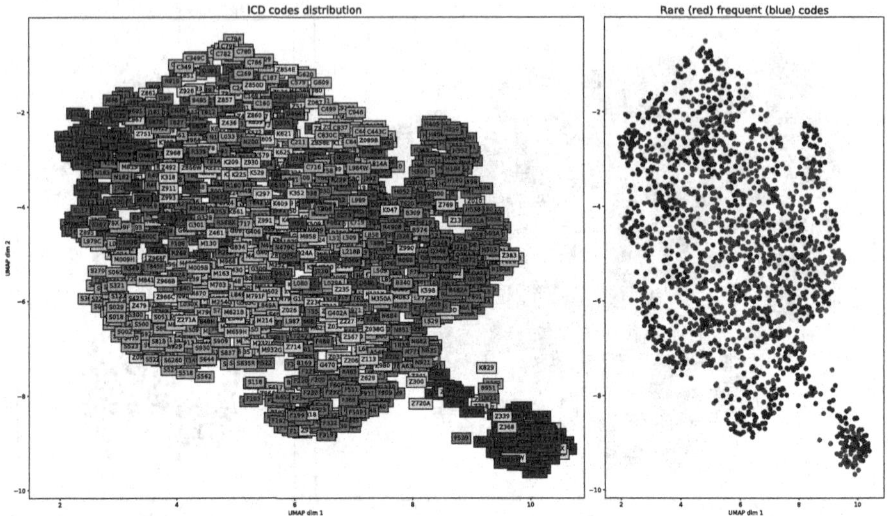

Fig. 5. KAFE embeddings

5 Conclusion and Future Work

Knowledge improves the embedding quality and word frequency (an unwanted bias) corrupts it. We proposed KAFE which leverages knowledge graph and adversarial principles to learn high-quality embeddings (exemplified through experiments) that are well-aligned with the domain knowledge and indifferent to word frequency.

While the proposed model has only been validated on clinical data, it can be applied further afield provided a knowledge hierarchy exists that encodes the relationship between words. Moreover, the adversarial property in KAFE can be leveraged to dissect other subtle biases in patient or document representations such as gender, colour, ethnicity etc. that negatively influence the output of prediction models [35]. In the future we aim to leverage KAFE to dissect Berkson's bias when representing and selecting patients for retrospective cohort studies [36]. We also aim to explore the possibility of applying KAFE in the context of zero-shot learning to predict the onset of rare and out-of-sample diseases.

References

1. Eisenstein, J.: Introduction to Natural Language Processing. MIT press, Cambridge (2019)
2. Mikolov, T., Chen, K., Corrado, G., Dean, J.: Efficient estimation of word representations in vector space. arXiv preprint arXiv:1301.3781 (2013)
3. Shickel, B., Tighe, P.J., Bihorac, A., Rashidi, P.: Deep EHR: a survey of recent advances in deep learning techniques for electronic health record (EHR) analysis. IEEE J. Biomed. Health Inform. **22**(5), 1589–1604 (2017)

4. World Health Organization: International classification of diseases (ICD) information sheet (2018)
5. Xiao, C., Choi, E., Sun, J.: Opportunities and challenges in developing deep learning models using electronic health records data: a systematic review. J. Am. Med. Inform. Assoc. **25**(10), 1419–1428 (2018)
6. Gong, C., He, D., Tan, X., Qin, T., Wang, L., Liu, T.Y.: FRAGE: frequency-agnostic word representation. In: Advances in Neural Information Processing Systems, pp. 1334–1345 (2018)
7. Mu, J., Bhat, S., Viswanath, P.: All-but-the-top: simple and effective postprocessing for word representations. arXiv preprint arXiv:1702.01417 (2017)
8. Ashfaq, A.: Predicting clinical outcomes via machine learning on electronic health records. Ph.D. thesis, Halmstad University Press (2019)
9. Ashfaq, A., Nowaczyk, S.: Machine learning in healthcare-a system's perspective. arXiv preprint arXiv:1909.07370 (2019)
10. Luong, M.T., Socher, R., Manning, C.D.: Better word representations with recursive neural networks for morphology. In: Proceedings of the Seventeenth Conference on Computational Natural Language Learning, pp. 104–113 (2013)
11. Creutz, M., Lagus, K.: Unsupervised models for morpheme segmentation and morphology learning. ACM Trans. Speech Lang. Process. (TSLP) **4**(1), 1–34 (2007)
12. Ling, W., et al.: Finding function in form: compositional character models for open vocabulary word representation. arXiv preprint arXiv:1508.02096 (2015)
13. Kim, Y., Jernite, Y., Sontag, D., Rush, A.M.: Character-aware neural language models. arXiv preprint arXiv:1508.06615 (2015)
14. Bojanowski, P., Grave, E., Joulin, A., Mikolov, T.: Enriching word vectors with subword information. Trans. Assoc. Comput. Linguist. **5**, 135–146 (2017)
15. Xu, C., et al.: RC-NET: a general framework for incorporating knowledge into word representations. In: Proceedings of the 23rd ACM International Conference on Information and Knowledge Management, pp. 1219–1228 (2014)
16. Miller, G.A.: Wordnet: a lexical database for English. Commun. ACM **38**(11), 39–41 (1995)
17. Bahdanau, D., Cho, K., Bengio, Y.: Neural machine translation by jointly learning to align and translate. arXiv preprint arXiv:1409.0473 (2014)
18. Choi, E., Bahadori, M.T., Song, L., Stewart, W.F., Sun, J.: Gram: graph-based attention model for healthcare representation learning. In: Proceedings of the 23rd ACM SIGKDD International Conference on Knowledge Discovery and Data Mining, pp. 787–795 (2017)
19. Ma, F., You, Q., Xiao, H., Chitta, R., Zhou, J., Gao, J.: KAME: knowledge-based attention model for diagnosis prediction in healthcare. In: Proceedings of the 27th ACM International Conference on Information and Knowledge Management, pp. 743–752 (2018)
20. Song, L., Cheong, C.W., Yin, K., Cheung, W.K., Fung, B.C., Poon, J.: Medical concept embedding with multiple ontological representations. IJCAI **19**, 4613–4619 (2019)
21. Ashfaq, A., et al.: Data resource profile: regional healthcare information platform in Halland, Sweden. Int. J. Epidemiol. **49**(3), 738–739f (2020)
22. Lazaridou, A., Marelli, M., Baroni, M.: Multimodal word meaning induction from minimal exposure to natural text. Cogn. Sci. **41**, 677–705 (2017)
23. Herbelot, A., Baroni, M.: High-risk learning: acquiring new word vectors from tiny data. arXiv preprint arXiv:1707.06556 (2017)
24. Schick, T., Schütze, H.: Attentive mimicking: Better word embeddings by attending to informative contexts. arXiv preprint arXiv:1904.01617 (2019)

25. Schick, T., Schütze, H.: Rare words: a major problem for contextualized embeddings and how to fix it by attentive mimicking. In: Proceedings of the AAAI Conference on Artificial Intelligence, vol. 34, pp. 8766–8774 (2020)
26. Mikolov, T., Sutskever, I., Chen, K., Corrado, G.S., Dean, J.: Distributed representations of words and phrases and their compositionality. In: Advances in Neural Information Processing Systems, pp. 3111–3119 (2013)
27. Quan, H., et al.: Coding algorithms for defining comorbidities in ICD-9-CM and ICD-10 administrative data. Med. Care **43**, 1130–1139 (2005)
28. Bai, T., Zhang, S., Egleston, B.L., Vucetic, S.: Interpretable representation learning for healthcare via capturing disease progression through time. In: Proceedings of the 24th ACM SIGKDD International Conference on Knowledge Discovery and Data Mining, pp. 43–51 (2018)
29. Alshargi, F., Shekarpour, S., Soru, T., Sheth, A.P.: Metrics for evaluating quality of embeddings for ontological concepts (2018)
30. Miotto, R., Li, L., Kidd, B.A., Dudley, J.T.: Deep patient: an unsupervised representation to predict the future of patients from the electronic health records. Sci. Rep. **6**(1), 1–10 (2016)
31. Choi, E., Xiao, C., Stewart, W.F., Sun, J.: MIME: multilevel medical embedding of electronic health records for predictive healthcare. arXiv preprint arXiv:1810.09593 (2018)
32. Ashfaq, A., Sant'Anna, A., Lingman, M., Nowaczyk, S.: Readmission prediction using deep learning on electronic health records. J. Biomed. Inform. **97**, 103256 (2019)
33. Wendlandt, L., Kummerfeld, J.K., Mihalcea, R.: Factors influencing the surprising instability of word embeddings. arXiv preprint arXiv:1804.09692 (2018)
34. McInnes, L., Healy, J., Melville, J.: UMAP: uniform manifold approximation and projection for dimension reduction. arXiv preprint arXiv:1802.03426 (2018)
35. Obermeyer, Z., Powers, B., Vogeli, C., Mullainathan, S.: Dissecting racial bias in an algorithm used to manage the health of populations. Science **366**(6464), 447–453 (2019)
36. Goldstein, B.A., Bhavsar, N.A., Phelan, M., Pencina, M.J.: Controlling for informed presence bias due to the number of health encounters in an electronic health record. Am. J. Epidemiol. **184**(11), 847–855 (2016)

Improved Update Rule and Sampling of Stochastic Gradient Descent with Extreme Early Stopping for Support Vector Machines

Marcin Orchel[1](✉) and Johan A. K. Suykens[2]

[1] Kraków, Poland
marcin@orchel.pl

[2] ESAT-STADIUS, KU Leuven, 3001 Leuven (Heverlee), Belgium
johan.suykens@esat.kuleuven.be

Abstract. We propose three techniques for improving accuracy and speed of margin stochastic gradient descent support vector machines (MSGDSVM). The first technique is to use sampling with full replacement. The second technique is to use the new update rule derived from the squared hinge loss function. The third technique is to limit the number of values for tuning of the margin hyperparameter M. We also provide theoretical analysis of a novel optimization problem for the proposed update rule. The first two techniques improve accuracy of MSGDSVM and the last one speed of tuning. Experiments show that the proposed method achieves superior accuracy compared to MSGDSVM for binary and multiclass classification, with similar generalization performance to sequential minimal optimization (SMO) and is faster than MSGDSVM.

Keywords: Support vector machines · Stochastic gradient descent · Early stopping · Squared hinge loss function

The support vector machines (SVM) is one of the best methods for classification [4]. Recently, the efficient implementation of stochastic gradient descent support vector machines (SGDSVM) has been proposed [10]. The remarkable property of this method is speed. The computational advantage over popular sequential minimal optimization (SMO) algorithm is achieved by extreme early stopping, in the neural network terms, stopping even before the first epoch will finish. This is a special case of early stopping used to prevent overfitting in neural networks [5]. The idea of single pass over a training set has been proposed much earlier in the context of online learning in [1]. The SMO method has a stopping criterion based on Karush-Kuhn-Tucker (KKT) condition and the number of iterations may exceed the number of examples. The number of iterations of SGDSVM is always smaller than the number of examples and it is equal to the number of support vectors due to each weight updated only one time. Another improvement for SGDSVM

M. Orchel—Independent Researcher.

G. Nicosia et al. (Eds.): LOD 2021, LNCS 13164, pp. 147–161, 2022.
https://doi.org/10.1007/978-3-030-95470-3_11

called margin stochastic gradient descent support vector machines (MSGDSVM) has been recently proposed, which speed up SGDSVM by computing all solutions for all selected values of a hyperparameter M during a single pass over a training set. This is related to having weights independent of the hyperparameter M (see Eq. 8 and 15 in [12]). For SGDSVM and SMO, for each value of the hyperparameter C, we must run the method separately. This improvement is similar to a solution path investigated for C-SVM [6]. The conceptual differences between SGDSVM and SMO are as follows. The SGDSVM updates each weight only one time. It leads to the algorithm which is based on optimization problem without a regularization term, which is replaced by regularization by extreme early stopping. The sparsity is achieved by initialization with 0s, updating each weight maximally one time and a stopping criterion that can stop the algorithm before updating all weights, while in SMO the sparsity may be additionally achieved by zeroing weights that were already set. Additionally, the MSGDSVM returns solutions for different values of the hyperparameter M that have the same particular terms and differ only in the number of terms. The general aim of this research is to improve accuracy and speed of solvers with extreme early stopping. There are two potential limiting factors for accuracy compared to SMO, the limited number of iterations, and the simplified structure of the solution for different values of M. The requirement for improvements is to preserve fast tuning of the hyperparameter M.

We propose two techniques for improving accuracy of MSGDSVM. The first is sampling with full replacement. Currently, MSGDSVM updates each weight maximally one time. The idea of the improvement is to allow a fixed number of updates per each example during the iteration process (see the p parameter in Sect. 2). We create a fixed number of copies per each training example, for instance 1 copy. It is similar to the bagging technique, however the goal here is to generate one data set, while in bagging, we generate multiple data sets, and we average the results [8]. Moreover, in bagging not all examples are replicated. Resampling with replicates is called oversampling and is frequently used for imbalanced data sets to create multiple copies of examples of the minority classes. The proposed technique is used with iterative methods that update each parameter (related to one training example) maximally one time. So the consequence of creating 1 copy of each example is that each parameter can be updated maximally 2 times. The alternative approach of using multiple epochs has the disadvantage that more complicated update rules than addition of weights may be needed for the following epochs. The second proposed technique is the new update rule, that is derived from an alternative formulation of SVM with a squared hinge loss function (see Sect. 3). The squared hinge loss function has been investigated in [3]. In this article, we use the squared hinge loss function with MSGDSVM. We expect that due to extreme early stopping, considered as aggressive regularization, a loss function with a quadratic term may be beneficial. We solve a problem of getting rid of the M hyperparameter from the update rule after incorporating the squared hinge loss function to MSGDSVM. This is a requirement for fast tuning of M proposed in [12]. We also propose a special type of loss functions based on a conditional operator. This work improves the update rule proposed in [12] and in [10]. Earlier work on using stochastic gradient descent (SGD) to SVM regards an algorithm that does not have heuristic for selection of parameters based on a loss function and has a learning rate in a stopping criterion [13]. For more references, see [12]. The third improvement is related to speeding up MSGDSVM by skipping tuning of an algorithm for some hyperparameter values without sacrificing accuracy (see Sect. 4).

The outline of the article is as follows. First, we define a problem, then we present three proposed techniques that are sampling with full replacement, the new update rule and improvement of speed of tuning. Then we analyze theoretical properties of the proposed techniques. After that, we present methods and experiments on real world data sets. Finally, we summarize the article and we include also appendix at the end.

1 Problem

We consider a classification problem for a given sample data $\boldsymbol{x}_i$ mapped respectively to $y_i \in \{-1, 1\}$ for $i = 1, \ldots, n$ with the following decision boundary

$$f(\boldsymbol{x}) = \boldsymbol{w} \cdot \varphi(\boldsymbol{x}) = 0, \tag{1}$$

where $\boldsymbol{w} \in \mathbb{R}^m$ with the feature map $\varphi(\cdot) \in \mathbb{R}^m$, $f(\cdot)$ is a functional margin. We classify data according to the sign of the $f(\boldsymbol{x})$. This is the standard decision boundary formulation used in SVM with a feature map and without a bias term b [14]. The primal optimization problem for M support vector classification (MSVC) [12] is

Optimization Problem (OP) 1.

$$\min_{\boldsymbol{w}} \frac{1}{2} \|\boldsymbol{w}\|^2 + \sum_{i=1}^{n} \max\{0, M - y_i f(\boldsymbol{x}_i)\}, \tag{2}$$

where $M > 0$ is a desired margin.

The first term in (2) is known as a *regularization term* (regularizer), the second term is an *error term*. We call the loss function $L(z; y_i, M) = \max\{0, M - y_i z\}$ the *margin hinge loss function*. The $\boldsymbol{w}$ can be written in the form

$$\boldsymbol{w} \equiv \sum_{j=1}^{n} \beta_j \varphi(\boldsymbol{x}_j), \tag{3}$$

where $\beta_j \in \mathbb{R}$. We usually substitute (3) to a decision boundary (1) and we get

$$\sum_{j=1}^{n} \beta_j \varphi(\boldsymbol{x}_j) \cdot \varphi(\boldsymbol{x}) = 0. \tag{4}$$

The optimization problem OP 1 is reformulated to find β_j parameters.

The SGD procedure for finding a solution of SVM in a version proposed in [12] is to update parameters as follows

$$\beta_t \leftarrow \eta_k \begin{cases} y_t, & \text{if } M - y_t f_{k-1}(\boldsymbol{x}_t) \geq 0 \\ 0, & \text{otherwise} \end{cases}, \tag{5}$$

where k is the iteration counter, t is an index of the selected example in the k-th iteration, $f_{k-1}(\boldsymbol{x})$ is a functional margin found in the previous iteration, y_t is a class of the t-th example, $\boldsymbol{x}_t$ is the t-th example. We use the worst violator technique for the selection, which selects the example with the maximal value of the loss function. The η_k is a learning rate set to $\eta_k = 1/\sqrt{k}$. All β_j are initialized with 0s, $f_0(\boldsymbol{x}) \equiv 0$. We always stop when the condition in (5) is violated for all remaining examples. The pseudocode of an updating procedure is in Algorithm 1. The number of iterations n_c is the number of terms in a final decision boundary and it is also the number of support vectors which are defined as examples which correspond to updated parameters. Each parameter β_k is updated maximally one time. The solution is sparse due to usually $n_c < n$. This property is called *extreme early stopping*. The procedure of fast tuning of M proposed in [12] is based on the idea that M is only present in the stopping criterion, but not in the update term. So there is one iteration process of a prototype solver for $M = M_{\max}$ value during which we gather all solutions for different values of M. Below, we propose three techniques, first two of them improve accuracy of MSGDSVM, the last one improves speed of training.

Algorithm 1. MSGDSVM

Input: M
Output: $\boldsymbol{\beta}$
$k, t = 1$, $P = \emptyset$, $\boldsymbol{\beta}, \boldsymbol{d} = \boldsymbol{0}$
while $k \le n$ **do**
  $\mu = 1/\sqrt{k}$, k++, $\Delta\beta_t = y_t\mu$
  $\beta_t = \Delta\beta_t$, P.put(t)
  for $j \notin P$ **do**
    $d[j]$ += $\Delta\beta_t \varphi(\boldsymbol{x}_j) \cdot \varphi(\boldsymbol{x}_t)$
  $s = \max$, $t =$ unset
  for $j \notin P$ **do**
    if $y_j d[j] - M < s$ **then**
      $s = y_j d[j] - M$; $t = j$
  if t is unset $\|$ $y_t d[t] - M \ge 0$ **then** break

Algorithm 2. MSGDSVM$_{\text{2f,x2}}$

Input: $M \le 1$, $p = 1$ or 2
Output: $\boldsymbol{\beta}$
$k, t = 1$, $P = \emptyset$, $\boldsymbol{\beta}, \boldsymbol{d}, \boldsymbol{c} = \boldsymbol{0}$
while $k \le pn$ **do**
  k++, $\Delta\beta_t = -(y_t d[t] - 1)$
  if $\Delta\beta_t < 0$ **then** $\Delta\beta_t = 0$
  β_t += $y_t \Delta\beta_t$, P.put(t), c[t]++
  for j **do** ▷ for $p = 1$, $j \notin P$
    $d[j]$ += $\Delta\beta_t \varphi(\boldsymbol{x}_j) \cdot \varphi(\boldsymbol{x}_t)$
  $s = \max$, $t =$ unset
  for j **do** ▷ for $p = 1$, $j \notin P$
    if c[j] $< p$ && $y_j d[j] - M < s$ **then**
      $s = y_j d[j] - M$; $t = j$
  if t is unset $\|$ $y_t d[t] - M \ge 0$ **then** break

2 Sampling with Full Replacement

In MSGDSVM each weight is updated maximally one time. We extend this procedure to allow to update each weight p times, where p is a hyperparameter. We propose to use sampling with full replacement. It is equivalent to say that we create $p - 1$ additional copies of each example, and we extend the data set. For example, for $p = 2$, we create one copy of each example, and we have $2n$ examples. Then we use exactly the same update rule as before (5) on the extended data set. This procedure is not the same as running p epochs with the update rule (5) as usually done in neural networks. We cannot use such procedure, because the idea of a fast hyperparameter tuning presented in [12] is that all solutions for all values of the M hyperparameter differ only in the number of terms. This property is preserved for sampling with full replacement, but not for multiple epochs, because the next epoch is started with a solution from the previous epoch depended on a value of M.

3 New Update Rule

The second contribution is the new update rule. We propose the following update rule

$$\beta_t \leftarrow \begin{cases} y_t \max\{0, 1 - y_t f_{k-1}(\boldsymbol{x}_t)\}, & \text{if } M - y_t f_{k-1}(\boldsymbol{x}_t) \geq 0 \\ 0, & \text{otherwise} \end{cases}. \tag{6}$$

We do not have a learning rate anymore. This means that learning speed is automatically adjusted using the previous solution. The update term is based on a functional margin f_{k-1} and can be written in terms of the loss function $y_t L(z; y_t, 1)$. The update term is similar to the stopping criterion which is also based on a loss function. The difference between the update term and the stopping criterion, is that there is no M in the update term. Thus, we can use fast tuning of M.

4 Improvement of Speed of Tuning

The third contribution is related to the new update rule (6). When, we have $M \leq 1$, the max function in the update term does not matter. For all $M \geq 1$, we have exactly the same model as for $M = 1$, so the idea is to skip tuning all values $1 < M < M_{\max}$. This speeds up the training process, because after this change $M_{\max} = 1$, so the prototype solver is executed for smaller value of $M_{\max}$ and it will stop earlier. After this change, the update rule (6) simplifies to

$$\beta_t \leftarrow \begin{cases} y_t (1 - y_t f_{k-1}(\boldsymbol{x}_t)), & \text{if } M - y_t f_{k-1}(\boldsymbol{x}_t) \geq 0 \\ 0, & \text{otherwise} \end{cases} \quad \text{for } M \leq 1. \tag{7}$$

The computational complexity of this improvement stays the same as for MSGDSVM and has been given in [12].

5 Theoretical Analysis

We derive the update rule (6) from the following optimization problem for MSVC with the squared hinge loss function with a desired margin M

OP 2.

$$\min_{\boldsymbol{w}} \frac{1}{2} \|\boldsymbol{w}\|^2 + \sum_{i=1}^{n} (\max\{0, M - y_i f(\boldsymbol{x}_i)\})^2, \text{ where } M > 0. \tag{8}$$

In the previous work, we used the hinge loss function [12]. The intuition behind the idea of introducing the squared hinge loss function is to give more weight for losses, especially for algorithms with strong regularization by extreme early stopping. The original C support vector classification (C-SVC) with a squared hinge loss function mentioned in [3], page 104 is

OP 3.

$$\min_{w} \frac{1}{2} \|\boldsymbol{w}\|^2 + C \sum_{i=1}^{n} \left(\max\left\{0, 1 - y_i f\left(\boldsymbol{x}_i\right)\right\}\right)^2 . \tag{9}$$

The squared hinge loss is convex and gives sparse solutions [7]. We have the following proposition.

Proposition 1. *The decision boundary for OP 2 is equivalent to the decision boundary for OP 3 with $C = 1$.*

Proof. The proof is similar as in [12]. We can write (8) as

$$\min_{w} \frac{1}{2} \|\boldsymbol{w}\|^2 + M^2 \sum_{i=1}^{n} \left(\max\left\{0, 1 - y_i \left(\boldsymbol{w}/M \cdot \varphi\left(\boldsymbol{x}_i\right)\right)\right\}\right)^2 .$$

When we substitute $\boldsymbol{w}' \rightarrow \boldsymbol{w}/M$, we get

$$\min_{w'} \frac{1}{2} \left\|\boldsymbol{w}'M\right\|^2 + M^2 \sum_{i=1}^{n} \left(\max\left\{0, 1 - y_i \left(\boldsymbol{w}' \cdot \varphi\left(\boldsymbol{x}_i\right)\right)\right\}\right)^2 ,$$

$$\min_{w'} \frac{1}{2} \left\|\boldsymbol{w}'\right\|^2 + \sum_{i=1}^{n} \left(\max\left\{0, 1 - y_i \left(\boldsymbol{w}' \cdot \varphi\left(\boldsymbol{x}_i\right)\right)\right\}\right)^2 .$$

The solution of the optimization problem differs from OP 3 with $C = 1$ by a multiplicative constant, so the decision boundaries in the form (1) are equivalent. □

The consequence of this proposition is that for any value of M, we get the same solution as for OP 3 with $C = 1$. It may seem as a limitation of the margin approach. However, we use the extreme early stopping, in which every weight is updated maximally one time. In this setting, the situation is reversed, and the version with C may be more limited.

Proposition 2. *When we update each parameter maximally one time and initialize parameters with 0s, the ordinary iteration method with simultaneous steps based on a dual problem for OP 3 returns the same solution regardless of C.*

Proof. The dual formulation of OP 3 is

$$\max_{\alpha} \sum_{i=1}^{n} \alpha_i - \frac{1}{2} \sum_{i=1}^{n} \sum_{j=1}^{n} y_i y_j \alpha_i \alpha_j \left(\varphi\left(\boldsymbol{x}_i\right) \cdot \varphi\left(\boldsymbol{x}_j\right) + \delta_{ij}/C\right) \tag{10}$$

$$\text{subject to } 0 \leq \alpha_i . \tag{11}$$

The δ_{ij} is a Kronecker delta, defined to be 1 if $i = j$, otherwise 0. See [3], page 104 for derivation. We do not have an upper bound for α_i in (11). We can derive an update rule for one weight α_i by computing a derivative of (10) similarly as in [11], page 93

$$\alpha_i^{\text{new}} = \alpha_i - \left(y_i \sum_{j=1}^{n} y_j \alpha_j \left(\varphi\left(\boldsymbol{x}_i\right) \cdot \varphi\left(\boldsymbol{x}_j\right) + \delta_{ij}/C\right) - 1\right) / \left(\varphi\left(\boldsymbol{x}_i\right) \cdot \varphi\left(\boldsymbol{x}_i\right)\right).$$

This should be nonnegative due to (11). The Kronecker delta can be nonzero only when $i = j$, but for this case α_j is zero due to updating each parameter only one time and initialization with 0s. So there will be no terms with C in the update rule. □

Proposition 3. *When we update each parameter maximally one time and initialize parameters with 0s, the ordinary iteration method with simultaneous steps based on a dual problem for OP 2 returns solutions that depend on M.*

Proof. The dual formulation of OP 2 is

$$\max_{\boldsymbol{\alpha}} M \sum_{i=1}^{n} \alpha_i - \frac{1}{2} \sum_{i=1}^{n} \sum_{j=1}^{n} y_i y_j \alpha_i \alpha_j \left(\varphi\left(\boldsymbol{x}_i\right) \cdot \varphi\left(\boldsymbol{x}_j\right) + \delta_{ij}\right) \tag{12}$$

$$\text{subject to } 0 \leq \alpha_i. \tag{13}$$

We can derive an update rule for one weight α_i by computing a derivative of (12)

$$\alpha_i^{\text{new}} = \alpha_i - \left(y_i \sum_{j=1}^{n} y_j \alpha_j \left(\varphi\left(\boldsymbol{x}_i\right) \cdot \varphi\left(\boldsymbol{x}_j\right) + \delta_{ij}\right) - M\right) / \left(\varphi\left(\boldsymbol{x}_i\right) \cdot \varphi\left(\boldsymbol{x}_i\right)\right). \tag{14}$$

This should be nonnegative due to (13). Because we update each parameter only one time and we initialize parameters with 0s, some α_j may be 0, but this will not cause M to disappear. So the update term depends on M. □

For the iteration method with sequential steps, both variants would depend on C and M respectively. The next idea is that for fast tuning of M, we cannot have the M hyperparameter in the update term as in (14), only in the stopping criterion. In order to achieve this, we need to reformulate the OP 2.

Proposition 4. *The OP 2 is equivalent to*

OP 4.

$$\min_{\boldsymbol{w}} W = \frac{1}{2} \left\|\boldsymbol{w}\right\|^2 + \sum_{i=1}^{n} \left(M - y_i f\left(\boldsymbol{x}_i\right) \geq 0 \text{ ? } \max\left\{0, 1 - y_i f\left(\boldsymbol{x}_i\right)\right\} : 0\right)^2. \tag{15}$$

The notation is based on a conditional (ternary) operator, similarly defined as in programming languages [2]. It takes three operands: a condition followed by a question mark (?), then the expression to evaluate if the condition is true, followed by a colon (:), and finally the expression to evaluate if the condition is false.

Proof. We can rewrite OP 2 as

$$\min_{\boldsymbol{w}} \frac{1}{2} \|\boldsymbol{w}\|^2 + \sum_{i=1}^{n} \left(M - y_i f(\boldsymbol{x}_i) \geq 0 \ ? \ M - y_i f(\boldsymbol{x}_i) : 0\right)^2 ,$$

$$\min_{\boldsymbol{w}} \frac{1}{2} \|\boldsymbol{w}\|^2 + \sum_{i=1}^{n} \left(M - y_i f(\boldsymbol{x}_i) \geq 0 \ ? \ \max\{0, M - y_i f(\boldsymbol{x}_i)\} : 0\right)^2 ,$$

$$\min_{\boldsymbol{w}} \frac{1}{2} \|\boldsymbol{w}\|^2 + M^2 \sum_{i=1}^{n} \left(M - y_i f(\boldsymbol{x}_i) \geq 0 \ ? \ \max\{0, 1 - y_i (\boldsymbol{w}/M \cdot \varphi(\boldsymbol{x}_i))\} : 0\right)^2 .$$

When we substitute $\boldsymbol{w}' \to \boldsymbol{w}/M$, we get

$$\min_{\boldsymbol{w}'} \frac{1}{2} \|\boldsymbol{w}' M\|^2 + M^2 \sum_{i=1}^{n} \left(M - y_i f(\boldsymbol{x}_i) \geq 0 \ ? \ \max\{0, 1 - y_i \boldsymbol{w}' \cdot \varphi(\boldsymbol{x}_i)\} : 0\right)^2 .$$

□

The OP 4 does not have a desired margin M in the update term, only in the condition of a ternary operator, which will be used in a stopping criterion. The reason of introducing max function in (15) is that we want to use similar representation as in a dual formulation of SVM, where α_i weights are nonzero, see below (18). The OP 4 can be written as

$$\min_{\boldsymbol{w}} \frac{1}{2} \|\boldsymbol{w}\|^2 + \sum_{i=1}^{n} \left(M - y_i f(\boldsymbol{x}_i) \geq 0 \ \& \ 1 - y_i f(\boldsymbol{x}_i) \geq 0 \ ? \ 1 - y_i f(\boldsymbol{x}_i) : 0\right)^2 . \tag{16}$$

For $M > 1$, we get the equivalent optimization problems to $M = 1$, so it simplifies to

$$\min_{\boldsymbol{w}} \frac{1}{2} \|\boldsymbol{w}\|^2 + \sum_{i=1}^{n} \left(M - y_i f(\boldsymbol{x}_i) \geq 0 \ ? \ 1 - y_i f(\boldsymbol{x}_i) : 0\right)^2 , \ \text{where } M \leq 1. \tag{17}$$

Before going to derivation of an update rule, we need to change a representation of $\boldsymbol{w}$ in (3) to

$$\boldsymbol{w} \equiv \sum_{j=1}^{n} y_j \alpha_j \varphi(\boldsymbol{x}_j), \ \text{where } \alpha_j \geq 0. \tag{18}$$

Such representation is similar to the representation used in a dual problem of OP 1. It is more strict, because we have a constraint on a sign for each term in a decision boundary depended on a class y_j. This change is necessary for introducing a max function in the update term (6). The full derivation of the update rule is in the appendix. We provide only derivation for the primal optimization problem due to space constraints. However, we also derived the update rule for the dual problem of OP 4.

6 Methods

We propose three main methods. The first one is MSGDSVM with the update rule (7) (MSGDSVM_{2f}). The pseudocode of MSGDSVM_{2f} is in Algorithm 2 for $p = 1$. The second one is MSGDSVM with 1 copy (MSGDSVM_{x2}). It implements MSGDSVM with sampling with full replacement for $p = 2$. The third one is MSGDSVM with the update rule (7) and 1 copy ($\text{MSGDSVM}_{2f,x2}$). It combines all improvements. The pseudocode of $\text{MSGDSVM}_{2f,x2}$ is in Algorithm 2 for $p = 2$. We do not report directly the pseudocode for MSGDSVM_{x2}, because it is Algorithm 2 for $p = 2$ with additional replacement of lines 3 and 4 with a line 3 from Algorithm 1. The line 4 is redundant for methods with the update rule (7). If one wants to implement the update rule (6) instead of (7), then this line would be necessary. It is the case for MSGDSVM with the update rule (6) (MSGDSVM_2).

For sampling with full replacement, we use more efficient implementation than copying examples. We have a counter for each example how many updates have been already processed with a limit being p (variable c in Algorithm 2). In sampling with full replacement, we update functional margins for all parameters (line 6 in Algorithm 2), while in MSGDSVM only for parameters which have not yet been processed (line 5 in Algorithm 1) which may slow down sampling with full replacement.

For improved speed of training due to the update rule (7), we cut margin values to tune from 30 values ($[2^{-19}, \ldots, 2^{10}]$) to 11 values ($[2^0, \ldots, 2^{10}]$).

7 Experiments

We compared proposed methods with MSGDSVM and SMO implementation of SVM. The SMO is an iterative method, for which we have a constraint on the number of iterations set to $100n$, where n is the training set size. For MSGDSVM and MSGDSVM_2, the maximal number of iterations is maximally equal to n, for MSGDSVM_{x2} and $\text{MSGDSVM}_{2f,x2}$, it is $2n$. For SMO, we do not have fast tuning of the C hyperparameter, while for remaining methods we use fast tuning of M proposed in [12]. Another difference is that for SMO, we use a bias term. We additionally compare all methods with a special version of SMO with reduced number of iterations to n, called SMO with n iterations (SMO_{x1}). Because SMO with a bias term updates two parameters in one iteration, the number of updates in SMO_{x1} is $2n$. We use our own implementation of all methods. We compared performance of all methods for real world data sets for binary and multiclass classification (binary data sets are from aa to wa in Table 1). We use a one-vs-all classifier for multiclass. More details about data sets are on the LibSVM site [9]. We use the same names for data sets as on the LibSVM site, except the data set *aa* for which we merged *a1a* to *a9a* data sets, *w1a* to *w9a* data sets have been merged to *wa*, *german* corresponds to *german.numer*, *ionosphere_s* to *ionosphere_scale*. We selected all data sets from this site which have not been compressed. We merged training and test data sets in the case if they were split.

For all data sets, we scaled every feature linearly to [0, 1]. We use the radial basis function (RBF) kernel in the form $K(\boldsymbol{x}, \boldsymbol{z}) = \exp(-\|\boldsymbol{x} - \boldsymbol{z}\|^2 / (2\sigma^2))$. The number of hyperparameters to tune is 2, σ and M for MSGDSVM and the proposed methods, σ and C for SMO and its variants. For sampling with full replacement, we generate 1 copy of each example, that is $p = 2$. For all hyperparameters, we use a grid search method for finding the best values. The values of hyperparameters to tune are selected as in [12]. We use the procedure similar to repeated double cross validation for performance comparison. For the outer loop, we run a modified k-fold cross validation for $k = 15$, with the optimal training set size set to 80% of all examples with the maximal training set size equal to 1000 examples. We limit a test data set to 1000 examples. We limit all read data to 35000. When it is not possible to create the next fold, we shuffle data and start from the beginning. We use the 5-fold cross validation for the inner loop for finding optimal values of the hyperparameters. After that, we run the method on training data, and we report average results for the outer cross validation.

The observations are as follows. In the experiment 1, we compared 4 methods. That are SMO, SMO_{x1}, MSGDSVM and $\text{MSGDSVM}_{\text{2f,x2}}$. The proposed method $\text{MSGDSVM}_{\text{2f,x2}}$ is the fastest one, with improved ranking for accuracy compared to MSGDSVM (see Table 1). There is little difference in overall rank between SMO and SMO_{x1}. For the number of support vectors the best method is SMO. There are some statistically significant results for accuracy for only a few data sets (see Table 2). One significant difference is that $\text{MSGDSVM}_{\text{2f,x2}}$ has statistically better accuracy for the Madelon, glass identification and Sensorless data sets. The Madelon data set is an artificial dataset containing data points grouped in 32 clusters placed on the vertices of a five dimensional hypercube and randomly labeled +1 or −1. The Sensorless data set is sensorless drive diagnosis with features extracted from motor current. The motor has intact and defective components. The $\text{MSGDSVM}_{\text{2f,x2}}$ has worse accuracy than SMO for the svmguide2 (bioinformatics), letter recognition, svmguide4 (traffic light signals) and vehicle data sets. The vehicle data set is a classification a given silhouette as one of four types of vehicle, using a set of features extracted from the silhouette. In the experiment 2, we compared additional methods, including SMO with reduced number of iterations (see Table 3). Each technique proposed in this article improves accuracy separately over MSGDSVM (see $\text{MSGDSVM}_{\text{x2}}$ and MSGDSVM_2). Combination of the proposed techniques improves further accuracy (see $\text{MSGDSVM}_{\text{2f,x2}}$). The fastest methods are $\text{MSGDSVM}_{\text{2f}}$ and $\text{MSGDSVM}_{\text{2f,x2}}$. The $\text{MSGDSVM}_{\text{2f}}$ has improved speed over MSGDSVM_2. Removing a bias term improves accuracy for methods based on SMO (see $\text{SMONOB}_{\text{x2}}$, $\text{SMONOB}_{\text{2,x2}}$, SMONOB). For the madelon data set (column $errM$), the proposed method $\text{MSGDSVM}_{\text{2f,x2}}$ achieves superior accuracy over SMO based methods, and almost the best accuracy for a glass data set. For the svmguide4 and vehicle data sets, there is an advantage for computing multiple epochs as for SMO (columns $errS4$ and $errV$).

Table 1. Experiment 1. The numbers in descriptions of the columns mean the methods: 1 - SMO, 2 - SMO_{x1}, 3 - MSGDSVM, 4 - $MSGDSVM_{2f,x2}$. Column descriptions: *dn* – data set, *err* – misclassification error, the last row is an average rank, *sv* – the number of support vectors, the last row is an average rank, *t* – average training time per outer fold in seconds, the best time is in bold, the last row is the sum of the training times.

dn	err1	err2	err3	err4	sv1	sv2	sv3	sv4	t1	t2	t3	t4
aa	0.158	**0.158**	0.165	0.159	381	384	**344**	398	245	23	**12**	24
australian	**0.149**	**0.149**	0.151	0.152	320	335	252	**203**	286	20	**4**	10
breast-cancer	0.033	0.033	**0.029**	0.03	77	77	87	**53**	44	5	**4**	6
cod-rna	0.049	**0.049**	0.058	0.052	**199**	211	373	387	354	19	**13**	26
diabetes	0.239	**0.235**	0.243	0.24	338	347	**335**	350	200	20	**5**	10
fourclass	0.0	0.0	0.001	**0.0**	225	225	398	**101**	168	9	**6**	12
german	0.243	**0.238**	0.243	0.244	460	453	**419**	460	539	39	**8**	20
heart	0.168	**0.159**	0.162	0.159	**96**	110	117	126	44	1	**0**	2
ionosphere_s	0.063	**0.059**	0.079	0.079	117	124	**86**	97	29	**1**	**1**	**1**
liver-disorders	**0.344**	0.371	0.358	0.364	199	**195**	201	211	180	**1**	**1**	2
madelon	0.363	0.364	0.33	**0.306**	1000	994	970	**881**	677	24	**13**	37
mushrooms	0.002	0.002	0.001	**0.001**	578	**546**	956	948	130	24	**12**	27
phishing	0.059	**0.057**	0.061	0.059	384	370	**284**	357	259	23	**12**	24
skin_nonskin	0.005	**0.005**	0.008	0.007	223	263	178	**135**	421	19	**13**	22
splice	0.12	0.122	0.126	**0.117**	536	**521**	611	661	329	24	**12**	31
sonar_scale	0.124	**0.119**	0.133	0.125	119	122	105	**86**	8	1	**0**	1
svmguide1	0.033	**0.033**	0.037	0.035	**112**	129	138	163	635	18	**12**	30
svmguide3	**0.163**	0.175	0.193	0.184	416	473	**396**	447	918	60	**13**	25
wa	**0.025**	**0.025**	0.026	0.025	**159**	**159**	423	387	171	24	13	**12**
connect-4	**0.266**	0.267	0.279	0.27	1398	1485	**1125**	1469	867	100	71	**60**
dna	**0.077**	0.077	0.089	0.084	1611	1605	**1003**	1412	233	159	86	**75**
glass	0.374	0.392	**0.318**	0.319	**447**	522	510	456	38	3	**2**	**2**
iris	**0.042**	**0.042**	0.047	0.049	**53**	55	124	94	4	**0**	**0**	**0**
letter	0.295	**0.295**	0.31	0.326	**3387**	3391	19918	9480	3278	377	346	**127**
pendigits	0.024	0.024	0.024	**0.022**	**697**	**697**	3101	2952	914	152	134	**82**
satimage	0.14	0.14	0.136	**0.132**	1024	**1023**	2108	2076	972	107	91	**89**
segment	0.055	0.056	0.056	**0.053**	**428**	643	2254	1892	780	215	106	**62**
Sensorless	0.131	0.131	**0.083**	0.087	3733	**3668**	8829	7623	3411	195	161	**81**
shuttle	0.005	**0.005**	0.012	0.012	155	161	**118**	146	399	44	65	**22**
svmguide2	0.197	0.194	**0.192**	0.194	412	424	**323**	356	67	13	**5**	**5**
svmguide4	**0.258**	0.347	0.422	0.402	**635**	889	900	1110	865	59	23	**17**
vehicle	**0.171**	0.202	0.252	0.235	**567**	843	707	816	660	79	30	**27**
vowel	0.023	**0.022**	0.025	0.027	1591	**1580**	6164	4367	2179	246	108	**59**
wine	0.024	0.024	0.024	**0.02**	120	120	126	**85**	3	**1**	**1**	**1**
All	2.19	**2.1**	3.1	2.6	**2.15**	2.57	2.62	2.66	8965	934	613	**459**

Table 2. Bayesian test for generalization performance for the experiment 1. The numbers in descriptions of the columns mean the methods: 1 - SMO, 2 - SMO$_{x1}$, 3 - MSGDSVM, 4 - MSGDSVM$_{2f,x2}$. Description of values: *eT* – the Bayesian correlated t-test for misclassification error; value greater than 0.9 is in bold. For example, eT12 means probability that the method 2 is better than the method 1. In the last row, there is a Bayesian signed-rank test for all data sets. We report only rows with at least one statistical significant result.

dn	eT12	eT21	eT13	eT31	eT14	eT41	eT23	eT32	eT24	eT42	eT34	eT43
madelon	0.0	0.0	**1.0**	0.0	**1.0**	0.0	**1.0**	0.0	**1.0**	0.0	**0.99**	0.0
svmguide3	0.01	0.57	0.0	**0.99**	0.0	**0.93**	0.01	0.78	0.01	0.46	0.45	0.03
glass	0.03	0.73	**0.94**	0.02	**0.95**	0.01	**0.98**	0.0	**0.98**	0.0	0.26	0.32
letter	0.0	0.0	0.02	0.67	0.0	**0.97**	0.02	0.68	0.0	**0.97**	0.07	0.63
Sensorless	0.0	0.0	**1.0**	0.0	**1.0**	0.0	**1.0**	0.0	**1.0**	0.0	0.0	0.04
svmguide4	0.0	**1.0**	0.0	**1.0**	0.0	**1.0**	0.0	**1.0**	0.0	**0.98**	0.77	0.02
vehicle	0.0	**0.99**	0.0	**1.0**	0.0	**1.0**	0.0	**1.0**	0.0	**0.97**	0.74	0.01
All	0.0	0.16	0.14	0.21	0.15	0.22	0.15	0.2	0.15	0.17	0.11	0.0

Table 3. Experiment 2. Methods are SMO, SMO$_{x1}$, MSGDSVM, MSGDSVM$_{2f,x2}$, MSGDSVM$_{x2}$, MSGDSVM$_2$, MSGDSVM$_{2f}$, SMO without a bias term with $2n$ iterations (SMONOB$_{x2}$), SMO with the squared hinge loss function and n iterations (SMO$_{2,x1}$), SMO without a bias term and with the squared hinge loss function and with $2n$ iterations (SMONOB$_{2,x2}$), SMO without a bias term (SMONOB), SMO with the squared hinge loss function (SMO$_2$), SMO without a bias term with the squared hinge loss function (SMONOB$_2$). Column descriptions: *method*, *errR* – the average rank for the misclassification error, *svR* – the average rank for the number of support vectors, *t* – a sum of average training times per outer fold in seconds, *errM*, *errS3*, *errG*, *errL*, *errS*, *errS4*, *errV* – the misclassification error for the Madelon, svmguide3, glass, letter, Sensorless, svmguide4, vehicle data sets respectively.

Method	errR	svR	t	errM	errS3	errG	errL	errS	errS4	errV
SMO	6.9	**3.76**	8965	0.363	0.163	0.374	0.295	0.131	0.258	0.171
SMO$_{x1}$	6.76	4.65	934	0.364	0.175	0.392	0.295	0.131	0.347	0.202
MSGDSVM	9.99	5.68	613	0.33	0.193	0.318	0.31	0.083	0.422	0.252
MSGDSVM$_{2f,x2}$	7.66	5.72	459	**0.306**	0.184	0.319	0.326	0.087	0.402	0.235
MSGDSVM$_{x2}$	8.74	5.47	1275	0.328	0.179	0.313	0.321	0.083	0.4	0.218
MSGDSVM$_2$	8.71	5.51	586	**0.306**	0.19	0.318	0.309	0.088	0.435	0.269
MSGDSVM$_{2f}$	8.71	5.51	**262**	**0.306**	0.19	0.318	0.309	0.088	0.435	0.269
SMONOB$_{x2}$	5.29	8.46	1479	0.328	0.174	0.318	**0.266**	0.078	0.333	0.204
SMO$_{2,x1}$	7.25	8.28	2056	0.365	0.173	0.409	0.278	0.13	0.35	0.198
SMONOB$_{2,x2}$	**4.57**	11.69	2213	0.328	0.17	**0.313**	0.271	0.078	0.313	0.189
SMONOB	5.51	7.06	26208	0.328	0.166	0.315	0.267	**0.078**	0.292	0.173
SMO$_2$	5.9	8.43	11114	0.364	**0.161**	0.394	0.278	0.132	**0.253**	**0.163**
SMONOB$_2$	5.01	10.78	24105	0.328	0.164	0.313	0.273	0.078	0.271	0.164

8 Summary

We proposed two techniques for improving accuracy of MSGDSVM: sampling with replacement and the new update rule derived from the squared hinge loss function. We also proposed one technique for improving speed of tuning of MSGDSVM. The combination of all improvements $\text{MSGDSVM}_{\text{2f,x2}}$ is the fastest method among all tested, and the accuracy is close to SMO and its variants with reduced number of iterations. We additionally showed that we are able to improve accuracy of SMO with statistical significance for some data sets using methods without a bias term, and further substantial improvement can be attributed to the proposed new update rule. For some data sets, SMO may still be beneficial over solvers with extreme early stopping.

Acknowledgments. The theoretical analysis of the method is supported by the National Science Centre in Poland, UMO-2015/17/D/ST6/04010, titled "Development of Models and Methods for Incorporating Knowledge to Support Vector Machines" and the data driven method is supported by the European Research Council under the European Union's Seventh Framework Programme. Johan Suykens acknowledges support by ERC Advanced Grant E-DUALITY (787960), KU Leuven C1, FWO G0A4917N. This paper reflects only the authors' views, the Union is not liable for any use that may be made of the contained information.

A Appendix

We provide a derivation of the update rule (6).

Proof. The proof is based on the primal problem OP 4. The technique of a proof is similar as presented in [10]. First, we compute the gradient of the objective function in (15) and we get

$$\begin{aligned}\frac{\partial W}{\partial \boldsymbol{w}} &= \boldsymbol{w} + \sum_{i=1}^{n} 2\left(M - y_i f\left(\boldsymbol{x}_i\right) \geq 0 \text{ ? } 1 - y_i \boldsymbol{w} \cdot \varphi\left(\boldsymbol{x}_i\right) : 0\right) \\ &\quad \cdot \left(M - y_i \boldsymbol{w} \cdot \varphi\left(\boldsymbol{x}_i\right) \geq 0 \text{ ? } -y_i \varphi\left(\boldsymbol{x}_i\right) \ : \ 0\right).\end{aligned}$$

Because we have the same condition in the last two factors, we can write

$$\frac{\partial W}{\partial \boldsymbol{w}} = \boldsymbol{w} + 2\sum_{i=1}^{n} M - y_i \boldsymbol{w} \cdot \varphi\left(\boldsymbol{x}_i\right) \geq 0 \text{ ? } -\left(1 - y_i \boldsymbol{w} \cdot \varphi\left(\boldsymbol{x}_i\right)\right) y_i \varphi\left(\boldsymbol{x}_i\right) \ : \ 0.$$

After substitution (18), we get

$$\begin{aligned}&\sum_{i=1}^{n} y_i \alpha_i \varphi\left(\boldsymbol{x}_i\right) + 2\sum_{i=1}^{n} M - y_i \sum_{j=1}^{n} y_j \alpha_j \varphi\left(\boldsymbol{x}_j\right) \cdot \varphi\left(\boldsymbol{x}_i\right) \geq 0 \text{ ?} \\ &\quad -\Big(1 - y_i \sum_{j=1}^{n} y_j \alpha_j \varphi\left(\boldsymbol{x}_j\right) \cdot \varphi\left(\boldsymbol{x}_i\right)\Big) y_i \varphi\left(\boldsymbol{x}_i\right) \ : \ 0.\end{aligned}$$

For the stochastic update, we approximate the above formula (which should be equal to 0 for the optimal solution), so we have

$$\begin{aligned} &ny_k\alpha_k\varphi\left(\boldsymbol{x_k}\right)+2n\Big(M-y_k\sum_{j=1}^{n}y_j\alpha_j\varphi\left(\boldsymbol{x_j}\right)\cdot\varphi\left(\boldsymbol{x_k}\right)\geq 0\Big)\ ?\\ &-\Big(1-y_k\sum_{j=1}^{n}y_j\alpha_j\varphi\left(\boldsymbol{x_j}\right)\cdot\varphi\left(\boldsymbol{x_k}\right)\Big)y_k\varphi\left(\boldsymbol{x_k}\right)\ :\ 0=0. \end{aligned}$$

We can generate an update term as for the ordinary iteration method by transforming the equation into a fixed point form for α_k by dividing by $ny_k\varphi\left(\boldsymbol{x_k}\right)$. We assume that $\varphi\left(\boldsymbol{x_k}\right)\neq 0$ for any coefficient. For the RBF kernel, it means that each component of $\boldsymbol{x}$ should be different from 0. We move all terms except the first one to the right, and we get

$$\begin{aligned} \alpha_k\leftarrow 2\Big(M-y_k\sum_{j=1}^{n}y_j\alpha_j\varphi\left(\boldsymbol{x_j}\right)\cdot\varphi\left(\boldsymbol{x_k}\right)\geq 0\Big)\ ?\\ 1-y_k\sum_{j=1}^{n}y_j\alpha_j\varphi\left(\boldsymbol{x_j}\right)\cdot\varphi\left(\boldsymbol{x_k}\right)\ :\ 0. \end{aligned}$$

We can skip multiplier 2, because it will not affect the final decision boundary. Assuming also initialization with zero and extreme early stopping (updating each parameter maximally one time), we get

$$\alpha_k\leftarrow\left(M-y_kf_{k-1}\left(\boldsymbol{x_k}\right)\geq 0\right)\ ?\ 1-y_kf_{k-1}\left(\boldsymbol{x_k}\right)\ :\ 0.$$

By incorporating the additional assumption that the weight is positive, we get

$$\alpha_k\leftarrow\left(M-y_kf_{k-1}\left(\boldsymbol{x_k}\right)\geq 0\right)\ ?\ \max\left\{0,1-y_kf_{k-1}\left(\boldsymbol{x_k}\right)\right\}\ :\ 0.$$

By returning to the original representation with β_j weights, we get (6).

References

1. Bordes, A., Ertekin, S., Weston, J., Bottou, L.: Fast Kernel classifiers with online and active learning. J. Mach. Learn. Res. **6**, 1579–1619 (2005). http://jmlr.org/papers/v6/bordes05a.html
2. Conditional (ternary) operator, July 2021. https://developer.mozilla.org/en-US/docs/Web/JavaScript/Reference/Operators/Conditional_Operator
3. Cristianini, N., Shawe-Taylor, J.: An Introduction to Support Vector Machines and Other Kernel-Based Learning Methods. Cambridge University Press, New York (2000). https://doi.org/10.1017/CBO9780511801389
4. Delgado, M.F., Cernadas, E., Barro, S., Amorim, D.G.: Do we need hundreds of classifiers to solve real world classification problems? J. Mach. Learn. Res. **15**(1), 3133–3181 (2014). http://dl.acm.org/citation.cfm?id=2697065

5. Goodfellow, I.J., Bengio, Y., Courville, A.C.: Deep Learning. Adaptive Computation and Machine Learning. MIT Press, Cambridge (2016). http://www.deeplearningbook.org/
6. Hastie, T., Rosset, S., Tibshirani, R., Zhu, J.: The entire regularization path for the support vector machine. J. Mach. Learn. Res. **5**, 1391–1415 (2004)
7. Huang, X., Shi, L., Suykens, J.A.K.: Ramp loss linear programming support vector machine. J. Mach. Learn. Res. **15**(1), 2185–2211 (2014). http://dl.acm.org/citation.cfm?id=2670321
8. Japkowicz, N., Shah, M. (eds.): Evaluating Learning Algorithms: A Classification Perspective. Cambridge University Press, New York (2011)
9. LIBSVM data sets, June 2011. www.csie.ntu.edu.tw/~cjlin/libsvmtools/datasets/
10. Melki, G., Kecman, V., Ventura, S., Cano, A.: OLLAWV: online learning algorithm using worst-violators. Appl. Soft Comput. **66**, 384–393 (2018). https://doi.org/10.1016/j.asoc.2018.02.040
11. Orchel, M.: Incorporating prior knowledge into SVM algorithms in analysis of multidimensional data. Ph.D. thesis, AGH University of Science and Technology (2013). https://doi.org/10.13140/RG.2.1.5004.9441
12. Orchel, M., Suykens, J.A.K.: Fast hyperparameter tuning for support vector machines with stochastic gradient descent. In: Nicosia, G., et al. (eds.) LOD 2020. LNCS, vol. 12566, pp. 481–493. Springer, Cham (2020). https://doi.org/10.1007/978-3-030-64580-9_40
13. Shalev-Shwartz, S., Singer, Y., Srebro, N., Cotter, A.: Pegasos: primal estimated sub-gradient solver for SVM. Math. Program. **127**(1), 3–30 (2011). https://doi.org/10.1007/s10107-010-0420-4
14. Steinwart, I., Hush, D.R., Scovel, C.: Training SVMs without offset. J. Mach. Learn. Res. **12**, 141–202 (2011). http://portal.acm.org/citation.cfm?id=1953054

A Hybrid Surrogate-Assisted Accelerated Random Search and Trust Region Approach for Constrained Black-Box Optimization

Rommel G. Regis(✉)

Saint Joseph's University, Philadelphia, PA 19131, USA
rregis@sju.edu

Abstract. This paper presents a hybrid surrogate-based approach for constrained expensive black-box optimization that combines RBF-assisted Constrained Accelerated Random Search (CARS-RBF) with the CONORBIT trust region method. Extensive numerical experiments have shown the effectiveness of the CARS-RBF and CONORBIT algorithms on many test problems and the hybrid algorithm combines the strengths of these methods. The proposed CARS-RBF-CONORBIT hybrid alternates between running CARS-RBF for global search and a series of local searches using the CONORBIT trust region algorithm. In particular, after each CARS-RBF run, a fraction of the best feasible sample points are clustered to identify potential basins of attraction. Then, CONORBIT is run several times using each cluster of sample points as initial points together with infeasible sample points within a certain radius of the centroid of each cluster. One advantage of this approach is that the CONORBIT runs reuse some of the feasible and infeasible sample points that were previously generated by CARS-RBF and other CONORBIT runs. Numerical experiments on the CEC 2010 benchmark problems showed promising results for the proposed hybrid in comparison with CARS-RBF or CONORBIT alone given a relatively limited computational budget.

Keywords: Constrained optimization · Expensive black-box optimization · Accelerated random search · Trust region method · Clustering · Surrogate models · Radial basis functions

1 Introduction

Surrogate-based and surrogate-assisted approaches have been proposed for computationally expensive black-box optimization (e.g., [3,6,11,26]) and some of these methods can handle black-box inequality constraints (e.g., [2,4,5,10,13,17–19,24]). Among the widely used surrogates include Kriging or Gaussian Process (GP) models and Radial Basis Function (RBF) models, and these are used to approximate the objective and constraint functions globally or locally, or both. Examples of global optimization methods for problems with expensive black-box

G. Nicosia et al. (Eds.): LOD 2021, LNCS 13164, pp. 162–177, 2022.
https://doi.org/10.1007/978-3-030-95470-3_12

objective and constraints include the KCGO algorithm [13], which employs Kriging models of the objective and constraint functions to calculate the probability of feasibility and a lower confidence bound when generating an infill point (i.e., a function evaluation point). The COBRA algorithm [23] and its extension SACOBRA [2] select an infill point that minimizes an RBF surrogate of the objective subject to RBF surrogates of the constraints along with some distance requirements from previous sample points. The SEGO-KPLS(+K) method [4] employ Kriging models combined with partial least squares and selects an infill point that minimizes the surrogate while also maximizing the expected improvement criterion. Moreover, CARS-RBF [18] is a surrogate-assisted constrained version of the Accelerated Random Search algorithm [1], which possesses some theoretical convergence guarantees to the global minimum. CARS-RBF was extensively tested on more than 30 test problems and compared with several alternative methods, including the Accelerated Particle Swarm Optimization (APSO) algorithm [28] and the surrogate-based ConstrLMSRBF method [22].

On the other hand, there are also local minimization algorithms for constrained expensive black-box optimization that use surrogates. For example, the derivative-free trust region method in [7] uses a quadratic interpolation model to approximate the objective function. Moreover, quadratic models have also been used with the direct search algorithm NOMAD [8]. CONORBIT [25] is a trust region method that uses RBF interpolation to model both the objectives and constraints. It is an extension of the ORBIT trust region algorithm [27], which has been shown to converge to a stationary point. Extensive numerical experiments on almost 30 test problems have also shown the effectiveness of CONORBIT in comparison with several alternative methods, including COBYLA [21], NOMAD [12] and a sequential penalty derivative-free method [14].

Formally, this paper focuses on solving the constrained optimization problem:

$$\begin{array}{ll} & \min_{x\in\mathbb{R}^d}\ f(x) \\ \text{s.t.} & \\ & g_i(x) \le 0,\ \ i = 1, \ldots, m, \\ & \ell \le x \le u, \end{array} \tag{1}$$

where $[\ell, u] \subset \mathbb{R}^d$ is the search space and $f, g_1, \ldots, g_m$ are black-box functions whose values are obtained from computationally expensive but deterministic simulations. Here, one simulation yields the values of $f(x), g_1(x), \ldots, g_m(x)$ at a given $x \in \mathbb{R}^d$. For now, assume that the feasible region of the problem has a nonempty interior. Moreover, assume also that there is no noise in the calculation of the f and g_i's. Future work will deal with the case where the interior of the feasible region is empty, which happens when there are equality constraints, and it will also address the issue of noise in the objective and constraint functions.

This paper proposes a hybrid surrogate-based approach for constrained expensive black-box optimization that combines the CARS-RBF global optimization method with the CONORBIT trust region method. Extensive experiments have shown the effectiveness of both algorithms on many test problems and the hybrid algorithm combines the strengths of these methods. The proposed CARS-RBF-CONORBIT hybrid alternates between running CARS-RBF for global search and a series of local searches using the CONORBIT trust region algorithm. In particular, after each CARS-RBF run, a clustering method is used on a fraction of the feasible sample points with the best objective function values to identify potential basins of attraction of local minima. Then, in each cluster, CONORBIT is run using the sample points of that cluster as initial points together with infeasible sample points within a certain radius of the cluster centroid. The hybrid approach is meant to combine the strength of the CARS-RBF algorithm for global optimization with the ability of the CONORBIT trust region algorithm to quickly converge to a local minimum. Moreover, the combination is meant to reduce the effort spent by CONORBIT on expensive function evaluations by reusing some of the feasible and infeasible sample points that were previously generated by CARS-RBF and by other CONORBIT runs. Numerical experiments on the well-known CEC 2010 benchmark problems [15] showed that the proposed hybrid generally performs better than CARS-RBF or CONORBIT alone given a relatively limited computational budget.

2 Global and Local Constrained Black-Box Optimization Using Radial Basis Functions

2.1 RBF-Assisted Constrained Accelerated Random Search

The Constrained Accelerated Random Search (CARS) algorithm for constrained black-box optimization [18] is an extension of the Finite Descent Accelerated Random Search algorithm [1] that was originally designed for bound-constrained optimization. CARS selects its iterates uniformly at random within a box-shaped search neighborhood around the current best solution. This search neighborhood is reduced whenever the iterate does not improve the current best solution. It is re-initialized to the entire search space whenever the iterate improves the current best solution or whenever the size of the neighborhood falls below a threshold. Under certain conditions, the sequence of best solutions generated by CARS converges to the global minimum in a probabilistic sense. More precisely, the following theorem was proved in [18]:

Theorem 1. *Consider the constrained black-box optimization problem in (1) where f is continuous and the feasible region $\mathcal{D} = \{x \in \mathbb{R}^d \mid \ell \leq x \leq u, g_i(x) \leq 0,\ i = 1, \ldots, m\}$ is compact. Moreover, assume that int($\mathcal{D}$) $\neq \emptyset$ and that every neighborhood of a boundary point of $\mathcal{D}$ intersects int($\mathcal{D}$). Let $\{X_n\}_{n \geq 1}$ be the sequence of best solutions produced by CARS. Then $f(X_n) \longrightarrow f^* := \min_{x \in \mathcal{D}} f(x)$ almost surely (a.s.).*

In the above theorem, the requirement that every neighborhood of a boundary point of $\mathcal{D}$ intersects $\mathrm{int}(\mathcal{D})$ is satisfied if $\mathrm{cl}(\mathrm{int}(\mathcal{D})) = \mathrm{cl}(\mathcal{D})$, where $\mathrm{cl}(\cdot)$ denotes the closure of a set in $\mathbb{R}^d$.

The CARS-RBF algorithm [18] was developed to improve the performance of CARS on computationally expensive constrained black-box optimization problems by using radial basis function (RBF) surrogates of the objective and constraint functions. In each iteration of CARS-RBF, a large number of trial points is generated in the current search neighborhood. Then, RBF surrogates of the constraints are used to identify the trial points that are predicted to be feasible or that have the minimum number of predicted constraint violations. From these trial points, the sample point where the simulation is run is chosen to be the one with the best predicted objective function value. Extensive numerical experiments have shown the effectiveness of CARS-RBF in comparison with CARS with no surrogates and with alternative methods such as Accelerated Particle Swarm Optimization (APSO) [28], a sequential penalty derivative-free method [14], and it is competitive with the ConstrLMSRBF algorithm [22] on many test problems. More details about CARS-RBF can be found in [18].

2.2 CONORBIT Trust Region Method

CONORBIT [25] is a derivative-free trust region algorithm for constrained black-box optimization that employs RBF interpolation models for the objective and constraint functions. It is an extension of the ORBIT trust region algorithm [27] for unconstrained local optimization. CONORBIT uses a small margin for the RBF models of the constraints to facilitate the generation of feasible iterates and this approach was shown to improve performance. Extensive numerical experiments on many test problems showed that CONORBIT outperformed many alternative methods including the trust region method COBYLA [21], the ConstrLMSRBF method [22], a sequential penalty derivative-free method [14], the direct search method NOMAD [12], and an augmented Lagrangian method.

In each iteration of CONORBIT, the following trust region subproblem is solved approximately to obtain the next sample point:

$$\begin{aligned} &\min\ s_n^f(x) \\ &\text{s.t.} \\ &s_n^{g_i}(x) + \xi \mathbb{I}([g_i(x_n) \leq -\xi]) \leq 0, \quad i = 1, \ldots, m \\ &\|x - x_n\| \leq \Delta_n \\ &x \in \mathbb{R}^d,\ \ell \leq x \leq u \end{aligned} \tag{2}$$

Here, s_n^f and $s_n^{g_i}$, $i = 1, \ldots, m$ are the RBF models of the objective and constraints, x_n is the trust region center, Δ_n is the trust region radius, ξ is the margin on the RBF inequality constraints, and $\mathbb{I}$ is an indicator function. The margin ξ is meant to facilitate the generation of feasible iterates and was found to be helpful in numerical experiments in [25]. The trust region radius is adjusted according to the feasibility of the current iterate, the ratio of the actual improvement provided by the iterate to the predicted improvement, and the validity of the RBF models. Moreover, CONORBIT uses a stopping criterion based on a criticality measure that involves the negative gradient of the RBF model of the objective and a projection set onto an approximate feasible region defined by RBF models of the constraints. More details can be found in [25].

2.3 Radial Basis Function Interpolation

The CARS-RBF and CONORBIT algorithms both use the RBF interpolation model in Powell [20], which differs from the Gaussian RBF network in the machine learning literature. In this RBF model, each data point is a center, the basis functions are not necessarily Gaussian, and training simply involves solving a linear system. This type of RBF model has been successfully used on high-dimensional constrained black-box optimization problems with hundreds of decision variables and many black-box constraints (e.g., see [2,23]).

To fit this RBF model, suppose we are given n distinct points $x_1, \ldots, x_n \in \mathbb{R}^d$ and their function values $u(x_1), \ldots, u(x_n)$, where $u(x)$ is either an objective or constraint function. This RBF interpolation model has the form:

$$s_n(x) = \sum_{i=1}^{n} \lambda_i \phi(\|x - x_i\|) + p(x), \ x \in \mathbb{R}^d,$$

where $\| \cdot \|$ is the Euclidean norm, $\lambda_i \in \mathbb{R}$ for $i = 1, \ldots, n$, and $p(x)$ is a linear polynomial in d variables. Here, ϕ has the *cubic* form ($\phi(r) = r^3$). However, ϕ can also take other forms such as the thin plate spline ($\phi(r) = r^2 \log r$) or the Gaussian form ($\phi(r) = \exp(-\gamma r^2)$, where γ is a hyperparameter). A cubic RBF model is used here because of its simplicity and its success in prior RBF methods (e.g., [23,25]). Fitting this model involves solving a linear system that possesses some nice mathematical properties. More details can be found in Powell [20].

3 A Hybrid Surrogate-Based Algorithm for Constrained Black-Box Optimization

The proposed method is a hybrid surrogate-based approach for solving the constrained black-box optimization problem in (1) that combines the strengths of the CARS-RBF global optimization method [18] with the CONORBIT trust region method [25]. This hybrid will be referred to as CARS-RBF-CONORBIT.

As with any surrogate-based optimization method, CARS-RBF-CONORBIT begins by evaluating the objective and constraint functions at the points of an initial space-filling design over the entire search space. For now, assume that a feasible point is included among the initial design points. Such an assumption is not unreasonable in practice where a feasible design is sometimes available and one would like to find an improved solution. Future work will consider the case where none of the initial design points are feasible by considering a two-phase approach similar to that in [23].

After obtaining the objective and constraint function values at the initial points, a fixed number n_{global} of CARS-RBF iterations are performed. Each CARS-RBF iteration builds/updates RBF surrogates for the objective and constraints, which are then used to select one new sample point where the objective and constraint functions are evaluated. From the sample points obtained so far, we select a certain fraction $0 < \alpha \leq 1$ of the feasible sample points with the best objective function values. Then, we run a clustering algorithm on these feasible sample points to identify a maximum of $k_{\max}$ clusters that represent potential basins of attraction for local minima.

Next, a series of CONORBIT runs each up to a maximum of n_{local} iterations are performed, one for each cluster. The feasible sample points in each cluster are augmented by infeasible sample points within a certain radius δ of the cluster centroid and all these sample points are reused as initial points for the CONORBIT iterations. As with CARS-RBF, each iteration of CONORBIT yields one new sample point and its objective and constraint function values. This sample point is typically a solution of the trust region subproblem (see Sect. 2.2) that involves minimizing an RBF model of the objective within the trust region and subject to RBF surrogates of the constraints within some margin. When there are many sample points in a given cluster, CONORBIT might terminate early, after less than n_{local} iterations, if it satisfies the stopping condition. After all the CONORBIT iterations are completed, we repeat the process of running CARS-RBF iterations followed by clustering and then a series of CONORBIT iterations until the computational budget is exhausted.

Below is a pseudo-code of the proposed CARS-RBF-CONORBIT method for solving the constrained black-box optimization problem in (1).

CARS-RBF-CONORBIT Algorithm

(1) *(Initial Simulations)* Perform simulations to obtain objective and constraint function values at initial set of points in the search space. (For now, assume that one of these initial points is feasible.)

(2) *(Perform Iterations)* While the computational budget has *not* been exhausted, do:

- **(a)** *(CARS-RBF Iterations)* Perform n_{global} iterations of CARS-RBF using all available sample points.
- **(b)** *(Gather Top Fraction of Feasible Points)* Gather the top α of the feasible sample points with the best objective function values.
- **(c)** *(Cluster Feasible Sample Points)* Cluster the feasible sample points obtained in 2(b) up to a maximum of k_{max} clusters.
- **(d)** *(CONORBIT Iterations)* For each cluster, perform n_{local} iterations of CONORBIT using as initial points the sample points in that cluster augmented by infeasible sample points that are within radius δ of the cluster centroid.

(3) *(Return Best Solution Found)* Return best feasible solution found and the corresponding objective and constraint function values.

4 Numerical Experiments

4.1 Experimental Setup

CARS-RBF-CONORBIT with two parameter settings for the maximum number of clusters used (k_{max}) are compared with its component algorithms, CARS-RBF and CONORBIT, on 10-D instances of 11 test problems from the CEC 2010 benchmark [15]. In particular, CARS-RBF-CONORBIT is run using $k_{\text{max}} = 3$ and $k_{\text{max}} = 5$. Since CARS-RBF-CONORBIT and its component algorithms are designed for problems with inequality constraints, some of the CEC 2010 problems are modified by replacing the equality with inequality constraints.

In particular, the test problems used include C01, C07, C08, C14 and C15, and modified versions of C02, C05, C06, C09, C10 and C17 where the = constraints are replaced with $\leq$ constraints. CARS-RBF and CONORBIT outperformed several alternative methods for constrained black-box optimization on many test problems [18,25], so no additional methods are included in the comparisons.

The experiments are performed using Matlab 9.4. Each algorithm is run for 30 trials on each test problem. To get a fair comparison, the different algorithms used the same space-filling design for each trial. This design is an approximate maximin Latin hypercube design with $2(d+1)$ points that contains a subset of $d+1$ affinely independent points to ensure that the RBF interpolation matrix is nonsingular. None of these initial points are guaranteed to be feasible, so a feasible point is included as the first point among the initial design points.

For the algorithm parameters, all four methods used the cubic RBF model with a linear polynomial tail as in [25]. CARS-RBF used the same parameter settings as in [18] while CONORBIT used the same parameter settings and options as [25]. For CARS-RBF-CONORBIT, the number of simulations allocated for each cycle of CARS-RBF iterations is $n_{\text{global}} = 5(d+1)$ and the maximum number of simulations for each CONORBIT run is $n_{\text{local}} = 5(d+1)$. For the CARS-RBF component of the hybrid method, the number of trial points generated in each iteration is the same as in [18], which is $n_{\text{trial}} = \min(1000 * d, 10000)$. After the CARS-RBF iterations, the fraction of the feasible sample points with the best objective function values used to estimate the basins of attractions is set to $\alpha = 0.5$. Moreover, k-means clustering as implemented in [9] is used to identify clusters among the best feasible sample points where the number of clusters is determined automatically using the elbow method up a maximum of $k_{\max}$ clusters. For the CONORBIT runs within the hybrid method, most of the parameter settings and options are set as in [25]. In particular, CONORBIT uses the fmincon solver from the Matlab Optimization Toolbox to solve the trust region subproblems where the solver is applied to the RBF surrogates of the objective and constraints. In addition, the feasible sample points in each cluster are used as initial points together with the infeasible sample points within radius δ from the cluster centroid where δ is twice the distance between the cluster centroid and the farthest cluster point.

4.2 Comparison Using Data Profiles

The CARS-RBF-CONORBIT hybrid is compared with the component algorithms using data profiles [16]. To create these profiles, define a problem p to be a pairing of a test problem and the given feasible initial point. Since there are 11 test problems and 30 feasible starting points, the total number of problems is $11 \times 30 = 330$. Given the set $\mathcal{P}$ of 330 problems, we run a set $\mathcal{S}$ of solvers, which are the CARS-RBF-CONORBIT algorithms, CARS-RBF and CONORBIT.

Now, given a solver $s \in \mathcal{S}$, the *data profile* of s on the set of problems $\mathcal{P}$ is the function

$$d_s(\alpha) = \frac{1}{|\mathcal{P}|} \left| \left\{ p \in \mathcal{P} \; : \; \frac{t_{p,s}}{n_p + 1} \leq \alpha \right\} \right|, \qquad \text{for } \alpha > 0,$$

where $t_{p,s}$ is the number of simulations required by solver s to satisfy a convergence test on problem p, and n_p is the number of variables in p. For a given solver s and any $\alpha > 0$, $d_s(\alpha)$ is the fraction of problems "solved" by s within $\alpha \cdot (n_p+1)$ simulations (equivalent to α simplex gradient estimates) [16]. Here, "solving" a problem means generating a point satisfying the convergence test below and one *simplex gradient estimate* is equivalent to (n_p+1) calls to the simulator that calculates the objective and constraint function values. This notion was introduced in [16] and it is based on the idea that (n_p+1) affinely independent points and their corresponding function values are needed to calculate a simplex gradient of the function in the fully determined case.

In expensive black-box optimization, algorithms are typically compared given a fixed and relatively limited computational budget in number of simulations. As in [25], a point x obtained by a solver "solves" the problem if it satisfies the following convergence test:

$$\max_{i=1,\ldots,q} g_i(x) \leq \epsilon, \quad \text{and} \quad f(x^{(0)}) - f(x) \geq (1-\tau)\left(f(x^{(0)}) - f_L^{\epsilon,\mathcal{S},\mu_f}\right). \tag{3}$$

where $\epsilon, \tau > 0$ are tolerances and $f_L^{\epsilon,\mathcal{S},\mu_f}$ is the minimum objective function value of all ϵ-feasible points obtained by *any* of the solvers within a given budget μ_f of simulations. This means that if x^{ϵ,s,μ_f} is the best ϵ-feasible point obtained by solver $s \in \mathcal{S}$ within μ_f simulations, then $f_L^{\epsilon,\mathcal{S},\mu_f} = \min_{s\in\mathcal{S}} f(x^{\epsilon,s,\mu_f})$.

Note that the starting point $x^{(0)}$ is feasible in all trial runs of the different solvers. Hence, at least one solver $s \in \mathcal{S}$ will satisfy the convergence test (3) for any given $\epsilon, \tau, \mu_f > 0$. If there are multiple points obtained by solver s that satisfy (3) on problem p, the performance measure $t_{p,s}$ is the minimum number of simulations needed to satisfy (3).

4.3 Results and Discussion

Figure 1 shows the data profiles of the algorithms up to 100 simplex gradients, where each simplex gradient is equivalent to $d+1$ simulations. Hence, all algorithms are run to a maximum of $\mu_f = 100(d+1)$ simulations, where each simulation yields the values of the objective and all constraint functions at a given input. The tolerances used for the data profiles are $\epsilon = 10^{-6}$ and $\tau = 0.01$.

Figure 1 shows that the CARS-RBF-CONORBIT algorithms are better than CARS-RBF and CONORBIT after 45 simplex gradient estimates (equivalent to $45(d+1)$ simulations). In particular, CARS-RBF-CONORBIT with $k_{\max} = 3$ and $k_{\max} = 5$ solve about 73% and 71% of the problems, respectively, within 75 simplex gradient estimates. On the other hand, CARS-RBF and CONORBIT solve about 68% and 56% of the problems, respectively, within the same computational budget. Recall that a *problem* corresponds to a particular combination of test problem and initial feasible point for a given trial. Also, the two CARS-RBF-CONORBIT algorithms solve about 78% of the problems within 100 simplex gradient estimates while CARS-RBF and CONORBIT solve about 74% and 65% of the problems, respectively, within the same budget. Between 45 and

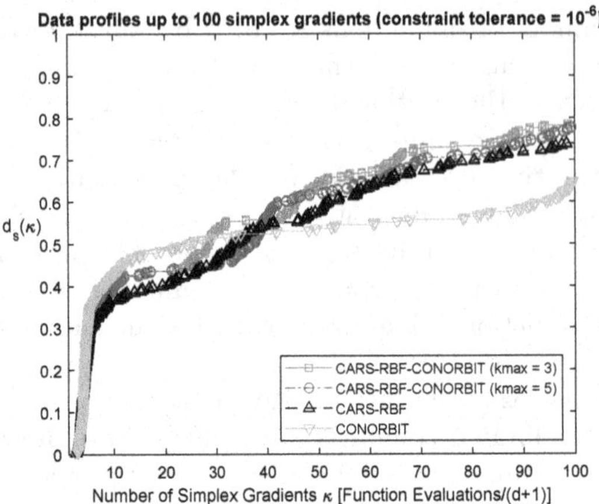

Fig. 1. Data profiles for 30 trials of CARS-RBF-CONORBIT and its component algorithms on 11 problems from the CEC 2010 benchmark.

100 simplex gradient estimates, the hybrid approach shows an advantage over the individual algorithms on the test problems used. The data profile for CONORBIT increases more quickly compared to that of the other methods within 20 simplex gradient estimates, but flattens out for a while because the method is meant for local optimization. CARS-RBF has more steady progress because it has better balance between global and local search, but the results show that it can be improved by running CONORBIT on sample points at promising basins of attractions found after a certain number of CARS-RBF iterations.

Figure 1 also shows that the performance of CARS-RBF-CONORBIT with $k_{\max} = 3$ is somewhat close to that of the hybrid with $k_{\max} = 5$, but with the former being slightly better than the latter after about 50 simplex gradient estimates. However, CARS-RBF-CONORBIT with $k_{\max} = 5$ caught up with CARS-RBF-CONORBIT with $k_{\max} = 3$ after about $100(d+1)$ simulations. In the long run, having a larger $k_{\max}$ allows CARS-RBF-CONORBIT to explore more local minima, which improves the chances of finding a better local minimum. However, having a smaller $k_{\max}$ allows the algorithm to quickly explore the most promising local minima in the earlier stages of the search.

Next, Table 1 shows the mean and standard error (over 30 trial runs) of the best feasible objective function values found by the CARS-RBF-CONORBIT, CARS-RBF and CONORBIT algorithms at three different computational budgets of $50(d+1)$, $75(d+1)$ and $100(d+1)$ simulations on the test problems. The best result is outlined by a solid box. The second best result is outlined by a box of dashes.

Table 1 indicates that the best feasible objective function values obtained by the CARS-RBF-CONORBIT algorithms are generally better than those obtained by CARS-RBF and CONORBIT at the different computational budgets. These results are consistent with the data profiles in Fig. 1. In particular,

Table 1. Mean and standard error (over 30 trials) of the best feasible objective function values found by CARS-RBF-CONORBIT and its component algorithms at different computational budgets on 11 problems from the CEC 2010 benchmark. The best is outlined by a solid box. The second best is outlined by a box of dashes.

Test Problem	No. of Simulations	CARS-RBF-CONORBIT ($k_{max} = 3$)	CARS-RBF-CONORBIT ($k_{max} = 5$)	CARS-RBF	CONORBIT
C01	$50(d+1)$	−0.4111 (0.0197)	−0.3840 (0.0190)	−0.4010 (0.0161)	−0.2931 (0.0117)
C01	$75(d+1)$	−0.4456 (0.0216)	−0.4148 (0.0210)	−0.4364 (0.0144)	−0.2931 (0.0117)
C01	$100(d+1)$	−0.4584 (0.0219)	−0.4489 (0.0235)	−0.4571 (0.0136)	−0.2931 (0.0117)
C02Mod	$50(d+1)$	−2.2086 (0.0141)	−2.2178 (0.0064)	−2.1860 (0.0343)	−1.4241 (0.1313)
C02Mod	$75(d+1)$	−2.2299 (0.0078)	−2.2351 (0.0036)	−2.1955 (0.0344)	−1.4258 (0.1317)
C02Mod	$100(d+1)$	−2.2376 (0.0043)	−2.2386 (0.0033)	−2.2114 (0.0320)	−1.4262 (0.1318)
C05Mod	$50(d+1)$	−454.04 (6.61)	−442.17 (7.18)	−455.52 (6.78)	−314.70 (25.52)
C05Mod	$75(d+1)$	−458.90 (6.42)	−454.63 (6.52)	−463.01 (6.16)	−318.89 (23.60)
C05Mod	$100(d+1)$	−462.67 (6.21)	−457.90 (6.48)	−471.65 (4.00)	−319.17 (23.64)
C06Mod	$50(d+1)$	−398.94 (23.58)	−399.00 (25.73)	−346.97 (17.86)	−238.29 (23.40)
C06Mod	$75(d+1)$	−436.20 (22.43)	−424.63 (26.35)	−358.72 (17.47)	−239.28 (23.35)
C06Mod	$100(d+1)$	−453.70 (22.60)	−441.66 (25.54)	−369.83 (17.47)	−239.52 (23.38)
C07	$50(d+1)$	21822.61 (14542.97)	13010.03 (5828.55)	1336867.68 (290091.08)	2385.22 (796.89)
C07	$75(d+1)$	8463.17 (5313.93)	4093.84 (1795.57)	640098.23 (133222.93)	2306.90 (797.59)
C07	$100(d+1)$	3578.09 (2075.19)	1526.48 (621.86)	360113.20 (74413.69)	2277.61 (795.72)
C08	$50(d+1)$	20436.79 (7812.00)	23337.31 (11619.16)	3048807.34 (830289.04)	8144.70 (4174.58)
C08	$75(d+1)$	9705.29 (4440.30)	13935.81 (7990.86)	1386161.70 (453408.52)	3878.31 (1604.32)
C08	$100(d+1)$	7208.80 (3472.04)	11392.69 (7256.83)	577530.53 (159798.76)	2637.64 (838.47)
C09Mod	$50(d+1)$	1.73E+06 (5.83E+05)	1.70E+06 (4.72E+05)	2.03E+08 (3.89E+07)	3.47E+06 (1.01E+06)
C09Mod	$75(d+1)$	1.60E+06 (5.76E+05)	1.53E+06 (4.54E+05)	7.66E+07 (1.55E+07)	3.46E+06 (1.01E+06)
C09Mod	$100(d+1)$	1.57E+06 (5.74E+05)	1.49E+06 (4.49E+05)	3.64E+07 (7.04E+06)	3.46E+06 (1.01E+06)
C10Mod	$50(d+1)$	1.16E+06 (6.33E+05)	1.20E+06 (6.62E+05)	1.85E+08 (2.59E+07)	3.85E+06 (1.28E+06)
C10Mod	$75(d+1)$	1.04E+06 (6.09E+05)	9.57E+05 (6.14E+05)	6.92E+07 (9.40E+06)	3.69E+06 (1.29E+06)
C10Mod	$100(d+1)$	9.74E+05 (5.98E+05)	8.99E+05 (5.97E+05)	3.52E+07 (5.31E+06)	3.68E+06 (1.29E+06)
C14	$50(d+1)$	9.19E+12 (3.38E+12)	1.27E+13 (3.33E+12)	1.02E+13 (2.33E+12)	1.13E+13 (4.44E+12)
C14	$75(d+1)$	1.73E+12 (8.68E+11)	3.86E+12 (1.26E+12)	1.12E+12 (5.87E+11)	4.09E+12 (1.89E+12)
C14	$100(d+1)$	1.18E+11 (5.33E+10)	7.45E+11 (3.09E+11)	8.89E+10 (7.27E+10)	1.06E+12 (4.61E+11)
C15	$50(d+1)$	1.611E+14 (1.43E+13)	1.610E+14 (1.44E+13)	1.67E+14 (1.48E+13)	1.26E+14 (1.19E+13)
C15	$75(d+1)$	1.59E+14 (1.42E+13)	1.55E+14 (1.35E+13)	1.66E+14 (1.48E+13)	1.11E+14 (1.1E+13)
C15	$100(d+1)$	1.56E+14 (1.39E+13)	1.54E+14 (1.35E+13)	1.65E+14 (1.48E+13)	1.01E+14 (1.08E+13)
C17Mod	$50(d+1)$	1.2914 (0.1865)	1.3297 (0.2389)	7.2946 (1.0405)	0.9025 (0.1870)
C17Mod	$75(d+1)$	1.0164 (0.1767)	0.7757 (0.1470)	5.6661 (0.6400)	0.6210 (0.1262)
C17Mod	$100(d+1)$	0.8101 (0.1674)	0.5459 (0.1180)	4.8892 (0.5690)	0.3792 (0.1182)

CARS-RBF-CONORBIT with $k_{\max} = 3$ is better than CARS-RBF on 9 of the 11 test problems (all except on C05Mod and C14) after $50(d+1)$, $75(d+1)$ and $100(d+1)$ simulations. Moreover, it is better than CONORBIT on 7 of the 11 test problems at the same computational budgets.

Table 1 also shows that CARS-RBF-CONORBIT with $k_{max} = 5$ is better than CARS-RBF on 8 of the 11 test problems (all except C01, C05Mod and C14) after $50(d+1)$, $75(d+1)$ and $100(d+1)$ simulations. It is also better than CONORBIT on 8 of the test problems (all except C08, C15 and C17Mod) after $100(d+1)$ simulations. Moreover, CARS-RBF-CONORBIT with $k_{max} = 5$ is better than CONORBIT on 7 and 6 of the test problems after $75(d+1)$ and $50(d+1)$ simulations, respectively.

Next, the two CARS-RBF-CONORBIT algorithms have comparable performances on the CEC 2010 problems. From Table 1, CARS-RBF-CONORBIT with $k_{max} = 5$ is better than the one with $k_{max} = 3$ on 6 of the 11 test problems at $75(d+1)$ and $100(d+1)$ simulations. Moreover, the latter is better than the former on 6 test problems at $50(d+1)$ simulations. This is also consistent with the data profiles in Fig. 1 where the graphs for the two algorithms are close to one another. This result suggests that CARS-RBF-CONORBIT is not very sensitive to k_{max} on the problems in the CEC 2010 benchmark.

It is worth noting that combining CARS-RBF and CONORBIT results in an algorithm that is more robust than either component method. For example, the results obtained by CARS-RBF on C07, C08, C09Mod and C10Mod are very poor in comparison with the other methods even after $100(d+1)$ simulations. However, the CARSRBF-CONORBIT algorithms never obtained such extremely poor results on the test problems. This indicates that the limited local search capability of CARS-RBF is not enough to quickly make progress on some of the CEC 2010 problems. CONORBIT provides the hybrid algorithm with better local search capability. On the other hand, CONORBIT is a local optimization method, so it is not expected to do well on problems with many local minima and this is evident from the data profiles in Fig. 1. The CARS-RBF component provides the hybrid with better global search capability.

Next, Table 2 shows the mean percent improvement and standard error (over 30 trials) in the best feasible objective function value found by each CARS-RBF-CONORBIT algorithm over the CARS-RBF global optimization method at computational budgets of $50(d+1)$, $75(d+1)$ and $100(d+1)$ simulations. Improvements over CONORBIT are not included since CONORBIT is only meant for local optimization. The few negative values in the table indicate the mean percent deterioration of a hybrid method over CARS-RBF. The table shows that CARS-RBF-CONORBIT with $k_{max} = 3$ obtained significant improvements over CARS-RBF on 9 of the 11 test problems (all except C05Mod and C14). It also shows that CARS-RBF-CONORBIT with $k_{max} = 5$ obtained significant improvements over CARS-RBF on 8 of the test problems. Moreover, the improvements obtained by the two CARS-RBF-CONORBIT algorithms over CARS-RBF are generally comparable, which is consistent with the results obtained from Fig. 1 and Table 1.

Table 2. Mean and standard error (over 30 trials) of the percent improvement of CARS-RBF-CONORBIT over CARS-RBF at different computational budgets on 11 problems from the CEC 2010 benchmark.

Test problem	Number of simulations	CARS-RBF-CONORBIT ($k_{max} = 3$)	CARS-RBF-CONORBIT ($k_{max} = 5$)
C01	$50(d+1)$	6.04 (6.07)	0.05 (6.11)
C01	$75(d+1)$	5.59 (6.26)	−2.23 (5.57)
C01	$100(d+1)$	3.41 (6.12)	1.37 (6.33)
C02Mod	$50(d+1)$	1.80 (1.74)	2.61 (2.58)
C02Mod	$75(d+1)$	2.47 (1.99)	2.95 (2.54)
C02Mod	$100(d+1)$	2.11 (2.21)	2.25 (2.44)
C05Mod	$50(d+1)$	0.50 (2.38)	−2.29 (2.17)
C05Mod	$75(d+1)$	−0.20 (2.24)	−1.37 (1.78)
C05Mod	$100(d+1)$	−1.61 (1.77)	−2.84 (1.27)
C06Mod	$50(d+1)$	19.65 (7.13)	17.44 (6.82)
C06Mod	$75(d+1)$	26.38 (7.23)	19.33 (5.94)
C06Mod	$100(d+1)$	26.45 (6.96)	21.07 (6.09)
C07	$50(d+1)$	98.59 (0.65)	99.04 (0.48)
C07	$75(d+1)$	98.79 (0.55)	99.50 (0.19)
C07	$100(d+1)$	98.86 (0.53)	99.63 (0.12)
C08	$50(d+1)$	98.29 (0.83)	98.61 (0.48)
C08	$75(d+1)$	98.32 (0.95)	97.80 (0.94)
C08	$100(d+1)$	96.46 (1.58)	96.85 (1.08)
C09Mod	$50(d+1)$	97.07 (1.22)	97.62 (0.80)
C09Mod	$75(d+1)$	93.91 (2.09)	94.33 (1.82)
C09Mod	$100(d+1)$	91.10 (2.74)	91.32 (2.56)
C10Mod	$50(d+1)$	98.50 (1.17)	98.33 (1.25)
C10Mod	$75(d+1)$	97.84 (1.43)	97.84 (1.46)
C10Mod	$100(d+1)$	96.68 (2.02)	96.71 (2.01)
C14	$50(d+1)$	−195.25 (143.85)	−231.99 (115.44)
C14	$75(d+1)$	−2701.94 (2182.42)	−4365.45 (2026.90)
C14	$100(d+1)$	−1031.47 (737.56)	−31688.43 (23823.24)
C15	$50(d+1)$	3.86 (0.58)	3.82 (0.44)
C15	$75(d+1)$	5.07 (0.76)	5.94 (1.25)
C15	$100(d+1)$	6.17 (0.92)	6.30 (1.33)
C17Mod	$50(d+1)$	66.86 (8.44)	74.00 (4.58)
C17Mod	$75(d+1)$	74.91 (4.73)	81.19 (3.74)
C17Mod	$100(d+1)$	79.15 (4.54)	83.52 (3.89)

Finally, Table 3 shows the mean and standard error (over 30 trials) of the run times (in sec) of the algorithms on the test problems at a computational budget of $100(d + 1)$ simulations. The experiments are carried out using an Intel(R) Core(TM) i7-7700T CPU @ 2.90 GHz, 2904 Mhz, 4 Core(s), 8 Logical Processor(s) Windows-based machine. Since the time to evaluate the objective and constraints of the test problems are negligible, these run times are mostly the overhead of the algorithms. The table shows that CARS-RBF tends to be

Table 3. Mean and standard error (over 30 trials) of the run times (in sec) of CARS-RBF-CONORBIT and its component algorithms at a computational budget of $100(d+1)$ simulations on 11 problems from the CEC 2010 benchmark.

Test problem	CARS-RBF-CONORBIT ($k_{max} = 3$)	CARS-RBF-CONORBIT ($k_{max} = 5$)	CARS-RBF	CONORBIT
C01	357.36 (8.46)	389.71 (12.36)	651.85 (6.80)	34.06 (3.21)
C02Mod	256.45 (5.02)	247.11 (3.54)	652.34 (6.68)	67.83 (8.38)
C05Mod	287.43 (18.03)	231.79 (15.20)	656.44 (10.23)	144.28 (10.03)
C06Mod	234.17 (5.53)	213.00 (3.77)	648.85 (11.38)	111.57 (9.21)
C07	325.55 (9.72)	334.57 (6.90)	651.83 (13.96)	86.41 (2.48)
C08	282.51 (6.43)	292.45 (5.53)	655.70 (8.58)	104.54 (7.37)
C09Mod	407.64 (9.80)	433.41 (8.40)	667.75 (7.91)	250.95 (43.00)
C10Mod	407.14 (11.05)	418.40 (11.72)	644.86 (12.26)	241.00 (42.62)
C14	1118.33 (33.23)	1307.08 (52.05)	656.99 (9.23)	672.74 (48.42)
C15	1454.65 (53.92)	1485.62 (55.49)	662.32 (9.51)	1735.72 (97.81)
C17Mod	524.60 (70.48)	572.46 (69.99)	671.25 (14.06)	1061.68 (171.67)

the slowest while CONORBIT tends to be the fastest on most of the problems. This is because in each iteration, CARS-RBF spends a significant amount of time using RBF surrogates to evaluate a large number of trial solutions while CONORBIT uses Matlab's fmincon to solve the trust region subproblem. These mean run times are all considerable for a computational budget of only $100(d+1)$ simulations because all the algorithms use RBF surrogates, which incur significant computing overhead. However, when the simulations are truly expensive (e.g., one hour per simulation), these run times only take up a tiny fraction of the computing time, so the differences shown in the table do not really matter.

5 Summary and Future Work

This paper proposed the surrogate-based CARS-RBF-CONORBIT hybrid for constrained expensive black-box optimization that combines the RBF-assisted CARS-RBF algorithm and the RBF-based CONORBIT trust region method. The hybrid method alternates between CARS-RBF iterations and a series of CONORBIT runs started at approximate basins of attractions of potential local minima. These approximate basins of attractions are obtained by applying k-means clustering on a fraction of the feasible sample points with the best objective function values. CARS-RBF is a surrogate-based global optimization method that performs both global and local search while CONORBIT is a trust region algorithm that focuses on finding the local minima that correspond to approximate basins of attractions. Numerical experiments on test problems from the CEC 2010 benchmark indicate that two hybrid algorithms that use up to a maximum of 3 and a maximum of 5 clusters perform better the component algorithms CARS-RBF and CONORBIT on these test problems. Hence, both CARS-RBF and CONORBIT have the potential to improve the performance of one another for constrained expensive black-box optimization by combining their strengths and achieving a better balance between global and local search.

The proposed CARS-RBF-CONORBIT assumes that a feasible initial point is given for each problem and allocates a fixed maximum number of simulations for each CARS-RBF run and CONORBIT run. Future work will consider the case when no feasible initial point is available and will explore adaptive procedures for switching between the component algorithms. Moreover, instead of using k-means clustering, other more suitable clustering methods that take into account the landscape of the multimodal objective function could be used. For example, some basins of attractions might correspond to narrow valleys that might be identified better using a different clustering technique. Also, future work will consider a wider variety of test problems, including problems with only bound constraints, from low to high dimensions and containing a few to many local minima. Although the proposed hybrid is for constrained optimization, a similar hybrid can be developed for bound-constrained problems. Finally, it would be best to also explore the performance of the proposed hybrid to real-world black-box optimization problems.

Acknowledgements. Thanks to Sebastien De Landtsheer for his Matlab code for k-means clustering with the elbow method to determine the optimal number of clusters.

References

1. Appel, M.J., LaBarre, R., Radulović, D.: On accelerated random search. SIAM J. Optim. **14**(3), 708–731 (2004)
2. Bagheri, S., Konen, W., Emmerich, M., Bäck, T.: Self-adjusting parameter control for surrogate-assisted constrained optimization under limited budgets. Appl. Soft Comput. **61**, 377–393 (2017)
3. Bartz-Beielstein, T., Zaefferer, M.: Model-based methods for continuous and discrete global optimization. Appl. Soft Comput. **55**, 154–167 (2017)
4. Bouhlel, M.A., Bartoli, N., Regis, R.G., Otsmane, A., Morlier, J.: Efficient global optimization for high-dimensional constrained problems by using the kriging models combined with the partial least squares method. Eng. Optim. **50**(12), 2038–2053 (2018)
5. Boukouvala, F., Hasan, M.M.F., Floudas, C.A.: Global optimization of general constrained grey-box models: new method and its application to constrained PDEs for pressure swing adsorption. J. Glob. Optim. **67**(1), 3–42 (2017)
6. Cheng, R., He, C., Jin, Y., Yao, X.: Model-based evolutionary algorithms: a short survey. Complex Intell. Syst. **4**(4), 283–292 (2018). https://doi.org/10.1007/s40747-018-0080-1
7. Conejo, P.D., Karas, E.W., Pedroso, L.G.: A trust-region derivative-free algorithm for constrained optimization. Optim. Meth. Softw. **30**(6), 1126–1145 (2015)
8. Conn, A.R., Le Digabel, S.: Use of quadratic models with mesh-adaptive direct search for constrained black box optimization. Optim. Meth. Softw. **28**(1), 139–158 (2013)

9. De Landtsheer, S.: kmeans_opt. MATLAB Central File Exchange (2021). (https://www.mathworks.com/matlabcentral/fileexchange/65823-kmeans_opt. Accessed 22 Jan 2021
10. Feliot, P., Bect, J., Vazquez, E.: A Bayesian approach to constrained single- and multi-objective optimization. J. Glob. Optim. **67**, 97–133 (2017)
11. Forrester, A.I.J., Sobester, A., Keane, A.J.: Engineering Design via Surrogate Modelling: A Practical Guide. Wiley, Hoboken (2008)
12. Le Digabel, S.: Algorithm 909: NOMAD: nonlinear optimization with the MADS algorithm. ACM Trans. Math. Softw. **37**(4), 44:1–44:15 (2011)
13. Li, Y., Wu, Y., Zhao, J., Chen, L.: A Kriging-based constrained global optimization algorithm for expensive black-box functions with infeasible initial points. J. Glob. Optim. **67**, 343–366 (2017)
14. Liuzzi, G., Lucidi, S., Sciandrone, M.: Sequential penalty derivative-free methods for nonlinear constrained optimization. SIAM J. Optim. **20**(5), 2614–2635 (2010)
15. Mallipeddi, R., Suganthan, P.N.: Problem definitions and evaluation criteria for the CEC 2010 competition on constrained real-parameter optimization. Technical Report. Nanyang Technological University, Singapore (2010)
16. Moré, J.J., Wild, S.M.: Benchmarking derivative-free optimization algorithms. SIAM J. Optim. **20**(1), 172–191 (2009)
17. Müller, J., Woodbury, J.D.: GOSAC: global optimization with surrogate approximation of constraints. J. Global Optim. **69**(1), 117–136 (2017). https://doi.org/10.1007/s10898-017-0496-y
18. Nuñez, L., Regis, R.G., Varela, K.: Accelerated random search for constrained global optimization assisted by radial basis function surrogates. J. Comput. Appl. Math. **340**, 276–295 (2018)
19. Palar, P.S., Dwianto, Y.B., Regis, R.G., Oyama, A., Zuhal, L.R.: Benchmarking constrained surrogate-based optimization on low speed airfoil design problems. In: GECCO'19: Proceedings of the Genetic and Evolutionary Computation Conference Companion, pp. 1990–1998. ACM, New York (2019)
20. Powell, M.J.D.: The theory of radial basis function approximation in 1990. In: Light, W. (ed.) Advances in Numerical Analysis, Volume 2: Wavelets, Subdivision Algorithms and Radial Basis Functions, pp. 105–210. Oxford University Press, Oxford (1992)
21. Powell, M.J.D.: A direct search optimization methods that models the objective and constraint functions by linear interpolation. In: Gomez, S., Hennart, J.P. (eds.) Advances in Optimization and Numerical Analysis, pp. 51–67. Kluwer, Dordrecht (1994)
22. Regis, R.G.: Stochastic radial basis function algorithms for large-scale optimization involving expensive black-box objective and constraint functions. Comput. Oper. Res. **38**(5), 837–853 (2011)
23. Regis, R.G.: Constrained optimization by radial basis function interpolation for high-dimensional expensive black-box problems with infeasible initial points. Eng. Optim. **46**(2), 218–243 (2014)
24. Regis, R.G.: A survey of surrogate approaches for expensive constrained black-box optimization. In: Le Thi, H.A., Le, H.M., Pham Dinh, T. (eds.) WCGO 2019. AISC, vol. 991, pp. 37–47. Springer, Cham (2020). https://doi.org/10.1007/978-3-030-21803-4_4
25. Regis, R.G., Wild, S.M.: CONORBIT: constrained optimization by radial basis function interpolation in trust regions. Optim. Meth. Softw. **32**(3), 552–580 (2017)
26. Vu, K.K., D'Ambrosio, C., Hamadi, Y., Liberti, L.: Surrogate-based methods for black-box optimization. Int. Trans. Oper. Res. **24**, 393–424 (2017)

27. Wild, S.M., Regis, R.G., Shoemaker, C.A.: ORBIT: optimization by radial basis function interpolation in trust-regions. SIAM J. Sci. Comput. **30**(6), 3197–3219 (2008)
28. Yang, X.-S.: Nature-Inspired Metaheuristic Algorithms, 2nd edn. Luniver Press (2010)

Health Change Detection Using Temporal Transductive Learning

Abhay Harpale(✉)

GE Global Research, New York, USA
harpale@ge.com
https://www.cs.cmu.edu/ aharpale

Abstract. Industrial equipment, devices and patients typically undergo change from a healthy state to an unhealthy state. We develop a novel approach to detect unhealthy entities and also discover the time of change to enable deeper investigation into the cause for change. In the absence of an engineering or medical intervention, health degradation only happens in one direction — healthy to unhealthy. Our transductive learning framework, known as max-margin temporal transduction (MMTT), leverages this chronology of observations for learning a superior model with minimal supervision. Temporal Transduction is achieved by incorporating chronological constraints in the conventional max-margin classifier — Support Vector Machines (SVM). We utilize stochastic gradient descent to solve the resulting optimization problem. We prove that with high probability, an ϵ-accurate solution for the proposed model can be achieved in $O\left(\frac{1}{\lambda\epsilon}\right)$ iterations. The runtime is $O\left(\frac{1}{\lambda\epsilon}\right)$ for the linear kernel and $O\left(\frac{n}{\lambda\epsilon}\right)$ for a non-linear Mercer kernel, where n is the number of observations from all entities — labeled and unlabeled. Our experiments on publicly available benchmark datasets demonstrate the effectiveness of our approach in accurately detecting unhealthy entities with less supervision as compared to other strong baselines — conventional and transductive SVM.

1 Introduction

Early detection of imminent disorders from routine measurements and checkups is an important challenge in the medical and prognostics research community. Multivariate observations from routine doctor visits are typically available for human patients. Similarly, sensor data is typically available through the life of industrial equipment, from the healthy to the unhealthy state. If a certain patient or equipment is diagnosed with a severe disorder, can we look back into the historical measurements to detect the earliest indicators of ensuing problems? Identification of such change-points from the past might lead to the discovery of external stimuli or operating conditions that were the root cause for the eventual failure. For example, the doctor may question the exposure of the individual to certain geographies and an engineer might investigate the operating conditions around the time of change.

Also, empowered by few such diagnoses, can we identify other individuals or machinery with incipient faults, to enable prevention? It should be noted that we seldom have examples of entities that have been deemed as healthy. A doctor might get

G. Nicosia et al. (Eds.): LOD 2021, LNCS 13164, pp. 178–192, 2022.
https://doi.org/10.1007/978-3-030-95470-3_13

visits from only people that are already symptomatic or ill, and others might hold on for their next routine visit. Similarly, industrial equipment may experience continued usage till the problem has already escalated, and customers might complain only about non-functioning units. Routine measurements are cheap, but expensive investigative analyses are usually not performed on asymptomatic entities. Thus, for most purposes, any asymptomatic entities are assumed to have an *unknown* health status.

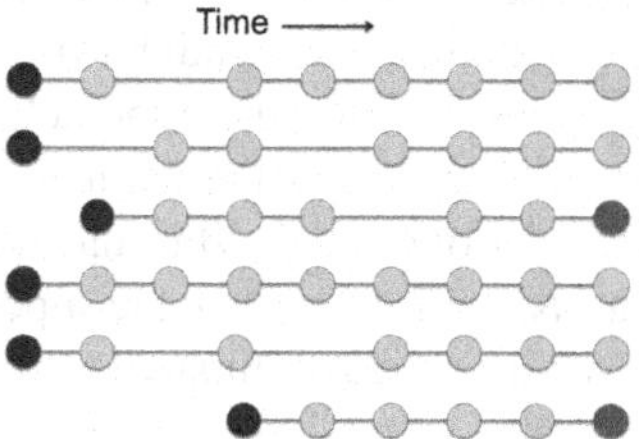

Fig. 1. The scenario for temporal transduction. Each row is a single entity (patient or equipment). Each circle is a multivariate measurement or observation from that entity. Along each row, the observations are chronologically ordered, with the one on left appearing first. Green indicates healthy state, red indicates unhealthy. Gray means measurements were made as part of routine checkups, but occurrence of disorder (health state) was unknown at that time. (Color figure online)

We depict this problem setting in Fig. 1. In this work, we assume that initially all the entities are healthy. For industrial equipment, this might be a safe assumption, since every new equipment is thoroughly inspected before being deployed. While doctors might perform a comprehensive examination before admitting a new patient, we understand that this assumption may not hold for disorders that were established before the first measurement. We will not address the setting of established (existed before the first measurement), disorder in this work. Additionally, in this setting, note that entities may not be monitored at the same intervals. Also, the time-span of available observations might be different. Only few entities may have known final unhealthy diagnoses, but the time of change is unknown.

We address this challenge using transductive learning [2,4,6] to model the problem setting with temporal information. Transductive learning is similar to semi-supervised learning, in its use of unlabeled data for inferring a better model with minimal supervision. It differs from semi-supervised learning by using test data as the unlabeled data, instead of any randomly chosen unlabeled data for model inference. The model is optimized for better performance on the unlabeled data used for training, not any other held out dataset. In our problem setting, the labeled set consists of two parts: the initial measurements from all entities (green circles) and the final measurements from the entities with known unhealthy states (red circles). The unlabeled set comprises of all the intermediate measurements and final observations from entities with unknown health status (gray circles). We are interested in classifying this unlabeled set for two reasons: Firstly, we wish to classify the final states of entities with unknown final diagnoses to identify those with disorders. Secondly, we wish to classify all the intermediate observations to identify change points — the time when the entity changed state from healthy to unhealthy. Thus, the unlabeled set is also the test set, hence the transductive learning setting.

In this paper, we propose a novel maximum margin classifier inspired by the popular SVM classifier [10] and its transductive counterpart [6]. We call this approach Max-Margin Temporal Transduction (MMTT). In addition to traditional constraints for maximum margin classifiers, MMTT incorporates a new constraint that penalizes violation in the chronology of events — the entity cannot go from unhealthy to healthy without intervention. We design a stochastic gradient descent approach to make our approach scalable. We also demonstrate that the approach can utilize not only linear, but also non-linear kernels, and in both cases, the number of iterations needed to infer an ϵ-accurate model are of the order of $O\left(\frac{1}{\lambda\epsilon}\right)$. Through experiments on multiple publicly available benchmark datasets, we demonstrate the superior predictive performance of our approach with minimal supervision as compared to conventional and transductive SVMs.

2 Notation and Background

Consider the set of entities $\mathbb{X} = \{\mathbf{x}_1, \ldots, \mathbf{x}_N\}$. From our examples earlier, each entity denotes the element being monitored - a patient or a jet engine. $\mathbf{x}_i \in \mathbb{R}^{T_i \times D}$, where D is the dimensionality of the multivariate time-series and T_i is the length of time-series of that particular entity i. Thus, in Fig. 1, i-th row depicts $\mathbf{x}_i$ and t-th circle in the i-th row depicts the observation $\mathbf{x}_{i_t}$. Let $\mathbb{Y} = \{\mathbf{y}_1, \ldots, \mathbf{y}_N\}$ be the health-status or labels for each of the entities, where $\mathbf{y}_i \in \{-1, 0, +1\}^{T_i}$. Without loss of generality, -1 denotes healthy (initial) state, $+1$ denotes unhealthy (changed) state and 0 indicates unknown diagnoses.

We denote the set of entities with known final diagnoses by $\mathbb{K} \subset \mathbb{X}$, and usually $|\mathbb{K}| \ll |\mathbb{X}|$. Most importantly, only entities with a changed final state comprise $\mathbb{K}$, thereby $\mathbf{y}_{i_{T_i}} = +1, \forall \mathbf{x}_i \in \mathbb{K}$. All entities start from the healthy state, thus, $\mathbf{y}_{i_1} = -1, \forall \mathbf{x}_i \in \mathbb{X}$. Thus, the *training set* consists of $\mathbb{X}$, $\mathbf{y}_{i_1} = -1, \forall \mathbf{x}_i \in \mathbb{X}$, and $\mathbf{y}_{i_{T_i}} = +1, \forall \mathbf{x}_i \in \mathbb{K}$.

The goal then is two-fold:

- **Final label prediction:** Identify the health statuses of entities not in $\mathbb{K}$. That means, we are interested in finding $\mathbf{y}_{i_{T_i}}, \forall \mathbf{x}_i \in \mathbb{K}^{\mathtt{C}}$, where $\mathbb{K}^{\mathtt{C}} = \mathbb{X} \setminus \mathbb{K}$. We denote these predictions by $\widehat{\mathbf{y}_{i_{T_i}}}, \forall \mathbf{x}_i \in \mathbb{K}^{\mathtt{C}}$, with the goal of minimizing the prediction error $\ell(\mathbf{y}_{i_{T_i}}, \widehat{\mathbf{y}_{i_{T_i}}})$.
- **Change-point detection:** Identify the earliest point of change from healthy to unhealthy state in an unhealthy entity.

Both goals can be addressed by classifying all observations along each entity, $\mathbf{x}_{i_t}, \forall \mathbf{x}_{i_t} \in \mathbf{x}_i, \forall \mathbf{x}_i \in \mathbb{X}$. In other words, we would like to predict $\widehat{\mathbf{y}_{i_t}}, \forall i_t$. We call this **Entire series prediction**. Analyzing the predictions on the entire series can enable the discovery of change-points in health state.

3 Our Approach

We propose a max-margin model that is an extension of the popular support vector machine (SVM) [10]. The predictive model is

$$\widehat{\mathbf{y}_{i_t}} = \langle \mathbf{w}, \phi(\mathbf{x}_{i_t}) \rangle \tag{1}$$

where $\mathbf{w} \in \mathbb{R}^D$ are parameters of the model.[1] $\phi(\cdot)$ is a feature transformation. For inferring $\mathbf{w}$, we solve the following minimization problem.

$$\underset{\mathbf{w}}{\text{minimize}} f(\mathbf{w}) := \frac{\lambda}{2}\|\mathbf{w}\|^2 + \frac{\ell_{\mathbb{L}}(\mathbf{w})}{|\mathbb{L}|} + \frac{\ell_{\mathbb{U}}(\mathbf{w})}{|\mathbb{U}|} + \frac{\ell_{\mathbb{C}}(\mathbf{w})}{|\mathbb{C}|} \tag{2}$$

where,

$$\begin{aligned}
\ell_{\mathbb{L}}(\mathbf{w}) &= \sum_{(\mathbf{x},y)\in\mathbb{L}} \max\{0, 1 - y\langle \mathbf{w}, \phi(\mathbf{x})\rangle\} \\
\ell_{\mathbb{U}}(\mathbf{w}) &= \sum_{(\mathbf{x},y)\in\mathbb{U}} \max\{0, 1 - |\langle \mathbf{w}, \phi(\mathbf{x})\rangle|\} \\
\ell_{\mathbb{C}}(\mathbf{w}) &= \sum_{(\mathbf{x}_{i_t},\mathbf{x}_{i_{t+1}})\in\mathbb{C}} \max\{0, \langle \mathbf{w}, \phi(\mathbf{x}_{i_t})\rangle - \langle \mathbf{w}, \phi(\mathbf{x}_{i_{t+1}})\rangle\} \\
\mathbb{L} &= \{(\mathbf{x}_{i_t}, \mathbf{y}_{i_t}) : \mathbf{x}_{i_t} \in \mathbb{X}, \mathbf{y}_{i_t} \in \mathbb{Y}, \mathbf{y}_{i_t} \neq 0, \forall\} \\
\mathbb{U} &= \{(\mathbf{x}_{i_t}, \mathbf{y}_{i_t}) : \mathbf{x}_{i_t} \in \mathbb{X}, \mathbf{y}_{i_t} \in \mathbb{Y}, \mathbf{y}_{i_t} = 0, \forall\} \\
\mathbb{A} &= \mathbb{L} \cup \mathbb{U} \\
\mathbb{C} &= \{(\mathbf{x}_{i_t}, \mathbf{x}_{i_{t+1}}) : \mathbf{x}_{i_t} \in \mathbb{A}, \mathbf{x}_{i_{t+1}} \in \mathbb{A}\}
\end{aligned}$$

$\ell_{\mathbb{L}}$, $\ell_{\mathbb{U}}$ and $\ell_{\mathbb{C}}$ refer to the constraints arising from the labeled, unlabeled and chronological considerations, respectively. $\ell_{\mathbb{L}}$ is the usual hinge-loss utilized in the context of supervised SVM classifiers [10]. $\ell_{\mathbb{U}}$ is the unlabeled loss that attempts to maximize the margin with the help of unlabeled data, a formulation very commonly used in transductive SVMs [6]. Finally, $\ell_{\mathbb{C}}$ is the *chronological* loss which penalizes if the chronological constraints are violated: that means, the transition from one state (e.g. healthy to unhealthy) is strictly a one-way transition and any violation results in a penalty that needs to be minimized. Note that a better chronological loss can be achieved by defining the set $\mathbb{C} = \{(\mathbf{x}_{i_t}, \mathbf{x}_{i_k}) : \mathbf{x}_{i_t} \in \mathbb{A}, \mathbf{x}_{i_k} \in \mathbb{A}, \forall t < k\}$. Thus, instead of requiring just the local chronological ordering of labels, we could instead require that the prediction at a particular observation is consistent with all the observations that follow it. While attractive for the linear kernel, this comprehensive loss is likely to be computationally expensive. By design, our localized constraint leads to an efficient streaming algorithm that requires just an observation and its immediate neighbor, not the entire series, making it attractive for scalability.

[1] We have ignored the bias b in this work, although it is likely to only further improve the results beyond those presented here. Detailed discussion on the bias term appears in [10].

To enable scalability, we propose a stochastic gradient descent based solution for the optimization problem. The sub-gradient[2] at step s of Eq. 2, w.r.t. $\mathbf{w}_s$, in the context of a single training example $(\mathbf{x}_{i_t}, \mathbf{y}_{i_t})$, is:

$$\begin{aligned} \nabla_s &= \lambda \mathbf{w}_s + \nabla_s(\mathbf{w}_s; \mathbf{x}_{i_t}, \mathbf{y}_{i_t}) \\ &\qquad\qquad \text{where, } \nabla_s(\mathbf{w}_s; \mathbf{x}_{i_t}, \mathbf{y}_{i_t}) = \\ &\mathbb{1}[\langle \mathbf{w}_s, \phi(\mathbf{x}_{i_t}) \rangle > \langle \mathbf{w}_s, \phi(\mathbf{x}_{i_{t+1}}) \rangle](\phi(\mathbf{x}_{i_t}) - \phi(\mathbf{x}_{i_{t+1}})) \\ &- \begin{cases} \mathbb{1}[\mathbf{y}_{i_t} \langle \mathbf{w}_s, \phi(\mathbf{x}_{i_t}) \rangle < 1] \mathbf{y}_{i_t} \phi(\mathbf{x}_{i_t}), \text{ if } \mathbf{y}_{i_t} \neq 0 \\ \mathbb{1}[|\langle \mathbf{w}_s, \phi(\mathbf{x}_{i_t}) \rangle| < 1] \operatorname{sign}(\langle \mathbf{w}_s, \phi(\mathbf{x}_{i_t}) \rangle) \phi(\mathbf{x}_{i_t}), \text{ otherwise} \end{cases} \end{aligned} \tag{3}$$

where $\mathbb{1}[\cdot]$ is the indicator function which takes value 1 only if $\cdot$ is true, else it is 0. Equipped with this sub-gradient, we can now iteratively learn $\mathbf{w}$ using the update $\mathbf{w}_{s+1} \leftarrow \mathbf{w}_s - \eta_s \nabla_s$. We use $\eta_s = \frac{1}{\lambda s}$, similar to Pegasos SVMs. This leads to a standard SGD algorithm, presented in Algorithm 1.

Algorithm 1. TTSGD algorithm

procedure TTSGD($\mathbb{X}, \lambda, S$)
  $\mathbf{w}_0 \leftarrow 0$
  for $s \leftarrow 1, 2, \cdots, S$ **do** ▷ SGD Iterations
    Choose $i_t \in \{1, \cdots, |\mathbb{A}|\}$ uniformly at random
    $\eta_s \leftarrow \frac{1}{\lambda s}$
    $\delta \leftarrow \lambda \mathbf{w}_s$
    if $\langle \mathbf{w}_s, \phi(\mathbf{x}_{i_t}) \rangle > \langle \mathbf{w}_s, \phi(\mathbf{x}_{i_{t+1}}) \rangle$ **then**
      $\delta \leftarrow \delta + (\phi(\mathbf{x}_{i_t}) - \phi(\mathbf{x}_{i_{t+1}}))$
    end if
    if $(\mathbf{x}_{i_t}, \mathbf{y}_{i_t}) \in \mathbb{L}$ AND $\mathbf{y}_{i_t} \langle \mathbf{w}_s, \phi(\mathbf{x}_{i_t}) \rangle < 1$ **then**
      $\delta \leftarrow \delta - \mathbf{y}_{i_t} \phi(\mathbf{x}_{i_t})$
    end if
    if $(\mathbf{x}_{i_t}, \mathbf{y}_{i_t}) \in \mathbb{U}$ AND $|\langle \mathbf{w}_s, \phi(\mathbf{x}_{i_t}) \rangle| < 1$ **then**
      $\delta \leftarrow \delta - \operatorname{sign}(\langle \mathbf{w}_s, \phi(\mathbf{x}_{i_t}) \rangle) \phi(\mathbf{x}_{i_t})$
    end if
    $\mathbf{w}_{s+1} \leftarrow \mathbf{w}_s - \eta_s \delta$
  end for
  return $\mathbf{w}$ ▷ The model parameters
end procedure

Note that TTSGD will only work for the linear kernel $\phi(\mathbf{x}_{i_t}) = \mathbf{x}_{i_t}$. Next, we design a representation of $\mathbf{w}_s$ that enables the use of non-linear Mercer kernels with our proposed model for achieving non-linear transforms.

Lemma 1. *The contribution of a sub-gradient $\nabla_k(\mathbf{w}_k; (\mathbf{x}_k, \mathbf{y}_k))$ used for update in the k-th round is $\frac{1}{\lambda s}$ in the s-th round, where $s > k$*

[2] Uses: $\frac{\partial |u|}{\partial v} = \frac{u}{|u|} \frac{\partial u}{\partial v}; u \neq 0$.

Proof. A typical update using a single example can be written as: $\mathbf{w}_{k+1} = (1 - \frac{1}{k})\mathbf{w}_k - \frac{1}{\lambda k}\nabla_k(\mathbf{w}_s; (\mathbf{x}_k, \mathbf{y}_k))$. With every upcoming update $j = k+1, \cdots, s$, the factor $\nabla_k(\mathbf{w}_k; (\mathbf{x}_k, \mathbf{y}_k))$ will be multiplied by $(1 - \frac{1}{j})$, starting with the initial weight of $\frac{1}{\lambda k}$. Thus, at the s-th round of SGD , the resulting overall weight of the factor $\nabla_k(\mathbf{w}_k; (\mathbf{x}_k, \mathbf{y}_k))$ will be

$$\frac{1}{\lambda k}\prod_{j=k+1}^{s}\left(1 - \frac{1}{j}\right) = \frac{1}{\lambda k}\prod_{j=k+1}^{s}\left(\frac{j-1}{j}\right) = \frac{1}{\lambda s}$$

Lemma 2. *During any round of SGD, $\mathbf{w}_{s+1}$ can be represented as a linear combination of $\phi(\mathbf{x})$'s, for $\mathbf{x} \in \mathbb{X}$.*

Proof. From Lemma 1, irrespective of the round in which $\nabla_k(\mathbf{w}_k; (\mathbf{x}_k, \mathbf{y}_k))$ is first used for SGD update, its contribution is weighed by the factor $\frac{1}{\lambda s}$ in the s-the iteration. Also, every time $(\mathbf{x}_{i_t}, \mathbf{y}_{i_t})$ leads to an update (if a corresponding indicator is true), it contributes $\mathbf{y}_{i_t}\phi(\mathbf{x}_{i_t}), \mathrm{sign}(\langle \mathbf{w}_k, \phi(\mathbf{x}_{i_t})\rangle)\phi(\mathbf{x}_{i_t})$, or $(\phi(\mathbf{x}_{i_{t+1}}) - \phi(\mathbf{x}_{i_t}))$ respectively for the three components of the loss. Thus, after s rounds, if $(\mathbf{x}_{i_t}, \mathbf{y}_{i_t})$ has resulted in updates $\mathbf{l}_{i_t}, \mathbf{u}_{i_t}, \mathbf{c}_{i_t}$-times (for the 3 components of loss), then cumulatively, $\mathbf{w}_{s+1}$ can be summarized in terms of the number of times each observation $(\mathbf{x}_{i_t}, \mathbf{y}_{i_t})$ contributes to the updates

$$\mathbf{w}_{s+1} = \frac{1}{\lambda s}\left(\sum_{(\mathbf{x}_{i_t}, \mathbf{y}_{i_t}) \in \mathbb{L}} \mathbf{l}_{i_t}\mathbf{y}_{i_t}\phi(\mathbf{x}_{i_t}) + \sum_{(\mathbf{x}_{i_t}, \mathbf{y}_{i_t}) \in \mathbb{U}} \mathbf{u}_{i_t}\phi(\mathbf{x}_{i_t}) + \sum_{(\mathbf{x}_{i_t}, \mathbf{x}_{i_{t+1}}) \in \mathbb{C}} \mathbf{c}_{i_t}\left(\phi(\mathbf{x}_{i_{t+1}}) - \phi(\mathbf{x}_{i_t})\right)\right) \quad (4)$$

From Lemma 2, the predictive model in Eq. 1, can be written as

$$\begin{aligned}\widehat{\mathbf{y}_{j_{t'}}} &= \mathbf{w}_S\phi(\mathbf{x}_{j_{t'}}) \\ &= \frac{1}{\lambda S}\left(\sum_{(\mathbf{x}_{i_t}, \mathbf{y}_{i_t}) \in \mathbb{L}} \mathbf{l}_{i_t}\mathbf{y}_{i_t}\mathcal{K}(\mathbf{x}_{i_t}, \mathbf{x}_{j_{t'}}) + \sum_{(\mathbf{x}_{i_t}, \mathbf{y}_{i_t}) \in \mathbb{U}} \mathbf{u}_{i_t}\mathcal{K}(\mathbf{x}_{i_t}, \mathbf{x}_{j_{t'}}) \right. \\ &\quad \left. + \sum_{(\mathbf{x}_{i_t}, \mathbf{x}_{i_{t+1}}) \in \mathbb{C}} \mathbf{c}_{i_t}(\mathcal{K}(\mathbf{x}_{i_{t+1}}, \mathbf{x}_{j_{t'}}) - \mathcal{K}(\mathbf{x}_{i_t}, \mathbf{x}_{j_{t'}}))\right) \quad (5)\end{aligned}$$

where, we have used the kernel representation $\mathcal{K}(a, b) = \langle \phi(a), \phi(b)\rangle$. Since the resulting predictive model is inner products on the feature transforms ϕ, we can easily utilize complex Mercer Kernels for non-linear transforms of the input feature space. This approach is described in Algorithm 2. Note that we are still optimizing for the primal, but due to the nature of the subgradient, we are able to utilize kernel products. In implementation, $\mathbf{w}_s$ is never explicitly calculated, but rather, $\langle \mathbf{w}_s, \phi(\mathbf{x}_{i_t})\rangle$ is estimated as $\widehat{\mathbf{y}_{i_t}}$, directly using Eq. 5.

Algorithm 2. TTSGD-Kernel algorithm

procedure TTSGDKERNEL($\mathbb{X}, \lambda, J$)
  $\mathbf{l}, \mathbf{u}, \mathbf{c} \leftarrow 0$
  for $s \leftarrow 1, 2, \cdots, S$ **do** ▷ SGD Iterations
    Choose $i_t \in \{1, \cdots, |\mathbb{A}|\}$ uniformly at random
    if $\langle \mathbf{w}_s, \phi(\mathbf{x}_{i_t}) \rangle > \langle \mathbf{w}_s, \phi(\mathbf{x}_{i_{t+1}}) \rangle$ **then**
      $\mathbf{c}_{i_t} \leftarrow \mathbf{c}_{i_t} + 1$
    end if
    if $(\mathbf{x}_{i_t}, \mathbf{y}_{i_t}) \in \mathbb{L}$ AND $\mathbf{y}_{i_t} \langle \mathbf{w}_s, \phi(\mathbf{x}_{i_t}) \rangle < 1$ **then**
      $\mathbf{l}_{i_t} \leftarrow \mathbf{l}_{i_t} + 1$
    end if
    if $(\mathbf{x}_{i_t}, \mathbf{y}_{i_t}) \in \mathbb{U}$ AND $|\langle \mathbf{w}_s, \phi(\mathbf{x}_{i_t}) \rangle| < 1$ **then**
      $\mathbf{u}_{i_t} \leftarrow \mathbf{u}_{i_t} + 1$
    end if
  end for
  return $\mathbf{l}, \mathbf{u}, \mathbf{c}$ ▷ The model parameters
end procedure

3.1 Analysis

Next, we analyze the convergence rate of the proposed optimization algorithm to the true parameter $\mathbf{w}^*$, and show that this rate is independent of the dataset size, labeled as well as unlabeled, but rather depends on the regularization parameter λ and the desired accuracy ϵ. We start with restating a key Lemma from [10].

Lemma 3 ***(Lemma 1 in*** **[10]).** *Let $f_1, \ldots, f_S$ be a sequence of λ-strongly convex functions*[3]*, and $\mathbf{w}_1, \ldots, \mathbf{w}_S$ be a sequence of vectors such that $\mathbf{w}_s \in \mathbb{R}^D, \forall s$. Also, let $\|\phi(\mathbf{x})\| \leq R$ for some $R \geq 1, R \in \mathbb{R}$. Let ∇_s belong to the subgradient set of f_s at $\mathbf{w}_s$, and $\mathbf{w}_{s+1} = \mathbf{w}_s - \eta_s \nabla_s$. If $\|\nabla_s\| \leq L, \forall s$, then for $\mathbf{w}^* = \arg\min_{\mathbf{w}} f(\mathbf{w})$ and $S > 3$,*

$$\sum_{s=1}^{S} f_s(\mathbf{w}_s) \leq \sum_{s=1}^{S} f_s(\mathbf{w}^*) + \frac{L^2 \ln(S)}{\lambda} \tag{6}$$

Lemma 4 ***(Theorem 2 in*** **[7]).** *If $f : \mathbb{R}^D \to [0, B]$, and $f(\mathbf{w})$ is a λ-strongly convex function with a Lipschitz constant L w.r.t the norm $\|\cdot\|$, i.e. $\forall \mathbf{w}_s \in \mathbb{R}^D, \|\nabla_s\| \leq L$, then, we have with probability at least $1 - 4\ln(S)\delta$,*

$$\frac{1}{S}\sum_{s=1}^{S} F(\mathbf{w}_s) - F(\mathbf{w}^*) \leq \frac{Reg_S}{S} + 4\sqrt{\frac{L^2 \ln(1/\delta)}{\lambda}}\frac{\sqrt{Reg_S}}{S} + \max\left\{\frac{16L^2}{\lambda}, 6B\right\}\frac{\ln(1/\delta)}{S} \tag{7}$$

where $Reg_S = \sum_{s=1}^{S} f(\mathbf{w}_s) - \sum_{s=1}^{S} f(\mathbf{w}^*)$, is the regret of the online learning algorithm.

[3] A function $f(\mathbf{w})$ is called λ-strongly convex if $f(\mathbf{w}) - \frac{\lambda}{2}\|\mathbf{w}\|^2$ is a convex function.

$F(\mathbf{w}_s) = E_{s-1}[f(\mathbf{w}_s; Z_s]$, *where* $E_{s-1}[\cdot]$ *indicates the conditional expectation w.r.t.* $Z_1, \ldots, Z_{s-1}$, *and* Z_s *is the data seen during iteration* s *of the online algorithm.*

Theorem 1. *For* $R = 1$, *small enough* λ *and* $S \geq 3$, *we have with probability at least* $1 - \delta$,

$$\frac{1}{S}\sum_{s=1}^{S} F(\mathbf{w}_s) - F(\mathbf{w}^*) = \mathrm{O}\left(\frac{\ln \frac{S}{\delta}}{\lambda S}\right)$$

Proof. It is trivial to show that our optimization function, $f(\mathbf{w})$, is a λ-strongly convex function. For a linear kernel, $\phi(\mathbf{x})$ is $\mathbf{x}$, so $\|\mathbf{x}\| \leq R$ can be enabled by normalizing $\mathbf{x}$ appropriately. In this work we utilize the non-linear Radial Basis Function kernel, and it is indeed the case[4] that $\|\phi(\mathbf{x})\| \leq R$. Using this and the Eq. 4, one can show that $\|\mathbf{w}_s\| \leq \frac{3R}{\lambda}$. This and the definition of ∇_s imply that $\|\nabla_s\| \leq 6R$. Thus we have shown $\|\nabla_s\| \leq L$, and in our case $L = 6R$, satisfying the Lipschitz criterion. Thus, our minimization problem satisfies all conditions for Lemma 3, and hence it is applicable.

Thus, from Lemma 3, we can see that for our problem $\text{Reg}_S \leq \frac{L^2 \ln(S)}{\lambda}$. Also, from the prior discussion, $\|\mathbf{w}_s\| \leq \frac{3R}{\lambda}$, and consequently, from Eq. 2, $f(\mathbf{w}) \in [0, \frac{9R^2}{2\lambda} + 4]$. Thus, our minimization problem satisfies the criteria for Lemma 4. Plugging in the appropriate values for L, Reg_S, and B, for $S \geq 3$, we arrive at the result.

For the TTSGD implementation in Algorithm 1, we are thus guaranteed to find an ϵ-accurate solution in $\mathrm{O}\left(\frac{1}{\lambda\epsilon}\right)$ iterations, with high probability. Just like the popular Pegasos [10] algorithm, the number of iterations is independent of the number of examples (labeled or unlabeled), but rather depends on the regularization and desired accuracy. For the TTSGD-Kernel approach in Algorithm 2, the runtime will depend on the number of observations, due to the $\min(s, |\mathbb{L}| + |\mathbb{U}| + |\mathbb{C}|)$ kernel evaluations at iteration s, bringing the overall runtime to $\mathrm{O}\left((|\mathbb{L}| + |\mathbb{U}| + |\mathbb{C}|)/\lambda\epsilon\right)$. The bounds derived above are for the average hypothesis, but in practice the performance of the final hypothesis is often better, as shown in [10]. For this work, we will use the final hypothesis for prediction.

4 Experiments

4.1 Baselines

In order to assess the incremental benefit of using temporal transduction, particularly the addition of the chronological constraint, we compare the proposed approach with related strong baselines:

- **SVM:** In this setting, the model is learnt merely from the available labeled set $\mathbb{L}$. We used the popular Pegasos SVM implementation.

[4] $\|\phi(\mathbf{x})\|^2 = \langle \phi(x)\phi(x) \rangle = \exp\left(\frac{-\|\phi(\mathbf{x}) - \phi(\mathbf{x})\|^2}{2\sigma^2}\right) = 1.$

– **TSVM:** In addition to the labeled set, we employ a transductive SVM (TSVM) on the observations from each entity. For TSVM, we used the transduction algorithm that is included with the popular SVM-light implementation.[5]

Chosen baselines utilize the same underlying predictive model as ours in Eq. 1. This choice is deliberate, as the goal is to compare the benefit of the chronological constraint on a similar model, not really compare predictive model families. Intuitively, the first baseline (SVM) uses the first constraint $\ell_{\mathbb{L}}$ in Eq. 2, TSVM uses $\ell_{\mathbb{L}}$ and $\ell_{\mathbb{U}}$, and our approach uses all three $\ell_{\mathbb{L}}, \ell_{\mathbb{U}}$, and $\ell_{\mathbb{C}}$. Additionally, as a straw-man baseline, we also use a *stratified*-dummy classifier that classifies randomly according to the distribution of the two categories in the labeled set. For this purpose, we use the *DummyClassifier(strategy = 'stratified')* routine from the scikit-learn library.

4.2 Experimental Setup

We developed our system in Python and used the scikit-learn[6] library of algorithms. All the reported results have been averaged over 100 runs, each starting with a randomized data-generation step based on Algorithm 3, and then proceeding to randomly hide the desired fraction of labels. Tuning the regularization parameter λ and penalization parameter C for the various approaches was performed as an inner cross-validation loop, using only the available training set. The experiments were run on a 12-core Mac-OSX server, with 64 GB of random access memory. For reproducibility, code and datasets are available for download.[7]

4.3 Datasets

The approach was designed on a real use case from the industry for equipment monitoring. However, we are unable to make the proprietary data available for public use. To facilitate reproducibility of results and continued research, we present results on a publicly available benchmark collection from the NASA Prognostics and Healthy Monitoring repository. Also, to enable a deeper investigation into the strengths and weaknesses of the algorithm, we propose an approach for transforming well-known binary classification datasets into the problem setting of interest.

4.4 Turbofan Engine Degradation

This dataset, released publicly by the NASA prognostics center of excellence, consists of run-to-failure observations of aircraft gas turbine engine [9]. In the seven years of its existence, it has been extensively used in over 70 publications for comparing performance of prognostics, classification and fault discovery. There are about 100 unique aircraft engines in the dataset and 21 sensor measurements are recorded at every cycle of engine operation. Engine lifespans are in the range 128–525 cycles per engine.

[5] http://svmlight.joachims.org/.

[6] http://scikit-learn.org.

[7] https://www.dropbox.com/s/z9kfii51nvc7izq/mmtt.zip?dl=0.

We use this dataset for the task of identifying engines that are near failure. Imagine an airline operator that is managing a fleet of several engines. One of them fails for an unforeseen reason (first unhealthy example), and immediately the operator is tasked with discovering others that are near failure. This enables the operator to control further damage, manage inventory, and minimize downtime. We model our experiment to this setting, by providing only one example of *unhealthy* engine in the training set to the compared methods. The data from each engine is randomly truncated at a randomly chosen cycle number. This ensures that lengths of time-series available for healthy estimation is different per engine. Class labels in the ground truth set are assigned based on remaining useful life of each engine. Engines that are failure free for the next 20 cycles are deemed *healthy* for further operation, while others are deemed unhealthy, and might need immediate repair. These settings accurately model a practical scenario observed in the industry.

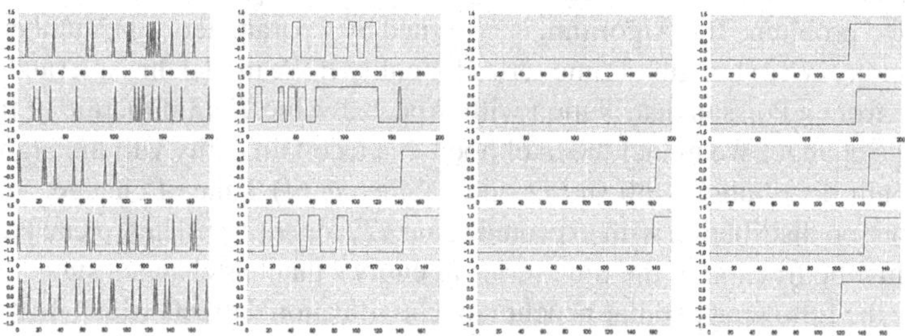

Fig. 2. Comparison of time-series predictions of health statuses of engines for each of the methods on 5 example unhealthy engines. From the left: Stratified, SVM, TSVM, and MMTT

In Fig. 2, we compare the full time-series predictions of the methods on few example unhealthy engines. Clearly, the fluctuation is significant in the predictions of the SVM classifier, but the MMTT method provides a one step change prediction, as is desirable in this problem setting. These average number of fluctuations in the predictions for a time-series have been summarized in our evaluation metric called *jitter*, which measures the average number of change-points in the predictions of a method along a time series. The results presented in Table 1 have been averaged over 100 random runs of this dataset. It can be observed that the performance of MMTT is significantly better than that of the other baselines. Although TSVM achieves the same jitter as that of MMTT, it is likely attributable to the failed detection of unhealthy engines, as evidenced in the accuracy and f1-score comparisons.

Note that the true time of change of state is unknown for this dataset, so claims about change-point detection are not possible. For these and further investigation, we present controlled experiments next.

Table 1. Engine dataset results. Best performance has been highlighted with a *

Metric	Stratified	SVM	TSVM	MMTT
Accuracy	0.13	0.80	0.72	0.85*
F1 score	0.35	0.74	0.65	0.78*
Jitter	24	6	1*	1*

4.5 Controlled Experiments

Intuitively, the time series of each entity can start from observations from one class (the healthy one), and then for a select group of entities change over to the other class (unhealthy one). Generating datasets in this manner enables the accurate identification of change-points, since ground truth about change is available for evaluation. Algorithm 3 describes the process of generating the state-change dataset given any binary classification problem. The Algorithm is governed by 4 parameters: (1) Number of entities to generate, N (2) Length of time series for entity i: We model this as a random variable drawn from a Poisson distribution with expected value T. (3) Fraction of instances that undergo change: We model the likelihood that a certain entity will undergo change as a Bernoulli distribution with success probability p. (4) Time of change: We model this as a Poisson distribution with expected value aT_i for entity i. Thus, a can be thought of as roughly the fraction of the time series of entity i that has changed state.

We use the following popular benchmark classification datasets which deal with the identification of onset of certain major disorders, and use Algorithm 3 to create custom datasets for controlled experiments. (a) Pima-Indian Diabetes [11]: Use glucose, plasma, serum and skin measurements from individuals of the Pima tribe to identify those which show signs of diabetes. (b) Parkinson Speech dataset [8]: Voice measurements from 31 people including 23 with Parkinson's disease, with roughly 6 measurements from each individual, to identify the diseased. (c) Ovarian Cancer [5]: Discover proteomic patterns in serum that distinguish ovarian cancer from non-cancer.

In addition to these datasets from the life-sciences community, we utilize some datasets that have been utilized for comparing SVM classifiers recently, namely the Adult, MNIST and USPS datasets used in the comparison of scalable SVM algorithms [1,10]. These are not life-sciences datasets, but are generally applicable as data that changes from one class to another over time, as enabled by Algorithm 3. We use the same version of these datasets as used in [1].

Table 2 describes the characteristics of the datasets. In our experiments, $T = 10, a = 0.5, p = 0.5$ and N was chosen to generate a wide variety of data. Note that the amount of data being classified by the classifier is approximately of the order of NT.

4.6 Results and Observations

In Fig. 3, we present the results comparing the accuracy of the various methods while increasing the number of known unhealthy entities. This is akin to the task of attempting to classify all monitored entities given a disease outbreak wherein, the information about few unhealthy entities becomes available at a time. The accuracy is reported on

Algorithm 3. Data generation process. N, T, p, a are pre-defined constants. N: total instances to generate. T: typical length of a series. p: fraction of instances that will change state. a: fraction of the time-series that denotes change. $\mathbb{C}_{-/+}$ is some input binary classification dataset

procedure DATAGEN($\mathbb{C}_-, \mathbb{C}_+$)
  $\mathbb{X} \leftarrow \{\}$
  $\mathbb{Y} \leftarrow \{\}$
  for $i \leftarrow 1$ to N **do**
    $T_i \sim \text{Poisson}(T)$ ▷ Length of time-series
    $\mathbf{y}_{i,T_i} \sim \text{Bernoulli}(p)$ ▷ Does $\mathbf{y}_i$ change state?
    if $\mathbf{y}_{i,T_i} \neq 0$ **then**
      $t_+ \sim \text{Poisson}(aT_i)$ ▷ Time of change
      $\mathbf{y}_{i,t} \leftarrow -1, \text{ if } 0 \leq t < t_+$
      $\mathbf{y}_{i,t} \leftarrow +1, \text{ if } t_+ \leq t \leq T_i$
      Draw $\mathbf{x}_{i,t}$ uniformly randomly from $\mathbb{C}_{y_{i,t}}$, with replacement
    else
      $\mathbf{y}_{i,t} \leftarrow -1, \text{ if } 0 \leq t \leq T_i$
      Draw $\mathbf{x}_{i,t}$ uniformly randomly from $\mathbb{C}_-$, with replacement
    end if
    $\mathbb{X} \leftarrow \mathbb{X} \cup \{\mathbf{x}_i\}$
    $\mathbb{Y} \leftarrow \mathbb{Y} \cup \{\mathbf{y}_i\}$
  end for
  return $\mathbb{X}, \mathbb{Y}$ ▷ The generated dataset
end procedure

being able to classify all entities at the final diagnoses, as well as all the intermediate predictions leading to the final state, the *entire series prediction* mentioned earlier in Sect. 2. It can be observed that MMTT outperforms all the other approaches, with minimal supervision. Barring the USPS dataset, the initial accuracy of MMTT is significantly superior. We attribute the particularly weak performance of svmlight, the conventional transductive baseline, to the need to maintain a balance of class proportions in the labeled and unlabeled sets. This assumption may not hold in our problem setting. The negative impact of this assumption is more pronounced in the straw-man baseline, the stratified classifier, which by definition, randomly assigns label by class proportions in the labeled

Table 2. Dataset characteristics

Dataset	# Observations	# Features	N
Diabetes	768	8	75
Parkinsons	195	22	20
Ovarian cancer	1200	5	10
Adult	48844	123	100
MNIST	70000	780	100
USPS	9298	256	100

set. Its performance worsens as more unhealthy instances are added to the labeled set, thereby skewing the class proportions. We discuss this challenge in detail in Sect. 4.7.

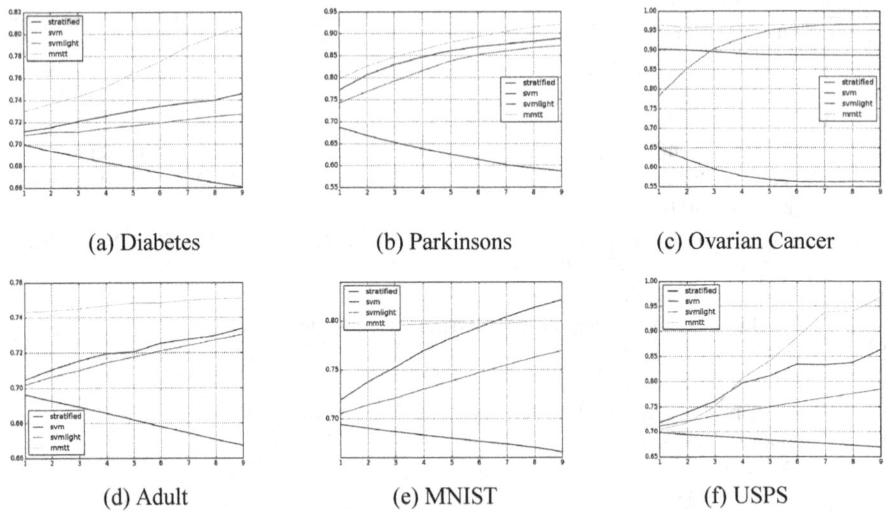

(a) Diabetes (b) Parkinsons (c) Ovarian Cancer

(d) Adult (e) MNIST (f) USPS

Fig. 3. Effect of increasing number of known unhealthy entities on accuracy.

Root-cause analysis is another use case where final diagnoses of many/all entities are available and the goal is to classify observations along the time-series for time-of-change identification, enabling investigation of external stimuli concurrent with time of change. In Fig. 4, we present the trends of accuracy on the various dataset when we have a knowledge of final diagnoses of entities of both kinds: those that end up healthy as well as unhealthy. The trends look similar to those in Fig. 3, albeit, the overall performance for all approaches has improved. Even with the knowledge of all final diagnoses, the performance of conventional transduction is sub-par compared to even the induction based simple SVM, and we again believe that the implicit strategy of attempting to maintain class proportions across labeled and unlabeled sets leads to poorer performance.

4.7 Need for a Balancing Constraint

To avoid problem of classifying all unlabeled instances to the majority class, balancing constraints have been used in most prior work on Transductive learning, for example by ensuring that the class proportions are maintained in the labeled and the unlabeled set [3,4,6]. To highlight the balancing problem with our approach, in Fig. 5, we present the predictions by the MMTT approach with increasing amounts of labeled information. The performance is very good when substantial supervision is available (2 rightmost images) in each chart. However, under minimalistic supervision (1 unhealthy known label), the entire dataset is predicted to be healthy. We believe this is because of the

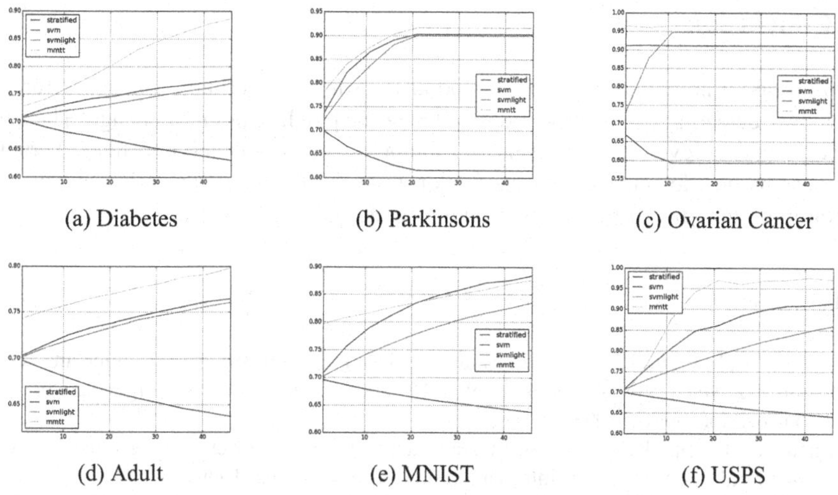

(a) Diabetes (b) Parkinsons (c) Ovarian Cancer

(d) Adult (e) MNIST (f) USPS

Fig. 4. Effect of increasing number of known entities (healthy and unhealthy) on accuracy

stringent chronological constraint that indirectly penalizes unhealthy predictions, but there is no penalty for predicting every observation as being healthy. While the problem is akin to the class balance situation highlighted in prior work on transductive learning, the remedy is not as simple as maintaining class proportions across labeled and unlabeled sets. The time-of-change is unknown (how many observations before and after change), and known unhealthy entities are disproportionate by design. At this stage, we want to highlight this as an open challenge for further exploration.

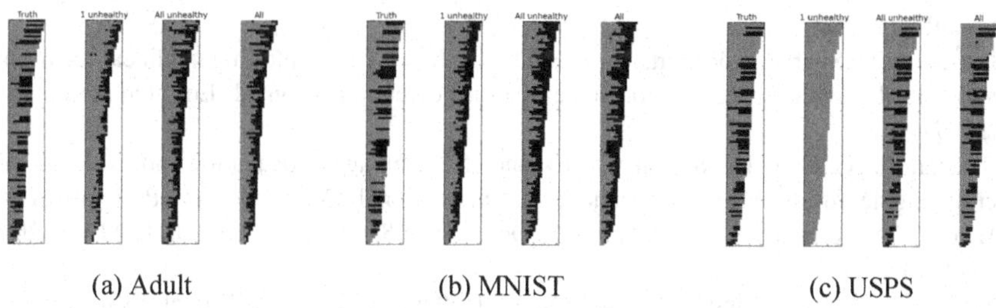

(a) Adult (b) MNIST (c) USPS

Fig. 5. The ground truth labels versus predictions at 3 stages: only 1 unhealthy entity known, all unhealthy entities known and final diagnoses of all entities are known. In each chart, each row is a time-series of health status (ground-truth or prediction), for a single entity. The entities in each dataset are sorted by length of time series T_i. Gray color indicates healthy, Black color indicates unhealthy, White is empty space (no recorded observations)

5 Conclusion

We proposed a novel approach for temporal transductive learning, especially for the problem of detecting changed entities and the corresponding point of change. The approach is scalable, even in the presence of temporal constraints and non-linear kernels. Experiments on life-science datasets demonstrate the potential for early detection of several important disorders such as cancer, diabetes and parkinsons.

References

1. Bordes, A., Ertekin, S., Weston, J., Bottou, L.: Fast kernel classifiers with online and active learning. J. Mach. Learn. Res. **6**, 1579–1619 (2005). http://leon.bottou.org/papers/bordes-ertekin-weston-bottou-2005
2. Chapelle, O., Vapnik, V., Weston, J.: Transductive inference for estimating values of functions. In: Advances in Neural Information Processing Systems (1999)
3. Chapelle, O., Zien, A.: Semi-supervised classification by low density separation. In: Proceedings of the Tenth International Workshop on Artificial Intelligence and Statistics, pp. 57–64 (2005)
4. Collobert, R., Sinz, F., Weston, J., Bottou, L.: Large scale transductive SVMs. J. Mach. Learn. Res. **7**, 1687–1712 (2006). http://dl.acm.org/citation.cfm?id=1248547.1248609
5. Petricoin, E.F., et al.: Use of proteomic patterns in serum to identify ovarian cancer. Lancet **359**, 572–577 (2002). https://doi.org/10.1016/S0140-6736(02)07746-2
6. Joachims, T.: Transductive inference for text classification using support vector machines. In: Proceedings of the Sixteenth International Conference on Machine Learning, ICML'99, pp. 200–209. Morgan Kaufmann Publishers Inc., San Francisco (1999). http://dl.acm.org/citation.cfm?id=645528.657646
7. Kakade, S.M., Tewari, A.: On the generalization ability of online strongly convex programming algorithms. In: Koller, D., Schuurmans, D., Bengio, Y., Bottou, L. (eds.) NIPS, pp. 801–808. Curran Associates, Inc. (2008). http://dblp.uni-trier.de/db/conf/nips/nips2008.html#KakadeT08
8. Little, M., McSharry, P., Roberts, S., Costello, D., Moroz, I.: Exploiting nonlinear recurrence and fractal scaling properties for voice disorder detection. Biomed. Eng. Online **6**(1), 23 (2007)
9. Saxena, A., Goebel, K., Simon, D., Eklund, N.: Damage propagation modeling for aircraft engine run-to-failure simulation. In: International Conference on Prognostics and Health Management, PHM 2008, pp. 1–9, October 2008. https://doi.org/10.1109/PHM.2008.4711414
10. Shalev-Shwartz, S., Singer, Y., Srebro, N., Cotter, A.: Pegasos: Primal Estimated sub-GrAdient SOlver for SVM. Math. Program. **127**(1), 3–30 (2011)
11. Smith, J.W., Everhart, J.E., Dickson, W.C., Knowler, W.C., Johannes, R.S.: Using the ADAP learning algorithm to forecast the onset of diabetes mellitus. J. Hopkins APL Tech. Dig. **10**, 262–266 (1988)

A Large Visual Question Answering Dataset for Cultural Heritage

Luigi Asprino[2], Luana Bulla[1], Ludovica Marinucci[1(✉)], Misael Mongiovì[1], and Valentina Presutti[1,2]

[1] ISTC - Consiglio Nazionale delle Ricerche, Rome and Catania, Italy
{luana.bulla,ludovica.marinucci,misael.mongiovi}@istc.cnr.it
[2] Università degli Studi di Bologna, Bologna, Italy
{luigi.asprino,valentina.presutti}@unibo.it

Abstract. Visual Question Answering (VQA) is gaining momentum for its ability of bridging Computer Vision and Natural Language Processing. VQA approaches mainly rely on Machine Learning algorithms that need to be trained on large annotated datasets. Once trained, a machine learning model is barely portable on a different domain. This calls for agile methodologies for building large annotated datasets from existing resources. The cultural heritage domain represents both a natural application of this task and an extensive source of data for training and validating VQA models. To this end, by using data and models from ArCo, the knowledge graph of the Italian cultural heritage, we generated a large dataset for VQA in Italian and English. We describe the results and the lessons learned by our semi-automatic process for the dataset generation and discuss the employed tools for data extraction and transformation.

Keywords: Visual Question Answering · Dataset · Cultural heritage

1 Introduction

Since its introduction a few years ago, Visual Question Answering (VQA) [6], the problem of generating the answer of a given question related to a certain image, has been intensively investigated by the machine learning community, achieving fascinating results (e.g., [10,11]). The necessity of accessing to training and validation data has stimulated the production of large datasets (e.g., [5]). Very recently, a raise of interest has been observed in developing computer science tools to engage people in museums and touristic sites, for instance automatic audio guides able to respond to user questions in natural language [9]. The application of VQA in this field presents some characteristics that differentiate it from the general VQA. The Cultural Heritage (CH) domain is particularly

This work was supported by the Italian PON project ARS01_00421: "IDEHA - Innovazioni per l'elaborazione dei dati nel settore del Patrimonio Culturale".
The authors are listed in alphabetical order.

G. Nicosia et al. (Eds.): LOD 2021, LNCS 13164, pp. 193–197, 2022.
https://doi.org/10.1007/978-3-030-95470-3_14

challenging due to the great variety of cultural assets that could potentially be considered. Moreover, if previous VQA approaches usually focus on answering questions that can be deduced by observing the image (e.g., how many people are depicted), in the CH domain answering meaningful questions related to an image generally requires background knowledge on the represented cultural property (e.g., who is the author of an artwork). Considering the peculiarity of the CH domain, models trained on general datasets barely adapt to this field. Specific datasets of question-answer pairs with associated images are limited to AQUA [4] and an annotated subset of Artpedia paintings [1].

The need for large datasets in the CH domain has motivated us to exploit the large amount of structured data in the ArCo Knowledge Graph [2] to produce a comprehensive VQA dataset, useful for training and evaluating VQA systems. ArCo consists of (i) a network of seven ontologies (in RDF/OWL) modeling the CH domain (with focus on cultural properties) at a fine-grained level of detail, and (ii) a Linked Open Data dataset counting ~200 M triples, which describe ~0.8 M cultural properties and their catalog records derived from the *General Catalog of Italian Cultural Heritage* (ICCD), i.e. the institutional database of the Italian CH, published by the *Italian Ministry of Culture* (MiC). ArCo ontology network is openly released with a CC-BY-SA 4.0 license both on *GitHub*[1] and on the official *MiC website*[2], where data can be browsed and acceded through the SPARQL query language.

Extracting information from ArCo to generate a dataset for VQA is not free of obstacles. First, ArCo does not give us a measure of which kind of questions might be interesting for average users in a real scenario. Second, ArCo data need to be suitably transformed and cleaned to produce answers in a usable form and questions need to be associated to corresponding answers. Third, the dataset we aim at generating is huge, and therefore manual validation of produced data cannot be performed.

We resorted to a semi-automatic approach that involves the collaboration of expert and non-expert users and the use of text processing and natural language processing techniques to obtain an accurate list of question-answer pairs. In the remainder of this paper, we discuss our methodology for the dataset generation and the characteristics of our resulting dataset.

2 Methodology

We considered a scenario where an image is associated to available knowledge either manually (e.g., artworks in a museum can be associated with their descriptions) or by object recognition (e.g., architectural properties identified by taking pictures), and generated a dataset as a list of question-answer pairs, each one associated to an image, a description and a set of available information. An instance of question-answer pair is: "Who is the author?" - "The author of the cultural asset is Pierre François Basan". A prediction model might build the

[1] https://github.com/ICCD-MiBACT/ArCo/tree/master/ArCo-release.

[2] http://dati.beniculturali.it/.

answer from the description, if available, or infer it by combining visual features (e.g., the painting style) with other information extracted from the description.

We adopted a semi-automatic approach consisting in two main steps. The first part of the process focused on generating a list of question templates with associated verbal forms by considering both expert and non-expert perspectives assessed by surveys. Then, for each question template, we automatically generated a list of question-answer pairs by combining question forms and associated answer templates with information from relevant cultural assets in ArCo, and accurately cleaning the results. This process was performed by an ad-hoc tool, developed following a build-and-evaluate iterative process. At each step we evaluated a sample of the produced dataset to propose new data cleaning rules for improving results. The process ended when the desired accuracy was achieved. Eventually, question-answer pairs from different question templates were combined. Next, we first detail our question templates generation process, then fully describe the question-answer pairs generation by drawing from question templates.

The *question templates generation* process was based on the following two perspectives carried out independently: a *domain experts' perspective*, represented by a selection of natural language competency questions (CQs) [7] previously considered to model the ArCo ontology network [2], and a *user-centered perspective*, represented by a set of questions from mostly non-expert (65 out of 104) users, collected through five questionnaires on a set of different images of cultural assets belonging to ArCo (five per questionnaire). In the questionnaires, the users were asked to formulate a number of questions (minimum 5, maximum 10) that they considered related to each image presented (questions they would ask if they were enjoying the cultural asset in a museum or a cultural site). In this way, we collected 2, 920 questions from a very heterogeneous group of users in terms of age (from 24 to 70 years old and 42 years average age), cultural background and interests. Then, the questions were semi-automatically analyzed and annotated in order to recognize their semantics, associate them (when possible) with ArCo's metadata, and create corresponding SPARQL queries for data extraction.

In the clustering process, we grouped user-produced questions into semantic clusters, named *question templates*, with the purpose of grouping together questions that ask for the same information. Clustering was first performed automatically by text analysis and sentence similarity, then validated and corrected manually. The automatic procedure consisted in the following steps. We initially aggregated sentences that resulted to be identical after tokenization, lemmatization and stop words removal. Then, for each question, we identified the most semantically similar one in the whole set by Sentence-BERT [8] and aggregated sentences whose similarity was above 84% (we found empirically that this value resulted in a low error rate). Eventually, we performed average linkage agglomerative clustering with a similarity threshold of 60%. To prepare for manual validation, we extracted a list of question forms, each one associated to a numerical ID representing the cluster it belongs to. Questions in the same cluster (e.g., "Who is the author?" and "Who made it?") were placed close to each other. After removing identical sentences, we obtained about 1659 questions, grouped in 126 clusters. Each question was then manually associated to a textual (human meaningful) ID (e.g., "AUTHOR") agreed by the annotators and a special "NODATA" ID (about 10%) was introduced for questions that refer to information that is not contained in ArCo. At the end of the process, after excluding clusters that ask for unavailable and unusable information in ArCo, we obtained 29 clusters, each of them representing a question template. Obtained question templates (labeled as "User") were aggregated with the ones from the domain experts (labeled as "Expert") obtaining 43 question templates, with 20 of them in common. Eventually, the experts defined an answer template and a SPARQL query for each question template.

We employed SparqlWrapper[3] for executing the SPARQL queries and extracting textual data and pictures from ArCo. We removed cultural assets that have zero or more than one associated pictures. For each record of the query results we generated a question-answer pair by randomly drawing a question form by the associated question cluster, with the same distribution of the results of the user questionnaires (frequently proposed questions are selected with higher probability), and building the associated answer from the answer template. In order to improve both the form of the answer itself and its rendering in its context, we adopted two approaches. We applied a set of cleaning rules, such as removing data with errors or marked as not validated and changing patterns of verbal forms (e.g., from "Baldin, Luigi" to "Luigi Baldin")[4], and employed pre-trained language models to improve the form of answers by adapting each sentence to its associated datum (e.g., Italian prepositions and articles have to be chosen according to the gender and number of corresponding nouns or adjectives). To solve this problem we applied the cloze task of BERT [3] on the generated answers, asking to infer words whose genre and number depend on the specific datum and cannot be previously determined. Furthermore, we applied a final grammar correction task by automatic translating the sentence from Italian to English and back to Italian by means of pre-trained language models for translation[5].

3 Results

The final dataset[6] contains 6.49M question-answer pairs covering cultural assets, 43 question templates and 282 verbal forms. The number of pairs per template ranges from 35 to 576 K. Each question-answer pair is associated with the corresponding cultural asset and its information, including its picture, a description and its URI in ArCo. In addition, on GitHub we provide two samples in Italian and English of 50 question-answer pairs per question template that we manually

[3] https://github.com/RDFLib/sparqlwrapper.

[4] A complete list is available on https://github.com/misael77/IDEHAdataset.

[5] https://huggingface.co/Helsinki-NLP/opus-mt-it-en and opus-mt-en-it.

[6] Available on GitHub https://github.com/misael77/IDEHAdataset.

evaluated. Results show an overall accuracy (percent of correct pairs) of 96,6% for the Italian sample, and of 93% for the English one. We also provide a table that reports, for each question template, its usage, the number of associated question forms, the number of question-answer pairs generated and the accuracy. Another table shows the breakdown of question-answer pair numbers by cultural asset type.

4 Conclusions and Future Work

The dataset we provide is the largest resource available for the training and validation of VQA models in the CH domain, which comprises 6.493.915 question-answer pairs, with associated visual, textual and structured information. We plan to apply this resource for training and evaluating a VQA system in the CH domain. We also plan to develop a second version of the dataset, which considers questions about restricted categories of cultural properties (e.g., paintings) in order to enhance the coverage of user questions related to specific types of cultural assets (e.g., who was it painted by?).

References

1. Bongini, P., Becattini, F., Bagdanov, A.D., Del Bimbo, A.: Visual question answering for cultural heritage. In: Proceeding of IOP Conference Series: Materials Science and Engineering (2020)
2. Carriero, V.A., et al.: ArCo: the Italian cultural heritage knowledge graph. In: Proceeding of ISWC, Part. II, pp. 36–52 (2019)
3. Devlin, J., Chang, M.W., Lee, K., Toutanova, K.: Bert: pre-training of deep bidirectional transformers for language understanding. In: Proceeding of NAACL-HLT, pp. 4171–4186 (2019)
4. Garcia, N., et al.: A dataset and baselines for visual question answering on art. In: Bartoli, A., Fusiello, A. (eds.) ECCV 2020. LNCS, vol. 12536, pp. 92–108. Springer, Cham (2020). https://doi.org/10.1007/978-3-030-66096-3_8
5. Krishna, R., et al.: Visual genome: connecting language and vision using crowdsourced dense image annotations. IJCV **123**(1), 32–73 (2017)
6. Malinowski, M., Fritz, M.: A multi-world approach to question answering about real-world scenes based on uncertain input. In: Proceedings of the NIPS, pp. 1682–1690 (2014)
7. Presutti, V., Blomqvist, E., Daga, E., Gangemi, A.: Pattern-based ontology design. In: Ontology Engineering in a Networked World, pp. 35–64 (2012)
8. Reimers, N., Gurevych, I.: Sentence-BERT: sentence embeddings using Siamese BERT-networks. In: Proceedings of the EMNLP (2019)
9. Seidenari, L., Baecchi, C., Uricchio, T., Ferracani, A., Bertini, M., Bimbo, A.D.: Deep artwork detection and retrieval for automatic context-aware audio guides. TOMM **13**(3s), 1–21 (2017)
10. Wang, P., Wu, Q., Shen, C., Hengel, A.V.D., Dick, A.: Explicit knowledge-based reasoning for visual question answering. In: Proceeding of IJCAI (2017)
11. Wu, Q., Teney, D., Wang, P., Shen, C., Dick, A., van den Hengel, A.: Visual question answering: a survey of methods and datasets. Comput. Vis. Image Underst. **163**, 21–40 (2017)

Expressive Graph Informer Networks

Jaak Simm(✉), Adam Arany, Edward De Brouwer, and Yves Moreau

ESAT-STADIUS, KU Leuven, 3001 Leuven, Belgium
{jaak.simm,adam.arany,edward.debrouwer,moreau}@esat.kuleuven.be

Abstract. Applying machine learning to molecules is challenging because of their natural representation as graphs rather than vectors. Several architectures have been recently proposed for deep learning from molecular graphs, but they suffer from information bottlenecks because they only pass information from a graph node to its direct neighbors. Here, we introduce a more expressive route-based multi-attention mechanism that incorporates features from routes between node pairs. We call the resulting method Graph Informer. A single network layer can therefore attend to nodes several steps away. We show empirically that the proposed method compares favorably against existing approaches in two prediction tasks: (1) 13C Nuclear Magnetic Resonance (NMR) spectra, improving the state-of-the-art with an MAE of 1.35 ppm and (2) predicting drug bioactivity and toxicity. Additionally, we develop a variant called injective Graph Informer that is provably more powerful than the Weisfeiler-Lehman test for graph isomorphism. We demonstrate that the route information allows the method to be informed about the *non-local topology* of the graph and, thus, it goes beyond the capabilities of the Weisfeiler-Lehman test. Our code is available at github.com/jaak-s/graphinformer.

Keywords: Graph neural networks · Graph attention · Self-attention · NMR · Bioactivity prediction

1 Introduction

Graphs are used as a natural representation for objects in many domains, such as compounds in computational chemistry or protein–protein interaction networks in systems biology. Machine learning approaches for graphs fall into two main lines of research: the *spectral* and the *spatial* approach. The spectral approach relies on the eigenvalue decomposition of the Laplacian of the graph and is well suited for problems involving a single fixed graph structure.

By contrast, *spatial* graph methods work directly on the nodes and edges of the graph. Convolutional neural networks (CNN) have inspired many spatial methods. Similarly to a local convolution filter running on the 2D grid of an image, spatial approaches update the hidden vectors of a graph node by aggregating the hidden vectors of its neighbors. Several spatial graph methods have

J. Simm and A. Arany—Contributed equally as first authors.

G. Nicosia et al. (Eds.): LOD 2021, LNCS 13164, pp. 198–212, 2022.
https://doi.org/10.1007/978-3-030-95470-3_15

been proposed, which vary in how to carry out this update. The most straightforward approach is to sum (or average) the hidden vectors of neighbors, and then transform the results with a linear layer and a nonlinearity (*e.g.*, as proposed in Neural Fingerprints [10] and Graph Convolutional Networks [19]). Instead of a simple dense feedforward layer, some methods use GRU-based gating [22], edge hidden vectors [17], and attention to the neighbors [29]. One of the advantages of spatial methods over spectral methods is that they can be straightforwardly applied to problems involving multiple graphs.

However, the update step of a node in spatial methods only has access to its own neighborhood, which limits the information flow throughout the whole graph. Increasing the accessible neighborhood of each node requires the stacking of several layers. The dilated filters in CNNs [32] are an example of solving this neighborhood limitation in images, by providing an efficient way to gather information over longer distances (see Fig. 1, (A) and (B)).

In this work, we propose a method that flexibly aggregates information over longer graph distances in one step, analogous to dilated filters. Our *Graph Informer* approach is inspired by the sequence-to-sequence transformer model [28]. The core contribution of our work is the introduction of *route-based multi-head self-attention* (RouteMHSA), which allows the attention mechanism to also access route information and, thus, base its attention scores both on the features of the nodes and the route between them. This enables Graph Informer to gather information from nodes that are not just direct neighbors. Furthermore, the proposed approach can straightforwardly use edge features, as they can be incorporated into the route features. In the case of chemical compounds, this allows using the annotation of the bonds between the atoms (single, double, aromatic, etc.). The central idea is illustrated in Fig. 1. Additionally, we investigate the expressiveness of Graph Informer, by describing a variant that is provably as powerful as the Weisfeiler-Lehman (WL) test and showing empirically that it can even go beyond the WL test.

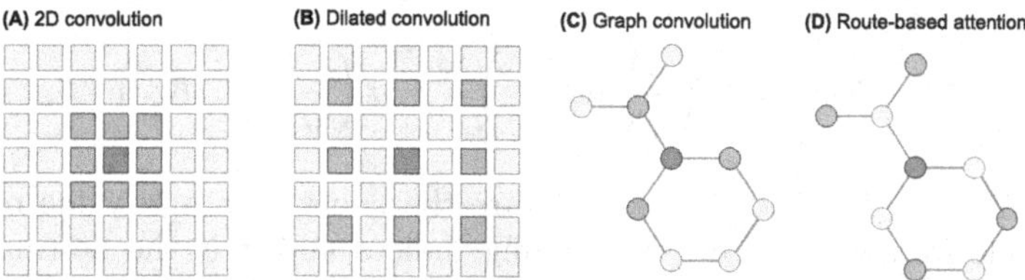

Fig. 1. Illustration of how route-based graph attention can mimic dilated filters in graphs. *Red* colored nodes are updated based on the vectors of the *blue* color nodes. (A) 2D convolutional filter, (B) 2D dilated filter, (C) graph convolution, (D) route-based graph attention. (Color figure online)

Our Graph Informer compares favorably with state-of-the-art graph-based neural networks on 13C NMR spectrum and drug bioactivity prediction tasks, which makes it suitable for both *node-level* and *graph-level* prediction tasks.

2 Proposed Approach

2.1 Setup

In this work, we consider learning problems on graphs where both nodes and edges are labeled. For a graph G with N nodes, we denote its node features by a matrix $X \in \mathbb{R}^{N \times F_{\text{nodes}}}$ and the route information by tensor $P \in \mathbb{R}^{N \times N \times F_{\text{route}}}$, F_{nodes} and F_{route} are the dimensions of the nodes and route features, respectively and the route information P is computed using the adjacency matrix of the graph and the edge labels. For example, for chemical compound graphs, where nodes are atoms, the vector between a pair of atoms in the route information tensor (*e.g.*, vector $P[i, j]$ for atoms i and j) can contain information on the type of route that connects the two atoms (rigid, flexible, aromatic) and how far they are from each other (shortest path).

We propose a method that works for both (1) *graph-level* tasks, such as graph classification where the whole graph G is assigned a label, and (2) *node-level* tasks where one should predict a value or a label for nodes (or subset of nodes) of the graph.

Due to the frequent use of tensors, in the following we adapt the Einstein summation convention, where the corresponding lower and upper indices are contracted (summed over). In case of matrices we use the matrix notation and the Einstein notation interchangeably for easier comparison to previous works, e.g. $AB = A_{ik}B_j^k$ corresponds to $\sum_k a_{ik}b_{kj}$. Note that $A_{ik}B_{jk}$ corresponds to a rank-3 tensor with elements $a_{ik}b_{jk}$ and not a matrix and that we consider the Euclidean metric tensor and hence use covariant (lower indices) and contravariant (upper indices) representations interchangeably.

2.2 Dot-Product Self-attention

Our approach is inspired by the encoder used in the transformer network [28], which proposed dot-product attention for sequence models formulated as

$$\text{Attn}(Q, K, V) = \sigma\left(\frac{1}{\sqrt{d_k}} QK^\top\right) V = \sigma\left(\frac{1}{\sqrt{d_k}} Q_{if} K_j^f\right) V_{f'}^j,$$

where $Q, K \in \mathbb{R}^{N \times d_k}$ and $V \in \mathbb{R}^{N \times d_v}$ are the query, key, and value matrices, i, j are node indices, f, f' are embedding indices and σ is the Softmax function, which makes sure that the attention probabilities for each node *sum to one*. This attention can be used as *self-attention* by projecting the input d-dimensional hidden vectors $H \in \mathbb{R}^{N \times d}$ into queries, keys and values. Q, K and V are computed from the input hidden vectors with corresponding weight matrices $W_Q \in \mathbb{R}^{d_k \times d}$, $W_K \in \mathbb{R}^{d_k \times d}$ and $W_V \in \mathbb{R}^{d_v \times d}$ (*i.e.*, $Q = HW_Q^\top$, $K = HW_K^\top$ and $V = HW_V^\top$). Intuitively, the i-th row of Q is the query vector of the i-th node (attending node), the j-th row of K is the key vector of node j (attended node) and the j-th row of V is the value vector for node j (value of the attended node).

To encode sequences with dot-product attention, the original transformer network [28] adds positional encodings to the hidden vectors H. In particular, sine waves with different frequencies and phases were used in the original paper to encode the position of the sequence elements. This allows the network to change attention probabilities based on the positions. However, as presented in the next section, such positional encoding is not directly applicable to graphs.

2.3 Route-Based Dot-Product Self-attention

One aspect to note is that in the case of graphs, there is no analogue for the global position of nodes. Therefore, we propose a novel *stable relative addressing* mechanism for graphs. The proposed mechanism adds a new component to the dot-attention that depends on the *features of the route* between two nodes in the graph. This route component is made up of a query and a key part. For the *route query* $Q_R \in \mathbb{R}^{N \times d_r}$ we map the hidden vectors of the nodes (H) with a matrix $W_Q^{\text{route}} \in \mathbb{R}^{d_r \times d}$

$$Q_R = H(W_Q^{\text{route}})^\top,$$

where d_r is the dimension of the route query and W_Q is the query weight matrix. Similarly, for the *route key* $K_R \in \mathbb{R}^{N \times N \times d_r}$ we map the route features P with a matrix $W_K^{\text{route}} \in \mathbb{R}^{d_r \times d}$

$$(K_R)_{ijf} = P_{ijh}(W_K^{\text{route}})_f^h,$$

where i, j refer to nodes dimensions, f to the embedding dimension and h to route features. The route information allows the attention mechanism to access the topology of the graph. Note that while Q_R is a matrix of route queries, one for each node of the graph, K_R is a tensor of rank 3, with each slice $K_R[i,j]$ corresponding to the route from node i to node j, and f indexing route embedding.

The route query and key can be now contracted into a $N \times N$ matrix of (logit) attention scores, which can be added to the original node-based scores $QK^\top$.

The intuition behind this is straight-forward: we compute the scalar product between the query vector $Q_R[i]$ of the node i and the key vector $K_R[i,j]$ (by contracting the index f) of the route from i to j.

Now the route-based attention probabilities $A \in [0,1]^{N \times N}$ are given by

$$A_{ij} = \sigma\left(\frac{1}{\sqrt{d_k + d_r}}\left(Q_{if}K_j^f + (Q_R)_{if}(K_R)_{ij}^f\right)\right), \tag{1}$$

where we also added the size of the route key d_r to the normalization for keeping the values within trainable (non-plateau) ranges of the softmax function. This enables the network to use both the information on the routes ($(Q_R)_{if}(K_R)_{ij}^f$), as well as the node hidden vectors ($Q_{if}K_j^f$), to decide the attention probabilities.

Analogously, we project the route information into the value component V of the dot-product attention. Firstly, we map the route data P into

$$(V_R)_{ijf} = P^h_{ij}(W^{\text{route}}_V)_{fh},$$

where $V_R \in \mathbb{R}^{N \times N \times d_v}$ is the tensor of rank 3 with slice $V_R[i,j]$ corresponding to the value vector for the route from node i to node j. The route value tensor V_R can then be weighted according to the attention probabilities A from Eq. 1 giving as the final route-based graph attention:

$$\text{RouteAttn}(Q, K, V, Q_R, K_R, V_R)_{if} = A_{ij}V^j_f + A_{i,j}(V_R)^j_{if}. \tag{2}$$

2.4 Locality-Constrained Attention

The route-based attention defined in Eq. 2 is general and allows unconstrained access throughout the graph. In some applications, localized access is preferred and this can be easily forced by adding a masking matrix to the attention scores:

$$(A_{\text{LC}})_{ij} = \sigma\left(\frac{1}{\sqrt{d_k + d_r}}\left(Q_{if}K^f_j + (Q_R)_{if}(K_R)^f_{ij}\right) + (M_{\text{route}})_{ij}\right), \tag{3}$$

where $M_{\text{route}} \in \{-\infty, 0\}^{N \times N}$ is 0 for unmasked routes and minus infinity for masked connections (in practice, we use a large negative value). This allows us, for example, to mask all nodes whose shortest distance from the current node is larger than 3 or some other number, thereby creating an attention ball around the current node. Similarly, it is possible to use the mask M_{route} to create an attention shell (*i.e.*, attending nodes within a given range of shortest distances).

In our experiments, we only used the attention ball with different values for the radius.

2.5 Implementation Details

As in the transformer architecture, we use multiple heads. Several route-based self-attentions are executed in parallel and their results are concatenated, allowing the network to attend to several neighbors at the same times.

Pool Node. Additionally, we introduce a *pool* node that has a different embedding vector than the graph nodes. The pool node has no edges to the graph nodes, but is unmasked by M_{route}, so that the attention mechanism can always attend and read, if required. This idea has two motivations. It allows the information to be easily shared across the graph and can be used as a "not found" answer when the attention mechanism does not find the queried node within the graph nodes.

Learnable Module. We use the attention probabilities A_{LC} for RouteAttn from Eq. 2, which takes in (1) hidden vectors of the nodes H, with size $(d, N_{\mathrm{nodes}}, N_{\mathrm{hidden}})$, (2) route information P, with size $(d, N_{\mathrm{nodes}}, N_{\mathrm{nodes}}, F_{\mathrm{route}})$, and (3) masks M_{node} and M_{route}. We use multiple heads and concatenate their results, giving us a route-based multi-head self-attention module (RouteMHSA):

$$\begin{aligned}\mathrm{RouteMHSA} &= \mathrm{Concat}(X_1, \ldots, X_{N_{\mathrm{head}}}) \\ \text{with } X_i &= \mathrm{RouteAttn}(HW_{i,Q}, HW_{i,K}, HW_{i,V}, \\ &\qquad HW_{i,Q}^{\mathrm{route}}, PW_{i,K}^{\mathrm{route}}, PW_{i,V}^{\mathrm{route}}),\end{aligned}$$

where W are the learnable parameters of the module.

The values for hidden size d, key size d_k, value size d_v, route key size d_r, and the number of heads N_{heads} can be chosen independently of each other. However, in our experiments we set d equal to $d_k N_{\mathrm{heads}}$, with $N_{\mathrm{heads}} \in \{6, 8\}$. We also use $d_v = d_k = d_r$.

Computational Complexity. The computational complexity of the RouteMHSA is quadratic with respect to the number of nodes in the graph, because the attention mechanism computes all pairwise attention probabilities. In practice, this means that the method can scale to graphs of a few hundred, maximally 1,000 nodes. This is more than enough for handling drug-like compounds, which typically have less than 100 (non-hydrogen) atoms.

3 Architecture of the Network

The architecture of Graph Informer is inspired by transformer networks [28], where the output of the multi-head attention is fed through a linear layer and added to its input, then a subsequent layer normalization [3] is applied. Additionally, a feedforward network (FFN) was applied to each hidden vector separately. However, in our graph experiments, for both node-level and graph-level tasks, we found this architecture to be quite difficult to train due to gradient flow issues.

Instead, we found that creating a residual-style network [14] with LayerNorm solved the gradient flow issue and was easy to train. The layer for this architecture can be expressed as

$$T = H + \mathrm{LayerNorm}(\mathrm{Linear}(\mathrm{RouteMHSA}(H))) \tag{4}$$

$$H' = T + \mathrm{LayerNorm}(\mathrm{FFN}(T)), \tag{5}$$

with H' being the output of the block (*i.e.*, updated hidden vectors). The architecture is depicted in Fig. 2. Graph Informer uses both neuron-level and channel-level dropout, with dropout rate 0.1.

4 Expressiveness of Graph Informer

Next, we investigate the general ability of Graph Informer to distinguish between distinct but similar graphs, also known as the graph isomorphism problem.

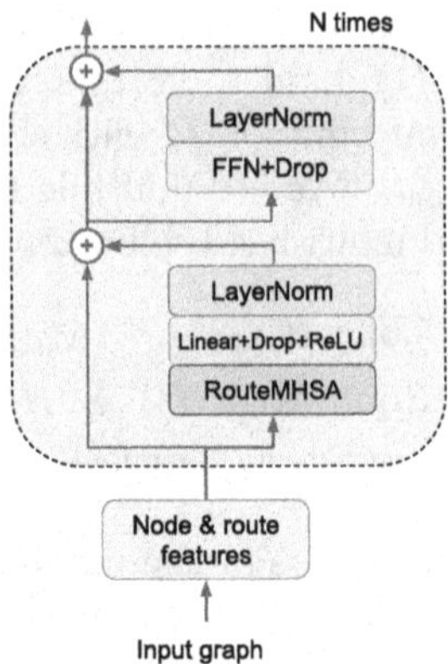

Fig. 2. Layer setup for the architectures of Graph Informer.

4.1 The Weisfeiler-Lehman Test

We propose a variant of our approach that is provably as powerful as the Weisfeiler-Lehman test of isomorphism of dimension 1 ($WL_{\text{dim}=1}$). Generalizing the notation of [31] to include route information, P, we write the update function of our model as

$$h_v^k = \phi\left(h_v^{k-1}, \left\{(h_u^{k-1}, P_{v,u}) : u \in \mathcal{N}(v)\right\}\right),$$

where $\left\{(h_u^{k-1}, P_{v,u}) : u \in \mathcal{N}(v)\right\}$ is the multiset of node vectors, h_u^{k-1}, and their route information vectors $P_{v,u}$, which can be attended by node v.

Such a network is provably as powerful as the Weisfeiler-Lehman test of isomorphism if the function $\phi(.)$ is injective[1] with respect to the hidden vectors h_v^{k-1} and h_u^{k-1} arguments, as follows straightforwardly from the theorem in [31].

RouteMHSA satisfies this condition with respect to the first argument (h_v^{k-1}) because one of the heads can attend to itself. However, to make it *provably* injective with respect to the second argument, we can replace the softmax mapping in Eq. 3 by an elementwise sigmoid. We call this version the *Injective* Graph Informer.

4.2 Beyond the Weisfeiler-Lehman Test

Several graph neural networks, including GIN [31] and NeuralFP [10], are limited in their expressiveness to the Weisfeiler-Lehman test of dim = 1. By contrast, the Graph Informer can aggregate route information and, therefore, has the capability to go beyond $WL_{\text{dim}=1}$ as the routes between the nodes reveal non-local topology. Indeed, we can show that Graph Informer has the following capability:

[1] We assume here that the output function is also injective.

Lemma 1. *Injective Graph Informer using the histogram of the lengths of all (possibly cyclic) routes between two nodes up to length N can distinguish any pair of graphs with different spectra.*

The proof of this lemma follows from the fact that injective Graph Informer is injective with respect to route features P and that graphs are cospectral (in terms of their adjacency matrix) if and only if $\mathrm{Tr}(A^r)$ are equal for both matrices for all values of r [2]. Notably, our method offers a more detailed view because it can also distinguish cospectral graphs (*e.g.*, the Tesseract (Q4) and Hoffman graphs, which are both 16-node and 4-regular).

We hypothesize that in practice this capability of Graph Informer gives strong additional expressiveness and makes it difficult to construct counterexamples because the route information allows the method to capture the non-local topology of the graph in detail.

5 Related Research

There are several neural network approaches that can directly work on the graph inputs. The key component in many of the works is to propagate hidden vectors of the nodes H using the adjacency matrix[2] A, either by summing the neighbor vectors, AH, or by averaging them as $\tilde{A}H$, where $\tilde{A} = A(D)^{-1}$ is the normalized adjacency matrix with D being the diagonal matrix of degrees. Neural Fingerprint [10] proposed to use the propagation AH, combined with a node-wise linear layer and a nonlinearity to compute fingerprints for chemical compounds. Based on the degree of the node, Neural Fingerprint uses a different set of linear weights. Similarly, Graph Convolutional Networks (GCN, [19]) propose to combine AH with a linear layer and a nonlinearity for semi-supervised learning, but by contrast with Neural Fingerprint, they use the same linear layer for all nodes, irrespective of their degree. Kearnes *et al.* [17] proposed the Weave architecture, where they use embedding vectors also for the edges of the graph and the propagation step updates nodes and edges in a coupled way.

In a separate line of research, [25] proposed Graph Neural Networks (GNN), which also use AH to propagate the hidden vectors between the nodes, but instead of passing the output to the next layer, the same layer is executed until convergence, which is known as the Almeida-Pineda algorithm [1]. A recent GNN-based approach is called Gated Graph Neural Networks (GGNN) [22], where at each iteration the neighborhood information AH is fed into a GRU [7]. An iterative graph network approach, motivated by the Message Passing algorithm in graphical model inference [24] was proposed by [8]. Gilmer et al. [11] generalized the existing approaches to the Neural Message Passing (NeuralMP) algorithm and showed the effectiveness of the approach for predicting simulated quantum properties based on 3D structures of compounds.

Hamilton et al. [13] propose a representation learning method, GraphSAGE, that does not require supervised signals and inductively embeds nodes in large graphs.

[2] The adjacency matrix here is assumed to include the self-connection (*i.e.*, $A_{ii} = 1$).

The method closest to ours is Graph Attention Networks (GAT) [29]. Similarly to us, they proposed to use multi-head attention to transform the hidden vectors of the graph nodes. However, GAT cannot integrate route features and its attention is only limited to the neighbors of every node. For graph isomorphism testing, this means it fails to separate any regular graphs, similar to $WL_{\text{dim}=1}$.

In contrast to the spatial methods reviewed above, spectral methods use the graph Fourier basis. One of the first spectral approaches to introduce this concept in the context of neural networks was SCNN [6]. To avoid the costly eigenvalue decomposition, Chebyshev polynomials were proposed to define localized filters [9]. Based on this work, proposed a neural network that combines spectral methods with metric learning for graph classification was proposed [20].

For sake of completeness, we note that another loosely related line of research has been working on the problem of transforming an input graph to a target graph and are known as *graph transformers* [5,33]. Those works usually do not use the attention mechanism used in the transformer network [28]. Yet, in a recent communication, also called graph transformers, [21] proposed a node-level attention mechanism for the same graph-to-graph task. In contrast, our model tackles node- or graph-level classification and regression tasks, and uses route-level attention mechanism.

In graph isomorphism research, as already mentioned, [31] showed that injective update rule and readout are crucial to guarantee expressiveness at the level of $WL_{\text{dim}=1}$. In a recent publication, [23] proposed a graph neural network as powerful as $WL_{\text{dim}=3}$.

In one of our main application areas, Nuclear Magnetic Resonance (NMR) chemical shift prediction (covered in Sect. 6.2), is an important element in structure elucidation. Recently, [15] proposed an imitation learning approach to solve this NMR inverse problem.

6 Evaluation

We compare our method against several baselines on two tasks. We first consider a *node-level* task where, given the graph of the compound, the goal is to predict 13C NMR peaks for all of the carbon atoms, see Sect. 6.2. For the *graph-level* tasks, we consider the drug–protein activity data set ChEMBL [4] and compound toxicity prediction from MoleculeNet [30], see Sect. 6.3. In all tasks and for all methods, we used the same feature set (see Appendix). For methods that could not use route features but supported edge features, we used the bond type (single, double, triple, aromatic). We consider an extensive set of baselines: GGNN [22], NeuralFP [10], GCN [19], Weave [17], GAT [29], NeuralMP [11] and GIN [31].

6.1 Model Selection

All methods used the validation set to choose the optimal hidden size from $\{96, 144, 192, 384\}$ and dropout from $\{0.0, 0.1\}$ [27]. For our method, attention radius from $\{1, 2, 3, 4\}$ and head count from $\{6, 8\}$ were also selected. All methods were trained with the Adam optimizer [18]. The validation set was used for early stopping (*i.e.*, we executed all methods for 100 epochs and picked the best model). At epochs 40 and 70, the learning rate was decreased by a factor of 0.3. Finally, the best early stopped model was used to measure the performance on the test set.

6.2 Node-Level Task: NMR 13C Spectrum Prediction

13C NMR is a cost-efficient spectroscopic technique in organic chemistry for molecular structure identification and structure elucidation. If 13C NMR measurements can be combined with a highly accurate prediction, based on the graph (structure) of the molecule, it is possible to validate the structure of the sample. It is crucial to have highly accurate predictions, because in many cases several atoms have peaks that are very close to each other. If such predictions are available, the chemist can avoid running additional expensive and time-consuming experiments for structure validation.

To create the data set, we used 13C NMR spectra from NMRShiftDB2.[3] After basic quality filtering 25,645 molecules remained with peak positions raging from –10 to 250 ppm (parts per million) with mean and standard deviation of 95 and 52 ppm.

In accordance to the most common Lorentzian peak shape assumption in NMR spectroscopy, we minimize the Mean Absolute Error (MAE) [12]. For evaluation, we split the data into train-validation-test sets containing 20,000, 2,500, and 3,145 molecules, respectively, and report results over 3 repeats with different splits. All methods use the same node-level output head, consisting of a tanh layer followed by a linear output layer. The MAE loss is minimized only for the atoms that have measurements for the 13C peaks (*i.e.*, the carbon atoms). For the results, see Table 1.

Our proposed Graph Informer reaches 1.35 MAE, which is also better than the state-of-the-art results from computational chemistry literature, which have reported 1.43 MAE for 13C for the NMRShiftDB2 [16]. At these levels of accuracy, many molecules with *densely packed* 13C NMR spectra can be resolved. This class of molecules with *densely packed* spectra are common for aromatic ring systems, which are prevalent in drug-like compounds. As an example, see Fig. 3a depicting 2-chloro-4-fluorophenol, which has three carbons, labelled 1, 3, 4 with peaks within the 115.2 to 116.6 ppm range. Figure 3b displays the predictions from the molecule when it was in the test set. The predictions match closely the true spectrum and, critically, predict the true ordering, which allows us to perfectly match peaks.

For a Graph Informer with 3 layers and 6 heads all having radius 2, we depict the attention probabilities in Fig. 4 for the bottom carbon (number 1 in Fig. 3a) of same 2-chloro-4-fluorophenol compound. We can see that all heads in Layer 1 strongly use the pool node as a reference. Also, we can see that in Layer 1 the

[3] Available here: https://nmrshiftdb.nmr.uni-koeln.de/.

Table 1. Test MAEs for NMR 13C spectrum prediction averaged over different folds.

Method	#layers	Test MAE
Mean model		46.894
GCN	2	6.284 ± 0.053
GCN	3	8.195 ± 0.385
GIN	2	5.015 ± 0.082
GIN	3	6.275 ± 0.306
GGNN	*na*	1.726 ± 0.055
NeuralMP	*na*	1.704 ± 0.025
GAT	1	2.889 ± 0.021
GAT	2	3.148 ± 0.067
GAT	3	5.193 ± 0.196
NeuralFP	1	3.839 ± 0.039
NeuralFP	2	2.248 ± 0.027
NeuralFP	3	2.040 ± 0.037
NeuralFP	4	1.966 ± 0.045
Graph Informer	1	1.827 ± 0.035
Graph Informer	2	1.427 ± 0.024
Graph Informer	3	1.375 ± 0.016
Graph Informer	4	**1.348** ± 0.007
State of the art [16]		1.43 ± NA

Table 2. AUC-ROC scores for multi-task datasets ChEMBL and Tox21 of MoleculeNet averaged over different folds.

Method	ChEMBL	Tox21
NeuralFP	0.826 ± 0.005	0.829
GCN	0.814 ± 0.005	0.809
GIN	0.816 ± 0.004	0.810
GGNN	0.756 ± 0.014	0.809
NeuralMP	0.707 ± 0.005	0.804
Weave	0.830 ± 0.006	0.822
Graph Informer	**0.839** ± 0.003	**0.848**

first head focuses on the atom itself, the second head attends to all neighbors within radius 2, and the third head focuses on the halogen atoms.

6.3 Results for Graph-Level Tasks

We benchmark the methods on two drug bioactivity data sets: ChEMBL v23 by [4] (94 tasks, 87,415 molecules, and 349,192 activity points) and Tox21 from MoleculeNet by [30] (12 tasks, 7,831 molecules, and 77,946 activity points). For ChEMBL, we used a repeated holdout-validation, and for Tox21 we used the single train-validation-test split given by MoleculeNet. In the case of ChEMBL, because of well-known series effect in drug design, allocating molecules randomly to the training and test sets results in overly optimistic and inconsistent performance estimates [26] because highly similar molecules end up in both training and test sets. Therefore, we apply *clustered* hold-out validation. We allocate the clusters (rather than individual molecules) to the training (60%), validation (20%), and test sets (20%). We repeat this split 5 times.

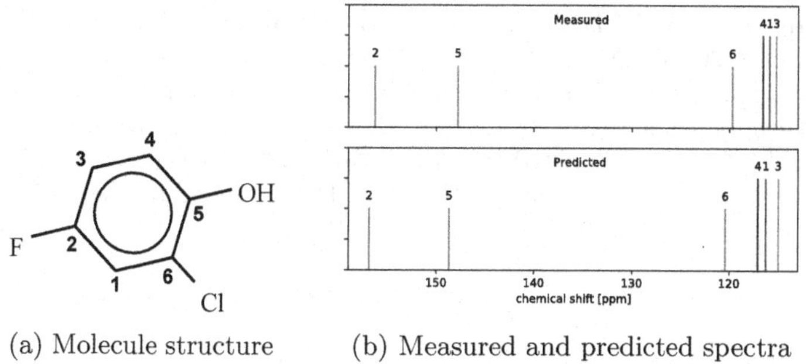

(a) Molecule structure (b) Measured and predicted spectra

Fig. 3. 13C NMR spectrum for 2-chloro-4-fluorophenol. The predicted NMR peaks are in the same order, even in the densely packed region (carbons 4, 1, 3).

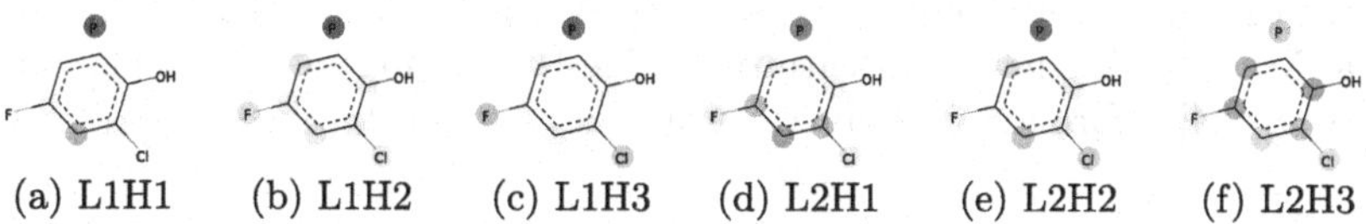

(a) L1H1 (b) L1H2 (c) L1H3 (d) L2H1 (e) L2H2 (f) L2H3

Fig. 4. Attention probabilities from carbon 1 (bottom in the figures) in the 2-chloro-4-fluorophenol molecule. We show selected three heads (H) from layer 1 and layer 2 (L).

Our and all adapted methods (GCN, GIN and GGNN) use a graph pooling layer consisting of a linear layer, ReLU, mean pooling, and a linear layer. We minimize the cross-entropy loss and Table 2 reports the AUC-ROC averages over tasks for each data set. Graph Informer outperforms all baseline methods in both ChEMBL and Tox21 data sets.

7 Conclusion

In this work, we proposed a route-based multi-head attention mechanism that allows the attention to use relative position and the type of connection between the pair of nodes in the graph. Because of its route features, our approach can incorporate information from non-direct neighbors efficiently—similar to the use of dilated filters in CNNs. In our theoretical analysis, we showed that a variant of Graph Informer is provably as powerful as the $WL_{\mathrm{dim}=1}$ test. Our empirical evaluation demonstrated that the proposed method is suitable both for node-level and graph-level prediction tasks, and delivers significant improvements over existing approaches in 13C NMR spectrum and drug bioactivity prediction.

Acknowledgments. YM is funded by Research Council KU Leuven: C14/18/092 SymBioSys3, C3/20/100 and CELSA/17/032; CELSA-HIDUCTION CELSA/17/032; FWO (Elixir Belgium I002819N and Infrastructure I002919N); Flemish AI Research Program; Authors are affiliated to Leuven.AI - KU Leuven institute for AI; VLAIO PM (HBC.2019.2528) and MaDeSMart (HBC.2018.2287); IMI initiative MELLODDY grant No 831472; EU H2020 grant No. 956832. EDB is funded by the FWO.

Appendix

Node and Route Features

In addition to the following features, we fed the one-hot encoding of the atom types by concatenating it to the node feature vector, see Table 3. The atom types that occurred in the two data sets are {C, N, O, Cl, F, S, I, Br, P, B, Zn, Si, Li, Na, Mg, K}. As the route features (Table 4) contain all edge labels (single, double, ...), all information about the graph topology is retained.

Table 3. Node features

Position	Description
0–2	Formal charge, one-hot encoded {−1, 0, +1}
3–7	Hybridization state, one-hot encoded {s, sp, sp2, sp3}
8–13	Explicit valence, one-hot encoded integer, between 0 and 5
14	Aromaticity, binary
15	Whether it is in a ring size 3, binary
16	Whether it is in a ring size 4, binary
17	Whether it is in a ring size 5, binary
18	Whether it is in a ring size 6, binary
19	Whether it is in any ring, binary
20	Partial charge, computed by Gasteiger method, real number
21	Whether it is a H-acceptor, binary
22	Whether it is a H-donor, binary
23	Whether it is an R stereo center, binary
22	Whether it is an S stereo center, binary

Table 4. Route features

Position	Description
0–8	Bond distance, binned, [0, 1, 2, 3, 4, $5 \leq d \leq 6$, $7 \leq d \leq 8$, $9 \leq d \leq 12$, $13 \leq d$]
9	Whether the shortest pure conjugated path containing at most 4 bonds, binary
10	Whether the shortest pure conjugated path containing at least 5 bonds, binary
11	Whether there is a route containing only up to 13 single bonds, binary
12	Whether there is a route containing only up to 13 double bonds, binary
13	Triple bond, binary
14	Whether there is a shortest path which is a pure conjugated path, binary
15	Whether the endpoints are in the same ring, binary
16	Single bond, binary
17	Double bond, binary
18	Aromatic bond, binary

References

1. Almeida, L.B.: A learning rule for asynchronous perceptrons with feedback in a combinatorial environment. In: Proceedings, 1st First International Conference on Neural Networks, vol. 2, pp. 609–618. IEEE (1987)
2. Alzaga, A., Iglesias, R., Pignol, R.: Spectra of symmetric powers of graphs and the Weisfeiler-Lehman refinements. arXiv preprint arXiv:0801.2322 (2008)
3. Ba, J.L., Kiros, J.R., Hinton, G.E.: Layer normalization. arXiv:1607.06450 [cs, stat], July 2016
4. Bento, A.P., et al.: The ChEMBL bioactivity database: an update. Nucleic Acids Res. **42**(D1), D1083–D1090 (2014)
5. Bottou, L., Bengio, Y., Le Cun, Y.: Global training of document processing systems using graph transformer networks. In: Proceedings of IEEE Computer Society Conference on Computer Vision and Pattern Recognition, pp. 489–494. IEEE (1997)
6. Bruna, J., Zaremba, W., Szlam, A., LeCun, Y.: Spectral networks and locally connected networks on graphs. arXiv preprint arXiv:1312.6203 (2013)
7. Cho, K., van Merrienboer, B., Gülçehre, Ç., Bougares, F., Schwenk, H., Bengio, Y.: Learning phrase representations using RNN encoder-decoder for statistical machine translation. CoRR arXiv:1406.1078 (2014)
8. Dai, H., Dai, B., Song, L.: Discriminative embeddings of latent variable models for structured data. In: International Conference on Machine Learning, pp. 2702–2711 (2016)
9. Defferrard, M., Bresson, X., Vandergheynst, P.: Convolutional neural networks on graphs with fast localized spectral filtering. In: Advances in Neural Information Processing Systems, pp. 3844–3852 (2016)
10. Duvenaud, D.K., et al.: Convolutional networks on graphs for learning molecular fingerprints. In: Advances in Neural Information Processing Systems, pp. 2224–2232 (2015)
11. Gilmer, J., Schoenholz, S.S., Riley, P.F., Vinyals, O., Dahl, G.E.: Neural message passing for quantum chemistry. In: Proceedings of the 34th International Conference on Machine Learning, vol. 70, pp. 1263–1272. JMLR. org (2017)

12. Gunther, H.: NMR Spectroscopy: An Introduction, vol. 81. Wiley, Hoboken (1980)
13. Hamilton, W., Ying, Z., Leskovec, J.: Inductive representation learning on large graphs. In: Advances in Neural Information Processing Systems, pp. 1024–1034 (2017)
14. He, K., Zhang, X., Ren, S., Sun, J.: Identity mappings in deep residual networks. In: Leibe, B., Matas, J., Sebe, N., Welling, M. (eds.) ECCV 2016. LNCS, vol. 9908, pp. 630–645. Springer, Cham (2016). https://doi.org/10.1007/978-3-319-46493-0_38
15. Jonas, E.: Deep imitation learning for molecular inverse problems. In: Advances in Neural Information Processing Systems, pp. 4991–5001 (2019)
16. Jonas, E., Kuhn, S.: Rapid prediction of NMR spectral properties with quantified uncertainty. J. Cheminform. **11**(1), 1–7 (2019)
17. Kearnes, S., McCloskey, K., Berndl, M., Pande, V., Riley, P.: Molecular graph convolutions: moving beyond fingerprints. J. Comput. Aided Mol. Des. **30**(8), 595–608 (2016). https://doi.org/10.1007/s10822-016-9938-8
18. Kingma, D.P., Ba, J.: Adam: a method for stochastic optimization. arXiv preprint arXiv:1412.6980 (2014)
19. Kipf, T.N., Welling, M.: Semi-supervised classification with graph convolutional networks. arXiv preprint arXiv:1609.02907 (2016)
20. Li, R., Wang, S., Zhu, F., Huang, J.: Adaptive graph convolutional neural networks. In: Thirty-Second AAAI Conference on Artificial Intelligence (2018)
21. Li, Y., Liang, X., Hu, Z., Chen, Y., Xing, E.P.: Graph transformer (2019). https://openreview.net/forum?id=HJei-2RcK7
22. Li, Y., Tarlow, D., Brockschmidt, M., Zemel, R.: Gated graph sequence neural networks. arXiv preprint arXiv:1511.05493 (2015)
23. Maron, H., Ben-Hamu, H., Serviansky, H., Lipman, Y.: Provably powerful graph networks. In: Advances in Neural Information Processing Systems, pp. 2153–2164 (2019)
24. Pearl, J.: Probabilistic Reasoning in Intelligent Systems: Networks of Plausible Inference. Elsevier, Amsterdam (2014)
25. Scarselli, F., Gori, M., Tsoi, A.C., Hagenbuchner, M., Monfardini, G.: The graph neural network model. IEEE Trans. Neural Netw. **20**(1), 61–80 (2008)
26. Simm, J., et al.: Repurposing high-throughput image assays enables biological activity prediction for drug discovery. Cell Chem. Biol. **25**(5), 611–618 (2018)
27. Srivastava, N., Hinton, G., Krizhevsky, A., Sutskever, I., Salakhutdinov, R.: Dropout: a simple way to prevent neural networks from overfitting. J. Mach. Learn. Res. **15**(1), 1929–1958 (2014)
28. Vaswani, A., et al.: Attention is all you need. In: Guyon, I., et al. (eds.) Advances in Neural Information Processing Systems, vol. 30, pp. 5998–6008. Curran Associates, Inc. (2017). http://papers.nips.cc/paper/7181-attention-is-all-you-need.pdf
29. Veličković, P., Cucurull, G., Casanova, A., Romero, A., Lio, P., Bengio, Y.: Graph attention networks. arXiv preprint arXiv:1710.10903 (2017)
30. Wu, Z., et al.: MoleculeNet: a benchmark for molecular machine learning. Chem. Sci. **9**(2), 513–530 (2018)
31. Xu, K., Hu, W., Leskovec, J., Jegelka, S.: How powerful are graph neural networks? In: 7th International Conference on Learning Representations, ICLR 2019, New Orleans, LA, USA, 6–9 May 2019 (2019). https://openreview.net/forum?id=ryGs6iA5Km
32. Yu, F., Koltun, V.: Multi-scale context aggregation by dilated convolutions. arXiv:1511.07122 [cs], November 2015
33. Yun, S., Jeong, M., Kim, R., Kang, J., Kim, H.J.: Graph transformer networks. In: Advances in Neural Information Processing Systems, pp. 11960–11970 (2019)

Zero-Shot Learning-Based Detection of Electric Insulators in the Wild

Ibraheem Azeem(✉) and Moayid Ali Zaidi(✉)

Østfold University College, Halden, Norway
ia@smartinspection.no, moayid.ali.zaidi@adonishr.com

Abstract. An electric insulator is an essential device for an electric power system. Therefore, maintenance of insulators on electric poles has vital importance. Unmanned Aerial Vehicles (UAV's) are used to inspect conditions of electric insulators placed in remote and hostile terrains where human inspection is not possible. Insulators vary in terms of physical appearance and hence the insulator detection technology present in the UAV in principle should be able to identify an insulator device in the wild, even though it has never seen that particular type of insulator before. To address this problem a Zero-Shot Learning-based technique is proposed that can detect an insulator device, that has never seen during the training phase. Different convolutional neural network models are used for feature extraction and are coupled with various signature attributes to detect an unseen insulator type. Experimental results show that inceptionsV3 has better performance on electric insulators dataset and basic signature attributes; "Color and number of plates" of the insulator is the best way to classify insulators dataset while the number of training classes doesn't have much effect on performance. Encouraging results were obtained.

Keywords: Zero-shot learning · Signature attribute · Electrical insulators · Object detection

1 Introduction

Deep learning has shown incredible performance in problems like image classification and object detection, yet this comes at the cost of a huge number of annotated samples for training. It is also not possible to correctly identify the unseen classes those are not used while training and hence in such situation we need to train the classifier from scratch again. In the past few years, special forms of neural networks "Convolutional Neural Networks (CNN's)" are advancing the state of the art of the many computer vision applications, and these are specifically designed to require advantage of the 2D structure of image data. However, CNNs typically require an enormous amount of training data, that is usually not available in many tasks such as power cables component detection and classification [10]. To the best of our knowledge, the very little approach

G. Nicosia et al. (Eds.): LOD 2021, LNCS 13164, pp. 213–225, 2022.
https://doi.org/10.1007/978-3-030-95470-3_16

is available that might efficaciously ankle these quandaries across many tasks. Some of the approaches to handle the dearth of training data problems are i. Utilizing synthetic data and data augmentation techniques to come up with more training data ii. Employing one-shot learning that aims to be told information about incipient object categories from just one or some training images [12] iii. Adopting zero-shot learning that seeks to ascertain information about incipient object categories from only descriptions [15].

Zero-shot learning is used to generalize the learned information to explore the unfamiliar object classes. During this context, both unseen and seen classes require to be cognate to the auxiliary semantic features. The setting within zero-shot learning considered the profound case of transfer learning. The model is trained to replicate human competency on unseen classes, which don't seem to be within the training stage [4]. The core in zero-shot learning is to ascertain a multi-model projection between semantics and visual features by utilizing the labeled optically discerned classes [8]. Besides this, the patrol transmission line of Unmanned Aerial Vehicle (UAV) has become very popular in transmission line inspection and research hot topic. It has shown efficiency, reliability, cost-effectiveness, and helped in inspections where the human approach is not easy. With the development of automation in UAV, it's the interest to identify the types of insulators during an inspection by capturing images of the insulator by UAV camera. For this purpose, a large number of images of insulators are required which is always a problem in deep learning. In this paper, we are addressing this problem by introducing a zero-shot model and use insulators dataset images captured by UAV.

Considering the objectives involved in the study, our research comes up with the subsequent questions: 1. Which deep architecture can generate the most powerful features? 2. What type of "Signature Attributes" can we propose to perform ZSL for our defined problem? 3. Does the number of training classes affect system performance?

2 Related Work

Researchers have used different approaches to machine learning to achieve a lack of data training problems. They have proposed multiple ways in their studies to apply zero-shot learning on the datasets.

Study in [8] introduces a multi-model explication model that categorically integrates three LSTMs (long short-term recollection models) and implemented on CUB, SUN, and AWA2 dataset. It shows the way to extract visual and textual explications. Furthermore, two incipient aspects are insights as well i.e. explanatory diversity and explanatory consistency respectively. Accuracy on both optically discerned and unseen classes was 38.4 and 35.6. In [9] zero-shot learning Hierarchical relegation approach for previously unseen classes by trading specificity for accuracy and mapping to semantic attributes of unseen classes from image features to human recognizable attributes are utilized. The direct method for semantic attributes discussed in [2] shows that the classifier is learned for every attribute from the examples within the training dataset. The study in [16] used binary embedding predicated zero-shot learning (BZSL) that apperceived an unseen instance by efficiently probing its most proximate class codes with minimum Hamming distance while incrementing the binary embedding dimension can inevitably improve the apperception precision. [5] has proposed the hybrid model consists of "Random Attribute Selection" (RAS) and conditional "Generative Adversarial Network" (GCN) to find out the realistic generation of attributes by their correlations in nature and improve discrimination for a large number of classes while [14] showed an approach for learning semantic-driven attributes using two different datasets i.e. AwA and APY. Human gaze embedding is used as auxiliary information to be told compatibility between image and label space for zero-shot learning [6]. The embedding framework that maps multiple text parts is joined with multiple semantic parts into a typical space in [1]. Moreover, a Fine-grained Caltech UCSD Birds-2011 dataset is used and improves state-of-the-art on the CUB dataset to 56.5 (from 50.2) within the supervised setting while improving the state-of-the-art also within the unsupervised setting to 33.9 (from 24.2). In [13] AwA dataset is used as in [14] also, contains 30,475 images from 50 animal classes while each class is annotated with 85 user-defined attributes. CUB dataset contains 11,788 images of 200 bird classes and each class is annotated with 312 attributes. User-defined semantic attributes jointly with discriminative and background lat1ent attributes approach are proposed in the study. The limitations of the proposed approach are that it takes a fixed feature representation as an input. A joint feature and attribute learning approach is suggested to overcome the limitations on some level. Another study [11] has used a semantic attention-based comparative network (SACN) to resolve the zero-shot visual recognition problem with the same dataset state above. ResNet and GoogleNet models are used as a backbone in it. The best results reported in the study are 86.5 on AWA using ResNet as a backbone while GoogleNet gives 84.5. For the CUB dataset 63.4 and 62.0 are reported results respectively. [18] has used both Nearest Neighbor + Self-Training (NN + ST) and data augmentation on video action recognition tasks. However, simple self-training and data augmentation strategies can address these challenges and achieve the state of the art results for zero-shot action recognition within the video. The study showed that Semantic embedding is similar to the state-of-the art low-level feature-based classification and better than the standard attribute-based intermediate representation. The possible reasons given are, attribute-space being less discriminative than semantic word space used, or because of the reliance on human annotation. This shows that some annotated signature attributes are might not be detectable or don't seem to be discriminative for class [18]. Objects are identified supported by a high-level description that's phrased in terms of semantic attributes, like the object's color or shape. NN isn't better than the attribute-based approaches in [7]. The zero-shot classification in [17] is used to classify target classes specifically supported learning a semantic mapping. It targets classes from feature space to semantic knowledge space. Considering the model as best, it can give the best accuracy in both the

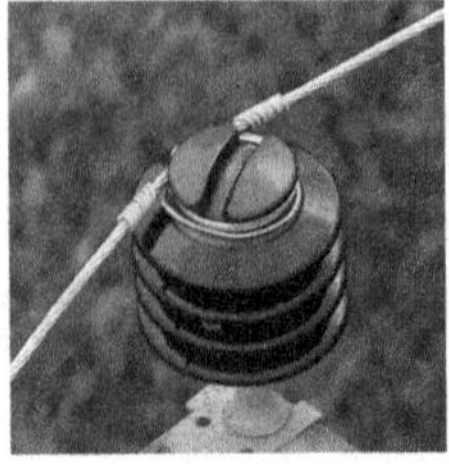

Fig. 1. Brown insulator

Fig. 2. Green insulator

AwA dataset and CUB datasets. In the study [3], the researcher has proposed a unified semi-supervised learning framework. This framework learns the attribute classifiers by exploring the correlations between images and utilizing multiple signature attributes. The usage of multiple features makes the attribute prediction more vigorous. An optimal graph is another choice to enhance the execution of zero-shot image categorization whether or not labeled images are less.

A literature study has shown that different types of approaches are proposed to implement zero-shot learning on an identical style of the dataset. Most of the researches has used datasets relevant to animals or birds. No study has used the dataset we have proposed for zero-shot learning. Due to the unusual attribute nature of our dataset, we are required to put it to the test and determine the most effective possibilities in it.

3 Dataset Details

The dataset contains different images of insulators that are captured against several types of complex background by UAV camera. The backgrounds of image scenes is with the vast majority of different other objects that include forest, crops, grass, poles, and wires etc. Using a UAV camera insulator in images was placed on a special frame so that the images can be obtained in conditions close to the real environment. The images of insulators are structured in folders in such a way that their basic attributes should reflect the difference between them. Figure 1 and Fig. 2 are example images of two different insulators.

4 Methodology

In the context of recognizing insulators, one must provide a large number of data that is not available when one uses a UAV for power inspection images. Moreover, a deep learning algorithm can classify a test data to any of its training classes, but fail when dealing with an "unseen" class. On the other hand, it cannot be ignored that deep learning feature extraction is discriminatory which is not acceptable in some cases. Here, the proposed method uses deep learned features along with a "Zero-shot Learning" framework to counter the problem of recognizing the "untrained" class of images.

A: Zero-Shot Learning

Starting from pattern recognition and the theory that computers can learn without instructions, researchers showed interest to see if computers could learn from data. To classify the samples where training examples are not available, we use zero-shot learning. Recognizing an object from an image without training the image using zero-shot learning allows us to recognize the unseen class of the object. It can give a high-level description of a new or unseen class and create a connection between it. Inspired by this human's ability, the interest of researchers in zero-shot learning is increasing for scaling up visual recognition.

B: Semantic Attributes

Recent advancements in object detection are directly learning a mapping from an image feature space to a semantic space. For that, the semantic attribute approach is the best way to go with it. Semantic attributes provide a bridge from automatically generated image features to human intuition. In our case, we have a dataset of electric insulators and there are several types of insulators. The most commonly used are the pin type, strain insulator, shackle insulator, and suspension type. We aim to differentiate the insulator based on their characteristics and signature attributes. To get information about the attribute we must focus on the naturally shows up region instead of the entire spatial domain. Different objects can have certain similar attributes but their category label is preserved. Attributes are also especially useful in problems that aim at modeling intra-category variations such as fine-grained classification. Even objects are encoded according to their semantic attributes or features and have become quite very practical now a day. An object can be encoded over a large set of exploratory attributes, while each attribute can be assigned a specific or multiple value/s that shows its probability, weight, or importance.

5 Experimental Results

The dataset consists of 38 total classes that we distributed to training and validation. We randomly took 30 classes from it to fine-tune the pertained models. The object of the experiments at this stage is to get to know which deep learning architecture can generate the most features from the dataset.

"Which Deep-Architecture Can Generate the Most Powerful Features?"

We already knew that we have a large number of classes in our dataset but images in each dataset are so less than models will not able to train them in a good manner. We decided to go for testing on the dataset and check, what we get in results. We tried to tune VGG16, VGG19, Xception, and inceptionV3 models on the dataset and got expected results as we assumed before. We did not get results better than 0.10, which we got from VGG16. We decided to go for data augmentation on this stag to increase the images in the dataset and tried to deceive the models. There are multiple approaches proposed by people for data augmentation. The first approach we used was changing the shape of the images. We increase the number of images in each class from 100 to 600 only one class contained 400 images because its original images were so less i.e. 50. The original image size in the dataset was mostly 1397 × 1397 but some were different as well. After augmentation sizes, all dataset images size, become 240 × 240 each.

Experiment#1:
We started tuning of models again on random 30 classes and got some different results from the previous experiment. We divided the dataset into training and validation with 0.80 for training and 0.20 for validation. The total number epoch was set as 100 with batch size 32 and the input image size is 240 × 240 × 3. We aimed to get the exact bounding box area/cropped of electric insulators so in the "ImageDataGenerator" function we used "shear_range" and "zoom_range" as 0.1. We also did "horizontal_flip" as True on the images. As a result, vgg16 gave 0.43 validation accuracy with 1.56 validation loss with a default learning rate of "Adam" as an optimizer. On decreasing the learning rate of vgg16 validation accuracy increases very slowly, for example, it changes in points 0.21, 0.219 then 0.22 so this is the learning progress of vgg16.

Experiment#2:
VGG16 achieved 0.52 accuracy on validation on changing the "shear_range" and "zoom_range" as 0.2. Inceptionv3 did not give any closer result to vgg16. It trained at all on 0.1, 0.01, 0.001 and 0.0001 learning rates. Vgg19 does not give any good results if we used sharp and zoom on the training set but if we removed it, it started learning and gave 0.52 accuracy at the end.

Experiment#3:
We have generated a new dataset using multiple augmentation approaches.

1. Translation
2. Rotation
3. Transformation

In these approaches, we have changed the values of each image degree, rows, and columns. After that, we have refined the dataset again manually and remove the black images containing a small description of the image generated from data augmentation. We split the data into the train and validation folder where mostly original data images were in the validation folder and augmented data were in the train folder. The first observation on tuning did not improve the results we got before but later we mix the randomly in training and validation folders again and then run the experiment. We also changed some of the parameters for the training model. Set the learning rates (0.0001, 0.001), beta_1 = 0.9, beta_2 = 0.999, epsilon = 1e−06 and amsgrad = True of Adam optimizer and set all zoom properties to 0.2. The results we got on different models are given in Table 1.

"What Type of "Signature Attributes" Can We Propose to Perform ZSL for Our Defined Problem?"
To classify the data for zero-shot learning, we arrange the data into their homogeneous/similar groups according to the common characteristics they have. Data

Table 1. CNN results

CNNs	val_accuracy
InceptionV3	75.5%
Xception	72.4%
VGG16	69.4%

Table 2. Signature attributes

No	Attributes
1	Color, Selected body part, No. of plates
2	Color, Selected body part No. of plates, Plates linked on body
3	Color, Selected body part, No. of plates Plates linked on body, Other Selected part, Shape of other selected part
4	Color, Selected body part, No. of plates Plates linked on body Other Selected part, Shape of other selected part, Surface
5	Color, Selected body part, No. of plates Selected part1, Shape of part1, Selected part2, Shape of part2, Surface

without a proper classification is not understandable by CNN and it is not valuable for further analysis and interpretation. The arrangement of data will helps CNN in training and provide ease to do comparison and analysis. We have different types of insulators in our dataset that can be defined concerning their properties, feature, and physical appearance, see Table 2.

Observations: We can create more combinations of attributes and define the objects with more details for classification. Some attributes are common in most of the classes, for-example our dataset contains one class of insulator that has brown color but there are certainly other classes as well that has the same property. We need can separate the attributes based on colors or the combination of other attributes as well. In general, our dataset is classified based on two attributes that are color and number of plates that give us 38 classes in total. On selection of certain attributes and going into more detail of objects we can either reduce or increase the number of classes, this was the zero-shot learning come. We are selecting one of the approaches where our data is not based on the number of plates but there are certainly other attributes that we have selected e.g. we have selected colors and how plates are linked on the body.

We have a total number of 11 classes that are created based on brown color and number of plates at the start but now we have selected the above given two signature attributes and the number of classes is reduced to two. In the same way, the total number of 38 classes is now 8 after the new classification of image objects. Now we have classified in such a way that we can use multiple signature attributes and distribution of dataset can show more details. We have chosen color brown, selected body part, number of plates, and selected part with shape and generated new classification by naming the classes like this:

1. Brown_round_top_seven_plate
2. white_long_top_four_plate
3. white_round_neck_eight_plate
4. brown_round_top_six_plate

Using this technique, we have generated 38 classes of the dataset, defining more details of each object is separating it from other objects for classification.

Preprocessing: Here we have used the class vector and image embedding technique (CVIE) to check the zero-shot learning score. We generated vectors of both train and zero-shot classes. We considered the word2vec Google news model to generate class vectors, which generated the vector of each class with 300 dimensions. For image embedding with classes, we have used the VGG16 pre-trained model for feature extraction. After that, we have created two files i.e. class vector and zero-shot data files. Class vector file contained an array of all classes whereas zero-shot data contained classes and subclass labels as signature attributes attached with images array. One file in txt format contained train classes that checked when the dataset was training based on their signature attributes. The dataset was partitioned into three i.e. training, validation, and testing. Training and validation contain the data of training classes while testing contained zero-shot classes that were not seen by the model before and it was used to evaluate the zero-shot score on the trained model.

Dataset Training Setup: We have used two different activation functions in training i.e. relu and softmax. We have used batch normalization that allows each layer of a NN to learn by itself independently. We have used a "categorical cross-entropy" loss to train the CNNs to output the probability over the classes for each image.

Evaluation of the Model: For evaluation of the model, we have used KDTree to query the zero-shot classes on trained classes in our trained model. We have categorized the results into three i.e. top 5, top 3, and top 1 accuracy of the zero-shot model.

Experiment#1: We have used on a large number of attributes for the classification i.e. Basic signature attributes with top details and surface attribute. E.g. 1. brown_round_top_five_plate_solid, 2. glass_pin_top_two_plate_transparent, 3. orange_round_top_thirty_plate_solid. The total number of classes we got based on the above-given classification is 38. We took 32 classes as seen and 6 for

Table 3. ZSL Results Phase 1

Seen classes	Unseen-classes	Top 5	Top 3	Top 1
25	13	0.40	0.23	0.11
33	5	0.47	0.33	0.23
16	22	0.41	0.26	0.13

Table 4. ZSL Results Phase 2

Seen classes	Unseen-classes	Top 5	Top 3	Top 1
25	13	0.52	0.33	0.13
31	5	0.55	0.43	0.29
16	22	0.47	0.28	0.10

unseen classes. To find out zero-shot learning possible score, we set the value "K = 5" to query the results and the best result we got with BATCH_SIZE = 32 and Epoch = 33 were Top_5 Accuracy: 0.37 Top_3 Accuracy: 0.29 Top_1 Accuracy: 0.13.

Experiment#2: We have classified the data based on attributes i.e. Basic signature attributes with top details and surface attributes. The total number of classes we got based on the above-given classification is 38. We took 32 classes as seen and 6 for unseen classes. E.g. 1. brown_round_top_five_plate, 2. glass_pin_top_two_plate, 3. orange_round_top_thirty_plate. To find out zero-shot learning possible score we set the value "K = 5" to query the results and the best result we got with BATCH_SIZE = 32 and Epoch = 53 were: Top_5 Accuracy: 0.38 Top_3 Accuracy: 0.31 Top_1 Accuracy: 0.14.

Experiment#3: We have classified the data based on attributes i.e. Basic signature attributes E.g. 1. brown_five_plate, 2. glass_two_plate, 3. orange_thirty_plate. The total number of classes we got on based on the above-given classification is 36. We took 26 classes as seen and 10 for unseen classes. To find out zero-shot learning possible score we set the value "K = 5" to query the results and the best result we got with BATCH_SIZE = 32 and Epoch = 53 were: Top_5 Accuracy: 0.54 Top_3 Accuracy: 0.43 Top_1 Accuracy: 0.26.

Does the Number of Training Classes Affect System Performance? We have done the number of experiments to check the performance of the model by increasing and decreasing the number of classes. In the first phase, we selected the signature attributes that were used in research question#2 i.e. experiment#2 to increase the performance of the model. We did three experiments and chose 23 unseen and 25 seen classes in the first experiment. As a result, we found a slight increase in accuracy for top-5 but a decrease in top-3 and 1. We increase the seen classes to 33 and decreases unseen classes to 5, we observed a good effect on all top-1, 3, and 5 accuracy scores. We reduced the seen classes to 16 and unseen classes to 22 to confirm the effect on performance. Below given table shows the experimental results of the first phase Table 3:

In the second phase, we again did three experiments using experimental attributes as used in research question#2 experiment#3, we choose 25 seen and 23 unseen classes. As a result, we found a slight increase in accuracy for top-5 but a decrease in top-3 and 1. We picked 31 seen classes and increases unseen classes to 6 and we observed little incremental effect on all top-1, and 5 accuracy scores but top-3 remained the same. We reduced the classed to seen classes to 16 and used 22 unseen classes to confirm the effect of classes. Below given table shows the experimental results of the second phase Table 4:

6 Discussion

Different CNN models are used in the study to find out the most powerful feature extraction model on the insulators dataset. On performing three different experiments, we can see that results differ every time because of certain reasons. In the beginning, we can observe that our dataset was so small and with numerous classes, none of the models trained it at all. It demonstrates that deep learning required a large number of data for training. Our first data augmentation technique that only changed the shape of the images and increased the number of images in the dataset was is not helpful to get good training results. On changing parameters like learning rate, batch size, epoch, and layers, training stopped after 0.43. From our second experiment, we can observe that the "ImageDataGenerator" parameters also affect training. Even, changing the value of shear_range and zoom_range hand effects on training and started increasing accuracy i.e. 0.52 on VGG16, but this is not with every model. Our first augmentation technique reflects that there is not much difference between the real and new images. The only resolution of images is changed and CNN models are not able to give us good results on them. To overcome this problem, we can consider multiple data augmentation techniques i.e. rotation, translation, and transformation of the images as a better way to regenerate the data. We can see CNN's works much better on new data generated in our last experiments. With new data generated, without changing any "ImageDataGererator" parameter from the default, the learning rate 0.0001 learning rate works properly on all architectures. Furthermore, beta_1 = 0.9, beta_2 = 0.999, epsilon=1e−06 and amsgrad = True of Adam optimizer also has effect on training. The use of "spatial categorical_crossentropy" has clear effects on CNN's training as well; "categorical_crossentropy" is used previously. We perceived inceptionv3 as the most powerful feature extraction model on our dataset as compared to VGG16, and Xception based on results but still, there was not much difference between Xception and inceptionv3.

For zero-shot learning (ZSL) selection of signature, attributes are one of the important tasks before we do classification. One straightforward and naive strategy for zero-shot to select attributes is to pick such attributes that give larger information. We can analyze that the number of classes also changes based on attributes in the electric insulator dataset. Although, it is hard to select attributes for electric insulators, as sometimes insulators are cover with some other type of objects that does not have any role to show uniqueness. The basic signature attributes for such types of the dataset are enough to classify them and extract a good amount of information. Based on the experimental results, we demonstrate that different attributes have a different level of information and predictability, thus we cannot treat them equally. After the experiments, we can say, on increasing the number of signature attributes on such a dataset decreases the number of classes and we lose the zero-shot accuracy score while a small number of signature attributes we were getting much better results. Insulators have different types of shapes, but the main difference between them we found out was the color and number of plates for classification. There is always an impact of choosing training and zero-shot classes on results as well. We analyze the performance of the model during testing by changing the number of classes. There is a change in results using a similar number of classes with a different selection of classes for training and zero-shot. The effects of reducing or increasing the number of classes is depending upon the data. In general, there is no guarantee that if we reduce or increase the number of classes with such type of data we can able to increase the classification accuracy. It will improve the performance if we combine similar classes and apply zero-shot on classes that are never seen by the model. Imagine that we are classifying brown color insulators into four classes based on the different numbers of plates. If we train these classes together and apply zero-shot on white color insulators classification, the overall ZSL score will improve.

7 Conclusion

In this study, we have tuned CNN's popular architecture and applied zero-shot learning on UAV-captured images of insulators. We have also checked the performance of the model based on some classes. The experimental results showed inceptionv3 is the powerful CNN model to extract features for us. We also introduced possible signature attributes for this type of dataset to state-of-the-art and applied zero-shot learning. We used the class and image embedding (CVIE) approach to test the possible zero-shot score. From the results, we can conclude that the insulator's dataset accuracy is directly dependent upon the classification of insulators based on semantic attributes. Increasing the number of signature attributes will losses the zero-shot accuracy score while a small number of signature attributes better results were achieved on this type of dataset. The best results we got are classification on basic attributes e.g. color and number of plates of insulators. The effects of reducing or increasing the number of classes were depending on the data inside each class. We cannot surely say that if we reduce or increases the number of classes we will able to increase the performance of the model.

Acknowledgment. The author would like to thank eSmart Systems for the support in the work with this paper.

References

1. Akata, Z., Malinowski, M., Fritz, M., Schiele, B.: Multi-cue zero-shot learning with strong supervision. In: 2016 IEEE Conference on Computer Vision and Pattern Recognition (CVPR), pp. 59–68 (2016)
2. Burlina, P.M., Schmidt, A.C., Wang, I.: Zero shot deep learning from semantic attributes. In: 2015 IEEE 14th International Conference on Machine Learning and Applications (ICMLA), pp. 871–876 (2015)
3. Gao, L., Song, J., Shao, J., Zhu, X., Shen, H.: Zero-shot image categorization by image correlation exploration. In: Proceedings of the 5th ACM on International Conference on Multimedia Retrieval, ICMR'15, pp. 487–490. Association for Computing Machinery, New York (2015)
4. Guo, J., Guo, S.: A novel perspective to zero-shot learning: towards an alignment of manifold structures via semantic feature expansion. ArXiv arXiv:2004.14795 (2020)
5. Zhang, L.L.L.S.H., Long, Y.: Adversarial unseen visual feature synthesis for zero-shot learning. Neurocomputing **329**(7), 12–20 (2019)
6. Karessli, N., Akata, Z., Schiele, B., Bulling, A.: Gaze embeddings for zero-shot image classification. In: 2017 IEEE Conference on Computer Vision and Pattern Recognition (CVPR), pp. 6412–6421 (2017)
7. Lampert, C.H., Nickisch, H., Harmeling, S.: Attribute-based classification for zero-shot visual object categorization. IEEE Trans. Pattern Anal. Mach. Intell. **36**(3), 453–465 (2014)
8. Liu, Y., Tuytelaars, T.: A deep multi-modal explanation model for zero-shot learning. IEEE Trans. Image Process. **29**, 4788–4803 (2020)
9. Markowitz, J., Schmidt, A.C., Burlina, P.M., Wang, I.: Hierarchical zero-shot classification with convolutional neural network features and semantic attribute learning. In: 2017 Fifteenth IAPR International Conference on Machine Vision Applications (MVA), pp. 194–197 (2017)
10. Nguyen, V.N., Jenssen, R., Roverso, D.: Automatic autonomous vision-based power line inspection: a review of current status and the potential role of deep learning. Int. J. Electr. Power Energy Syst. **99**, 107–120 (2018). https://doi.org/10.1016/j.ijepes.2017.12.016
11. Nian, F., Sheng, Y., Wang, J., Li, T.: Zero-shot visual recognition via semantic attention-based compare network. IEEE Access **8**, 26002–26011 (2020)
12. Vinyals, O., Blundell, C., Lillicrap, T.P., Kavukcuoglu, K., Wierstra, D.: Matching networks for one shot learning. NIPS (2016)
13. Peng, P., Tian, Y., Xiang, T., Wang, Y., Pontil, M., Huang, T.: Joint semantic and latent attribute modelling for cross-class transfer learning. IEEE Trans. Pattern Anal. Mach. Intell. **40**(7), 1625–1638 (2018)
14. Qin, J., Wang, Y., Liu, L., Chen, J., Shao, L.: Beyond semantic attributes: discrete latent attributes learning for zero-shot recognition. IEEE Signal Process. Lett. **23**(11), 1667–1671 (2016)
15. Manning, C.D., Ng, A.Y., Socher, R., Ganjoo, M.: Zero-shot learning through cross-modal transfer (2013)

16. Shen, F., Zhou, X., Yu, J., Yang, Y., Liu, L., Shen, H.T.: Scalable zero-shot learning via binary visual-semantic embeddings. IEEE Trans. Image Process. **28**(7), 3662–3674 (2019)
17. Wang, K., Wu, S., Gao, G., Zhou, Q., Jing, X.: Learning autoencoder of attribute constraint for zero-shot classification. In: 2017 4th IAPR Asian Conference on Pattern Recognition (ACPR), pp. 605–610 (2017)
18. Xu, X., Hospedales, T., Gong, S.: Semantic embedding space for zero-shot action recognition. In: 2015 IEEE International Conference on Image Processing (ICIP), pp. 63–67 (2015)

Randomized Iterative Methods for Matrix Approximation

Joy Azzam, Benjamin W. Ong, and Allan A. Struthers(✉)

Department of Mathematical Sciences, Michigan Technological University, Houghton, USA
{atazzam,ongbw,struther}@mtu.edu

Abstract. Standard tools to update approximations to a matrix A (for example, Quasi-Newton Hessian approximations in optimization) incorporate computationally expensive one-sided samples AV. This article develops randomized algorithms to efficiently approximate A by iteratively incorporating cheaper two-sided samples $U^\top AV$. Theoretical convergence rates are proved and realized in numerical experiments. A heuristic accelerated variant is developed and shown to be competitive with existing methods based on one-sided samples.

Keywords: Matrix approximation · Randomized algorithms · Two-sided samples · Quasi-Newton

1 Introduction and Motivation from Optimization

Effective nonlinear optimization algorithms require 1st derivative information, $\nabla f(x)$, while superlinear convergence requires some 2nd derivative approximation [12]. For example, standard Quasi-Newton (QN) methods such as BFGS (complete gradient and an approximate Hessian generated from gradient differences) have superlinear terminal convergence. Limited-Memory (LM) QN methods [11], such as LBFGS, which approximate the Hessian efficiently by storing only the most recent gradient differences, are widely used in large-scale optimization. This article formulates randomized QN like algorithms which can be used to approximate Hessians (as well as general matrices) with reduced cost.

Alternatively, consider Stochastic Gradient Descent (SGD) [14], which is a common dimension reduction technique in statistics and Machine learning. SGD minimizes the average of cost functions $f_i : \mathbb{R}^m \to \mathbb{R}$,

$$f(x) = \frac{1}{n}\sum_{i=1}^{n} f_i(x),$$

by approximating $\nabla f \approx s^{-1}\sum_{i=1}^{s} \nabla f_i$. Here, f_i is associated with the i-th entry of a large (n entry) data or training set. The SGD approximation is simply $\nabla f \approx F\,p$, where F is the matrix with ith column ∇f_i and the sparse vector p has

G. Nicosia et al. (Eds.): LOD 2021, LNCS 13164, pp. 226–240, 2022.
https://doi.org/10.1007/978-3-030-95470-3_17

s non-zero entries of $1/s$ at sampled indices. Our algorithms generalize SGD by incorporating flexible sampling in $\mathbb{R}^n$ along with sampling in the parameter space $\mathbb{R}^m$ to provide flexibility to tune our algorithms to computational hardware.

Section 2 introduces the fundamental problem, two-sided samples and terminology. Section 3 reviews randomized one-sided Quasi-Newton algorithms while Sect. 4 develops our randomized, two-sided Quasi-Newton algorithms. Section 5 provides probabilistic convergence rates and error estimates. Section 6 numerically demonstrates the convergence of the algorithms. Section 7 incorporates an inner block power iteration to accelerate our two-sided algorithms and compares the result to one-sided algorithms based on similar heuristics.

2 Fundamental Problem, Samples, and Terminology

The fundamental problem our algorithms addresses is how to efficiently construct a sequence of approximations to a matrix, $A \in \mathbb{R}^{m\times n}$, from a stream of incomplete and possibly noisy data. Specifically, we develop and analyze algorithms to *iteratively* embed aggregate information from

$$U^\top A V \in \mathbb{R}^{s_1\times s_2}, \quad U \in \mathbb{R}^{m\times s_1}, \quad V \in \mathbb{R}^{n\times s_2}.$$

These weighted linear combinations of the rows and columns of the data A are called two-sided samples. This is in contrast to weighted linear combinations of the rows, $U^\top A$, or weighted linear combinations of the columns, $A V$, which we refer to as one-sided samples. Two-sided samples have been used before in non-iterative algorithms: [9] compares Schatten-p norm estimates (pth root of the sum of the pth power of the singular values) using two-sided samples, $U^\top A V$, (termed a bi-linear sketch) to estimates using one-sided samples, $A V$. Large eigenvalues estimates using two-sided random projectors are examined in [1]; and two-sided samples are used in [2] to tighten bounds on low-rank approximations. We follow their lead by simply counting sample entries to estimate data cost: $m\, s_2$ for the one-sided samples, $A V$, and $s_1\, s_2$ for the two-sided sample, $U^\top A V$. Algorithms using two-sided samples are a subset of randomized numerical linear algebra. An overview of existing algorithms and applications is provided by the extensive list of articles citing the comprehensive review [7]. The algorithms in [1,7] expend significant up front effort computing projections Ω and Ψ so that the projected matrix, $\Omega^\top A \Psi$, approximates A on dominant eigenvalues with the goal that uniform random sampling of $\Omega^\top A \Psi$ yields good approximations to A. In contrast, our algorithms produce an improving sequence of approximations by iteratively embedding small randomized two-sided samples, $U^\top A V$, with sample dimensions $s_1 \times s_2$ that can be chosen to suit available hardware. Throughout we compare algorithms using the cost estimates (respectively $s_1\, s_2$ and $m\, s_2$ for the samples $U^\top A V$ and $A V$) from [9].

We use notation motivated by QN algorithms in non-linear optimization. SPD means symmetric positive definite and W is an SPD weight matrix. X^+ denotes the Moore-Penrose pseudo-inverse of X; $\langle X, Y\rangle_F = \mathrm{Tr}\left[\mathrm{X}^\top \mathrm{Y}\right]$ and

$\|X\|_F^2 = \langle X, X\rangle_F$ are the Frobenius inner product and norm. For conforming SPD weights W_1 and W_2 the weighted Frobenius norm (with special case $X = X^\top$ and $W = W_1 = W_2$ written $F(W^{-1})$) is

$$\|X\|^2_{F(W_1^{-1}, W_2^{-1})} = \|W_1^{-1/2} X\, W_2^{-1/2}\|_F^2.$$

The W-weighted projector $\mathcal{P}$, which projects onto the column space of $W\,U$,

$$\mathcal{P} = P_{W^{-1},U} = W\,U(U^\top\,W\,U)^{-1}U^\top, \tag{1}$$

satisfies $\mathcal{P}\,W = W\,\mathcal{P}^\top = \mathcal{P}\,W\,\mathcal{P}^\top$ and $W^{-1}\mathcal{P} = \mathcal{P}^\top W^{-1} = \mathcal{P}^\top\,W^{-1}\,\mathcal{P}$.

3 Randomized One-Sided Quasi-Newton Algorithms

Our iterative approximations to A using two-sided samples are motivated by the one-sided sampled algorithms in [6] and QN optimization algorithms. Classical QN schemes for SPD matrices A are formulated as constrained minimum change updates for $B \approx A$ or $H \approx A^{-1}$ in weighted Frobenius norms [12]: the constraint enforces the new information while the minimum change condition stabilizes the update. The one-sided sampled update algorithms in [5,6] are given by the KKT [8,12] equations (with particular choices of weight W) for the quadratic programs

$$B_{k+1} = \arg\min_B \left\{\frac{1}{2}\|B - B_k\|^2_{F(W^{-1})} \mid B\,U_k = A\,U_k \text{ and } B = B^\top\right\}, \tag{2}$$

$$H_{k+1} = \arg\min_H \left\{\frac{1}{2}\|H - H_k\|^2_{F(W^{-1})} \mid U_k = H\,A\,U_k \text{ and } H = H^\top\right\}. \tag{3}$$

The analytical updates defined by Eqs. (2) and (3), are

$$B_{k+1} = B_k + \mathcal{P}_B(A - B_k) + (A - B_k)\mathcal{P}_B^\top - \mathcal{P}_B(A - B_k)\mathcal{P}_B^\top, \tag{4}$$

$$H_{k+1} = H_k + \mathcal{P}_H(A^{-1} - H_k) + (A^{-1} - H_k)\mathcal{P}_H^\top - \mathcal{P}_H(A^{-1} - H_k)\mathcal{P}_H^\top, \tag{5}$$

where the weighted projectors $\mathcal{P}_B$ and $\mathcal{P}_H$ defined by Eq. (1) are

$$\mathcal{P}_B = P_{W^{-1},U_k} = W\,U_k(U_k^\top\,W\,U_k)^{-1}U_k^\top,$$
$$\mathcal{P}_H = P_{W^{-1},A\,U_k} = W\,A\,U_k(U_k^\top\,A\,W\,A\,U_k)^{-1}U_k^\top\,A.$$

Note, these are two different updates using the same one-sided sample $A\,U_k$ which are not simply connected by the Sherman-Morrison-Woodbury (SMW) formula. In Eq. (4), B_{k+1} is an improved approximation to A while in Eq. (5), H_{k+1} is an improved approximation to A^{-1}. Familiar algorithms are obtained by selecting different weights, W. Block DFP [15] is Eq. (4) with $W = A$

$$B_{k+1} = (I_n - \mathcal{P}_{\text{DFP}})\,B_k(I_n - \mathcal{P}_{\text{DFP}}^\top) + \mathcal{P}_{\text{DFP}}\,A,$$

where $\mathcal{P}_{\text{DFP}} = P_{A^{-1},U_k} = A\,U_k(U_k^\top\,A\,U_k)^{-1}U_k^\top$. Block BFGS [5,6] is the result of inverting Eq. (4) with $W = A^{-1}$ using the SMW formula

$$B_{k+1} = B_k - B_kU_k\left(U_k^\top B_kU_k\right)^{-1}U_k^\top B_k + A\,U_k\left(U_k^\top AU_k\right)^{-1}U_k^\top\,A.$$

QN algorithms are commonly initialized with multiples of the identity.

4 Randomized Two-Sided Quasi-Newton Algorithms

A general algorithm (defined by SPD weight matrices W_1 and W_2) to approximate non-square matrices and two distinct algorithms specialized to symmetric matrices are developed. As with one-sided sampled algorithms, different weights give different algorithms. Algorithms and theorems are developed for a generic initialization B_0.

4.1 General Two-Sided Sampled Update

Analogous to Eq. (2), our first algorithm is defined by the minimization

$$B_{k+1} = \arg\min_{B} \left\{ \frac{1}{2} \|B - B_k\|^2_{F(W_1^{-1}, W_2^{-1})} \mid U_k^\top B\, V_k = U_k^\top A\, V_k \right\}. \tag{6}$$

Solving the KKT equations for Eq. (6) gives the self-correcting update

$$B_{k+1} = B_k + P_{W_1^{-1}, U_k} (A - B_k) P^\top_{W_2^{-1}, V_k}. \tag{7}$$

Since this update explicitly corrects the projected residual sample $R_k = U_k^\top (A - B_k)\, V_k$, it decreases the weighted Frobenius norm $\|A - B_k\|^2_{F(W_1^{-1}, W_2^{-1})}$ unless the approximation is correct on the sampled spaces, i.e., $U_k^\top (A - B_k)\, V_k = 0$.

Given $A \in \mathbb{R}^{m\times n}$, initial approximation $B_0 \in \mathbb{R}^{m\times n}$, two-sided sample sizes $\{s_1, s_2\}$, and SPD weights $\{W_1, W_2\}$, Eq. (7) generates a sequence $\{B_k\}$ that converges monotonically to A in the appropriate weighted Frobenius norm. Pseudocode is provided in Algorithm 1: boxed values give the two-sided sample size per iteration; double boxed values the total for all iterations. For symmetric A, the independent left and right hand sampling fails to preserve symmetry.

Require: $B_0 \in \mathbb{R}^{m\times n}$, SPD $W_1 \in \mathbb{R}^{m\times m}$, $W_2 \in \mathbb{R}^{n\times n}$, $\{s_1, s_2\} \in \mathbb{N}$.
repeat $\{k = 0, 1, \ldots\}$
  Sample $U_k \sim N(0,1)^{m\times s_1}$ and $V_k \sim N(0,1)^{n\times s_2}$
  Compute $R_k = U_k^\top A\, V_k - U_k^\top B_k V_k \in \mathbb{R}^{s_1\times s_2}$ .... $\boxed{s_1\, s_2}$
  Update $B_{k+1} = B_k + W_1 U_k (U_k^\top W_1 U_k)^{-1} R_k (V_k^\top W_2 V_k)^{-1} V_k^\top W_2$
until convergence
return B_{k+1} .... $\boxed{\boxed{(k+1)\,(s_1\, s_2)}}$

Algorithm 1: NS: Non-Symmetric Two-Sided Sampling

4.2 Symmetric Update

Unsurprisingly, the fully symmetrized general algorithm ($A = A^\top, B_0 = B_0^\top$, $V_k = U_k$ and $W = W_1 = W_2$) give symmetric approximations. Pseudocode is provided in Algorithm 2 with sample counts boxed as before.

Require: $B_0 \in \mathbb{R}^{n\times n}$ satisfying $B_0^\top = B_0$, SPD $W \in \mathbb{R}^{n\times n}$, $s \in \mathbb{N}$.
repeat $\{k = 0, 1, \ldots\}$
  Sample $U_k \sim \mathcal{N}(0,1)^{n\times s_1}$
  Compute $R_k = U_k^\top A U_k - U_k^\top B_k U_k \in \mathbb{R}^{s_1\times s_1}$ .... $\boxed{s_1^2}$
  Compute $\tilde{P}_k = W U_k (U_k^\top W U_k)^{-1}$
  Update $B_{k+1} = B_k + \tilde{P}_k R_k \tilde{P}_k^\top$
until convergence
return B_{k+1} .... $\boxed{\boxed{(k+1)\,(s^2)}}$

Algorithm 2: SS1: Symmetric Two-sided Sampling

Remark 1. Algorithm 2 can give non-SPD updates from SPD input *e.g.*

$$W = \begin{bmatrix} 1 & 0 \\ 0 & 1 \end{bmatrix}, \quad A = \begin{bmatrix} 1 & 0 \\ 0 & 1 \end{bmatrix}, \quad B = \begin{bmatrix} 1 & 0 \\ 0 & 9 \end{bmatrix}, \quad \text{and} \quad U = \frac{1}{\sqrt{2}} \begin{bmatrix} 1 \\ 1 \end{bmatrix}.$$

4.3 Multi-step Symmetric Updates

Enforcing symmetry for $A = A^\top$ and $B_0 = B_0^\top$ with an internal step

$$\begin{aligned} B_{k+1/2} &= B_k + P_{W^{-1},U_k}(A - B_k)P_{W^{-1},V_k}^\top \\ B_{k+1} &= \frac{1}{2}\left(B_{k+1/2} + B_{k+1/2}^\top\right) \end{aligned}$$

gives convergence comparable to Algorithm 2. However, the two-step algorithm

$$\begin{aligned} B_{k+1/3} &= B_k + P_{W_1^{-1},U_k}(A - B_k)P_{W_2^{-1},V_k}^\top \\ B_{k+2/3} &= B_{k+1/3} + P_{W_2^{-1},V_k}(A - B_{k+1/3}^\top)P_{W_1^{-1},U_k}^\top \\ B_{k+1} &= \frac{1}{2}\left(B_{k+2/3} + B_{k+2/3}^\top\right), \end{aligned} \tag{8}$$

has superior convergence properties and requires no additional data since

$$P_{W_2^{-1},V_k} A\, P_{W_1^{-1},U_k}^\top = \left(P_{W_1^{-1},U_k} A\, P_{W_2^{-1},V_k}^\top\right)^\top.$$

Pseudocode is provided in Algorithm 3 with sample counts boxed as before.

5 Convergence Analysis

Our convergence results rely on properties of randomly generated projectors. In our experiments, we orthogonalize square matrices with entries drawn from $N(0,1)$ to generate rotations from a rotationally invariant distribution [16]. Our algorithms use symmetric rank s projectors defined by an SPD weight W

$$\hat{z} = W^{1/2} U (U^\top W U)^{-1} U^\top W^{1/2},$$

Require: $B_0 \in \mathbb{R}^{n \times n}$ satisfying $B_0 = B_0^\top$, SPD $W \in \mathbb{R}^{m \times m}$, $\{s_1, s_2\} \in \mathbb{N}$.

repeat $\{k = 0, 1, \ldots\}$

  Sample $U_k \sim N(0,1)^{n \times s_1}$ and $V_k \sim N(0,1)^{n \times s_2}$

  Compute residual, $R_k = U_k^\top A V_k - U_k^\top B_k V_k \in \mathbb{R}^{s_1 \times s_2}$ .... $\boxed{s_1\, s_2}$

  Compute $B_{k+1/3} = B_k + W U_k (U_k^\top W U_k)^{-1} R_k (V_k^\top W V_k)^{-1} V_k^\top W$

  Compute $R_{k+1/3} = (U_k^\top A V_k)^\top - V_k^\top B_{k+1/3} U_k \in \mathbb{R}^{s_2 \times s_1}$

  Compute $B_{k+2/3} = B_{k+1/3} + W V_k (V_k^\top W V_k)^{-1} R_{k+1/3} (U_k^\top W U_k)^{-1} U_k^\top W$

  Update $B_{k+1} = \frac{1}{2}(B_{k+2/3} + B_{k+2/3}^\top)$

until convergence

return B_{k+1} .... $\boxed{(k+1)\,(s_1\, s_2)}$

Algorithm 3: SS2: Two-Step Symmetric Two-Sided Sampling

where U is simply the first s columns of a random rotation. The smallest and largest eigenvalues λ_1 and λ_n of the expectation $\mathbf{E}[\hat{z}]$ of these random projections determines convergence of our algorithms with optimal rates when $\lambda_1 = \lambda_n$.

Definition 1. *A random matrix, $\hat{X} \in \mathbb{R}^{m \times n}$, is rotationally invariant if the distribution of $Q_m \hat{X} Q_n$ is the same for all rotations $Q_i \in \mathcal{O}(i)$.*

Proposition 1. *Let $\mathcal{Z}$ be any distribution of real, rank s projectors in $\mathbb{R}^n$. Then,*

$$0 \leq \lambda_{\min}(\boldsymbol{E}[\hat{z}]) \leq \frac{s}{n} \leq \lambda_{\max}(\boldsymbol{E}[\hat{z}]) \leq 1, \quad \hat{z} \in \mathcal{Z}.$$

Further, if $\hat{z}$ is rotationally invariant, then $\boldsymbol{E}[\hat{z}] = \frac{s}{n} I_n$.

Proposition 2. *For $R \in \mathbb{R}^{m \times n}$ and conforming symmetric projections $\hat{y}, \hat{z}$,*

$$\begin{aligned} \langle R\,\hat{z}, R\,\hat{z}\rangle_F &= \langle R, R\,\hat{z}\rangle_F \\ \langle \hat{y}\,R\,\hat{z}, \hat{y}\,R\,\hat{z}\rangle_F &= \langle \hat{y}\,R\,\hat{z}, R\,\hat{z}\rangle_F = \langle \hat{y}\,R\,\hat{z}, R\rangle_F \end{aligned} \tag{9}$$

Proposition 3. *For any $R \in \mathbb{R}^{m \times n}$ and conforming symmetric positive semi-definite matrices S_1, S_2, and (in the special case $m = n$) S we have the bounds:*

$$\begin{aligned} \lambda_{\min}(S_1)\langle R, R\rangle_F &\leq \langle S_1\, R, R\rangle_F \leq \lambda_{\max}(S_1)\langle R, R\rangle_F, \\ \lambda_{\min}(S_2)\langle R, R\rangle_F &\leq \langle R, R\, S_2\rangle_F \leq \lambda_{\max}(S_2)\langle R, R\rangle_F, \\ \lambda_{\min}(S)^2\langle R, R\rangle_F &\leq \langle S\, R, R\, S\rangle_F \leq \lambda_{\max}(S)^2\langle R, R\rangle_F. \end{aligned}$$

Remark 2. Convergence results for Algorithms 1 to 3. are for $\mathbf{E}[\|B - A\|_F^2]$. Such results dominate similar results for $\|\mathbf{E}[B - A]\|_F^2$ since

$$\|\mathbf{E}\left[B - A\right]\|_F^2 = \mathbf{E}\left[\|B - A\|_F^2\right] - \mathbf{E}\left[\|B - \mathbf{E}\left[B\right]\|_F^2\right].$$

Theorem 1 (Convergence of Algorithm 1 - NS). *For $A \in \mathbb{R}^{m \times n}$ and $B_0 \in \mathbb{R}^{m \times n}$ with $W_1 \in \mathbb{R}^{m \times m}$ and $W_2 \in \mathbb{R}^{n \times n}$ fixed SPD weights. If $U_k \in \mathbb{R}^{m \times s_1}$*

and $V_k \in \mathbb{R}^{n\times s_2}$ *are random, independently selected orthogonal matrices with full column rank (with probability one), then* B_k *from Algorithm 1 satisfies*

$$\boldsymbol{E}\left[\|B_{k+1}-A\|^2_{F(W_1^{-1},W_2^{-1})}\right] \le (\rho_{NS})^k \boldsymbol{E}\left[\|B_0-A\|^2_{F(W_1^{-1},W_2^{-1})}\right],$$

where $\rho_{NS} = 1-\lambda_{\min}(\boldsymbol{E}[\hat{y}])\,\lambda_{\min}(\boldsymbol{E}[\hat{z}])$, *with*

$$\hat{y}_k = W_1^{1/2}U_k(U_k^\top W_1 U_k)^{-1}U_k^\top W_1^{1/2}, \quad \hat{z}_k = W_2^{1/2}V_k(V_k^\top W_2 V_k)^{-1}V_k^\top W_2^{1/2}.$$

Proof. Define the kth residual as $R_k := W_1^{-1/2}(B_k - A)W_2^{-1/2}$. With some algebraic manipulation, Eq. (7) can be re-written as $R_{k+1} = R_k - \hat{y}_k R_k \hat{z}_k$. Computing the squared Frobenius norm of both sides,

$$\begin{aligned}\langle R_{k+1}, R_{k+1}\rangle_F &= \langle R_k - \hat{y}_k R_k\hat{z}_k, R_k - \hat{y}_k R_k \hat{z}_k\rangle_F\\ &= \langle R_k, R_k\rangle_F - \langle R_k, \hat{y}_k R_k \hat{z}_k\rangle_F - \langle \hat{y}_k R_k\hat{z}_k, R_k\rangle_F + \langle \hat{y}_k R_k\hat{z}_k, \hat{y}_k R_k\hat{z}_k\rangle_F\\ &= \langle R_k, R_k\rangle_F - \langle \hat{y}_k R_k\hat{z}_k, R_k\hat{z}_k\rangle_F,\end{aligned}$$

where we have made use of Proposition 2. Taking the expected value with respect to independent samples U_k (leaving V_k and R_k fixed) gives

$$\begin{aligned}&\mathbf{E}\left[\|R_{k+1}\|_F^2 \mid V_k, R_k\right] = \langle R_k, R_k\rangle_F - \langle \mathbf{E}[\hat{y}_k]R_k\hat{z}_k, R_k\hat{z}_k\rangle_F\\ &\le \langle R_k, R_k\rangle_F - \lambda_{\min}(\mathbf{E}[\hat{y}_k])\langle R_k\hat{z}_k, R_k\hat{z}_k\rangle_F \le \langle R_k, R_k\rangle_F - \lambda_{\min}(\mathbf{E}[\hat{y}_k])\langle R_k, R_k\hat{z}_k\rangle_F,\end{aligned}$$

where we applied Proposition 3 to the symmetric positive semi-definite matrix $\mathbf{E}[\hat{y}_k]$, and used Eq. (9). Taking the expected value with respect to independent samples V_k and leaving R_k fixed gives

$$\begin{aligned}\mathbf{E}[\|R_{k+1}\|_F^2 \mid R_k] &\le \langle R_k, R_k\rangle_F - \lambda_{\min}(\mathbf{E}[\hat{y}_k])\langle R_k, R_k\mathbf{E}[\hat{z}_k]\rangle_F\\ &\le \langle R_k, R_k\rangle_F - \lambda_{\min}(\mathbf{E}[\hat{y}_k])\lambda_{\min}(\mathbf{E}[\hat{z}_k])\langle R_k, R_k\rangle_F.\end{aligned}$$

Taking the full expectation and noting $\mathbf{E}[\|R_{k+1}\|_F^2] = \mathbf{E}[\|B_k - A\|^2_{F(W_1^{-1},W_2^{-1})}$

$$\begin{aligned}\mathbf{E}[\|R_{k+1}\|_F^2] &\le \mathbf{E}\left[\langle R_k, R_k\rangle_F\right] - \lambda_{\min}(\mathbf{E}[\hat{y}_k])\lambda_{\min}(\mathbf{E}[\hat{z}_k])\mathbf{E}\left[\langle R_k, R_k\rangle_F\right]\\ &= (1-\lambda_{\min}(\mathbf{E}[\hat{y}_k])\lambda_{\min}(\mathbf{E}[\hat{z}_k]))\mathbf{E}[\langle R_k, R_k\rangle_F]\end{aligned}$$

gives the result by unrolling the recurrence. Note, independence of U_k and V_k justifies $\mathbf{E}[\langle \hat{y}_k R_k\hat{z}_k, R_k\hat{z}_k\rangle_F] = \langle \mathbf{E}[\hat{y}_k]R_k\hat{z}_k, R_k\hat{z}_k\rangle_F$.

Theorem 2 (Convergence of Algorithm 2 - SS1). *Let* $A, W \in \mathbb{R}^{n\times n}$ *be fixed SPD matrices and* $U_k \in \mathbb{R}^{n\times s}$ *be a randomly selected matrix having full column rank with probability* 1. *If* $B_0 \in \mathbb{R}^{n\times n}$ *is an initial guess for* A *with* $B_0 = B_0^\top$, *then* B_k *from Algorithm 2 satisfies*

$$\boldsymbol{E}[\|B_{k+1}-A\|^2_{F(W^{-1})}] \le (\rho_{SS1})^k \boldsymbol{E}[\|B_0 - A\|^2_{F(W^{-1})}],$$

where $\rho_{SS1} = 1-\lambda_{\min}(\boldsymbol{E}[\hat{z}])^2$ *and* $\hat{z}_k = W^{1/2}U_k(U_k^\top W U_k)^{-1}U_k^\top W^{1/2}$.

Proof. Following similar steps outlined in the proof in Theorem 1, we arrive at

$$\langle R_{k+1}, R_{k+1}\rangle_F = \langle R_k, R_k\rangle_F - \langle R_k, \hat{z}_k R_k \hat{z}_k\rangle_F.$$

Taking the expected value with respect to U_k leaving R_k fixed we have

$$\begin{aligned}
&\mathbf{E}\left[\|R_{k+1}\|_F^2 \mid R_k\right] = \langle R_k, R_k\rangle_F - \mathbf{E}\left[\langle R_k, \hat{z}_k R_k \hat{z}_k\rangle_F\right] \\
&= \langle R_k, R_k\rangle_F - \mathbf{E}\left[\mathrm{Tr}[\mathrm{R}_k^\top \hat{\mathrm{z}}_k \mathrm{R}_k \hat{\mathrm{z}}_k]\right] \\
&= \langle R_k, R_k\rangle_F - \mathrm{Tr}\left[\mathbf{E}\left[\mathrm{R}_k \hat{\mathrm{z}}_k \mathrm{R}_k \hat{\mathrm{z}}_k\right]\right] \le \langle \mathrm{R}_k, \mathrm{R}_k\rangle_\mathrm{F} - \mathrm{Tr}\left[\mathbf{E}\left[\mathrm{R}_k \hat{\mathrm{z}}_k\right]^2\right],
\end{aligned}$$

where the inequality arises from application of Jensen's Inequality. Simplifying and applying Proposition 3,

$$\begin{aligned}
&\mathbf{E}[\|R_{k+1}\|_{F(W^{-1})}^2 \mid R_k] \le \langle R_k, R_k\rangle_F - \mathrm{Tr}\left[\mathbf{E}\left[\mathrm{R}_k \hat{\mathrm{z}}_k\right]^2\right] \\
&= \langle R_k, R_k\rangle_F - \mathrm{Tr}\left[\mathrm{R}_k \mathbf{E}\left[\hat{\mathrm{z}}_k\right] \mathrm{R}_k \mathbf{E}\left[\hat{\mathrm{z}}_k\right]\right] \\
&= \langle R_k, R_k\rangle_F - \langle \mathbf{E}[\hat{z}_k] R_k, R_k \mathbf{E}[\hat{z}_k]\rangle_F \le \langle R_k, R_k\rangle_F - \lambda_{\min}(\mathbf{E}[\hat{z}_k])^2 \langle R_k, R_k\rangle_F.
\end{aligned}$$

Taking the full expectation and un-rolling the recurrence yields the desired result.

Theorem 3 (Convergence of Algorithm 3 - SS2). *Let A, U_k, V_k and B_0 be defined as in Theorem 1, and let W be a fixed SPD matrix then B_k from Algorithm 3 satisfies*

$$\boldsymbol{E}\left[\|B_k - A\|_{F(W^{-1})}^2\right] \le (\rho_{SS2})^k \boldsymbol{E}\left[\|B_0 - A\|_{F(W^{-1})}^2\right],$$

where

$$\rho_{SS2} = 1 - 2\lambda_{\min}(\boldsymbol{E}[\hat{y}])\lambda_{\min}(\boldsymbol{E}[\hat{z}]) + \lambda_{\min}(\boldsymbol{E}[\hat{y}])^2 \lambda_{\min}(\boldsymbol{E}[\hat{z}])^2.$$

Proof. Let R_k be the kth residual R_k, and $\hat{y}_k, \hat{z}_k$ be projectors as in Theorem 1 with $W = W_1 = W_2$. Eq. (8) can be re-written in terms of R_k as follows.

$$R_{k+1/3} = R_k - \hat{y}_k R_k \hat{z}_k, \quad R_{k+2/3}^\top = R_{k+1/3}^\top - \hat{z}_k R_{k+1/3}^\top \hat{y}_k,$$
$$R_{k+1} = \frac{1}{2}\left(R_{k+2/3} + R_{k+2/3}^\top\right).$$

Theorem 1 gives

$$\mathbf{E}\left[\left\|R_{k+1/3}\right\|_F^2\right] \le (\rho_{\mathrm{NS}})\mathbf{E}\left[\|R_k\|_F^2\right],$$

and a repeated application of Theorem 1 gives

$$\mathbf{E}\left[\left\|R_{k+2/3}\right\|_F^2\right] \le (\rho_{\mathrm{NS}})\mathbf{E}\left[\|R_{k+1/3}\|_F^2\right] \le (\rho_{\mathrm{NS}})^2\mathbf{E}\left[\|R_k\|_F^2\right].$$

Lastly, we observe via the triangle inequality that

$$\begin{aligned}
&\mathbf{E}\left[\|R_{k+1}\|_F^2\right] = \mathbf{E}\left[\left\|\frac{1}{2}\left(R_{k+2/3} + R_{k+2/3}^\top\right)\right\|_F^2\right] \\
&\le \frac{1}{2}\mathbf{E}\left[\left\|R_{k+2/3}\right\|_F^2\right] + \frac{1}{2}\mathbf{E}\left[\left\|R_{k+2/3}^\top\right\|_F^2\right] = (\rho_{\mathrm{NS}})^2\mathbf{E}\left[\|R_k\|_F^2\right],
\end{aligned}$$

Un-rolling the loop for k iterations gives the desired result.

With the relevant rate ρ below error bounds for Algorithms 1 to 3 are

$$\|R_{k+1}\|^2_{F(W_1^{-1},W_2^{-1})} \leq \rho \|R_k\|^2_{F(W_1^{-1},W_2^{-1})} \tag{10}$$

where $y_1 = \lambda_{\min}(\mathbf{E}[\hat{y}])$, $z_1 = \lambda_{\min}(\mathbf{E}[\hat{z}])$, and

$$\rho_{\text{NS}}(y_1, z_1) = 1 - y_1 z_1, \quad \rho_{\text{SS1}}(z_1) = 1 - z_1^2, \quad \rho_{\text{SS2}}(y_1, z_1) = (1 - y_1 z_1)^2. \tag{11}$$

Since any symmetric rank s random projection $\hat{z}$ on $\mathbb{R}^n$ satisfies $0 \leq z_1 \leq \frac{s}{n} \leq z_n \leq 1$ and rotationally invariant distributions, e.g. UU^+ with $U \sim N(0,1)^{n\times s}$, further satisfy $\mathbf{E}[\hat{z}] = \frac{s}{n}$, minimizing the various convergence rates ρ over the appropriate domains gives the following optimal rates.

Corollary 1. *The optimal convergence rates for Algorithms 1 to 3 are obtained attained for U_k and V_k sampled from rotationally invariant distributions,*

$$\rho_{NS}^{opt} = 1 - \frac{s_1}{m}\frac{s_2}{n}, \quad \rho_{SS1}^{opt} = 1 - \left(\frac{s_2}{n}\right)^2, \quad \rho_{SS2}^{opt} = \left(1 - \frac{s_1}{m}\frac{s_2}{n}\right)^2. \tag{12}$$

Remark 3. Theorems 1 to 3 all assume the weight matrix W and distributions are fixed. All our non-accelerated numerical experiments use fixed weights and sample from fixed rotationally invariant distributions.

Remark 4. Corollary 1 is an extremely strong result. Consider for simplicity $s_1 = s_2 = s$. Although the convergence rates are $\sim 1 - \left(\frac{s}{n}\right)^2$, only $s \times s$ aggregated pieces of information are used each iteration. If a one-sided sampled algorithm uses $s \times n$ pieces of information, e.g. [6], our algorithm can take $\frac{n}{s}$ iterations with the *same amount of information*. Consequently the error decrease after $\frac{n}{s}$ iterations, is comparable to convergence rates of one-sided sampled QN methods.

$$\left(1 - \frac{s^2}{n^2}\right)^{n/s} \approx 1 - \frac{n}{s} \cdot \frac{s^2}{n^2},$$

Lower bounds on the convergence rates (analogous to the upper bounds in Theorems 1 to 3 but using the upper bounds in Proposition 3) are easily derived. For example, the two-sided error bound for Algorithm 1 is

$$\rho_{\text{NS}}(y_m, z_n)\mathbf{E}[\|R_k\|_F^2] \leq \mathbf{E}[\|R_{k+1}\|_F^2] \leq \rho_{\text{NS}}(y_1, z_1)\mathbf{E}[\|R_k\|_F^2],$$

where as before $y_1 \leq y_2 \leq \cdots \leq y_m$ is the spectrum of $\mathbf{E}[\hat{y}]$, $z_1 \leq z_2 \leq \cdots \leq z_n$ is the spectrum of $\mathbf{E}[\hat{z}]$ and the explicit form for ρ_{NS} is in Eq. (11). We collect the similar results for Algorithms 1 to 3 in Corollary 2.

Corollary 2 (Two-Sided Convergence Rates). *Given the assumptions of Theorems 1 to 3 the explicit formulas Eq.* (11) *for* ρ *give two-sided bounds,*

$$\rho_{NS}(y_m, z_n)^k \leq \frac{\boldsymbol{E}\left[\|B_{k+1} - A\|^2_{F(W_1^{-1}, W_2^{-1})}\right]}{\|B_0 - A\|^2_{F(W_1^{-1}, W_2^{-1})}} \leq \rho_{NS}(y_1, z_1)^k$$

$$\rho_{SS1}(z_n)^k \leq \frac{\boldsymbol{E}\left[\|B_{k+1} - A\|^2_{F(W^{-1})}\right]}{\|B_0 - A\|^2_{F(W^{-1})}} \leq \rho_{SS1}(z_1)^k$$

$$\rho_{SS2}(y_n, z_n)^k \leq \frac{\boldsymbol{E}\left[\|B_{k+1} - A\|^2_{F(W^{-1})}\right]}{\|B_0 - A\|^2_{F(W^{-1})}} \leq \rho_{SS2}(y_1, z_1)^k$$

where y_1, y_m, z_1, z_n *are the extreme eigenvalues of* $\boldsymbol{E}[\hat{y}]$ *and* $\boldsymbol{E}[\hat{z}]$.

Remark 5. If $\hat{y}$ and $\hat{z}$ are rotationally invariant, the upper and lower probabilistic bounds in Corollary 2 coincide since $z_1 = z_n = \frac{s_1}{n}$ and $y_1 = y_m = \frac{s_2}{m}$. Algorithms 1 to 3 all use rotationally invariant distributions and converge predictably at the expected rate. The algorithms still converge with other distributions provided the smallest eigenvalue of the expectation is positive.

6 Numerical Results

Algorithm 1 to 3 were implemented in the MATLAB framework from [6] and tested on representative SPD matrices from the same article: $A = XX^\top$ with $X \sim \mathcal{N}(0,1)^{n \times n}$; the Gisette-Scale ridge regression matrix from [3]; and the NASA matrix from [4]. The author's website [13] contains MATLAB scripts and similar results for all matrices from [6]. Computations were performed on Superior, the HPC facility at Michigan Technological University.

Many metrics can be used to objectively compare algorithmic costs. Common metrics include number of FLOPS, total memory used, communication overhead, and for matrix-free black-box procedures the number of individual matrix-vector products Av. As noted by [9] the analogous metric for a black box procedure to compute the matrix product required for our $s_1 \times s_2$ two-sided sample $U^\top A V$ is the number of sampled entries, $s_1 s_2$. As an explicit example, for $f : \mathbb{R}^m \to \mathbb{R}$, Forward-Forward mode [10] Algorithmic Differentiation (AD) simultaneously computes $f(x)$ and a directional 2nd derivative $u^\top \nabla^2 f(x) v$. With sufficient shared-memory processors, AD can efficiently compute $f(x)$ and the two-sided sample $U^\top \nabla^2 f(x) V$ of the Hessian with cost $s_1 s_2$.

Algorithms in [6] are for symmetric matrices. We compare the convergence of (unweighted i.e. $W = I$) Algorithms 1 to 3 (sample size $s = \lceil\sqrt{n}\rceil$ matching [6]) to the one sided algorithms in [6]: Fig. 1a for $A = XX^\top$ with $X \sim \mathcal{N}(0,1)^{5000 \times 5000}$; Fig. 1b for Gisette-Scale; and Fig. 1c) for NASA. Our algorithms achieve the theoretical convergence rates from Eq. (12) (dotted lines). Weighted algorithms DFP and BFGS use $B_0 = I$, un-weighted Algorithms 1 to 3 use $B_0 = 0$. Runs were terminated after $5n^2$ iterations or when the relative

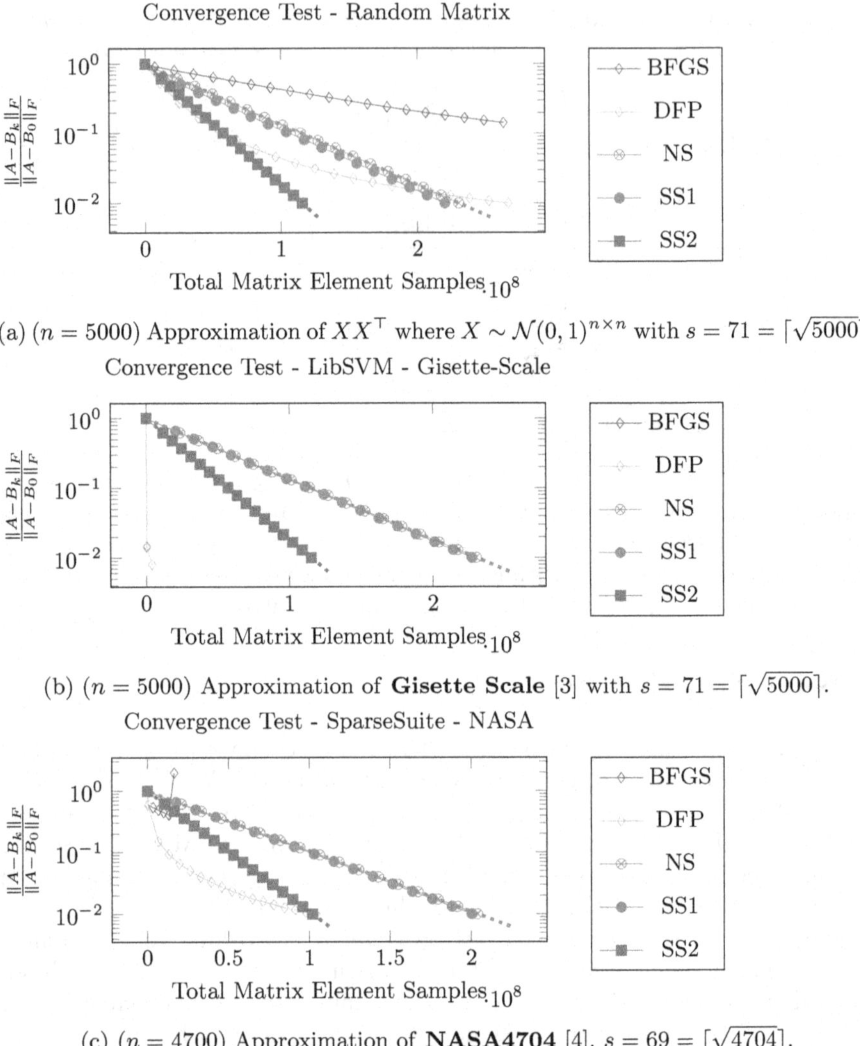

(a) ($n = 5000$) Approximation of $XX^\top$ where $X \sim \mathcal{N}(0,1)^{n\times n}$ with $s = 71 = \lceil\sqrt{5000}\rceil$.

(b) ($n = 5000$) Approximation of **Gisette Scale** [3] with $s = 71 = \lceil\sqrt{5000}\rceil$.

(c) ($n = 4700$) Approximation of **NASA4704** [4]. $s = 69 = \lceil\sqrt{4704}\rceil$.

Fig. 1. Two-sided sampled algorithm performance with theoretical rate as dots. BFGS and DFP can be: a) comparable; b) superior; or c) stall/diverge.

residual norm fell below 0.01. The one-sided algorithms DFP and BFGS have target dependent weight matrices: DFP is Eq. (4) with weight $W = A$ while BFGS is the Sherman-Morrison-Woodbury inversion of Eq. (5) with $W = A^{-1}$. Figure 1a shows our algorithms outperforming both BFGS and DFP for small tolerances. Figure 1b shows enhanced initial convergence for DFP and BFGS, but Fig. 1c demonstrates that BFGS may not converge. In contrast, our two-sided algorithms converge consistently, achieving the theoretical convergence rates.

7 Heuristic Accelerated Schemes

Algorithm 2 corrects the symmetric projected residual $U_k^\top(A - B_k)U_k$ at each stage; significant corrections occur if U_k aligns with large eigenvalues of R_k. Block power iteration is a standard heuristic [7] to enhance alignment.

Incorporating p steps of a block power iteration to enrich U_k produces the hybrid algorithm in Algorithm 2: the loop from line 4 to line 9 enriches a random U by multiplying by the residual and re-orthogonalizing p times. As before, work estimates are boxed on the right (p block power iterations require $p\,n\,s$ and the square symmetric sample requires s^2) with the total double boxed. Although the inner iteration requires significantly more matrix samples per iteration, conventional wisdom [7] suggests one or two inner iterations are likely to be beneficial. Our experiments show Algorithm 2 is competitive for $p = 2$.

Require: $B_0 \in \mathbb{R}^{n\times n}$ satisfying $B_0^\top = B_0$, SPD $W \in \mathbb{R}^{n\times n}$, $s \in \mathbb{N}$.
1: **repeat** $\{k = 0, 1, \ldots\}$
2: Sample $U_{0,k} \sim \mathcal{N}(0,1)^{n\times s}$
3: $B_{0,k} = B_k$
4: **loop** $\{i = 1, 2, \ldots, p\}$
5: $\Lambda = AU_{i-1,k} - B_{i-1,k}U_{i-1,k}$
6: $\Sigma = \Lambda(U_{i-1,k}^\top W\, U_{i-1,k})^{-1}U_{i-1,k}^\top W$
7: $B_{i,k} = B_{i-1,k} + \Sigma + \Sigma^\top - W\, U_{i-1,k}(U_{i-1,k}^\top W\, U_{i-1,k})^{-1}U_{i-1,k}^\top\Sigma$
8: $U_{i,k} = \Lambda$
9: **end loop** .... $\boxed{p\,n\,s}$
10: Compute $R_k = U_{p,k}^\top A\, U_{p,k} - U_{p,k}^\top B_{p,k}U_{m,k} \in \mathbb{R}^{s\times s}$ .... $\boxed{s^2}$
11: Compute $\tilde{P}_k = W\, U_{p,k}(U_{p,k}^\top W\, U_{p,k})^{-1}$
12: Update $B_{k+1} = B_k + \tilde{P}_k\, R_k \tilde{P}_k^\top$
13: **until** convergence
14: **return** B_{k+1} .... $\boxed{\boxed{(k+1)(p\,n\,s + s^2)}}$

Algorithm 4: SS1A: Accelerated Symmetric Approximation

Acceleration Convergence Results

Algorithm 4 (rotationally invariant samples with $p = 2$) is compared to BFGS-A (the result of applying the SMW formula to the accelerated method AdaRBFGS from [6]) and the three one-sided, non-accelerated algorithms: S1 and DFP defined by Eq. (4) with weights $W = I$ and $W = A$ (respectively), and BFGS defined by applying the SMW formula to Eq. (5) with weight $W = A^{-1}$. We use the same test matrices, initialization, and termination conditions described in Sect. 6: Fig. 2a shows results for $A = XX^\top$; Fig. 2b shows results for the Hessian matrix Gisette Scale [3]; and Fig. 2c shows results for NASA4704 [4]. SS1A matches or outperforms all other algorithms for the three matrices. As before [13] contains MATLAB scripts and results for all matrices from [6]. Both

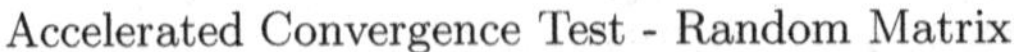

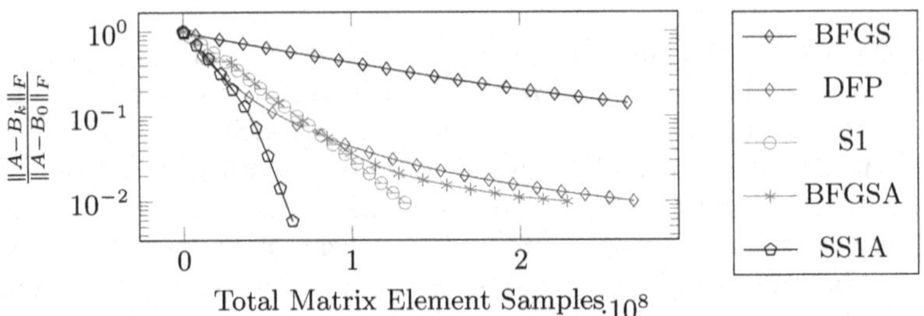

(a) ($n = 5000$) Approximation of $XX^\top$ where $X \sim \mathcal{N}(0,1)^{n\times n}$

Accelerated Convergence Test - LibSVM - Gisette-Scale

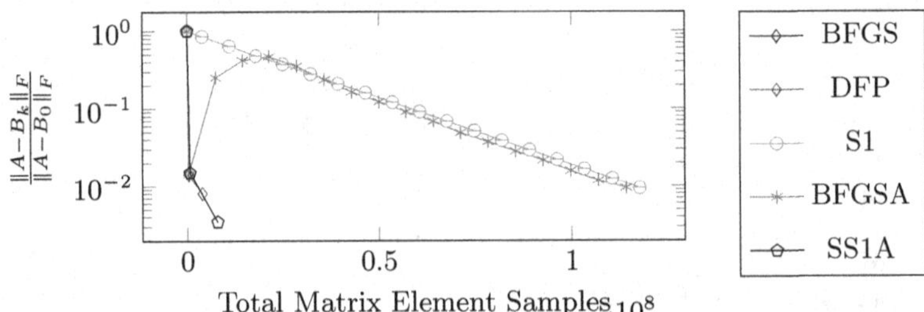

(b) ($n = 5000$) Hessian approximation of Hessian of **GisetteScale**

Accelerated Convergence Test - SparseSuite - NASA

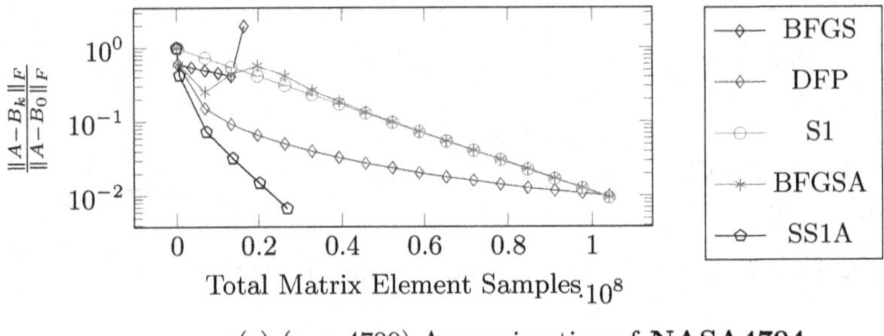

(c) ($n = 4700$) Approximation of **NASA4704**

Fig. 2. Accelerated SS1A algorithm outperforms accelerated BFGS algorithms.

BFGS-A and SS1A adaptively sample their update spaces. BFGS-A samples from the columns of the Cholesky decomposition of B_k while SS1-A effectively samples from a small power of the residual $A - B_k$. Comparing BFGS to BFGS-A and S1 to SS1-A shows the benefits of adaptivity. The hybrid SS1A performs consistently well.

8 Conclusions and Future Work

Algorithms 1 to 3 iteratively generate matrix approximations to a fixed target from two-sided samples. Rotationally invariant sampling gives optimal theoretical convergence in general and predicted convergence rates are experimentally verified for several real world matrices, with comparable performance to existing one-sided algorithms. A hybrid method combining simultaneous iteration (to enrich a subspace) with the two-sided sampled update is developed and shown to be competitive with existing one-sided accelerated schemes.

The algorithms systematically make minimal changes and drive weighted residual norms for a fixed A monotonically to zero. Such self-correcting algorithms can potentially approximate slowly changing matrices, $A(x)$. For example, QN optimization algorithms have a slowly changing Hessian target $\nabla_x^2 f(x_k)$ while solvers for stiff ODEs $y'(x) = f(y(x))$ have a slowly changing Jacobian target $\nabla_y f(y(x_k))$. The two-sided sampled matrix approximation algorithms and theory presented in the article provide a general foundation for these and other applications. Efficient factorized updates, compact low rank approximations, inverse approximation, and sparse matrix sampling are all planned.

References

1. Andoni, A., Nguyen, H.L.: Eigenvalues of a matrix in the streaming model, pp. 1729–1737. SIAM (2013). https://doi.org/10.1137/1.9781611973105.124
2. Avron, H., Clarkson, K.L., Woodruff, D.P.: Sharper bounds for regularized data fitting. In: Jansen, K., Rolim, J.D.P., Williamson, D., Vempala, S.S. (eds.) Approximation, Randomization, and Combinatorial Optimization. Algorithms and Techniques, Leibniz International Proceedings in Informatics (LIPIcs), vol. 81, pp. 27:1–27:22. Schloss Dagstuhl-Leibniz-Zentrum fuer Informatik, Dagstuhl (2017). https://doi.org/10.4230/LIPIcs.APPROX-RANDOM.2017.27
3. Chang, C.C., Lin, C.J.: LIBSVM: a library for support vector machines. ACM Trans. Intell. Syst. Technol. **2**, 27:1–27:27 (2011)
4. Davis, T.A., Hu, Y.: The university of florida sparse matrix collection. ACM Trans. Math. Softw. **38**(1), 1:1–1:25 (2011). https://doi.org/10.1145/2049662.2049663
5. Gao, W., Goldfarb, D.: Block BFGS methods. SIAM J. Optim. **28**(2), 1205–1231 (2018). https://doi.org/10.1137/16M1092106
6. Gower, R.M., Richtárik, P.: Randomized Quasi-Newton updates are linearly convergent matrix inversion algorithms. SIAM J. Matrix Anal. Appl. **38**(4), 1380–1409 (2017). https://doi.org/10.1137/16M1062053
7. Halko, N., Martinsson, P., Tropp, J.: Finding structure with randomness: probabilistic algorithms for constructing approximate matrix decompositions. SIAM Rev. **53**(2), 217–288 (2011). https://doi.org/10.1137/090771806
8. Karush, W.: Minima of Functions of Several Variables with Inequalities as Side Conditions. ProQuest LLC, Ann Arbor, MI (1939). Thesis (SM)-The University of Chicago
9. Li, Y., Nguyen, H.L., Woodruff, D.P.: On Sketching Matrix Norms and the Top Singular Vector, pp. 1562–1581. SIAM (2014). https://doi.org/10.1137/1.9781611973402.114

10. Naumann, U.: The Art of Differentiating Computer Programs. Society for Industrial and Applied Mathematics (2011). https://doi.org/10.1137/1.9781611972078
11. Nocedal, J.: Updating Quasi-Newton matrices with limited storage. Math. Comput. **35**(151), 773–782 (1980). https://doi.org/10.2307/2006193
12. Nocedal, J., Wright, S.J.: Numerical Optimization. Springer Series in Operations Research and Financial Engineering, 2nd edn. Springer, New York (2006). https://doi.org/10.1007/978-0-387-40065-5
13. Ong, B., Azzam, J., Struthers, A.: Randomized iterative methods for matrix approximation - supplementary material and software repository (2021). https://www.mathgeek.us/publications.html
14. Robbins, H., Monro, S.: A stochastic approximation method. Ann. Math. Stat. **22**(3), 400–407 (1951). https://doi.org/10.1214/aoms/1177729586
15. Schnabel, R.: Quasi-newton methods using multiple secant equations. Computer Science Technical Reports 244, 41 (1983). https://scholar.colorado.edu/csci_techreports/244/
16. Stewart, G.: The efficient generation of random orthogonal matrices with an application to condition estimators. SIAM J. Numer. Anal. **17**(3), 403–409 (1980). https://doi.org/10.1137/0717034

Improved Migrating Birds Optimization Algorithm to Solve Hybrid Flowshop Scheduling Problem with Lot-Streaming of Random Breakdown

Ping Wang[1,2], Renato De Leone[1(✉)], and Hongyan Sang[2]

[1] University of Camerino, Camerino, MC 62022, Italy
renato.deleone@unicam.it
[2] Liaocheng University, Liaocheng, LC 252000, China

Abstract. An improved migrating birds optimization (IMBO) algorithm is proposed to solve the hybrid flowshop scheduling problem with lot-streaming of random breakdown (RBHLFS) with the aim of minimizing the total flow time. To ensure the diversity of the initial population, a Nawaz-Enscore-Ham (NEH) heuristic algorithm is used. A greedy algorithm is used to construct a combined neighborhood search structure. An effective local search procedure is utilized to explore potential promising neighborhoods. In addition, a reset mechanism is added to avoid falling into local optimum. Extensive experiments and comparisons demonstrate the feasibility and effectiveness of the proposed algorithm.

Keywords: Hybrid flowshop · Lot-streaming · MBO algorithm · Random breakdown · NEH algorithm · Greedy algorithm

1 Introduction

Nowadays, most of the researches on hybrid flowshop scheduling problem with lot-streaming (HLFS) are from the perspective of a static environment [1]. However, the problem isintrinsically a dynamic constraint problem. The research on hybrid flowshop scheduling problem with lot-streaming of random breakdown (RBHLFS) has received much less attention. Nie et al. [2] proposed a improved Genetic Algorithm to solve the flexible scheduling problem subject to machine breakdown. Tian et al. [3] consider a semi-online scheduling problem on a single machine with an unexpected breakdown period.

The migrating birds optimization (MBO) algorithm was proposed by Duman in 2012 [4]. Zhang et al. [5] proposed an discrete MBO algorithm to solve the no-wait flowshop scheduling problem. Zhang et al. [6] proposed a variation of the MBO algorithm to solve the batch splitting scheduling flexible flowshop problem.

G. Nicosia et al. (Eds.): LOD 2021, LNCS 13164, pp. 241–245, 2022.
https://doi.org/10.1007/978-3-030-95470-3_18

2 Problem Statement

The RBHLFS can be described as follows. There are j jobs, which need to go through n processing stages, and each stage requires m (m>1) machines. At each stage, a job is processed according to his specific job sequence. Each job can be further divided into several sublots, and the sublots of different jobs are not allowed to be mixed on the same machine. One machine can only process one job at the time. Moreover, at any stage, only one machine can be selected for a job. During the processing, a machine can have a breakdown at any time. The objective of RBHLFS is to determine a permutation π^* of the set of all the jobs for which the total flow time is minimum. Take the ith sublot of job j in the stage n as an example, $CT_{n,j,i}$ is the completion time; $ST_{n,j,i}$ is the start time; $P_{n,j,i}$ is the processing time. The repair time of the malfunctioning machine is indicated by r. The RBHLFS can be stated as:

$$C(\pi^*) = min\ C(\pi) = min \sum_{j=1}^{n} CT_{n,j,i}, \quad \pi \in \Pi \tag{1}$$

Due to the uncertainty of the time when a machine breakdown occurs, we consider three situations (see Fig. 1).

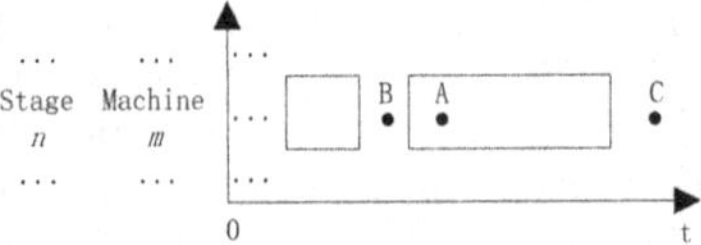

Fig. 1. Gantt chart of machine breakdown classification

A: If $ST_{n,j,i} \leq$ A $< CT_{n,j,i}$

$$ST_{n,j,i} = ST_{n,j,i}, CT_{n,j,i} = ST_{n,j,i} + P_{n,j,i} + r \tag{2}$$

B: If $CT_{n,j,i-1} \leq$ B $< ST_{n,j,i}$ or $CT_{n,j-1,i} \leq$ B $< ST_{n,j,i}$

$$ST_{n,j,i} = max\left\{ST_{n,j,i}, B + r\right\}, CT_{n,j,i} = ST_{n,j,i} + P_{n,j,i} \tag{3}$$

C: If C $\geq CT_{n,j,i}$

$$ST_{n,j,i} = ST_{n,j,i}, CT_{n,j,i} = CT_{n,j,i} \tag{4}$$

3 The IMBO Algorithm for RBHLFS

3.1 Population Initialization

For the flowshop scheduling problem, the NEH (Nawaz-Enscore-Ham) algorithm is currently one of the most effective heuristic algorithms [7]. A good initial population can make the algorithm quickly converge to the approximate optimal solution, and improve the efficiency and accuracy of the algorithm. Therefore, the NEH algorithm is used here to construct an initial solution, and other solutions are randomly generated.

3.2 Neighborhood Structure

A Greedy Algorithm is an algorithm that searches for the optimal solution locally, optimizing the scheduling goal. However, it cannot guarantee that the solution is a global optimal solution [8]. In this paper, a greedy algorithm is utilized to design two neighborhood structures N_1 and N_2, which are respectively denoted as insert greedy (IG) and exchange greedy (EG).

3.3 Local Search and Reset Mechanism

In order to improve the current solution, a local search is performed. A new solution is generated through insertion and exchange operations. The number of iterations is fixed a priori. If the new solution is better than the original solution, the current optimal solution is updated. If the maximum number of iterations is reached, the local search ends.

A reset mechanism is utilized in the IMBO algorithm. For each individual, the initial age is set to 1. If the individual evolves and the optimal solution is not updated, the individual's solution age is increased by 1. If the age exceeds a limit value and there is no change, the current individual is discarded, a local search is performed to generate a new individual, and the current optimal solution is updated.

3.4 The Proposed Algorithm

The complete process the IMBO algorithm can be described as follows:

Step 1: Algorithm initialization. Set the algorithm parameters and termination conditions, and set *flag*=1. The leader is generated by the NEH algorithm, the followers are randomly generated.

Step 2: Evolving of the leader. The k neighbourhood solutions of the leader are generated by the neighborhood structure N_1 and N_2, and the best neighboring solution is selected to update the original solution. Among the remaining solutions, x solutions are selected, and added to the shared neighborhood solution sets P_l and P_r, for the evolution phase as followers.

Step 3: Evolving of the followers. Based on the neighbourhood structures, k-x neighbourhood solutions are generated for each follower in the left lists, and are denoted by S. If the optimal solution in $S \cup P_l$ is better than the current solution, the new solution becomes the next follower. Then $P_l = \emptyset$. Add the other neighbourhood solutions in S to P_l for the evolution of another follower. The right list performs the same operation.

Step 4: Update the current solution and its age, perform the reset mechanism and a local search if no change occurs for L generations.

Step 5: Determine whether the maximum number of iteratives G has been reached. If not, go to step 2, otherwise, proceed to step 6.

Step 6: Leader change. If *flag*=1(0), the leader solution will move to the end of the left (right) list to become a follower. The first follower in the left (right) list moves forward to become the new leader set *flag*=0(1).

Step 7: If T has been reached, the algorithm ends; otherwise, go to step 2.

4 Experimental Results

The algorithms has been encoded in C++, and all experiments have been performed on an AMD A12-9700P CPU @3.4 GHz with 4.0 GB main memory with Windows 10 OS. In our experiments, we set the number of jobs between 20 and 100, while the number of stages has been fixed to 5 and 10. The frequency of machine breakdown is 1 and 30, and the machine repair time is r=2. We generate 10 instances, and performed 100 independent experiments on each instance. The relevant parameters of the algorithm are set as follows: s=51, G=10, k=3, x=1, L=50.

The relative percentage increase (RPI) is computed as follows. Let C represents the total flow time generated by each algorithm in the experiment, and let C^* be the minimum total flow time among all algorithms in the experiment. Then,

$$RPI = \frac{C - C^*}{C^*} \times 100\% \tag{5}$$

We compared the IMBO algorithm with other intelligent algorithms: Particle Swarm Optimization (PSO) [9], Genetic Algorithm (GA) [10], Invasive Weed Optimization (IWO) [11], and Artificial Bee Colony (ABC) [12]. The results are reported in Table 1. In all instances, the IMBO algorithm mostly achieves the minimum RPI value, and even surpassed other algorithms by a considerable amount. The average RPI value for the IMBO algorithm is the smallest, showing the effectiveness of the proposed IMBO algorithm.

Table 1. RPI values of different algorithm for RBHLFS

Instance	IWO	ABC	GA	PSO	IMBO
20× 5	0.24	0.32	0.24	0.28	**0.23**
20× 10	0.76	0.47	0.51	0.45	**0.43**
40× 5	0.64	0.66	0.32	0.29	**0.29**
40× 10	0.24	0.87	0.37	**0.16**	0.49
60× 5	0.95	0.85	**0.30**	0.69	0.49
60× 10	0.57	0.84	0.89	**0.29**	0.60
80× 5	0.57	0.64	0.68	1.02	**0.50**
80× 10	0.81	0.93	0.96	0.84	**0.74**
100× 5	0.88	0.91	0.83	0.69	**0.74**
100× 10	1.02	1.30	1.07	0.94	**0.91**
Mean	0.67	0.78	0.63	0.57	**0.54**

5 Conclusions

In order to solve the hybrid flowshop scheduling problem with lot-streaming of random breakdown (RBHLFS), an improved migrating birds optimization

(IMBO) algorithm is proposed to optimize the total flow time. The Nawaz-Enscore-Ham (NEH) heuristic algorithm is used to generate the initial solution. A greedy algorithm is instead used to construct a neighborhood search structure. Then a local search method is utilized. In addition, a reset mechanism is added to avoid falling into local optima. In randomly generated example of RBHLFS, computational experiments show that the IMBO algorithm is more effective and shows better performance than other algorithms from the literature.

References

1. Peng, K., Pan, Q., Zhang, B.: An improved artificial bee colony algorithm for steelmaking-refining-continuous casting scheduling problem. Chin. J. Chem. Eng. **26**(8), 1727–1735 (2018)
2. Nie, L., Wang, X., Liu, K., Bai, Y.: A rescheduling approach based on genetic algorithm for flexible scheduling problem subject to machine breakdown. J. Phys. Conf. Series, 1453-012018 (2020)
3. Tian, J., Zhou, Y., Fu, R.: An improved semi-online algorithm for scheduling on a single machine with unexpected breakdown. J. Combin. Optim. **40**(1), 170–180 (2020). https://doi.org/10.1007/s10878-020-00572-6
4. Duman, E., Uysal, M., and Alkaya, A.F.: Migrating birds optimization: a new metaheuristic approach and its performance on quadratic assignment problem. Inf. Sci., (217) (2012)
5. Zhang, S.J., Gu, X.S., and Zhou, F.N.: An improved discrete migrating birds optimization algorithm for the no-wait flow shop scheduling problem. IEEE Access (99), 1 (2020)
6. Zhang, M., Tan, Y., Zhu, J., Chen, Y., and Chen, Z.: A competitive and cooperative migrating birds optimization algorithm for vary-sized batch splitting scheduling problem of flexible job-shop with setup time. Simul. Modell. Practice Theory (100), 102065(2019)
7. Wang, P., Sang, H., Tao, Q., Guo, H., Han, Y.: Improved migrating birds optimization algorithm to solve hybrid flowshop scheduling problem with lot-streaming. IEEE Access (99), 1 (2020)
8. Mao, J., Hu, X.L., Pan, Q.K., Miao, Z., Tasgetiren, M.F.: An iterated greedy algorithm for the distributed permutation flowshop scheduling problem with preventive maintenance to minimize total flowtime. In: 2020 39th Chinese Control Conference. https://doi.org/10.23919/CCC50068.2020.9189642
9. Marichelvam, M.K., Geetha, M., Tosun, M.: An improved particle swarm optimization algorithm to solve hybrid flowshop scheduling problems with the effect of human factors-a case study. Comput. Oper. Res. (114), 104812 (2019)
10. Chen, T.L., Cheng, C.Y., Chou, Y.H.: Multi-objective genetic algorithm for energy-efficient hybrid flow shop scheduling with lot streaming. Ann. Oper. Res. **290**(4), 813–836(2020)
11. Sang, H.-Y., Pan, Q.-K., Duan, P.-Y., Li, J.-Q.: An effective discrete invasive weed optimization algorithm for lot-streaming flowshop scheduling problems. J. Intell. Manuf. **29**(6), 1337–1349 (2015). https://doi.org/10.1007/s10845-015-1182-x
12. Huang, J.P., Pan, Q.K., Miao, Z.H., Gao, L.: Effective constructive heuristics and discrete bee colony optimization for distributed flowshop with setup times. Eng. Appl. Artif. Intell. (97), 104016(2021)

Building Knowledge Base for the Domain of Economic Mobility of Older Workers

Ying Li(✉), Vitalii Zakhozhyi, Yu Fu, Joy He-Yueya, Vishwa Pardeshi, and Luis J. Salazar

Giving Tech Labs, 721 4th Avenue #699, Kirkland, WA 98033, USA
yingli@evanalysiscorp.com, luis@giving.tech

Abstract. This paper presents the work of building a knowledge base for the domain of economic mobility for older workers. To extract high-quality entities and relations that are important to the specific domain, domain specificity scores for entities and relations are designed and applied. To assist human-in-the-loop ontology construction, a novel topic modeling method, named "description guided topic modeling", is developed. It clusters domain entities based on their embedding and organizes those clusters according to descriptions of potential topics important to the domain. To demonstrate feasibility, these methods are applied to a collection of knowledge sources related to economic mobility for older workers. These methods are further tested through a case study on one specific barrier for economic mobility, i.e., limited broadband access for older workers, to show the potential of these methods.

Keywords: Knowledge base · Domain specific lexicon · Knowledge graph · Social impact · Ontology · Economic mobility · Older workers

1 Introduction

The social and economic security of older Americans has been disrupted by the fast-paced changes in the work environment, automation, and artificial intelligence trends. Women and men over 65 experience many barriers to economic mobility that create uncertainties with their employment and workforce development. Moreover, older Americans' plight has been made even starker by the COVID-19 pandemic, which deteriorated the lives of people of age 65+ [1,2,8].

The urgency and importance of issues related to older workers' economic mobility call for serious actions to develop and implement comprehensive policies and new opportunities for people of age 65+. The starting point is the need for relevant and trustworthy information. However, this information is not well-organized and presented as useful insights; instead, the search for knowledge is often filled with noise and disinformation. Moreover, even if there are trusted information sources on a particular social issue, gathering insights on the issue often relies on a manual process that is inefficient, lengthy, and costly.

Funding for this research was partially provided by CWI Labs, a wholly-owned subsidiary of the Center for Workforce Inclusion, a national nonprofit organization.

G. Nicosia et al. (Eds.): LOD 2021, LNCS 13164, pp. 246–260, 2022.
https://doi.org/10.1007/978-3-030-95470-3_19

The purpose of the work presented in this paper is to address the need for organized knowledge through building knowledge bases using knowledge graph methodologies. A knowledge graph describes entities and their interrelations, organized in a graph, with a schema that defines possible entities and relations between entities and allows for potentially interrelating arbitrary entities [7,20]. Knowledge graphs have been used with demonstrated efficacy in solving problems addressing complex social challenges, for instance, understanding the impact of the opioid crisis in the U.S. [11], combating human trafficking [12].

The challenges in building knowledge bases for social issues are manifold. The *lack of trusted knowledge sources* makes the relevant domain contents extremely sparse and continually drowned out by irrelevant contents that add large amount of noise for knowledge extraction. The *lack of established domain information architecture* means *data cannot be systematically annotated*, limiting the applicability of supervised ML methods even when enough raw data are gathered.

Our approach for addressing these intertwined challenges is a semi-automated solution with human-in-the-loop [6,13] for scoping the domain and learning domain information architecture, outlined in Fig. 1. The tool loop enables iterative execution of entity and relation extraction, phrase clustering, and topic modeling. The human loop supports domain experts to provide feedback for tuning parameters such as thresholds for entity and relation extraction and cluster grouping for topic modeling.

We tested the feasibility on three sets of knowledge sources, listed in Table 1. These knowledge sources are gathered from domain experts, including nonprofit representatives and academic researchers, in the forms of white papers, verified web pages, academic research articles, official documents, etc. The diverse sets of knowledge sources ensure sufficient coverage over the domain for bootstrapping topic modeling and relation extraction.

We further conducted validation of discovered knowledge on the issue of limited broadband access for older workers, against knowledge sources of *Broadband Access and Affordability*, the full results of which are presented in a white paper [24] due to space limit of this paper. Positive validation demonstrates the potential of our methods in tackling the challenges, whereas lack of benchmark data for automated objective evaluation remains a challenge.

Our main contributions are:

1. methods for domain specific entity extraction, phrase extraction, and relation extraction utilizing *domain specificity scores*, resulting in a domain lexicon built for the domain of Economic Mobility of Older Workers (Sect. 2)
2. a novel application of topic modeling (Sect. 3), we named *description guided topic modeling*, an iterative refinement of large number of topic clusters according to brief topic descriptions from human input. This enabled a human loop, the result of which is the domain ontology (Sect. 4)
3. a case study focusing on the issue of broadband access for older workers that prevent them from being active in workforce (Sect. 5)

Throughout this paper, references to concepts and algorithms are provided in context instead of a separate section, to maximize the benefit for practitioners.

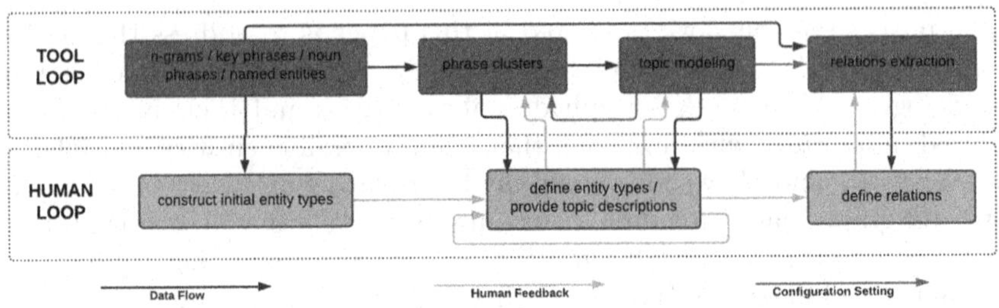

Fig. 1. High level overview of data flow between components and support for human-in-the-loop to build domain lexicon and construct domain ontology.

Table 1. Knowledge sources

Knowledge Source	Description
Economic mobility & Barriers for older workers	Our main knowledge source, on the topics of older adults and aging, economic mobility and barriers, evolving workspaces, workforce development, and existing policies and opportunities landscape for improving economic mobility of older workers
Broadband access and affordability	Our knowledge source for case study (Sect. 5) on the topics of broadband access, impact of broadband on the economic security and health of older people, broadband infrastructure, digital divide, and connectivity policies on both federal and state level
X4Impact Social Challenges https://x4i.org	Our test knowledge sources for running our tool loop on multiple domains simultaneously, contains documented social challenges introduced from nonprofits, academia, and private sectors, covers social issues related to all 17 United Nations Sustainable Development Goals

2 Building Domain Lexicon

A domain lexicon consists of key phrases, noun phrases, named entities, and relational verbs that are important for properly describing the domain. Early work in phrase mining and entity recognition generally followed supervised approach. More methods started to utilize corpus with human annotation, such as wikipedia, whereas unsupervised methods for phrase mining and entity and relation extraction typically rely on heuristic rules and example templates, as well as basic syntactic knowledge and corpus statistics [5,14]. The work in [23] represents one recent state of the art in phrase mining and provides a comprehensive overview of some established methods. It also pointed out that human feedback is necessary for tuning the algorithm parameters, such as thresholds, not only for unsupervised methods but also for well developed supervised methods.

Below, we first introduce the concept of domain specificity score in Sect. 2.1. It utilizes domain statistics for measuring individual term and phrase's importance to a domain in comparison with a general domain, such as wikipedia. Experiment details for entity extraction and relation extraction are presented in Sect. 2.2 and Sect. 2.3.

2.1 Domain Specificity Score

We regard the notion of a phrase being of importance or interest to a domain as *domain specificity.* The domain corpus may contain domain specific phrases as well as phrases commonly occurring in a general domain or a domain different from the current domain of interest. Although useful for understanding the domain, general words and phrases may crowd out the domain-specific ones. For instance, word *geriatric* may be important for describing the domain of older workers, it may still be in the long tail of the domain corpus. This calls for a method to measure the domain specificity of each word and phrase.

Domain Specificity Scores for Unigrams. We adopt the concept of *term frequency - inverse sentence frequency* (TF-ISF) [15], which is an adaptation of the conventional TF-IDF by replacing *document* with *sentence* in the computation. For a given word w, we first compute its tf-isf for each sentence s and sum up its tf-isf values across all sentences. Formally, we calculate $\textit{tf-isf}(w)$ as:

$$\textit{tf-isf}(w) = \sum_{s \in \mathcal{S}} \textit{tf-isf}(w, s) \tag{1}$$

where $\mathcal{S}$ is the set of all sentences in the given corpus.

Then, we multiply $\textit{tf-isf}(w)$ by the log of ratio of document-level word density in the domain corpus over the raw word density in the general corpus to get the final domain specificity score for w:

$$\text{domain-specificity}(w) = \textit{tf-isf}(w) \times \log\left(\frac{\frac{dc_{\mathcal{D}}(w)}{DN_{\mathcal{D}}}}{\frac{c_{\mathcal{G}}(w)}{N_{\mathcal{G}}}}\right) \tag{2}$$

where $dc_{\mathcal{D}}(w)$ is the document-level word frequency for w in the domain corpus $\mathcal{D}$ (i.e., the number of documents in $\mathcal{D}$ that contain w), $c_{\mathcal{G}}(w)$ is the raw word frequency for w in the general corpus $\mathcal{G}$ (i.e., the number of times w appears in $\mathcal{G}$), $DN_{\mathcal{D}}$ is the total number of documents in $\mathcal{D}$, and $N_{\mathcal{G}}$ is the total number of words in $\mathcal{G}$.

Domain Specificity Scores for Phrases. Similarly, for a given n-gram p that consists of multiple words, we define its domain specificity score as:

$$\text{domain-specificity}(p) = \textit{tf-isf}(p) \times \max_{w \in \mathcal{P}} \log\left(\frac{\frac{dc_{\mathcal{D}}(w)}{DN_{\mathcal{D}}}}{\frac{c_{\mathcal{G}}(w)}{N_{\mathcal{G}}}}\right) \tag{3}$$

where $\mathcal{P}$ is the set of words in the phrase p, and $\textit{tf-isf}(\text{p})$ is computed the same as Eq. 1.

Our metric for domain specificity uses the tf-isf value to up-weight terms that appear in more sentences but not too many sentences, which helps us filter out misspellings or invalid words and words that appear more frequently in general. In the meantime, the ratio term up-weights words or phrases that have relatively high density in the domain corpus versus the general corpus. Thus, higher domain specificity scores indicate that the words or phrases are more domain-specific.

In our computation of the domain specificity scores, we used the Wikipedia Download from March 2019 as the general corpus. We also experimented with other commonly used general corpora, such as the Reuters Corpora, the New York Times Annotated Corpus, and Google Web Trillion Word Corpus. Compared with Wikipedia March 2019, these earlier corpora do not perform well due to less coverage on words related to topics and issues emerging more recently.

2.2 Phrase Extraction and Term Recognition

We scraped the data from trusted knowledge sources, parsed and retained relevant chunks of text, and conducted conventional NLP pre-processing before we applied domain specificity scores for extracting key phrases, noun phrases, and named entities.

Domain-specific Key Phrases Extraction. We extracted n-grams (n = 1, 2, 3, and 4) and computed their domain specificity scores as defined in Sect. 2.1 to identify the top domain-specific keywords in our knowledge sources. Since some stop-words might have high tf-isf values and crowd out the domain specific terms, we removed all terms with a tf-isf value that is greater than 5. Table 2 shows some examples of extracted n-grams ranked by their domain specificity scores that shows the difference between high and low domain specificity scores. With human input, we can set the threshold as 20 for selecting domain specific key phrases.

Domain-specific Noun Phrase Extraction. We used regular expression of POS (Part-Of-Speech) tagging to extract noun phrases from knowledge sources. Since knowledge sources consist of many different web pages and PDF documents of diverse characteristics, there is inevitably noise in the results of automated text content scraping and parsing. Heuristic rules concerning length outliers or frequency outliers are applied to remove repetitive text elements from headers, footers, table of contents, organization taglines, or article titles that are usually not proper noun phrases and contribute little to lexicon building. Domain specificity score for each extracted noun phrase is calculated using Eq. 3. In Table 3, we present some examples of domain-specific noun phrases that are above threshold 20 as we did for key phrase extractions.

Domain-specific named Entity Recognition. We used Microsoft Azure Text Analytics API for Named Entity Recognition (NER). The entities identified belong to various pre-defined classes such as Organization, Skill, Event, Product etc. We then calculated the domain specificity scores for the extracted entities using Eq. 3.

2.3 Relation Extraction

Relation extraction has been an active topic in research with available methods that are supervised, unsupervised, or distance supervised. Supervised methods are generally not feasible for a domain with unknown information structure.

Table 2. Examples of extracted phrases with high and low domain specificity scores.

Key Phrase	Domain Specificity Score
An aging workforce	64.0
Telemedicine	53.8
Aging in america	50.8
Geriatrics	46.6
Broadband connectivity	42.4
Alaska native elders american	41.9
Effects of disability discrimination	38.4
People living with dementia	35.3
Counseling	5.0
Engage in meaningful activities	0.1

Table 3. Examples of domain-specific noun phrases with high domain specificity scores.

Noun Phrase	Domain Specificity Score
Effects of disability discrimination laws	39.0
The deployment of broadband	38.7
Growing demand of soft skills	35.0
Disability discrimination laws on hiring	30.0
COVID in nursing homes	25.6
Broadband service in unserved areas	21.6

Strictly unsupervised methods may extract large amount of relations that are hard to be mapped to important types of relations needed for the given domain. Distance supervised methods [18] utilize large datasets that already have semantic information, such as Freebase or Wikipedia annotation, and achieved good performance without having to incur high cost of obtaining labeled data. Recent works such as [22] on inter-sentence relation extraction using graph convolution network have produced state-of-the-art results but still require basic labels on the types of relations in the graph.

Our relation extraction makes use of domain specificity scores for selecting phrases and entities for candidate mention, and conducts entity linking to enable extraction of inter-sentence relations. We then conduct post-processing to cluster relation verbs to make the results consumable in the human loop (see Fig. 1).

Entity Linking. Entity linking enables replacement of references of entities by standard entity form. We implemented entity linking through three sub-tasks: 1) *abbreviation resolution* where entities are resolved via common abbreviation; 2) *co-reference resolution* utilizing *NeuralCoref*; 3) *deduplication* through fuzzy-matching using tf-idf, n-grams, and cosine similarity.

Domain-specific Relation Extraction. Relations are extracted between two entities or noun phrases which are of the form Entity—Verb—Entity or Noun Phrase—Verb—Noun Phrase. We applied POS tagging and entity-linking to

Table 4. Examples of extracted relations.

Entity/Noun phrase 1 (Type)	Relation	Entity/Noun phrase 2 (Type)
Senate aging committee (Organization)	Hosted by	AARP (Organization)
SCAN Foundation (Organization)	Led by	National Association of Area Agencies on Aging (Organization)
Workplace injuries (Barrier)	Compared with	Younger counterparts (Beneficiary)
Younger counterparts (Beneficiary)	Contend with	Discrimination against older workers (Barrier)

Table 5. Examples of clustered verb phrases.

Verbs in cluster	Avg Domain Specificity Score	Entity—Relation—Entity
'improving', 'improved by', 'improve with', 'improved with', 'improved in', 'improved at', 'improve', 'improves'	39.5	broadband communications—improve—increase broadband availability at affordable costs
'prioritize', 'prioritized', 'prioritising', 'prioritize for', 'prioritizing'	17.0	changing hiring policies—prioritize—hiring skilled workers
'minimizing', 'understates', 'minimizes', 'understate', 'minimize'	7.9	the impact on older workers—understate—the pandemic

identify direct relations, parallel relations, and inter-sentence relations. Table 4 shows some examples of extracted relations.

Such extracted relations can be very noisy and in large numbers. To make them useful, we performed post-processing steps: 1) generate clusters of semantically similar relation-verbs; 2) compute the domain specificity score for each relation-verb cluster as the average of the domain specificity score for each verb/web phrase by Eq. 2 and Eq. 3; 3) use the domain specificity score for the verb cluster to differentiate relevant and important relations from generic and less important ones. Table 5 presents some examples of verb clusters, their scores and sample relations in the cluster. We can observe that higher domain specificity scores correspond to verb clusters that are more specific to the domain of broadband access.

3 Description Guided Topic Modeling

Topic modeling helps tackle the challenges of sparse data, unlabeled data, and an unknown domain structure [10]. To learn the key concepts that are critical for constructing the domain ontology, we developed a novel application of clustering and word embedding that we named *description guided topic modeling*. The goal of description guided topic modeling is to group unstructured noun phrases from a corpus into a domain knowledge framework, and eventually map them to entity types to form the basis for ontology construction.

Recall our overall workflow as depicted in Fig. 1, the tool loop takes input from a domain expert in the form of keyword descriptions about potential domain topics, runs clustering with respect to the descriptions, and provides evaluation metrics back to the expert. Based on the evaluation metrics, the expert can decide to refine the topic descriptions and run a new iteration of the topic modeling. The iterative process stops when the expert is satisfied with the collection of topics along with their corresponding entities.

3.1 Algorithm Details

Shown in Algorithm 1, our algorithm for description guided topic modeling is a semi-supervised, bottom up, two-step adaptive clustering algorithm that assigns all extracted noun phrases to one of the predefined topics.

1. Clustering step: an unsupervised clustering of the extracted noun phrases. We adopted the spherical k-means as it is an appropriate method for clustering of text embedding spaces [16]. The number of unsupervised clusters should be significantly bigger than the number of predefined topics.
2. Topic Assignment step: semi-supervised assignment of clusters to topics via topic descriptions provided by the human expert. Each unsupervised cluster is mapped to the topic closest to it in the distance from cluster center and the keyword descriptions of the topics.

3.2 Experimentation Settings and Results

As input to Algorithm 1, the noun phrases and topic keyword descriptions are pre-processed and mapped into a text embedding space by the pre-trained FastText model [4]. We designed and conducted experiments to search for the optimal configurations regarding: 1) the number of clusters to run k-means; 2) metrics to use for assigning clusters to topics; 3) the embeddings to use.

Number of Clusters in k-Means. The desired clustering results are compact clusters that contain semantically similar noun phrases. We use the skewness of the distribution of average within-cluster distances across all the clusters for inspecting clustering results. The average within-cluster distance is calculated by taking the average of the cosine distances between the centroid and all noun phrases of the cluster. When the skewness is high, the distribution skews to the left, and more clusters have lower average within-cluster distances. This indicates that the algorithm generates more compact clusters. Therefore, we are primarily looking for the number of clusters that give high skewness. Additionally, we also check for small clusters because we do not want to generate clusters with only one or too few noun phrases.

The number of clusters in the k-means algorithm for description guided topic modeling is determined by combining the Elbow Method and our interpretation

Algorithm 1: DESCRIPTION GUIDED TOPIC MODELING

Input: Noun phrases $NP_1, ..., NP_N$
Topic descriptions $TD_1, ..., TD_M$
Number of clusters for k-means algorithm K, $K > M$
Text embedding function $\mathscr{F}$
Output: Topic assignment $\mathbf{T}$ such that $T_i = j$, meaning noun phrase NP_i belongs to topic j
$\mathbf{X} \leftarrow [x_1, x_2, ..., x_N] \leftarrow [\mathscr{F}(NP_1), \mathscr{F}(NP_2), ..., \mathscr{F}(NP_N)]$
$\mathbf{Y} \leftarrow [y_1, y_2, ..., y_N] \leftarrow [\mathscr{F}(TD_1), \mathscr{F}(TD_2), ..., \mathscr{F}(TD_M)]$
// c_i is the cluster assignment of NP_i
// μ_k is the center of cluster k
$\mathbf{c}, \mu \leftarrow$ **Spherical-KMeans**(X, K)
for k *from 1 to* K **do**
    // a_k is the assignment of cluster k to the topic with the closest topic description
    $a_k \leftarrow \arg\min_{j \in [1,M]}$ **Cosine-Dist**(y_j, μ_k)
end
for i *from 1 to* N **do**
    // The topic assignment of NP_i is the same topic assignment of NP_i's cluster
    $T_i \leftarrow a_{c_i}$
end
return $\mathbf{T} \leftarrow [T_1, T_2, ..., T_N]$

of the number of small clusters, and the skewness of the distribution of average within-cluster distances. The commonly used Elbow Method for k-means chooses the number of clusters by looking at the inertia plot's turning corner. The inertia, or within-cluster sum of squares, is defined as the squared sum of within-cluster distances from all the noun phrases in a cluster to the cluster's centroid, and summed across all the clusters.

We plot the skewness versus the number of clusters in Fig. 2. The chosen value for number of clusters is marked with red ($K = 264$) with high skewness.

Topic Assignment Metrics and Topic Description Refinement. Recall the topic assignment step in Algorithm 1, clusters are assigned to topics, hence a topic can be viewed as a group of clusters.

We define the linkage criterion L between two topics as the median of all pairwise cosine distances between noun phrases from each topic:

$$L(T_1, T_2) = \mathbf{Median}\{D(\mathscr{F}(NP_i), \mathscr{F}(NP_j))\}$$
$$\forall NP_i \in T_1, \forall NP_j \in T_2$$

This linkage criterion reflects the distance between two topics as large clusters of noun phrases. Higher values of the linkage criterion indicate that the corresponding clusters are better at separating topics in the embedding space.

Topic descriptions are small sets of keywords that human expert provided initially and then refined after each round of topic assignment according to: 1)

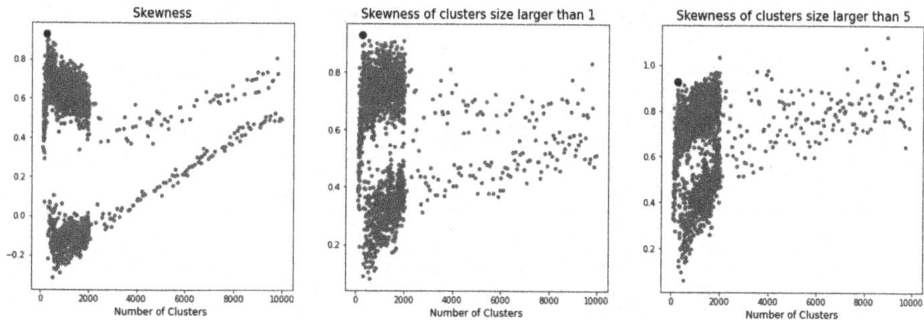

Fig. 2. Skewness over spherical k-means clusters: over all clusters (left), over clusters of size great than 1 (center), and over clusters of size greater than 5 (right).

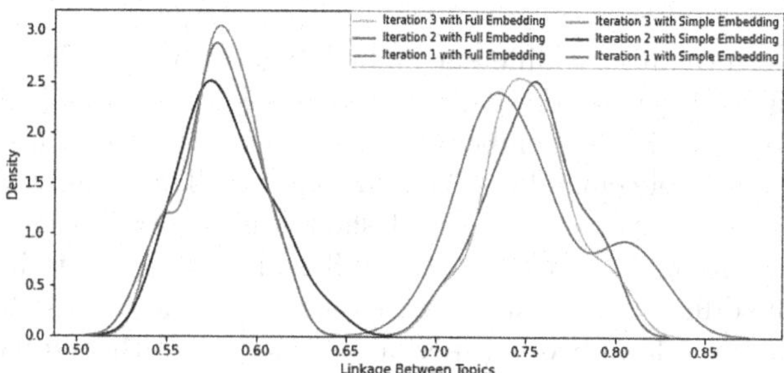

Fig. 3. Comparison of linkage criterion of topic modeling between iterations of topic description refinements and embeddings.

specificity of the words belonging to the topic with minimum overlap between the topics descriptions; 2) *avoidance* of general or contextual words for simple yet comprehensive definitions of the topics; 3) *iterative learning* from each iteration to design topics and description for the next iteration.

Three iterations of topic description refinement were conducted. Figure 3 shows the distributions of linkage criterion for all pairs of topics for each iteration under different embedding. From the plot, we can see that the distribution of linkage criterion of iteration 2 and 3 shifts towards the higher end compared to the distribution of iteration 1. Also, the distribution of iteration 3 converges better compared to that of iteration 2.

Text Embeddings. We chose FastText embedding because it captures subword semantic information better when compared to other methods like Word2Vec [17] and GloVe [21]. We experimented with two different FastText embeddings. The first embedding is trained on English Wikipedia (referred to as "full") and the second embedding is trained on Simple English Wikipedia (referred to as "simple"). The simple English Wikipedia has less pages and is limited to 850 English words and simple sentence structures. Our experiments

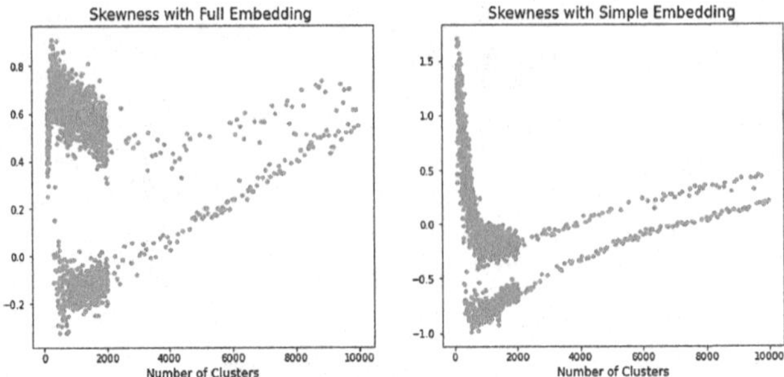

Fig. 4. Comparison of the skewness of the distribution of average within-cluster distances between full embeddings (left) and simple embedding (right).

compare the performance of different embeddings in both K-Means and topic assignment. Figure 4 depicts the K-Means skewness for both embeddings.

We can see that the full embedding converges with the large number of clusters while the simple embedding fails to converge. The comparison of linkage criterion for topic assignments in Fig. 3 shows the lower linkage criterion for the simple embedding than that of the full embeddings. This indicates that the simple embedding is less effective in separating the topics. We can draw conclusion that the full embedding performs better than the simple embedding in both K-means and topic assignment.

4 Constructing Domain Ontology

A domain ontology is a formal specification of concepts and relationships between them in a particular domain of discourse [9]. Developing an ontology, as outlined in [19], includes: 1) defining entities (concepts, classes) in the ontology; 2) defining relations between the entities; 3) defining properties allowed or restricted for the entities and relations. Ontology development process utilizes *competency questions* to help scoping a domain and specifying focus areas of interest. Below are some competency questions for the domain of economic mobility of older workers:

- What barriers prevent older people from improving their economic mobility?
- What organizations focus on a specific barrier to economic mobility? What organizations work to support older workers' economic mobility?
- What population groups are impacted by an issue and who are the beneficiary groups of the programs?
- What is the status quo on economic mobility and what are the measurements and indicators that describe a specific barrier?

Due to the scope and heterogeneity of the domain, learning about the actual concepts (entities), attributes, and relation values can be too costly and hard to scale if it is done fully manually with just human experts defining all necessary

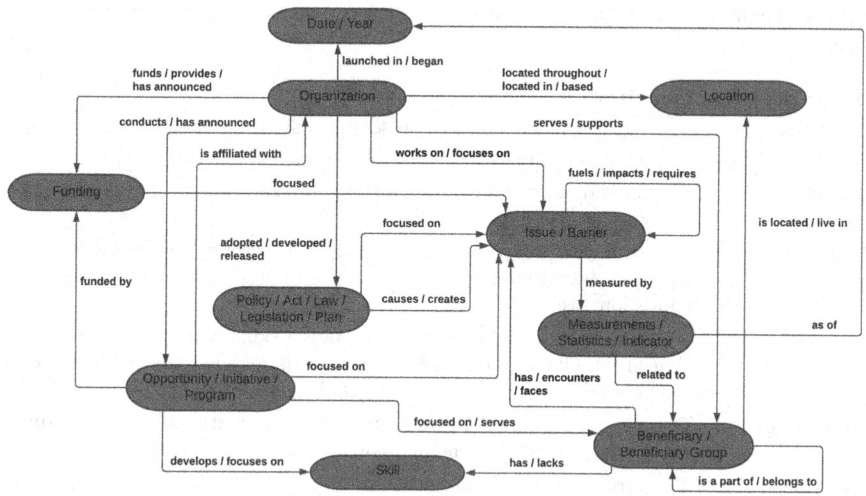

Fig. 5. Ontology for a domain-specific knowledge graph for economic mobility of older workers.

knowledge. At the same time, fully automated learning and ontology mining can often be characterized as low quality because of excessive noise [3].

We used a semi-automated human-in-the-loop approach for ontology construction. As described in Fig. 1, the results from tool loop (detailed in Sect. 2 and Sect. 3) provide input to human loop, where human intervention takes on the form of continuous input-feedback loop. Human interventions include: adjusting data corpus, setting term frequency cut-off thresholds, providing topic descriptions, adjusting number of clusters for topic modeling, selecting appropriate evaluation metrics and benchmarks, reviewing and verifying experiment results. The tool loop takes human expert's input and returns next iteration of results and evaluation metrics back to the human expert who in turn learns more about the domain.

Figure 5 presents the final constructed ontology for the domain of economic mobility of older workers, resulted from the above mentioned iterative human-in-the-loop learning process. The definitions of the entity types are presented in Table 6. With this ontology, we can populate the knowledge base with instances and values for the entities, relations, and their properties.

5 Case Study on the Issue of Broadband Access

To validate the results of the above knowledge discovery process, we conducted a case study on the specific topic of broadband access. We applied our methods against knowledge sources of *Broadband Access and Affordability* (see Table 1), which provide a good coverage on this issue in our corpus and our domain experts can provide quality feedback during results testing and validation. Below

Table 6. Definition of entity types of the Ontology on economic mobility of older workers

Entity type	Definition
Issue/Barrier	Central part of the ontology. It is a problem that is often discussed or argued about, and that affects the interests of an older worker
Measurement	Specific indicators and rates which can characterize or describe a certain instance of the issue entity
Beneficiary group	A person or a group of people with certain specific characteristics like age, race and ethnicity, socioeconomic status, income, profession, education, etc.
Skill	Abilities or expertise in some area. Assign skills as a separate entity type elevate its importance in bridging existing gaps that prevent older workers from being an active part of the workforce
Organization	Organizations (national private or public, international) which conduct their activity within economic mobility/older worker arenas
Opportunity	Programs, initiatives, and other training opportunities which can help older workers gain new skills or new knowledge
Policy	Laws, plans, or interventions which can either address one of the issues or create a challenge related to older workers
Funding	Grants and other financial resources and tools aimed at solving or supporting a certain issue
Date	Time related entity
Location	Geographical area

is a brief summary of the case study, the full results are presented in a white paper [24] due to space limit of this paper.

After unsupervised extraction of a domain lexicon from the sources, we aggregated and analyzed the extraction results of key phrases, noun phrases, named entities, and relation verbs as well as the results of the description guided topic modeling method. We applied domain specificity scores to extract entities and relationships that are more specific to the domain. These results were validated and approved by the domain experts. The positive validation demonstrates the potential of our methods in tackling the challenges that we outlined in Sect. 1.

We evaluated the knowledge discovery process through testing against the competency questions and validate the results with domain experts. The competency questions presented in the Sect. 4 are addressed through showcasing extracted results. The white paper explores deeper the issue of the broadband access gap for older Americans and provides findings that can be used to promote a better policy agenda to help solve the issue.

6 Conclusions and Future Work

In this paper we presented a set of tools and methods we developed for building a knowledge base for the domain of economic mobility of older workers. We demonstrated that the domain specificity score is effective for building domain lexicon from raw data when there is no labeled data nor categorical knowledge about the domain. Our description guided topic modeling method helps the learning of the domain ontology through an iterative human-in-the-loop approach. The resulting ontology and domain scoping were successfully applied in a case study for insight generation.

We recognize the need for objective evaluation and reproducibility in face of lacking benchmark ground truth data fitting for the social domains we work with. Building on top of this paper's work with the basic domain lexicon and ontology constructed, one can start to experiment with semi-automated annotation tools for building up ground truth dataset for robust evaluation. Another future work can be to populate a knowledge graph based on the domain ontology with a continuous process that layers data from many more diverse sources.

References

1. Policy brief: the impact of covid-19 on older persons (2020). unsdg.un.org/sites/default/files/2020-05/Policy-Brief-The-Impact-of-COVID-19-on-Older-Persons.pdf
2. Akinola, S.: Covid-19 has worsened ageism. here's how to help older adults thrive. (2020). www.weforum.org/agenda/2020/10/covid-19-has-worsened-ageism-here-s-how-to-help-older-adults-thrive//
3. Alexopoulos, P.: Building a large knowledge graph for the recruitment domain with text kernel's ontology. www.textkernel.com/newsroom/building-a-large-knowledge-graph-for-the-recruitment-domain-with-textkernels-ontology/
4. Bojanowski, P., Grave, E., Joulin, A., Mikolov, T.: Enriching word vectors with subword information. arXiv preprint arXiv:1607.04606 (2016)
5. Deane, P.: A nonparametric method for extraction of candidate phrasal terms. In: Proceedings of the 43rd Annual Meeting of the Association for Computational Linguistics (ACL'05), pp. 605–613. Association for Computational Linguistics, Ann Arbor, Michigan. 1219840.1219915 (2005). www.aclweb.org/anthology/P05-1075
6. Doan, A.: Human-in-the-loop data analysis: a personal perspective, pp. 1–6, 3209900.3209913 (2018)
7. Ehrlinger, L., Woss, W.: Towards a definition of knowledge graphs. In: SEMANTiCS (2016)
8. Goger, A.: For millions of low-income seniors, coronavirus is a food-security issue (2020). www.brookings.edu/blog/the-avenue/2020/03/16/for-millions-of-low-income-seniors-coronavirus-is-a-food-security-issue/
9. Gruber, T.: Toward principles for the design of ontologies used for knowledge sharing. Int. J. Hum. Comput. Stud. **43**, 907–928 (1995)
10. Hall, D., Jurafsky, D., Manning, C.D.: Studying the history of ideas using topic models. In: Proceedings of the Conference on Empirical Methods in Natural Language Processing. EMNLP '08, Association for Computational Linguistics, pp. 363–371. USA (2008)
11. Kamdar, M.R., Hamamsy, T., Shelton, S., Vala, A., Eftimov, T., Zou, J., Tamang, S.: A knowledge graph-based approach for exploring the u.s. opioid epidemic (2019). arxiv.org/abs/1905.11513
12. Kejriwal, M., Szekely, P.: Knowledge graphs for social good: an entity-centric search engine for the human trafficking domain. IEEE Tran. Big Data, 1 (2017). TBDATA.2017.2763164

13. Kejriwal, M., Shao, R., Szekely, P.: Expert-guided entity extraction using expressive rules. pp. 1353–1356(072019). 3331184.3331392
14. Kim, S.N., Baldwin, T., Kan, M.Y.: Extracting domain-specific words - a statistical approach. In: Proceedings of the Australasian Language Technology Association Workshop 2009, pp. 94–98. Sydney, Australia (2009). www.aclweb.org/anthology/U09-1013
15. Larocca Neto, J., Alexandre, N., Santos, D., Kaestner, C., Freitas, A.: Document clustering and text summarization. In: Proceedings of 4th International Conference Practical Applications of Knowledge Discovery and Data Mining (PADD-2000), pp. 41–55. London (2000)
16. Meng, Y., et al.: Discriminative topic mining via category-name guided text embedding. In: Proceedings of The Web Conference 2020. WWW'20, Association for Computing Machinery, pp. 2121–2132. New York, NY, USA. 3366423.3380278 (2020)
17. Mikolov, T., Chen, K., Corrado, G.S., Dean, J.: Efficient estimation of word representations in vector space (2013). arXiv:1301.3781
18. Mintz, M., Bills, S., Snow, R., Jurafsky, D.: Distant supervision for relation extraction without labeled data. In: Proceedings of the Joint Conference of the 47th Annual Meeting of the ACL and the 4th International Joint Conference on Natural Language Processing of the AFNLP. Association for Computational Linguistics, pp. 1003–1011. Suntec, Singapore (2009). www.aclweb.org/anthology/P09-1113
19. Noy, N., Mcguinness, D.: Ontology development 101: a guide to creating your first ontology. Knowl. Syst. Laboratory **32** (2001)
20. Paulheim, H.: Knowledge graph refinement: a survey of approaches and evaluation methods. Semantic Web **8**, 489–508 (2016)
21. Pennington, J., Socher, R., Manning, C.D.: Glove: Global vectors for word representation. In: Empirical Methods in Natural Language Processing (EMNLP), pp. 1532–1543 (2014). www.aclweb.org/anthology/D14-1162
22. Sahu, S.K., Christopoulou, F., Miwa, M., Ananiadou, S.: Inter-sentence relation extraction with document-level graph convolutional neural network. In: Proceedings of the 57th Annual Meeting of the Association for Computational Linguistics. Association for Computational Linguistics, pp. 4309–4316. Florence, Italy, pp. 19–1423 (2019). www.aclweb.org/anthology/P19-1423
23. Shang, J., Liu, J., Jiang, M., Ren, X., Voss, C.R., Han, J.: Automated phrase mining from massive text corpora. CoRR arXiv:1702.04457 (2017). arXiv:1702.04457
24. CWI Labs, Giving Tech Labs, and X4Impact. Limited broadband access as a barrier to economic mobility of older americans knowledge extraction and data analysis (2021). giving.tech/wp-content/uploads/2021/01/CWI-Labs-Giving-Tech-Labs-X4Impact-White-Paper-Limited-Broadband-Access-as-a-Barrier-to-Economic-Mobility-of-Older-Americans.pdf

Optimisation of a Workpiece Clamping Position with Reinforcement Learning for Complex Milling Applications

Chrismarie Enslin[1], Vladimir Samsonov[1(✉)], Hans-Georg Köpken[2], Schirin Bär[1], and Daniel Lütticke[1]

[1] Institute for Information Management in Mechanical Engineering, RWTH Aachen University, Aachen, Germany
vladimir.samsonov@ima.rwth-aachen.de
[2] Digital Industries, Siemens AG, Erlangen, Germany

Abstract. Fine-tuning and optimisation of production processes in manufacturing are often conducted with the help of algorithms from the field of Operations Research (OR) or directly by human experts. Machine Learning (ML) methods demonstrate outstanding results in tackling optimisation tasks within the research field referred to as Neural Combinatorial Optimisation (NCO). This opens multiple opportunities in manufacturing for learning-based optimisation solutions. In this work, we show a successful application of Reinforcement Learning (RL) to the task of workpiece (WP) clamping position and orientation optimisation for milling processes. A carefully selected clamping position and orientation of a WP are essential for minimising machine tool wear and energy consumption. With the example of 3- and 5-axis milling, we demonstrate that a trained RL agent can successfully find a near-optimal orientation and positioning for new, previously unseen WPs. The achieved solution quality is comparable to alternative optimisation solutions relying on Simulated Annealing (SA) and Genetic Algorithms (GA) while requiring orders of magnitude fewer optimisation iterations.

Keywords: Reinforcement Learning · Supervised learning · Manufacturing · Process optimisation · Milling optimisation · Tool path

1 Introduction

This study looks into the adaptation of learning-based methods to an optimisation task in the context of mechanical engineering. The object of investigation is applying RL for optimising WP clamping position and orientation in a Computer Numerical Control (CNC) milling machine. The milling process involves the removal of material from the WP with a rotary cutting tool. The considered CNC milling machine belongs to a widely used type of milling machines with a rotary table and a swivelling spindle head allowing complex movements of the

G. Nicosia et al. (Eds.): LOD 2021, LNCS 13164, pp. 261–276, 2022.
https://doi.org/10.1007/978-3-030-95470-3_20

cutting tool related to the WP. This enables the production of a wide variety of complex WP geometries on a single CNC milling machine.

Designing a new CNC milling process is a laborious task relying heavily on human expertise. Firstly, a Numerical Control (NC) program needs to be created based on the WP geometry, the chosen processing technology and the type of CNC machine. The resulting NC program defines the toolpath relative to the WP. A second step is the definition of a WP clamping position and orientation in the working space of the CNC machine. Different WP clamping positions and orientations result in different movements of the machine axes. Therefore, certain clamping positions and orientations will require higher accelerations of the heavy machine axes and a higher number of axes movements, directly influencing machine wear, machining accuracy and energy efficiency.

The standard approach to determine a suitable WP clamping position and orientation relies solely on human expertise gathered through experience. The preceding work [18] demonstrates the concept of formalising the task of finding the optimal WP clamping position and orientation for the milling process as an optimisation problem with subsequent use of RL. In this study, we considerably enhance the proposed approach and demonstrate the capability of the proposed method in addressing more challenging milling tasks while improving the quality of solutions with fewer search iterations. A direct comparison to alternative optimisation approaches, such as GA [14] and SA [20], demonstrates the capability of the proposed RL-solution to yield comparable results for WPs not seen during training, while requiring considerably fewer optimisation iterations.

2 ML Applications in Mechanical Engineering

Learning-based and data-driven methods are widely deployed to drive progress in the field of smart production and manufacturing [4,15]. ML is considered the key enabling technology for further cost savings, quality improvement, and minimisation of waste in applications relying on the use of cutting processes along with heuristic optimisation approaches. At the same time, milling and turning receive the most attention from the research community [5].

Multiple studies concentrate on adopting learning-based methods for condition monitoring and machine tool diagnosis to enhance cutting processes. Wu et al. [22] demonstrate the applicability of simple regression models based on Random Forest (RF), Support Vector Machine (SVM) or Multilayer Perceptron (MLP) for the prediction of tool wear. Kothuru et al. [11] investigate the possibility of using SVM prediction models to estimate the condition of a cutting tool using only audible signals.

More advanced supervised ML models capable of processing sequential data demonstrated their efficiency for condition monitoring and quality prediction. Wang et al. [21] utilise recurrent predictive models on time series for tool wear monitoring. In this study, a Gated Recurrent Unit (GRU) predictive model demonstrates superior performance compared to conventional methods. For a similar Use Case (UC), Serin et al. [19] introduce the use of RL in combination with an LSTM-based (Long Short-term Memory) control system. The LSTM-model is used as a memory base and can suggest optimal cutting parameters to the RL agent. The study of Yuan et al. [24] emphasises the advantages of enhancing recurrent predictive models with attention mechanisms for quality prediction in complex production processes.

RL methods are often used in manufacturing engineering for planning, controlling and iterative optimisation of the production process. Xanthopoulos et al. [23] deploy an RL agent for learning joint production control and maintenance strategies on a deteriorating production system. A trained RL agent can suggest maintenance schedules and production plans that are superior to methods currently implemented in practice. Pol et al. [16] use decentralised RL agents to perform online scheduling in flexible production systems with the advantage of generalising to uncertain situations that deviate from the plan. Meyes et al. [13] demonstrate the use of an RL agent for sample-efficient optimisation of a heavy plate rolling process. The application UC considered in the study contains many process-, material- and machine parameters interacting with each other. These different interactions of parameters produce different product quality levels, which relate to the height and the grain size of the finished products. A trained RL agent can estimate suitable pass schedules for heavy plate rolling to achieve the desired material characteristics.

Bhinge et al. [1] demonstrate the importance of the tool path for the total energy consumption of the machine tool. Rangarajan et al. [17] emphasise the importance of the WP orientation in milling to minimise the processing time and drive loads. Campatelli et al. [3] propose a mathematical model to reduce energy consumption through optimal WP placement in the milling process. However, the proposed approach is applicable only for the finishing operations and is not capable of axis collisions avoidance. To the best of our knowledge, only the study from Samsonov et al. [18] demonstrates the application of RL methods to the task of optimal WP positioning in a machine tool.

3 Problem Statement

This study aims to find a near-optimal clamping position and orientation for a previously unseen WP in a CNC machine with RL. An optimal position refers to a placement that minimises the acceleration and the distance travelled in the axis directions of the machine while milling the WP. This placement should also accommodate the avoidance of all possible collisions of CNC machine parts caused by the movements of the machine axes. The task of finding WP position and orientation can be formalised as a continuous optimisation problem maximising the objective function equal to the reward function discussed in Sect. 4.1.

Two UCs of CNC milling are investigated in this work. Firstly, a simple WP geometry is considered with a groove along the perimeter of the part and the milling slot on top of it (see Fig. 1). Given shape requires a 3-axis milling process with possible movements in a front-to-back (X-axis), side-to-side (Y-axis), as well as up-and-down (Z-axis) directions by the cutting tool. A milling slot is a widespread feature involving multiple changes of the tool movement direction. Having the grove along the perimeter of the WP increases the chance of axes collisions if the WP position and orientation are not selected correctly. A further element of simplicity is that the machine coordinate system and the WP coordinate systems coincide. Changing the orientation of the slot (slot angle) allows the generation of a set of WP geometries for training and evaluation of the proposed RL solution. This WP geometry is considered to a certain extent in [18] and is used in this work to demonstrate the achieved improvements on the search efficiency of the solution space.

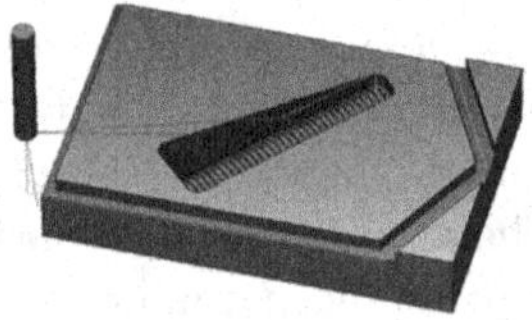

Fig. 1. Visualisation of a 3-axis WP, slot angle = 45°

(a) Slot angle = 0°, Tower Z = 60mm, Tower X = 55mm (b) Slot angle = 45°, Tower Z = 60mm, Tower X = 100mm (c) Slot angle = 90°, Tower Z = 235mm, Tower X = 130mm (d) Slot angle = 210°, Tower Z = 235mm, Tower X = 55mm

Fig. 2. Visualisations of 5-axis milling WPs

The second UC, and focus of the current study, is based on a 5-axis milling process and decoupling of the WP and machine coordinate systems. Apart from the milling slot and groove along the WP perimeter included in the first UC, a tower-shaped surface is added on the top of the WP (see Fig. 2 for various examples of WP geometries). The introduced tower feature is a complex spiral-formed shape with inclined sides towards the center. This makes the feature fairly representative for complex milling operations, requiring coordinated movements of multiple axes and constant change in the moving speed/accelerations. The milling process of considered 5-axis WP geometry involves a front-to-back movement (Z-axis), a side-to-side movement (X-axis), an up-and-down movement (Y-axis), a rotation movement along the Y-axis and pitch movement from side-to-side of the cutting tool relative to the WP. In this case, the RL agent needs to learn how to handle non-trivial axis collision avoidance patterns and a different WP design. To vary the shape of the WP, the slot angle can be rotated 360° and the tower position can be moved continuously along the top and bottom of the WP. The NC program created to mill the WP is calibrated such that the slot and the tower will not overlap. Modifications of the slot and tower features change the energy exertion and wear on the main axes of the milling machine and the optimal placement is influenced.

The 3-axis UC provides a good framework to test modifications of the RL solution, with the goal of finally transferring and testing the results on the more complicated 5-axis UC.

4 RL Experiment Setup

The WP clamping position and orientation optimisation task is formalised as a fully observable Markov Decision Process (MDP). The RL agent is allowed to iteratively move the WP in the machine's working space to improve the reward from the WP clamping position and orientation. In the following sub-chapters the state space, action space and reward function for the RL agent are discussed, along with search space efficiency implementations. The training and validation scheme for the RL agent is outlined and the generation of WP data and an efficient way of using the milling simulation data by training an ML model is explained.

4.1 State Space, Action Space and Reward Function

The position of the WP in the machine space is described by axes coordinates (X, Y, Z) and the rotation angle of the WP with respect to the orientation of the machine. The RL agent can choose the location and orientation of the WP every time an action is required. For the 3-axis machine, the only possible variation in the shape of the WP is in terms of the milling slot angle. Therefore, each WP position and shape is uniquely captured by (X, Y, *Rotation Angle*, *Slot Angle*). The 5-axis machine includes the possibility of milling a tower, as well as a slot, and a combination of these two features uniquely specifies the WP. The parameters to describe the WP placement and shape to an RL agent in this case are (Z, X, Y, *Rotation Angle*, *Tower Z*, *Tower X*, *Slot Angle*). For the RL agent to know what kind of WP is currently being processed, it is important to provide this unique parameter set explicitly or to determine a proxy for the WP shape that could be provided.

After placement of the WP in the machine's working space at the specified coordinates, with the specified orientation and including the unique features of the WP, a milling process is carried out. From this milling process, the sum of the squared accelerations (e_Z, e_X, e_Y) and distances travelled along every machine axis (d_Z, d_X, d_Y) are recorded for further evaluation of the current WP placement and orientation.

The tuples for the state-action-reward representation found in the 5-axis UC are briefly summarised in Table 1. The state space includes the unique WP parameters, the location and orientation parameters in the working space of the milling machine, the resulting squared accelerations and distances travelled, the collision switch (*limit Z*) and the number of *steps left* before reaching the step limit in the episode. The latter parameter helps the agent with planning its search strategy and choosing step sizes while exploring the solution space. The action space consists of the changes to the WP location in terms of the machine coordinates, as well as the rotation angle of the WP and a decision to terminate the episode or not.

Table 1. Summary of the main parameters of the optimisation task.

State	*(Z, X, Y, Rotation Angle, Tower X, Tower Z, Slot Angle,* d_Z, d_X, e_Z, e_X, *limit Z, steps left)*
Action	*(ΔZ, ΔX, ΔY, ΔRotation Angle, Stop)*
Reward	$R = 0.7e + 0.3d$

The reward function provides feedback on the quality of the suggested placement of the WP in the machine space by the RL agent. Maximising the reward ensures WP positions and orientations close to the optimum. The 3-axis and 5-axis milling processes have similar reward functions, as the concept of minimising wear on the machine and energy consumption remains the same. Therefore, the reward function includes a component representing the sum of squared acceleration (e) and the sum of distance travelled (d) on the main axes. For 3-axis milling, the directions of travel are the X- and Y-axis, with the X-axis representing the heavier axis. For 5-axis milling the directions of travel include an X-axis, a Y-axis and a Z-axis. Here the Z-axis is the heaviest axis and the movement along the Y-axis is not included in the reward function, as the movement is minimal in this UC. To limit the wear of the machine as much as possible, the movement of the heavier axes should be one of the components minimised by the RL agent.

The squared acceleration (e) is given higher importance than distance travelled (d) by domain experts in finding a near-optimal position of the WP and therefore the reward function is formulated as:

$$R = \begin{cases} 0.7e + 0.3d & \text{no axis collision} \\ -1 & \text{axis collision.} \end{cases} \tag{1}$$

The weights for the optimisation function are also chosen by domain experts and reflect specific industry needs. If these weights were to be changed the RL agent would have to be retrained, but the optimisation problem would be formulated in the same way.

The terms concerning acceleration (e) and distance (d) are both accumulated terms across the different machine axes and over the entire WP milling run.

The components e_Z and d_Z are weighted twice as heavy as e_X and d_X in the combined reward components e and d, as advised by domain experts because the movement of the Z-axis should be avoided. The combined components are:

$$e = 2e_Z + e_X \tag{2}$$

$$d = 2d_Z + d_X, \tag{3}$$

where e_Z, e_X, d_Z and d_X represent normalised terms over the minimum and maximum observed values from multiple experiments. The normalisation function negates the e- and d-values, such that lower values of e and d will result in a higher reward. If any axis collisions are encountered, the lowest possible reward is returned. The intention of the optimisation algorithm should firstly be to learn collision avoidance, and thereafter an optimal WP placement.

4.2 Search Efficiency Modifications

To reduce the number of optimisation iterations, we introduce several changes to the search process established in [18]. Firstly, for the 3-axis UC, we **extend the maximum range of the WP position change per step** (step size) along the X-axis, Y-axis and the rotation angle for orientation, from 40 mm, 40 mm and 35°, respectively, to 800 mm, 800 mm and 360°. This larger step size allows the RL agent to move the WP into any position within the working space of the milling machine. Similarly, for the 5-axis UC, we allow the RL agent to step into any position in the work space. Therefore, the step sizes are 300 mm, 200 mm and 50 mm for the Z-, X- and Y-axis, respectively, as well as a full 360° WP rotation.

Secondly, instead of assuming that the RL agent has to complete an entire episode before continuing to the next one, we introduce an additional action dimension referred to as **early stopping**. Early stopping allows the RL agent to stop the search process at any iteration step as soon as it assumes the current WP position is suitable. This updates the action space to *(*ΔZ*,* ΔX*,* ΔY*,* Δ*Rotation Angle, Stop)*.

Thirdly, we switch from a dense reward calculated after every iteration of the search process to a **sparse reward**, returned only when the RL agent chooses to stop the search process or when the maximum allowed number of steps per search episode is reached. A sparse reward forces the RL agent to step to the optimum and terminate the episode as quickly as possible, to start collecting rewards. The sparse reward is a prerequisite for the early stopping implementation. The dense reward is not compatible with early stopping, since longer episodes generate higher cumulative rewards, incentifying the RL agent not to interrupt the episodes earlier.

4.3 RL Agent Training and Validation

The RL training routine is represented in Fig. 3. For every training episode in the 5-axis UC, a new WP is generated with a random slot angle, tower position Z

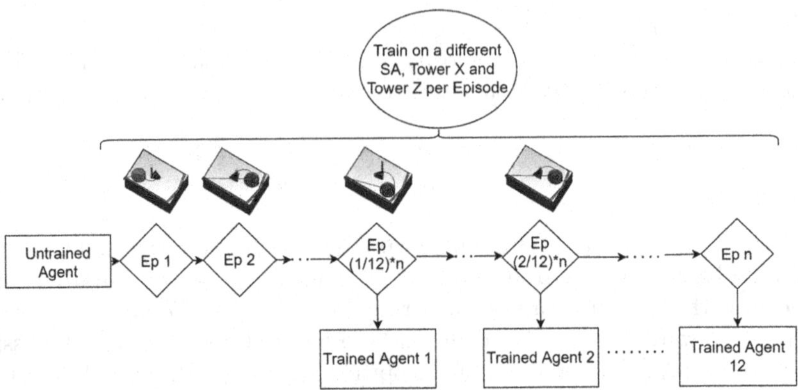

Fig. 3. Training setup for the RL agent

and tower position X. For the 3-axis UC, this is reduced to only varying the slot angle. The restrictions on these training WPs are as follows: slot angle between 0° and 180°, tower X position between 55 mm and 145 mm along the X-axis and tower Z position either 60 mm or 235 mm along the Z-axis. The decision to keep tower position Z fixed at only 60 mm or 235 mm is made to keep the tower and the slot from overlapping. To keep some WPs aside, purely for validation, the slot angles between 40° and 50° and tower X position between 90 mm and 100 mm are disallowed for training. The allowance of only two tower Z positions makes it impossible to keep certain tower Z positions aside for validation, therefore both positions are available for training and validation.

In the training procedure, episodes of different maximal lengths are investigated. The combination of larger step sizes and a sparse reward leads to a reduction in the maximum episode length required to find a near-optimal positioning of a WP. This drastically reduces the step count from 110 steps per episode to two steps per episode for 3-axis and one step per episode for 5-axis. These episodes have the option to terminate early if the agent believes it has reached the optimum. Therefore, an episode can terminate immediately after initialisation, if the agent finds that the initialisation position is an optimal position. All 5-axis RL agents are trained for 300.000 steps and the 3-axis RL agents are trained for 100.000 steps and this is repeated for three different random seeds in each case.

Throughout the training process 12 evaluation phases are equally spaced between training episodes, to account for the possibility that an intermediate version of the RL agent might be superior in performance to the RL agent in a later stage of training. During each evaluation phase, the current version of the trained RL agent for a given run solves the WP positioning task for 20 different initialisation points (WP positions) for each WP involved in the evaluation. This allows for the testing of the overall robustness and consistency of the evaluation scheme.

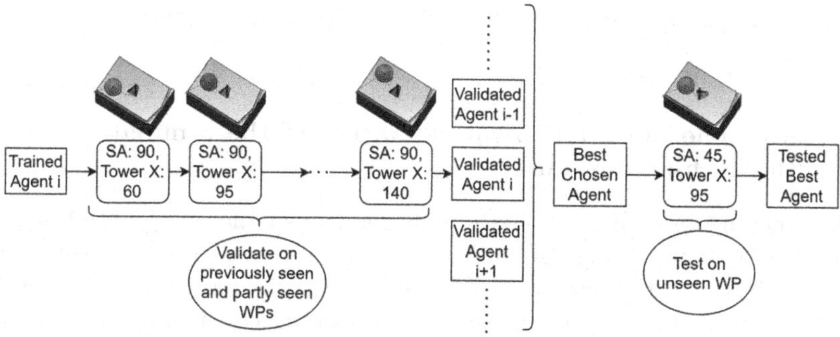

Fig. 4. Evaluation setup for the RL agent

The evaluation scenario for monitoring of the intermediate RL performance for the 3-axis UC is based on 11 WPs with slot angles that have been **seen** during training and are equally spaced throughout the 180° of allowed slot angles. The version of the RL agent from the evaluation phase with the best average reward is selected as a trained RL agent and is finally validated on the completely **unseen** WP with a slot angle of 45°. The rewards achieved on the unseen WP are used for the final estimation of the quality and generalisation ability of the RL agent, but it is not involved in choosing the best RL agent, as this can be seen as data leakage.

The 5-axis RL agent is regularly evaluated on 94 WPs with a combination of 6 different slot angles, 8 different tower X positions and the two available Z positions. These validation WPs are **seen**, or **partially seen**, during training, but the WP with a 45° slot angle and a tower located at the X coordinate of 95mm is kept away from the evaluation process. Similar to the 3-axis approach, the version of the RL agent from the evaluation phase with the best average reward is selected for the final validation on the **unseen** WP with slot angle 45° and tower X position of 95mm. The validation scheme for the 5-axis UC is summarised in Fig. 4.

The meta-heuristics GA [14] and SA [20] are used as baseline models for comparison to the RL agents. These heuristics are popular methods for solving optimisation problems and GA is often used in the context of production process optimisation as seen in [5]. Each meta-heuristic run is given 100.000 iterations to find the solution to the WP positioning task. Analogous to the RL evaluation, 20 independent initialisations and three random seeds are used. This results in 60 independent evaluation runs for each solution approach. The reward function is used as a cost function for the optimisation heuristics.

The RL agent used for the 3-axis and 5-axis UCs is the Soft Actor-Critic (SAC) introduced in [8]. In our study the stable baselines [10] implementation is used, to make it easily comparable to other RL implementations. The experiments are all performed in docker containers [12] for full reproducibility of experiments. The entire implementation of the milling simulation is done as an

OpenAI Gym environment [2]. The meta-heuristics are implementations from the mlrose package [9].

4.4 Data Generation and Approximation of the Simulation with Machine Learning

Experiments in this investigation are performed on simulations created by SinuTrain software, reproducing the 3-axis and 5-axis milling processes performed on a CNC machine. For the purpose of RL training, the simulation is time intensive. It can take up to 4 s to generate the output of a successful run for the 3-axis environment, with failed runs taking up to a second to generate outputs. We develop a set of ML models closely mimicking the behaviour of the simulation. ML models predict the output of a simulation run in a fraction of a second. As a result the development process of the RL solution is considerably accelerated. It is important to note that there are two layers of separation from the CNC machine and what the RL agent is trained on, namely the Sinutrain simulation and the ML models. Therefore the accuracy of the ML models are thoroughly validated to confirm the accuracy of this method, as summarised in Table 2 and Table 3. The details of the data generation process and the 3-axis ML model is stipulated in Samsonov et al. [18].

As this study also shifts to the more complex 5-axis milling process, another ML model is required to mimic the behaviour of the SinuTrain machine simulation. The ML models are even more justified in the 5-axis UC as a successful run can take up to 15 s to generate the required outcomes. For the purpose of ML model training, 83.252 data points are generated from the SinuTrain simulation. The ML model ensemble in this study consists of five gradient boosting models (a success/fail classifier and a regression model for each of e_Z, e_X, d_Z and d_X), fitted with the LightGBM package [7]. LightGBM is known for providing fast and efficient training of gradient boosting models. The training process, together with the ML models and the inputs and outputs of all components are outlined in Fig. 5.

The first model in the ensemble is a classification model, which distinguishes between input combinations that lead to a successful milling process or to an axis collision. The only axis collision possible in the training data is on the Z-axis. Table 2 summarises the F1-Score and the overall accuracy of the classifier model, as well as the size of the training datasets in each class, to demonstrate the support behind the different accuracy values.

The data points that are classified as successful milling process runs continue to the four regression models. These models produce estimates for e_Z, e_X, d_Z and d_X, respectively. As preprocessing of the data, the offset between the Z coordinate and the Tower Z position, as well as the offset between the X coordinate and the tower X position, is calculated and used as input values. The accuracy of the four models is summarised in Table 3, where the accuracy measure is chosen to be the R^2-values.

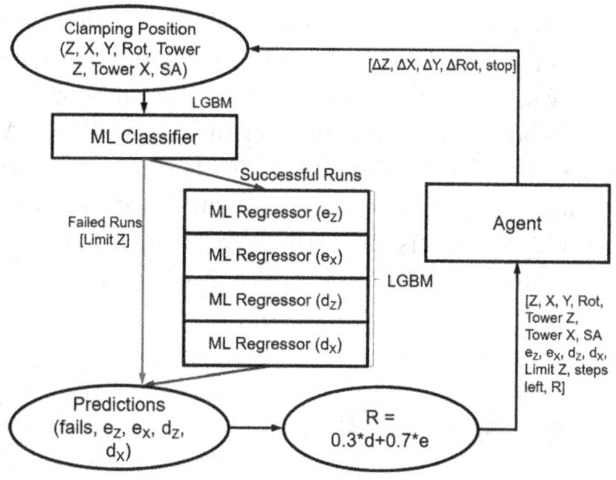

Fig. 5. ML model architecture

Table 2. Classifier accuracy

	F1-Score	Totals
Success	0.981	34943
Limit Z	0.987	48859
Weighted average	0.984	83252

Table 3. Regression models accuracy

	R^2-values
e_Z	0.995
e_X	0.984
d_Z	0.997
d_X	0.995

5 Experimental Results

With a set of experiments, it is demonstrated to what extent the original WP positioning optimisation approach proposed by Samsonov et al. [18] can be enhanced and transferred to a more challenging 5-axis milling tasks. The use of the increased step size, early stopping and sparse reward, as described in Sect. 4.2, considerably improves the search efficiency of the trained RL agent. Averaged over all evaluation runs and random seeds, a trained RL agent solves the task with just one step in 95% of all cases for the 5-axis UC. For the 3-axis UC, the RL agent prefers to use two steps for solving the task in 90% of the observations. This allows us to introduce a hybrid approach, augmenting the best trained RL agent with a simple search heuristic to improve the absolute performance. For each WP positioning task, instead of solving the task once with an arbitrary initialisation point, the RL agent is allowed to conduct the search multiple times with different initialisation points and the solution with the best reward is selected. In our work this approach is referred to as Hybrid RL and the trained RL agent is given 20 attempts to find a near-optimal solution.

The runs of all three random seeds demonstrate the capability of the RL agent to generalise to unseen WP geometries, to consistently avoid axis collisions and to find good WP clamping positions, as seen in Fig. 6 and Fig. 7. The RL training with random seed 4821 results in a noticeably lower performance across all 3- and 5-axis runs. Given the observed performance, the decrease is possibly related to bad initialisation weights of the policy- and value networks. The hybrid RL approach considerably boosts the absolute performance. The GA meta-heuristic demonstrates the best results, both in terms of the result consistency and the absolute reward value. The SA meta-heuristic could not match the performance of the RL-based or GA solutions in both 3-axis and 5-axis UCs. To conserve space, further evaluations of SA meta-heuristic are omitted.

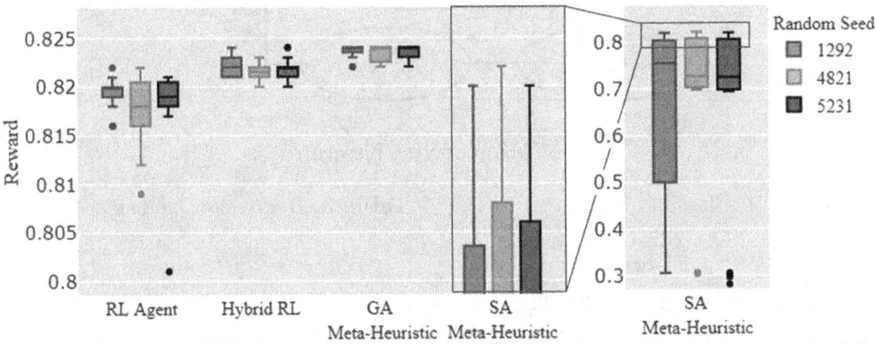

Fig. 6. The achieved rewards for the 3-axis WP positioning search guided by the trained RL agent, hybrid RL approach, GA and SA meta-heuristics

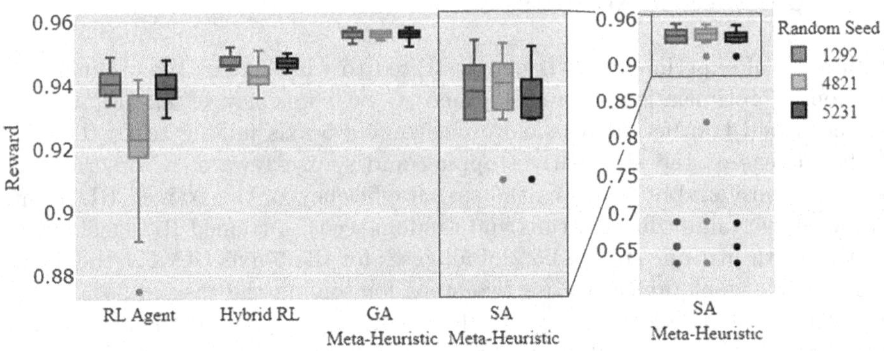

Fig. 7. The achieved rewards for the 5-axis WP positioning search guided by the trained RL agent, hybrid RL approach, GA and SA meta-heuristics

The conducted comparison of the absolute performance demonstrates that a well-tuned GA solver can surpass the proposed RL-based approaches. However, not only the absolute performance is essential for practical applications, the time required to find a viable solution is often a critical viability factor. A trained RL agent needs between one to two steps per task to find a near-optimal positioning of a WP. The hybrid RL requires on average 21 and 36 steps per task for the 3-axis and 5-axis UC correspondingly.

To investigate how fast the GA meta-heuristic achieves results comparable to solutions found with the RL-based approaches, we track the performance at each interaction with the simulation environment during the meta-heuristic search. The pairwise comparison is always conducted between the GA and RL methods with identical random seeds used for the search/training. Figure 8a and Fig. 8b demonstrate that the GA meta-heuristic needs on average 660 steps for the 3-axis UC and 615 steps for the 5-axis UC to match the performance of the corresponding RL agent. GA meta-heuristic takes on average 7842 and 1756 steps on for the 3- and 5-axis UC correspondingly to surpass the performance of the hybrid RL approach (see Fig. 8c and Fig. 8d).

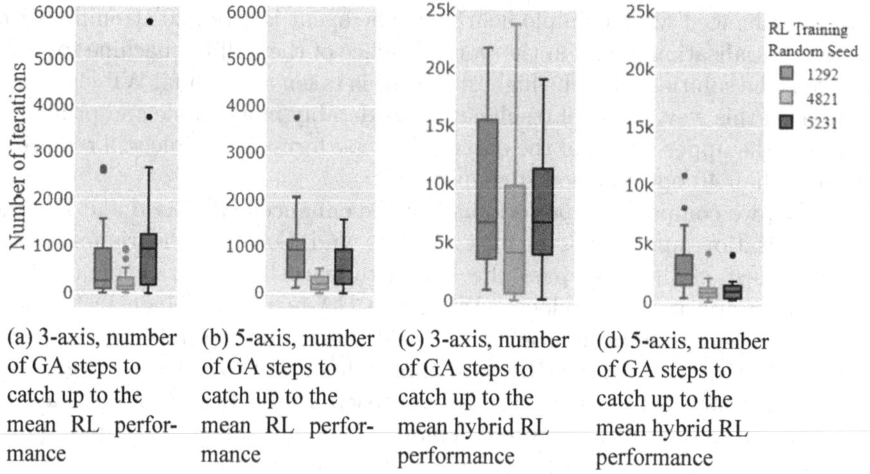

(a) 3-axis, number of GA steps to catch up to the mean RL performance

(b) 5-axis, number of GA steps to catch up to the mean RL performance

(c) 3-axis, number of GA steps to catch up to the mean hybrid RL performance

(d) 5-axis, number of GA steps to catch up to the mean hybrid RL performance

Fig. 8. The number of search steps the GA heuristic requires to match the mean performance of the trained RL agents and the hybrid RL approach

The considered evaluations are conducted on a fast ML-based environment to make extensive testing computationally viable. However, these ML models are only intended for the training of the RL agent. Building such an ML-based environment for RL training covers a finite number of WP geometries, thus keeping the general effort limited. During the deployment of a new WP in production, the search for optimal clamping parameters needs to be conducted directly in the SinuTrain simulation environment. A GA search involving 7842 steps could require 8,3 h for 3-axis and similarly, a GA search involving 1756 steps could take 4,8 h for 5-axis. This is in direct contrast to the possible 2 s for 3-axis and 10 s for 5-axis needed for the hybrid RL approach with matching performance.

6 Conclusion and Future Work

In this work, we extend the conceptual RL-based approach to optimise a WP clamping position in a CNC milling machine tool. We introduce several significant additions to the original approach [18]. The **first contribution** is the demonstration of the direct transfer to a more complex UC involving 5-axis milling, using the same hyperparameter set and training scheme. The RL agent successfully learns and avoids more elaborate collision patterns related to the 5-axis milling scenario, as well as consistently finds near-optimal WP clamping positions.

The **second contribution** is the introduction of early stopping, combined with larger action ranges and sparse rewards. As a result, a trained RL agent needs between one and two optimisation steps to find a near-optimal WP clamping position. Improved search efficiency allows us to make the **third contribution** by introducing the Hybrid RL approach where RL-guided optimisation search is enhanced with a simple heuristic. The agent is given 20 attempts with different initialisation points in the working space of the milling machine to solve one task. The solution with the highest reward is taken as the final WP clamping position. In this way, the final results are considerably more stable and primarily located at the upper bound of the observed RL performance, while still requiring not more than 40 optimisation steps per task.

Finally, we compare the performance of the enhanced RL-based and Hybrid RL optimisation approaches against the SA and GA meta-heuristics. Both learning-based methods surpass the SA meta-heuristic in terms of absolute performance and search efficiency. While the GA-heuristic demonstrates better absolute performance, it needs about 300 times more optimisation steps to match the performance of the RL-based approach and is two orders of magnitude less efficient compared to the Hybrid RL approach. In practice, it means that the RL-based optimisation methods are capable of solving the WP clamping position task in seconds instead of half-days of runtime in the SinuTrain simulation required by the considered meta-heuristics.

While demonstrating significant improvements, the proposed RL-based methods still remain in a prototype phase. In future work, we plan to address the current need for a handcrafted set of features describing the WP. A handcraft-free WP description can be achieved by representing the WP milling process as a set of vectors covering the change of the relative position, speed and acceleration of the milling tool tip point related to the WP during the milling process. A compact WP representation can be learned in an unsupervised manner, using an autoencoder, which feeds into the state space in an RL-based optimisation task. An additional direction of work is enhancing the RL-based optimisation methods by combining them with more advanced heuristics. Avoiding local optima is a common challenge while designing and applying meta-heuristics [6]. If a near-optimal WP position, determined by an RL-based method, is used as a starting point for an additional heuristic search, better absolute performance can be achieved while still maintaining the overall acceptable computation time.

References

1. Bhinge, R., et al.: An intelligent machine monitoring system for energy prediction using a Gaussian Process regression. In: Lin, J. (ed.) 2014 IEEE International Conference on Big Data (Big Data 2014), pp. 978–986. IEEE, Piscataway (2014). https://doi.org/10.1109/BigData.2014.7004331
2. Brockman, G., et al.: OpenAI Gym (2016). https://arxiv.org/pdf/1606.01540
3. Campatelli, G., Scippa, A., Lorenzini, L., Sato, R.: Optimal workpiece orientation to reduce the energy consumption of a milling process. Int. J. Precis. Eng. Manuf. Green Technol. **2**(1), 5–13 (2015)
4. Cioffi, R., Travaglioni, M., Piscitelli, G., Petrillo, A., de Felice, F.: Artificial intelligence and machine learning applications in smart production: progress, trends, and directions. Sustainability **12**(2), 492 (2020). https://doi.org/10.3390/su12020492
5. Du Preez, A., Oosthuizen, G.A.: Machine learning in cutting processes as enabler for smart sustainable manufacturing. Procedia Manuf. **33**, 810–817 (2019). https://doi.org/10.1016/j.promfg.2019.04.102
6. Gandomi, A.H. (ed.): Metaheuristic Applications in Structures and Infrastructures, 1st edn. Elsevier Insights, Elsevier, London (2013)
7. Ke, G., et al.: LightGBM: A Highly Efficient Gradient Boosting Decision Tree (2017, undefined)
8. Haarnoja, T., Zhou, A., Abbeel, P., Levine, S.: Soft Actor-Critic: Off-Policy Maximum Entropy Deep Reinforcement Learning with a Stochastic Actor (2018). http://arxiv.org/pdf/1801.01290v2
9. Hayes, G.: mlrose: Machine Learning, Randomized Optimization and SEarch package for Python (2019). https://github.com/gkhayes/mlrose
10. Hill, A., et al.: Stable Baselines (2018)
11. Kothuru, A., Nooka, S.P., Liu, R.: Application of audible sound signals for tool wear monitoring using machine learning techniques in end milling. Int. J. Adv. Manuf. Technol. **95**, 3797–3808 (2017). https://doi.org/10.1007/s00170-017-1460-1
12. Merkel, D.: Docker: lightweight linux containers for consistent development and deployment. Linux J. **2014**(239), 2 (2014)
13. Meyes, R., et al.: Interdisciplinary data driven production process analysis for the internet of production. Procedia Manufa. **26**, 1065–1076 (2018). https://doi.org/10.1016/j.promfg.2018.07.143
14. Mitchell, M.: An Introduction to Genetic Algorithms. Complex Adaptive Systems. MIT, Cambridge and London (1996)
15. Nti, I.K., Adekoya, A.F., Weyori, B.A., Nyarko-Boateng, O.: Applications of artificial intelligence in engineering and manufacturing: a systematic review. J. Intell. Manuf. 1–21 (2021). https://doi.org/10.1007/s10845-021-01771-6
16. Pol, S., Baer, S., Turner, D., Samsonov, V., Meisen, T.: Global reward design for cooperative agents to achieve flexible production control under real-time constraints. In: Proceedings of the 23rd International Conference on Enterprise Information Systems. SCITEPRESS - Science and Technology Publications (2021). https://doi.org/10.5220/0010455805150526
17. Rangarajan, A., Dornfeld, D.: Efficient tool paths and part orientation for face milling. CIRP Ann. **53**(1), 73–76 (2004). https://doi.org/10.1016/S0007-8506(07)60648-9

18. Samsonov, V., Enslin, C., Köpken, H.G., Baer, S., Lütticke, D.: Using reinforcement learning for optimization of a workpiece clamping position in a machine tool. In: Proceedings of the 22nd International Conference on Enterprise Information Systems, pp. 506–514. SCITEPRESS - Science and Technology Publications (2020). https://doi.org/10.5220/0009354105060514
19. Serin, G., Sener, B., Ozbayoglu, A.M., Unver, H.O.: Review of tool condition monitoring in machining and opportunities for deep learning. Int. J. Adv. Manuf. Technol. **109**(3–4), 953–974 (2020)
20. van Laarhoven, P.J.M., Aarts, E.H.L.: Simulated annealing. In: van Laarhoven, P.J.M., Aarts, E.H.L. (eds.) Simulated Annealing: Theory and Applications, pp. 7–15. Springer, Dordrecht (1987). https://doi.org/10.1007/978-94-015-7744-1_2
21. Wang, J., Yan, J., Li, C., Gao, R.X., Zhao, R.: Deep heterogeneous GRU model for predictive analytics in smart manufacturing: application to tool wear prediction. Comput. Ind. **111**, 1–14 (2019). https://doi.org/10.1016/j.compind.2019.06.001
22. Wu, D., Jennings, C., Terpenny, J., Gao, R.X., Kumara, S.: A comparative study on machine learning algorithms for smart manufacturing: tool wear prediction using random forests. J. Manuf. Sci. Eng. **139**(7) (2017). https://doi.org/10.1115/1.4036350
23. Xanthopoulos, A.S., Kiatipis, A., Koulouriotis, D.E., Stieger, S.: Reinforcement learning-based and parametric production-maintenance control policies for a deteriorating manufacturing system. IEEE Access **6**, 576–588 (2018). https://doi.org/10.1109/ACCESS.2017.2771827
24. Yuan, X., Li, L., Wang, Y., Yang, C., Gui, W.: Deep learning for quality prediction of nonlinear dynamic processes with variable attention-based long short-term memory network. Can. J. Chem. Eng. **98**(6), 1377–1389 (2020). https://doi.org/10.1002/cjce.23665

Thresholding Procedure via Barzilai-Borwein Rules for the Steplength Selection in Stochastic Gradient Methods

Giorgia Franchini(✉), Valeria Ruggiero, and Ilaria Trombini

Department of Mathematics and Computer Science, University of Ferrara, Ferrara, Italy
{giorgia.franchini,valeria.ruggiero}@unife.it,
{ilaria.trombini}@edu.unife.it

Abstract. A crucial aspect in designing a learning algorithm is the selection of the hyperparameters (parameters that are not trained during the learning process). In particular the effectiveness of the stochastic gradient methods strongly depends on the steplength selection. In recent papers [9,10], Franchini et al. propose to adopt an adaptive selection rule borrowed from the full-gradient scheme known as Limited Memory Steepest Descent method [8] and appropriately tailored to the stochastic framework. This strategy is based on the computation of the eigenvalues (Ritz-like values) of a suitable matrix obtained from the gradients of the most recent iterations, and it enables to give an estimation of the local Lipschitz constant of the current gradient of the objective function, without introducing line-search techniques. The possible increase of the size of the sub-sample used to compute the stochastic gradient is driven by means of an augmented inner product test approach [3]. The whole procedure makes the tuning of the parameters less expensive than the selection of a fixed steplength, although it remains dependent on the choice of threshold values bounding the variability of the steplength sequences. The contribution of this paper is to exploit a stochastic version of the Barzilai-Borwein formulas [1] to adaptively select the endpoints range for the Ritz-like values. A numerical experimentation for some convex loss functions highlights that the proposed procedure remains stable as well as the tuning of the hyperparameters appears less expensive.

Keywords: Stochastic gradient methods · Learning rate selection rule · Barzilai-Borwein rules · Adaptive sub-sampling strategies · Reduction variance techniques

1 Introduction

This work aims to consider the most expensive phase in terms of power and time of the Machine Learning (ML) methodologies, i.e. the training phase [4],

This work has been partially supported by the INdAM research group GNCS and by POR-FSE 2014–2020 funds of Emilia-Romagna region.

G. Nicosia et al. (Eds.): LOD 2021, LNCS 13164, pp. 277–282, 2022.
https://doi.org/10.1007/978-3-030-95470-3_21

where it is crucial to correctly set the hyperparameters connected to the optimizer, in particular the steplength and the mini-batch size. Starting from an idea developed in the deterministic field [8] and following an approach similar to that in [12–14], it is possible to tailor this steplength selection strategy for the stochastic gradient (SG) method. Furthermore, in the basic SG iteration, the adaptive steplength selection can be combined with an adaptive sub-sampling strategy (based on the augmented inner product test [3]) with the aim to assure the descent features in expectation of the stochastic gradient directions [9,10]. The aim of this paper is to introduce in this scheme a novel technique to adaptively set the bounds of the range in which the steplengths can move so that the method is made almost automatic.

This work is essentially structured in three sections. Section 2 resumes the main features of the method in [10] and introduces the novel contribution. Section 3 presents a set of numerical experiments, aimed to evaluate the effectiveness of the proposed approach and, finally, Sect. 4 concludes the paper.

2 Novel Contribution in Steplength Selection via Ritz-like Values

Among the state-of-the-art steplength selection strategies for deterministic gradient methods, the Limited Memory Steepest Descent (LMSD) rule proposed in [8] is one of the most effective ideas for capturing second-order information on the objective function from the gradients of few consecutive iterations. The LMSD rule is based on collecting the gradients of a group of $m \geq 1$ successive iterations, called *sweep*, where m is a small number (generally not larger than 7). After each sweep, these gradients enable to compute by a very inexpensive procedure a symmetric tridiagonal $m \times m$ matrix, whose eigenvalues λ_i, known as Ritz values, are interpreted as approximations of the eigenvalues of the Hessian of the objective function at the current iteration. Then, their inverses are used as the steplengths for the new sweep of iterations. The LMSD method is suitable for the minimization of non-linear and non-convex objective functions [7,8]. A variant of LMSD is based on the harmonic Ritz values (see [5] for details).

The idea proposed in [10] is to tailor LMSD method to the stochastic case by combining it with a procedure for adaptively increasing the mini-batch size. The main difference with respect to the deterministic case consists in building the tridiagonal matrix from a set of stochastic gradients instead of full gradients. Its eigenvalues are named Ritz-like values. With a similar strategy, the harmonic Ritz-like values can also obtained. The inverses of the Ritz-like values (or the harmonic Ritz-like values), appropriately thresholded within a prefixed interval $(\alpha_{min}, \alpha_{max})$, are used as steplengths in the iterates of the next sweep. In [10] the authors adopt an alternation of the Ritz-like and harmonic Ritz-like values, by evaluating two different versions. In particular, an effective strategy consists of replacing the Ritz-like values with the shorter harmonic Ritz-like ones for the next sweep when the size of the current sub-sample is increased (Adaptive Alternation of Ritz-like values or **AA-R**). Indeed, at any iteration, the variance of the descent condition is checked, by verifying that the stochastic gradient related to a sub-sample with the current mini-batch size satisfies the *augmented inner product test* proposed in [3]. When this does not arise, the size of sub-sample is increased so that the new stochastic gradient related to this new sub-sample satisfies the augmented inner product test (for details see [10]). Despite the approach just described being less expensive in terms of time and resources than finding an optimal steplength through trial and error, it remains dependent on the values $(\alpha_{min}, \alpha_{max})$ and on a safeguard value $\overline{\alpha}$. This value is used to start a new sweep when all the inverses of the Ritz-like values are out of the interval $(\alpha_{min}, \alpha_{max})$ and they are discarded. Starting from **AA-R** method, the new contribution is finalized to avoid dependence from α_{min}, α_{max} and $\overline{\alpha}$. In particular, we propose to exploit the properties of the Barzilai-Borwein (BB) formulas to approximate the endpoints of the spectrum of the average Hessian matrix between two iterates (see for example [2]). In practice, at the end of the computation of the Ritz-like (or harmonic Ritz-like) values, from the last iterates and the collected gradients, we determine $1/\alpha_k^{BB1}$ and $1/\alpha_k^{BB2}$ by the standard rule for α_k^{BB1} and the variant suggested in [6] for α_k^{BB2}:

$$\alpha_k^{BB1} = \frac{s_{k-1}^T s_{k-1}}{s_{k-1}^T y_{k-1}}, \quad \alpha_k^{BB2} = \frac{\| s_{k-1} \|}{\| y_{k-1} \|}, \tag{1}$$

where y_{k-1} is the difference between the last two stochastic gradients and s_{k-1} is the difference between the last two iterates. Then, unless the case when $s_{k-1}^T y_{k-1} < 0$, $\frac{1}{\alpha_k^{BB1}}$ and $\frac{1}{\alpha_k^{BB2}}$ are used as $\frac{1}{\alpha_{max}}$ and $\frac{1}{\alpha_{min}}$ respectively, i.e., the eigenvalues out of the interval $(\frac{1}{\alpha_k^{BB1}}, \frac{1}{\alpha_k^{BB2}})$ are discarded. Furthermore, after the first iterations, $\overline{\alpha}$ is replaced by the minimum value of the steplengths employed in the last sweep. We refer to this method as **AA-R-BB**.

3 Numerical Experiments

We consider the optimization problems arising in training a binary classifier on the well-known datasets *MNIST*, *w8a* and *china0* equipped with 60000, 49749 and 16033 examples, respectively. The objective function $F_n(x)$ is the finite average of Logistic Regression (LR) or Smooth Hinge (SH) loss functions. The numerical experiments are aimed to evaluate the behaviour of the proposed **AA-R-BB** method and its previous version **AA-R** with respect to the prefixed bounds of the range $(\alpha_{min}, \alpha_{max})$ for the steplengths. Both methods are compared also to a stochastic algorithm exhibiting a good performance only for a set of hyperparameters carefully selected. In particular, we consider the mini-batch stochastic gradient method **SG-mini**. For this scheme, the steplength is set equal to $50 \cdot \alpha_{OPT}$, where 50 is the fixed mini-batch size and α_{OPT} is the best-tuned steplength value for the basic stochastic gradient method. These values are those related to the best results obtained by a trial process with different mini-batch size and several steplength values.

The numerical experiments are carried out in Matlab® on Intel Core i7 processor. We decided to conduct the experiments on a maximum of 10 epochs because this is sufficient to reach an accuracy comparable with other popular learning methodology on the considered datasets.

Table 1. Numerical results of the considered methods after 10 epochs for $F_n(x)$ given by the SH loss function.

Method	*MNIST*		*w8a*		*china0*	
	$F_n(\bar{x}) - F_*$	$A(\bar{x})$	$F_n(\bar{x}) - F_*$	$A(\bar{x})$	$F_n(\bar{x}) - F_*$	$A(\bar{x})$
$\alpha_{OPT} \cdot (10^{-2}, 5 \cdot 10^2)$						
SG-mini	0.0058	0.8998	0.0023	0.9065	0.0123	0.9155
AA-R	0.0299	0.8819	0.0095	0.9031	0.0709	0.8803
AA-R-BB	0.0138	0.8947	0.0027	0.9066	0.0210	0.9160
$\alpha_{OPT} \cdot (10^{-3}, 10^3)$						
SG-mini	0.0058	0.8998	0.0023	0.9065	0.0123	0.9155
AA-R	0.0302	0.8817	0.0094	0.9031	0.0735	0.8793
AA-R-BB	0.0138	0.8949	0.0027	0.9063	0.0172	0.9157

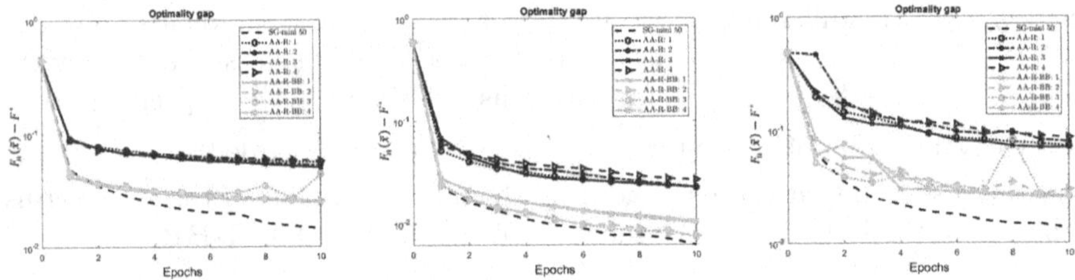

Fig. 1. Behaviour of the optimality gap in 10 epochs with LR on *MNIST* dataset (on the left panel), *w8a* dataset (on the center panel) and *china0* dataset (on the right panel).

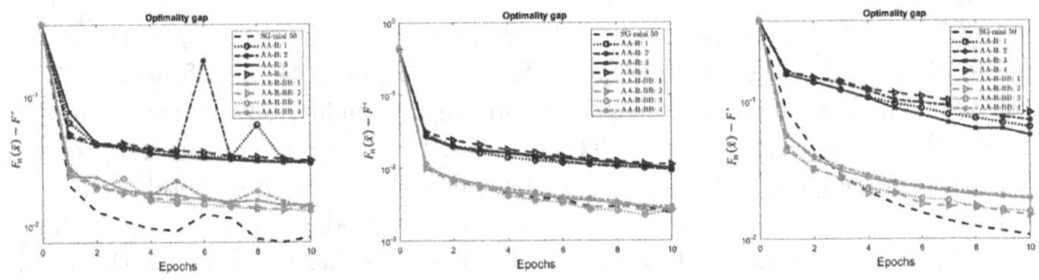

Fig. 2. Behaviour of the optimality gap in 10 epochs with SH on *MNIST* dataset (on the left panel), *w8a* dataset (on the center panel) and *china0* dataset (on the right panel).

As in [10], 4 different settings of the range $(\alpha_{min}, \alpha_{max})$ are used in **AA-R** and **AA-R-BB** methods: 1) $\alpha_{OPT} \cdot (10^{-2}, 5 \cdot 10^2)$, 2) $\alpha_{OPT} \cdot (10^{-3}, 10^3)$, 3) $\alpha_{OPT} \cdot (10^{-2}, 10^3)$, 4) $\alpha_{OPT} \cdot (10^{-3}, 5 \cdot 10^2)$. For any numerical simulation, 10 runs are executed, leaving the possibility to the random number generator to vary and measuring the average of the results. In particular, in Table 1, $A(\bar{x})$ denotes the average accuracy for the test set (examples never seen during the training phase) and $F_n(\bar{x}) - F_*$ is the average value of the *optimality gap* for the train set at the end of 10 epochs. Here $\bar{x}$ is the iterate after 10 epochs and F^* denotes a ground-truth value for the loss function minimum, obtained with a huge number of iterations of a deterministic full gradient method. The results shown in Figs. 1–2 highlight that the **AA-R-BB** strategy leads to a more effective steplength selection than that of **AA-R**. Indeed the effectiveness of **AA-R-BB** method is comparable with the one of **SG-mini** equipped with a best-tuned steplength. These results are confirmed by further numerical experiments, showing that the adaptive steplength rules in **AA-R-BB** method make the choice of a suitable steplength a less difficult task with respect to the setting of a carefully selected fixed value in standard methods. Furthermore, Table 1 shows as the accuracy obtained by **AA-R-BB** seems to be less dependent on the values of α_{max} and α_{min} with respect to **AA-R** case, since **AA-R-BB** adaptively adjusts the bounds (for the sake of synthesis only the results for the SH loss function are reported). Indeed, while the values of $A(\bar{x})$ decreases for the wider range $\alpha_{OPT}(10^{-3}, 10^3)$ for **AA-R** method, in the case of **AA-R-BB** a greater stability is observed.

4 Conclusions and Future Works

In this work the numerical experimentation highlighted that the steplength selection rules as inverses of Ritz-like values coupled with the threshold procedure based on the BB formulas enable to obtain an accuracy similar to the one obtained by a method equipped with hyperparameters carefully selected by a trial process, as for example **SG-mini**. In conclusion, the proposed technique seems provide a guidance on the learning rate selection and it allows to perform similarly to standard approaches equipped with the best-tuned steplength.

Future works will concern the possibility of combining the proposed steplength selection rule with other methods, as Momentum and Adam methods, following the analysis in [11] and involving other loss functions, in particular non-convex ones, as, for instance, those involved in deep learning framework.

References

1. Barzilai, J., Borwein, J.M.: Two-point step size gradient methods. IMA J. Numer. Anal. **8**(1), 141–148 (1988)
2. Birgin, E.G., Martínez, J.M., Raydan, M.: Spectral projected gradient methods: review and perspectives. J. Stat. Softw. **60**(3), 1–21 (2014)

3. Bollapragada, R., Byrd, R., Nocedal, J.: Adaptive sampling strategies for stochastic optimization. SIAM J. Optim. **28**(4), 3312–3343 (2018)
4. Bottou, L., Curtis, F.E., Nocedal, J.: Optimization methods for large-scale machine learning. SIAM Rev. **60**(2), 223–311 (2018)
5. Curtis, F.E., Guo, W.: Handling nonpositive curvature in a limited memory steepest descent method. IMA J. Numer. Anal. **36**(2), 717–742 (2016). https://doi.org/10.1093/imanum/drv034
6. Al-Baali, M., Grandinetti, L., Purnama, A. (eds.): Numerical Analysis and Optimization. SPMS, vol. 134. Springer, Cham (2015). https://doi.org/10.1007/978-3-319-17689-5
7. Di Serafino, D., Ruggiero, V., Toraldo, G., Zanni, L.: On the steplength selection in gradient methods for unconstrained optimization. Appl. Math. Comput. **318**, 176–195 (2018)
8. Fletcher, R.: A limited memory steepest descent method. Math. Program. Ser. A **135**, 413–436 (2012)
9. Sergeyev, Y.D., Kvasov, D.E. (eds.): NUMTA 2019. LNCS, vol. 11973. Springer, Cham (2020). https://doi.org/10.1007/978-3-030-39081-5
10. Franchini, G., Ruggiero, V., Zanni, L.: Ritz-like values in steplength selections for stochastic gradient methods. Soft. Comput. **24**(23), 17573–17588 (2020). https://doi.org/10.1007/s00500-020-05219-6
11. Franchini, G., Ruggiero, V., Zanni, L.: Steplength and mini-batch size selection in stochastic gradient methods. In: Nicosia, G., et al. (eds.) Machine Learning, Optimization, and Data Science, pp. 259–263. Springer International Publishing, Cham (2020)
12. Liang, J., Xu, Y., Bao, C., Quan, Y., Ji, H.: Barzilai-Borwein-based adaptive learning rate for deep learning. Pattern Recogn. Lett. **128**, 197–203 (2019). https://doi.org/10.1016/j.patrec.2019.08.029
13. Tan, C., Ma, S., Dai, Y., Qian, Y.: Barzilai-Borwein step size for SGD. In: Lee, D., Sugiyama, M., Luxburg, U., Guyon, I., Garnett, R. (eds.) Advances in Neural Information Processing Systems 29 (NIPS) (2016)
14. Yang, Z., Wang, C., Zang, Y., Li, J.: Mini-batch algorithms with Barzilai-Borwein update step. Neurocomputing **314**, 177–185 (2018)

Learning Beam Search: Utilizing Machine Learning to Guide Beam Search for Solving Combinatorial Optimization Problems

Marc Huber(✉) and Günther R. Raidl

Institute of Logic and Computation, Algorithms and Complexity Group, TU Wien, Vienna, Austria
{mhuber,raidl}@ac.tuwien.ac.at

Abstract. Beam search (BS) is a well-known incomplete breadth-first-search variant frequently used to find heuristic solutions to hard combinatorial optimization problems. Its key ingredient is a guidance heuristic that estimates the expected length (cost) to complete a partial solution. While this function is usually developed manually for a specific problem, we propose a more general Learning Beam Search (LBS) that uses a machine learning model for guidance. Learning is performed by utilizing principles of reinforcement learning: LBS generates training data on its own by performing nested BS calls on many representative randomly created problem instances. The general approach is tested on two specific problems, the longest common subsequence problem and the constrained variant thereof. Results on established sets of benchmark instances indicate that the BS with models trained via LBS is highly competitive. On many instances new so far best solutions could be obtained, making the approach a new state-of-the-art method for these problems and documenting the high potential of this general framework.

Keywords: Beam search · Combinatorial optimization · Machine learning · Longest common subsequence problem

1 Introduction

Beam search (BS) is a prominent graph search algorithm frequently applied to heuristically solve hard planning and discrete optimization problems in limited time. In this context, it traverses a state graph from a root node, representing an initial state, in a breadth-first-search manner to find a best path to a target node. To keep the computational effort within limits, BS evaluates the reached nodes at each level and selects a subset of only up to β most promising nodes to continue with; the other nodes will not be pursued further, making BS an

This project is partially funded by the Doctoral Program "Vienna Graduate School on Computational Optimization", Austrian Science Foundation (FWF), grant W1260-N35.

G. Nicosia et al. (Eds.): LOD 2021, LNCS 13164, pp. 283–298, 2022.
https://doi.org/10.1007/978-3-030-95470-3_22

incomplete search. The subset of selected nodes at a current level is called *beam*, and parameter β *beam width*. In this way, BS continues level by level until there are no nodes to further expand. A shortest or longest path from the root node to a target node is finally returned as solution. As we consider here maximization problems, we assume w.l.o.g. that the goal is to find a longest path.

Clearly, the way how nodes are evaluated and selected for the beam plays a crucial role for the solution quality. Typically, the length of the longest path to a node so far is considered, and a heuristic value that estimates the maximum further length to go in order to reach a target node is added. This latter heuristic value is calculated by a function also called *guidance function* or guidance heuristic. It is typically developed in a manual, highly problem-specific way, frequently involving many computational experiments and comparisons of different options. Finding a promising guidance function is often challenging as the function not only needs to deliver good estimates but also needs to be fast as it is evaluated for each node in the BS.

The key idea of this work is to use a machine learning (ML) model as guidance function in BS, more specifically a neural network (NN), to approximate the maximum further length to go from a current node to reach a target node. In such an approach, it is a challenge to train the ML model appropriately. Classical supervised learning would mean that labeled training data is available in the form of problem-specific nodes (states) plus *real/exact* maximum path lengths to target nodes. Such data would only be obtainable with huge computational effort and for smaller problem instances. With the BS, however, we primarily want to address large problem instances that cannot practically be solved exactly. Concepts from *reinforcement learning* come to our rescue: In our *Learning Beam Search* (LBS) we start with a randomly or naively initialized ML model and create training data on the fly by performing the search many times on representative, randomly created problem instances. Better estimates for the maximum lengths to go than the ML model usually delivers are determined for subsets of reached nodes by means of *nested BS calls.* This generated training data is buffered in a FIFO replay buffer and used to continuously train the ML model, intertwined with the LBS's further training data production.

While the general principle of this LBS is quite generic, we consider here two well-known NP-hard problems as specific case studies: the Longest Common Subsequence (LCS) problem and the Constrained Longest Common Subsequence (CLCS) problem. Our experimental results show that for both problems, LBS automatically trained on independent random instances is able to compete with the so far leading approaches and in many cases obtains better solutions in comparable runtimes.

Section 2 reviews related work. In Sect. 3, we present the new LBS in a problem-independent way. The LCS and CLCS problems are introduced in Sect. 4. The problem-specific state graphs and how the guidance functions are specifically realized by NNs are described in Sects. 5 and 6, respectively. Results of computational experiments are discussed in Sect. 7. Finally, we conclude in Sect. 8, where we also outline promising future work.

2 Related Work

The increasing popularity of ML also affected classical combinatorial optimization. There is a growing interest in utilizing ML to better solve hard discrete problems. While end-to-end ML approaches to combinatorial optimization also have been attempted by a number of researchers and appear promising, see, e.g., [5], these approaches are usually still not competitive with state-of-the-art problem-solving techniques. However, a broader range of approaches has been suggested to improve classical optimization methods with ML components.

In the context of tree search techniques, one approach is *imitation learning*, i.e., to learn a heuristic by imitating an expert's behavior. In this direction, He et al. [11] proposed to speed up a branch-and-bound by learning a node selection and pruning policy from solving training problems given by an oracle that knows optimal solutions. Concerning variable branching in mixed integer programming, Khalil et al. [15] suggested a ML framework that attempts to mimic the decisions made by strong branching through solving a learning to rank problem. Moreover, Khalil et al. [16] introduced a framework for learning a binary classifier to predict the probability of whether a heuristic will succeed at a given node of a search tree. Training data is collected by running a heuristic at every node at the search tree, gathering the binary classification labels. A general learning to search framework that uses a retrospective oracle to generate feedback by querying the environment on roll-out search traces to improve itself after initial training by an expert was suggested by Song et al. [26].

AlphaGo and its successor AlphaZero gained broader recognition as agents excelling in the games of Go, chess, and shogi [25]. They are based on Monte Carlo tree search in which a deep NN is used to evaluate game states, i.e., to estimate their values in terms of the probabilities to win or lose. Additionally, the NN provides a policy in terms of a probability distribution over the next possible moves. Training is done via reinforcement learning by self-play. Thus, training data is continuously produced by simulating many games against itself, stored in a replay buffer, and used to continuously improve the NN. We apply a similar principle also in our LBS. Several researchers adapted AlphaZero to address combinatorial optimization problems: For example, Laterre et al. [18] applied it to a 3D packing problem, Abe et al. [1] to problems on graphs including minimum vertex cover and maximum cut, and Huang et al. [12] to graph coloring. The latter two approaches used different kinds of graph neural networks as ML models. Mittal et al. [21] suggested another form of heuristic tree search for various graph problems that is guided by a graph neural network. Here, training is done on the basis of smaller instances with known solutions in a supervised fashion, but results indicate that the approach generalizes well to larger instances not seen during training. In the more general context of metaheuristics, a recent survey on utilizing ML can be found in [14].

Beam search was originally proposed in the context of speech recognition [19]. Since then it has been applied in a variety of areas including machine translation [27] and syntactic parsing [29]. Concerning combinatorial optimization problems, many applications exist in particular in the domains of scheduling, see, e.g., [3,8,23], and string-related problems originating in bioinformatics, see, e.g., [6,7,13], but also packing [2].

Concerning specifically the guidance of BS by a ML model, we are only aware of the work by Negrinho et al. [22], who examined this topic from a pure theoretical point of view. They formulated the approach as learning a policy for an abstract structured prediction problem to traverse the combinatorial search space of beams and presented a unifying meta-algorithm as well as novel no-regret guarantees for learning beam search policies using imitation learning.

3 Learning Beam Search

We consider a discrete maximization problem that can be expressed as a longest path problem on a (possibly huge) directed acyclic state graph $G = (V, A)$ with nodes V and arcs A. Each node $v \in V$ represents a problem-specific state, for example, the partial assignment of values to the decision variables in a solution. An arc $(u, v) \in A$ exists between nodes $u, v \in V$ if and only if state v can be obtained from state u by a valid problem-specific action, such as the assignment of a specific feasible value to a so far unassigned decision variable in state u. Let label $\ell(u, v)$ denote this action transitioning from state u to state v. There is one dedicated root node $r \in V$ representing the initial state, in which typically all decision variables are unassigned. Moreover, there are one or more target nodes $T \subset V$, which have no outgoing arcs and represent valid final states, e.g., in which all decision variables have feasible values. Note that this definition of the state graph also covers classical branching trees. Each arc $(u, v) \in A$ has associated a length (or cost) $c(u, v)$. Any path from the root node r to a target node $t \in T$ represents a feasible solution, and we assume that its length, which is the sum of the path's edge lengths, corresponds to the objective value of the solution. As we consider a maximization problem, we seek a longest r–t path, over all $t \in T$.

Our LBS builds upon classical BS, i.e., a breadth-first-search in which at each level a subset of at most β nodes, called the beam B, is selected and pursued further. This selection is performed by evaluating each node u of the current level with the evaluation function $f(v) = g(v) + h(v)$, where $g(v)$ corresponds to the length of a longest so far identified path from the root r to node v, and $h(v)$ is a heuristic guidance function estimating the maximum further length to go to some target node. Note that in an implementation values $g(v)$ are stored with each node v as well as a reference to a predecessor node $u = pred(v)$ on a maximum length path, and thus, $g(v) = g(u) + c(u, v)$; only the root node has no predecessor. In this way, once a target node t is reached, a maximum length r–t path within the investigated part of graph G can be efficiently identified, and the corresponding solution is obtained via the respective arc labels.

As already stated in the introduction, the heuristic guidance function $h(v)$ estimating the length to go is usually crafted manually in a problem-specific way. In our LBS, however, we use an ML model. Still, problem-specific aspects will play a role in the choice of the specific model, in particular, which features are

Algorithm 1. Learning Beam Search (LBS)

Input: nr. of iterations z, beam width β, exp. nr. of training samples per instance α, NBS beam width β', replay buffer size ρ, min. buffer size for training γ
Output: trained guidance function h
$h \leftarrow$ untrained guidance function h (ML regression model)
$R \leftarrow \emptyset$ // replay buffer: FIFO of max. size ρ
for z iterations **do**
    $I \leftarrow$ create representative random problem instance
    Beam Search with training data generation (I, β, R, α, β')
    if $|R| \geq \gamma$ **then**
        (re-)train h with data from R
    end if
end for
return h

derived from a problem-specific state and which kind of ML is actually used. But for now, it is enough to assume that $h(v)$ is a learnable function mapping a state to a scalar value in $\mathbb{R}$.

The core idea of LBS is to train function h via self-learning by iterated application on many random instances generated according to the properties of the instances expected in the future application. The principle is comparable to how learning takes place in AlphaZero [25]. A pseudocode for the main part of the LBS is shown in Algorithm 1. It maintains an initially empty replay buffer R which will contain the training data. This buffer is realized as a first-in first-out (FIFO) queue of maximum size ρ. The idea hereby is to also remove older, outdated training samples when the guidance function has already been improved. A certain number (z) of iterations is then performed. In each iteration, a new independent random problem instance is created and the actual BS applied. This BS, however, is extended by a training data generation that adds in the expected case α new training samples with labels to the replay buffer R; details on this data generation will follow below. After each BS run, a check is performed to determine if the buffer R already contains a minimum number of samples γ, and if this is the case, the guidance function is (re-)trained with data from R. As this training is performed in each iteration, it is usually enough to do a small incremental form of training if the ML model provides this possibility. More specifically, we will use a neural network and train for one epoch over R with mini-batches of size 32. The improved guidance function is then immediately used in the next BS call.

Algorithm 2 shows the actual BS, which is enhanced by the optional training data generation via *nested beam search* (NBS) calls. It receives as input parameters a specific problem instance I to solve, the beam width β, and when training data should be generated the replay buffer R to which the new samples will be added, the expected number of samples to generate α, and a possibly different beam width β' for the NBS. The procedure starts by initializing the beam B with the single root node created for the problem instance I. The outer while-loop performs the BS level by level until B becomes empty. In each iteration, each node in the beam

Algorithm 2. Beam Search with optional training data generation

1: **Input:** problem instance I, beam width β,
2: only when training data should be generated: replay buffer R, exp. nr. of samples α, NBS beam width β'
3: **Output:** best found target node t
4: $B \leftarrow \{r\}$ with r being a root node for problem instance I
5: $t \leftarrow$ none // so far best target node
6: **while** $B \neq \emptyset$ **do**
7: $\quad V_{\text{ext}} \leftarrow \emptyset$
8: $\quad$ **for** $v \in B$ **do**
9: $\quad\quad$ expand v by considering all valid actions, add obtained new nodes to V_{ext}
10: $\quad$ **end for**
11: $\quad$ **for** $v \in V_{\text{ext}}$ **do**
12: $\quad\quad$ evaluate node by $f(v) = g(v) + h(v)$
13: $\quad\quad$ filter dominated nodes (optional, problem-specific)
14: $\quad\quad$ **if** $v \in T \wedge t =$ none $\vee\, g(t) < g(v)$ **then**
15: $\quad\quad\quad$ // new best terminal node encountered
16: $\quad\quad\quad$ $t \leftarrow v$
17: $\quad\quad$ **end if**
18: $\quad\quad$ **if** R given $\wedge$ rand() $< \alpha/n_{\text{nodes}}$ **then** // generate training sample?
19: $\quad\quad\quad$ $t' \leftarrow$ Beam Search $(I(v), \beta')$ // NBS call
20: $\quad\quad\quad$ add training sample $(v, g(t'))$ to R
21: $\quad\quad$ **end if**
22: $\quad$ **end for**
23: $\quad$ $B \leftarrow$ select (up to) β nodes with largest f-values from V_{ext}
24: **end while**
25: **return** t

is expanded by considering all feasible actions for the state the node represents and creating respective successor nodes. These are added to set V_{ext}. Each node in V_{ext} is then evaluated by calculating $g(v)$, $h(v)$ as well as the sum $f(v)$. Optionally and depending on the specific problem, domination checks and filtering can be applied to reduce V_{ext} to only meaningful nodes. Next, line 14 checks if a training sample should be created from the current node v, which is done with probability α/n_{nodes} when the replay buffer R has been provided. Hereby, n_{nodes} is an estimate of the total number of (non-dominated) nodes a whole BS run creates so that we can expect to obtain about α samples. More specifically, in our implementation we initially set $n_{\text{nodes}} = 0$ for the very first LBS iteration, actually producing no training data but counting the number of overall produced nodes, and update n_{nodes} for each successive iteration by the average number of nodes produced over all so far performed LBS iterations. Thus, n_{nodes} is adaptively adjusted. To actually obtain a training sample for a current node v, the sub-problem instance $I(v)$ to which state v corresponds is determined, and an independent NBS call is performed for this subproblem with beam width β'. This NBS returns the target node t' of a longest identified path from node v onward, and thus $g(t')$ will typically be a better approximation to the real maximum path length than $h(v)$. State v and value

$g(t')$ are therefore together added as training sample and respective label (target value) to the replay buffer.

Computational Complexity. Let us assume that the expansion and evaluation of one node takes the problem-specific time T_{node} and the maximum height of the BS tree is H. One NBS call then requires time $O(\beta' \cdot H \cdot T_{\text{node}})$. Considering that LBS performs z iterations and in each makes in the expected case α NBS calls, we obtain that LBS runs in $O(z \cdot (\beta + \alpha \cdot \beta') \cdot H \cdot T_{\text{node}})$ total time.

4 Case Studies

We test the general LBS approach specifically on the following two problems.

The Longest Common Subsequence (LCS) Problem. A string is a sequence of symbols from an alphabet Σ. A subsequence of a string s is a sequence derived by deleting zero or more symbols from that string without changing the order of the remaining symbols. A *common subsequence* of a set of m non-empty strings $S = \{s_1, \ldots, s_m\}$ is a subsequence that all these strings have in common. The LCS problem seeks a common subsequence of maximum length for S. For example, the LCS of strings AGACT, GTAAC, and GTACT is GAC.

The LCS problem is well-studied and has many applications in particular in bioinformatics, where it is used to find relationships among DNA, RNA, or protein sequences. For $m = 2$ strings the problem can be solved efficiently [10], while for general m it is NP-hard [20]. Many heuristics have been proposed for the general LCS problem, and most so far leading ones rely on BS. See [7] for a state-of-the-art method and a rigorous comparison of methods. The BS proposed in [7] utilizes a sophisticated guidance function that approximates the expected LCS length for the remaining input string lengths assuming uniform random strings.

Notations. For a string s, we denote its length by $|s|$. Let $n = \max_{s_i \in S} |s_i|$ be the maximum input string length. The j-th letter of a string s is $s[j]$, with $j = 1, \ldots, |s|$. By $s[j, j']$ we refer to the substring of s starting with $s[j]$ and ending with $s[j']$ if $j \leq j'$ or the empty string ε else. Let $|s|_a$ be the number of occurrences of letter $a \in \Sigma$ in string s.

As in previous work [7], we prepare the following data structure in preprocessing to allow an efficient "forward stepping" in the strings. For each $i = 1, \ldots, m$, $j = 1, \ldots, |s_i|$, and $c \in \Sigma$, $succ[i, j, c]$ stores the minimal position j' such that $j' \geq j \wedge s_i[j'] = c$ or 0 if c does not occur in s_i from position j onward.

The Constrained Longest Common Subsequence (CLCS) Problem. This problem extends the LCS problem on m input strings by additionally considering a *pattern string* P that *must* appear as subsequence in a solution.

For $m = 2$ input strings besides the pattern string, this problem can again be solved efficiently, see, e.g., [28], but for general m the problem also is NP-hard.

Concerning heuristics to address large instances of this general variant, there exists an approximation algorithm [9], which, however, is in practice clearly outperformed by the BS approaches in [6]. One of these BSs is of similar nature as the above-mentioned BS for the LCS problem [7] as it also utilizes an expected length calculation, however, it required a careful extension to consider the pattern string.

As additional data structure, a table $embed[i, j]$, $i = 1, \ldots, m$, $j = 1, \ldots, |P|$ that stores the right-most position j' in input string s_i such that $P[j, |P|]$ is a subsequence of $s_i[j', |s_i|]$ is prepared here during preprocessing.

5 State Graphs for the LCS and CLCS Problems

The state graph $G = (V, A)$ searched by our LBS for solving the LCS problem corresponds to the one used in former work [7]. We therefore only briefly summarize the main facts. A state (node) v is represented by a *position vector* $p^v = (p_i^v)_{i=1,\ldots,m}$ with $p_i^v \in 1, \ldots, |s_i| + 1$, indicating the still relevant substrings of the input strings $s_i[p_i^v, |s_i|]$, i=1,. . . ,m. Note that these substrings form the LCS subproblem instance $I(v)$ induced by node v, for which LBS may perform an independent NBS call to obtain a target value for training. The root node $r \in V$ has position vector $p^r = (1, \ldots, 1)$, and thus, $I(r)$ corresponds to the original LCS instance. An arc $(u, v) \in A$ refers to transitioning from state u to state v by appending a valid letter $a \in \Sigma$ to a partial solution, and thus, arc (u, v) is labeled by this letter, i.e., $\ell(u, v) = a$. In other words, appending letter $a \in \Sigma$ to a partial solution at state u only is feasible if $succ[i, p_i^v, a] > 0$ for $i = 1, \ldots, m$, and yields in this case state v with $p_i^v = succ[i, p_i^v, a] + 1, \ i = 1, \ldots, m$. States that allow no feasible extension are jointly represented by the single terminal node $t \in V$ with $p^t = (|s_i| + 1)_{i=1,\ldots,m}$. As the objective is to find a maximum length string, and with each arc always one letter is appended to a partial solution, the length (cost) of each arc $(u, v) \in A$ is here $c(u, v) = 1$, and thus, $g(v)$ corresponds to the number of arcs of the longest identified r–v path.

In case of the CLCS problem, we also need to consider pattern string P. The position vector is therefore extended by an additional value p_{m+1}^v indicating the position from which on P is not yet covered by the partial solutions leading to state v. A letter $a \in \Sigma$ is only feasible as extension, if the state that would be obtained by it still allows to cover the remaining pattern string, i.e., if $succ[i, p_i^v, c(a)] + 1 \leq embed[i, p_{m+1}^v]$ for $i = 1, \ldots, m$.

For both, the LCS and the CLCS problem, dominance checks and filtering are performed in our LBS exactly as described in [7] and [6], respectively.

6 ML Models for the LCS and CLCS Problems

In principle, any ML regression model may be considered for LBS as guidance function $h(v)$. Clearly, the model needs to be flexible enough, and providing the possibility of incremental learning is a particular advantage in the context of the LBS. Therefore, we consider here for both of our test problems a simple dense feedforward NN with two hidden layers, both equipped with ReLU activation functions. The output layer consists of a single neuron without activation function only – remember that its value is supposed to approximate the maximum further length to go from state v.

Formally, we have defined $h(v)$ to directly receive a state v as input. However, it makes sense to consider the actual input used for the NN more carefully. As an intermediate step, we transform the raw state (and problem instance) information into a more meaningful *feature vector*, which is then actually provided to the NN.

The well-working guidance heuristic from [7] is based on the *remaining string lengths* $|s_i| - p_i^v + 1$, $i = 1, \ldots, m$, only. Therefore, we also use them as features for our NN. Note that the order of the strings and therefore also these values are irrelevant. To avoid possible difficulties in learning these symmetries, we avoid them by always sorting the remaining string lengths before providing them as input to the NN.

In case of the CLCS problem, we additionally have the position p_{m+1}^v in the pattern string P as part of the state, and consequently, we also provide $|P| - p_{m+1}^v + 1$ as an additional feature.

Moreover, earlier guidance heuristics for the LCS problem rely on the *minimum numbers of letter appearances* $\min_{i=1,\ldots,m} |s_i[p_i^v, |s_i|]|_c$, $c \in \Sigma$, from which also a (usually weak) lower bound on the solution length may be calculated. Therefore, we also provide these values for both problems as further features to the NN.

The NN is initialized with random weights. Once the replay buffer has reached the minimum fill level of γ samples, incremental training is done in each LBS iteration by sampling mini-batches of size 32 for one epoch from the replay buffer and applying the ADAM optimizer with step size 0.001 and exponential decay rates for the moment estimates 0.9 and 0.999 as recommended in [17]. As loss function we use the mean squared error.

7 Experimental Evaluation

We implemented LBS in Julia 1.6 using the Flux package for the NN. All experiments were performed in single-threaded mode on a machine with an Intel Xeon E5-2640 processor with 2.40 GHz and a memory limit of 20 GB. Benchmark instances are grouped by the alphabet size $|\Sigma|$, the number of input strings m, and the maximum string length n. LBS was applied to train a NN for each combination of $|\Sigma|$, m, and n. Remember that this learning takes place on the basis of independent random instances that LBS creates on its own. Finally, the benchmark instances are used to evaluate the performance of the BS using the correspondingly trained NN as guidance function. Preliminary tests led to the following LBS configuration that turned out to be suitable for all our benchmarks unless stated otherwise: no. of LBS iterations $z = 1000$, min. buffer size for learning $\gamma = 3000$, LBS and NBS beam widths $\beta = \beta' = 50$, max. buffer size $\rho = 5000$, and exp. nr. of training samples generated per instance $\alpha = 60$. In the

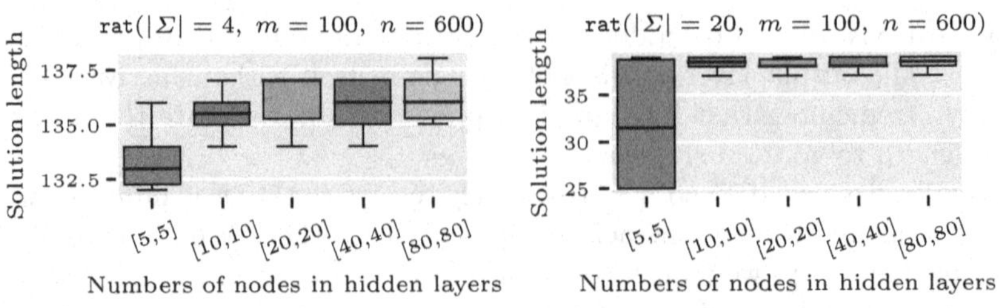

Fig. 1. Impact of the numbers of nodes in the hidden layers on the solution length of LBS on `rat` benchmark instances.

following, we first discuss the experiments for the LCS problem and then those for the CLCS problem.

7.1 LCS Experiments

For the LCS problem, two frequently used benchmark sets are considered. The first one denoted as `rat` was introduced in [24] and consists of 20 instances composed of sequences from rat genomes. The sequences of these instances are close to independent random strings, each sequence has length $n = 600$, but all instances differ in their combinations of values for $|\Sigma|$ and m. The second benchmark set BB from [4] consists of 80 random instances for eight different combinations of $|\Sigma|$ and m (ten instances per combination) and all have string lengths up to $n = 1000$. These instances stand out in that the strings of each exhibit large similarities. Consequently, we generated the random instances within the LBS in the same manner, to obtain suitable NNs specifically for this kind of instances.

A parameter of major importance is the number of hidden nodes in the NN. Clearly, the network size has a direct impact on the computation times of the BS as the guidance function needs to be evaluated for each non-terminal node. Thus, we want to make the NN as small as possible, but at the same time, large enough to get high-quality predictions. In order to examine this aspect, we made tests with different NN configurations. Figure 1 shows exemplary box plots for final LCS lengths obtained from ten LBS runs per NN configuration on selected `rat` instances. We conclude that 20 nodes in both hidden layers are a robust choice. Smaller NNs are sometimes too restrictive, occasionally implying significantly worse results. Therefore, we use this configuration in all further experiments.

Next, we investigate the impact of the beam width on the solution quality, performing again ten runs per configuration. Figure 2 shows respective boxplots. The same beam width has been used for the LBS (β), for the NBS calls (β'), as well as for the final testing on the `rat` benchmark instances. As one may expect, larger beam widths in general yield better results. In particular, using NBS beam widths of $\beta' \leq 30$ turned out to yield clearly inferior results. Therefore, we set $\beta = \beta' = 50$ in all further experiments if not indicated otherwise.

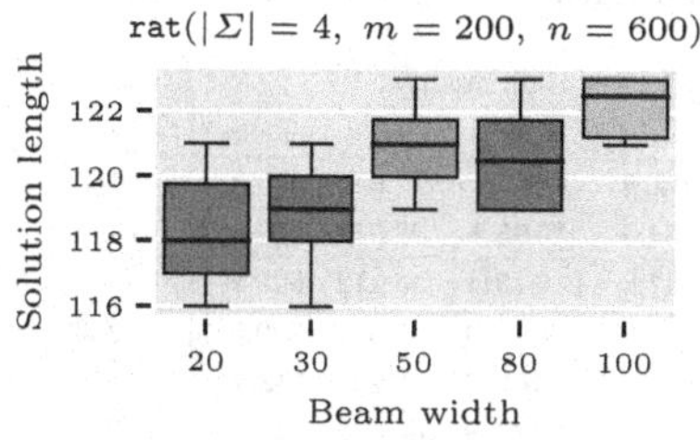

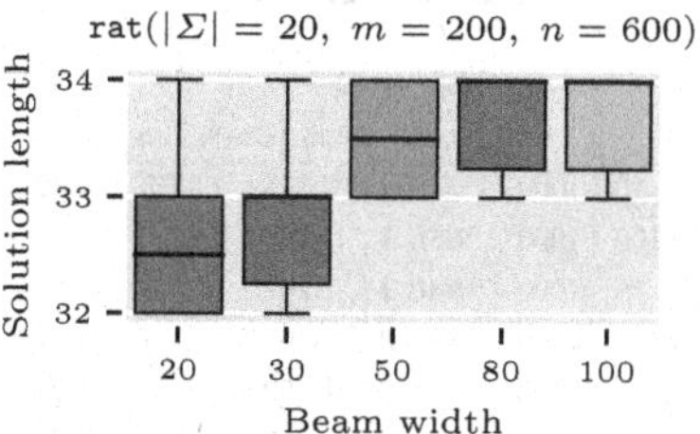

Fig. 2. Impact of beam width $\beta = \beta'$ in training and testing on `rat` instances.

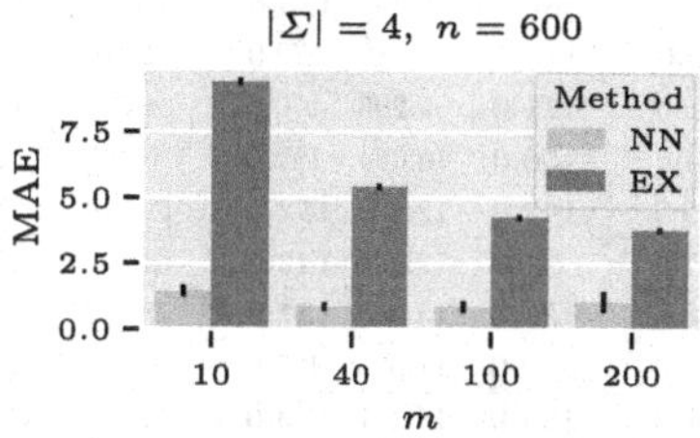

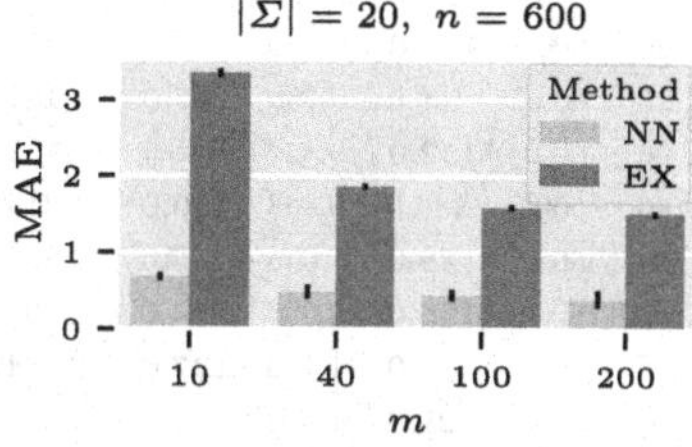

Fig. 3. Mean absolute error of the trained NNs and EX on test samples created by a BS with EX guidance function.

Of interest also is how well a NN trained by LBS actually approximates the real LCS length. As we cannot obtain exact LCS lengths for instances of interesting size, we approximate them by applying the so far leading BS with the approximate expected length (EX) guidance function from [7]. More specifically, to consider instances with a broad range of different input string lengths, we generated 10000 labeled test samples by LBS using EX as guidance function in the outer BS as well as in the NBS calls instead of an NN. This was done for $|\Sigma| \in \{4, 20\}$, $m \in \{10, 40, 100, 200\}$, and $n = 600$. Ten NNs were then trained by LBS for each configuration, and these NNs as well as EX were tested on the generated data sets. Figure 3 shows obtained Mean Absolute Errors (MAEs); standard deviations are indicated by the small black lines. We can observe that the NNs approximate the LCS lengths much better than EX, and differences are particularly large for smaller m. The MAE of EX is about four to six times as large as the MAE of the NNs.

Finally, we compare our approach to the state-of-the-art methods from the literature. While all training with LBS was done with $\beta = \beta' = 50$, we do the tests on the benchmark instances following [7] with two different beam widths: aiming for *low (computation) time* with $\beta = 50$ and aiming for *high quality* with $\beta = 600$. Table 1 shows obtained results. For our LBS, the average solution length $|s_{\mathrm{LBS}}|$ and the runtime of the BS with the trained NN t_{LBS} are listed for each instance group. Columns $|s_{\text{BS-EX}}|$ and $t_{\text{BS-EX}}$ show the respective solution qualities and runtimes for the BS from [7] with the EX guidance function. For a fair time comparison, we re-implemented this approach in our Julia-framework and list the times measured by us, while the solution lengths correspond to those

Table 1. LCS results on benchmark sets BB and rat.

Set	$\|\Sigma\|$	m	n	Low times					High quality				
				$\|s_{\text{LBS}}\|$	t_{LBS} [s]	$\|s_{\text{BS-EX}}\|$	$t_{\text{BS-EX}}$ [s]	$\|s_{\text{lit-best}}\|$	$\|s_{\text{LBS}}\|$	t_{LBS} [s]	$\|s_{\text{BS-EX}}\|$	$t_{\text{BS-EX}}$ [s]	$\|s_{\text{lit-best}}\|$
BB	2	10	1000	651.2	0.855	635.1	0.824	**662.9**	673.1	12.044	673.5	9.180	**676.5**
BB	2	100	1000	***556.1**	1.550	525.1	1.765	551.0	***565.8**	22.979	536.6	18.368	560.7
BB	4	10	1000	***540.1**	1.262	453.0	0.954	537.8	**545.4**	18.112	545.2	12.467	**545.4**
BB	4	100	1000	***381.3**	2.591	318.6	2.174	371.2	***392.9**	35.331	329.5	24.233	388.8
BB	8	10	1000	462.4	1.452	338.8	1.270	**462.6**	**462.7**	28.232	**462.7**	19.155	**462.7**
BB	8	100	1000	***267.4**	4.319	198.0	3.257	260.9	***274.8**	60.682	210.6	36.785	272.1
BB	24	10	1000	**385.6**	5.430	**385.6**	4.172	**385.6**	**385.6**	67.455	**385.6**	48.177	**385.6**
BB	24	100	1000	***148.2**	10.314	95.8	9.399	147.0	**149.5**	153.194	113.3	138.174	**149.5**
rat	4	10	600	199.0	0.550	198.0	1.138	**201.0**	**205.0**	8.591	**205.0**	4.240	204.0
rat	4	15	600	***184.0**	0.660	182.0	1.134	182.0	**185.0**	9.097	**185.0**	7.276	184.0
rat	4	20	600	**169.0**	0.620	168.0	2.082	**169.0**	***173.0**	8.082	172.0	4.120	170.0
rat	4	25	600	166.0	0.766	**167.0**	1.182	166.0	***171.0**	9.295	170.0	4.766	168.0
rat	4	40	600	***152.0**	0.844	146.0	1.172	151.0	***156.0**	10.064	152.0	5.265	150.0
rat	4	60	600	149.0	0.868	**150.0**	1.315	149.0	**152.0**	12.129	**152.0**	12.016	151.0
rat	4	80	600	***138.0**	1.056	137.0	1.368	137.0	140.0	12.564	**142.0**	13.292	139.0
rat	4	100	600	***135.0**	0.483	131.0	1.408	133.0	**137.0**	13.650	**137.0**	7.739	135.0
rat	4	150	600	**127.0**	1.176	**127.0**	2.734	125.0	***130.0**	11.625	129.0	16.841	126.0
rat	4	200	600	**121.0**	1.572	**121.0**	1.733	**121.0**	**123.0**	14.117	**123.0**	19.567	**123.0**
rat	20	10	600	**70.0**	1.108	**70.0**	2.501	**70.0**	**71.0**	10.104	**71.0**	7.579	**71.0**
rat	20	15	600	**62.0**	1.117	**62.0**	2.660	61.0	**63.0**	12.048	**63.0**	13.448	62.0
rat	20	20	600	***54.0**	1.059	53.0	2.553	53.0	**54.0**	13.704	**54.0**	7.970	**54.0**
rat	20	25	600	***51.0**	1.152	50.0	2.545	50.0	**52.0**	13.073	**52.0**	13.573	51.0
rat	20	40	600	***49.0**	0.529	47.0	2.872	48.0	**49.0**	16.005	**49.0**	8.801	**49.0**
rat	20	60	600	**46.0**	1.945	**46.0**	3.234	**46.0**	**47.0**	19.734	46.0	13.413	**47.0**
rat	20	80	600	42.0	1.953	41.0	2.236	**43.0**	43.0	24.741	43.0	23.051	**44.0**
rat	20	100	600	**38.0**	2.007	**38.0**	3.932	**38.0**	39.0	24.441	**40.0**	25.239	39.0
rat	20	150	600	***37.0**	2.457	36.0	2.481	36.0	**37.0**	28.719	**37.0**	29.312	**37.0**
rat	20	200	600	**34.0**	2.048	**34.0**	3.189	**34.0**	**34.0**	32.118	**34.0**	26.838	**34.0**

reported in [7]. Last but not least, so far best known solution lengths from other approaches, as also reported in [7], are shown in column $|s_{\text{lit-best}}|$. Best solution lengths are printed bold, and new best ones obtained by LBS are additionally marked with an asterisk. In 13 out of 28 cases from the low time experiments and in 7 out of 28 cases from the high quality experiments, new best results could be achieved by LBS. In the remaining cases, the quality of the LBS solutions either matched so far best results or were only by a small amount behind. Concerning runtimes, we can conclude that they are very similar to those of BS-EX.

7.2 CLCS Experiments

For the CLCS problem, we use the benchmark set from [6]: ten instances for each combination of $|\Sigma| \in \{4, 20\}$, $m \in \{10, 50, 100\}$, and $n \in \{100, 500, 1000\}$, and ratios of $\frac{n}{|P|} \in \{4, 10\}$ concerning the pattern strings. Note that it is guaranteed that the pattern string appears in the input strings in the way the instances were created, for details on the creation see [6]. We compare the results of the following seven methods from the literature with those obtained by the LBS: the approximation algorithm from [9] (Approx), and Greedy, Random, BS-UB,

Table 2. Results for the CLCS problem on benchmark instances from [6].

$\frac{n}{\lvert P\rvert}$	$\lvert\Sigma\rvert$	m	n	$\lvert s_{\text{LBS}}\rvert$	t_{LBS} [s]	$\lvert s_{\text{Approx}}\rvert$	$\lvert s_{\text{Greedy}}\rvert$	$\lvert s_{\text{Random}}\rvert$	$\lvert s_{\text{BS-UB}}\rvert$	$\lvert s_{\text{BS-Prob}}\rvert$	$\lvert s_{\text{BS-EX}}\rvert$	$\lvert s_{\text{BS-Pat}}\rvert$
4	4	10	100	**34.5**	0.198	28.6	32.2	31.4	**34.5**	**34.5**	**34.5**	**34.5**
4	4	50	100	**27.5**	0.009	26.4	26.9	26.9	**27.5**	**27.5**	**27.5**	**27.5**
4	4	100	100	**26.5**	0.006	25.9	26.2	26.1	**26.5**	**26.5**	**26.5**	**26.5**
4	4	10	500	***183.2**	12.080	134.3	160.4	153.8	179.3	182.4	181.1	168.6
4	4	50	500	147.9	9.912	130.1	139.5	138.1	146.2	**148.3**	146.3	142.7
4	4	100	500	140.6	10.387	128.9	135.8	134.5	140.4	**140.8**	140.3	137.3
4	4	10	1000	***366.5**	27.570	264.7	317.4	308.1	350.3	361.7	361.4	330.8
4	4	50	1000	***296.6**	23.909	257.4	277.3	274.5	291.9	296.4	289.5	284.2
4	4	100	1000	**282.5**	27.273	256.4	270.7	268.1	279.7	**282.5**	279.0	273.3
4	20	10	100	**25.0**	0.005	**25.0**	**25.0**	**25.0**	**25.0**	**25.0**	**25.0**	**25.0**
4	20	50	100	**25.0**	0.005	**25.0**	**25.0**	**25.0**	**25.0**	**25.0**	**25.0**	**25.0**
4	20	100	100	**25.0**	0.005	**25.0**	**25.0**	**25.0**	**25.0**	**25.0**	**25.0**	**25.0**
4	20	10	500	**125.0**	0.006	**125.0**	**125.0**	**125.0**	**125.0**	**125.0**	**125.0**	**125.0**
4	20	50	500	**125.0**	0.009	**125.0**	**125.0**	**125.0**	**125.0**	**125.0**	**125.0**	**125.0**
4	20	100	500	**125.0**	0.012	**125.0**	**125.0**	**125.0**	**125.0**	**125.0**	**125.0**	**125.0**
4	20	10	1000	**250.0**	0.012	**250.0**	**250.0**	**250.0**	**250.0**	**250.0**	**250.0**	**250.0**
4	20	50	1000	**250.0**	0.014	**250.0**	**250.0**	**250.0**	**250.0**	**250.0**	**250.0**	**250.0**
4	20	100	1000	**250.0**	0.031	**250.0**	**250.0**	**250.0**	**250.0**	**250.0**	**250.0**	**250.0**
10	4	10	100	**34.6**	2.628	22.9	29.6	26.5	**34.6**	**34.6**	34.3	32.1
10	4	50	100	***25.1**	3.307	19.8	21.8	21.0	24.9	25.0	24.3	23.5
10	4	100	100	**23.0**	3.668	18.9	20.8	19.6	**23.0**	**23.0**	21.9	21.5
10	4	10	500	***186.3**	21.081	121.4	163.7	147.9	182.2	185.0	184.8	165.9
10	4	50	500	***143.4**	26.723	114.2	129.5	123.6	138.7	142.9	141.8	131.2
10	4	100	500	***133.8**	37.620	111.3	122.0	118.3	129.2	133.3	132.0	124.3
10	4	10	1000	***377.2**	43.442	245.5	329.1	294.8	365.0	375.8	376.3	330.4
10	4	50	1000	***290.9**	56.531	233.5	266.5	254.9	279.6	289.2	290.4	266.0
10	4	100	1000	***272.7**	79.228	230.3	253.2	246.8	262.3	270.9	272.1	255.2
10	20	10	100	**10.2**	0.004	**10.2**	10.1	**10.2**	**10.2**	**10.2**	**10.2**	**10.2**
10	20	50	100	**10.0**	0.005	**10.0**	**10.0**	**10.0**	**10.0**	**10.0**	**10.0**	**10.0**
10	20	100	100	**10.0**	0.005	**10.0**	**10.0**	**10.0**	**10.0**	**10.0**	**10.0**	**10.0**
10	20	10	500	**53.1**	0.007	51.0	52.5	52.7	**53.1**	**53.1**	**53.1**	**53.1**
10	20	50	500	**50.0**	0.006	**50.0**	**50.0**	**50.0**	**50.0**	**50.0**	**50.0**	**50.0**
10	20	100	500	**50.0**	0.008	**50.0**	**50.0**	**50.0**	**50.0**	**50.0**	**50.0**	**50.0**
10	20	10	1000	**105.4**	0.011	101.0	103.9	104.6	**105.4**	**105.4**	**105.4**	**105.4**
10	20	50	1000	**100.0**	0.009	**100.0**	**100.0**	**100.0**	**100.0**	**100.0**	**100.0**	**100.0**
10	20	100	1000	**100.0**	0.012	**100.0**	**100.0**	**100.0**	**100.0**	**100.0**	**100.0**	**100.0**

BS-Prob, BS-EX, and BS-Pat from [6]. In all BS approaches, the same beam width $\beta = 2000$ was used for the tests on the benchmark instances. Results are shown in Table 2. Here, in ten out of 36 cases, new best results could be achieved by LBS, and it scores worse in only two out of 36 cases; in the remaining cases, the solutions values from LBS are equal to the so far best known ones.

8 Conclusions and Future Work

We presented a general learning beam search framework to solve combinatorial optimization problems for which the solution space can be represented by a state graph. Instead of the frequently challenging manual design of a meaningful guidance function, we train a regression model to approximate the real length to go from a state and use this model thereafter in a BS. Training is done in the spirit of reinforcement learning by performing many BS runs on randomly created instances and calling a nested beam search to obtain labeled training data. Our case studies on the LCS and the CLCS problems clearly show that this learning approach can be highly effective. On many benchmark instances new best solutions could be obtained, making this approach a new state-of-the-art method for the considered two problems.

Clearly, the proposed LBS is not entirely problem-agnostic: Still, it is important to use a suitable state space, to derive meaningful features from states, and to choose an appropriate ML model for a problem at hand. Moreover, note that in our implementation for the LCS and CLCS, individual models need to be trained for specific choices of $|\Sigma|$, m, and n. In future work specifically for the LCS and CLCS problems, we aim at relying on different features that just describe the distribution of remaining input string lengths, in order to learn models that are independent of m and possibly also n.

General improvement potential for LBS lies in the fact that the guidance function actually does not need to approximate the length to go well, but only needs to provide scores for ranking the solutions in the beam. Can this flexibility be used to come up with alternative optimization targets and loss functions for the training, yielding overall better results? Parallelization and the utilization of GPUs are further natural possibilities to speed up in particular the learning. Last but not least, we aim at applying LBS to further problems and to also investigate other ML models than NNs.

References

1. Abe, K., Xu, Z., Sato, I., Sugiyama, M.: Solving NP-hard problems on graphs with extended AlphaGo Zero. arXiv:1905.11623 [cs, stat] (2020)
2. Akeba, H., Hifib, M., Mhallah, R.: A beam search algorithm for the circular packing problem. Comput. Oper. Res. **36**(5), 1513–1528 (2009)
3. Blum, C., Miralles, C.: On solving the assembly line worker assignment and balancing problem via beam search. Comput. Oper. Res. **38**(1), 328–339 (2011)
4. Blum, C., Blesa, M.J.: Probabilistic beam search for the longest common subsequence problem. In: Stützle, T., Birattari, M., Hoos, H.H. (eds.) SLS 2007. LNCS, vol. 4638, pp. 150–161. Springer, Heidelberg (2007). https://doi.org/10.1007/978-3-540-74446-7_11
5. Dai, H., Khalil, E.B., Zhang, Y., Dilkina, B., Song, L.: Learning combinatorial optimization algorithms over graphs. In: Advances in Neural Information Processing Systems, vol. 31, pp. 6348–6358. Curran Associates, Inc. (2017)

6. Djukanovic, M., Berger, C., Raidl, G.R., Blum, C.: On Solving a generalized constrained longest common subsequence problem. In: Olenev, N., Evtushenko, Y., Khachay, M., Malkova, V. (eds.) OPTIMA 2020. LNCS, vol. 12422, pp. 55–70. Springer, Cham (2020). https://doi.org/10.1007/978-3-030-62867-3_5
7. Djukanovic, M., Raidl, G.R., Blum, C.: A beam search for the longest common subsequence problem guided by a novel approximate expected length calculation. In: Nicosia, G., Pardalos, P., Umeton, R., Giuffrida, G., Sciacca, V. (eds.) LOD 2019. LNCS, vol. 11943, pp. 154–167. Springer, Cham (2019). https://doi.org/10.1007/978-3-030-37599-7_14
8. Ghirardi, M., Potts, C.N.: Makespan minimization for scheduling unrelated parallel machines: a recovering beam search approach. Eur. J. Oper. Res. **165**(2), 457–467 (2005)
9. Gotthilf, Z., Hermelin, D., Lewenstein, M.: Constrained LCS: hardness and approximation. In: Ferragina, P., Landau, G.M. (eds.) CPM 2008. LNCS, vol. 5029, pp. 255–262. Springer, Heidelberg (2008). https://doi.org/10.1007/978-3-540-69068-9_24
10. Gusfield, D.: Algorithms on Strings, Trees, and Sequences: Computer Science and Computational Biology. Cambridge University Press, New York (1997)
11. He, H., Daumé, H.C., Eisner, J.M.: Learning to search in branch-and-bound algorithms. In: Ghahramani, Z., et al. (eds.) Advances in Neural Information Processing Systems, vol. 27. Curran Associates, Inc. (2014)
12. Huang, J., Patwary, M., Diamos, G.: Coloring big graphs with AlphaGo Zero. arXiv:1902.10162 [cs] (2019)
13. Huang, L., et al.: LinearFold: linear-time approximate RNA folding by 5'-to-3' dynamic programming and beam search. Bioinformatics **35**(14), i295–i304 (2019)
14. Karimi-Mamaghan, M., Mohammadi, M., Meyer, P., Karimi-Mamaghan, A.M., Talbi, E.G.: Machine learning at the service of meta-heuristics for solving combinatorial optimization problems: a state-of-the-art. Eur. J. Oper. Res. (2021). https://doi.org/10.1016/j.ejor.2021.04.032
15. Khalil, E.B., Bodic, P.L., Song, L., Nemhauser, G., Dilkina, B.: Learning to branch in mixed integer programming. In: Proceedings of the 26th International Joint Conference on Artificial Intelligence, pp. 724–731. AAAI Press (2016)
16. Khalil, E.B., Dilkina, B., Nemhauser, G.L., Ahmed, S., Shao, Y.: Learning to run heuristics in tree search. In: Proceedings of the 26th International Joint Conference on Artificial Intelligence, pp. 659–666. Melbourne, Australia (2017)
17. Kingma, D.P., Ba, J.: Adam: a method for stochastic optimization. In: Proceedings of the 3rd International Conference on Learning Representations, San Diego, CA (2015)
18. Laterre, A., et al.: Ranked reward: enabling self-play reinforcement learning for combinatorial optimization. In: AAAI 2019 Workshop on Reinforcement Learning on Games. AAAI Press (2018)
19. Lowerre, B.: The harpy speech recognition system. Ph.D. thesis, Carnegie Mellon University, Pittsburgh, PA (1976)
20. Maier, D.: The complexity of some problems on subsequences and supersequences. J. ACM **25**(2), 322–336 (1978)
21. Mittal, A., Dhawan, A., Manchanda, S., Medya, S., Ranu, S., Singh, A.: Learning heuristics over large graphs via deep reinforcement learning. arXiv:1903.03332 [cs, stat] (2019)
22. Negrinho, R., Gormley, M., Gordon, G.J.: Learning beam search policies via imitation learning. In: Bengio, S., et al. (eds.) Advances in Neural Information Processing Systems, vol. 31, pp. 10652–10661. Curran Associates, Inc. (2018)

23. Ow, P.S., Morton, T.E.: Filtered beam search in scheduling. Int. J. Prod. Res. **26**, 297–307 (1988)
24. Shyu, S.J., Tsai, C.Y.: Finding the longest common subsequence for multiple biological sequences by ant colony optimization. Comput. Oper. Res. **36**(1), 73–91 (2009)
25. Silver, D., et al.: A general reinforcement learning algorithm that masters chess, shogi, and Go through self-play. Science **362**(6419), 1140–1144 (2018)
26. Song, J., Lanka, R., Zhao, A., Bhatnagar, A., Yue, Y., Ono, M.: Learning to search via retrospective imitation. arXiv:1804.00846 [cs, stat] (2019)
27. Sutskever, I., Vinyals, O., Le, Q.V.: Sequence to sequence learning with neural networks. In: Advances in Neural Information Processing Systems, vol. 27. Curran Associates, Inc. (2014)
28. Tsai, Y.: The constrained longest common subsequence problem. Inf. Process. Lett. **88**, 173–176 (2003)
29. Weiss, D., Alberti, C., Collins, M., Petrov, S.: Structured training for neural network transition-based parsing (2015)

Modular Networks Prevent Catastrophic Interference in Model-Based Multi-task Reinforcement Learning

Robin Schiewer(✉) and Laurenz Wiskott

Institute for Neural Computation, Ruhr-University Bochum, Bochum, Germany
{robin.schiewer,laurenz.wiskott}@ini.rub.de
https://www.ini.rub.de/

Abstract. In a multi-task reinforcement learning setting, the learner commonly benefits from training on multiple related tasks by exploiting similarities among them. At the same time, the trained agent is able to solve a wider range of different problems. While this effect is well documented for model-free multi-task methods, we demonstrate a detrimental effect when using a single learned dynamics model for multiple tasks. Thus, we address the fundamental question of whether model-based multi-task reinforcement learning benefits from shared dynamics models in a similar way model-free methods do from shared policy networks. Using a single dynamics model, we see clear evidence of task confusion and reduced performance. As a remedy, enforcing an internal structure for the learned dynamics model by training isolated sub-networks for each task notably improves performance while using the same amount of parameters. We illustrate our findings by comparing both methods on a simple gridworld and a more complex vizdoom multi-task experiment.

Keywords: Model-based reinforcement learning · Multi-task reinforcement learning · Latent space models · Catastrophic interference · Task confusion

1 Introduction

In recent years, deep reinforcement learning (RL) has shown impressive results in problem domains such as robotics and game playing [12,15,19,20,27]. However, sample inefficiency is still a major shortcoming of many of the methods. To achieve superhuman performance e.g. in video games, the required number of interactions for complex tasks lies in the tenths of millions. Model-based approaches mitigate this problem by integrating the collected sample information into a coherent model of the environmental dynamics [30]. Those learned models are used for direct policy learning [11], planning [7,13] or to augment existing model-free approaches [8,21,24]. In the presence of a readily available analytical model of the environment, that does not have to be learned, perfect planning can be used to learn a policy with great success [29,31].

G. Nicosia et al. (Eds.): LOD 2021, LNCS 13164, pp. 299–313, 2022.
https://doi.org/10.1007/978-3-030-95470-3_23

Despite the success of deep RL across a wide range of problem domains, most state of the art approaches have to be re-trained for every new task. Since similar problems might very well share a common underlying structure, this is a potential waste of resources [3]. Furthermore, an agent can also encounter a different mixture of tasks in a single environment. If the agent is created without the concept of distinct tasks in mind, an alteration of the task mixture may negatively impact the agent's performance. While greater sample efficiency is one reason to engage in multi-task learning, it also produces a more capable agent that can solve a wider range of problems. Whereas there exists a vast amount of research around multi-task RL in general [6,9,10,26], the combination of model-based latent space RL and multiple tasks is still largely unexplored. Since the former has brought impressive improvements for sample efficiency and the latter promises greater flexibility and reusability, we combine latent space models with multi-task RL in this paper and investigate its usefulness.

We focus on model-based multi-task deep RL for complex observation domains like images. The main question we address is if training the same dynamics model on multiple similar tasks helps performance through knowledge transfer or if catastrophic interference outweighs the benefits. If so, how can a model-based multi-task agent be structured to avoid catastrophic interference? Our contributions in answering these questions are the following:

- In our experiments we show a detrimental effect on performance through catastrophic interference when training a single dynamics model on multiple tasks simultaneously. To mitigate the effect, we propose a world model that uses multiple distinct latent space dynamics models which are activated through a context-based task classification network.
- By strict separation of the dynamics networks, the probability that dynamics of different tasks interfere with each other is minimized. As a side effect, retraining individual tasks and recombining learned tasks is straightforward.
- We demonstrate the performance difference between a single dynamics model and our method in a 3-task gridworld and a more complex 2-task 3D environment. To evaluate both approaches, we perform planning on the learned dynamics models and measure the obtained reward.

2 Related Work

Latent Space Models: Previous work on learning latent space models from high dimensional inputs has produced impressive results especially in the atari learning environment and continuous control benchmarks [14,17,22]. The architectures share a similar structure where a convolutional encoder embeds inputs to a latent representation which is fed into a recurrent prediction network for states and rewards. By stacking predictions on previous predictions, imaginary rollout trajectories based on the learned dynamics are produced. Those are used for planning [13], training an actor-critic agent [13,14] or to drive exploration [28]. To mitigate accumulating errors caused by imperfect dynamics models, probabilistic prediction models and long rollout trajectories during training are used. While the above mentioned methods demonstrate impressive performance of latent space models in RL, we specifically focus on resource efficiency in a multi-task setting.

Multi-task RL: To overcome catastrophic interference and pave the way for lifelong learning, in [26] a policy network from multiple modules called columns is built. Each column is a mini network with individual weights while the same architecture is shared among all columns. For each task, a new column is added and layer-wise laterally connected to all previous columns. Since the connections are learned, this enables new columns to selectively benefit from previously acquired behavior. The authors train their method using A3C [19] and demonstrate performance improvements by comparing it to various versions of a single-column baseline network.

By learning task-specific pathways of active subsections within one large neural network, [9] use a similar concept than [26]. A subsection corresponds to a small, localized region in the larger network. However, instead of hardwiring the connections between subsections, they are learned by an evolutionary strategy. Given an active path, training of the involved subsections is done via gradient descent. At the same time, the remaining weights are frozen so that change only occurs in the currently active modules. The authors describe their approach as an evolutionary version of dropout, where thinned out networks are evolved instead of randomly generated (as in regular dropout).

A significant difference between the above mentioned approaches and our method is that our algorithm learns modular dynamics models instead of modular policies. Furthermore, in our proposed method lateral connections between the models are absent since they would introduce dependencies which prevent arbitrary recombination of the learned task models.

3 Method Description

The foundation of our approach lies in the combination of multiple recurrent dynamics models (RDMs), a vector-quantized variational autoencoder (VQ-VAE) and a task classification network (TCN) for online task detection. The VQ-VAE is used to encode the image observation stream from the environment to a latent space representation which we simply refer to as embedding from now on. The TCN receives the stream of embeddings and predicts the probability of each RDM to be responsible for the task at hand. The most probable RDM is then used to generate predictions in latent space. A graphical overview of the method is presented in Fig. 1. In the following sections, we will explain each of the components in more detail. While we discuss the hyperparameters we use for some parts of the architecture, we do not mention all of them and refer to the link to our code repository in the conclusion for an exhaustive list.

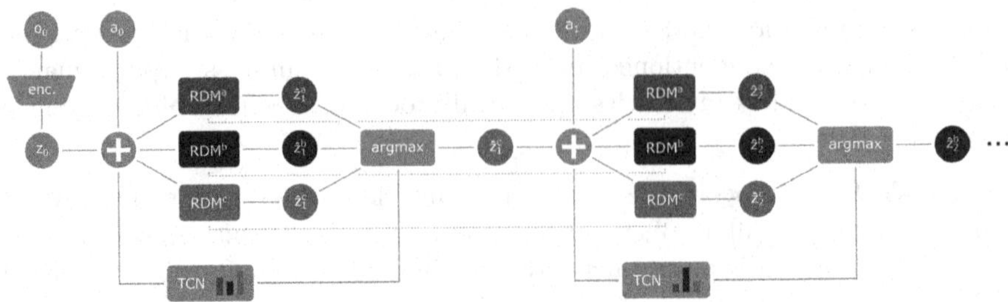

Fig. 1. Overview of the multi RDM architecture. In the first timestep, observation o_0 is embedded by the encoder of the VQ-VAE. The resulting embedding is combined with the action and presented to the RDMs and the TCN. The latter chooses one of the RDMs (visualized by the argmax node), which is consequently used to predict the embedding for the next timestep. For simplicity, reward and terminal state probability outputs are not visualized here. Dashed lines indicate recurrent components that maintain information persistent between the individual time steps.

3.1 Vector-Quantized Variational Autoencoder

To embed an image observation into a latent space, we use the Vector-Quantized Variational Autoencoder (VQ-VAE) from [23]. It consists of a convolutional encoder block, a quantization layer and a deconvolutional decoder block. In the following, we summarize the parameters of all VQ-VAE components with the parameter vector ϕ. The encoder transforms an image observation $\mathbf{o}$ into a 3D tensor $E_\phi(\mathbf{o})$ of size $w \times h \times c$. It can be interpreted as a more compact image of size w times h with c channels. This image tensor is passed to the quantization layer that replaces each pixel along the channel dimension with the closest from a set of codebook vectors. The resulting tensor $\mathbf{z}$ is a per-pixel quantized version of $E_\phi(\mathbf{o})$. Conceptually equivalent to a regular VAE, the deconvolutional decoder D transforms $\mathbf{z}$ back to the (reconstructed) original input image $\hat{\mathbf{o}} = D_\phi(\mathbf{z})$. The VQ-VAE is trained by minimizing the following loss function:

$$\mathcal{L}_\phi(\mathbf{o}, D_\phi(\mathbf{z})) = \underbrace{||\mathbf{o} - D_\phi(\mathbf{z})||_2^2}_{\text{decoder}} + \underbrace{||sg[E_\phi(\mathbf{o})] - \mathbf{z}||_2^2}_{\text{codebook}} + \beta \underbrace{||sg[\mathbf{z}] - E_\phi(\mathbf{o})||_2^2}_{\text{encoder}} \quad (1)$$

Whereas the parameter β trades off encoder vs. decoder loss and the $sg[\cdot]$ operator indicates the stop of gradient backpropagation. We used $\beta = 0.5$ in all our experiments, since it performed best among a collection of tested values between

0.25 and 0.75. According to the recommendations of [23], we swap the codebook loss in Eq. 1 for an exponentially moving average update of the codebook vectors e_i:

$$N_i^{(t)} := N_i^{(t-1)}\gamma + n_i^{(t)}(1-\gamma), \quad m_i^{(t)} := m_i^{(t-1)}\gamma + \sum_j^{n_i^{(t)}} E(o)_{i,j}^{(t)}(1-\gamma), \quad e_i^{(t)} := \frac{m_i^{(t)}}{N_i^{(t)}}$$

where $n_i^{(t)}$ is for the current batch the number of pixels/vectors in $E(\mathbf{o})$ that will be replaced with codebook vector e_i and γ is a decay parameter between 0 and 1. We used the default $\gamma = 0.99$ for all experiments and a Keras [2,4] port of the VQ-VAE implementation in the Sonnet library [1]. The autoencoder architecture of our choice naturally forms clusters in embedding space, which has been reported to increase prediction performance of dynamics models like ours [23] and potentially helps the TCN in differentiating between tasks.

3.2 Recurrent Dynamics Models

For the design of our RDMs, we follow ideas from [13] and combine probabilistic and deterministic model components to increase predictive performance and prevent overfitting [5]. As previously explained, following the work in [22] we additionally embed environment observations into latent space representations z and operate entirely in that latent space from there on. Since z is a 3D tensor, the architecture of our RDM contains mainly convolutional layers (conv) to carry through any spatial information contained in the image observations. An RDM is comprised of four submodules: The belief-state network (conv), the embedding network (conv), the reward network (dense) and the terminal state probability network (dense).

As already briefly mentioned, to make the prediction of z_t more robust to uncertainty, we combine a stochastic and a deterministic prediction path. The deterministic path is realized through a convolutional LSTM network which computes the belief state h_t from the previous belief state h_{t-1}, action a_{t-1} and embedding z_{t-1}. Given h_t, the stochastic prediction path is a categorical distribution from which z_t is sampled. This way, information about past observations, actions and belief states can be accumulated in h_t and passed on to future time steps. At the same time, uncertainty regarding the next z_t can be expressed through the sampling process. Because the embeddings are defined completely by the indices of their codebook vectors, we use a categorical distribution to sample those indices, namely the Gumbel Softmax [16,18], which is a reparameterized approximation to a categorical one-hot distribution. It can be used with backpropagation and at the same time offers almost discrete sampling of the required one-hot vectors. The rewards are sampled from a diagonal Gaussian and the terminal transition probabilities are sampled from a Bernoulli distribution. The structure of the RDM is summarized as follows:

$$\text{belief state:} h_t = f_\theta(h_{t-1}, z_{t-1}, a_{t-1}) \tag{2}$$

$$\text{embedding:} z_t \sim q_\theta(z_t|h_t) \tag{3}$$

$$\text{reward:} r_t \sim q_\theta(r_t|h_t) \tag{4}$$

$$\text{terminal transition:} \gamma_t \sim q_\theta(\gamma_t|h_t) \tag{5}$$

Note that we summarize all RDM component's parameters into one vector θ for simplicity. To train the RDM, the following loss function is minimized per time step:

$$\mathcal{L}_\theta = \mathcal{L}_\theta^e + \mathcal{L}_\theta^r + \mathcal{L}_\theta^\gamma \tag{6}$$

whereas the individual prediction loss terms are the parameter maximum likelihood solutions given the transition data:

$$\text{embedding:}\mathcal{L}_\theta^e = \mathbb{E}_{q_\phi(z_t|o_t)}\left[-\ln q_\theta(z_t|h_t)\right] \tag{7}$$

$$\text{reward:}\mathcal{L}_\theta^r = \mathbb{E}_{q_\theta(z_t|h_t),p(r_t|o_t,a_t,o_{t+1})}\left[-\ln q_\theta(r_t|h_t,z_t)\right] \tag{8}$$

$$\text{terminal:}\mathcal{L}_\theta^\gamma = \mathbb{E}_{q_\theta(z_t|h_t),p(\gamma_t|o_t,a_t,o_{t+1})}\left[-\ln q_\theta(\gamma_t|h_t,z_t)\right] \tag{9}$$

while we denote all learned distributions by $q_\theta(\cdot)$ or $q_\phi(\cdot)$ and the true environment dynamics by $p(\cdot)$.

3.3 Context Detection

In contrast to the algorithms presented in [11,13,22], we use distinct dynamics models to combat catastrophic interference when training on multiple tasks. In theory, a single (monolithic) dynamics model can exploit similarities in the tasks it learns, which positively influences learning speed and predictive performance. But since weight updates for different tasks may interfere with each other negatively, a monolithic dynamics model could instead suffer from reduced performance. This effect has been shown by [33] for policy networks in multi-task reinforcement learning. To prevent this catastrophic interference, we isolate knowledge about different task dynamics in separate RDMs and use the TCN to orchestrate them. This essentially makes the TCN a context detector that learns to classify an embedding-action-stream and choose the correct RDM for the task at hand. In every prediction step, the TCN receives the current observation embedding and action as input. While it is theoretically able to choose a different RDM in every time step, we did not observe this behavior in practice. Because different tasks can have locally similar or equivalent embeddings, the TCN contains convolutional LSTM layers and is thus able to remember past embeddings and actions. It is thereby capable to disambiguate temporarily similar embedding streams. The TCN is trained in a supervised manner via categorical crossentropy loss.

3.4 Planning

We directly use our learned RDMs for planning in the environments. Thereby, we follow the crossentropy method that has originally been introduced by [25] and is also employed by [11] and [5]. A planning procedure consists of n iterations. At the beginning of the first iteration, the current observation is embedded into latent space by the VQ-VAE and serves as a starting point for k rollouts of length T. To generate the action sequences required for the rollouts, we initialize one categorical distribution per time step at random from which we sample k actions. This results in a $k \times T$ matrix of actions where each row represents the action sequence for one rollout trajectory. The resulting trajectories produced by the monolithic or multi RDM are ranked w.r.t. their discounted return and the top ρ percent are chosen as winners. In the next iteration, we use the maximum likelihood parameters of the previous winners for the action distributions and again sample $k \times T$ actions to generate new rollouts. For the gridworld setting, we use $n = 20$ iterations, $k = 500$ rollout trajectories and $T = 100$ planning steps, because especially in the third task the reward is far away from the starting location. We take the top $\rho = 10/500$ of all rollout trajectories in each iteration. For the 3D environment, we use $n = 5$ iterations, $k = 400$ rollout trajectories and $T = 60$ time steps due to technical limitations regarding the capacity and of the used GPUs (the algorithm is running on a single Tesla v100 with 16 GB of VRAM). We take the top $\rho = 10/400$ of all rollout trajectories in each iteration. For both the gridworld and the 3D environment, we add noise of 0.05 to the maximum likelihood parameters during planning because this effectively improved the results.

3.5 Training

Presented with a batch of encoded trajectories from m tasks, the TCN outputs the probabilities for each of the m RDMs per sample (i.e. per trajectory per time step). Since the individual prediction errors of the RDMs are weighted with their probability of being chosen, only those with a high probability receive a strong learning signal and consequently adapt their weights. To discourage the TCN from switching around RDMs during a task, we provide the task ID as a supervised learning signal during training. The TCN produces a less concentrated probability distribution in the beginning of the training when the prediction error of the RDMs dominates the loss. At this time, multiple RDMs can be chosen for one task without hurting the loss significantly, which means that most RDMs are exposed to more than one task during the course of the training. This increases variety in training samples for each of the RDMs and encourages them to learn characteristics shared across tasks. However, this situation only lasts for the first few training epochs. As soon as the prediction error decreases, the TCN has to allocate the correct models to their respective tasks in order to further minimize the loss. The resulting system is then able to activate a task-specific RDM for each task it was trained on. As a consequence of using distinct networks for each task, individual RDMs may be re-trained or trained longer on tasks where the performance is not yet as desired without the risk of sacrificing performance in other tasks.

We compare the multi RDM model with a larger, monolithic RDM in two different experiments: A simple 3-task gridworld and a more demanding 2-task setting comprised of two vizdoom [32] environments. The latter setting can be considered more demanding since it features a dynamic 3D environment, more diverse observations and a more complex reward function. We want to emphasize that per experiment, both the multi RDM and the monolithic RDM have the same amount of trainable parameters. Since the monolithic RDM does not have a TCN and to maintain a fair comparison, the architecture of the monolithic

RDM is slightly modified. Another network head is added and trained to output the current task ID similarly to the TCN in the multi RDM architecture. This way, the monolithic architecture gets exactly the same training information as the multi RDM architecture. Training is summarized as follows:

1. Collect around 30 000 (gridworld)/200 000 (vizdoom) transitions from each task by random action sampling. Since only complete trajectories are collected, the final number of collected transitions might end up slightly higher.
2. Train the VQ-VAE on the collected samples for 200 (both) epochs.
3. Train either the monolithic or the multi RDM predictors on the trajectories using the VQ-VAE to embed observations. Training is done for 150 (gridworld)/500 (vizdoom) epochs.
4. Perform planning utilizing the trained predictors.

When fully trained, we use the VQ-VAE and RDMs to perform imaginary rollouts as described in Sect. 3.4. Since we specifically want to assess the quality of the learned dynamics for both the monolithic and the multi RDM, we do not re-plan actions after each step. This way, groundtruth observation data would be injected into the system continuously, which potentially helps a suboptimal model. Instead, in the gridworld experiment our algorithm uses only the starting observation of the agent to perform a full planning routine until the end of the episode. This has the beneficial side effect of being considerably less time consuming than re-planning after every step. Note that the starting observations of the gridworld tasks are unique, so they should in theory provide sufficient information for the algorithm. In the vizdoom experiment, one of the two tasks features a dynamic environment that can change without interaction of the agent. Thus, periodic re-planning is necessary to incorporate new information from the environment. In an effort to find a good balance between computational cost and planning accuracy, we use the first 30 actions from the planning procedure and then re-plans from there on.

To obtain the gridworld experiment data discussed in the next section, we repeated the above mentioned four-step process three times with different random seeds and averaged the results appropriately. To obtain the vizdoom experiment data, we did a single iteration of the above mentioned four-step process. We want to clarify that we trained one pair of dynamics models (a monolithic and a multi RDM) per experiment. This means, the dynamics models trained on the gridworld experiment did not receive any data from the vizdoom experiment and vice versa.

4 Experiments

To answer the question of whether dynamics models can benefit from training on multiple tasks at once, we use a simple gridworld experiment (cf. Fig. 2) and a more complex vizdoom experiment (cf. Fig. 3). For both experiments, observations from different tasks can easily be confused with each other. Yet, the system

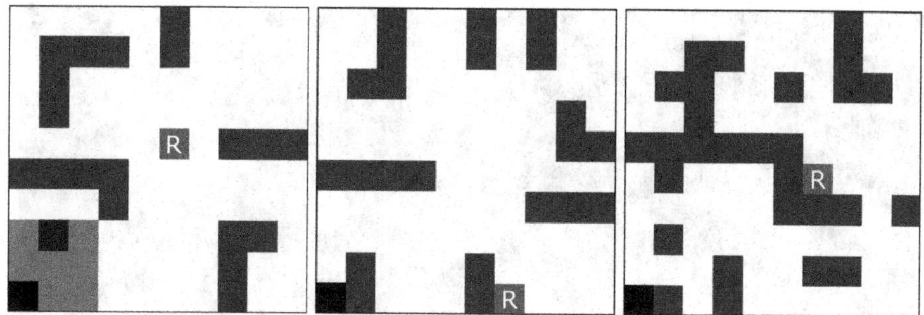

Fig. 2. The three training tasks of the gridworld experiment. The agent (green square in the lower left corner) has to find the reward (red square marked with "R"). Every action moves the agent one grid cell into the respective direction if no obstacles are met. Walls (black squares) block movement without reward penalty. The shaded blue area in the first environment (left panel) shows an example of the top-down viewport of the agent. If part of the viewport is outside of the environment at any time, the respective part is filled with zeros (i.e. it is colored black). (Color figure online)

has to learn to identify the tasks and memorize task dynamics without confusing them. This is especially difficult for the gridworld experiment, since although having very different global structures, the individual image observations look similar across all three tasks. On the one hand, this makes it easier to learn shared properties across them, e.g. that all objects in the viewport are shifted down one cell if an "up" move is performed. We expect this to be advantageous for the monolithic RDM. On the other hand, in the absence of striking visual cues it becomes harder to distinguish between different tasks, which should put the multi RDM at an advantage because it uses the specialized TCN for that. In case of the vizdoom experiment, distinguishing the tasks is easier because there are salient optical cues in every observation (e.g. the gun and HUD indicating the "VidzoomBasic" task). We expect this to benefit the monolithic RDM. Yet, although the environment dynamics are partially similar for both tasks (agent movement), we see quite different observations and reward functions. We expect this reduced degree of common structure among the vizdoom tasks to put the multi RDM at an advantage.

Gridworld Experiment. All mazes have a size of 10 by 10 cells, a fixed layout per maze for reproducibility and a maximum episode length of 100 steps. Per environment, there exists one reward of value 1.0 at a fixed position. Additionally, the agent receives a -0.01 penalty per step, resulting in a positive reward only if the agent is able to find the goal state fast enough. During data collection, the agent starts at a random, unobstructed cell somewhere in the maze and performs random actions. This takes exploration strategies, which exceed the scope of this work, out of the equation. To assess the performance of the learned dynamics models in a comparable way, the agent's starting position is fixed to the lower left corner of every maze during control. The action space is discrete

Fig. 3. The two training tasks of the vizdoom experiment. In both cases, the agent is situated in a rectangular room with its back against one wall, facing the opposite wall. The agent can move sideways left, sideways right and attack in both tasks. In the "VizdoomBasic" task (left image), the agent has to shoot the monster on the opposite wall. In the "TakeCover" task (right image), the agent has to dodge incoming fireballs shot by enemies that randomly spawn on the opposite wall.

and provides four choices: Up, down, left and right. The observation emitted by the environment is an image of a 5×5 cells top-down view centered on the agent position, visualized in Fig. 2 in the leftmost panel. Although we conduct our experiments on a gridworld, we emphasize that the input space consists of proper image observations.

Vizdoom Experiment. In both tasks the agent is positioned in a rectangular 3D room viewed from the ego perspective (c.f. Fig. 3). Standing with its back against one wall, the agent faces the opposite wall and can choose from three actions: Move sideways left, move sideways right and attack. Since the agent does not hold a weapon in the second task, the attack action here simply advances the environment one time step. In the first task ("VizdoomBasic"), the agent has to shoot the monster at the opposite wall for a reward of 100. Every move action results in a reward penalty of -1 and every fired shot in a reward penalty of -5. The maximum number of allowed time steps is 300. In the second task ("Take-Cover") the agent has to evade fireballs shot by enemies randomly spawning at the opposite side of the room. Since every time step the agent is alive results in a reward of 1, the goal is to survive as long as possible. Although there is formally no time limit for the second task, more and more monsters will spawn over time. This makes evading all fireballs increasingly difficult and ultimately impossible. For both tasks, the agent and monster starting positions are randomized during training data collection and fixed for the control experiments. Both, monolithic and multi RDM get the same sequence of starting positions.

4.1 Evaluation

To evaluate the quality of the trained monolithic as well as multi RDM, we use them to directly generate agent behavior via planning. This way, the performance

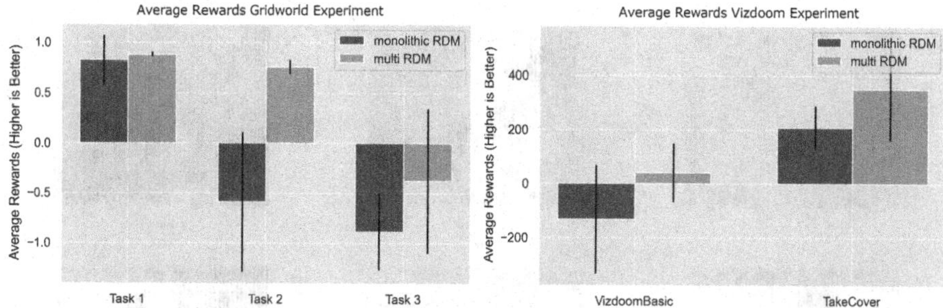

Fig. 4. Control results for the gridworld and the vizdoom experiments. Each task is depicted as a column on the x-axis, in turn separated into two columns representing the monolithic (blue) or multi (orange) RDM. The y-axis shows the average episode rewards. **Left:** Results for the gridworld experiment obtained via planning rollouts averaged over 60 trials per task. Note that the 60 trials originate from 3 full training runs as explained in Sect. 3.5 with 20 trials per run. **Right:** Results for the vizdoom experiment obtained via planning rollouts averaged over 35 trials per task. Note that the 35 trials originate from 1 full training run as explained in Sect. 3.5. (Color figure online)

of the approaches can be assessed with minimal additional complexity. The results obtained with our trained RDMs for the gridworld experiment are shown in Fig. 4 (left panel). The overall difficulty increase from task one to three is reflected in the reward decrease for both approaches. While the first task can be completed with a comparable performance by both architectures, the second and third task show significant differences between the two approaches, the split architecture outperforming the monolithic network ($p < 0.0001$, Mann-Whitney-U-test). The third task is generally the hardest, since the way to the reward is most obstructed and longer than for the other tasks. The results for the vizdoom experiment are shown in Fig. 4 (right panel) and follow a similar pattern as those of the gridworld experiment. The multi RDM architecture manages to deliver significantly higher average rewards than the monolithic architecture in both tasks ($p < 0.002$, Mann-Whitney-U-test). Note that the maximum reward that can be obtained in the first task is capped at around 85 (depending on the agent and monster starting positions) while it is potentially infinite for the second task.

With these findings in mind, the question is why the monolithic RDM performs worse than the multi RDM in planning. To shed light on possible reasons for the performance difference, we carefully inspected the generated rollout samples of both approaches in the gridworld experiment setting. Per task and architecture, we generated 64 trajectories of length 50 from 64 arbitrary starting observations. The actions were randomly sampled from the tasks' action spaces. Figure 5 shows the most likely environment of the produced latent space embedding $z_{i,t}$ for every trajectory i and time step t. To find the most likely environment, each $z_{i,t}$ was decoded into an image using the VQ-VAE decoder and compared to a list of representative images. This list contains one image for

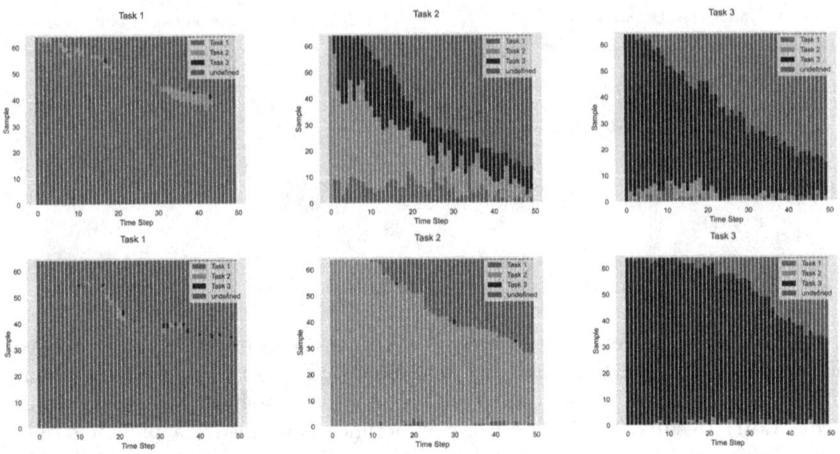

Fig. 5. For each plot, 64 rollouts of length 50 with randomly chosen initial observations and random actions were performed. On the y-axis all samples of a single time step are listed, color-coded depending on their most probable class. Samples belonging to task (1, 2, 3) are colored in (blue, orange, green) and samples that could not be assigned to any task are colored in red. On the x-axis, the distinct time steps are shown. For visual clarity, data obtained exclusively from one of the three training runs is shown. **Top row:** Results of the monolithic architecture. **Bottom row:** Results of the multi RDM. (Color figure online)

every possible agent location in the three tasks, which means that if the decoded $z_{i,t}$ resembles any valid observation, this was detected with absolute certainty. If the average per-pixel distance to the closest match was above a certain threshold (here 0.01), the $z_{i,t}$ was classified as "undefined" (red in the plots). Otherwise, it was classified to belong to the environment of the closest match (blue for task 1, orange for task 2, green for task 3). Note that this analysis is feasible only for environments with a relatively small amount of states, which is another reason why we decided to use the gridworld tasks for this analysis.

It is clearly visible that the monolithic RDM produces trajectories of lower overall quality (cf. Fig. 5, top row). First of all, compared to the multi RDM the fraction of undefined observations is notably higher in tasks 2 and 3. Even more importantly, the monolithic RDM has a higher chance to generate z that belong to a different task, most evidently reflected in task 2 (cf. Fig. 5, top row, middle panel). This can lead to malformed rollout trajectories which miss rewards or report rewards where there are none, harming the planning process in general. In the worst case, the monolithic RDM switches tasks in the middle of a rollout trajectory. By visual inspection of the rollout trajectories, we observed this to happen regularly for the monolithic RDM in both the gridworld and the vizdoom experiments. Through separation of the task dynamics in the multi RDM, considerably less confusion among the tasks arises and the fraction of undefined samples is reduced as well. This stands in line with measurably better planning performance for the multi RDM architecture. We conclude from our analysis that catastrophic interference outweighs possible advantages of transfer learning in the monolithic RDM. In our experiments, the simple use of isolated dynamics models mitigates the harmful interference to a large degree.

5 Conclusion

In this work, we demonstrate an unintuitive effect in model-based multi-task RL. Contrary to expectations fueled by model-free multi-task RL, using a single monolithic RDM for all tasks can harm performance in model-based multi-task RL instead of improving it. We conclusively assume that the positive effect of transfer learning between similar tasks is either absent for dynamics models or outweighed by catastrophic interference. Moreover, while using the same parameter budget for both approaches we show that imposing an internal structure to the dynamics model can lead to notably improved learned dynamics. We show this by using both approaches for control through planning, where the multi RDM scores measurably higher rewards than the monolithic RDM. By analyzing the rollout trajectories of both the monolithic and the multi RDM in the gridworld setting, we find cleaner, more task-specific trajectories and less overall wrong predictions in case of the multi RDM. We conclude that separating task recognition from task dynamics effectively prevents catastrophic interference (or task confusion) to a large degree, leading to the measured performance improvement. The code for reproducing our experiments, including all hyperparameters, can be found at https://github.com/rschiewer/lrdm.

Acknowledgements. We thank Jan Bollenbacher, Dr. Anand Subramoney and Prof. Dr. Tobias Glasmachers for their feedback and help, which greatly influenced this work.

References

1. Sonnet (2017). https://github.com/deepmind/sonnet
2. Abadi, M., et al.: TensorFlow: large-scale machine learning on heterogeneous systems (2015). https://www.tensorflow.org/
3. Caruana, R.: Multitask learning. Mach. Learn. **28**, 41–75 (1997). https://doi.org/10.1023/A:1007379606734
4. Chollet, F., et al.: Keras (2015). https://keras.io
5. Chua, K., Calandra, R., McAllister, R., Levine, S.: Deep reinforcement learning in a handful of trials using probabilistic dynamics models. In: Advances in Neural Information Processing Systems (NIPS), December 2018, pp. 4754–4765, May 2018. http://arxiv.org/abs/1805.12114
6. Duan, Y., Schulman, J., Chen, X., Bartlett, P.L., Sutskever, I., Abbeel, P.: Rl^2: fast reinforcement learning via slow reinforcement learning. arXiv preprint arXiv:1611.02779 abs/1611.02779 (2016). http://arxiv.org/abs/1611.02779
7. Ebert, F., Finn, C., Dasari, S., Xie, A., Lee, A., Levine, S.: Visual foresight: model-based deep reinforcement learning for vision-based robotic control. arXiv preprint arXiv:1812.00568 (2018)

8. Feinberg, V., Wan, A., Stoica, I., Jordan, M.I., Gonzalez, J.E., Levine, S.: Model-based value expansion for efficient model-free reinforcement learning. In: Proceedings of the 35th International Conference on Machine Learning (ICML 2018) (2018)
9. Fernando, C., et al.: PathNet: evolution channels gradient descent in super neural networks. arXiv preprint arXiv:1701.08734 (2017)
10. Finn, C., Abbeel, P., Levine, S.: Model-agnostic meta-learning for fast adaptation of deep networks. In: International Conference on Machine Learning, pp. 1126–1135. PMLR (2017)
11. Ha, D., Schmidhuber, J.: Recurrent world models facilitate policy evolution. arXiv preprint arXiv:1809.01999 abs/1809.01999 (2018)
12. Haarnoja, T., Zhou, A., Abbeel, P., Levine, S.: Soft actor-critic: off-policy maximum entropy deep reinforcement learning with a stochastic actor. In: International Conference on Machine Learning, pp. 1861–1870. PMLR (2018)
13. Hafner, D., et al.: Learning latent dynamics for planning from pixels. In: International Conference on Machine Learning, pp. 2555–2565. PMLR (2019)
14. Hafner, D., Lillicrap, T.P., Norouzi, M., Ba, J.: Mastering atari with discrete world models. arXiv preprint arXiv:2010.02193 abs/2010.02193 (2020)
15. Hessel, M., et al.: Rainbow: combining improvements in deep reinforcement learning. In: AAAI, pp. 3215–3222 (2018). https://www.aaai.org/ocs/index.php/AAAI/AAAI18/paper/view/17204
16. Jang, E., Gu, S., Poole, B.: Categorical reparameterization with Gumbel-Softmax. In: 5th International Conference on Learning Representations, ICLR 2017, Toulon, France, 24–26 April 2017, Conference Track Proceedings. OpenReview.net (2017)
17. Kaiser, L., et al.: Model-based reinforcement learning for Atari. arXiv preprint arXiv:1903.00374, March 2019. http://arxiv.org/abs/1903.00374
18. Maddison, C.J., Mnih, A., Teh, Y.W.: The concrete distribution: a continuous relaxation of discrete random variables. In: 5th International Conference on Learning Representations, ICLR 2017, Toulon, France, 24–26 April 2017, Conference Track Proceedings. OpenReview.net (2017)
19. Mnih, V., et al.: Asynchronous methods for deep reinforcement learning. In: International Conference on Machine Learning, pp. 1928–1937. PMLR (2016)
20. Mnih, V., et al.: Playing Atari with deep reinforcement learning. arXiv preprint arXiv:1312.5602 (2013)
21. Nagabandi, A., Kahn, G., Fearing, R.S., Levine, S.: Neural network dynamics for model-based deep reinforcement learning with model-free fine-tuning. In: 2018 IEEE International Conference on Robotics and Automation (ICRA), pp. 7559–7566 (2018). https://doi.org/10.1109/ICRA.2018.8463189
22. Oh, J., Guo, X., Lee, H., Lewis, R., Singh, S.: Action-conditional video prediction using deep networks in Atari games. CoRR **10**(1), 52–55 (2015). http://papers.nips.cc/paper/5859-action-conditional-video-prediction-using-deep-networks-in-atari-games.pdf, http://arxiv.org/abs/1507.08750
23. van den Oord, A., Vinyals, O., Kavukcuoglu, K.: Neural discrete representation learning. In: Guyon, I., et al. (eds.) Advances in Neural Information Processing Systems, vol. 30. Curran Associates, Inc. (2017). https://proceedings.neurips.cc/paper/2017/file/7a98af17e63a0ac09ce2e96d03992fbc-Paper.pdf
24. Racanière, S., et al.: Imagination-augmented agents for deep reinforcement learning. In: Proceedings of the 31st International Conference on Neural Information Processing Systems, pp. 5694–5705 (2017)
25. Rubinstein, R.Y.: Optimization of computer simulation models with rare events. Eur. J. Oper. Res. **99**(1), 89–112 (1997)

26. Rusu, A.A., et al.: Progressive neural networks. arXiv preprint arXiv:1606.04671 abs/1606.04671 (2016)
27. Schulman, J., Wolski, F., Dhariwal, P., Radford, A., Klimov, O.: Proximal policy optimization algorithms. arXiv preprint arXiv:1707.06347 (2017)
28. Sekar, R., Rybkin, O., Daniilidis, K., Abbeel, P., Hafner, D., Pathak, D.: Planning to explore via self-supervised world models. In: International Conference on Machine Learning, pp. 8583–8592. PMLR (2020)
29. Silver, D., et al.: A general reinforcement learning algorithm that masters chess, shogi, and Go through self-play. Science **362**(6419), 1140–1144 (2018). https://doi.org/10.1126/science.aar6404,http://science.sciencemag.org/
30. Sutton, R.S.: Dyna, an integrated architecture for learning, planning, and reacting. SIGART Bull. **2**(4), 160–163 (1991)
31. Tesauro, G.: Temporal difference learning and TD-Gammon. Commun. ACM **38**(3), 58–68 (1995). https://doi.org/10.1145/203330.203343,https://dl.acm.org/doi/abs/10.1145/203330.203343
32. Wydmuch, M., Kempka, M., Jaśkowski, W.: VIZDoom competitions: playing doom from pixels. IEEE Trans. Games **11**, 248–259 (2018)
33. Yu, T., Kumar, S., Gupta, A., Levine, S., Hausman, K., Finn, C.: Multi-task reinforcement learning without interference. In: NeurIPS, pp. 1–9 (2019)

A New Nash-Probit Model for Binary Classification

Mihai-Alexandru Suciu and Rodica Ioana Lung(✉)

Centre for the Study of Complexity, Babeş-Bolyai University, Cluj-Napoca, Romania
rodica.lung@econ.ubbcluj.ro

Abstract. The Nash equilibrium is used to estimate the parameters of a Probit binary classification model transformed into a multiplayer game. Each training data instance is a player of the game aiming to maximize its own log likelihood function. The Nash equilibrium of this game is approximated by modifying the Covariance Matrix Adaptation Evolution Strategy to search for the Nash equilibrium by using tournament selection with a Nash ascendancy relation based fitness assignment. The Nash ascendancy relation allows the comparison of two strategy profiles of the game. The purpose of the approach is to explore the Nash equilibrium as an alternate solution concept to the maximization of the log likelihood function. Numerical experiments illustrate the behavior of this approach, showing that for some instances the Nash equilibrium based solution can be better than the one offered by the baseline Probit model.

Keywords: Binary classification · Probit model · Nash equilibrium

1 Introduction

Machine learning techniques have evolved to cope with large amounts of data, partly due to the increase in computational power available. However, there are still issues related to the fine classification even of small data sets, with many paths to explore. Methods that are based on the optimization of a loss or likelihood function may be limited by the way these functions are constructed, i.e. reaching the optimum value of a function may not provide the best classification model for the data. In this sense, there is room for exploring different solution concepts, such as those provided by game theory.

Game theory models interactions among different agents aiming to maximize their payoffs. These payoffs are computed based on their strategies as well as on all the others'. Various real-world situations led to different kind of games such as non-cooperative and cooperative, normal form games and extensive form games, with perfect and imperfect information, etc. For each type of game solution concepts with appealing theoretical and practical properties are devised [8]. Non-cooperative games model the situation in which agents do not cooperate or

G. Nicosia et al. (Eds.): LOD 2021, LNCS 13164, pp. 314–324, 2022.
https://doi.org/10.1007/978-3-030-95470-3_24

communicate to each other. One of the most popular solution concepts for non-cooperative games is the Nash equilibrium (NE): a situation of the game in which no player has an incentive for unilateral deviation [9].

In this paper we explore the use of the Nash equilibrium concept in conjunction with the Probit classification model: instead of maximizing the log likelihood function, a game in which each instance chooses the model parameters that maximize its own log likelihood is devised. To ensure that each players' payoff depends on its choice as well as on the others', the mean values of the strategies of all players are used within the payoff function. The NE of this game is approximated by a modified version of the Covariance Matrix Adaptation Evolution Strategy (CMA-ES) algorithm [2] - called Covariance Matrix Adaptation Nash Evolutionary Strategy (CMA-NES) - that computes game equilibria instead of optima of an objective function. The game is not directly converted into an optimization problem, the search of CMA-ES is diverted towards the equilibrium by ranking individuals based on a Nash ascendancy relation [5].

1.1 The Nash-Probit Game

A binary classification problem can be defined as follows: given a set of data $\mathcal{X} \subset \mathbb{R}^{N\times p}$ and corresponding labels $Y = (y_1, \ldots, y_N)^\top$, with $y_i \in \{0, 1\}$, such that label $y_i \in Y$ corresponds to instance $x_i \in \mathcal{X}$, $i = 1, \ldots, N$, find a model that is able to make a good prediction of Y from $\mathcal{X}$ [3, p. 11]. A probabilistic model will provide some function $\phi : \mathbb{R}^p \times \mathbb{R}^p \to [0, 1]$ that can assign to an instance $x \in \mathbb{R}^p$ the probability $\phi(x; \beta)$ that it has a certain label, usually 1. The expression of the function ϕ is chosen by the decision maker, and parameter β is computed most often by some optimization procedure.

Probit Classification. Within this model the probability that an instance x has a label 1 is estimated by using the cumulative distribution function Φ of the standard normal distribution:

$$\phi(x; \beta) = \Phi(x\beta) \tag{1}$$

where $x\beta$ denotes the dot product of $x = (x_1, \ldots, x_p)$ and $\beta \in \mathbb{R}^p$. Parameter β is computed by maximizing the log likelihood function:

$$\log \mathcal{L}(\mathcal{X}; \beta) = \sum_{i=1}^{N} \left(y_i(\log \Phi(x_i\beta)) + (1 - y_i) \log(1 - \Phi(x_i\beta))\right). \tag{2}$$

The Probit classification model is well known and extensively used in many applications. Most current research focuses on improving the Probit model, and often for a particular problem. To the best of our knowledge, an attempt to use the Nash equilibrium concept with the Probit classification model similar to the one presented in what follows has not been previously made.

The Nash-Probit Game. In the Probit model the aim is to estimate β such that for each instance $x \in \mathcal{X}$ with label 1 the probability $\Phi(x;\beta)$ that x is classified as having label 1 is maximized and for each $x \in \mathcal{X}$ having label 0 the probability $1 - \Phi(x;\beta)$ of x being classified having label 0 is maximized as well. The maximization of these probabilities for all $x \in \mathcal{X}$ can be formulated as a non-cooperative game among instances that try choose β values in order to maximize their corresponding probabilities. Thus, we can define a game $\Gamma = (\mathcal{N}, \mathbb{R}^p, U|\mathcal{X}, Y)$ in the following manner:

- $\mathcal{N}$ is the set of players composed of instances in $\mathcal{X}$: instance $x_i \in \mathcal{X}$ is player i in $\mathcal{N}$;
- the strategy of a player i is to choose $\beta_i \in \mathbb{R}^p$; a strategy profile of the game would be $\beta = (\beta_1, \ldots, \beta_N)$ consisting of the strategies of all players $i \in \mathcal{N}$;
- the payoff $u_i : \mathbb{R}^p \times \mathbb{R}^p \to \mathbb{R}$ of player i is defined as:

$$u_i(x_i;\beta) = y_i \log \Phi(x_i\tilde{\beta}) + (1 - y_i)\log(1 - \Phi(x_i\tilde{\beta})), \tag{3}$$

where $\tilde{\beta} = (\tilde{\beta}^1, \ldots, \tilde{\beta}^p)$, and $\tilde{\beta}^j$ is the average of β_i^j for all $i \in \{1, \ldots, N\}$, $j \in \{1, \ldots, p\}$.

Averages in Eq. (3) ensure that the payoff of each player depends on their strategies as well as on the strategies of all other players, while maximizing its part of the log likelihood function (Eq. (2)).

A possible solution of a game is the Nash equilibrium: a strategy profile such that no player has an incentive for unilateral deviation. In the case of game Γ a Nash equilibrium would be a strategy profile $\beta^* = (\beta_1, \ldots, \beta_N)$ such that no unilateral change of a β_i could improve the payoff function u_i of player i, i.e. it reflects the maximum probability to be classified with the right label, while all other player maintain their strategies unchanged. The corresponding $\tilde{\beta}^*$ can be then used as parameter in the Probit model.

2 Covariance Matrix Adaptation - Nash - Evolution Strategy

CMA-ES is an efficient evolution strategy designed for nonlinear optimization problems [2] that evolves a population of size λ randomly generated each iteration following a normal multivariate distribution with an adaptive covariance matrix. The mean of the population is updated each iteration based on the position of the best μ individuals in the population. In a game theoretic context, CMA-ES has been previously used only to compute mixed Nash equilibria by converting the game into an optimization problem [7,10,13]. However, in the case of game Γ we are looking for pure Nash equilibria for continuous payoff functions, and thus we need an entirely different approach to equilibria approximation. CMA-NES, the *Covariance Matrix Adaptation - Nash Evolution Strategy* is the proposed adaptation of the well known CMA-ES optimizer for Nash equilibria detection of such games.

Consider a noncooperative game (N, S, U), where N is the set of players, $N = \{1, \ldots, n\}$, and n is the number of players, $S = S_1 \times \ldots \times S_n$ is the set of strategy profiles of the game, $S_i \subset \mathbb{R}$ the strategy set of player i, and $U = (u_1, \ldots, u_n)$ is the payoff function, with $u_i : S \rightarrow \mathbb{R}$ the payoff function of player i, $i \in N$. Elements $s = (s_1, \ldots, s_n)$ in S are called strategy profiles, or situations of the game in which player $i \in N$ has chosen strategy s_i.

CMA-NES encodes in an individual a strategy profile $s \in S$ and searches for the equilibrium strategy, which represents a situation of a game such that no player can improve its payoff by unilateral deviation. The search is guided by replacing the selection strategy of CMA-ES with tournament selection. During the tournament individuals are compared by using a fitness based on a domination relation known to direct the search towards Nash equilibria [5]. The Nash ascendancy relation used is described in Alg. 1. Two strategy profiles of the game are compared with respect to the Nash equilibrium concept by counting how many players would improve their payoffs by unilateral deviation from one strategy profile to the other. The strategy/individual with a lower number of such players is considered better with respect to this relation than the other. If there is an equal number of players that can improve their payoff by unilateral switching in this way, the two strategies are considered indifferent.

Algorithm 1. Nash ascendancy test to compare individuals θ and β

```
k1 = k2 = 0;
for j = 0; j < p; j = j + 1 do
    if θj <> βj then
        θ' = θ, β' = β;
        θ'j = βj, β'j = θj;
        if uj(θ') > uj(θ) then
            k1 ++;
        end if
        if uj(β'j) > uj(β) then
            k2 ++;
        end if
    end if
end for
if k1 < k2 then
    return θ Nash ascends β is TRUE;
else if k2 < k1 then
    return θ Nash ascends β is FALSE;
else
    return θ and β are indifferent to each other;
end if
```

CMA-NES uses tournament selection to choose μ individuals and to assign them a fitness based on the tournament results in order to rank them among each other. Comparisons within the tournament are performed using the Nash

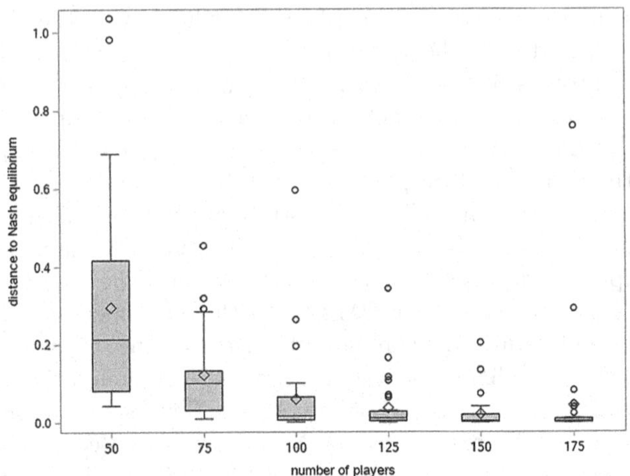

Fig. 1. Distances to NE reported by CMA-NES for the oligopoly presented in Example 1 for games with different number of players.

ascendancy relation. The fitness is computed as the number t of tournaments an individual has won. Because CMA-NES minimizes the fitness function, this number is converted to $\frac{\mu-t}{\mu}$. Thus, each individual that has won a tournament during selection has a fitness of $\frac{\mu-t}{\mu}$, while all the others that will not contribute to the updated mean will have a fitness of 1.

Example 1. Consider the following game inspired from the Cournot oligopoly: we have a set of n players, the strategy profiles set is $\mathbb{R}^n$ and the payoff function for player i is $u_i(s) = s_i(100 - \sum_{j=1}^{n} s_j)$, where $s = (s_1, \ldots, s_n) \in \mathbb{R}^n$, and s_i is the strategy of player i, $i = \{1, \ldots, n\}$. This game has one NE, with all components equal to $\frac{100}{n+1}$. CMA-NES was run 30 times on instances of this game with 50–175 players. Figure 1 presents boxplots of distances to the NE by the mean reported by CMA-NES. It is an interesting behavior of CMA-NES that increasing the number of players also increases its accuracy. Similar behaviour is displayed by NEO [6]. Even without modifying any other aspects of CMA-ES, results reported here are better than those obtained by using Differential Evolution and Extremal optimization [6].

CMA-NES for Binary Classification

The main purpose of our endeavour is to use CMA-NES to approximate an equilibrium strategy for the classification game Γ. Due to the nature of the problem we add additional information during the search: the accuracy score of an individual is added to the tournament selection based fitness. In order to direct the search towards smaller β values the norm of an individual is also added to the fitness minimized by CMA-NES.

Iinitialization. CMA-ES requires an initial value for the population mean. CMA-NES starts the search with the values estimated by the Probit classification model, assuming that the game equilibrium might be found in the same region of the search space, and in order to speed up the search process.

Extensions of the Game. CMA-NES approximates the Nash equilibrium for any payoff functions. Game Γ is defined with the Probit method as baseline. Other classification models can be considered. To illustrate this, we will also use CMA-NES with a logistic regression game model for which in the payoff function in (3) we replace the normal cumulative distribution function with the sigmoid function

$$\sigma(x;\beta) = \frac{1}{1+e^{-x_i\tilde{\beta}}}.$$

We will denote by CMA-NES-P the model that estimates Probit Nash parameters and CMA-NES-LR the model that estimates logistic regression parameters.

3 Numerical Examples

Numerical examples are used to illustrate the potential of the game theoretic approach. We use a set of synthetic and real-world data. We use a set of small datasets with various difficulties in order to assess if the game theoretic model is capable to enhance the baseline models used before extending our study to larger datasets.

Data. Synthetic data were generated with the *make_classification* function from the *scikit-learn*[1] python package [11]. For testing the performance and generalization characteristics of the proposed method we generate five synthetic data sets with various degrees of difficulty for the binary classification problem. We control the difficulty of a test set by varying the degree of overlap between classes and the number of instances in each class (the balance of the class). For reproducibility we report the parameters that we use to generate the test data: class separator $\in \{0.1, 0.5\}$, class weight $\in \{0.5, 0.7, 0.8\}$, the seed used to generate the data $\{5, 10\}$ and 100 instances for each data set. For the real-world data[2] we use a binary version of the *iris* data set (we remove the *Iris Setosa* instances from the data to make the problem a binary classification one, the remaining 100 instances are overlapping thus making the problem harder), the *Statlog (heart disease)* data set with 270 instances and 13 attributes and the *Somerville happiness survey* data set with 143 instances and 7 attributes.

Performance Evaluation. For estimating the prediction error we use the Stratified k-Fold Cross Validation approach [4], with $k = 10$. We repeat the Stratified 10-fold cross validation 14 times, each time we generate the folds using the *StratifiedKFold* function from *scikit-learn* with a different seed. We report the area

[1] Version 0.23.1.

[2] Data available at UCI machine learning repository [1].

under the receiver operating characteristic curve (AUC) [12] and the accuracy (ACC) [4] for each fold. Significance of differences in results is evaluated by using a paired t-test with $\alpha = 0.05$. Results reported by CMA-NES are compared with those reported by the corresponding baseline model.

Results and Discussions. Tables 1 and 2 present results reported by the two CMA-NES variants and the corresponding baseline models for the synthetic datasets. Results for the real datasets are presented in Tables 3 and 4. Boxplots representing the distribution of the values of AUC and ACC are represented in Figs. 2 and 3.

Table 1. AUC and ACC values for the synthetic data sets for different values of the class separators (s) and weights (w), and different tournament sizes (ts) for CMA-NES that runs the game based on Probit and on Logistic Regression.

s	w	ts	CMA-NES-P				CMA-NES-LR			
			AUC		ACC		AUC		ACC	
			Mean	Std	Mean	Std	Mean	Std	Mean	Std
0.1	0.5	2	0.54	0.05	0.52	0.04	0.53	0.08	0.51	0.06
		3	0.56	0.03	0.53	0.04	0.56	0.03	0.54	0.04
		4	0.56	0.04	0.53	0.04	0.56	0.03	0.54	0.03
		5	0.56	0.03	0.54	0.04	0.56	0.03	0.54	0.03
	0.7	2	0.54	0.06	0.51	0.04	0.54	0.07	0.52	0.05
		3	0.61	0.05	0.54	0.03	0.61	0.05	0.54	0.04
		4	0.61	0.04	0.54	0.03	0.59	0.05	0.52	0.03
		5	0.6	0.04	0.52	0.04	0.59	0.06	0.52	0.03
0.5	0.5	2	0.83	0.03	0.77	0.02	0.84	0.02	0.78	0.02
		3	0.87	0.02	0.82	0.03	0.87	0.02	0.82	0.03
		4	0.87	0.01	0.82	0.03	0.87	0.02	0.82	0.03
		5	0.86	0.02	0.82	0.03	0.87	0.02	0.82	0.03
	0.7	2	0.83	0.04	0.72	0.03	0.83	0.04	0.71	0.02
		3	0.85	0.03	0.79	0.02	0.85	0.03	0.79	0.01
		4	0.85	0.03	0.79	0.02	0.85	0.03	0.78	0.02
		5	0.85	0.03	0.78	0.02	0.85	0.03	0.78	0.02
	0.8	2	0.76	0.09	0.63	0.04	0.76	0.09	0.62	0.04
		3	0.86	0.02	0.77	0.02	0.78	0.07	0.69	0.07
		4	0.86	0.02	0.77	0.01	0.78	0.07	0.69	0.07
		5	0.86	0.02	0.76	0.01	0.78	0.08	0.7	0.07

Table 2. AUC and ACC values for the synthetic data sets for different values of the class separators (s) and weights (w) for the Probit and Logistic Regression classifiers.

s	w	Probit (AUC)		Probit (ACC)		LogRegr (AUC)		LogReg (ACC)	
		Mean	Std	Mean	Std	Mean	Std	Mean	Std
0.10	0.50	0.60	0.02	0.54	0.02	0.60	0.02	0.54	0.02
	0.70	0.62	0.03	0.56	0.03	0.62	0.03	0.56	0.03
0.50	0.50	0.87	0.02	0.82	0.03	0.87	0.02	0.82	0.03
	0.70	0.85	0.03	0.77	0.03	0.85	0.03	0.77	0.03
	0.80	0.81	0.06	0.66	0.04	0.79	0.06	0.64	0.04

Table 3. AUC and ACC values for the real data sets for different tournament sizes (ts) of CMA-NES and data sets: 1 - Iris, 7 - Statlog, 8 - Sommerville

Data set	ts	CMA-NES-P				CMA-NES-LR			
		AUC		ACC		AUC		ACC	
		Mean	Std	Mean	Std	Mean	Std	Mean	Std
1	2	0.96	0.07	0.91	0.07	0.60	0.12	0.50	0.02
	3	0.96	0.07	0.91	0.07	0.69	0.14	0.54	0.03
	4	0.96	0.07	0.9	0.07	0.71	0.10	0.55	0.05
	5	0.96	0.07	0.9	0.07	0.78	0.12	0.55	0.04
7	2	0.62	0.05	0.51	0.02	0.50	0.02	0.53	0.02
	3	0.62	0.04	0.51	0.02	0.52	0.02	0.53	0.02
	4	0.59	0.06	0.52	0.02	0.53	0.03	0.55	0.01
	5	0.57	0.04	0.54	0.02	0.51	0.03	0.54	0.01
8	2	0.56	0.04	0.53	0.02	0.54	0.05	0.51	0.02
	3	0.56	0.02	0.54	0.01	0.55	0.05	0.52	0.02
	4	0.54	0.02	0.54	0.02	0.58	0.05	0.52	0.02
	5	0.54	0.02	0.54	0.02	0.57	0.04	0.53	0.02

We find that CMA-NES improves upon Probit for the synthetic data sets with class weight 0.7 and 0.8 (in 7 cases accuracy results are significantly better and in 8 cases they are as good as those reported by Probit). While also an improvement in AUC levels is observed for the data set with 0.5/0.8, the logit version only improves the accuracy of the search for this dataset, with similar results for the other instances of the data.

Table 4. AUC and ACC values for the real data sets for Probit and Logistic regression classifiers and data sets: 1 - Iris, 7 - Statlog, 8 - Sommerville

Data set	Probit (ROC)		Probit (ACC)		LogRegr (ROC)		LogRegr (ACC)	
	Mean	Std	Mean	Std	Mean	Std	Mean	Std
1	0.96	0.07	0.91	0.06	0.96	0.07	0.91	0.06
7	0.90	0.00	0.84	0.00	0.90	0.00	0.84	0.00
8	0.51	0.02	0.53	0.02	0.51	0.02	0.53	0.02

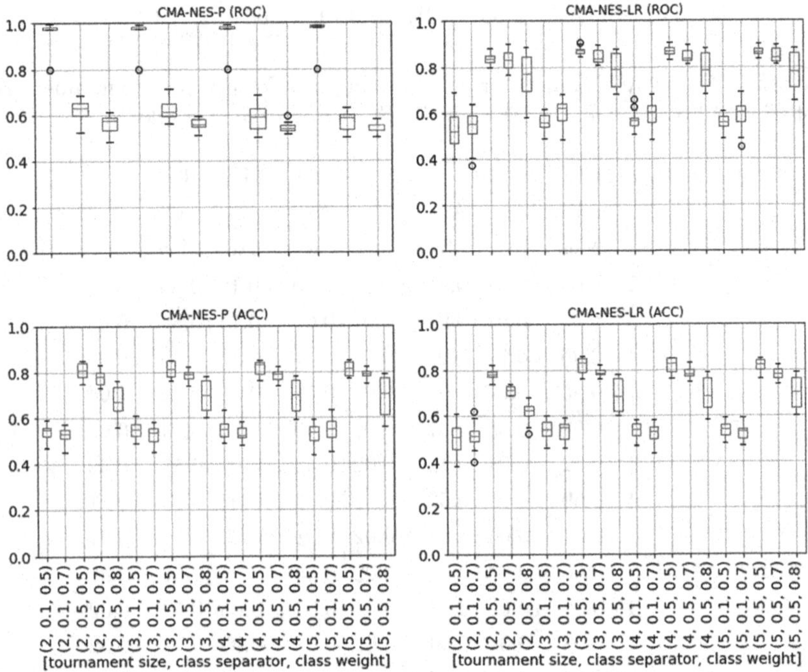

Fig. 2. Box plots of AUC (ROC) and ACC values for the synthetic data sets for different values of the class separators and weights, and different tournament sizes for CMA-NES.

The tournament size does not influence the search results significantly, indicating the robustness of CMA-NES with respect to this parameter. In a similar manner Fig. 3 illustrates results reported for the real-datasets. The accuracy reported by CMA-NES is as good as that of Probit for the *Iris* data-set, significantly better for *Sommerville* data set and worse for *Statlog*. The tournament size does not influence the outcome of CMA-NES significantly for these data also. The Logistic regression game model also improves results for the *Sommerville* data-set but they are significantly worst for the other two.

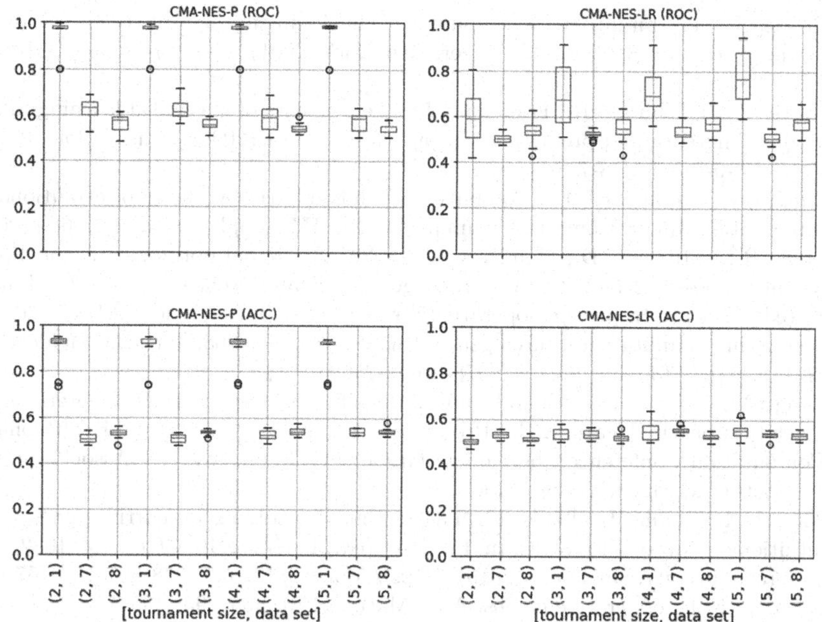

Fig. 3. Box plots of AUC (ROC) and ACC values for the real data sets for different tournament sizes of CMA-ES and data sets: 1 - Iris, 7 - Statlog, 8 - Sommerville

4 Conclusions

The use of the Nash equilibrium concept as a solution for the binary classification problem is explored in this paper. It is shown how an optimization heuristic can be adapted to compute game equilibria and used to estimate model parameters that approximate the equilibrium, and hopefully its appealing properties. While results illustrate the potential of the approach, the main contribution of the paper is that it opens the path to actually embed game theory solution concepts into classification models.

Acknowledgements. This work was supported by a grant of the Ministry of Research, Innovation and Digitization, CNCS/CCCDI - UEFISCDI, project number 194/2021 within PNCDI III.

References

1. Dua, D., Graff, C.: UCI machine learning repository (2017). http://archive.ics.uci.edu/ml
2. Hansen, N., Ostermeier, A.: Completely derandomized self-adaptation in evolution strategies. Evol. Comput. **9**(2), 159–195 (2001). https://doi.org/10.1162/106365601750190398

3. Hastie, T., Tibshirani, R., Friedman, J.: The Elements of Statistical Learning. Springer Series in Statistics. Springer, New York (2001). https://doi.org/10.1007/978-0-387-84858-7
4. Hastie, T., Tibshirani, R., Friedman, J.: The elements of statistical learning: data mining, inference and prediction. Springer, New York (2009). https://doi.org/10.1007/978-0-387-84858-7
5. Lung, R.I., Dumitrescu, D.: Computing Nash equilibria by means of evolutionary computation. Int. J. Comput. Commun. Control **III**(Suppl. Issue), 364–368 (2008)
6. Lung, R.I., Mihoc, T.D., Dumitrescu, D.: Nash extremal optimization and large cournot games. In: Pelta, D.A., Krasnogor, N., Dumitrescu, D., Chira, C., Lung, R. (eds.) Nature Inspired Cooperative Strategies for Optimization (NICSO 2011). Studies in Computational Intelligence, vol. 387, pp. 195–203. Springer, Heidelberg (2011). https://doi.org/10.1007/978-3-642-24094-2_14
7. Lung, R.I., Suciu, M.A.: Equilibrium in classification: a new game theoretic approach to supervised learning. In: Proceedings of the 2020 Genetic and Evolutionary Computation Conference Companion, GECCO 2020, pp. 137–138. Association for Computing Machinery, New York (2020)
8. Nan, J.-X., Zhang, L., Li, D.-F.: The method for solving bi-matrix games with intuitionistic fuzzy set payoffs. In: Li, D.-F. (ed.) EAGT 2019. CCIS, vol. 1082, pp. 131–150. Springer, Singapore (2019). https://doi.org/10.1007/978-981-15-0657-4_9
9. Nash, J.: Non-cooperative games. Ann. Math. **54**(2), 286–295 (1951)
10. Pavlidis, N., Parsopoulos, K., Vrahatis, M.: Computing Nash equilibria through computational intelligence methods. J. Comput. Appl. Math. **175**(1), 113–136 (2005). Selected Papers of the International Conference on Computational Methods in Sciences and Engineering
11. Pedregosa, F., et al.: Scikit-learn: machine learning in Python. J. Mach. Learn. Res. **12**, 2825–2830 (2011)
12. Rosset, S.: Model selection via the AUC. In: Proceedings of the Twenty-First International Conference on Machine Learning, ICML 2004, p. 89. Association for Computing Machinery, New York (2004). https://doi.org/10.1145/1015330.1015400
13. Suciu, M.-A., Lung, R.I.: Nash equilibrium as a solution in supervised classification. In: Bäck, T., et al. (eds.) PPSN 2020. LNCS, vol. 12269, pp. 539–551. Springer, Cham (2020). https://doi.org/10.1007/978-3-030-58112-1_37

An Optimization Method for Accurate Nonparametric Regressions on Stiefel Manifolds

Ines Adouani[1] and Chafik Samir[2(✉)]

[1] University of Sousse, ISSATSO, Sousse, Tunisia
[2] CNRS UCA LIMOS UMR 6158, Aubière, France
chafik.samir@uca.fr

Abstract. We consider the problem of regularized nonlinear regression on Riemannian Stiefel manifolds when only few observations are available. In this paper, we introduce a novel geometric method to estimate missing data using continuous and smooth temporal trajectories to overcome the discrete nature of observations. This approach is important in many applications and representation spaces where nonparametric regression for data that verify orthogonality constraints is crucial. To illustrate the behavior of the proposed approach, we give all numerical details and provide geometric tools for computational efficiency.

Keywords: Nonlinear regression · Optimization · Stiefel manifolds

1 Introduction

The most widely used statistical learning models are regression and classification [1–4]. Regression is a supervised learning technique that translates the relationship between an input t_i (predictors) and the corresponding output X_i (outcome variable) for $i = 0, \ldots, n$. Generally, we write $X_i = r(t_i) + \epsilon_i$ where ϵ_i denotes the noise term and r the regression function. When r is a linear function, the supervised learning algorithm is called linear regression. Using this simplest predictive model, one can infer for the unseen data. However, it is largely admitted that linear regression models are not efficient when it comes to real data, high dimension, nonlinear spaces, etc.

Another popular family of solutions are polynomials. The standard method to extend linear regression to a non-linear relationship between X and t. This approach has been widely used in different fields and wide applications. In fact, this technique seems to be easy to implement, can be extended to infinite functional solutions when used to form orthogonal bases and allow more nicely solutions, e.g. splines. The latter will be at the heart of our motivations. Unluckily, polynomial regression has some limitations: i) First of all they can not be extended to nonlinear spaces, ii) the complexity increases with the number of data points,

Supported by CNRS France.

G. Nicosia et al. (Eds.): LOD 2021, LNCS 13164, pp. 325–337, 2022.
https://doi.org/10.1007/978-3-030-95470-3_25

which is hard to handle and iii) polynomial-based regression models are known to drastically over-fit, even for one dimensional observations. To overcome the limitations of polynomial-based regression, we consider spline-based regression.

Spline-based regression models are among the most important non-linear regression techniques. Indeed, this approach is locally adaptive for both functional time series and spatial data. It can be extended to more complicated spaces and can use an improved regression technique which, instead of building one model for the entire dataset, divides the dataset into multiple bins and fits each bin with a separate model. For example, with standard polynomials we need the global structure on the underlying space. This is a hard constraint. To avoid such constraints, many methods have been proposed to divide data to partitions and solve the problem locally. The points where the division occurs are called knots and local solutions lead to piecewise functions.

In order to construct more sophisticated solutions, one can add extra conditions on regression functions such as continuity and differentiability. Note that when we formulate the problem as spline interpolation or fitting, we should place the knots in an ordered manner. Many techniques exist and choices are made depending on the applications at hand. While other configurations may have potential advantages, it is very common to use equally-spaced observation times $t_i, i = 1, \ldots, N$. Indeed, searching for optimal knots is not a straightforward task on nonlinear manifolds and consequently out of the scope of this paper. More importantly, we focus on the search space of smooth regression splines where data points verify orthogonality constraints and t_i are distinct and ordered time instants. We show that this problem occurs in many real situations and leads to specific optimization methods, usually called optimizations on Riemannian manifolds [5,6].

Data points on Riemannian manifolds are fundamental objects in many fields including, subspace filtering, machine learning, signal and image-video processing, communication and medical imaging [2,3,7–9]. To cite but few examples, tracking, face and action recognition and statistical shape analysis [4,5,10–12]. In many real-world applications, Stiefel manifold and Grassmann manifold are most commonly preferred as representation Riemannian manifolds. A common limitation in many of these applications has been the geometric structure on some manifolds, e.g. Grassmann and Stiefel manifolds [5]. As increasingly real-world applications have to deal with non-vector data, a great number of algorithms for manifold embedding and manifold learning have been introduced. Recently, many efforts have been made to develop important geometric and statistical tools: Riemannian exponential map and its inverse, means, distributions, geodesic arcs, etc. [1,3,13].

Motivated by the success of these approaches, we consider problems for which the finite set of data points $X_0, X_1, \ldots, X_N$ consist in observations on Stiefel manifold $\mathcal{M} = St(n, k)$, the manifold of k-tuples of pairwise orthogonal unit vectors in $\mathbb{R}^n$, associated to a set of observation times $(0 = t_0 < t_1 < ... < t_N = 1)$ such that the goal is to seek a spline $\gamma : [t_0, t_N] \rightarrow \mathcal{M}$ that minimizes the functional

$$E(\gamma) = \frac{\lambda}{2} \int_{t_0}^{t_N} < \frac{D^2\gamma(t)}{Dt^2}, \frac{D^2\gamma(t)}{Dt^2} >_{\mathcal{M}} + \frac{1}{2} \sum_{i=0}^{N} d^2_{\mathcal{M}}(\gamma(t_i), X_i) \tag{1}$$

where $< ., . >_{\mathcal{M}}$ and $d_{\mathcal{M}}$ denote the canonical metric and geodesic distance on $\mathcal{M}$ [6].

A vast number of methods for solving the optimization problem on Riemannian manifolds have appeared in the literature [2,14–22]. In this paper, we propose a geometric algorithm that generates a solution of the problem (1). This resulting solution takes the form of a spline and satisfies the following properties:

1. $\gamma(t_i) = X_i$.
2. γ is of class C^1.

In short, the task is that of regression on a homogeneous space in the purpose to estimate/predict missing data from few available observations. By observations we mean any data points that can be obtained from temporal acquisitions. For example, medical images at different time instants are usually used to analyze the evolution of a disease. In this context and due to logistic and time constraints, it is very common to store few discrete moments only. Then at each time instant, we have a data point that is represented as an element of a manifold [1]. So there is a need to estimate missing data points on such manifold at non observed time instants. Several discrete-time models on smooth manifolds and Lie groups have been studied in the literature [23,24]. Here, we consider a continuous-time model and will address the problem of regularized non-linear regression from finite observations [9]. The main contributions of the proposed methods shows that: (i) the formulation leads to a generalization of splines on Stiefel manifolds, (ii) the solution exists and can be efficiently solved and (iii) the geometric algorithm is simpler and can be applied in a lot of fields with various data types.

Various works have been developed to construct an interpolating Bézier spline on Riemannian manifolds with some degree of smoothness. Without being exhaustive, we mention [8,25–29] for an account of important theoretical contributions in this area. However, only a little has been done in this direction for Stiefel manifold [30,31]. Indeed, Riemannian optimization problem on Stiefel manifold is generally extremely hard to solve due to the geometric structure and to the orthogonality constraints that represent this manifold. In [24,32], the authors propose an iterative algorithm to compute the Riemannian log map equipped with the canonical metric which allow to find the geodesic between two given points in the Stiefel manifold. In this paper, we make use of this algorithm to present a novel approach to construct an interpolating Bézier spline of class C^1 on $\mathcal{M}$. The results presented here extend our results introduced in [8]. Accurately, we follow the same idea from the energy minimization formulation of least-squares fitting in Euclidean spaces and generalize this concept to this manifold. The proposed method is geometrically simpler, extensible and easy-to-implement. In fact, we demonstrate the utility of the proposed solution on different vision problems.

The paper is organized as follows. In Sect. 2 we review the basic differential geometry of the Stiefel manifolds that will be used to derive our main results. In Sect. 3 we address the fitting problem on the Stiefel manifolds and we describe our method to construct a solution. Section 4 shows numerical results and potential applications and Sect. 5 concludes the paper.

2 Preliminaries

The goal of this section is to recall the basic facts about the Stiefel manifold $\mathcal{M} = St(n, k)$, with $k \leq n$, that will allow us to build a simple but consistent geometric algorithm to generate interpolating Bézier spline. We describe a few computational tools (namely geodesics, the Riemannian Exp and Log maps) derived from a chosen Riemannian metric. For a detailed exposition on these concepts, we refer the reader to [5,6].

The Stiefel manifold $\mathcal{M}$ is a compact matrix manifold of k-dimensional orthonormal frames in $\mathbb{R}^n$. A point in the Stiefel manifold is represented by an n-by-k matrix as following:

$$\mathcal{M} = \{X \in \mathbb{R}^{n \times k} \mid X^T X = I_k\}.$$

When $k = 1$, we simply have the sphere S^{n-1}, while when $k = n$, we have the orthogonal Lie group $O(n)$ and when $k = n - 1$, we obtain the special orthogonal group $SO(n)$. The Stiefel manifold can also be viewed as a quotient manifold of the orthogonal Lie group $O(n)$. Consequently, the matrix manifold $\mathcal{M}$ is diffeomorphic to $O(n)/O(n-k)$ which turns it into a homogeneous space. For any matrix representative $X \in \mathcal{M}$, the tangent space of $\mathcal{M}$ at X is defined as

$$T_X\mathcal{M} = \{Z \in \mathbb{R}^{n \times k} \mid X^T Z + Z^T X = 0\}.$$

Hence the dimension of both $T_X\mathcal{M}$ and $\mathcal{M}$ is $nk - \frac{1}{2}k(k+1)$.

We can endow the Stiefel manifold with a different Riemannian metrics: the Euclidean metric, and the canonical metric. In the two special cases when $k = 1$ and $k = n$, these two Riemannian metrics are equal. Otherwise they differ, and yield different formulas for geodesics and parallel translation. For the purpose of this paper, we endow $\mathcal{M}$ with the canonical metric. In fact, let $X \in \mathcal{M}$, and $Z_1, Z_2 \in T_X\mathcal{M}$, then we define the canonical metric on $T_X\mathcal{M}$ by,

$$\langle Z_1, Z_2 \rangle_X = \text{trace}(Z_1^T(1 - \frac{1}{2}XX^T)Z_2) \tag{2}$$

Geodesics on a Riemannian manifold are locally shortest curves that are parametrized by the arc length. For a curve γ on $\mathcal{M}$, they satisfy the second order differential equation:

$$\ddot{\gamma} + \dot{\gamma}\dot{\gamma}^t\gamma + \gamma\left((\gamma^t\dot{\gamma})^2 + \dot{\gamma}^t\dot{\gamma}\right) = 0. \tag{3}$$

It is clear that this equation is numerically difficult to solve. Hopefully, the canonical structure allows a practical decomposition of the tangent space that simplify the characterization of geodesics. In [6], the authors have included a general formula for geodesics on Stiefel manifolds. A geodesic is then determined by a starting point $X \in \mathcal{M}$ and a direction Z in $T_X\mathcal{M}$. In fact, the idea starts with a decomposition of Z into its horizontal and vertical components with respect to X, $Z = XX^TZ + (I - XX^T)Z$. Then, by letting $A = X^TZ$ a skew symmetric matrix and by means of a QR decomposition of $(I - XX^T)Z$, we have $Z = XA + QR$. Therefore, the geodesic γ on the Stiefel manifold starting from X with direction Z is given by

$$\gamma(t) = XM(t) + QN(t), \tag{4}$$

$M(t)$ and $N(t)$ are k-by-k matrices given by the matrix exponential

$$\begin{bmatrix} M(t) \\ N(t) \end{bmatrix} = \exp\left(t\begin{pmatrix} A & -R^t \\ R & 0 \end{pmatrix}\right)\begin{bmatrix} I_k \\ 0 \end{bmatrix}. \tag{5}$$

We can also describe geodesics on the Stiefel manifold by the orthogonal Lie group $O(n)$ actions, i-e by multiplying the orthogonal matrices to the initial point X [6].

Now let $\gamma : [0, 1] \to \mathcal{M}$ be a geodesic curve such that $\gamma(0) = X_1$ and $\dot{\gamma}(0) = Z \in T_{X_1}\mathcal{M}$. In terms of the canonical metric, the Riemannian exponential map $\text{Exp}_{X_1} : T_{X_1}\mathcal{M} \to \mathcal{M}$ that sends a Stiefel tangent vector Z to the endpoint $\gamma(1) = X_2$ of the geodesic γ is given explicitly by:

$$\text{Exp}_{X_1}(Z) = X_1M + QN = X_2 \in \mathcal{M}, \tag{6}$$

where M, Q and N are the same as described in the above setting. Conversely, given two data points X_1 and $X_2 \in \mathcal{M}$, the inverse exponential map $\text{Exp}_{X_1}^{-1}$ (also known as the logarithmic map Log_{X_1}) allows the recovery of the tangent vector Z between them. Formulas to compute the Riemannian log map on the Stiefel manifold relative to the Euclidean metric are provided in [33]. To the best of our knowledge, up to now, there exist two different approaches for evaluating the log map on the Stiefel manifold with respect to the canonical metric [24,32]. In this paper, we adopt the method provided in [32] and we suppose that each two data points belong to a geodesic ball with an injectivity radius determined in [24]. Consequently, the geodesic arc joining X_1 to X_2 in $\mathcal{M}$ can be parameterized explicitly by:

$$\gamma(t, X_1, X_2) = \text{Exp}_{X_1}\left(t\text{Log}_{X_1}(X_2)\right), \; t \in [0, 1], \tag{7}$$

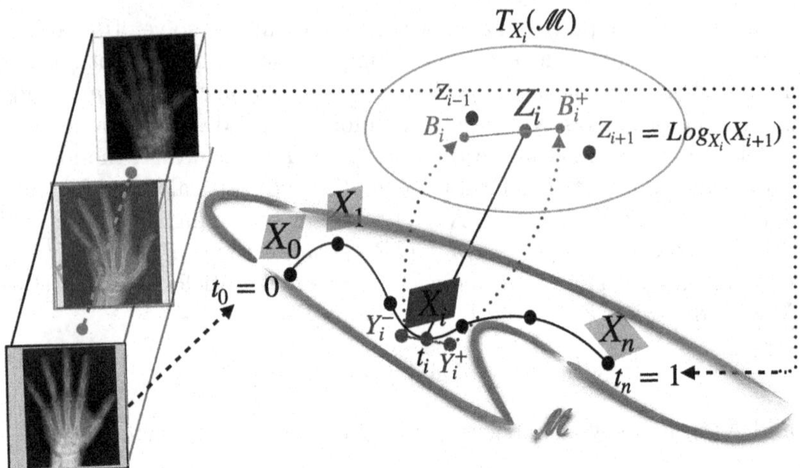

Fig. 1. Geometrical illustration of the Riemannian manifold $\mathcal{M}$ and its tangent space $T_{X_i}\mathcal{M}$ at $X_i \in \mathcal{M}$. γ is an interpolating spline with $X_t = \gamma(t)$ for $t \in I$ which verifies $\gamma(t)^T\gamma(t) = I_k$ for all $t \in I$.

3 Regression on Stiefel Manifolds

In this section, a method to construct a C^1 interpolating Bèzier curve for smoothing data that are constrained to live in the Stiefel Manifold $\mathcal{M}$ equipped with its canonical Riemannian metric is proposed. More precisely, given $X_0, ..., X_N$ a set of $(N+1)$ distinct points in $\mathcal{M}$ and $t_0 < t_1 < ... < t_N$ an increasing sequence of time instants. For simplicity of the exposition we will assume that the time instants are $t_i = i$. Our main objective is to construct an interpolating Bézier spline $\alpha : [0, N] \to \mathcal{M}$ of class C^1 and minimizing the cost functional Eq. (1).

In Euclidean spaces, Bézier curves are defined as a combination of linear polynomial functions, whose coefficients are called control points. The first and last control points are the endpoints of the curve, but the intermediate control points are in general not on the curve. The number of control points determines the degree of the polynomial spline. Moreover, Bézier curves can be constructed by the classical de Casteljau algorithm [34]. By trading linear interpolation by geodesic interpolation, the Casteljau algorithm was generalized to accommodate Bézier curve for Riemannian manifolds.

In what follows, $\alpha_j : [t_0, t_N] \to \mathcal{M}$ denotes the Bézier curve of order j on $\mathcal{M}$ determined by control points V_j, for $j = 0, ..., N$, and defined by a generalization of the classical de Casteljau algorithm. More precisely, let $V_i^0(t) = V_i$, and for $i = 0, ..., j - m, m = 1, ..., j$,

$$V_i^m(t) = \alpha^m(t, V_i, ..., V_{i+m}) = \gamma(t, V_i^{m-1}, V_{i+1}^{m-1}), \tag{8}$$

where $\gamma(t,.,.)$ represent the geodesic on $\mathcal{M}$ defined by Eq. (7). In this framework, the Bézier spline $\alpha : [0, N] \rightarrow \mathcal{M}$ will be constructed by a sequence of N Bézier curve of order two and three such that the segment joining X_0 and X_1, as well as the segment joining X_{N-1} and X_N are Bézier curves of order two, while all the other segments are Bézier curves of order three. Furthermore, we will suppose that there exists two artificial control points (Y_i^-, Y_i^+) on the left and on the right hand side of the interpolation point X_i for $i = 1, ..., (N-1)$. Hence, the Bézier spline $\alpha : [0, N] \longrightarrow \mathcal{M}$ is given by

$$\alpha(t) = \begin{cases} \alpha_2^0(t; X_0, Y_1^-, X_1), & 0 \leq t \leq 1, \\ \alpha_3^i(t-i; X_i, Y_i^+, Y_{i+1}^-, X_{i+1}), & i-1 \leq t \leq i \\ \alpha_2^{N-1}(t-(N-1); X_{N-1}, Y_{N-1}^+, X_N), & N-1 \leq t \leq N \end{cases}$$

Since the Bézier spline α interpolates the first and the last control points of each Bézier curve α_j^i, $j \in \{2, 3\}, 0 \leq i \leq n-1$, therefore the continuity of α at joint points is well satisfied. However, as mentioned in the last section, we would ideally like that α be of class C^1. It is immediate, by construction, that the spline α is C^∞ on $]t_i, t_{i+1}[$, for $i = 1, ..N-1$. We are thus looking to describe the best positions of the unknown control points that ensure the differentiability condition at the knot points.

Our main idea to handle this issue is to treat the fitting problem on different tangent space. Explicitly, let $X_0, ..., X_N$ be a set of distinct given points in $\mathcal{M}$ with X_l being in the cut locus of X_i, $i \neq l$. By means of the algorithm of the logarithm map developed in [24,32] for Stiefel manifold, we transport data points $X_0, ..., X_N$ to $T_{X_i}\mathcal{M}$ at a point $X_i \in \mathcal{M}$, $i = 1, ..., N-1$. Let us denote the mapped data by $Z^i = (Z_0^i, ..., Z_N^i)$ with $Z_m^i = \mathrm{Log}_{X_i}(Z_m)$ for $m = 0, ..., N$. Now our next concern is to search for the control points of a C^1 Bézier spline on $T_{X_i}\mathcal{M}$, $i = 1, ..., N-1$.

From this tangential solution, the Riemannian exponential map Exp_{X_i} defined on $\mathcal{M}$ by Eq. (6) will bring back the solution to the matrix manifold $\mathcal{M}$. The resulting Bézier spline α is then reconstructed with De Casteljau algorithm and we prove that is optimal. So let $\beta : [0, N] \rightarrow T_{X_i}\mathcal{M}$ denote the Bézier spline on $T_{X_i}\mathcal{M}$, $i = 1, ..., N-1$ defined identically to the Bézier spline on $\mathcal{M}$ by N Bézier curves β_j^i, $j \in \{2, 3\}, 0 \leq i \leq N-1$. And let $(B_m^i)^-$ and $(B_m^i)^+$ denote control points on the left and on the right hand side of the interpolation point Z_m, for $m = 1, ..., (N-1)$.

Interestingly, we observe that the optimization problem Eq. (1) which is not easy to solve directly on $\mathcal{M}$ became a simplified euclidean cost function. In [8] we treat in details the Euclidean case and we found an explicit relation between control points $(B_m^i)^-$ and $(B_m^i)^+$. Therefore control points of the Bézier curve β are exactly solution of the problem of minimization of the mean square acceleration expressed uniquely in terms of $(B_m^i)^-$. Specifically, we have on $T_{X_i}\mathcal{M}$:

$$\begin{aligned}
&\min_{(B_1^i)^-,\dots,(B_{N-1}^i)^-} E((B_1^i)^-,\dots,(B_{N-1}^i)^-) \\
&:= \min_{(B_1^i)^-,\dots,(B_{N-1}^i)^-} \int_0^1 \|\ddot{\beta}_2^0(t;Z_0^i,(B_1^i)^-,Z_1^i)\|^2 \\
&+ \sum_{m=1}^{N-2} \int_0^1 \|\ddot{\beta}_3^i(t;Z_m^i,(B_m^i)^+,(B_{m+1}^i)^+,Z_{m+1}^i)\|^2 \\
&+ \int_0^1 \|\ddot{\beta}_2^{N-1}(t;Z_{N-1},(B_{N-1}^i)^+,Z_N^i)\|^2.
\end{aligned} \tag{9}$$

where $||.||$ represent the canonical norm on the tangent space $T_{X_i}\mathcal{M}$. It suffices now to compute the inner product of the acceleration and then evaluate the integral of each term of (9). Similar to [8] we obtain a simple PDE in terms of $(B_m^i)^-$ and Z_m. Hence, to solve this optimization problem, it remains to search for the critical points of the gradient of the energy function, see [8] for more details. For the convenience of the reader, we give a geometrical illustration (sketch) in Fig. 1.

Theorem 31. *Given $X_0,\dots,X_N$ be a set of data points in $\mathcal{M}$ and a sequence of time instants $t_0 < \dots < t_N$. The Bézier spline $\alpha : [0,N] \to \mathcal{M}$ interpolating points X_i at $t_i = i$ on $\mathcal{M}$, for $i = 0,\dots,N$ satisfies the following properties:*

(i) α is uniquely defined by $(N-1)$ control points $Y = [Y_1^-,\dots,Y_{N-1}^-]^T \in \mathbb{R}^{n(N-1)\times n}$ where the rows of Y are given by:

$$Y_i^- = Exp_{X_i}((B_i^i)^-). \tag{10}$$

(ii) α is C^1 on $\mathcal{M}$.

Proof. The proof runs as Corollary 3.3. in [8]. Detailed steps are summarized in Algorithm 1.

4 Experimental Results

In this section, we show the performance of the proposed nonlinear regression method via two experiments. In all experiments we have considered a finite set of data points $X_0, X_1, \dots, X_N$ on a Stiefel manifold $\mathcal{M}$. Each data point X_i is given at a fixed time instant t_i with $i = 0, 1, \dots, N$ and $(t_0, t_N) = (0, 1)$. In the first setting, we consider a very common situation where data points are elements on $St(n, 1) = S^{n-1}$. This situation is standard in many applications.

Algorithm 1. C^1 Bézier spline on $\mathcal{M}$.

Input: $N \geq 3$, $(X_0, ..., X_N)$ as data points on $\mathcal{M}$ at time instants $(t_0, ..., t_N)$.
Output: $\widehat{Y}$

for $i = 1 : N - 1$ **do**
  Compute $Z = [Z_0^i, ..., Z_N^i]^T$ on $T_{X_i}\mathcal{M}$:
  for $m = 0 : N$ **do**
    $Z_m^i = \text{Log}_{X_i}(X_m)$
    Compute $B = [(B_1^i)^-, ..., (B_{N-1}^i)^-]^T$ on $T_{X_i}\mathcal{M}$, using Algorithm (1) in [8].
    Compute control point $Y_i^- = \text{Exp}_{X_i}(B_i^i)^-)$.
  end for
end for
return $\widehat{Y}$

To cite but few popular ones: normalized directions, longitudinal data, path of rotations is equivalent on $SO(n)$ to the path of a point on $St(n, 1)$ when rotations applied to a unit vector. In the second regression setting, we consider a popular example from morphometric analysis where data points are landmark curves and are considered as elements on $St(n, p)$ where $p = 2$ for planar curves and $p = 3$ for spatial curves.

4.1 Special Case with Directions and Rotations

These experiments concern the nonlinear regression problem where observations are on the finite unit sphere. It is very well known that the geodesics exist and are unique for non antipodal points. Both the Riemannian exponential and its inverse are diffeomorphisms inside a ball of radius π. In all examples we display the resulting path using Algorithm 1 detailed in the previous section and not a specific formulation for spherical data.

We remind that the problem of regression with cubic splines can efficiently be solved in this case, and the solution may be better. However, our strategy is different: demonstrate that the proposed method produces good estimators when the Stiefel manifold coincides with the sphere. Thus, and for a visualization purpose, we consider the case $(n, p) = (3, 1)$ and show original data points and the optimal α with different values of N. The time instants are uniformly spaced in $[0, 1]$. We display different examples in Fig. 2.

4.2 Special Case with Procrustes Process

These experiments concern the nonlinear regression problem where observations are curves with orthogonal constraint. So, they can be considered as elements on the unit sphere, usually denoted Σ_n^p and called the Kendall space [11]. They were largely studied for analyzing biological data [9]. It is also common to perform other transformations on curves during Procrustes analysis. We remind that our main objective is to show that the proposed method is successful in this

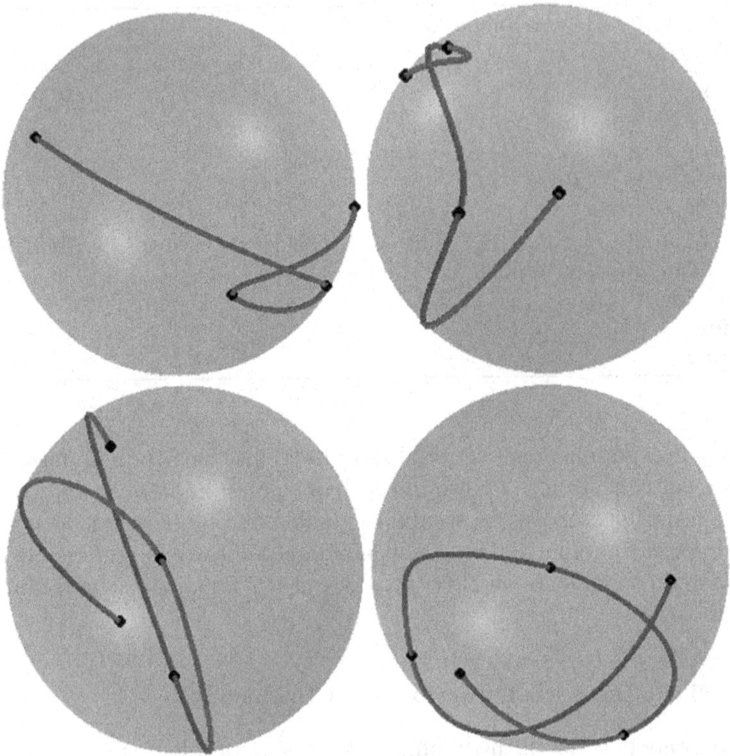

Fig. 2. Examples with different data points on $\mathcal{M} = St(3,1)$. The original data are black dotted and the optimal solution α in red. (Color figure online)

particular case and remains more general for other cases. Otherwise, geodesics, the exponential map, log map are detailed in [9].

In these experiments, we observe four data points X_0, X_1, X_2, X_3 on (n,p) at time instants $(t = 0, t_1 = \frac{1}{3}, t_2 = \frac{2}{3}, t_3 = 1)$. In Fig. 3, observation are displayed on the diagonal with red boxes. So, the goal is to estimate the missing points X_t for $t \notin \{0,1,2,3\}$. We display the resulting solution using Algorithm 1 detailed in the previous section for this manifold. Our is more general but we restrict ourselves to show a good estimator when the Stiefel manifold coincides with this manifold. Thus, and for a visualization purpose, we consider the case $(n,p) = (3,2)$ and show original data points (red boxes) and the optimal $\alpha : t \mapsto \alpha(t) = X_t$. We remind that the time instants are uniformly spaced in $[0,1]$. We generate 4 intermediate points (estimated) between any two successive observations at different time instants, e.g. $t = \frac{1}{15}, \frac{2}{15}, \frac{3}{15}, \frac{4}{15}$. Note that we can generate as many intermediate points as we want but we choose 4 to maintain a good quality of illustrations. Following the same principle, we present two different examples in Fig. 3.

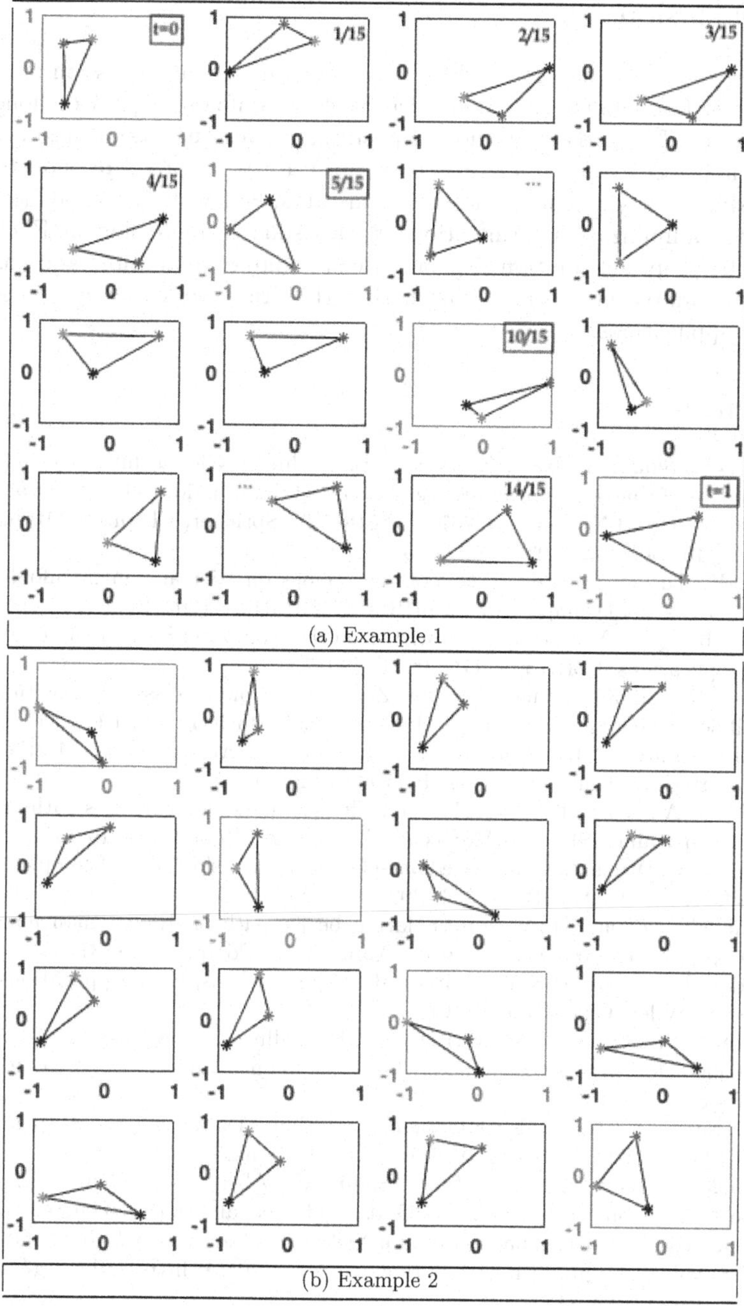

Fig. 3. Two examples with 4 different data points on $\mathcal{M} = St(3, 2)$. The original observations are given on the diagonal: Time instants ($t = 0, \dots, \frac{i}{15}, \dots, 1$) are different and uniformly spaced in $[0, 1]$. The $x - y$ axes display the second components. (Color figure online)

5 Conclusion

We have presented a new method for regularized nonlinear regression on a special class of Riemannian manifolds: the Stiefel manifolds $\mathcal{M}$ of k-orthonormal frames in $\mathbb{R}^n$. Using the definition of geodesic curves and taking into account the rich and nice structure of this space, we use the proposed method to predict missing data points at specified time instants. Following the same idea from the energy minimization formulation of least-squares regression in Euclidean spaces, the proposed solution is geometrically simpler, extensible and easy-to-implement. Moreover, we have shown that this framework can be applied for different applications.

References

1. Lee, D.S., Sahib, A., Narr, K., Nunez, E., Joshi, S.: Global diffeomorphic phase alignment of time-series from resting-state fMRI data. In: Martel, A.L., et al. (eds.) MICCAI 2020. LNCS, vol. 12267, pp. 518–527. Springer, Cham (2020). https://doi.org/10.1007/978-3-030-59728-3_51
2. Kim, K., Dryden, I., Le, H.: Smoothing splines on Riemannian manifolds, with applications to 3D shape space. CoRR abs/1801.04978 (2020)
3. Chakraborty, R., Vemuri, B.C.: Statistics on the (compact) Stiefel manifold: theory and applications. CoRR abs/1708.00045 (2017)
4. Zhang, R., Li, X., Zhang, H., Jiao, Z.: Geodesic multi-class SVM with Stiefel manifold embedding. IEEE Trans. Pattern Anal. Mach. Intell., 1 (2021)
5. Absil, P., Mahony, R., Sepulchre, R.: Optimization Algorithms on Matrix Manifolds. Princeton University Press, Princeton (2008)
6. Edelman, A., Arias, T.A., Smith, S.T.: The geometry of algorithms with orthogonality constraints. SIAM J. Matrix Anal. Appl. **20**(2), 303–353 (1998)
7. Cherian, A., Wang, J.: Generalized one-class learning using pairs of complementary classifiers. CoRR abs/2106.13272 (2021)
8. Samir, C., Adouani, I.: C^1 interpolating bézier path on Riemannian manifolds, with applications to 3D shape space. Appl. Math. Comput. **348**, 371–384 (2019)
9. Dryden, I.L., Mardia, K.V.: Statistical Shape Analysis, with Applications in R, 2nd edn. Wiley, Chichester (2016)
10. Rentmeesters, Q., Absil, P., Van Dooren, P., Gallivan, K., Srivastava, A.: An efficient particle filtering technique on the Grassmann manifold. In: IEEE ICASSP, pp. 3838–3841 (2010)
11. Srivastava, A., Klassen, E.: Functional and Shape Data Analysis. Springer, New York (2016). https://doi.org/10.1007/978-1-4939-4020-2
12. Arnould, A., Gousenbourger, P.-Y., Samir, C., Absil, P.-A., Canis, M.: Fitting smooth paths on Riemannian manifolds: endometrial surface reconstruction and preoperative MRI-based navigation. In: Nielsen, F., Barbaresco, F. (eds.) GSI 2015. LNCS, vol. 9389, pp. 491–498. Springer, Cham (2015). https://doi.org/10.1007/978-3-319-25040-3_53
13. Pealat, C., Bouleux, G., Cheutet, V.: Improved time-series clustering with UMAP dimension reduction method. In: 2020 25th International Conference on Pattern Recognition (ICPR), pp. 5658–5665 (2021)
14. Silva Leite, F., Machado, L.: Fitting smooth paths on Riemannian manifolds. In. J. Appl. Math. Stat. **06**(4), 25–53 (2006)

15. Samir, C., Absil, P.A., Srivastava, A., Klassen, E.: A gradient-descent method for curve fitting on Riemannian manifolds. Found. Comput. Math. **12**, 49–73 (2012)
16. Dyn, N.: Linear and nonlinear subdivision schemes in geometric modeling. Found. Comput. Math. Hong Kong 2008 London Math. Soc. Lecture Note Ser. **363**, 68–92 (2009)
17. Wallner, J., Nava Yazdani, E., Grohs, P.: Smoothness properties of lie group subdivision schemes. Multiscale Model. Simul. **6**(2), 493–505 (2007)
18. Mustafa, G., Hameed, R.: Families of non-linear subdivision schemes for scattered data fitting and their non-tensor product extensions. Appl. Math. Comput. **359**, 214–240 (2019)
19. Shingel, T.: Interpolation in special orthogonal groups. IMA J. Numer. Anal. **29**(3), 731–745 (2009)
20. Hinkle, J., Fletcher, P.T., Joshi, S.: Intrinsic polynomials for regression on Riemannian manifolds. J. Math. Imaging Vis. **50**, 32–52 (2014)
21. Lin, L., St. Thomas, B., Zhu, H., Dunson, D.: Extrinsic local regression on manifold-valued data. J. Amer. Statist. Assoc. **112**(519), 1261–1273 (2017)
22. Petersen, A., Müller, H.G.: Fréchet regression for random objects with Euclidean predictors. Ann. Statist. **47**(2), 691–719 (2019)
23. Boumal, N., Absil, P.A.: Discrete regression methods on the cone of positive-definite matrices. In: IEEE ICASSP (2011)
24. Rentmeesters, Q.: A gradient method for geodesic data fitting on some symmetric Riemannian manifolds, pp. 7141–7146 (2011)
25. Gousenberger, P.Y., Massart, E., Absil, P.-A.: Data fitting on manifolds with composite bézier-like curves and blended cubic splines. J. Math. Imaging Vis. **61**, 645–671 (2019)
26. Gousenbourger, P., Samir, C., Absil, P.: Piecewise-Bézier c^1 interpolation on Riemannian manifolds with application to 2D shape morphing. In: 2014 22nd International Conference on Pattern Recognition (ICPR) (2014)
27. Popiel, T., Noakes, L.: Bézier curves and c^2 interpolation in Riemannian manifolds. J. Approx. Theory **148**(2), 111–127 (2007)
28. Geir, B., Klas, M., Olivier, V.: Numerical algorithm for c^2-splines on symmetric spaces. SIAM J. Numer. Anal. **56**(4), 2623–2647 (2018)
29. Adouani, I., Samir, C.: A constructive approximation of interpolating Bézier curves on Rimannian symmetric spaces. J. Optim. Theory Appl. **187**, 1–23 (2020)
30. Huper, K., Kleinsteuber, M., Silva Leite, F.: Rolling Stiefel manifolds. Int. J. Syst. Sci. **39**, 881–887 (2008)
31. Krakowski, K., Machado, L., Silva Leite, F., Batista, J.: A modified Casteljau algorithm to solve interpolation problems on Stiefel manifolds. J. Comput. Appl. Math. **311**, 84–99 (2017)
32. Zimmermann, R.: A matrix-algebraic algorithm for the Riemannian logarithm on the Stiefel manifold under the canonical metric. SIAM J. Matrix Anal. Appl. **38**(2), 322–342 (2017)
33. Bryner, D.: Endpoint geodesics on the Stiefel manifold embedded in Euclidean space. SIAM J. Matrix Anal. Appl. **38**(4), 1139–1159 (2017)
34. de Casteljau, P.: Outillages méthodes de calcul. Technical report, André Citroën Automobiles, Paris (1959)

Using Statistical and Artificial Neural Networks Meta-learning Approaches for Uncertainty Isolation in Face Recognition by the Established Convolutional Models

Stanislav Selitskiy(✉), Nikolaos Christou, and Natalya Selitskaya

School of Computer Science, University of Bedfordshire, Luton LU1 3JU, UK
stanislav.selitskiy@study.beds.ac.uk
https://www.beds.ac.uk/computing

Abstract. We investigate ways of increasing trust in verdicts of the established Convolutional Neural Network models for the face recognition task. In the mission-critical application settings, additional metrics of the models' uncertainty in their verdicts can be used for isolating low-trust verdicts in the additional 'uncertain' class, thus increasing trusted accuracy of the model at the expense of the sheer number of the 'certain' verdicts. In the study, six established Convolutional Neural Network models are tested on the makeup and occlusions data set partitioned to emulate and exaggerate the usual for real-life conditions training and test set disparity. Simple A/B and meta-learning supervisor Artificial Neural Network solutions are tested to learn the error patterns of the underlying Convolutional Neural Networks.

Keywords: Uncertainty · Meta-learning · Face recognition · Makeup · Occlusion · Paper with code

1 Introduction

When Machine Learning (ML) algorithms leave laboratory settings and enter the real-life application, especially in the mission-critical tasks with the high cost of errors or rewards, questions of the explainability or, at least, the trustworthiness of the ML-generated decisions become no less important than the proper prediction verdict. Frequently, when ML algorithms' training data do not fully represent real-life data, the very notion of the 'proper' verdict becomes blurry. If the correct prediction scored a tiny bit more than the alternative wrong one, should we consider it a true success? Can we, and if the answer is 'yes', how we can trust ML algorithms in their decisions?

The domain of mission-critical applications has been entered by the ML solutions first in military and medical areas [12], and later in the autonomous driving and the robot operation field that experiences explosive development

G. Nicosia et al. (Eds.): LOD 2021, LNCS 13164, pp. 338–352, 2022.
https://doi.org/10.1007/978-3-030-95470-3_26

[26]. Concerns about how ML algorithms handle real-life uncertainty and to what degree it could be trusted have also entered general scientific, engineering, and public discourse [1], frequently as a part of the Artificial Intelligence (AI) perspectives debate.

The very founders of the contemporary field of AI scientific research were sceptical about the worthiness of the attempts to answer the question of what AI is. Instead, they suggested answering the question of how well AI can emulate human intelligence and finding ways of quantifying the success of that imitation [17,24]. N. Chomsky, in numerous lectures and publications (f.e. [5]), even more categorically elaborated that AI is a human linguistic concept rather than an independent phenomenon. Trust, also an abstract human concept, in the end, should be left for humans to decide. However, metrics of the uncertainty of the ML and AI algorithms in their decisions and algorithms that can produce those metrics may supply humans with quantified data to make decisions on the confidence level suitable for the task and audience.

Being a complex system itself, ML or AI (whatever it is, at least it exists as a human concept) operation can be the subject of study or modelling by ML itself. The idea of learning the ML processes was introduced in the 90's [23], and recently gained traction in various flavours of meta-learning [25]. They can be viewed either as a narrower 'learning to learn' approaches such as an extension of transfer learning [2,6,19], or model hyper-parameter optimization [3,21], or a wider horizon 'learning about learning' approach conjoint to the explainable learning [13,15].

Of course, the generalized topic of meta-learning of the real-life uncertainty handling by ML algorithms, or even more specifically ANN models, to increase trust in and explainability of ANN verdicts is too big for this paper scope. Therefore, the study concentrates on limited aspects of the task in terms of the application domain, kinds of the training-test data disparity, underlying types of ANN models, their parameters and behaviour patterns to learn, and meta-learning methods.

This research builds upon the previous pilot study of the makeup and occlusions effects on face recognition (FR) accuracy of the established convolutional neural network (CNN) models such as AlexNet that was trained on the non-disguised images only [22]. On the one hand, biometrics applications in general, and FR in particular, may seem not mission-critical applications in which confidence in positive or negative verdicts is essential. However, in legal and security applications, where the lives of individuals or a large number of people may be at stake, false-positive errors in the former case or false-negative errors in the latter may have huge societal costs and impacts. On the other hand, solutions for trustworthy ML found for the FR task may be generalized for other image processing applications or applications dealing with other kinds of data or analysis.

In this study, performance in FR of the additional state-of-the-art (SOTA) CNN models of various complexity and computational heaviness, such as VGG19, GoogLeNet, ResNet50, Inception v.3, InceptionResnet v.2, is investigated in the presence of unexpected makeup and occlusions in test images. Instead of analyzing test data to identify samples not represented by the training data, which may produce unreliable prediction verdicts [7], we propose to analyze the behaviour of the underlying CNNs to identify verdicts about which CNNs are 'uncertain' and separate them into additional class.

Three metrics, derived from the softmax activations distribution of the underlying CNNs, are investigated whether they may be used to generate the trusted verdict at the desired confidence level. The first one is suggested in the original work: the highest CNN softmax activation to which a simple A/B test is applied. The A and B class distributions here are the CNN softmax distributions of the wrongly identified and correctly identified images, with a threshold separating those classes at the desired confidence level. The hypothesis is that even such a simple test would produce consistent results ensuring false-positive or false-negative errors on the given confidence level or better across all models for makeup and occluded images.

Another two metrics are produced by the meta-learning supervising ANN (SNN) attached to the softmax output layers of the ensembles of the underlying CNN models. The hypothesis is that SNN can learn 'uncertainty' patterns in the underlying CNN softmax activation distributions. The SNN output metrics are predictions of the 'certainty' class and the very SNN's softmax activation for that class. The number of 'certainty' classes is proportional to the underlying CNN ensemble size. SNN is not yet trained to meet the forehead set confidence level at this stage of the research. We assess what level of confidence in false positive or false negative errors a simple meta-learning supervising ANN version of the fully connected neural network (FNN) consisting of three dense layers is capable of achieving.

The paper is organized as follows. Section 2 briefly introduces ML and uncertainty concepts used in developing the proposed solution and outlines intuition and a general idea of what the supervisor ANN may learn. Section 3 describes the data set used for experiments; Sect. 4 outlines experimental algorithms in detail; Sect. 5 presents the obtained results, and Sect. 6 discusses the results, draws practical conclusions, and states directions of the research of not yet answered questions.

2 Machine Learning and Uncertainty

2.1 High-Level View on Classification with ANN

Each layer of the simple fully-connected artificial neural network (FNN) can be represented as a composite linear transformation, which can be expressed via multiplication of the input vector $\mathbf{x}_i$ by a matrix W_i, and non-linear transformation by an activation function a_i or c.

$$\pi(\mathbf{x}) = c \circ f_k \dots a_i \circ f_i \dots a_1 \circ f_1(\mathbf{x}), \forall \mathbf{x} \in \mathcal{X} \subset \mathbb{R}^m \tag{1}$$

where $\mathbf{y}_i = f_1(\mathbf{x}_i) = \mathsf{W}_i \mathbf{x}_i$, for example $a_i(y_{ij}) = y_{ij}^+$.

The point estimate classification c based on the class scores is frequently implemented via a normalized mass function, for example, softmax, and possibly following argmax classification functions: $i_{max} = c(\mathbf{y}_k) = arg\,max(\dots, \frac{e^{y_{kj}}}{\sum_j e^{y_{kj}}}, \dots)$.

However, for the uncertainty-aware classification, the range space should be not just a one-dimensional integer one $c : \mathcal{Y} \subset \mathbb{R}^n \mapsto \mathcal{C} \subset \mathbb{I}$, but rather a product with either binary trusted/not-trusted flag dimension:

$$c : \mathcal{Y} \subset \mathbb{R}^n \mapsto \mathcal{C} \subset \mathbb{I} \times \mathbb{B} \tag{2}$$

or a product with the real one-dimensional space that can be given probabilistic interpretation $c : \mathcal{Y} \subset \mathbb{R}^n \mapsto \mathcal{C} \subset \mathbb{I} \times \mathbb{R}$, or a product with the real n-dimensional space that can be interpreted as a probability distribution over all classes $c : \mathcal{Y} \subset \mathbb{R}^n \mapsto \mathcal{C} \subset \mathbb{I} \times \mathbb{R}^n$, or, if not normalized, may be given the fuzzy logic interpretation.

2.2 Trusted Accuracy Metrics

If only the classification verdict is used as a final result of the ANN model then accuracy of the target CNN model can be calculated only as the ratio of the number of correctly identified test images by the CNN model, to the number of all test images $Accuracy = \frac{N_{correct}}{N_{all}}$.

When additional dimension in classification is used, for example softmax activation value of the target CNN, or amending verdict of the meta-learning supervisor ANN, (see Formula 2), and $\pi(\mathbf{x}) = \mathbf{c}$, where $\forall \mathbf{x} \in \mathcal{X}$, $\forall \mathbf{c} \in \mathcal{C} = \{(1, f), (2, f) \dots (n, f)\}$, $\forall f \in \mathbb{B} = \{True, False\}$, then the trusted accuracy and other trusted quality metrics can be calculated as:

$$Accuracy_t = \frac{N_{correct:f=T} + N_{wrong:f \neq T}}{N_{all}} \tag{3}$$

As a mapping to a more usual notations, $N_{correct:f=T}$ can be as the True Positive (TP) number, $N_{wrong:f \neq T}$ - True Negative (TN), $N_{wrong:f=T}$ - False Positive (FP), and $N_{correct:f \neq T}$ - False Negative (FN).

Trusted precision, as a measure of the 'pollution' of the true positive verdicts by the false positive errors:

$$Precision_t = \frac{N_{correct:f=T}}{N_{correct:f=T} + N_{wrong:f=T}} \tag{4}$$

Trusted recall, as a measure of the true positive verdicts 'loss' due to false negative errors:

$$Recall_t = \frac{N_{correct:f=T}}{N_{correct:f=T} + N_{correct:f \neq T}} \tag{5}$$

Or, trusted specificity, as a measure of the true-negative verdicts 'loss' by false-positive errors, or, in the context of A/B testing, is equal to the confidence level percentage of the wrongly identified images that are recognized of such.

$$Specificity_t = \frac{N_{wrong:f \neq T}}{N_{wrong:f \neq T} + N_{wrong:f=T}} \tag{6}$$

Where $N_{correct}$ and N_{wrong} number of correctly and incorrectly identified test images by the CNN model.

2.3 Bayesian View on ANN Classification

Mapping softmax activations $\mathbf{y}_k$ to the desired probabilistic classification space is not enough because they capture only uncertainty of the model about image observations, but not the uncertainty of the already trained model about its parameters. Using the Bayesian learning rule with ANNs, which allows capturing the latter uncertainty, was introduced in the mid-1990s [16,18] and remains a popular probabilistic framework for uncertainty representation.

The Bayesian learning rule can be derived and used even without accepting the Bayesian view on probability. However, conclusions made from the results or even merits of the rule use may differ if one shares frequentist views on probability.

$$P(c|x) = \frac{P(x|c)P(c)}{P(x)} = \frac{P(x|c)P(c)}{\sum_{c \in C} P(x,c)} \tag{7}$$

where $\forall x \in \mathcal{X}$, $\forall y \in \mathcal{Y}$, $P(x)$ is a probability of observing entity x, $P(y)$ is a probability of observing entity y, $P(x,y)$ - probability of observing entity x and y together, and $P(x|y)$ - probability of observing entity x given observation of y [9].

One does not have to be the Bayesian to use Formula 7 if the right-hand side parameters are more easily obtainable than the left-hand side ones. For example, in the ML model comparison in the controlled environment of the training and test data distribution $P(x)$. However, if the test images' real-life distribution is unknown or significantly differs from the training distribution, a frequentist would refuse to make any conclusions without collecting sufficient statistics on the new test data. However, Bayesian view on probability, unlike the frequentist one, allows it to exist, or at least a belief about it, without or with few observations, suggest to use the Bayes rule as the learning tool that refines the prior beliefs or imperfect knowledge at the moment t by using the new information D at the moment $t+1$:

$$P(c_{t+1}|x, D_{t+1}) = \frac{P(x, D_{t+1}|c_t)P(c_t)}{P(x, D_{t+1})} \tag{8}$$

Therefore, one can start learning the real-life test environment probabilities, starting with the old training data or any other beliefs in the 'reasonable' probability distributions. Gradually approaching 'real' distributions, which nicely fits the continuous, reinforcement, and flavours of the active learning paradigms [4].

One can introduce ML parameters, for example weights W of the ANN, as such a new information. ANN weights and class verdicts, of course, are conditioned by the training data set D_{tr}, however, if ANN is not retrained on the test stage, data set term may be dropped for simplicity:

$$p(c|x) = \int_W p(c|x, W)p(W)dW \tag{9}$$

or, for approximation:

$$p(c|x) = \sum_{i \in \{1,\dots,N\}} p(c|x, W_i)p(W_i)\Delta W \tag{10}$$

where i indexes N samples of the CNN weight sets $W_i \in \mathcal{W} \subset \mathbb{R}^k$, out of the k-dimensional weight space.

The first term can be viewed as responsible for so-called *aleatoric* uncertainty, or uncertainty related to handling data by the model. In contrast, the second term - *epistemic* uncertainty or uncertainty related to the model stability [11].

While Bayesian formulas are straightforward, in practice, their analytical solution is unrealistic due to multi-dimensionality, multi-modality, and non-linearity of the activation functions. Therefore, practical solutions for Bayesian neural networks (BNN) include approximation via various sampling methods of either *aleatoric* or *epistemic* uncertainties or both, such as Variational Inference [10], Markov Chain Monte Carlo [18], Monte Carlo Dropout [8], Deep Ensembles [14]. As alternative for the *aleatoric* uncertainty estimations Direct Modeling and Error Propagation approaches, which expect particular parametric distribution of parameters and their variance propagation model, were suggested [20].

2.4 Proposed Solution: Supervisor ANN

We propose to investigate the feasibility of using meta-learning supervisor neural networks (SNN) as means of the implicit approximation of the Bayesian integral Formula 9 as follows:

$$P(c_t|x, M, D) = \int_W P(c_t|x, W, M, D)p(W|M, D)dW \approx \sum_{c_s \in C^T} y_s|c_s, D_2 \tag{11}$$

where

$$\mathbf{y}_s = snn((\mathbf{y}_{ti}|m_i)_{i \in 1 \dots |M|}|x, M, D_1) \tag{12}$$

Where $P(c|x, \dots)$ is a conditional probability of the image x being classified by the CNN model ensemble M trained on the data set D, and set of the network weights W as an additional parameter, $p(W|M, D)$ - probability distribution of the weights W across the layers of the model ensemble. Training data set $D = D_1 + D_2$ is partitioned into data set D_1 for the target CNN ensemble training, and data set D_2 - for the SNN training. Sum of the softmax activations of the supervisor network $y_s|c_s$ of those classes that associate with the 'trustworthy' state of the target network image classification $C^T_s = \{c_{si} \in C_s : c_{si} = \mathit{true\ verdict}\}$ may be viewed as the Bayesian integral approximation. For practical purposes, there may be just two SNN verdicts: 'trusted' or 'untrusted', or categorization may be more granular, depending on the ensemble vote.

Transformation *snn* represents mapping that the supervising neural network performs from the softmax activations of the target network ensemble into softmax activations of the 'un/trusted' classes. The intuition of the proposed solution goes as that the structure of the fully connected network transformation (Formula 1) is similar to the Bayesian integral (Formula 9) approximation (Formula 10): a softmax-normalized piecewise linear combination (with ReLU dropouts) of the softmax activations of the target networks $(\mathbf{y}_{ti}|m_i)$ (which carry the same information about the image variations uncertainty as the first term of the Bayesian integral) multiplied by the SNN weights. With proper training on the variations in the softmax vectors of different models of the ensemble, those weights can learn information about models' uncertainty in their parameters and become analogous to the second term of the Bayesian integral.

3 Data Set

The BookClub artistic makeup data set contains images of $|C| = 21$ subjects. Each subject's data may contain a photo-session series of photos with no-makeup, various makeup, and images with other obstacles for facial recognition, such as wigs, glasses, jewellery, face masks, or various types of headdress. Overall, the data set features 37 photo sessions without makeup or occlusions, 40 makeup sessions, and 17 sessions with occlusions. Each photo session contains circa 168 JPEG images of the 1072×712 resolution of six basic emotional expressions (sadness, happiness, surprise, fear, anger, disgust), a neutral expression, and the closed eyes photo-shoots taken with seven head rotations at three exposure times on the off-white background. The photos were taken over two months, and several subjects were posed at multiple sessions over several weeks in various clothing with changed hairstyles, downloadable from https://data.mendeley.com/datasets/yfx9h649wz/3. All subjects have given written consent for using their anonymous images in public scientific research.

A high in-session variety BookClub data set provides *aleatoric* uncertainty for the following experiments. The non-makeup training and makeup test data partition was applied to boost the *epistemic* uncertainty.

4 Experiments

The experiments were designed to maximize the uncertainty the established SOTA CNN models have to face with in the realistic life-scenario fashion. For the natural way of increasing uncertainties, neither face detectors nor other segmentation techniques were used to include varying hairstyles and closes into the models' input.

The experiments were run on the Linux (Ubuntu 18.04) operating system with three GeForce GTX 1070 Ti GPUs (with 8 GB GDDR5 memory each), X299 chipset motherboard, 128 GB DDR4 RAM, and i7-7800X CPU. Experiments were run using MATLAB 2020b.

Experiments verifying the accuracy and uncertainty handling by the established SOTA CNN were done on MATLAB implementations of the models (as a part of the Deep Learning Toolbox) with the last two layers resized to match the number of classes in the BookClub data set (21), and retrained using 'adam' learning algorithm with 0.001 initial learning coefficient, 'piecewise' learning rate drop schedule with 5 iterations drop interval, and 0.9 drop coefficient, mini-batch size 128, and $10-20-30$ epochs parameters for various models with different convergence rate to ensure at least 95% learning accuracy. The following CNN models were experimented with: AlexNet consisting of 25 elementary layers and taking input images scaled to 277×277 dimension, VGG19 model having 47 elementary layers, GoogLeNet - 144, Resnet50 - 177, and they take 224×224 scaled images as input. The Inception v.3 contains 315 elementary layers, and InceptionResnet v.2 - 824, both taking 299×299 scaled images as input. All experimented with models, but AlexNet and VGG19 models, have the Directed Acyclic Graph (DAG) architecture. The training set for CNN models contained $|D_1| = 4529$ non-makeup and non-occluded images.

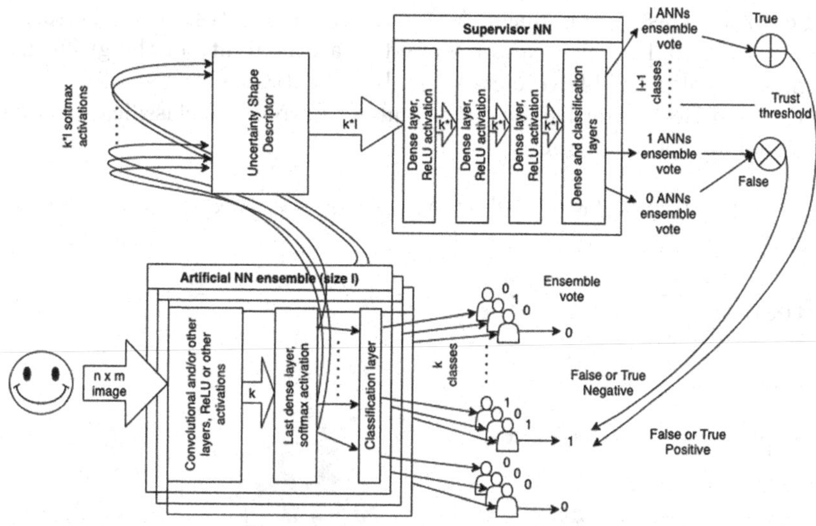

Fig. 1. Supervisor (SNN) and being supervised ANN block schema.

For experiments with A/B testing and SNN, for each subject with more than one non-makeup session, one non-makeup session was set aside and was not used for the target CNN training. These images were used for A/B threshold determination and as SNN's training set $|D_2| = 1653$. Each of these images was run against the target CNN, and then, activations of the target CNN were fed as a training input into SNN.

The SNN transformation, mentioned in Formula 11 that was used in the experiments can be represented as a composite function of the 'uncertainty shape descriptor' function presented in the Algorithm 1 and FNN transformation Formula 1. The SNN's architecture, Fig. 1, in MATLAB implementation consist of three fully connected layers with ReLU activation function and number of neurons $nLayer1 = nLayer2 = nLayer3 = |C| * |M|$ and final dense layer with softmax activation followed by the classification layer, both with neuron number $nVerdicts = |M| + 1$. It can be presented as:

$$\mathbf{y}_s = snn((\mathbf{y}_{ti}|m_i)_{i \in 1...|M|}|x, M, D_1) = fnn \circ usd((\mathbf{y}_{ti}|m_i)_{i \in 1...|M|}|x, M, D_1) \tag{13}$$

where $(\mathbf{y}_{ti}|m_i)_{i \in 1...|M|}$ is a vector composed of the softmax activation vectors of all CNN models m in the ensemble M trained on the data set D_1, x is a given test image, and $\mathbf{y}_s$ is the vector of SNN softmax activations.

The SNN was trained using 'adam' learning algorithm with 0.01 initial learning coefficient, 'piecewise' learning rate drop schedule with 5 iterations drop interval, and 0.9 drop coefficient, mini-batch size 64, and 200 epochs parameters.

Number of classes the SNN was trained on was equal to the number of models in ensemble plus one. Meaning of each class is an ensemble vote verdict $ev \in \{0, \frac{1}{N_{models}}, \dots \frac{N}{N_{models}}, \dots, 1\}$, where N is a predicted number of models that would correctly classify the image, $N \leq N_{models}$. At the test time, the softmax vote score $vs \in [0, 1]$, which can be viewed as a probability of the given image belonging to that ensemble vote class, can be also attached to the target CNN's verdict c_t and SNN's verdict ev_s, this making CNN-SNN classification multi-dimensional and probabilistic:

$$c : \mathcal{X} \subset \mathbb{R}^m \mapsto \mathcal{C} \subset \mathbb{I} \times \mathbb{R}^2,\ c(\mathbf{x}) = (c_t, ev_s, vs_s),\ \forall \mathbf{x} \in \mathcal{X},\ \forall \mathbf{c} \in \mathcal{C} \tag{14}$$

5 Results

Subj.7, Sess.MK2; Subj.10, Sess.MK4; Subj.14; Sess.HD1; Subj.21, Sess.MK1

Fig. 2. Image class examples misidentified or identified with low accuracy by all CNN models.

When retrained on the BookClub training set comprised of only non-makeup and non-occluded sessions, the well-known SOTA CNN models correctly identified a significant part of the makeup and occluded sessions. However, a number of the session posed a difficulty for all or majority of the CNN models, Fig. 2

Algorithm 1. The 'uncertainty shape descriptor' building function

Input: List of softmax activations $(\mathbf{y}_{ti}|m_i)_{i\in 1\ldots|M|}$ of the models m in the ensemble M

Parameters: Number of models $|M|$

Output: Flattened softmax activation vector $\mathbf{y}_t$ a.k.a. 'uncertainty shape descriptor'

for all models m_i, $i \in \{1,\ldots|M|\}$ **do**
  $\mathbf{y}_{ci}|m_i \leftarrow sort(\mathbf{y}_{ti}|m_i)\, by\, 'descending'$; Sort softmax activations inside each model vector
end for
$M_{index} \leftarrow sort(M)\, by\, y_c(1)$; Order model vectors by the largest softmax activation and extract index of the new order
$\mathbf{y}_{c1\,index} \leftarrow index(\mathbf{y}_{ci}|m_{M_{index}(1)})$; Extract the order of activations in the vector with largest softmax activation (first in M_{index})
for all models m_i, $i \in M_{index}$ **do**
  $\mathbf{y}_{ti}|m_i \leftarrow sort(\mathbf{y}_{ti}|m_i)\, by\, \mathbf{y}_{c1\,index}$; Rearrange order of activations in each vector to the order of activations in the vector with largest softmax activation (first in M_{index})
end for
$\mathbf{y}_t \leftarrow flatten(\mathbf{y}_{t1}|m_1 \ldots \mathbf{y}_{ti}|m_i \ldots \mathbf{y}_{t|M|}|m_{|M|})$
return solution

and Table 1. On top of that, those incorrect classifications were made with a high confidence level (for raw results, source code, and pre-trained models, see https://github.com/Selitskiy/LOD2021).

Table 1. Image classes misidentified or identified with low accuracy by all CNN models

Session	AlexNet	VGG19	GoogLeNet	Resnet50	Inception3	InceptRes2
S1HD1	0.0000	0.0000	0.0000	0.0482	0.0000	0.0000
S1MK2	0.0060	0.0476	0.0000	0.0000	0.0000	0.0714
S1MK3	0.0000	0.0398	0.0000	0.4034	0.0000	0.3921
S7FM1	0.0000	0.0000	0.0000	0.0000	0.3642	0.0989
S10MK2	0.0479	0.0659	0.0000	0.0000	0.0299	0.0060
S10MK3	0.0000	0.0000	0.0000	0.2426	0.0000	0.0059
S10MK4	0.0060	0.0000	0.0000	0.0240	0.0000	0.0599
S21MK1	0.0000	0.2927	0.0000	0.0427	0.0000	0.0000

The models also demonstrated specialization in handling various types of makeup and occlusions. Particularly, the VGG19 model had problems with the artificial white wig that other models easily recognized. Inception v.3, which was the most accurate and reliable model overall, had few 'blinders' on simple makeups, easily recognizable by even simpler models. Resnet50 failed to recognize painted realistic human faces. Furthermore, GoogLeNet failed on face masks which other DAG models solved recognition. Inception v.3 and InceptionResnet v.2 solved recognition of the light theatrical type makeup, which was problematic for other models. VGG19, while failing on many easy cases, uniquely recognized heavily painted over faces with contrast pigments. GoogLeNet and Resnet50 were particularly successful with recognition in the presence of wigs and dark glasses, and Inception v.3 - for face mask recognition.

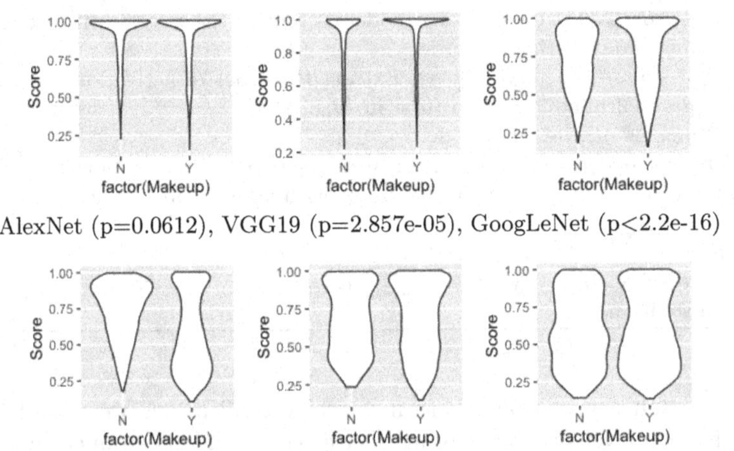

AlexNet (p=0.0612), VGG19 (p=2.857e-05), GoogLeNet (p<2.2e-16)

Resnet50 (p<2.2e-16), Inception v.3 (p=0.0567), InceptionResnet v.2 (p=0.1807)

Fig. 3. Violin plots and two-sided Kolmogorov-Smirnov p-value for prior (selected non-makeup sessions) and posterior (sessions with makeup and occlusions) distributions of the soft-max activation score that drove the classification verdict for the wrongly identified images for AlexNet, VGG19, GoogLeNet, Resnet50, Inception v.3, InceptionResnet v.2 models.

The similarity of the highest softmax activation distributions for selected non-makeup images, which were not a part of the underlying CNN training, and makeup and occluded images can be seen in Fig. 3. That suggests that such non-makeup images may be used for finding out the softmax activation threshold to ensure exclusion of the low-activation wrongly identified on the test makeup set, as well, on the desired confidence level. However, rigorous goodness of fit non-parametric tests, such as Kolmogorov-Smirnov, have shown a relatively low probability that those sample distributions were drawn from the same one. Still, for practical purposes, the hypothesis was relaxed that a simple A/B test might give reasonable results even on the single highest activation input, see Table 2. Therefore, a higher-dimensional input and more non-linear ANN approach can be proceeded with.

The meta-learning SNN experiments for multiple ensemble sizes were conducted on the Inception v.3 - the most accurate and robust CNN model out of

Table 2. Unconstrained accuracy, trusted accuracy and precision of the makeup and occluded test images calculated based on the confidence level (CL) thresholds of the prior distributions of the wrongly identified non-makeup training images.

Metric	AlexNet	VGG19	GoogLeNet	Resnet50	Incept	IncRes2
Accuracy	0.3712	0.2009	0.4534	0.5250	0.5553	0.5675
75% CL Trust. accuracy	0.6648	0.6714	0.7498	0.7555	0.7390	0.8021
90% CL Trust. accuracy	0.6104	0.7859	0.7829	0.7336	0.7203	0.8046
75% CL Trust. precision	0.5756	0.2977	0.6781	0.8437	0.8342	0.8337
90% CL Trust. precision	–	–	0.7466	0.8809	0.9265	0.9111
75% CL Trust. recall	0.5311	0.3937	0.8961	0.7256	0.7243	0.8482
90% CL Trust. recall	0.0000	0.0000	0.8184	0.6412	0.5978	0.7568
75% CL Trust. specificity	0.7501	0.7471	0.6186	0.8000	0.7630	0.7282
90% CL Trust. specificity	1.0000	1.0000	0.7510	0.8710	0.9219	0.8814

all the models being investigated. Number of images in the test set was 11145, and model ensemble size $N_{models} \in \{1, 2, 4, 8, 16\}$.

Table 3. Training time [s], accuracy, loss of the final mini-batch, and testing time [s] of the session of 170 images of the meta-learning SNN attached to the Inception v.3 model ensemble of the various size

Ens. size	1	2	4	8	16
Train. time	251	264	283	443	595
Train. acc.	86%	100%	100%	100%	100%
Train. loss	0.2920	0.0107	0.0022	0.0021	0.0044
Test time	14.76	15.90	18.17	34.44	48.65

The training time, final mini-batch accuracy and loss, Table 3 show similar results as in [14], - ensemble of 4 models reaches the accuracy parameters at the level that changes slightly with the model number increase, Table 4. A single model does not give the SNN enough information on the uncertainty in its parameters and fails to train above the circa 70–90% accuracy rate.

For the trusted accuracy metric parameters calculation (Formulas 3–6), two schemes were used: considering the ensemble vote ev only, and considering both ensemble vote ev and vote score vs, from Formula 14. Such that the combined CNN-SNN supervised verdict $\mathbf{sc}_i = (c_i, ev_i) \in \mathcal{TC} \subset \mathbb{I} \times \mathbb{R}$ is considered trusted if $ev_i \geq ev_t = 0.5$ for the former schema, see results in Tables 4 and 5. And if $ev_i \geq ev_t = 0.5 \vee vs_i \geq vs_t = 0.95$ - for the latter schema, see results on GitHub.

For the practically 'good enough' ensemble size of 4 models, experiments with makeup and occlusions sessions were conducted to test the trusted accuracy and other metrics improvement, Table 5.

6 Discussion, Conclusions, and Future Work

The observed success and failure 'specialization' of the SOTA models on different types of makeup and occlusions is a fascinating and prospective phenomenon. The most successful last 4 CNN models use DAG architecture with 'Inception'-type cascade cells implanted in various numbers and deepness on the networks. Those cells with parallel convolutional layers of various filter sizes are meant to self-tune for the most prospective one in that place of the network. Yet, these models fail or succeed differently. A natural direction of the future work is to use explainable learning solutions to identify architectural reasons for this variability and extend the depth and width of these cells in the proper direction and improve placement of the cells.

Table 4. Accuracy metrics of the Inception v.3 CNN model ensemble improved by the ensemble vote verdict only of the proposed SNN for ensemble sizes 1, 2, 4, 8, 16

Ens. size	1	2	4	8	16
Accuracy	0.6221	0.6046	0.6784	0.6567	0.6668
Trusted accuracy	0.7514	0.7744	0.7624	0.7344	0.7437
Trusted precision	0.8040	0.8860	0.8732	0.8186	0.8141
Trusted recall	0.7939	0.7195	0.7602	0.7651	0.7978
Trusted specificity	1.000	0.8585	0.7670	0.6757	0.6355

The proposed simple and flexible SNN appears to learn uncertainty of the verdicts of the underlying target SOTA CNN ensembles even for small ensemble sizes. Performance of the homogeneous models' ensemble converges to the improved accuracy results at a low number of models, thus keeping the time and resource overheads reasonably low. When the underlying CNN model fails to recognize most images in the test session, SNN still produces high trusted negative metrics.

Table 5. Accuracy metrics of the SOTA CNN model ensembles of size 4 improved by the ensemble vote verdict only of the proposed SNN

Model	AlexNet	VGG19	GoogLeNet	Resnet50	Inception3	InceptRes2
Accuracy	0.4128	0.4352	0.5087	0.6464	0.6785	0.6571
Trusted accuracy	0.6697	0.6944	0.4973	0.8188	0.7624	0.7828
Trusted precision	0.9214	0.7446	0.6408	0.9477	0.8732	0.8282
Trusted recall	0.2185	0.4532	0.0273	0.7617	0.7602	0.8446
Trusted specificity	0.9869	0.8802	0.9841	0.9231	0.7670	0.6642

However, the straightforward approach of attaching CNN ensemble output to SNN input without Algorithm 1 does not generalize, and SNN training does not converge at all. In that case, CNNs' error patterns, even if learned, are learned specifically for each subject or even specifically to photo sessions. Algorithm 1 allows the creation of subject-invariant softmax distribution shape descriptors. There are variations of the Algorithm, and determining more effective is another direction for further investigation. The intuition that SNN implicitly captures *epistemic* uncertainty, i.e. dispersion of the weights of the CNN ensemble, requires quantitative confirmation as another future work item.

The proposed 'uncertainty' metrics produced by the statistical A/B test and meta-learning SNN allow achieving trusted accuracy in the circa 70–80% range on BookClub data set in the described training/test set partition. A natural continuation will be investigating the algorithms' performance on other data set. The ultimate goal of achieving trusted prediction verdicts in its true sense will be to use the proposed 'uncertainty' metrics to produce results with errors at the confidence level forehead acceptable by the human users for a given application domain. Incorporating these confidence levels into SNN's feedback parameters and loss functions is also the future work direction.

References

1. Amodei, D., Olah, C., Steinhardt, J., Christiano, P., Schulman, J., Mané, D.: Concrete problems in AI safety (2016)
2. Andrychowicz, M., et al.: Learning to learn by gradient descent by gradient descent. In: Proceedings of the 30th International Conference on Neural Information Processing Systems, NIPS'16, pp. 3988–3996, Curran Associates Inc., Red Hook (2016)
3. Bergstra, J., Bardenet, R., Bengio, Y., Kégl, B.: Algorithms for hyper-parameter optimization. In: Advances in Neural Information Processing Systems, vol. 24. Curran Associates, Inc. (2011). https://proceedings.neurips.cc/paper/2011/file/86e8f7ab32cfd12577bc2619bc635690-Paper.pdf
4. Briggs, W.: Uncertainty: The Soul of Modeling, Probability & Statistics. Springer, Cham (2016). https://doi.org/10.1007/978-3-319-39756-6
5. Chomsky, N.: Powers and Prospects: Reflections on Human Nature and the Social Order. South End Press, Boston (1996)
6. Finn, C., Abbeel, P., Levine, S.: Model-agnostic meta-learning for fast adaptation of deep networks. In: Proceedings of the 34th International Conference on Machine Learning. Proceedings of Machine Learning Research, vol. 70, pp. 1126–1135. PMLR, August 2017. http://proceedings.mlr.press/v70/finn17a.html
7. Fort, S., Ren, J., Lakshminarayanan, B.: Exploring the limits of out-of-distribution detection. CoRR arXiv:2106.03004 (2021)
8. Gal, Y., Ghahramani, Z.: Dropout as a Bayesian approximation: representing model uncertainty in deep learning. In: Proceedings of the 33rd International Conference on International Conference on Machine Learning, ICML'16, vol. 48, pp. 1050–1059. JMLR.org (2016)
9. Ghahramani, Z.: Probabilistic machine learning and artificial intelligence. Nature **521**(7553), 452–459 (2015). https://doi.org/10.1038/nature14541
10. Graves, A.: Practical variational inference for neural networks. In: Proceedings of the 24th International Conference on Neural Information Processing Systems, NIPS'11, pp. 2348–2356. Curran Associates Inc., Red Hook (2011)

11. Kendall, A., Gal, Y.: What uncertainties do we need in Bayesian deep learning for computer vision? arXiv:1703.04977 (2017)
12. Kurd, Z., Kelly, T.: Establishing safety criteria for artificial neural networks. In: Palade, V., Howlett, R.J., Jain, L. (eds.) KES 2003. LNCS (LNAI), vol. 2773, pp. 163–169. Springer, Heidelberg (2003). https://doi.org/10.1007/978-3-540-45224-9_24
13. Lake, B.M., Ullman, T.D., Tenenbaum, J.B., Gershman, S.J.: Building machines that learn and think like people. Behav. Brain Sci. **40**, e253 (2017). https://doi.org/10.1017/S0140525X16001837
14. Lakshminarayanan, B., Pritzel, A., Blundell, C.: Simple and scalable predictive uncertainty estimation using deep ensembles. In: Proceedings of the 31st International Conference on Neural Information Processing Systems, NIPS'17, pp. 6405–6416. Curran Associates Inc., Red Hook (2017)
15. Liu, X., Wang, X., Matwin, S.: Interpretable deep convolutional neural networks via meta-learning. In: 2018 International Joint Conference on Neural Networks (IJCNN), pp. 1–9 (2018). https://doi.org/10.1109/IJCNN.2018.8489172
16. MacKay, D.J.C.: A practical Bayesian framework for backpropagation networks. Neural Comput. **4**(3), 448–472 (1992). https://doi.org/10.1162/neco.1992.4.3.448
17. McCarthy, J., Minsky, M.L., Rochester, N., Shannon, C.E.: A proposal for the Dartmouth summer research project on artificial intelligence, August 31, 1955. AI Mag. **27**(4), 12 (2006)
18. Neal, R.M.: Bayesian Learning for Neural Networks. Lecture Notes in Statistics, vol. 118. Springer, New York (1996). https://doi.org/10.1007/978-1-4612-0745-0
19. Nichol, A., Achiam, J., Schulman, J.: On first-order meta-learning algorithms. ArXiv arXiv:1803.02999 (2018)
20. Postels, J., Ferroni, F., Coskun, H., Navab, N., Tombari, F.: Sampling-free epistemic uncertainty estimation using approximated variance propagation. CoRR arXiv:1908.00598 (2019)
21. Ram, R., Müller, S., Pfreundt, F., Gauger, N., Keuper, J.: Scalable hyperparameter optimization with lazy Gaussian processes. In: 2019 IEEE/ACM Workshop on Machine Learning in High Performance Computing Environments (MLHPC), pp. 56–65 (2019)
22. Selitskaya, N., Sielicki, S., Christou, N.: Challenges in real-life face recognition with heavy makeup and occlusions using deep learning algorithms. In: Nicosia, G., et al. (eds.) LOD 2020. LNCS, vol. 12566, pp. 600–611. Springer, Cham (2020). https://doi.org/10.1007/978-3-030-64580-9_49
23. Thrun, S., Pratt, L.: Learning to Learn. Springer, Boston (1998). https://doi.org/10.1007/978-1-4615-5529-2
24. Turing, A.M.: I.-Computing machinery and intelligence. Mind **LIX**(236), 433–460 (1950). https://doi.org/10.1093/mind/LIX.236.433
25. Vanschoren, J.: Meta-learning: a survey. ArXiv arXiv:1810.03548 (2018)
26. Casimiro, A., Ortmeier, F., Schoitsch, E., Bitsch, F., Ferreira, P. (eds.): SAFECOMP 2020. LNCS, vol. 12235. Springer, Cham (2020). https://doi.org/10.1007/978-3-030-55583-2

Multi-Asset Market Making via Multi-Task Deep Reinforcement Learning

Abbas Haider(✉), Glenn I. Hawe, Hui Wang, and Bryan Scotney

School of Computing, Ulster University, Newtownabbey, UK
{haider-a,gi.hawe,h.wang,bw.scotney}@ulster.ac.uk

Abstract. Market making (MM) is a trading activity by an individual market participant or a member firm of an exchange that buys and sells same securities with the primary goal of profiting on the bid-ask spread, which contributes to the market liquidity. Reinforcement learning (RL) is emerging as a quite popular method for automated market making, in addition to many other financial problems. The current state of the art in MM based on RL includes two recent benchmarks which use temporal-difference learning with Tile-Codings and Deep Q Networks (DQN). These two benchmark approaches focus on single-asset modelling, limiting their applicability in realistic scenarios, where the MM agents are required to trade on a collection of assets. Moreover, the Multi-Asset trading reduces the risk associated with the returns. Therefore, we design a Multi-Asset Market Making (MAMM) model, known as MTDRLMM, based on Multi-Task Deep RL. From a Multi-Task Learning perspective, multiple assets are considered as multiple tasks of the same nature. These assets share common characteristics among them, along with their individual traits. The experimental results show that the MAMM is more profitable than Single-Asset MM, in general. Moreover, the MTDRLMM model achieves the state-of-the-art in terms of investment return in a collection of assets.

Keywords: Multi-Asset Market Making · Multi-Task Deep Reinforcement Learning · Tile-Codings · Deep Q Networks

1 Introduction

Market Making (MM) is a well known financial problem, the goal of which is to provide liquidity to traders in the market. The market maker (MMer) is a trader who is under obligation of standing ready to buy or sell assets from other traders. The objective of the MMer is to place ask and bid quotes in the market (order books), simultaneously. The difference between the ask quote price and the bid quote price denotes bid-ask spread. The earning of the MMer generates from the bid-ask spread, the higher the spread the higher will be the return. The return can be positive representing profit or negative incurring loss on a trade.

G. Nicosia et al. (Eds.): LOD 2021, LNCS 13164, pp. 353–364, 2022.
https://doi.org/10.1007/978-3-030-95470-3_27

The necessary condition for obtaining a profit is the execution of both ask and bid quotes placed by the MMer in the market.

Reinforcement Learning (RL) is a well-known approach of solving sequential decision making problems [21]. The decision maker, formally known as the agent in RL paradigm, derives a mapping from state space to the action space, known as a policy. The RL agent's policy represents the experience of mapping situations to the actions through assessing the scalar reward signals. RL has widely been used in solving the MM problem. The first benchmark by [19] uses temporal-difference learning RL algorithm with linear function approximation (FA) i.e. Tile-Codings (TC). Another state-of-the-art solution by [16] uses DQN (a non-linear FA) based MM agent. There are some more recent work in Multi-Asset Market Making (MAMM) such as [14] and [6], however these belong to the non-empirical approaches and provides non-ML based solutions. [19] and [16] are the two most recent practical RL solutions with realistic LOB simulations, however these solutions focus on Single-Asset Market Making (SAMM). In fact, [6] states that most of the MM solutions, present in the literature, focus on SAMM only. The single asset MM models cannot be used for MM portfolios. There is a need to design and develop new RL algorithms which can understand the complexity of markets and are able to generate profitable MM portfolios. Hence, the objective of this paper is to provide first empirical RL based solution of MAMM.

The assets traded in the market have some common and some specific characteristics. Therefore, these SAMM based RL agents fail to obtain a combined optimal policy for MAMM. In other words, these RL agents are only capable of providing liquidity for a single asset rather than the multiple assets, simultaneously. The MMer usually trades on a collection of assets rather than on a single asset in the real market, e.g. a stock exchange. Therefore, these RL agents have limited applicability in real markets, and the MAMM reduces the risk through the investment diversification across multiple assets rather than single asset. We refer to the method in [19] as SAMM-TC, and to the method in [16] as SAMM-DQN throughout the paper.

Multi-Task Learning (MTL) is a machine learning method, and known for improving generalization of machine learning (ML) models. The ML models are trained on multiple datasets representing different tasks in the RL domain. With the advancement in deep learning (DL), the automated feature extraction capability of neural networks (NNs) greatly improves the generalization of RL models in comparison to traditional feature engineering method. Deep RL (DRL) has become quite popular after human-level performance of DRL agents in automated video game playing. Notably, DRL helped Deepmind's researchers in developing the AlphaGo program which has beaten the top-ranked GO players.

This work is motivated by the achievements of DRL, capabilities of MTL and the lack of RL based MAMM methods. The proposed method is the first attempt to fill this gap, and facilitates MAMM which is currently not addressed by any existing RL based MM methods. We design a MAMM RL model inspired by *hard parameter sharing*, a well-known MTL method. The proposed MAMM model constitutes two DRL models, namely the *Shared* model and the *Asset-specific* model(s) connected with each other. From the concepts of MTL, the assets are treated as multiple related tasks with some common and some specific features. The *Shared* model is responsible for automatic extraction of common features when trained on multiple assets, simultaneously. However, the *Asset-specific* model learns the specific characteristics of a particular asset. The designed model can trade on a collection of assets, unlike the existing SAMM based RL models. We use quote driven limit order book (LOB) to simulate the real market as the RL environment. LOBs are widely accepted and used by the majority of the financial markets [13]. The *trade* and *quote* (TAQ) data is collected from an open source data provider[1] of Chicago Board of Options Exchange (CBOE) to populate the LOB.

An Exchange Traded Fund (ETF), is a marketable asset which tracks an index or an asset. ETFs are associated with a price, and they can be traded like regular stocks in the market. ETFs provide diversification among different kinds of assets, and therefore are quite popular among traders these days. We choose three ETFs and two Options for comparison among the proposed framework and the benchmarks. We call our proposed model "MTDRLMM" throughout the paper. [16] argued that non-linear FA in RL such as DQN is better than linear FAs due to the obvious advantages including automated feature extraction and compact representations in high dimensional data. From the inspection of literature, we are aware that DRL based MAMM does not exist. Hence, this paper aims to bridge the gap between MAMM and DRL through MTDRLMM model. The main contributions are: 1) the novel MTDRLMM model based on MTL for MAMM; 2) the first empirical study on RL based MAMM.

The remainder of the paper is structured as follows, Sect. 2 reviews the relevant literature. Section 3 describes the proposed method, and Sect. 4 focusses on experiments and results. Finally, Sect. 5 concludes the paper.

2 Related Work

There is a long line of work in ML based MM including RL. The first practical RL model of MM was developed by [10]. They evaluated the impact of uninformed market participants on the behaviour of market makers and argued that MM agents derive RL policies while balancing profit and spread, successfully. [4] solved MM problem through deriving optimal bid and ask quotes for LOB. Moreover, the order arrivals and executions process uses probabilistic knowledge. [11] solved exploration versus exploitation problem as price discovery versus profit earning, and studied the effect of MM on price formation. This work was focussed on price prediction and stability, and does not improve market fluidity (measured by market spread). [1] uses convex hull optimization to design an automated SAMM framework. [2] assumes the market has sufficient liquidity while evaluating their online learning based MM model. [3] designed a high frequency MM model that takes advantage of speed in placing quotes and also argues that speed and profit are positively correlated. Then, [19] developed a state-of-the-art model based on RL which is the first attempt to study this problem on realistic

[1] https://www.cboe.com/us/equities/market_data_services/.

grounds taking into consideration all market phenomena that affect MM. Most recently, [16] proposed a DQN based MM agent which outperforms the traditional RL model of MM [19]. [20] studied the MM problem from the game theory point of view. They designed a robust MM agent using existing Adversarial RL method. [18] showed that DQN based MM proves to be an optimal solution. They maximise the profit obtained from bid-ask spreads and minimise inventory through a reward function. [7] designed a stock trader using an ensemble of multiple DRL models.

In quantitative finance domain, [8] argued that if LOs are forced to stay in LOB for a minimum time then liquidity increases. [5] proposes a simple analytical extension to the MM problem, and the flexibility of existing algorithms used in designing MM strategies increases. [6] proposed a closed form approximation to the existing MAMM solution by [4]. However, this method solves the problem theoretically, and has assumptions of [4]. There is a recent mathematical solution by [14] which solves the MM for a large number of bonds. They design a model-based actor-critic like algorithm to provide numerical approximation of traditional SAMM model by [4]. [22] designed a SAMM strategy using Q learning algorithm, and they claim that the strategy can be implemented using lookup table. They use state aggregation, similar to this work and [19] for state discretization. [17] designed a Multi-Agent stock market simulator which reproduce the market microstructure metrics. The RL agents learn to trade autonomously. Contrary to all, [15] studied multi-asset trading, and used DQN algorithm for long-short strategy. They backtested the strategy on S&P 500 futures contract dataset. They simply trained an RL agent on an index dataset, similar to the SAMM benchmarks (SAMM-TC and SAMM-DQN). However, we aim to exploit the common and individual traits of multiple assets, simultaneously via MTL, to improve the RL policy through return enhancement. Hence, the RL based solution of MAMM is missing from the literature. Our work aims to fill this gap, and provide a road ahead for further research and development towards more complex MAMM RL algorithms for MM portfolios.

3 The Proposed Method

MTL has now become quite popular in ML domain since recent advancements including stock prediction, object detection, and natural language processing. The aim is to jointly learn multiple related tasks instead of learning them separate. These multiple related tasks are considered to have common characteristics which can be exploited to improve the model generalization for each of the individual task [12]. There are many application domains where the amount of training data is small, due to various reasons including data is not freely available e.g. historical TAQ data in stock markets. The idea behind MTL is to combine multiple similar tasks and learn a model. By similar tasks or the tasks of same nature means the objective of the problem remains the same for each task. For instance, if a bot learns the task of walking, then the bot can be further trained to learn the task of running on top of the learned experience of walking. The same

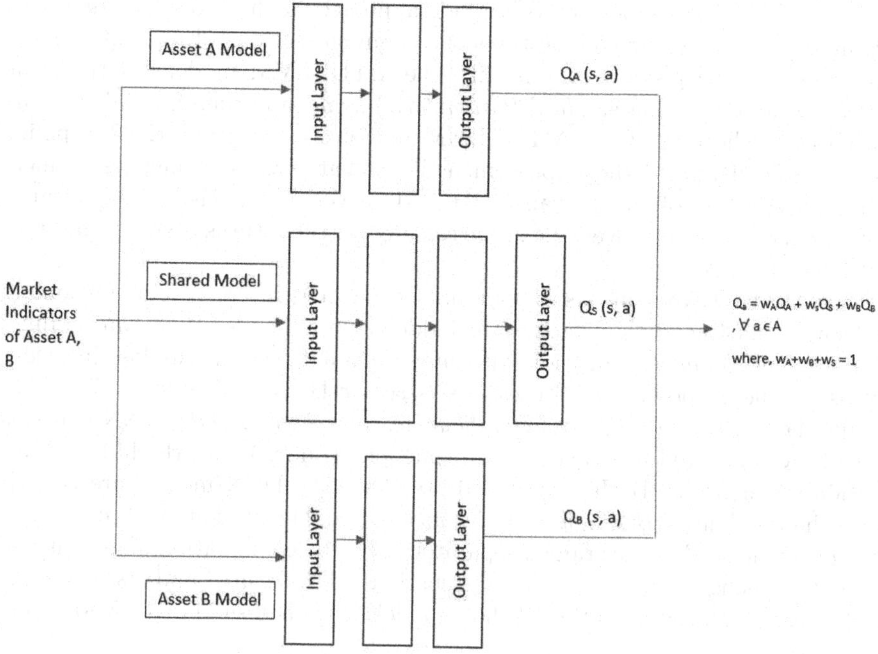

Fig. 1. Schematic diagram of the MTDRLMM

rationale is extended here, a MMer is a trader which has to trade on a collection of assets rather than single asset in actual environment, e.g. a stock exchange. These multiple assets are treated as multiple similar tasks, as the structure of the LOB remains the same. Each asset has its own dynamics which governs the interest of traders and market response. A collection of assets containing assets, spanning multiple sectors and organisations, has some common and individual features or traits.

We propose a DRL solution to the problem of MAMM using MTL, known as MTDRLMM. The proposed model employs two types of trained DNN models, as shown in Fig. 1. The proposed model comprises of one *shared* DNN and multiple *task-specific* DNNs, trained using backpropagation algorithm. Both *shared* and *task-specific* or *Asset-specific* DNNs are fully-connected DNNs. Each asset in the collection has an individual *Asset-specific* NN, trained separately. From Fig. 1, two assets A and B have their respective *Asset-specific* DNNs, the Q value function is the output of each DNN. Q_A represents the Q function for Asset A, whereas Q_B is the Q function for Asset B, and Q_S denotes the Q function of *shared* DNN, trained on both assets A and B.

The proposed model is inspired from the well-known MTL method, namely *Hard Parameter Sharing.* The method is quite popular in the MTL literature and dates back to [9]. In this method, a DNN learns several related tasks by sharing some layers, and each task is associated with an individual layer as well. The major advantage of this MTL approach is the reduction in model overfitting. The DNN model captures the common features of multiple related tasks and tries to produce a good fit. The MTDRLMM model has the same architectural foundation as *Hard Parameter Sharing*, and therefore inherits the advantages. The aim of the MTDRLMM model is to improve the RL policy through MTDRL, and the improvement is evaluated via two metrics, namely returns and total volume of transactions. Moreover, the statistical significance test shows that the improvement in the earnings or the returns is significant with p value equal 0.05.

The *shared* DNN consists of four fully connected layers (Input, Output, 2 Hidden) with 64 neurons in each of the hidden layers. The *reLU* activation function, and the *adam* optimization with *mean squared error* as the loss function, are used. The proposed model uses *Asset-specific* DNN models instead of asset specific layers, like *Hard Parameter Sharing*. The *Asset-specific* DNN contains three fully connected layers (Input, Output, one Hidden), and the hidden layer contains 64 neurons. Both *shared* and *Asset-specific* DNN models are trained using the well-known *Bellman recursive update* (Eq. 1) equation in RL paradigm. Moreover, the models are trained separately, and the combination of output layers is done using weighted mean of Q functions of the *shared* and *Asset-specific* DNN models. The combination of these models is termed as MTDRLMM model based MMer.

$$Q^*(s,a) = E[r + \gamma max Q^*(s',a')], a' \in A \tag{1}$$

In Eq. 1, r is the reward (using Eq. 2 by [19], where ϕ is profit/loss, I is the inventory accumulated, and λ is the dampening factor), γ is the discount factor, and $Q^*(s,a)$ denotes the optimal action-value function with s as state and a as action.

$$r = \phi - \lambda max(I, 0) \tag{2}$$

The fundamental technical indicators used to simulate the environment for MTDRMM agent are described below:

- volatility: the dispersion of returns
- volume imbalance: the ratio of ask and bid volumes
- relative strength index: measures the recent fluctuation in asset price
- market spread: the difference between lowest ask and the highest bid of an asset
- mid-price movement: the change in the midpoint of best ask and best bid prices
- ask distance: distance between the best open order in the ask book and the best ask price
- bid distance: distance between the best open order in the bid book and the best bid price
- inventory: holdings of assets

The input layers receive these market technical indicators computed from the LOB simulation data for the MTDRLMM agent. The output layers outputs the action-values, also known as Q values in RL paradigm, of all the discrete actions.

The lookback window (length is 60) method is used to compute volatility, relative strength index and market spread from LOB TAQ data. The *mean squared error* loss function computes the *mean squared error* between the actual (calculated from *Bellman recursive update* equation) and observed Q values. Then, the input layer weights get updated using backpropagation algorithm. The *shared* model is trained on multiple assets through simultaneous feeding of market environment variables randomly. A *uniform-distribution* picks the training example, also known as a state tuple, for the input to the *shared* model. The trained MTDRLMM model conducts out-of-sample backtesting on historical LOB time-series data. The final output is the weighted mean of the Q values from both *shared* and *Asset-specific* models. The Q_a is weighted Q function (refer Eq. 3), of n number of assets, which computes the action-values or Q values of each discrete action in the action space. The action space consists of nine discrete actions, directly taken from [19]. The action computes the ask and the bid quotes. These quotes are then placed in the LOB on both sides (ask and bid). Each action has some quoted-spread (the difference between the ask and the bid quote), and the quoted-spread multiplied by the volume yields the net profit or net earning. The Q function of ith *Asset-specific* DNN is associated with the corresponding weight w_i. Q_S represents the Q function of *shared* DNN with the corresponding weight w_{n+1}

$$\begin{aligned} Q_a &= w_1 Q_1 + w_2 Q_2 + w_3 Q_3 + \cdots + w_n Q_n + w_{n+1} Q_S, \\ \sum_{i=1}^{n+1} w_i &= 1 \end{aligned} \tag{3}$$

The Q_a denotes the output Q function which generates the Q values of each discrete action. In this empirical study, the unity is uniformly distributed among the weights shown in Eq. 3. In simple words, each constituent model (*Asset-specific* and *shared*) of MTDRMM is allocated equal weight. However, the weights can be skewed to incorporate the weightage of a particular asset in the basket or portfolio. In simple words, one asset can be given more importance over others by skewing weights in a portfolio.

4 Experiments

4.1 Data and Settings

We collect level 2 order book Trade and Quote (TAQ) historical *Intraday* tick-by-tick data, with the frequency of approximately five seconds, from CBOE. The training datasets contains 6500 training examples, whereas test datasets comprises of 1600 test examples. Each example constitutes a LOB snapshot and an executed ask and bid order associated with price and volume. Investors use ETFs to construct a well diversified portfolio. They are designed to track indices rather than individual assets. Another benefit of ETFs is they provide risk management to the investors by allowing them to trade in futures and options.

Table 1. MTDRLMM hyperparameters

Serial no.	Hyperparameter	Value	Range
1	Dimension of state space	8	
2	Dimension of action space	9	
3	RL algorithm	Sarsa	
4	RL policy	ϵ-greedy	
5	Training dataset (split %)	80	
6	Validation dataset (split %)	20	
7	Test dataset (split %)	20	
8	Training episodes	20	
9	Testing epsiodes	20	
10	Tilings in tile-codings	8	
11	Inventory range	[−100, 100]	
12	Order size	10	
13	Discount factor (γ)	0.95	0–1
14	Exploration rate (ϵ)	0.5	0–1
15	Learning rate (α)	0.001	0–0.1
16	Activation function	Linear	
17	Loss function	Mean squared error	
18	Optimizing function	Adam	

We prepare a basket of five randomly picked assets (from top 100 listed on Yahoo finance), namely SPY, DIA, XLF, TXN and UPS, for the empirical analysis. The scraped data requires preprocessing such as eliminating redundancy and NULLs. Data cleaning was done via python scripts[2]). The github link to the code[3] and data[4] is provided to ensure replicability. The hyperparameters of MTDRLMM are specified in Table 1.

4.2 Results and Discussion

Empirically, we compare the two versions of MM, namely SAMM and MAMM, on both linear and non-linear FA RL methods. Both SAMM-TC and MAMM-TC use the Tile-Codings FA to train the RL model of MM. In MAMM-TC, a function approximator is trained on all five assets collectively. In MAMM-DQN, *shared* and *Asset-specific* DNNs are trained collectively and individually, respectively. Whereas, SAMM-TC and SAMM-DQN train a function approximator for every asset. These two versions of MM, in combination with two types of FAs, are then backtested and their returns are compared, as shown in Table 2.

[2] https://github.com/Haider93/mm_data/blob/main/preprocess_data.py.
[3] https://github.com/Haider93/MTDRLMM.
[4] https://github.com/Haider93/MTDRLMM/data.

Table 2. Out-of-Sample backtesting averaged PnL obtained over twenty episodes

Asset	SAMM-TC ($)	MAMM-TC ($)	SAMM-DQN ($)	MAMM-DQN ($)	SAMM-AS ($)
SPY	40.9792	46.7382	8.5623	39.7740	−3.6079
DIA	161.3019	160.1599	68.5310	70.1787	−4.4794
XLF	−8.9907	−8.7499	−0.1800	3.7599	−4.5947
UPS	−28.7761	−6.8000	−3.6998	−3.6998	−4.5583
TXN	−18.8399	−15.5601	−13.8057	−12.4257	−4.7656

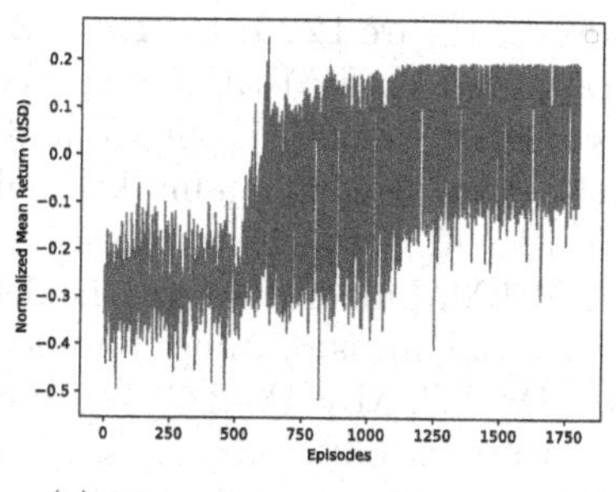

(a) Mean Normalized Return ($)

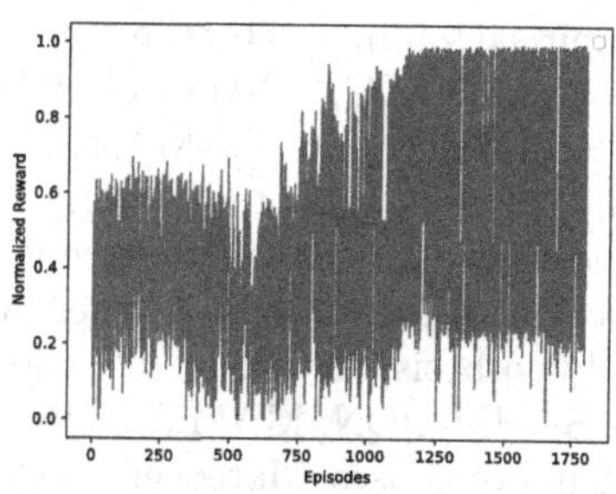

(b) Normalized Reward

Fig. 2. Training curves of the MTDRLMM model.

These two FA methods generate different returns whether used in SAMM or MAMM, and neither of them outperform the other completely in all five assets. This could be attributed to the fact that the trading strategies differ in their returns. The MAMM-TC obtains significantly higher returns than SAMM-TC in four out of five assets. Moreover, the same pattern is observed in DQN FA, where SAMM-DQN yields lower return than MAMM-DQN, except UPS. The MAMM-DQN represents the proposed MTDRLMM model. In the basket of these five assets, SAMM-TC obtains profit in SPY and DIA ETFs and incurs loss in others. MAMM-TC, if treated as a MM strategy, is profitable in both SPY and DIA, while the other three saw negative returns. In the case of DQN FA, SAMM-DQN yields profit in SPY and DIA only, however MAMM-DQN is able to make profit in SPY, DIA and XLF. Moreover, the returns of MAMM-DQN in all five assets are higher than SAMM-TC, MAMM-TC and SAMM-DQN, except in DIA. The MAMM reduces the risk associated with profit through appropriate diversification of the capital investment into a collection of assets. Moreover, the return over the investment is also observed higher, except in DIA while using TC. This could be attributed to the fact that stock markets are quite volatile, and a single asset can yield higher return than an improper and naive diversified collection of assets, due to the higher market liquidity. In general MAMM proves to be more profitable and less risky than SAMM in both FA methods. Upon observation across assets in four columns of Table 2, MAMM-DQN return profit in three out of five assets higher than others. Hence, the MTDRLMM model proves to be more profit yielding and less volatile, empirically. The training curves, shown in Fig. 2, represent the mean normalized PnL value (refer Fig. 2(a)) and the normalized reward (refer Fig. 2(a)) observed when training the MTDRLMM model. The graphs clearly shows that the MTDRLMM model based RL agent successfully maximizes the PnL and the accumulated reward throughout the training phase.

We add one more evaluation criterion, namely total number of transactions (Ask limit orders + bid limit orders + market buys + market sells) in the LOB. The greater the number of total transactions the higher will be the added liquidity to the market by the market maker. We now show learning curves (Normalized PnL vs training episodes, and Cumulative Reward vs training episodes). The empirical findings of total number of transactions are **1226, 1081, 1318 and 1566** for SAMM-TC, MAMM-TC, SAMM-DQN and MAMM-DQN, respectively. From the overall results, the MAMM-DQN is found to be more profitable and provided highest amounts of market liquidity (the primary goal of a market maker). We conduct statistical significance t test with equal variances to prove the significance of the returns enhancement via MTDRLMM model. We use the default NULL hypothesis, H0: $\mu 1 = \mu 2$. The t-test one tail statistic values between the groups SAMM-DQN, MAMM-DQN and SAMM-TC, MAMM-TC are same i.e. **1.8595** in both cases. Moreover, the two tail values for both pairs are same again i.e. **2.306**. The tabular t statistic value for degrees of freedom 8 one-tail is **1.86** and two-tail is **2.306**, with default significance level $\alpha = 0.05$. The p value for both groups is equal to **0.05**, hence the we reject the NULL hypothesis. There is a significant difference between means, hence the empirical results are statistically significant.

5 Conclusion

The MM is a well-known fundamental trading problem which obligates the MMer to continuously buy and sell assets during market operational hours. RL has been quite popular in solving MM problem. The existing RL solutions focus on single asset modelling of the problem. However, the applicability of SAMM RL models in real scenarios is practically limited. From the concepts and advantages of MTL, multiple assets are treated as multiple related tasks, and therefore we propose a MAMM RL model based on MTL. The proposed MTDRLMM model proves to obtain significantly higher percentage return in a basket of assets. In the empirical study, the SAMM RL benchmarks and MTDRLMM model are trained first and then backtested on historical LOB TAQ data. The MTDRLMM model is shown to be a practical MAMM method, and provides a more profitable MM RL policy than SAMM RL benchmarks. Moreover, the return improvement is found to be statistically significant via statistical t test, with default significance level of 0.05 and the p value of 0.05 is obtained. The research findings presented opens new doors to the further research and development of more flexible and complex RL algorithms for large MM portfolios.

References

1. Abernethy, J., Chen, Y., Vaughan, J.W.: An optimization-based framework for automated market-making. In: EC'11: Proceedings of the 12th ACM Conference on Electronic Commerce, pp. 297–306 (2011)
2. Abernethy, J., Kale, S.: Adaptive market making via online learning. In: Proceedings of the 26th International Conference on Neural Information Processing Systems (NIPS'13), vol. 2, pp. 2058–2066. Curran Associates Inc. (2013)
3. Ait-Sahalia, Y., Saglam, M.: High frequency market making: optimal quoting (2017). Available at SSRN: https://ssrn.com/abstract=2331613 or https://doi.org/10.2139/ssrn.2331613
4. Avellaneda, M., Stoikov, S.: High-frequency trading in a limit order book. Quant. Finance **8**(3), 217–224 (2008)
5. Baldacci, B., Derchu, J., Manziuk, I.: An approximate solution for options market-making in high dimension (2020). https://arxiv.org/pdf/2009.00907.pdf
6. Bergault, P., Evangelista, D., Gueant, O., Vieira, D.: Closed-form approximations in multi-asset market making, September 2020. https://arxiv.org/pdf/1810.04383.pdf
7. Carta, S., Corriga, A., Ferreira, A., Podda, A.S., Recupero, D.R.: A multi-layer and multi-ensemble stock trader using deep learning and deep reinforcement learning. Appl. Intell. **51**(2), 889–905 (2020). https://doi.org/10.1007/s10489-020-01839-5
8. Cartea, A., Wang, Y.: Market making with minimum resting times. Quant. Finance **19**, 903–920 (2019)
9. Caruana, R.: Multitask learning: a knowledge-based source of inductive bias. In: Proceedings of the Tenth International Conference on Machine Learning (1993)
10. Chan, N.T., Shelton, C.: An electronic market-maker. Technical report. MIT (2001)
11. Das, S.: The effects of market-making on price dynamics. In: Proceedings of the 7th International Joint Conference on Autonomous Agents and Multiagent Systems (AAMAS'08), vol. 2, pp. 887–894. International Foundation for Autonomous Agents and Multiagent Systems (2008)
12. Eramo, C.D., Tateo, D., Bonarini, A., Restelli, M., Peters, J.: Sharing knowledge in multi-task deep reinforcement learning. In: International Conference on Learning Representations (ICLR) (2020)
13. Gould, M.D., Porter, M.A., Williams, S., McDonald, M., Fenn, D.J., Howison, S.D.: Limit order books. Quant. Finance **13**(11), 1709–1742 (2013)
14. Guéant, O., Manziuk, I.: Deep reinforcement learning for market making in corporate bonds: beating the curse of dimensionality. Appl. Math. Finance **26**, 387–452 (2020)
15. Hirsa, A., Osterrieder, J., Hadji-Misheva, B., Posth, J.A.: Deep reinforcement learning on a multi-asset environment for trading. arXiv:2106.08437, June 2021
16. Kumar, P.: Deep reinforcement learning for market making. In: Proceedings of the 19th International Conference on Autonomous Agents and MultiAgent Systems, pp. 1892–1894 (2020)
17. Lussange, J., Lazarevich, I., Bourgeois-Gironde, S., Palminteri, S., Gutkin, B.: Modelling stock markets by multi-agent reinforcement learning. Comput. Econ. **57**(1), 113–147 (2020). https://doi.org/10.1007/s10614-020-10038-w
18. Selser, M., Kreiner, J., Maurette, M.: Optimal market making by reinforcement learning (2021)

19. Spooner, T., Fearnley, J., Savani, R., Koukorinis, A.: Market making via reinforcement learning. In: Proceedings of the 17th International Conference on Autonomous Agents and MultiAgent Systems (AAMAS'18), pp. 434–442. International Foundation for Autonomous Agents and Multiagent Systems (2018)
20. Spooner, T., Savani, R.: Robust market making via adversarial reinforcement learning. In: Proceedings of the Twenty-Ninth International Joint Conference on Artificial Intelligence (IJCAI-20) Special Track on AI in FinTech (2020)
21. Sutton, R., Barto, A.: Reinforcement Learning: An Introduction, 2nd edn. MIT Press, Cambridge (2018)
22. Zhong, Y., Bergstrom, Y., Ward, A.: Data driven market making via model free learning. In: Proceedings of the Twenty-Ninth International Joint Conference on Artificial Intelligence Special Track on AI in FinTech, pp. 4461–4468 (2020)

Evaluating Hebbian Learning in a Semi-supervised Setting

Gabriele Lagani[1(✉)], Fabrizio Falchi[2], Claudio Gennaro[2], and Giuseppe Amato[2]

[1] Department of Computer Science, University of Pisa, 56127 Pisa, Italy
gabriele.lagani@phd.unipi.it
[2] ISTI-CNR, 56124 Pisa, Italy
{fabrizio.falchi,claudio.gennaro,giuseppe.amato}@cnr.it

Abstract. We propose a semi-supervised learning strategy for deep Convolutional Neural Networks (CNNs) in which an unsupervised pre-training stage, performed using biologically inspired Hebbian learning algorithms, is followed by supervised end-to-end backprop fine-tuning. We explored two Hebbian learning rules for the unsupervised pre-training stage: soft-Winner-Takes-All (soft-WTA) and nonlinear Hebbian Principal Component Analysis (HPCA). Our approach was applied in sample efficiency scenarios, where the amount of available labeled training samples is very limited, and unsupervised pre-training is therefore beneficial. We performed experiments on CIFAR10, CIFAR100, and Tiny ImageNet datasets. Our results show that Hebbian outperforms Variational Auto-Encoder (VAE) pre-training in almost all the cases, with HPCA generally performing better than soft-WTA.

Keywords: Hebbian learning · Deep learning · Semi-supervised · Sample efficiency · Neural networks · Bio-inspired

1 Introduction

While deep learning has achieved outstanding results in a variety of domains, ranging from computer vision [15] to language processing [10], and reinforcement learning [41], learning algorithms are typically based on supervised end-to-end Stochastic Gradient Descent (SGD) training with error backpropagation (*backprop*), which needs a large number of labeled training samples in order to achieve high results. However, gathering labeled samples is expensive, as it requires a significant amount of human work. On the other hand, gathering unlabeled samples is relatively simple. Therefore, researchers started to investigate learning strategies to exploit large amounts of unlabeled data, in addition to the fewer labeled data, for sample efficient learning [5–7,9,18,21,27,37,40,45,47]. This led to the *semi-supervised* learning approach, in which an unsupervised pre-training stage

This work was partially supported by the H2020 projects AI4EU (GA 825619) and AI4Media (GA 951911).

G. Nicosia et al. (Eds.): LOD 2021, LNCS 13164, pp. 365–379, 2022.
https://doi.org/10.1007/978-3-030-95470-3_28

is performed on all the available samples (but without using label information), and then it is followed by a supervised fine-tuning stage on the few labeled samples only (in this case, the label information is used for supervision).

Note that backprop is not considered to be biologically plausible from the neuroscientific community [34]. On the other hand, a biologically motivated learning principle is represented by the Hebbian learning paradigm [12,14]. This approach does not require supervision, nor backpropagation. Since biological brains appear to be able to generalize from few samples, research on Hebbian learning algorithms seems a promising direction.

In this work, we propose a semi-supervised learning approach, in which the unsupervised pre-training step is performed by means of the Hebbian learning paradigm. Two Hebbian learning variants are considered: soft-Winner-Takes-All (soft-WTA) [31], and nonlinear Hebbian Principal Component Analysis (HPCA) [19]. We test our approach in sample efficiency scenarios, performing experiments on CIFAR10, CIFAR100 [23], and Tiny ImageNet [46] datasets. Different regimes of sample efficiency are considered, comparing the results with another popular unsupervised pre-training method, namely the Variational Auto-Encoder (VAE) [20]. The results show that our approach outperforms VAE pre-training in almost all the cases, especially when the number of labeled samples available for the successive supervised fine-tuning stage is low. Moreover, HPCA generally performs better than soft-WTA.

Integration of Hebbian learning and deep learning is still an emerging topic. However, our results are encouraging, motivating further interest in this direction.

The main contributions of this paper are the following:

- For the first time, Hebbian learning approaches are applied in a semi-supervised scenario, in which an unsupervised pre-training stage, based on Hebbian approach, is followed by a supervised end-to-end fine-tuning stage based on SGD and backprop;
- We provide extensive experimental evaluation of the approaches, from a sample efficiency perspective, on different object recognition datasets;

The remainder of this paper is structured as follows: Sect. 2 gives an overview on related work concerning semi-supervised training and Hebbian learning; Sect. 3 introduces the various Hebbian learning strategies that we explored; Sect. 4 illustrates the sample efficiency problem and defines our semi-supervised approach based on Hebbian learning; Sect. 5 delves into the details of our experimental setup; In Sect. 6, the results of our simulations are illustrated; Finally, Sect. 7 presents our conclusions and outlines possible future developments.

2 Related Work

In this section, we present an overview of related work concerning both semi-supervised training and Hebbian learning.

2.1 Semi-supervised Training and Sample Efficiency

Early work on deep learning had to face problems related to convergence to poor local minima during the training process. This led researchers to exploit a pre-training phase that allowed them to initialize network weights in a region near a good local optimum [5,27]. In these studies, greedy layerwise pre-training was performed by applying unsupervised autoencoder models, layer by layer . It was shown that such pre-training was indeed helpful to obtain a good initialization for a successive supervised training stage.

In successive works, the idea of enhancing neural network training with an unsupervised learning objective was considered [21,37,45,47]. In [21], Variational Auto-Encoders (VAE) were considered, in order to perform an unsupervised pre-training phase using a limited amount of labeled samples. Also [37] and [47] relied on autoencoding architectures to augment supervised training with unsupervised reconstruction objectives, showing that joint optimization of supervised and unsupervised losses helped to regularize the learning process. In [44], joint supervised and unsupervised training was again considered, but the unsupervised learning part was based on manifold learning techniques.

Another approach, SimCLR [9], used a Contrastive Loss to perform the unsupervised learning part. The approach relied on data augmentation, in order to produce transformed variants of a given input. The unsupervised loss basically encouraged hidden representations to match for transformed variants generated from the same input.

In this paper, we focus our comparisons on similar methods based on unsupervised pre-training, using VAE pre-training as baseline. Nonetheless, it is also worth mentioning that different approaches to semi-supervised learning were also proposed. For example, in [18,40], graph-based methods were used to generate *pseudo-labels* for unlabeled samples, which were then used as target during training. In [6,7,40], the *mixup* approach was also used: convex combinations of pairs of input samples were generated, and a consistency criterion was imposed that pushed the prediction for the combination to match the corresponding combination of predictions. It should be noticed that our method is not in contrast with these other approaches, but rather they can be integrated together, as also suggested in Sect. 7.

2.2 Hebbian Learning

Several variants of Hebbian learning rules were developed over the years. Some examples are: Hebbian learning with Winner-Takes-All (WTA) competition [13], Hebbian learning for Principal Component Analysis (PCA) [4,14,19,39], Hebbian/anti-Hebbian learning [35,36]. A brief overview is given in Sect. 3. However, it was only recently that Hebbian learning started gaining attention in the context of DNN training [2,3,25,26,42,43].

In [25], a Hebbian learning rule based on inhibitory competition was used to train a neural network composed of fully connected layers. The approach was validated on object recognition tasks. Instead, the Hebbian/anti-Hebbian learning rule developed in [36] was applied in [3] to train convolutional feature extractors. The resulting features were shown to be effective for classification. Convolutional layers were also considered in [42,43], where a Hebbian approach based on WTA competition was employed in this case.

However, the previous approaches were based on relatively shallow network architectures (2–3 layers). A further step was taken in [2,26], where a Hebbian WTA learning rule was considered. The learning rule was applied for training a 6-layer Convolutional Neural Network (CNN). The results suggested that Hebbian learning is suitable for training early feature detectors, as well as higher network layers, but not very effective for training intermediate network layers. Furthermore, Hebbian learning was successfully used to retrain the higher layers of a pre-trained network, achieving results comparable to backprop. The advantage was that Hebbian learning required fewer training epochs, thus suggesting potential applications in the context of transfer learning (see also [8,28,29]).

The novelty of our contribution w.r.t. previous work is that, for the first time, we investigate unsupervised Hebbian learning in combination with supervised backprop training, in a semi-supervised fashion. In addition, extensive experimental evaluation is performed.

3 Hebbian Learning Strategies

Consider a single neuron with weight vector $\mathbf{w}$ and input $\mathbf{x}$. Call $y = \mathbf{w}^T \mathbf{x}$ the neuron output. A learning rule defines a weight update as follows:

$$\mathbf{w}_{new} = \mathbf{w}_{old} + \Delta\mathbf{w} \tag{1}$$

where $\mathbf{w}_{new}$ is the updated weight vector, $\mathbf{w}_{old}$ is the old weight vector, and $\Delta\mathbf{w}$ is the weight update.

The Hebbian learning rule, in its simplest form, can be expressed as $\Delta\mathbf{w} = \eta\, y\, \mathbf{x}$ (where η is the learning rate) [12,14]. Basically, this rule states that the weight on a given synapse is reinforced when the input on that synapse and the output of the neuron are simultaneously high. Therefore, connections between neurons whose activations are correlated are reinforced. In order to prevent weights from growing unbounded, a weight decay term is generally added. In the context of competitive learning [13], this is obtained as follows:

$$\Delta\mathbf{w}_\mathrm{i} = \eta\, y_i\, \mathbf{x} - \eta\, y_i\, \mathbf{w}_\mathrm{i} = \eta\, y_i\, (\mathbf{x} - \mathbf{w}_\mathrm{i}) \tag{2}$$

where the subscript i refers to the i'th neuron in a given network layer. Moreover, the output y_i can be replaced with the result r_i of a competitive nonlinearity, which allows to decorrelate the activity of different neurons. In the Winner-Takes-All (WTA) approach [13], at each training step, the neuron which produces the strongest activation for a given input is called the *winner*. In this case, $r_i = 1$ if the i'th neuron is the winner and 0 otherwise. In other words, only the winner is allowed to perform the weight update, so that it will be more likely for the same neuron to win again if a similar input is presented again in the future.

In this way different neurons are induced to specialize on different patterns. In soft-WTA [31], r_i is computed as $r_i = \frac{y_i}{\sum_j y_j}$. We found this formulation to work poorly in practice, because there is no tunable parameter to cope with the variance of activations. For this reason, we introduce a variant of this approach that uses a *softmax* operation in order to compute r_i:

$$r_i = \frac{e^{y_i/T}}{\sum_j e^{y_j/T}} \tag{3}$$

where T is called the *temperature* hyperparameter (the name comes from statistical mechanics, where this function was first introduced) [11]. The advantage of this formulation is that we can tune the temperature in order to obtain the best performance on a given task, depending on the distribution of the activations.

The Hebbian Principal Component Analysis (HPCA) learning rule, in the case of nonlinear neurons, is obtained by minimizing the so-called *representation error* [4,14,39]:

$$L(\mathbf{w_i}) = E[(x - \sum_{j=1}^{i} f(y_j)\, \mathbf{w_j})^2] \tag{4}$$

where $f()$ is the neuron activation function. Minimization of this objective leads to the nonlinear HPCA rule [19]:

$$\Delta \mathbf{w_i} = \eta f(y_i)(x - \sum_{j=1}^{i} f(y_j)\mathbf{w_j}) \tag{5}$$

It can be noticed that these learning rules do not require supervision, and they are *local* for each network layer, i.e. they do not require backpropagation. In the next section, we discuss how Hebbian learning is integrated with backprop in a semi-supervised training approach.

4 Sample Efficiency Scenario and Semi-supervised Approach Based on Hebbian Learning

Let's define the *labeled set* $\mathcal{T}_L$ as a collection of elements for which the corresponding label is known. Conversely, the *unlabeled set* $\mathcal{T}_U$ is a collection of elements whose labels are unknown. The whole *training set* $\mathcal{T}$ is given by the union of $\mathcal{T}_L$ and $\mathcal{T}_U$. All the samples from $\mathcal{T}$ are assumed to be drawn from the same statistical distribution. In a *sample efficiency* scenario, the number of samples in $\mathcal{T}_L$ is typically much smaller than the total number of samples in $\mathcal{T}$. In particular, an s%-sample efficiency *regime* is characterized by $|\mathcal{T}_L| = \frac{s}{100}|\mathcal{T}|$ (where $|\cdot|$ denotes the cardinality of a set, i.e. the number of elements inside the set). In other words, the size of the labeled set is s % that of the whole training set (labeled plus unlabeled).

Traditional supervised approaches based on SGD and backprop work well provided that the size of the labeled set is sufficiently large, but they do not exploit the unlabeled set. To tackle this limitation, we consider a semi-supervised approach in two phases. During the first phase, latent representations are obtained from hidden layers of a DCNN, which are trained using unsupervised Hebbian learning. Such approach, inspired by biology, has the advantage of being able to learn representations without requiring label information, nor backpropagation. This unsupervised pre-training is performed on all the available training samples, unlabeled and labeled (but without using label information in the latter case). During the second phase, a final linear classifier is placed on top of the features extracted from deep network layers. Classifier and deep layers are fine-tuned in a supervised training fashion, by running an end-to-end SGD optimization procedure using only the few labeled samples at our disposal (with the corresponding labels).

5 Experimental Setup

In the following, we describe the details of our experiments and comparisons, discussing the network architecture and the training procedure[1].

5.1 Datasets Used for the Experiments

The experiments were performed on the following datasets: CIFAR10, CIFAR100 [23] and TinyImageNet [46].

The CIFAR10 dataset contains 50,000 training images and 10,000 test images, belonging to 10 classes. Moreover, the training images were randomly split into a training set of 40,000 images and a validation set of 10,000 images.

The CIFAR100 dataset also contains 50,000 training images and 10,000 test images, belonging to 100 classes. Also in this case, the training images were randomly split into a training set of 40,000 images and a validation set of 10,000 images.

The TinyImageNet dataset contains 100,000 training images and 10,000 test images, belonging to 200 classes. Moreover, the training images were randomly split into a training set of 90,000 images and a validation set of 10,000 images.

We considered sample efficiency regimes in which the amount of labeled samples was respectively 1%, 2%, 3%, 4%, 5%, 10%, 25% and 100% of the whole training set.

5.2 Network Architecture and Training

We considered a six layer neural network as shown in Fig. 1: five deep layers plus a final linear classifier. The various layers were interleaved with other processing stages (such as ReLU nonlinearities, max-pooling, etc.). The architecture was inspired by AlexNet [24], but with slight modifications in order to reduce the

[1] The code to reproduce the experiments described in this paper is available at: https://github.com/GabrieleLagani/HebbianPCA/tree/hebbpca.

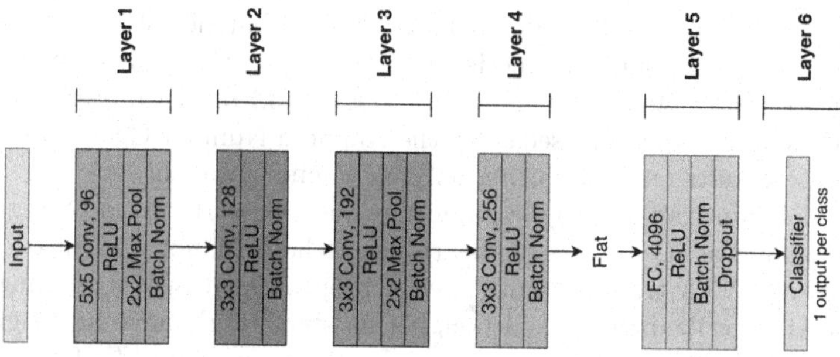

Fig. 1. The neural network used for the experiments.

overall computational cost of training. We decided to adopt a simple network model in order to be able to evaluate the effects of different learning approaches on a layer-by-layer basis. This choice also makes it more practical for other researchers to reproduce the experiments.

For each sample efficiency regime, we trained the network with our semi-supervised approach. First, we used the soft-WTA and the HPCA unsupervised pre-training in the internal layers. This was followed by the fine tuning stage with SGD training, involving the final classifier as well as the previous layers, in an end-to-end fashion.

For each sample efficiency configuration we also created a baseline for comparison. In this case, we used another popular unsupervised method, namely the Variational Auto-Encoder (VAE) [20], for the unsupervised pre-training stage. This was again followed by the supervised end-to-end fine tuning based on SGD. VAE-based semi-supervised learning was also the approach considered in [21].

5.3 Testing Sample Efficiency at Different Layer Depths

In our experiments, in addition to evaluating the entire network trained as discussed above, we also evaluated the sample efficiency capability on the various internal layers of the trained models. To this end, we cut the networks in correspondence of the output of the various layers and we trained a new linear classifier on top of each already pre-trained layer. For each configuration, the supervised SGD training stage was performed using the labeled samples, thus fine tuning the classifier, as well as the previous network layers. Then, the resulting accuracy was evaluated. This process was done both for the Hebbian trained networks, and the VAE trained network, used as baseline, in order to make comparisons.

5.4 Details of Training

We implemented our experiments using PyTorch. All the hyperparameters mentioned below resulted from a parameter search aimed at maximizing the validation accuracy on the respective datasets, following the Coordinate Descent (CD) approach [22].

Training was performed in 20 epochs using mini-batches of size 64. No more epochs were necessary, since the models had already reached convergence at that point. Networks were fed input images of size 32×32 pixels. Experiments were performed using five different seeds for the Random Number Generator (RNG), averaging the results and computing 95% confidence intervals.

In the Hebbian training, the learning rate was set to 10^{-3}. No L2 regularization or dropout was used, since the learning method did not present overfitting issues. For soft-WTA training, images were preprocessed by a whitening transformation as described in [23], although this step didn't have any significant effect for other training methods. The *temperature* parameter T of the softmax operation used in soft-WTA was set to $T = 0.02$.

For VAE training, the network in Fig. 1, up to layer 5, acted as encoder, with an extra layer mapping layer 5 output to 256 gaussian latent variables, while a specular network branch acted as decoder. VAE training was performed without supervision, in an end-to-end encoding-decoding task, optimizing the β-VAE Variational Lower Bound [16], with coefficient $\beta = 0.5$.

For the supervised training stage, based on SGD, the initial learning rate was set to 10^{-3} and kept constant for the first ten epochs, while it was halved every two epochs for the remaining ten epochs. We also used momentum coefficient 0.9, Nesterov correction, dropout rate 0.5 and L2 weight decay penalty coefficient set to $5 \cdot 10^{-2}$ for CIFAR10, 10^{-2} for CIFAR100 and $5 \cdot 10^{-3}$ for TinyImageNet. Cross-entropy loss was used as optimization metric.

To obtain the best possible generalization, *early stopping* was used in each training session, i.e. we chose as final trained model the state of the network at the epoch when the highest validation accuracy was recorded.

6 Results and Discussion

In this section, the experimental results obtained with each dataset are presented and analyzed. We report the classification accuracy, along with the 95% confidence intervals, in the various sample efficiency regimes, for the CIFAR10, CIFAR100 and Tiny ImageNet datasets.

6.1 CIFAR10

Table 1 reports the top-1 accuracy results obtained on the CIFAR10 dataset. We only report top-1 accuracy, given that CIFAR10 contains only 10 classes.

As we can observe, Hebbian approaches perform better than VAE in almost all the cases. In particular, when low sample efficiency regimes are considered (between 1% and 5%) Hebbian approaches achieve significantly higher results than VAE. Only when the number of available labeled samples increases (beyond 10%), VAE pre-training starts to become competitive, obtaining results comparable to Hebbian training. Overall, Hebbian pre-training appears to be more effective than VAE, in particular when the number of available labeled samples is relatively low (5% or less). The maximum improvement of Hebbian approaches over VAE is achieved in the 5% sample efficiency regime, in correspondence of

Table 1. CIFAR10 accuracy (top-1) and 95% confidence intervals, obtained with a linear classifier on top of various layers, for the various sample efficiency regimes. Results obtained with VAE and Hebbian pre-training are compared.

Regime	Pre-Train	L1	L2	L3	L4	L5
1%	VAE	33.54 ±0.27	34.41 ±0.84	29.92 ±1.25	24.91 ±0.66	22.54 ±0.60
	soft-WTA	35.47 ±0.19	35.75 ±0.65	36.09 ±0.27	30.57 ±0.36	30.23 ±0.37
	HPCA	**37.01** ±0.42	**37.65** ±0.19	**41.88** ±0.53	**40.06** ±0.65	**39.75** ±0.50
2%	VAE	37.65 ±0.35	39.13 ±0.40	36.52 ±0.47	29.39 ±0.32	26.78 ±0.72
	soft-WTA	41.05 ±0.39	42.09 ±0.34	43.48 ±0.36	37.85 ±0.28	36.59 ±0.23
	HPCA	**41.60** ±0.28	**42.12** ±0.24	**46.56** ±0.38	**45.61** ±0.19	**45.51** ±0.43
3%	VAE	41.22 ±0.27	43.16 ±0.44	42.60 ±0.87	31.91 ±0.44	29.00 ±0.33
	soft-WTA	44.67 ±0.37	**46.12** ±0.27	48.08 ±0.42	43.22 ±0.31	41.54 ±0.50
	HPCA	**44.74** ±0.08	45.61 ±0.28	**49.75** ±0.41	**48.94** ±0.45	**48.80** ±0.27
4%	VAE	44.39 ±0.30	45.88 ±0.39	46.01 ±0.40	34.26 ±0.21	31.15 ±0.35
	soft-WTA	46.77 ±0.36	**49.24** ±0.40	51.23 ±0.37	46.90 ±0.27	45.31 ±0.18
	HPCA	**47.10** ±0.25	48.26 ±0.09	**52.00** ±0.16	**51.05** ±0.29	**51.28** ±0.28
5%	VAE	46.31 ±0.39	48.21 ±0.21	48.98 ±0.34	36.32 ±0.35	32.75 ±0.32
	soft-WTA	48.34 ±0.27	**52.90** ±0.28	**54.01** ±0.24	49.80 ±0.16	48.35 ±0.26
	HPCA	**48.49** ±0.44	50.14 ±0.46	53.33 ±0.52	**52.49** ±0.16	**52.20** ±0.37
10%	VAE	53.83 ±0.26	56.33 ±0.22	57.85 ±0.22	52.26 ±1.08	45.67 ±1.15
	soft-WTA	54.23 ±0.18	**59.40** ±0.20	**61.27** ±0.24	**58.33** ±0.35	**58.00** ±0.26
	HPCA	**54.36** ±0.32	56.08 ±0.28	58.46 ±0.15	56.54 ±0.23	57.35 ±0.18
25%	VAE	**62.51** ±0.24	67.26 ±0.32	68.48 ±0.21	68.79 ±0.29	68.70 ±0.15
	soft-WTA	61.29 ±0.23	**68.23** ±0.31	**70.09** ±0.41	**70.01** ±0.17	**69.85** ±0.37
	HPCA	61.45 ±0.26	65.25 ±0.16	64.71 ±0.17	62.43 ±0.13	64.77 ±0.22
100%	VAE	**67.53** ±0.22	75.83 ±0.31	80.78 ±0.28	84.27 ±0.35	85.23 ±0.26
	soft-WTA	67.37 ±0.16	**77.39** ±0.04	**81.83** ±0.47	**84.42** ±0.15	**85.37** ±0.03
	HPCA	66.76 ±0.13	75.16 ±0.20	79.90 ±0.18	83.55 ±0.33	84.38 ±0.22

network layer 5, where a gap of almost 16% points is observed between VAE and soft-WTA, and a gap of almost 20% points is observed between VAE and HPCA. Moreover, in low sample efficiency regimes (10% or less) it is possible to notice that VAE and soft-WTA approaches suffer from a decrease in performance when going deeper with the number of layers. This issue is common with unsupervised methods, because the absence of a supervision signal (or still its scarcity, in case of semi-supervised training) makes it harder to develop task-specific features on higher layers, which is essential to achieve higher performances, as it emerges from previous studies on deep CNNs [1]. With HPCA, this problem seems to alleviate, and the accuracy remains pretty much constant with the number of layers, again in the low sample efficiency regimes (10% or less), meaning that the features produced by this approach are more meaningful for the classification task.

Table 2. CIFAR100 accuracy (top-5) and 95% confidence intervals obtained with a linear classifier on top of various layers, for the various sample efficiency regimes. Results obtained with VAE and Hebbian pre-training are compared.

Regime	Pre-Train	L1	L2	L3	L4	L5
1%	VAE	21.69 ±0.10	21.70 ±0.30	17.61 ±0.54	13.45 ±0.54	12.28 ±0.50
	soft-WTA	19.60 ±0.23	20.08 ±0.32	18.50 ±0.44	15.21 ±0.36	15.30 ±0.28
	HPCA	**22.30** ±0.38	**22.28** ±0.63	**23.58** ±0.21	**21.70** ±0.61	**22.63** ±0.55
2%	VAE	28.24 ±0.13	**28.42** ±0.31	23.56 ±0.73	17.01 ±0.37	15.25 ±0.63
	soft-WTA	26.73 ±0.34	26.67 ±0.20	25.48 ±0.20	20.22 ±0.32	20.76 ±0.24
	HPCA	**29.65** ±0.52	26.57 ±0.26	**33.20** ±0.20	**30.21** ±0.54	**30.83** ±0.35
3%	VAE	31.28 ±0.54	31.71 ±0.27	27.46 ±1.23	18.26 ±0.24	16.44 ±0.12
	soft-WTA	30.53 ±0.37	30.81 ±0.52	29.99 ±0.44	23.22 ±0.25	23.69 ±0.49
	HPCA	**32.81** ±0.18	**33.08** ±0.55	**37.75** ±0.38	**35.02** ±0.36	**35.04** ±0.17
4%	VAE	34.60 ±0.10	35.44 ±0.31	32.34 ±0.79	19.68 ±0.32	17.89 ±0.27
	soft-WTA	33.51 ±0.26	34.15 ±0.21	32.85 ±0.18	25.78 ±0.21	26.91 ±0.24
	HPCA	**36.13** ±0.39	**36.23** ±0.20	**41.21** ±0.39	**39.16** ±0.32	**38.89** ±0.15
5%	VAE	36.68 ±0.17	37.26 ±0.26	35.33 ±0.81	20.55 ±0.44	18.48 ±0.26
	soft-WTA	35.71 ±0.29	36.83 ±0.37	35.80 ±0.18	28.39 ±0.43	29.57 ±0.13
	HPCA	**38.03** ±0.20	**38.02** ±0.25	**43.76** ±0.33	**41.66** ±0.20	**41.42** ±0.23
10%	VAE	42.64 ±0.34	44.84 ±0.48	46.04 ±0.44	27.81 ±0.13	23.80 ±0.60
	soft-WTA	41.91 ±0.27	**45.61** ±0.29	44.98 ±0.28	36.39 ±0.27	38.26 ±0.46
	HPCA	**43.51** ±0.34	44.84 ±0.26	**50.84** ±0.22	**49.53** ±0.19	**48.93** ±0.38
25%	VAE	**53.53** ±0.12	57.63 ±0.52	**62.16** ±0.57	55.29 ±0.68	52.59 ±1.02
	soft-WTA	50.60 ±0.34	**57.84** ±0.26	59.94 ±0.15	51.26 ±0.41	56.26 ±0.34
	HPCA	51.51 ±0.31	54.22 ±0.23	59.60 ±0.44	**58.29** ±0.29	**58.70** ±0.18
100%	VAE	**67.51** ±0.11	**73.83** ±0.30	**78.70** ±0.23	**79.58** ±0.18	**79.97** ±0.14
	soft-WTA	64.00 ±0.23	73.06 ±0.20	76.39 ±0.12	76.07 ±0.12	79.80 ±0.11
	HPCA	65.61 ±0.12	70.38 ±0.23	74.10 ±0.12	73.38 ±0.18	74.42 ±0.14

Furthermore, HPCA seems to perform generally better than soft-WTA, especially on higher layers, when low sample efficiency regimes are considered (5% or less). The maximum improvement of HPCA over soft-WTA is achieved in the 1% sample efficiency regime, in correspondence of network layers 5 and 4, where a gap of almost 9–10% points is observed.

6.2 CIFAR100

Since CIFAR10 contained just 10 different classes, to validate our observations with a similar, yet more difficult scenario, we also performed tests with CIFAR100, containing 100 classes. In Table 2 the top-5 accuracy results obtained on the CIFAR100 dataset are shown.

As we can observe, in low sample efficiency regimes (10% or less), Hebbian approaches perform better than VAE in almost all the cases. In particular, soft-WTA generally performs better than VAE on higher network layers, and HPCA generally performs better than both soft-WTA and VAE on all network layers. Only when the number of available labeled samples increases (beyond 10%), VAE pre-training starts to really kick in, obtaining higher results than Hebbian training in almost all the cases. The maximum improvement of Hebbian approaches over VAE is achieved in the 10% sample efficiency regime, in correspondence of network layer 5, where a gap of almost 15% points is observed between VAE and soft-WTA, and a gap of over 25% points is observed between VAE and HPCA. Moreover, in low sample efficiency regimes (10% or less) it is possible to notice that VAE and soft-WTA approaches suffer from a decrease in performance when going deeper with the number of layers. As already observed on the previous dataset, this is likely due to the lack of task-specificity of higher layer features provided by unsupervised training. With HPCA, this problem seems to alleviate, and the accuracy remains pretty much constant with the number of layers, again in the low sample efficiency regimes (10% or less), meaning that the features produced by this approach are more meaningful for the classification task. Furthermore, HPCA seems to perform generally better than soft-WTA, especially on higher layers (except for the 100% regime). The maximum improvement of HPCA over soft-WTA is achieved in the 4–5% sample efficiency regimes, in correspondence of network layers 5 and 4, where a gap of almost 12–13% points is observed. Overall, the results suggest that HPCA scales better than other approaches with the complexity of the dataset, especially for low sample efficiency regimes (10% or less), while VAE is generally preferable in regimes when more labeled samples are available (25% or higher).

6.3 Tiny ImageNet

Further experiments on Tiny ImageNet allowed us to validate the scalability of our previous observations to larger datasets. Tiny ImageNet has 200 classes and the training set consists of 100,000 samples (90,000 of which are used for training and 10,000 for validation). In Table 3 the top-5 accuracy results obtained on the Tiny ImageNet dataset are shown.

As we can observe, in low sample efficiency regimes (10% or less), Hebbian approaches perform better than VAE. In particular, soft-WTA generally performs better than VAE on higher network layers, and HPCA performs better than both soft-WTA and VAE on all network layers. Only when the number of available labeled samples increases (beyond 10%), VAE pre-training starts to really kick in, obtaining higher results than Hebbian training. The maximum improvement of Hebbian approaches over VAE is achieved in the 10% sample efficiency regime, in correspondence of network layer 5, where a gap of over 3% points is observed between VAE and soft-WTA, and a gap of almost 15% points is observed between VAE and HPCA. Moreover, in low sample efficiency regimes (10% or less) it is possible to notice that VAE and soft-WTA approaches

Table 3. TinyImageNet accuracy (top-5) and 95% confidence intervals obtained with a linear classifier on top of various layers, for the various sample efficiency regimes. Results obtained with VAE and Hebbian pre-training are compared.

Regime	Pre-Train	L1	L2	L3	L4	L5
1%	VAE	9.63 ±0.26	9.49 ±0.39	7.58 ±0.28	5.99 ±0.19	5.55 ±0.23
	soft-WTA	9.25 ±0.20	9.63 ±0.29	8.71 ±0.12	6.54 ±0.29	6.20 ±0.20
	HPCA	**10.81** ±0.27	**10.99** ±0.36	**12.15** ±0.46	**11.05** ±0.27	**11.38** ±0.41
2%	VAE	12.94 ±0.37	13.06 ±0.23	10.86 ±0.28	7.40 ±0.27	6.74 ±0.20
	soft-WTA	12.67 ±0.26	12.56 ±0.30	11.36 ±0.18	8.57 ±0.21	8.56 ±0.29
	HPCA	**14.12** ±0.23	**14.32** ±0.31	**16.89** ±0.61	**15.28** ±0.28	**15.71** ±0.47
3%	VAE	14.31 ±0.18	15.17 ±0.20	13.67 ±0.36	8.35 ±0.29	7.74 ±0.19
	soft-WTA	14.66 ±0.17	14.50 ±0.33	13.71 ±0.18	9.95 ±0.25	10.26 ±0.18
	HPCA	**16.25** ±0.21	**16.54** ±0.28	**19.78** ±0.47	**18.31** ±0.24	**18.23** ±0.33
4%	VAE	16.09 ±0.20	17.05 ±0.20	16.83 ±0.51	8.86 ±0.11	8.45 ±0.21
	soft-WTA	16.20 ±0.31	16.51 ±0.26	15.70 ±0.17	11.04 ±0.29	11.52 ±0.07
	HPCA	**17.70** ±0.44	**18.33** ±0.24	**21.95** ±0.57	**20.86** ±0.32	**20.55** ±0.28
5%	VAE	17.44 ±0.26	18.62 ±0.32	19.16 ±0.52	9.92 ±0.24	9.29 ±0.17
	soft-WTA	17.72 ±0.17	18.06 ±0.49	17.03 ±0.30	12.15 ±0.19	12.55 ±0.15
	HPCA	**19.26** ±0.41	**19.93** ±0.41	**23.97** ±0.52	**22.95** ±0.26	**22.46** ±0.17
10%	VAE	21.62 ±0.25	23.83 ±0.19	27.42 ±0.18	16.69 ±0.18	13.51 ±0.34
	soft-WTA	21.22 ±0.43	23.08 ±0.21	21.90 ±0.15	16.21 ±0.27	16.70 ±0.17
	HPCA	**22.82** ±0.33	**24.34** ±0.29	**28.69** ±0.36	**28.79** ±0.26	**28.13** ±0.38
25%	VAE	**29.40** ±0.31	**32.42** ±0.29	**39.93** ±0.31	**37.97** ±0.62	**37.89** ±0.54
	soft-WTA	26.36 ±0.48	31.31 ±0.28	32.54 ±0.13	22.39 ±0.11	24.96 ±0.23
	HPCA	28.01 ±0.75	30.63 ±0.16	35.87 ±0.53	36.98 ±0.26	37.10 ±0.23
100%	VAE	**42.32** ±0.16	**48.54** ±0.53	**58.31** ±0.12	**59.60** ±0.23	**60.23** ±0.65
	soft-WTA	38.55 ±0.20	46.82 ±0.33	48.91 ±0.24	42.35 ±0.24	54.94 ±0.10
	HPCA	40.34 ±0.31	45.00 ±0.40	53.12 ±0.26	52.95 ±0.28	53.96 ±0.43

suffer from a decrease in performance when going deeper with the number of layers. As already observed on previous datasets, this is likely due to the lack of task-specificity of higher layer features provided by unsupervised training. With HPCA, this problem seems to alleviate, and the accuracy remains pretty much constant or slightly increases with the number of layers, again in the low sample efficiency regimes (10% or less), meaning that the features produced by this approach are more meaningful for the classification task. Furthermore, HPCA seems to perform generally better than soft-WTA, especially on higher layers (except for the 100% regime). The maximum improvement of HPCA over soft-WTA is achieved in the 25% sample efficiency regime, in correspondence of network layers 5 and 4, where a gap of almost 13–14% points is observed. Overall, the results suggest that HPCA scales better than other approaches with the complexity of the dataset, especially for low sample efficiency regimes (10% or less), while VAE is preferable in regimes when more labeled samples are available (25% or higher).

7 Conclusions and Future Work

In summary, our results suggest that our semi-supervised approach based on unsupervised Hebbian pre-training performs generally better than VAE pre-training, especially in low sample efficiency regimes, in which only a small portion of the training set (between 1% and 10%) is assumed to be labeled. In particular, the HPCA approach appears to perform generally better than soft-WTA. Moreover, HPCA seems to scale better than other approaches when the complexity of the dataset increases, especially when low sample efficiency regimes are considered. On the other hand, VAE pre-training seems to become more effective in regimes where a larger portion of the training set (25% or higher) is labeled. Therefore, our method is preferable in scenarios in which manually labeling a large number of training samples would be too expensive, while gathering unlabeled samples is relatively cheap.

In future works, further improvements might come from exploring more complex feature extraction strategies, which can also be formulated as Hebbian learning variants, such as Independent Component Analysis (ICA) [17] and sparse coding [32,33,38]. Moreover, Hebbian approaches can also be combined with pseudo-labeling and consistency methods mentioned in Sect. 2 [6,7,18,40]. In addition to the semi-supervised learning scenario considered in this paper, it would also be interesting to investigate Hebbian approaches in a meta-learning scenario. Hebbian learning already found application in the context of meta-learning, with the *differentiable plasticity* model [30]. In this case, the simple Hebbian learning rule, $\Delta \mathbf{w} = \eta \, y \, \mathbf{x}$, was used, but further improvements might come from applying more advanced Hebbian rules, such as those studied in this paper. Finally, an exploration on the behavior of such algorithms w.r.t. adversarial examples also deserves attention.

References

1. Agrawal, P., Girshick, R., Malik, J.: Analyzing the performance of multilayer neural networks for object recognition. arXiv preprint arXiv:1407.1610 (2014)
2. Amato, G., Carrara, F., Falchi, F., Gennaro, C., Lagani, G.: Hebbian learning meets deep convolutional neural networks. In: Ricci, E., Rota Bulla, S., Snoek, C., Lanz, O., Messelodi, S., Sebe, N. (eds.) Image Analysis and Processing. LNCS, vol. 11751. Springer, Cham (2019). https://doi.org/10.1007/978-3-030-30642-7_29
3. Bahroun, Y., Soltoggio, A.: Online representation learning with single and multi-layer hebbian networks for image classification. In: Lintas, A., Rovetta, S., Verschure, P., Villa, A. (eds.) Artificial Neural Networks and Machine Learning. LNCS, vol. 10613. Springer, Cham (2017). https://doi.org/10.1007/978-3-319-68600-4_41
4. Becker, S., Plumbley, M.: Unsupervised neural network learning procedures for feature extraction and classification. Appl. Intell. **6**(3), 185–203 (1996)
5. Bengio, Y., Lamblin, P., Popovici, D., Larochelle, H.: Greedy layer-wise training of deep networks. In: Advances Neural Information Processing Systems, pp. 153–160 (2007)
6. Berthelot, D., et al.: Remixmatch: semi-supervised learning with distribution alignment and augmentation anchoring. arXiv preprint arXiv:1911.09785 (2019)

7. Berthelot, D., Carlini, N., Goodfellow, I., Papernot, N., Oliver, A., Raffel, C.: Mixmatch: A holistic approach to semi-supervised learning. arXiv preprint arXiv:1905.02249 (2019)
8. Aguilar Canto, F.J.: Convolutional neural networks with Hebbian-based rules in online transfer learning. In: Martinez-Villasenor, L., Herrera-Alcantara, O., Ponce, H., Castro-Espinoza, F.A. (eds.) Advances in Soft Computing. MICAI 2020. LNCS, vol. 12468. Springer, Cham (2020). https://doi.org/10.1007/978-3-030-60884-2_3
9. Chen, T., Kornblith, S., Norouzi, M., Hinton, G.: A simple framework for contrastive learning of visual representations. In: International Conference on Machine Learning, pp. 1597–1607. PMLR (2020)
10. Devlin, J., Chang, M.W., Lee, K., Toutanova, K.: Bert: Pre-training of deep bidirectional transformers for language understanding. arXiv preprint arXiv:1810.04805 (2018)
11. Gao, B., Pavel, L.: On the properties of the softmax function with application in game theory and reinforcement learning. arXiv preprint arXiv:1704.00805 (2017)
12. Gerstner, W., Kistler, W.M.: Spiking neuron models: single neurons, populations, plasticity. Cambridge University Press (2002)
13. Grossberg, S.: Adaptive pattern classification and universal recoding: I. parallel development and coding of neural feature detectors. Biological cybernetics **23**(3), 121–134 (1976)
14. Haykin, S.: Neural networks and learning machines. Pearson, 3rd edn. (2009)
15. He, K., Zhang, X., Ren, S., Sun, J.: Deep residual learning for image recognition. In: Proceedings of the IEEE Conference on Computer Vision and Pattern Recognition, pp. 770–778 (2016)
16. Higgins, I., et al.: beta-vae: learning basic visual concepts with a constrained variational framework (2016)
17. Hyvarinen, A., Karhunen, J., Oja, E.: Independent component analysis. Stud. Inf. Control **11**(2), 205–207 (2002)
18. Iscen, A., Tolias, G., Avrithis, Y., Chum, O.: Label propagation for deep semi-supervised learning. In: Proceedings of the IEEE/CVF Conference on Computer Vision and Pattern Recognition, pp. 5070–5079 (2019)
19. Karhunen, J., Joutsensalo, J.: Generalizations of principal component analysis, optimization problems, and neural networks. Neural Netw. **8**(4), 549–562 (1995)
20. Kingma, D.P., Welling, M.: Auto-encoding variational bayes. arXiv preprint arXiv:1312.6114 (2013)
21. Kingma, D.P., Mohamed, S., Jimenez Rezende, D., Welling, M.: Semi-supervised learning with deep generative models. Adv. Neural Inf. Proc. Syst. **27**, 3581–3589 (2014)
22. Kolda, T.G., Lewis, R.M., Torczon, V.: Optimization by direct search: new perspectives on some classical and modern methods. SIAM Rev. **45**(3), 385–482 (2003)
23. Krizhevsky, A., Hinton, G.: Learning multiple layers of features from tiny images (2009)
24. Krizhevsky, A., Sutskever, I., Hinton, G.E.: Imagenet classification with deep convolutional neural networks. Adv. Neural Inf. Proc. Syst. **25**, 1097–1105 (2012)
25. Krotov, D., Hopfield, J.J.: Unsupervised learning by competing hidden units. Proc. Nat. Acad. Sci. **116**(16), 7723–7731 (2019)
26. Lagani, G.: Hebbian learning algorithms for training convolutional neural networks. Master's thesis, School of Engineering, University of Pisa, Italy (2019). etd.adm.unipi.it/theses/available/etd-03292019-220853/
27. Larochelle, H., Bengio, Y., Louradour, J., Lamblin, P.: Exploring strategies for training deep neural networks. J. Mach. Learn. Res. **10**(1) (2009)

28. Magotra, A., kim, J.: Transfer learning for image classification using hebbian plasticity principles. In: Proceedings of the 2019 3rd International Conference on Computer Science and Artificial Intelligence, pp. 233–238 (2019)
29. Magotra, A., Kim, J.: Improvement of heterogeneous transfer learning efficiency by using hebbian learning principle. Appl. Sci. **10**(16), 5631 (2020)
30. Miconi, T., Clune, J., Stanley, K.O.: Differentiable plasticity: training plastic neural networks with backpropagation. arXiv preprint arXiv:1804.02464 (2018)
31. Nowlan, S.J.: Maximum likelihood competitive learning. In: Advances in neural information processing systems, pp. 574–582 (1990)
32. Olshausen, B.A.: Learning linear, sparse, factorial codes. Massachusetts Institute of Technology, AIM-1580 (1996)
33. Olshausen, B.A., Field, D.J.: Emergence of simple-cell receptive field properties by learning a sparse code for natural images. Nature **381**(6583), 607 (1996)
34. O'Reilly, R.C., Munakata, Y.: Computational explorations in cognitive neuroscience: understanding the mind by simulating the brain. MIT press (2000)
35. Pehlevan, C., Chklovskii, D.B.: Optimization theory of hebbian/anti-hebbian networks for pca and whitening. In: 2015 53rd Annual Allerton Conference on Communication, Control, and Computing (Allerton), pp. 1458–1465. IEEE (2015)
36. Pehlevan, C., Hu, T., Chklovskii, D.B.: A hebbian/anti-hebbian neural network for linear subspace learning: a derivation from multidimensional scaling of streaming data. Neural Comput. **27**(7), 1461–1495 (2015)
37. Rasmus, A., Berglund, M., Honkala, M., Valpola, H., Raiko, T.: Semi-supervised learning with ladder networks. In: Advances in Neural Information Processing Systems, pp. 3546–3554 (2015)
38. Rozell, C.J., Johnson, D.H., Baraniuk, R.G., Olshausen, B.A.: Sparse coding via thresholding and local competition in neural circuits. Neural Comput. **20**(10), 2526–2563 (2008)
39. Sanger, T.D.: Optimal unsupervised learning in a single-layer linear feedforward neural network. Neural Netw. **2**(6), 459–473 (1989)
40. Sellars, P., Aviles-Rivero, A.I., Schönlieb, C.B.: Laplacenet: a hybrid energy-neural model for deep semi-supervised classification. arXiv preprint arXiv:2106.04527 (2021)
41. Silver, D., et al.: Mastering the game of go with deep neural networks and tree search. Nature **529**(7587), 484 (2016)
42. Wadhwa, A., Madhow, U.: Bottom-up deep learning using the hebbian principle (2016)
43. Wadhwa, A., Madhow, U.: Learning sparse, distributed representations using the hebbian principle. arXiv preprint arXiv:1611.04228 (2016)
44. Weston, J., Chopra, S., Bordes, A.: Memory networks. arXiv preprint arXiv:1410.3916 (2014)
45. Weston, J., Ratle, F., Mobahi, H., Collobert, R.: Deep Learning via Semi-supervised Embedding. In: Montavon, G., Orr, G.B., Muller, K.R. (eds.) Neural Networks: Tricks of the Trade. LNCS, vol. 7700. Springer, Heidelberg (2012). https://doi.org/10.1007/978-3-642-35289-8_34
46. Wu, J., Zhang, Q., Xu, G.: Tiny imagenet challenge. Technical report, Stanford University (2017)
47. Zhang, Y., Lee, K., Lee, H.: Augmenting supervised neural networks with unsupervised objectives for large-scale image classification. In: International Conference on Machine Learning, pp. 612–621 (2016)

Experiments on Properties of Hidden Structures of Sparse Neural Networks

Julian Stier(✉), Harshil Darji, and Michael Granitzer

University of Passau, Passau, Germany
julian.stier@uni-passau.de, darji01@ads.uni-passau.de
https://www.fim.uni-passau.de/en/data-science/

Abstract. Sparsity in the structure of Neural Networks can lead to less energy consumption, less memory usage, faster computation times on convenient hardware, and automated machine learning. If sparsity gives rise to certain kinds of structure, it can explain automatically obtained features during learning.

We provide insights into experiments in which we show how sparsity can be achieved through prior initialization, pruning, and during learning, and answer questions on the relationship between the structure of Neural Networks and their performance. This includes the first work of inducing priors from network theory into Recurrent Neural Networks and an architectural performance prediction during a Neural Architecture Search. Within our experiments, we show how magnitude class blinded pruning achieves 97.5% on MNIST with 80% compression and re-training, which is 0.5 points more than without compression, that magnitude class uniform pruning is significantly inferior to it and how a genetic search enhanced with performance prediction achieves 82.4% on CIFAR10. Further, performance prediction for Recurrent Networks learning the Reber grammar shows an R^2 of up to 0.81 given only structural information.

Keywords: Sparse recurrent neural networks · Pruning · Hidden structural prior · Neural architecture search · Architecture performance prediction

1 Introduction

Understanding the structure of deep neural networks promises advances across many open problems such as energy-efficient hardware, computation times, and domain-specific performance improvements. The structure is coupled with sparsity on different levels of the neural architecture, and if there is no sparsity, then there is also no structure: a single hidden layered neural network is capable of universal approximation [14], but as soon as there exists a deeper structure, there naturally occurs sparsity.

Clearly, the structure between the input domain and the first hidden layer is tightly coupled with the structure within the data – correlations between

G. Nicosia et al. (Eds.): LOD 2021, LNCS 13164, pp. 380–394, 2022.
https://doi.org/10.1007/978-3-030-95470-3_29

the underlying random variables such as the spatial correlation of images or correlation in windows of time series data. In theory and with perfectly fitting functions, that should be all there is, but in practice, neural architectures got deeper and deeper, and hidden structures seem to have an effect when neural networks are not just measured by their goodness of fit but also, e.g., on hardware efficiency or robustness [3]. Assuming such hidden structures exist for the better, we wonder how we can automatically find them, how they can be controlled during learning, and whether we can exploit given knowledge about them.

We give our definition for sparse neural networks and show experiments on automatic methods to obtain hidden structures: pruning, neural architecture search, and prior initialization. With structural performance prediction, we also show experiments on exploiting structural information to speed up neural architecture search methods.

Our **contributions** comprise a pytorch tool called *deepstruct*[1] which provides models and tools for Sparse Neural Networks, a *genetic neural architecture search* enhanced with structural performance prediction, a *comparison of magnitude-based pruning* on feed-forward and recurrent networks, an **original** correlation analysis on *recurrent networks* with different biologically plausible **structural priors** from social network theory, and *performance prediction* results on these recurrent networks. Details on the experiments and code for reproducibility can be found at github.com/innvariant/sparsity-experiments-2021.

2 Sparse Neural Networks

Sparse Neural Networks (SNNs) are deep neural networks f with a low proportion of connectivity $\xi(f)$ with respect to all possible connections.

Sparsity. Given a vector $x \in \mathbb{R}^d$ with $d \in \mathbb{N}$, its sparsity is $\xi(x) = \frac{||x||_0}{d} = \frac{1}{d} \cdot \sum_{i=0}^{d} |x_i|^0$, given the cardinality function $|| \cdot ||_0$ (of which 0 refers to the case of $p = 0$ of a $\mathcal{L}_p$ norm) and the size of the vector. Density is defined as its complement with $1 - \xi(x)$. The definition extends naturally to tensors and simply provides the proportion of non-zero elements in a tensor compared to the total number of its elements. A tensor can be considered as sparse as soon as its sparsity is below a given threshold value, e.g., $\xi(x) < 0.5$ – as soon as more than 50% of its elements are zero.

What is the **motivation** for sparsity at all? First, more sparsity implies a lower number of parameters which is desirable if the approximation and generalization capabilities are not heavily affected. In theory, it also implies a lower number of computations. From a technical perspective, sparse structures could lead to specialized hardware. Further, sparsity means that there is space for compression that can affect the overall model memory footprint. Memory requirements are an important aspect for limited capacity devices such as in mobile deployment. In the feature transformation layers, sparsity explains data dependencies and provides room for explainability.

[1] http://github.com/innvariant/deepstruct.

Neural Networks. A neural network is a function composed of non-linear transformation layers $\sigma(Wx + B)$ extended with transformations for skip-layer connections such that $z^l = \sum_{s=1}^{l-1} W^{s\to l} \cdot a^s + B^l$ with $a^l = \sigma(z^l)$ being the activation of layer l with σ being e.g. *tanh* or *max(x, 0)*. $W^{s\to l}$ describes the weights from layer $l-1$ to l for a network with $l \in \{1, \ldots, L\}$. The input to the function a^0 is $x \in \mathbb{R}^{d_x}$ from the input domain. Consecutive sizes of weight matrices W need to be aligned and define the layer sizes. The final weight matrices $W^{s\to L}$ map to the output domain $\mathbb{R}^{d_y}$ with $B^L \in \mathbb{R}^{d_y}$.

Given the weights of a neural network f as a set of grouped vectors W, we overload ξ such that we obtain the sparsity of a neural network $\xi(f) = \frac{1}{|W|}\sum_{x\in W} \xi(x)$. A **Sparse Neural Network** is a neural network f with low sparsity, e.g. $\xi(f) < 0.5$. The set of grouped vectors could, e.g., be all neurons with their weights from all possible incoming connections.

Sparse Recurrent Neural Networks (SRNN). Recurrent Neural Networks additionally have recurrent connections which unfold over time. These recurrent connections are initialized as hidden states, h. At any sequence t, $h_t = \sigma(Wx_t + Uh_{t-1} + b_h), t \in \{1, ..., T\}$ with W and U being input-to-hidden weigths and hidden-to-hidden weights, respectively. We refer to x_t as the input at sequence t and h_{t-1} as the hidden state value from the previous step.

Similar to SNNs, SRNNs also consists of extended non-linear transformation layers with skip-layers such that $h_t^l = \sum_{s=1}^{l-1} W^{s\to l} h_t^s + Uh_{t-1}^l + b_h^l$ with $t \in \{1, ..., T\}$.

In SRNNs, directed acyclic graphs are not associated with recurrent connections, only with consecutive transformation layers. Therefore, to reflect a graph in an SRNN, only input-to-hidden weights $W^{s\to l}$ are multiplied with masks $M^{s\to l}$ through the Hadamard product $\odot$ such that h_t^l can be formulated as:

$$\sum_{s=1}^{l-1} \left(W^{s\to l} \odot M^{s\to l}\right) h_t^s + Uh_{t-1}^l + b_h^l$$

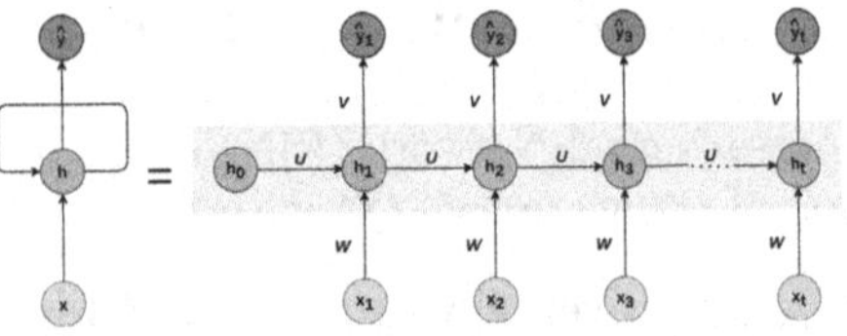

Fig. 1. A simple **Recurrent Neural Network** unrolled with t sequences. Here, W represents *input-to-hidden weights*, U represents *hidden-to-hidden weights*, and V represents *hidden-to-output weights*.

Achieving Sparsity. Sparsity refers to a structural property of Neural Networks which can be desirable due to reasons such as model size or hardware acceleration. There exist multiple ways to achieve sparsity, e.g., through regularization, pruning, constraints, or by prior initialization.

Regularization affects the optimization objective such that not only a target loss but also a parameter norm is minimized. As such, regularization takes effect during training and can force weights to be of small magnitude. Under sufficient conditions, e.g., with an L1-norm and rectified linear units as activation functions, sparsity in the trained network can be achieved in an end-to-end fashion during learning.

Pruning refers to removing elements of the network outside of the end-to-end training phase. Based on a selection criterion such as the magnitude of a weight, one or multiple weights can be set to zero. A pruning scheme decides on what sets the criterion is applied or how often the pruning is repeated. Sparsity is enforced based on this selection criterion and pruning scheme.

Prior design is a constraint on the overall search space of the optimization procedure. More generally, prior structure to a neural network is restricting the hypothesis space of all possible obtainable functions during learning to a smaller space. Convolutions as feed-forward neural networks with local spatial restrictions can be understood as such a prior design.

Pruning Neural Networks. Pruning is a top-down method to derive a model from an originally larger one by iteratively removing elements of the neural network. The motivation to prune is manifold: 1) finding high-performing network structures can be faster in comparison to other search methods such as grid- or random-search, 2) pruning can improve metrics such as error, generalization, fault tolerance, or robustness or 3) reduce computational costs in terms of speed, memory requirements and energy efficiency or 4) support the interpretation of neural networks.

Pruning consists of a selection method and a strategy. The selection method decides which elements to choose based on a criteria, e.g., the magnitude of a weight. The strategy applies the selection method repeatedly on a model until some stopping criterion is reached, e.g., a certain number of iterations are conducted.

One-shot or single pruning refers to applying the pruning method once. After pruning, often a certain number of re-training cycles are conducted. Fixed-size pruning refers to selecting a fixed number of elements based on the ranking obtained through the pruning selection method. In each step, the same number of elements are removed. Relative or percentage pruning refers to selecting a percentage of remaining elements to be pruned. This results in fewer numbers to be removed in sudden decays of performance. Bucket pruning holds a bucket value which is filled by, e.g., the weight magnitude or the saliency measure of the pruning selection method, and as many elements as the bucket can hold are removed per step.

A naive method for pruning is the random selection of k components. Differences can be made by defining the granularity, e.g., whether to prune weights, neurons, or even channels or layers. Random pruning often serves as a baseline for pruning methods to show their general effectiveness, and it has been shown in various articles that most magnitude- and error-based methods outperform their random baseline, see Fig. 2a. Usually, models drop in performance after pruning but recover within a re-training phase of few epochs.

For magnitude-based pruning, good explanations can be found in [11,22] and in recent surveys such as [8,19]. Class-blinded selects weights based on their magnitude regardless of their class, i.e., their layer, class-uniform selects the same amount of weights from each class, and class-distributed selects elements in proportion to the standard deviation of weight magnitudes in the respected class.

Prior Design. Restricting the search space of the neural network architecture fosters faster convergence, and domain-specific improved performance can be achieved. Convolutions with kernels applied over spatially related inputs are a good example for such a prior. Similarly, realizations as the MLP-Mixer [28] show that even multi-layer perceptrons with additional imposed structure and final poolings can achieve state-of-the-art performance.

3 Related Work

Pruning. Recent work on pruning has been conducted by Han et al. [11] using different magnitude-based pruning methods for deep neural networks in the context of compression or by Dong et al. [4]. A survey by Liang et al. [19] provides extensive insights into pruning and quantization for deep neural networks. In 2018, authors of the Lottery Ticket Hypothesis reported on finding sparse subnetworks after iteratively pruning, re-setting to the original weight initialization, and training it from scratch to a comparable performance [7]. We collected over 300 articles on pruning just up until 2019.

First pruning experiments on Recurrent Network Network were performed by Lee et al. [9]. Han et al. [10] proposed recurrent self-organising neural networks, adding or pruning the hidden neurons based on their competitiveness during training. A Baidu research group [25] could reduce the network size by 8× while managing the near-original performance. In 2019, Zhang et al. [33] proposed one-shot pruning for Recurrent Neural Networks using the recurrent Jacobian spectrum. Using this technique, the authors confirmed that their network, even with 95% sparsity, performed better than fully dense networks.

Sparse Training. Besides L_1- or L_2-regularizations which can lead to real-zeros with appropriate activation functions such as ReLUs, there are also training methods outside the optimization objective to enforce sparsity such as Sparse Evolutionary Training (SET) by Mocanu et al. [23], Dynamic Sparse Reparamterization (DSR) [24] or the Rigged Lottery (RigL) [6]. For RNNs, there exists Selfish sparse RNN training [21].

Neural Architecture Search. On Neural Architecture Search (NAS), there are two notable recent surveys by Elsken et al. [5] and Wistuba et al. [31] providing an overview and dividing NAS into the definition of a **search space**, a **search strategy** over this space and the **performance estimation** strategy. Differentiable architecture search [20] is a notable method for finding sparse neural networks on a high-level graph based search space by allowing to choose among paths in a categorical and differentiable manner. While there exist hundreds of variations in the definition of search spaces and methods, the field recently came up with benchmarks and comparable metrics [32].

Structural Performance Prediction. Structural Performance Prediction refers to using structural features of a neural network to predict a performance estimate without any or only partial training. Such performance prediction was already conducted by Baker et al. [1] on two structurally simple features, namely the total number of weights and the number of layers. But they mostly focused on prediction based on hyperparameters and time-series information. Klein et al. also did performance prediction based on time-series information. They conducted "learning curve prediction with Bayesian Neural Networks" [15]. In [27] more extensive graph properties of randomly induced structures were used to predict the performance of neural networks for image classification. A related work on a "genetic programming approach to design convolutional neural network architectures" [30] included an acceleration study for accuracy prediction based on path depth, breadth-first-search level width, layer out height and channels, and connection type counts. The performance prediction during a NAS yields a 1.69× speed-up. Similar to [27] in [3] we used structural properties to predict the robustness of recurrent neural networks.

4 Experiments

We conducted four experiments: First, pruning feed-forward neural networks to investigate the effect of different pruning methods, namely random pruning, magnitude class-blinded, magnitude class-uniform, magnitude class-distributed, and Optimal Brain Damage [18]. Second, pruning recurrent neural networks to investigate whether we observe similar compression rates and to have a baseline comparison for recurrent models in the subsequent experiment. Third, inducing random graphs as structural priors into recurrent neural networks, based on the biological motivation that biological neural networks are also connected like small-world networks [12]. And fourth, conducting a genetic neural architecture search with architectural performance prediction.

4.1 Pruning Feed-Forward Networks

On `MNIST` [17] we used two different feed-forward architectures with rectified linear units with 100 and 300–100 neurons in the hidden layers, trained up to 200 epochs with cross-entropy, a batch size of 64, and a learning rate of 0.01 and 0.0001. With the five pruning methods, we conducted several repeated experiments with iterative fixed-size pruning or iterative relative pruning of the number of weights.

We found that magnitude class blinded pruning clearly outperforms the other methods and Optimal Brain Damage, surprisingly, performs nearly the same although it uses second-order derivative information for pruning. Pruning in general can dramatically reduce overfitting in the examined network and can even outperform other regularization techniques.

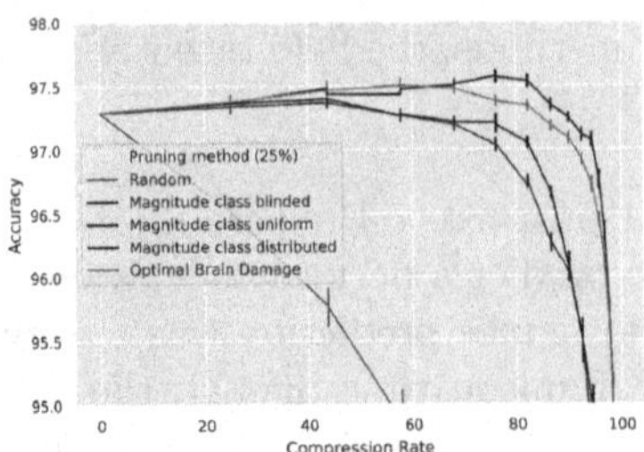

(a) Five pruning methods on feed-forward networks. Choosing weight-magnitude over random selections clearly has advantages. Optimal Brain Damage is expensive and worth in a low-parameter regime.

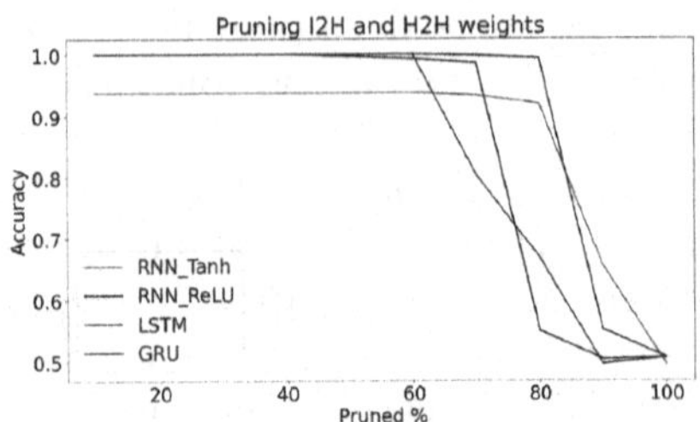

(b) Pruning both input-to-hidden and hidden-to-hidden weights on a recurrent neural network.

Fig. 2. Pruning performance in feed-forward networks (Fig. 2a) and recurrent networks (Fig. 2b).

4.2 Pruning Recurrent Networks

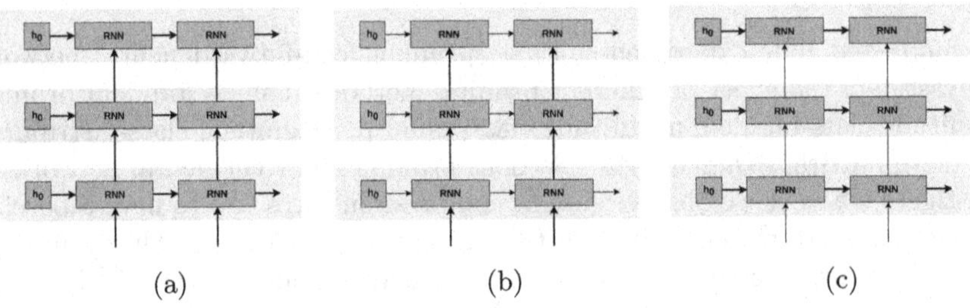

Fig. 3. The red right arrow (⟶) resembles pruning of hidden-to-hidden weights, the red up-arrow (↑) pruning of input-to-hidden weights. Figure 3a shows pruning both, Fig. 3b only i2h weights, and Fig. 3c only h2h weights.

In this experiment, we prune input-to-hidden "i2h" and hidden-to-hidden "h2h" weights individually and simultaneously as depicted in Fig. 3 on a pre-trained base recurrent model for the `Reber` grammar [26], trained for 50 epochs. The base recurrent model consists of an embedding layer that accepts input in the form of ASCII values of each character in the input `Reber` sequence. Three recurrent layers of 50 neurons follow the embedding layer and a linear layer predicts the final scores. TanH and ReLU are used as non-linearities. The models are trained with a learning rate of 0.001 and a batch size of 32.

From the `Reber` grammar, we generated 25000 sequences, out of which 12500 are true `Reber` grammar sequences, and the remaining are false. The dataset resembles a binary classification task in which a model has to predict whether a sequence is in the Reber grammar or not. Logically, a baseline performance from random guessing is accuracy of 50%. We then split this dataset into a train-test split of 75%-25%, with 18750 sequences in the training set and 6250 sequences in the test set.

The threshold based on which we prune weights is calculated based on the percent of weights to prune. Therefore, for example, to prune $p = 10\%$ of weights for a given layer, the threshold is the 10th percentile of all the **absolute** weights in that layer. In our experiment, we go from percent $p = 10$ to 100 while incrementing p by 10 after each round.

We considered LSTM, GRU, and vanilla RNNs as architectures for comparison. Base models were trained separately for each to get the base performance. Then, we pruned i2h and h2h weights; simultaneously and individually. Based on these results, we can identify the effect pruning has on the performance of RNNs and the amount of re-training required to regain the original performance.

The base models for RNN_ReLU, LSTM, and GRU achieve perfect accuracies of 100% on the test set within the first two epochs. RNN_TanH achieved 90% after six epochs and showed a drop in accuracy between epoch three and five down to 50%, which we observed over multiple repetitions of the experiment.

Pruning both i2h and h2h weights simultaneously, about 80% of weights in RNN_Tanh, 70% of weights in RNN_ReLU, 60% of weights in LSTM, and 80% of weights in GRU can be safely reduced as can be observed in Fig. 2b. After pruning above the safe threshold, we re-trained each pruned model and found that it only takes one epoch to regain the original performance. Pruning 100% weights, the model never recovers.

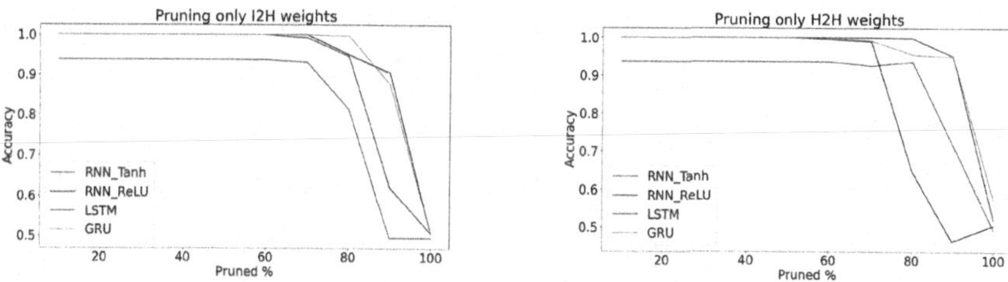

(a) Pruning only input-to-hidden weights. (b) Pruning only hidden-to-hidden weights.

Fig. 4. Accuracies of RNN_Tanh, RNN_ReLU, LSTM, and GRU after applying iterative magnitude percentage pruning on a common base model.

Pruning only i2h weights, results showed that we safely prune about 70% for RNN_Tanh, RNN_ReLU, and LSTM. For GRU, we prune 80% of i2h weights without noticing a significant reduction in performance, see Fig. 4a. As in the case of pruning both i2h and h2h weights simultaneously, our pruned model still recovers only after one re-training epoch with up to 90% of pruning i2h weights in RNN_ReLU, LSTM, and GRU. RNN_Tanh takes about two re-training epochs to recover after 90% of i2h weight pruning. Finally, as expected, this pruned model never recovers with 100% of i2h weights pruning.

Subsequently, we prune only h2h weights of each recurrent layer in our trained base model. Results showed that we could safely prune about 70% of h2h weights for RNN_ReLU and LSTM, while 80% of h2h weights for RNN_Tanh and GRU, see Fig. 4b. Like pruning only i2h weights, models still recover after one re-training epoch with up to 90% of pruning h2h weights. Pruning 100% h2h weights, RNN_Tanh and RNN_ReLU never recover, but GRU and LSTM still retain the original performance with just one re-training epoch.

4.3 Random Structural Priors for Recurrent Neural Networks

Another method than pruning to induce sparsity in a recurrent network is by applying prior structures by design. We use random structures that are generated by converting random graphs into neural architectures, similar as in [27]. For this, we begin with a random graph and calculate the layer indexing of each vertex. A layer index is obtained recursively by $v \mapsto max(\{ind^l(s)|(s,v) \in E_v^{in}\} \cup \{-1\})+1$. This layer indexing helps to identify the layer of a neural architecture a vertex belongs to.

Such a graph is used to generate randomly structured ANNs by embedding it between an input and an output layer, as in Fig. 5b. RNNs can be understood as a sequence of neural networks, in which a network model at sequence t accepts outputs from a model at sequence $t-1$. Introducing recurrent connections as in Fig. 5c provides us with Sparse RNNs with random structure.

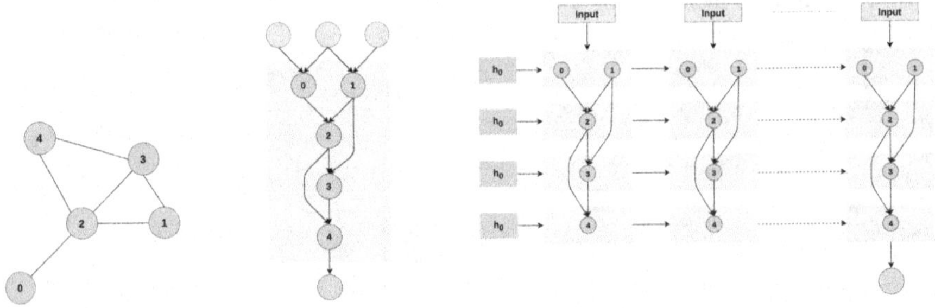

(a) Initial graph (b) Sparse Neural Network with induced structural prior. (c) A sparse RNN with a structural prior based on the initial graph of five vertices.

Fig. 5. We select one directed version of the graph from Fig. 5a, compute its topological ordering based on the described layer indexing and embed it into a neural network as a structural prior as shown in Fig. 5b. This randomly structured ANN can then be converted into a randomly structured RNN by introducing recurrent connections, as in Fig. 5c.

We generate 100 connected Watts–Strogatz [29] and 100 Barabási–Albert [2] graphs using the graph generators provided by NetworkX. The graphs are transformed into recurrent networks and trained on the `Reber` grammar dataset.

Analogue to pruning, this experiment is also conducted with RNN with Tanh nonlinearity, RNN with ReLU nonlinearity, LSTM, and GRU.

To identify essential graph properties that correlate with the performance, we calculated the Pearson correlation of each graph property to its corresponding performance results. Table 1 shows the Pearson correlation between test accuracy and different graph properties.

Table 1. Pearson correlation between the test accuracy of an architecture and different graph properties.

Property	Correlation with test accuracy			
	RNN_Tanh	RNN_ReLU	LSTM	GRU
Layers	0.25	0.30	0.28	0.34
Nodes	**0.40**	**0.44**	**0.44**	**0.49**
Edges	0.38	**0.43**	**0.42**	**0.49**
Source_nodes	0.35	**0.47**	**0.57**	**0.74**
Diameter	−0.23	−0.27	−0.32	−0.20
Density	0.29	0.15	0.29	0.34
Average_shortest_path_length	−0.27	−0.25	−0.36	−0.23
Eccentricity_var	−0.22	−0.24	−0.30	−0.21
Degree_var	−0.28	−0.26	−0.39	**−0.58**
Closeness_var	**−0.46**	−0.39	**−0.51**	**−0.67**
Nodes_betweenness_var	**−0.49**	**−0.41**	**−0.56**	**−0.52**
Edge_betweenness_var	−0.34	−0.30	**−0.44**	−0.26

Based on this correlation, we found closeness_var, nodes_betweenness_var, and the number of nodes to be essential properties for randomly structured RNN_Tanh. For randomly structured RNN_ReLU, the essential properties are the number of nodes, the number of edges, the number of source nodes, and nodes_betweenness_var. In the case of randomly structured LSTM, we found six essential properties, i.e., the number of nodes, the number of edges, the number of source nodes, closeness_var, nodes_betweenness_var, and edge_betweenness_var. Similarly, we found six essential properties for randomly structured GRU, namely, the number of nodes, the number of edges, the number of source nodes, degree_var, closeness_var, and nodes_betweenness_var.

By storing the graph properties and their corresponding performance during the training of randomly structured recurrent networks, we create a small dataset of 200 rows for each RNN variant. We then train three different regression algorithms, namely Bayesian Ridge, Random Forest, and AdaBoost, on this dataset and report an R-squared value for each.

Performance of **RNN_TANH** was best predicted with Bayesian Ridge (BR) Regression with an R^2 of 0.47919, while Random Forest (RF) achieved 0.43163 and AdaBoost (AB) 0.35698. All regressors have an R^2 of below 0.5, from which we conclude only a weak fit and predictability based on the used structural features. For **RNN_RELU**, RF was best with an R^2 of 0.61504, followed by AB with 0.53469 and BR with 0.36075. Structural features on **LSTM** predicted performance with AB with 0.59514, with RF with 0.57933, and with BR with 0.37206. We found a moderate fit for random forests, similar as in [27]. **GRU** accuracies were predicted with AB with 0.78313 and with BR with 0.67224, and RF achieved an R^2 of **0.87635**. This indicates a strong fit and a good predictability that we interpreted carefully as potentially coming from a skewed underlying distribution of the overall dataset of Sparse Neural Networks but also an indication of possible strong predictability in larger settings in which structural properties have even more impact.

4.4 Architectural Performance Prediction in Neural Architecture Search

We investigated a Genetic Neural Architecture Search to find correlations between structural properties of sparse priors and neural networks on a more coarse level and analysed the predictive capabilities for performances of Sparse Neural Networks when having just architectural information available.

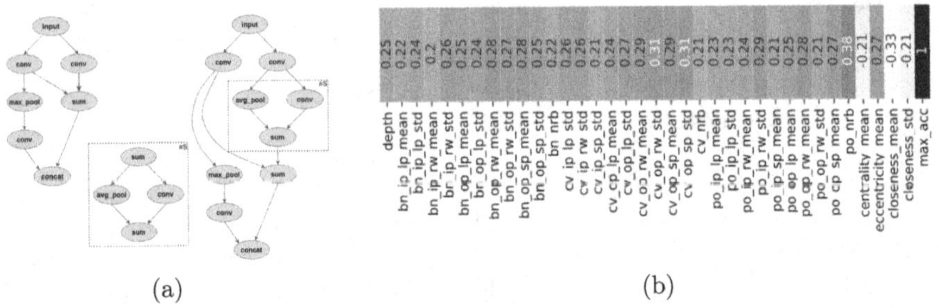

Fig. 6. Figure 6a shows an exemplary graph from our search space with a mutation through an inserted sub-graph of depth two. Figure 6b shows the correlations between structural properties and the maximum validation accuracy.

Our **search space** is based on directed acyclic graphs (DAG) and follows Irwin-Harris [13] to represent CNN architectures as depicted in Fig. 6a. Each vertex of the DAG is labelled with an operation: a convolution, max or average pooling, a summation, or concatenation or a skip connection. A convolution can have kernel sizes of either 3×3, 5×5 or 7×7.

In total, five **genotypical operations** were used on the search space: two mutations and three crossover operations. The first mutation remaps vertex operations in a genotype, e.g. it could replace *max_pool* in the left genotype of Fig. 6a with an *avg_pool*. Figure 6a depicts the result of the second mutation operation by inserting a smaller sub-graph #S into a randomly selected edge.

The first crossover operation considers the longest simple sub-paths and swaps them between both parent genotypes. Bridge-crossover searches for bridges in both parent genotypes and swaps the succeeding child graphs after both found bridges. In a third crossover operation, matching layers with feature maps are searched. The number of feature maps within all vertices of both genotypes for matching layers are averaged.

We use a minimum depth of 6 and a maximum of 12 for all DAGs. Due to our final choice of mutation operation that increases the depth size with every generation, we set the minimum and maximum depths for the random search to 10 and 36. The hyperparameter search uses a population size of 30, a mutation probability of 0.5, a crossover probability of 0, a probability of removing the worst candidates of 0.1, and an architectural estimation fraction of 0.5.

Architectural Performance Prediction. The experiment is conducted on `cifar10` [16] and the meta dataset to investigate architectural performance prediction consists of 56 features with a total of 2,472 data points - we split it into 70% training and 30% testing. The resulting meta-dataset constitutes a new supervised learning task containing graph-based features and the estimated performance of each candidate evaluated on `cifar10`. Three categories, namely layer masses, path lengths, and remaining graph properties, make up the meta-dataset.

The layer mass is the number of channels of the current vertex times the sum of all channels of preceding vertices with an incoming connection to the current vertex. The average and standard deviation of this layer mass are used as a graph-based feature in the meta-dataset. For path lengths, we consider the shortest, longest and expected random walk length from a source to a target vertex. Again, we take the average and standard deviation of these properties over all vertices in a graph and obtain 36 features over all possible vertex operations. Further, we include the depth, edge betweenness, eccentricity, closeness, average neighborhood degree, and vertex degree as graph properties to the meta-dataset features.

Six of the ten most important features relate to standard deviations of path lengths regarding convolutional blocks or pooling layers. When combining this result with the correlations of features and the maximum accuracy, which shows positive correlations for all these six features and the maximum accuracy, it seems that an even distribution of pooling and convolution layers benefits the performance of an architecture. This assumption is further backed by the observation that handcrafted models like DenseNet use pooling vertices to connect architectural cells. Mean centrality is also included in the ten most important features and shows a negative correlation to the maximum accuracy. Centrality is an inverted measurement of the farness of one vertex to other vertices in the network. Thus, we interpret these findings as evidence that deeper architectures perform on average better than shallower ones and that varying path lengths might support a similar effect as model ensembles. Compare Table 1 and Fig. 6b for correlations between graph properties and accuracy estimations across different experiments.

5 Discussion and Conclusion and Future Work

We presented experimental results on different methods for optimizing structures of neural networks: pruning, neural architecture search, and prior initialization. These methods unite under joint questions of how structure influences the performance of neural networks given data. Structural performance prediction is an emerging method to exploit this fact to speed up search procedures or to control bias towards desirable properties such as low memory or energy consumption or computational speed for specialized hardware.

We compared five pruning techniques for pruning feed-forward networks, namely random pruning, magnitude class blinded, magnitude class uniform, magnitude class distributed, and optimal brain damage. Out of these five, random pruning immediately showed an accuracy drop, while the other four performed consistently near original performance for over 60% compression rate. In the end, magnitude class blinded outperformed the remaining four.

While applying pruning on recurrent networks, we found models to perform consistently better for up to 60% pruning. All the pruned models regain the original performance in just one to two epochs of re-training. This means these recurrent networks can achieve the same results as any dense recurrent network with almost 60% fewer weights. As opposed to our expectations, LSTM and GRU recovered even after 100% pruning of hidden-to-hidden weights. In LSTM, this might be due to a separate cell state that acts as long-term memory.

Our experiment with random structural priors for recurrent networks aimed to find essential graph properties and use them for performance prediction. Similar to the results of [27], three of the essential features are the number of edges, vertices, and source nodes. Although the construction and properties of Watts-Strogatz and Barabási-Albert random graphs are different, recurrent networks based on this two performed equally with RNN_Tanh and RNN_ReLU. Barabási-Albert based recurrent networks perform better than Watts-Strogatz based with LSTM and GRU.

Correlation analyses between structural properties and the performance of an untrained network reveal themselves to be difficult – after all, the mere structure is, at first sight, and ignoring the structural dependencies on the input feature space, independent of data. All the more promising is if such a relationship between structure of models and application domains can be found. The idea that structures contain relevant information implies that architectural priors or search strategies over architectures can be heavily biased and influenced. To what extend this bias takes shape is difficult to understand, and we hope to foster more research towards its impact.

Acknowledgements. Paul Häusner and Jerome Würf contributed to this work during their research at the University of Passau, and we thank them for their contributions and valuable discussions.

References

1. Baker, B., Gupta, O., Naik, N., Raskar, R.: Accelerating neural architecture search using performance prediction (2018)
2. Barabási, A., Albert, R.: Emergence of scaling in random networks. Science, **286**(5439), 509–512 (1999)
3. Amor, M.B., Stier, J., Granitzer, M.: Correlation analysis between the robustness of sparse neural networks and their random hidden structural priors. arXiv preprint arxiv:2107.06158 (2021)
4. Dong, X., Chen, S., Pan, S.J.: Learning to prune deep neural networks via layer-wise optimal brain surgeon. In: Advances in Neural Information Processing Systems, pp. 4860–4874 (2017)
5. Elsken, T., Metzen, J.H., Hutter, F.: Neural architecture search: a survey. J. Mach. Learn. Res. **20**(1), 1997–2017 (2019)
6. Evci, U., Gale, T., Menick, J., Castro, P.S., Elsen, E.: Rigging the lottery: making all tickets winners. In: International Conference on Machine Learning, pp. 2943–2952. PMLR (2020)
7. Frankle, J., Carbin, M.: The lottery ticket hypothesis: finding sparse, trainable neural networks. In: International Conference on Learning Representations (2019).openreview.net/forum?id=rJl-b3RcF7
8. Gale, T., Elsen, E., Hooker, S.: The state of sparsity in deep neural networks. arXiv preprint arXiv:1902.09574 (2019)
9. Omlin, C. W., Giles, C. L.: Pruning recurrent neural networks for improved generalization performance. IEEE Trans. Neural Netw. **5**(5), 848–851 (1994)
10. Han, H.-G., Zhang, S., Qiao, J.-F.: An adaptive growing and pruning algorithm for designing recurrent neural network. Neurocomputing **242**, 51–62 (2017)
11. Han, S., Pool, J., Tran, J., Dally, W. J.: Learning both weights and connections for efficient neural network. In: Advances in Neural Information Processing Systems, pp. 1135–1143 (2015)
12. Hilgetag, C.C., Goulas, A.: Is the brain really a small-world network? Brain Structure and Function **221**(4), 2361–2366 (2015). https://doi.org/10.1007/s00429-015-1035-6
13. Irwin-Harris, W., Sun, Y., Xue, B., Zhang, M.: A graph-based encoding for evolutionary convolutional neural network architecture design. In: 2019 IEEE Congress on Evolutionary Computation (CEC), pp. 546–553. IEEE (2019)
14. Kidger, P., Lyons, T.: Universal approximation with deep narrow networks. In: Conference on Learning Theory, pp. 2306–2327 (2020)
15. Klein, A., Falkner, S., Hutter, F.: Learning curve prediction with bayesian neural networks (2016)
16. Krizhevsky, A., Sutskever, I., Hinton, G.E., et al.: Learning multiple layers of features from tiny images. (2009)
17. LeCun, Y., Bottou, L., Bengio, Y., Haffner, P.: Gradient-based learning applied to document recognition. Proc. IEEE **86**(11), 2278–2324 (1998)
18. LeCun, Y., Denker, J.S., Solla, S.A.: Optimal brain damage. In: Advances in Neural Information Processing Systems, pp. 598–605 (1990)
19. Liang, T., Glossner, J., Wang, L., Shi, S., Zhang, X.: Pruning and quantization for deep neural network acceleration: a survey. arXiv preprint arXiv:2101.09671 (2021)
20. Liu, H., Simonyan, K., Yang, Y.: DARTS: differentiable architecture search. In: International Conference on Learning Representations (2019). openreview.net/forum?id=S1eYHoC5FX

21. Liu, S., Mocanu, D.C., Matavalam, A.R.R., Pei, Y., Pechenizkiy, M.: Sparse evolutionary deep learning with over one million artificial neurons on commodity hardware. selfish sparse rnn training. In: International Conference on Machine Learning, Neural Comput. Appl. **33**(7), 2589-2604 PMLR (2021)
22. Marchisio, A., Hanif, M. A., Martina, M., Shafique, M.: Class-blind pruning method for deep neural networks. In: 2018 International Joint Conference on Neural Networks (IJCNN), pp. 1–8. IEEE (2018)
23. Mocanu, D.C., et al.: Scalable training of artificial neural networks with adaptive sparse connectivity inspired by network science. Nat. Commun. **9**(1), 1–12 (2018)
24. Mostafa, H., Wang, X.: Parameter efficient training of deep convolutional neural networks by dynamic sparse reparameterization. In: International Conference on Machine Learning, pp. 4646–4655. PMLR (2019)
25. Narang, S., Elsen, E., Diamos, G., Sengupta, S.: Exploring sparsity in recurrent neural networks. arXiv preprint arXiv:1704.05119 (2017)
26. Reber, A.S.: Implicit learning of synthetic languages: the role of instructional set. J. Exper. Psychol. Human Learn. Memory, **2**(1), 88 (1976)
27. Stier, J., Granitzer, M.: Structural analysis of sparse neural networks. Proc. Comput. Sci. **159**, 107–116 (2019)
28. Tolstikhin, I., et al.: Mlp-mixer: an all-mlp architecture for vision. arXiv preprint arXiv:2105.01601, 2021
29. Duncan, J.W., Steven, H.S.: Collective dynamics of 'small-world'networks. Nature, **393**(6684), 440–442 (1998)
30. Wendlinger, L., Stier, J., Granitzer, M.: Evofficient: reproducing a cartesian genetic programming method. In: EuroGP Genetic Programming, Cham, Springer International Publishing, pp. 162–178 (2021). ISBN 978-3-030-72812-0
31. Wistuba, M., Rawat, A., Pedapati, T.: A survey on neural architecture search. arXiv preprint arXiv:1905.01392 (2019)
32. Ying, C., Klein, A., Christiansen, E., Real, E., Murphy, K., Hutter, F.: Nas-bench-101: towards reproducible neural architecture search. In: International Conference on Machine Learning, pp. 7105–7114. PMLR (2019)
33. Zhang, M.S., Stadie, B.: One-shot pruning of recurrent neural networks by jacobian spectrum evaluation. arXiv preprint arXiv:1912.00120 (2019)

Active Learning for Capturing Human Decision Policies in a Data Frugal Context

Loïc Grossetête[1(✉)], Alexandre Marois[2], Bénédicte Chatelais[2], Christian Gagné[3], and Daniel Lafond[2]

[1] Enseirb-Matmeca Engineering School, Bordeaux, France
lgrossetete@enseirb-matmeca.fr
[2] Thales Research and Technology Canada, Toronto, Canada
{alexandre.marois,benedicte.chatelais,
daniel.lafond}@thalesgroup.com
[3] Université Laval, Quebec, Canada
christian.gagne@gel.ulaval.ca

Abstract. Modeling human expert decision patterns can potentially help create training and decision support systems when no ground truth data is available. A cognitive modeling approach presented herein uses a combination of supervised learning methods to mimic expert strategies. Yet without historical data logs on human expert judgments in a given domain, training machine learning algorithms with new examples to be labelled one by one by human experts can be time-consuming and costly. This paper investigates the use of active learning methods for example selection in policy capturing sessions with an oracle in order to optimize frugal learning efficiency. It also introduces a new hybrid method aimed at improving predictive accuracy based on a better management of the exploration/exploitation tradeoff. Analyses on three datasets evaluated data exploration, data exploitation and finally hybrid methods. Results highlight different tradeoffs of those methods and show the benefits of using a hybrid approach.

Keywords: Active learning · Frugal learning · Policy capturing · Cognitive systems engineering · Cognitive modeling · Decision support

1 Introduction

Active learning is gaining momentum as a way to maintain good machine learning performances even with scarce amounts of data [1]. This approach requires the learner to have access to a consequent unlabeled dataset (of potential cases), who then queries an oracle (generally assumed to provide the ground truth but in the discussed context it would in practice be replaced by a human expert) to get the label of one instance. In active learning application contexts, queries are considered to have a prohibitive cost (obtaining each labeled example costs time, resources or efforts) so it is important that the queried instance gives the model as much information as possible.

G. Nicosia et al. (Eds.): LOD 2021, LNCS 13164, pp. 395–407, 2022.
https://doi.org/10.1007/978-3-030-95470-3_30

The Cognitive Shadow system, developed by Thales Research and Technology Canada [2–5], is a data-driven tool making use of frugal learning techniques to capture the judgement policies of human experts and provide online decision support. In the current paper, we aim to assess the potential of such a technique under specific constraints such as the need to operate in real time and propose a new hybrid data exploration/exploitation method to balance potential tradeoffs. We first present an overview of the Cognitive Shadow prototype, the frugal learning problem and active learning methods. Section 2 details the experimental methodology used to compare the effectiveness of different approaches. Section 3 reports the analyses and results. Section 4 provides a discussion on our findings and their implications.

1.1 Cognitive Shadow

The Cognitive Shadow [2–6], is an AI-based knowledge capture and decision support system that can be integrated into various mission systems. It automatically learns an operator's decision pattern and provides real-time warnings to prevent potential errors when a mismatch is detected between the predicted decision and the user's. To do so, the Cognitive Shadow relies on seven supervised machine learning techniques that use the decision outcome as the predicted variable and the contextual attributes as the predictors. This online learning system continually improves itself over time, meaning that models are automatically retrained following new decision outcomes received. Model predictions are combined using a voting rule based on an accuracy metric automatically computed at each training iteration using a 10-fold cross-validation procedure. The prototype system is implemented as a web service with a representational state transfer application programming interface (REST API) and a relational (PostgreSQL) database. Python scripts are used to train the machine learning algorithms available in Scikit-Learn. The system allows configuring new classification tasks (e.g., triage, medical diagnosis, risk assessment) using the web interface and deploying this capability either on premise or in the cloud to enable interactions with different computerized task environments.

Unlike data-mining which allows finding hidden patterns in large datasets that even experts fail to recognize, policy capturing aims to model human expert judgments at either the individual or group-of-experts level. While this approach does not require historical data logs with ground truth labels, it does require human experts to take part in knowledge capture sessions. Judgmental bootstrapping [7] occurs when those models outperform human performance (not being prone to human errors due to factors such as fatigue, distraction, stress, mental overload). The Cognitive Shadow does not need large datasets to bootstrap human judgments because it exploits frugal learning methods shown to decrease the amount of data necessary to reach good prediction accuracy [2].

1.2 Frugal Data Machine Learning

If machine learning tends to be more powerful when a large quantity of data is available, it is not always the case. Because of this, some machine learning techniques aim to be efficient even with a low number of examples. To achieve such goals, three main methods are used. The first one is simply to use models inherently better on small amount of data such as Random Forest, Linear Regression or Support Vector Machines (SVM) [8]. Secondly, it is also sometimes possible to use transfer learning [9, 10] by reusing models trained on a task similar to the current task with a consequent number of data and retraining them using a small amount of data from the current task. This method is only usable when working on small datasets with strong similarity to a larger available one. Lastly, it is possible to use active learning methods (or adaptive optimal experimental design) [1]. This type of method aims at reducing the number of uninformative examples presented to the learner by selecting the data to be labeled by a user (often called the oracle). It is especially useful when large unlabeled datasets are available but it is expensive to label the data. Three main active learning strategies are generally used: a) Pool-Based Sampling [11], where the algorithm selects the data to be labeled from a finite pool of unlabeled data; b) Membership Query Synthesis [12], where the algorithm generates instances to be labeled from a predefined distribution; and c) Stream-Based Selective Sampling [13, 14], where the algorithm receives a flow of unlabeled data and for each decide if it asks for the data labeling or not.

There are two main factors impacting the efficiency of an active learning method resulting in the exploration/exploitation tradeoff [15]. First it should explore the unlabeled data space, meaning that the extracted samples to be queried should be a good approximation of the unlabeled data space and cover well the space for having all representative clusters to be properly labeled. Second, it should exploit the model's predictions to query the samples containing the most crucial information for the models.

Random Sampling is the main method used for data exploration. Each instance presented to the oracle is taken from the unlabeled dataset at random. This method has many desirable properties since the data distribution of the labeled instances sample converges to the full dataset distribution.

Density Clustering Sampling [16] is an exploration method that rather uses clustering techniques to separate the dataset into p-clusters. Then, a representative instance of each cluster is selected to be labeled.

Many active learning strategies focusing on data exploitation were developed and can be grouped into different categories based on the information metric they aim to maximize:

- Query by committee [17]: Query the data for which the prediction differs the most across multiple models obtained from different learning algorithms.
- Uncertainty sampling [18]: Query the data for which it is the least confident according to the confidence measures over the models' predictions.
- Expected model change [19]: Query the data that is maximizing the model change measured effect.
- Expected error reduction [20]: Query the data maximizing the model's generalization error reduction.

Most exploitation methods tend to perform better than exploration ones such as a random selection with a sufficient number of data [1]. However, they may sometimes get worse performances when the first examples selected are not representative since they would need the committee to be retrained at each decision to perform best and could miss small clusters of data if they are not detected early.. To solve this issue, we propose to use a hybrid of exploration and exploitation which will start by mostly exploring the datasets and gradually switch to exploitation to refine the decision boundaries similar to [15].

1.3 Active Learning for Frugal Human Policy Capturing

Our aim is to use active learning to strengthen the prediction of the Cognitive Shadow when a large pool of unlabeled data is available but only a scarce amount can be labeled by an oracle with the objective of the oracle being replaced by a human in future experiments. Therefore our study of active learning is made under the following conditions: a) it is limited to classification problems; b) it is limited to pool-based sampling since it cannot be assumed that a method to generate potential queries, excluding membership query synthesis, and stream-based is not adapted to the context; c) the active learning methods should be time-efficient so no model retraining is allowed during the query selection phase; and d) the active learning method should be model-agnostic since any kind of model could be used. The only assumption on the model is that it can give a prediction and an estimation of its confidence for each possible class for a given example. These constraints prevent us from using methods such as expected model change or expected error reduction, since they induce retraining, and, as well, any non-model-agnostic methods.

To explore the data space, it is possible to select the data at random but we also tested selecting the most distant instance to any previously selected data using the Euclidean norm, other norms have been considered but there was no apparent benefit to prefer one over the one used except computing time. Considering that this study is focused on low amounts of data and the relatively weak time complexity of the method, it was deemed unnecessary to use another norm. The method ensures that the selected data is diverse by taking points far from each other. However, this technique is very sensitive to outliers and potentially generates a reduction of accuracy when applied to datasets where most data points are in clusters and a few outliers are isolated, as it does not take the data distribution into account, plus one-hot encoded data have two distinct features with a distance of 1 so have twice the maximal impact of other features. One possibility we studied to enhance random exploration was to do first a K-means clustering on the dataset, dividing it into 10 groups, and then taking at random an example from each group. Doing so allows us to maintain a mostly random exploration while avoiding the potential risks to get very similar data in the firsts queries.

We thus developed two hybrid methods to compare their performances with pure exploration and pure exploitation methods. They are method-agnostic and use two valuation vector ($v_{exploration}$ and $v_{exploitation}$) containing respectively the valuation of each data in the dataset (the valuations have to be in [0,1[) for the exploration and exploitation methods and produces a result vector equal to:

$$\omega \times v_{exploration} + (1 - \omega) \times v_{exploitation} \tag{1}$$

For the first method, ω is a fixed constant manually defined. For the second, ω starts at 1 so that only exploration is used for the first samples and decreases linearly to reach 0, i.e. when the number of data queried reaches a fixed value. Both are easy to implement, fast and should mitigate the potential pitfalls of complex datasets.

2 Methods

2.1 Rival Strategies

The exploitation methods satisfying the constraints posed by the Cognitive Shadow and retained for comparison were modified versions of the *ModAL* library [21]. These three original methods were: a) max disagreement b) vote entropy, and c) consensus entropy. Max disagreement calculates the Kullback-Leibler divergence of each learner to the consensus prediction for each instance and scales it using min/max in [0,1]. Vote entropy divides the number of learners predicting each class for each instance to produce a probability distribution vector. Then it computes the entropy of the given probability vector and scale it using min/max in [0,1]. Consensus entropy does the same as vote entropy but calculates the average of the class probabilities of each classifier using the *Scikit-learn* method *predict_proba* to get the probability distribution.

The exploration methods satisfying the constraints posed by Cognitive Shadow and retained for comparison were Random, Farthest and K-means. The methods to balance exploration and exploitation considered and retained for comparison were Fixed omega 0.5, and Dynamic omega (reach $\omega = 0$ after 100 queries) both using Random as the exploratory method and consensus entropy as the exploitation one. The Random method was used since the two others are new and could fail to generalize to other datasets.

2.2 Classifiers

In order to use committee sampling methods, we needed different learners. Given our goal to enhance the Cognitive Shadow algorithms, the classifiers it uses were selected with their default parameters to avoid the time cost of hyperparameter optimization. All of them are from Scikit-learn 0.24.1. These algorithms are: Logistic Regression (*LogisticRegression(multiclass="ovr")*), Decision Tree (*DecisionTreeClassifier*), Naive Bayes (*GaussianNB*), SVC (*SVC(probability=True)*), Random Forest (*RandomForestClassifier*), K Neighbors Classifier (*KNeighborsClassifier*), and NN (*MLPClassifier(alpha=1e−5)*). To produce a prediction based on all the classifiers individual predictions on a single example, we used a simple voting aggregation method. The prediction on which most algorithms agreed on was chosen and the tie breaker was random.

2.3 Datasets

The first dataset type used in our evaluation was the AMASCOS dataset. This dataset refers to a simulation based on the Thales Airborne Surveillance Mission System (AMASCOS [6, 22]). This simulation was done in a context of a maritime patrol task where tactical coordinators had to classify radar contacts. Surface vessels monitored are characterized by 14 attributes and can be classified as either "Allied", "Neutral", or "Suspect". The list of attributes represents a mix of categorical and numerical variables (e.g., platform type, speed, length, and sea lane deviation). A one-hot encoding method was used for the 11 categorical variables. To allow us to perform statistical analyses on the results, 20 datasets of this type containing 200 labels each with balanced classes were generated using a specific set of rules. For this dataset type, a dataset with 10,000 examples was used as test set.

The second dataset type used was a modified version of AMASCOS with 50 randomly-generated outliers, each having a randomly attributed category to be incoherent with the rules. This second dataset was privileged to evaluate the resilience against outliers since some datasets might have some and it would be beneficial to maintain a certain level of accuracy. Again, to permit statistical analyses on the results, 20 datasets of this type containing 250 labels, 200 with balanced classes were generated using a specific set of rules and 50 randomly generated outliers. For this dataset type, the same dataset as for the first dataset type with 10,000 examples was used as test set.

The last dataset type used was targeting issues of exploitation methods highlighted by the study appearing when faced with complex data patterns embedded in simple ones. Indeed, most exploitation methods are selecting examples near first established decision boundaries but can miss small data clusters due to overconfidence in prediction. They are also over-focusing on specific regions if the models are not retrained between queries and it should be more evident on low dimensionality dataset. To test this, the dataset used took the form presented in Fig. 1.

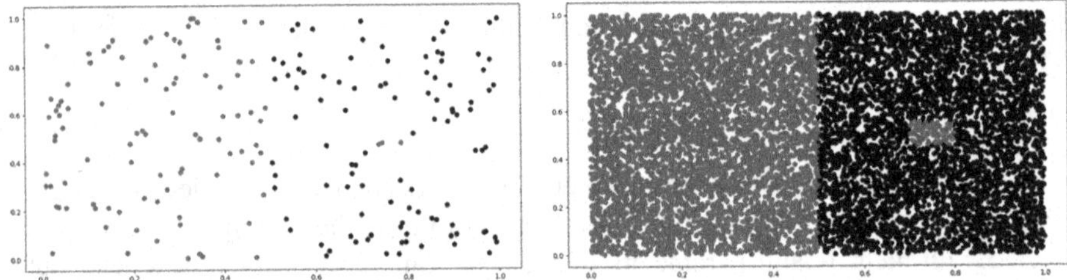

Fig. 1. Complex pattern dataset presentation: train set example (left) and test set example (right)

The goal here was to show that exploitation centered methods would be sometimes overconfident and assume that the only separation between the two class is in the middle and focus on refining the separation boundary while missing the small red cluster embedded in the black region. To allow us to perform statistical analysis on the results, 100 datasets of this type containing 200 labels each with 2 almost-balanced classes were generated using the following rule on a uniform distribution on $[0,1[^2$: $\forall (x_1, x_2) \in [0,1[^2$ the class is *Red* if $x_1 < 0.5 \cup (x_1, x_2) \in [0.7,0.8] \times [0.45,0.55]$ else it is *Black*. For this dataset type, a dataset with 10,000 examples was used as test set.

2.4 Evaluation Methods

For each dataset, we were interested in comparing the performances of each selected method. We therefore designed the following set of experiments each based on the following algorithm. For each training dataset, we made a loop where examples were chosen 10 by 10 by the desired selection method from the unlabeled dataset and added to a training pool. Each time, we retrained from scratch all the Cognitive Shadow algorithms on the training pool and used the aggregation method on the test dataset with the newly trained algorithms. We then gathered the results for 150 decisions for each method on each dataset. The first experiment was designed to compare the methods among themselves on the AMASCOS dataset. The second was made to do the same with the AMASCOS dataset containing additional outliers. The last experiment aimed at doing the same on the complex pattern dataset.

3 Analysis and Results

For all datasets, we generated a graph depicting the mean prediction accuracy of each method on a [0–1] scale as a function of the number of decisions (queries). To compare performance of the methods, we generated the average of the first three query sets (namely 10, 20 and 30 queries) and did the same for the three last query sets (i.e. 130, 140 and 150). The goal here was to compare the relationships between all methods in the first queries (with a small data sample) and in the last queries (with many data). Shapiro-Wilks tests and the number of data points raised issues pertaining to normality especially for the Complex Pattern dataset. Nonparametric Kruskal-Wallis H tests with the factor Methods (8 levels: Exploration random, Exploration farthest, Exploration K-means, Exploitation max, Exploitation vote, Exploitation consensus entropy, Hybrid fixed omega, and Hybrid dynamic omega) were thus privileged to identify whereas for both first and last queries groups there was at least a significant difference on the prediction accuracy among the different exploration, exploitation and hybrid methods. Bonferroni-corrected Mann-Whitney U tests were then executed among the methods to identify specific differences. Measures of effect size were also reported in the form of eta-squared measures (η^2) for the Kruskal-Wallis H tests and Mann-Whitney U tests.

3.1 AMASCOS Dataset

The results of the experiments comparing all exploration, exploitation and hybrid methods for the AMASCOS dataset are given in Fig. 2. The Kruskal-Wallis test ran on the first queries (queries 10, 20 and 30) raised at least one significant different among the methods, $\chi^2(7) = 31.75$, $p < .001$, $\eta^2 = .16$. Within the first queries, the Exploitation vote method significantly outperformed the Exploitation max method ($U = 60.23$, $p = .001$, $\eta^2 = .42$). The same pattern of results was observed in regards of the Exploration farthest method ($U = 51.60$, $p = .012$, $\eta^2 = .31$) and the Hybrid fixed omega method ($U = 60.75$, $p = .001$, $\eta^2 = .43$). All other differences failed to reach significance ($ps > .05$), suggesting that the other methods were not significantly different from the best method (Exploitation vote) but also from those with the lowest prediction accuracy (i.e. Exploitation max, Exploration farthest and Hybrid fixed omega).

The test performed on the last queries (queries 130, 140 and 150) of the AMASCOS dataset also raised at least one significant difference among the methods, $\chi^2(7) = 115.85$, $p < .001$, $\eta^2 = .72$. Consistent with the first queries, the Exploitation max method reached the lowest prediction accuracy, significantly differing from the two other exploitation methods, from the Exploration farthest method and from the two hybrid methods ($Us > 73.64$, $ps < .001$, $\eta^2 s > .63$). Coherent with the hypothesis that exploration methods can be less efficient with large datasets, the Exploration random and Exploration *K*-mean methods were outperformed by all the same methods than the Exploitation max method ($Us > 57.54$, $ps < .003$, $\eta^2 s > .39$). Globally, within the last queries (i.e. with larger datasets), both hybrid methods, the Exploration farthest method, the Exploitation consensus entropy method and the Exploitation vote methods led to the highest predictive accuracy.

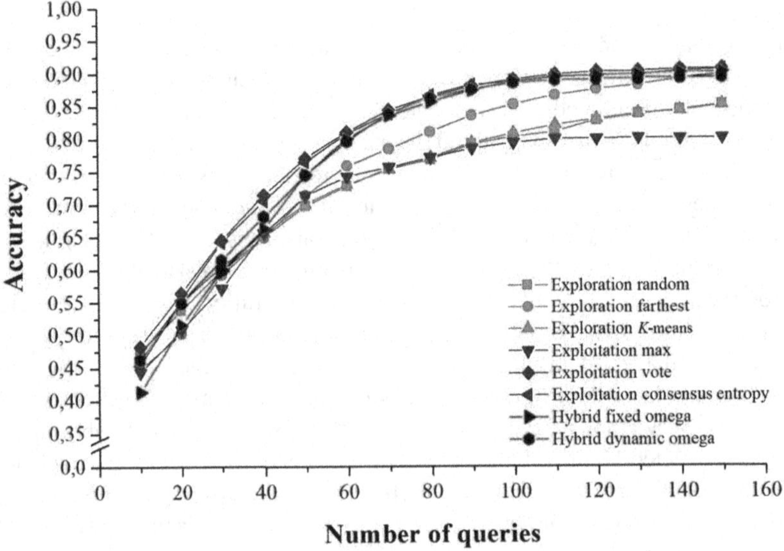

Fig. 2. AMASCOS dataset mean prediction accuracy for each exploration, exploitation and hybrid methods as a function of the number of queries

3.2 AMASCOS Dataset with Outliers

The different exploration, exploitation and hybrid methods were also compared for the AMASCOS dataset that contained outliers. Figure 3 depicts the mean predictive accuracy of all these methods for this dataset as a function of the number of queries. The Kruskal-Wallis test ran on the prediction accuracy of the first queries (10, 20 and 30) raised at least one significant difference among the methods, $\chi^2(7) = 61.24$, $p < .001$, $\eta^2 = .36$. The presence of outliers slightly changed the pattern of results as compared with the previous dataset. Indeed, the Exploration farthest method now became the worst method and was outperformed by all other methods, except for the Hybrid fixed omega method ($Us > 73.06$, $ps < .001$, $\eta^2 s > .62$). This latter method also generated accuracies that were significantly inferior to the Exploration random method ($U = 48.40$, $p = .027$, $\eta^2 = .27$) and the Exploitation vote method ($U = 52.20$, $p = .010$, $\eta^2 = .32$). All other differences did not reach significance ($ps > .05$). Overall, the best methods for these first queries were the Exploration random, Exploitation vote and Hybrid dynamic omega, but only the two first significantly differed from the two worst methods.

Regarding the last queries (queries 130, 140 and 150), the Kruskal-Wallis test executed on the prediction accuracies of the AMASCOS dataset with outliers also reached significance, $\chi^2(7) = 114.31$, $p < .001$, $\eta^2 = .71$. Multiple comparisons uncovered that the Exploitation max method, as for the AMASCOS dataset without any outlier, remained the worst method significantly inferior to all the others with the exception of the Exploration *K*-means method and the Exploration random method (Us > 69.37, ps < .001, η^2s > .56). In opposition with the pattern of results observed in the first queries, the Exploration random and the Exploration *K*-means methods dropped in rank and were outperformed by all other methods except for the Exploration farthest and the Exploitation max methods (Us > 52.59, ps < .010, η^2s > .32). Hence, with a higher number of queries, Hybrid fixed omega, Hybrid dynamic omega, Exploitation vote and Exploitation consensus entropy reached the best accuracies and differed from all the others but the Exploration farthest method.

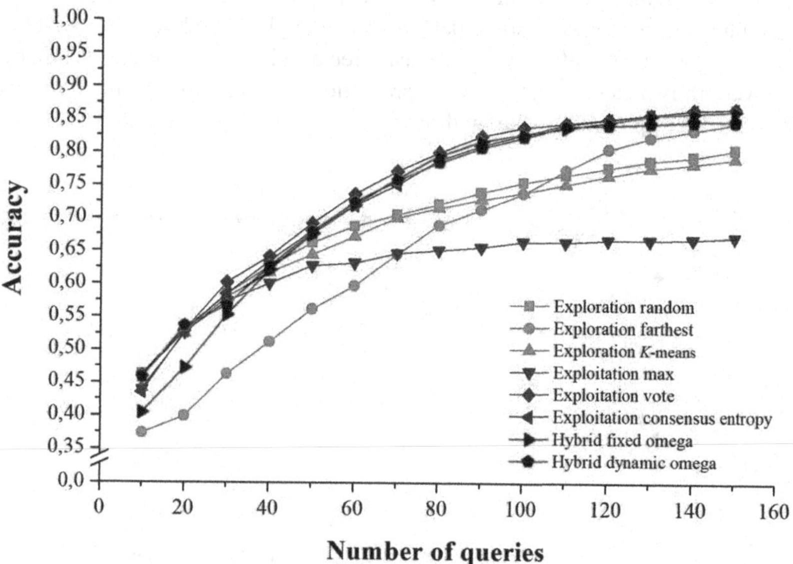

Fig. 3. AMASCOS dataset with outliers mean prediction accuracy for each exploration, exploitation and hybrid methods as a function of the number of queries

3.3 Complex Pattern Dataset

Mean prediction accuracies of the Complex pattern dataset for all the exploration, exploitation and hybrid methods were computed (see Fig. 4). At least one significant difference among the first queries (10, 20 and 30) could be found across the different methods with the Kruskal-Wallis omnibus test, $\chi^2(7) = 539.67$, $p < .001$, $\eta^2 = .67$. Further tests allowed to pinpoint that the three exploitation methods reached the poorest prediction accuracies and differed significantly from all the other methods, except with the Hybrid fixed omega method for the Exploitation consensus entropy and Exploitation vote methods (Us > 112.08, ps < .018, η^2s > .06). The Hybrid fixed omega method was also significantly outperformed by all the exploration methods and the Hybrid dynamic omega method (Us > 240.06, ps < .001, η^2s > .27). A significant advantage could also specifically be found for the Exploration farthest method compared with the Exploration random method ($U = 134.83, p = .001, \eta^2 = .09$). The three exploration methods and the Hybrid dynamic omega method were thus best suited for prediction with small amounts of queries.

A main effect of Method was also found for the last queries (130, 140 and 150) as the Kruskal-Wallis test reached significance level, $\chi^2(7) = 168.28, p < .001, \eta^2 = .20$. Decomposition of the effect with Mann-Whitney U tests uncovered that the Exploration random and Exploration K-means methods reached inferior prediction accuracy than all the other methods (Us > 139.87, ps < .002, η^2s > .09). The Exploration farthest method was also significantly inferior than the Exploitation max method ($U = 107.16$, $p = .029, \eta^2 = .05$) and the Exploitation vote method ($U = 165.22, p < .001, \eta^2 = .13$). The Exploitation vote method also outperformed both hybrid methods (Us > 109.63, ps < .008, η^2s > .06). Generally, among the last three queries, the exploitation and hybrid methods were thus better at reaching high prediction accuracy, though the best method seemed to be the Exploitation vote method.

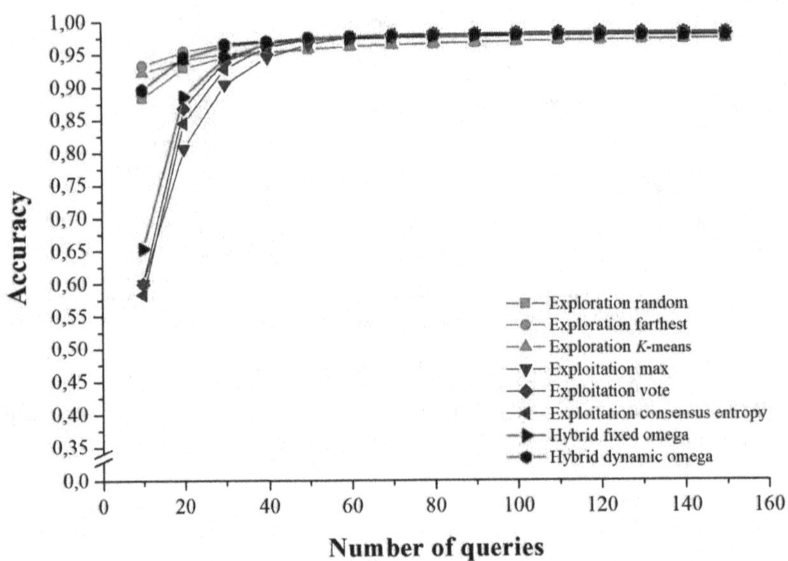

Fig. 4. Complex pattern dataset mean prediction accuracy for each exploration, exploitation and hybrid methods as a function of the number of queries

4 Discussion

This study highlighted some problems for active learning methods relying only on exploitation strategies with very scarce amounts of data when the committee is not retrained after each decision. Indeed, exploration methods showed an overall more robust pattern of results for the initial 30 learning trials (especially for the complex pattern dataset). The Exploration farthest method showed good overall performances compared to other exploration methods, but seems to be very sensitive to the type of outliers used in the study. Yet, it still manages to get decent performances after exhausting the outliers. Conversely, exploitation methods showed an overall greater potential once more data (130–150 examples) was collected, except for Exploitation Max method which yielded poor results compared to the other two exploitation methods.

This work introduced a potential solution to alleviate the problem by using a hybrid method allowing to transit gradually from an exploration phase to an exploitation phase. As expected, a hybrid approach combining exploitation and exploration methods produced the best overall pattern of results, but only when using a dynamic omega parameter (fully focusing on exploration at first and progressively switching to a full exploitation mode at 150 trials). The Hybrid static omega method fails to significantly increase overall performances. This is probably due to its lack of ability to transition from the exploration phase to the exploitation phase. The method proposed herein could be enhanced in many ways by modulating how the two strategies are weighed dynamically depending on performance variations. Furthermore, the two hybrid methods tested here relied on consensus entropy for exploitation and random sampling for exploration. Other exploration and exploitation strategy combinations should be tested as well in future works. This approach could also potentially be further improved by using other state of the art active learning methods such as [15, 23] for the exploratory phase. Directions for future works include testing this methodology on a greater variety of datasets ((including multiple complex-pattern datasets and public datasets) to assess generalizability of these results.

This study closely relates to previous work from [15] on balancing the exploration/exploitation trade-off, with a few key differences in methodology (mainly switching vs. combining methods, and query synthesis vs. query by committee). This work would also benefit from a sensitivity and specificity analysis which could highlight some potential advantages or disadvantages of some methods used [15].

Advances in frugal learning could prove beneficial in many data science applications since it allows a significant gain in accuracy for very low amounts of data without significantly degrading performances at other stages of learning overall. The methodology presented herein has the benefit of being modular and can be adapted for use with different types of methods.

This work contributed to the general issue of frugal learning since it is desirable to attain good performances even when working with very scarce amounts of data. This is particularly true for cognitive sciences and applied human factors since these fields often require humans to provide judgment labels which is generally time consuming and potentially costly.

The Cognitive Shadow and other future cognitive modelling technologies will benefit from these findings by improving the efficiency of interactive knowledge capture sessions with human experts. Indeed, capturing human expert decision policies may help create cognitive assistants, training agents, or safety nets for human decision making in different domains. Some of the many potential areas of application include medical triage and clinical making, maritime surveillance, environmental risk assessment, situation assessment, pilot decision making and human-machine teaming.

References

1. Settles, B.: Active Learning Literature Survey, vol. 52. University of Wisconsin Madison (2010)
2. Lafond, D., Roberge-Vallières, B., Vachon, F., Tremblay, S.: Judgment analysis in a dynamic multitask environment: capturing nonlinear policies using decision trees. J. Cogn. Eng. Decis. Mak. **11**, 122–135 (2017)
3. Labonté, K., Lafond, D., Hunter, A., Neyedli, H.F., Tremblay, S.: Comparing two decision support modes using the cognitive shadow online policy-capturing system. In: Proceedings of the Human Factors and Ergonomics Society Annual Meeting, vol. 64, pp. 1125–1129 (2020)
4. Lafond, D., Tremblay, S., Banbury, S.: Cognitive shadow: a policy capturing tool to support naturalistic decision making. Presented at the IEEE International Multi-Disciplinary Conference on Cognitive Methods in Situation Awareness and Decision Support (CogSIMA), San Diego, 1 February 2013 (2013)
5. Lafond, D., Labonté, K., Hunter, A., Neyedli, H.F., Tremblay, S.: Judgment analysis for real-time decision support using the cognitive shadow policy-capturing system. In: Ahram, T., Taiar, R., Colson, S., Choplin, A. (eds.) IHIET 2019. AISC, vol. 1018, pp. 78–83. Springer, Cham (2020). https://doi.org/10.1007/978-3-030-25629-6_13
6. Chatelais, B., Lafond, D., Hains, A., Gagné, C.: Improving policy-capturing with active learning for real-time decision support. In: Ahram, T., Karwowski, W., Vergnano, A., Leali, F., Taiar, R. (eds.) IHSI 2020. AISC, vol. 1131, pp. 177–182. Springer, Cham (2020). https://doi.org/10.1007/978-3-030-39512-4_28
7. Armstrong, J.S.: Judgmental bootstrapping: inferring experts' rules for forecasting. In: Armstrong, J.S. (ed.) Principles of Forecasting, pp. 171–192. Springer, Boston (2001). https://doi.org/10.1007/978-0-306-47630-3_9
8. Couronné, R., Probst, P., Boulesteix, A.-L.: Random forest versus logistic regression: a large-scale benchmark experiment. BMC Bioinformatics **19**, 270 (2018)
9. Pan, S.J., Yang, Q.: A survey on transfer learning. IEEE Trans. Knowl. Data Eng. **22**, 1345–1359 (2010)
10. Weiss, K., Khoshgoftaar, T.M., Wang, D.: A survey of transfer learning. J. Big Data **3**(1), 1–40 (2016). https://doi.org/10.1186/s40537-016-0043-6
11. Yang, Y.-Y., Lee, S.-C., Chung, Y.-A., Wu, T.-E., Chen, S.-A., Lin, H.-T.: libact: pool-based active learning in python (2017). https://arxiv.org/abs/1710.00379
12. Wang, L., Hu, X., Yuan, B., Lu, J.: Active learning via query synthesis and nearest neighbour search. Neurocomputing **147**, 426–434 (2015)
13. Atlas, L., et al.: Training connectionist networks with queries and selective sampling. In: Proceedings of the 2nd International Conference on Neural Information Processing Systems, pp. 566–573. MIT Press, Cambridge (1989)
14. Smailović, J., Grčar, M., Lavrač, N., Žnidaršič, M.: Stream-based active learning for sentiment analysis in the financial domain. Inf. Sci. **285**, 181–203 (2014)

15. Tharwat, A., Schenck, W.: Balancing exploration and exploitation: a novel active learner for imbalanced data. Knowl. Based Syst. **210**, 106500 (2020)
16. Wang, M., Min, F., Zhang, Z.-H., Wu, Y.-X.: Active learning through density clustering. Expert Syst. Appl. **85**, 305–317 (2017)
17. Seung, H.S., Opper, M., Sompolinsky, H.: Query by committee. In: Proceedings of the Fifth Annual Workshop on Computational Learning Theory - COLT '92, pp. 287–294. ACM Press, Pittsburgh (1992)
18. Lewis, D., Catlett, J., Cohen, W., Hirsh, H.: Heterogeneous Uncertainty Sampling for Supervised Learning (1996)
19. Cai, W., Zhang, Y., Zhou, J.: Maximizing expected model change for active learning in regression. In: 2013 IEEE 13th International Conference on Data Mining, pp. 51–60. IEEE, Dallas (2013)
20. Roy, N., Mccallum, A.: Toward optimal active learning through sampling estimation of error reduction. In: Proceedings of the 18th International Conference on Machine Learning (2001)
21. Danka, T., Horvath, P.: modAL: a modular active learning framework for python (2018). https://arxiv.org/abs/1805.00979
22. Marois, A., Chatelais, B., Grossetête, L., Lafond, D.: Evaluation of evolutionary algorithms under frugal learning constraints for online policy capturing. Presented at the IEEE International Multi-disciplinary Conference on Cognitive Methods in Situation Awareness and Decision Support (CogSIMA), Conference Presented Virtually 21 April (2021)
23. Bouneffouf, D., Laroche, R., Urvoy, T., Féraud, R., Allesiardo, R.: Contextual Bandit for Active Learning: Active Thompson Sampling (2014). https://hal.archives-ouvertes.fr/hal-01069802

Adversarial Perturbations for Evolutionary Optimization

Unai Garciarena, Jon Vadillo, Alexander Mendiburu, and Roberto Santana(✉)

Intelligent Systems Group, University of the Basque Country, San Sebastian, Spain
roberto.santana@ehu.eus

Abstract. Sampling methods are a critical step for model-based evolutionary algorithms, their goal being the generation of new and promising individuals based on the information provided by the model. Adversarial perturbations have been proposed as a way to create samples that deceive neural networks. In this paper we introduce the idea of creating adversarial perturbations that correspond to promising solutions of the search space. A surrogate neural network is "fooled" by an adversarial perturbation algorithm until it produces solutions that are likely to be of higher fitness than the present ones. Using a benchmark of functions with varying levels of difficulty, we investigate the performance of a number of adversarial perturbation techniques as sampling methods. The paper also proposes a technique to enhance the effect that adversarial perturbations produce in the network. While adversarial perturbations on their own are not able to produce evolutionary algorithms that compete with state of the art methods, they provide a novel and promising way to combine local optimizers with evolutionary algorithms.

Keywords: Adversarial perturbations · Model-based EAs · Neural networks · Sampling methods · EDAs

1 Introduction

For years, there has been an active cross-fertilization between the fields of neural networks (NNs) and evolutionary algorithms (EAs). While evolutionary methods are the most used algorithms for neural architecture search, neural networks have also been used to guide the search in model-based EAs [1]. One key question in the design of NN-EA approaches is how to combine the excellent efficiency of gradient based optimizers with the power of evolutionary operators to conduct global search. In this paper we investigate the effect that methods conceived for extracting information of NN have in the behavior of EAs. In particular, we focus on the study of adversarial perturbations applied to surrogates of fitness functions.

Given a machine learning model that classifies examples, an adversarial attack [20] aims to make imperceptible changes to the examples in such a way

G. Nicosia et al. (Eds.): LOD 2021, LNCS 13164, pp. 408–422, 2022.
https://doi.org/10.1007/978-3-030-95470-3_31

that they are incorrectly classified by the model. The effectiveness of the adversarial perturbation added to the example is measured both in terms of its capacity to fool the model and in terms of the amount of distortion required to make the adversarial attack successful. In this paper we investigate the use of adversarial perturbations as sampling method and model enhancer within an evolutionary algorithm. Using solutions already evaluated, we train a NN to classify whether a solution will have a high-fitness or a low-fitness. Then we apply adversarial perturbations to low-fitness solutions until the classifier is "fooled" to classify the solution as high-fitness. The assumption made is that, since the NN model captures information about the features that make solutions poor or good, the perturbations made to the solutions will indeed improve the fitness of solutions. Therefore, in order to make the network classify a poor solution as good, the adversarial perturbation will actually improve the solution in terms of the objective function being optimized.

Our work is in line with ongoing research on the use of different types of models to learn the most characteristic patterns of promising solutions in evolutionary algorithms [1,3,16]. There exists an apparent paradox between the high accuracy that a neural network can achieve at the time of capturing the characteristics of the data and the difficulty of effectively exploiting that information at the time of generating new solutions. None of the current approaches provides a general, satisfactory and efficient solution to sampling from neural networks within the evolutionary search scenario. Therefore, this paper introduces adversarial perturbations as a completely new approach for exploiting this information. This is a promising research direction because there is an active research on the theoretical basics of adversarial perturbations and a variety of approaches have been introduced with this goal.

The goal of our paper is not to introduce a new state of the art evolutionary algorithm for numerical optimization. Our objective is to analyze whether adversarial perturbations can be used as a new way to exploit the information learned by models of the best solutions opening the possibility of creating new hybrid optimization methods. We compare 15 different strategies used to create adversarial examples, most of which use gradient information about the model to create the perturbations. We identify those strategies that have a better performance within the context of evolutionary algorithms, and investigate whether gradient optimizers produce any advantage over black-box attacks (those that do not exploit information about the models to create the perturbations). Our preliminary results do not show a significant improvement in efficiency due to the use of adversarial perturbations, the results of the adversarial perturbation methods vary significantly depending on the problem. However, we identify methods that exploit the gradients of the models as the most effective.

The rest of this paper is organized as follows. In the following section we present adversarial examples and adversarial perturbations with a focus on those methods that will be later investigated in more detail. Section 3 introduces the elements of our proposal to use adversarial perturbation as part of the evolutionary process. Section 4 describes the algorithm and analyzes different aspects of the interaction between the EA, the NN and the technique for creating the adversarial perturbations that need to be taken into account for enhancing the search. Related work is reviewed in Sect. 5. Section 6 presents the experiments conceived to evaluate the performance of the different variants of adversarial attacks. Finally, conclusions drawn from the whole process are summarized in Sect. 7.

2 Adversarial Examples and Adversarial Perturbations

In this section we present a brief introduction of adversarial examples and adversarial perturbations. Our introduction is focused on the concepts and methods used in this paper. For a more detailed introduction to adversarial examples, we encourage the reader to examine [23].

Adversarial examples are inputs deliberately perturbed in order to produce a wrong response in a target deep neural network (DNN), while keeping the perturbation hardly detectable. The existence of slightly perturbed inputs able to fool state-of-the-art DNNs were first reported by [20], for image classification tasks.

Adversarial attacks can be classified according to different characteristics [23]. First of all, we denominate *targeted* adversarial examples to those inputs modified in order to produce a particular output class, or *untargeted* adversarial examples if the objective is to change the output of the model without fixing the output class we want to obtain. Moreover, we consider that an attack is *individual* if it has effect on just one input, or *universal* if it is able to fool the model for a large proportion of input samples. Regarding the information that an adversary has access to from the target model we can also differentiate between *white-box* or *black-box* attacks. In the former, which is the most common attack type, the adversary has full access to the model, including its weights, logits or gradients. In the latter, contrarily, we assume that it is not possible to obtain any information about the target DNN.

Goodfellow et al. [5] introduce the Fast Gradient Sign Method (FGSM) attack, in which the input x of ground-truth class y is perturbed in a single-step, in the same direction of the gradients of the loss function with respect to x:

$$x' \leftarrow x + \epsilon \text{sign}(\bigtriangledown \mathcal{L}(\theta, x, y)), \tag{1}$$

where $\mathcal{L}$ represents the loss function, θ the parameters of the target model and ϵ the maximum distortion allowed for each value in x. A straightforward extension of the FGSM approach is to perform more than one step, which is known as Basic Iterative Method (BIM) [8]. The *Momentum Iterative Attack* (MI) [2] is a further extension of this attack strategy, in which the momentum [14] of the gradients are considered in order to achieve a more effective attack. Thus, at each step the input sample is perturbed according to the following update rule:

$$x_t \leftarrow x_{t-1} + \epsilon \text{sign}(g_t), \tag{2}$$

where g_t represents the accumulated gradients until step t, with a decay factor μ:

$$g_t \leftarrow \mu \cdot g_{t-1} + \frac{\nabla \mathcal{L}(\theta, x_{t-1}, y)}{|| \nabla \mathcal{L}(\theta, x_{t-1}, y)||_1} \tag{3}$$

The Decoupled Direction and Norm attack [17], which also relies on iteratively perturbing the input sample in the direction determined by the gradients of the loss function, provides a different strategy in order to minimize the distortion amount of the perturbation. At each step t, the perturbation δ is updated with a step size α:

$$\delta_t \leftarrow \delta_{t-1} + \alpha \frac{\nabla \mathcal{L}(\theta, x_{t-1}, y)}{|| \nabla \mathcal{L}(\theta, x_{t-1}, y)||_2}, \tag{4}$$

and it is projected in the sphere of radius ϵ and centered in x, which constrains the ℓ_2 norm. However, at each step the value of ϵ is decreased by a factor of μ if x_{t-1} is able to fool the model or increased if it is not. In this way, after many updates,

$$x_t \leftarrow x + \epsilon_t \frac{\delta_t}{||\delta_t||_2}, \tag{5}$$

the solution is expected to converge to a valid adversarial example, while minimizing the ℓ_2 norm of the perturbation.

The attack rationale of the DeepFool algorithm [12], another iterative gradient-based approach, is to push an input sample x of class y towards the closest decision boundary of the target model.

$$r* = arg\ min\ ||r||_2 \text{ s.t. } f(x+r) \neq f(x). \tag{6}$$

To approximate $r*$, at step t, the decision boundaries of the model are linearly approximated in the vicinity of x, and for each $k \neq y$, the perturbation needed to reach (under the linear approximation) the decision boundary corresponding to the k-th class is determined by:

$$\delta_t^k \leftarrow \frac{f_k(x_t) - f_y(x_t)}{|| \nabla f_k(x_t) - \nabla f_y(x_t)||_2^2} (\nabla f_k(x_t) - \nabla f_y(x_t)) \tag{7}$$

being f_k the logits of f corresponding to the k-th class. Finally, the input is pushed towards that direction δ_t^* that requires the minimum distance: $x_{t+1} \leftarrow x_t + \delta_t^*$. This is done until the input finally reaches another decision region, that is, until the output of the model is changed.

Finally, the last method we focus on, the *Blend Uniform Noise Attack* method, relies on the addition of uniform noise to the input sample until it is able to produce a wrong prediction in the target model. Although this strategy may require a higher distortion amount to fool the model than the previous approaches, it can be suitable for black-box scenarios.

3 Adversarial Examples as Promising Solutions

In this section we define the components of the model that eventually is used for generating individuals for, and thus guide, the evolutionary algorithm. The three main components interacting in the whole evolutionary process are:

1. Sub-populations (S): Solutions grouped in classes.
2. Model (M): A supervised (neural network) classifier to classify a solutions.
3. Attack (A): A method for creating adversarial perturbations using the model.

3.1 Learning to Discriminate the Quality of the Solutions

The first step of the algorithm is to group the solutions into three groups according to their fitness values. Solutions are sorted according to their fitness and divided into three groups of similar size $\left(\frac{N}{3}\right)$. These groups are called: *Best*, *Middle*, and *Worst* solutions.

We could also split the population into two parts and consider only *Best* and *Worst* solutions. However, by creating an additional class grouping solutions with an intermediate value of the fitness, we intend to emphasize the difference between the groups comprising the best and worst performing solutions.

We define a prediction task that consists of, given a solution, determine whether it belongs to the *Best* or *Worst* groups (in this part of the procedure, the *Middle* set is ignored). This is a typical binary classification problem where the predictor variables or features are the decision variables of the optimization problem and the binary label can be interpreted as: 1) The solution belongs to the class of *Best* solutions. 0) The solution belongs to the class of *Worst* solutions.

Due to their capacity to fit any problem type, the model chosen for testing the adversarial methods is a multi-layer perceptron (MLP) [13]. In an MLP, every neuron in layer $l-1$ is exclusively connected to every other neuron in the next layer l. These dense connections can be represented as matrix operations:

$$n_l = \sigma_l(w_l \times n_{l-1} + b_l) \tag{8}$$

where, n_l represents data computed in the l-th layer, and $w_l, b_l \in \theta$ are the parameters for that layer. The first layer takes input data ($n_0 = x$), and for the output layer $n_l = \hat{y}$. Commonly, the outputs of the layers are activated by non-linear functions, in this equation represented by σ.

The neural network classifier is learned from scratch each generation using the individuals of the *Best* and *Worst* datasets from the previous generation. We also considered the possibility of starting the learning process from the network weights learned in the previous generation, but this approach led to a decreased ability of the network to learn from new solutions. Learning a new neural network in each generation can increase the computational load for problems with many variables. To cope with this question, starting the learning process from random weights could be triggered only at certain generations during the evolution.

3.2 Generating Promising Solutions with Adversarial Attacks

If the model has been successfully trained and it is able to generalize to unseen data, it will predict with a high accuracy whether solutions are good or poor. Therefore, we can start from a poor solution, which is correctly classified by the model as belonging to the class *Worst* and make adversarial perturbations on it until the model classifies it as belonging to the class *Best*. The assumption here is that the perturbations required for the model to be fooled are indeed perturbations that improve the quality of the solution in terms of the fitness.

Instead of using a low-fitness solution from the previous population, it is also possible to use a random solution, or a solution from the *Middle* class, as long as the model initially predicts it as belonging to the *Worst* class, because otherwise it will not be able to create an adversarial perturbation since the model can be fooled only if it initially predicts the class correctly.

3.3 Weaker Models Make More Adversarial Examples

As discussed in the previous section, it is not possible to improve solutions that belong to the *Best* class and are correctly classified by the model as such. This is a drawback because it means that the model can not further improve the best found solutions.

One partial remedy for this problem is, once the model has been trained, to partially modify it in order to make it incorrectly predict at least some of the good solutions as poor, but keeping the prediction of the poor solutions correct. Such modification will increase the pool of solutions to which adversarial perturbations can be applied, and will likely increase the gain in fitness when solutions in the *Best* class can be perturbed multiple times.

However, we would like the modifications made to the network not to distort the relevant information that it captures about the features that differentiate between good and poor solutions. Therefore, we constrain the modifications made to the network to the weights that connect the last hidden layer with the output layer. These weights can be represented by a matrix of dimension $m \times 2$, where m is the number of units in the last hidden layer. The two units in the output layer correspond to the values generated for the two classes.

Since in the output layer a `softmax` function is applied, and the output of the function is proportional to the two inputs values, we can expect that if we increase the weights that are connected to the unit of the *Worst* class then there will be more examples predicted as *Worst*. That is the type of modification made to the network. Therefore, we increase the weights connected to this unit in a parsimonious way until at least one of the good solutions used for training is incorrectly classified as *Worst*, or a maximum number of trials is reached.

This modification to the network can be beneficial not only because it increases the number of solutions to which adversarial perturbation could be applied to, but also because it can determine an increment in the amount of perturbation added to all solutions modified. The assumption is that, in order to reverse the classification given by the modified network to examples from the

Worst class, the adversarial perturbation method will require to add a higher amount of perturbation to the solution. Otherwise, it will not be able to balance the effect that the change of weights in the last connected layer of the modified network produced in the classification.

4 Surrogate Assisted EA with Adversarial Sampling

In model-based evolutionary algorithms (MEAs) [18], a model is used to discover latent dependencies between variables. In this type of algorithms, the synergism between the model and the method for sampling is one of the keys of the correct optimization workflow. Algorithm 1 shows a basic common MEA in pseudocode form.

Algorithm 1: Pseudo-code for a generic MEA.

```
pop = generate_population();
while halting condition is not met do
    fitness = evaluate_population(pop);
    selected_pop, selected_fit = select_solutions(pop, fitness);
    model = create_model(selected_pop, selected_fit);
    offs = sample_model(model);
    pop = offs + best(pop, fitness);
end
```

Several model types can be used to guide a MEA. A common choice are probably probabilistic graphical models, among which some of the most popular choices are Bayesian or Gaussian networks. MEAs driven by these models are known as estimation of distribution algorithms (EDAs). Nevertheless, a number of researchers [11,15,16] have also proposed the usage of different neural networks for guiding the search in EAs.

When inserted as part of Algorithm 1, adversarial perturbations can be seen as a method for sampling solutions generated by an MLP model of the best solutions. The algorithm can be also considered as a hybrid between local optimization methods used to create the adversarial perturbation and global search as implemented by the EA.

4.1 Inserting Adversarial Perturbation Methods in EAs

We can consider the methods for creating the adversarial perturbations as sampling procedures since we assume that they will produce solutions similar to the ones from the class *Best* used for training. However, it is not possible to know in advance which of the strategies discussed in Sect. 2 will be more effective to improve more the fitness of solutions. Some of the methods for creating the adversarial perturbations, such as FGSM are well known local optimizers, but other algorithms such as Deepfool exploit completely different strategies that depend on the information contained in the neural network representation.

We expect that adversarial methods, such as the `Blend Attack`, that do not exploit information about the network will be less effective than methods that exploit the gradient or the information about the class decision borders of the network, but this does not have to be the case.

Furthermore, evaluating the effect of the adversarial perturbations in the context of evolutionary optimization is also a difficult task since different criteria can be considered. For example, methods that increase the fitness the most in a single application can be desirable, but the speed of the method is another relevant criterion, as is the capacity of the algorithm to create "diverse" adversarial examples to accomplish a more exploratory search. None of these criteria are usually considered when evaluating the adversarial examples for attacking machine learning models.

5 Related Work

Although we did not find any previous application of adversarial perturbations as a component of an evolutionary algorithm, neural networks have been extensively investigated as surrogates in EAs [6]. They have been applied for estimating the values of the fitness functions and to assist in the application of mutation and crossover operators [7]. The main difference between our proposal and previous applications of neural networks as surrogates is that in our approach, the network information is used at the time of generating the sample, not to predict the quality of the sample.

In EDAs [9], sampling methods play an important role to generate new solutions. There are a number of papers that implement sampling methods to generate solutions from a neural network. Perhaps the best known examples are those algorithms based on variants of the Restricted Boltzmann Machines (RBM) [16,21] and Deep Boltzmann Machines [15]. This type of models keep a latent representation of the data, and exploit this information using Gibbs sampling or other Markov Chain Monte Carlo methods.

A closer relationship to our proposal exists with the method of ANN inversion [10], introduced by Baluja [1] to sample an MLP as part of an evolutionary algorithm (Deep-Opt-GA). The goal of ANN inversion (or back-drive) is, given the possible output value of a network, determine which input values produce that output. It is usually applied for neural networks that solve a regression problem [1,3].

There are important differences between the use of back-drive within the context of evolutionary search for generating solutions as in [1] and our proposal. First, back-drive is a single method to recover the inputs while adversarial perturbation is a diverse set of techniques of different nature. Second, back-drive assumes that the neural network has been trained to solve a regression task and therefore requires more detailed information about the solutions. We use a neural network that solves a classification task and therefore our method only requires a way to separate promising from poor solutions. Partial evaluation of solutions and other strategies can be used for this goal.

6 Experiments

In this experimental section we address different questions in order to evaluate the performance of the adversarial methods for guiding the evolutionary algorithm. More specifically, the goals of our experiments are:

- To determine whether the strategy used to initialize the solutions before applying the adversarial perturbations has an influence in the performance of the method.
- To investigate the behavior of the different adversarial perturbation methods within the evolutionary computation framework.
- To investigate whether the modification of the network improves the frequency in which adversarial perturbations can be applied.

The experiments were divided into three parts according to the questions stated above. In the first part of the experiments (Sect. 6.2), we investigate the influence of the initialization schemes for a reduced number of functions and a large set of adversarial perturbations methods. In the second part of the experiments (Sect. 6.3), we focus on the analysis of a reduced set of adversarial perturbation methods for a larger set of functions. In the third group of experiments, we investigate the effect of adversarial perturbations when networks have been tricked.

6.1 Problem Benchmark and Parameters of the Algorithm

The extensive problem suite of CEC-2005 [19], as implemented in [22], has been used to determine the performance of the adversarial methods driving an evolutionary algorithm. We select functions F1 to F17, except function F7 that has been discarded because of the range of possible values of the variables being not fixed.

For the first and third groups of experiments, the reduced set of functions comprises F1, F2, F6, F8, F13, and F14. These functions have been chosen as representative of three different levels of difficulty; univariate (F1 and F2), basic multimodal (F6 and F8), and expanded multimodal (F13 and F14).

All the experiments involve the same DNN structure: 2 hidden layers; sigmoid activation function after the two layers; softmax activation function after the output layer; weights (w_l) are initialized employing Xavier initialization [4]; and biases (b_l) are initialized to 0. The network is trained using the Adam optimizer, with a batch size of 50. Population size was $N = 1.000$ and we used truncation selection $T = 0.3$ with the *best elitism* method in which all the selected solutions are kept for the next population. The number of generations was $n_{gen} = 50$. For each combination and function, 5 independent executions have been performed. The algorithm was implemented in Python and the code is available from the authors upon request.

6.2 Initialization Schemes for Adversarial Perturbations

As explained in Sect. 3, the solutions are divided into three groups: *Best*, *Middle*, and *Worst*. The adversarial perturbations are applied to solutions in the third group and some solutions in the *Middle* class. The question we want to investigate is whether using these *poor* solutions as initial values can be beneficial for the workflow of the evolution. To test this hypothesis, we consider three scenarios: 1) Use the worst solutions to initialize the adversarial methods, 2) Use mutated variants of the solutions in the *Middle* class. 3) Use randomly initialized solutions. These scenarios are tested for all combinations of adversarial methods and functions.

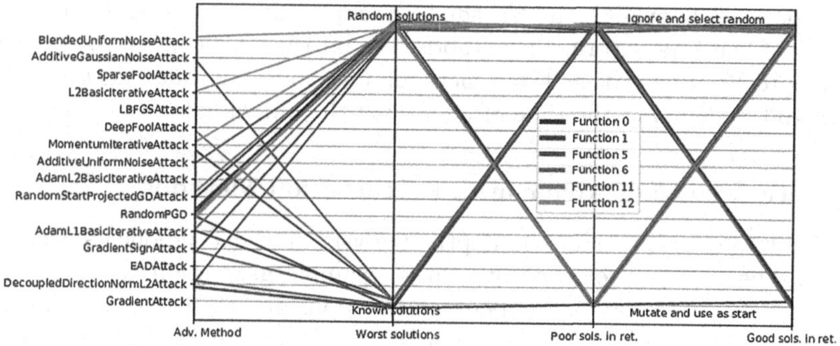

Fig. 1. Parallel coordinates showing the relation between the different components in the preliminary experimentation. For each problem, the top 3 runs are shown, one line for each one.

Figure 1 shows a summary of the results. In the figure, we represent, for each function, which have been the four best combinations of initialization and adversarial perturbation methods out of all the runs executed. It serves as way to identify top-performing configurations. In the figure, each configuration of adversarial method and individual initialization technique is represented by a line.

The analysis of Fig. 1 reveals that the effect of the initialization method is, to some extent, problem-dependent. Simpler problems can be better optimized initializing from the worst solutions, the more complex ones benefit almost exclusively from random solutions. This effect is less noticeable for solutions in the *Middle* set. In this case, initialization from solutions in the *Middle* class can also produce top results for complex functions. Runs that do not use random initialization were only able to achieve a top-3 performance once, for a simple function. These results indicate that the diversity component introduced by the random initialization is, overall, a positive feature.

In addition to clarifying the effects of the different initialization methods, the experiments show that the performance of the adversarial methods is in general problem-dependent. For example, some of the best results obtained for the expanded multimodal problems (the *clearest* lines), have been obtained when the EA uses adversarial perturbation methods unable to produce top results for any of the basic multimodal or unimodal problems (`BlendedUniformNoise` or `DeepFool`). Regarding these other problems, EAs incorporating adversarial methods that use the gradients and the `DecoupledDirectionNormL2` were the ones producing the best results. This finding is relevant because it points to a possible way to characterize the behavior of adversarial perturbation methods using different classes of optimization problems.

Methods which make use of the gradients (in this case `RandomPGD`, `DecoupledDirectionNormL2`, and `RandomStartProjectedGD`) outperform the others for the simplest type of functions. Other methods which showed good performances in the most difficult functions while still performing considerably well in the simpler ones are `DeepFool` and `BlendedUniformNoise`. As can be observed, the majority of the top-performing methods rely on the gradients of the network. This was to be expected, as these methods take more informed decisions with respect to the model at the time of modifying the individuals.

6.3 Performance of Adversarial Perturbation Methods

In order to perform a more in-depth analysis of the components that can affect the performance of the adversarial example generation technique as a guide of EAs, we enlarge the pool of functions in which the algorithm is tested. We also constrain the set of adversarial perturbation methods to: A method which does not rely on the gradients of the network (and therefore requires no information about the model): `BlendedUniformNoise`, and other four which do `DeepFool`, `DecoupledDirectionNormL2`, `RandomPGD`, and finally `RandomStartProjectedGD`. We evaluate two scenarios for applying the adversarial perturbation methods to solutions in the *Middle* class: R) random initialization and M) mutation of a solution in class *Middle*.

Figure 2 shows a heatmap which encompasses the results of this experiment. All fitness values obtained by every method and combination were scaled to [0, 1] in order to improve the interpretability of the results. For each problem (in the y axis), and method-variant combination (in the x axis), the color represents the mean of the best fitness values, computed from the five runs. The darker (and therefore, lower) the better. Additionally, each of the cells in the heatmap has an overlaid digit. This number is the negative logarithm in base 10 of the variance obtained across all five runs. In other words, the number is the positive version of the exponent of the variance. In this case, a larger number denotes less uncertainty about the final result.

Analyzing this figure row-wise, it is possible to observe that some of the problems have considerably brighter colors across all method combinations. The runs for these functions have produced largely diverse best fitness values (e.g., one *good* run and other four *very poor* runs), which explains the mean being high in the [0, 1] interval and the high variance.

Analyzing the figure as a whole, it becomes apparent that, again, there is not an absolute winner among the adversarial methods. For example, for problems

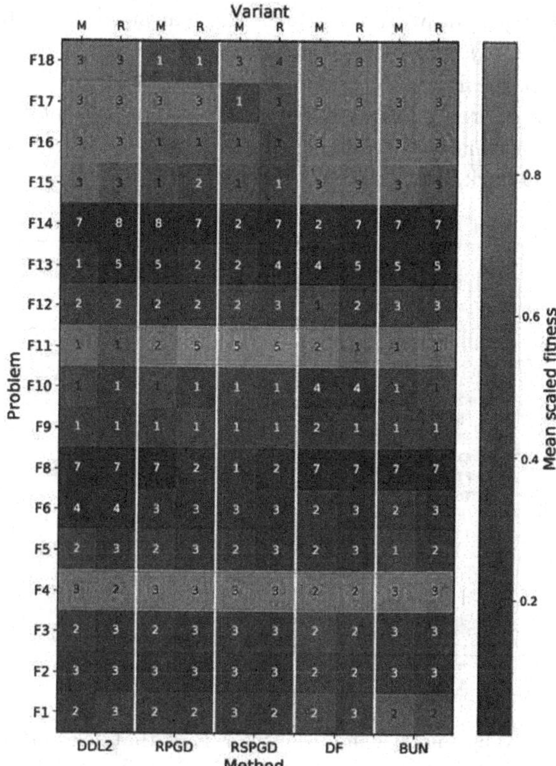

Fig. 2. Heatmap showing the mean values obtained by each method with different treatment of the *poor* solutions in the middle set, for each function. The numbers inside the figures represent the variance of all runs performed. Variants for starting solutions: Mutated versions of the solutions classified as poor in the middle set (M), or random solutions as (R). Methods: Decoupled Direction Norm L2 (DDL2), RandomPGD (RPGD), Random Start Projected GD (RSPGD), DeepFool (DF), Blended Uniform Noise (BUN).

F1-F3, DDL2, RPGD, and RSPGD produce the best mean fitness values. For F5-F8, DDL2 and DF offer the best performance, whereas for F9 and F10, only DF has a good performance. RSPGD and RPGD exhibit the lowest means for F15-F18, and for F4 and F11, DDL2 is the most consistent one. Almost in all cases, using a random initialization is better than applying mutation.

6.4 Network Tricking

Finally, we focus on trying to determine whether tricking the network in such a way that it is more difficult for it to predict a solution as *good* can be beneficial for the EA. If the adversarial perturbation method is forced to further modify the solutions so that the network recognizes them as good, then there is a chance of the solutions actually improving even more.

To test this hypothesis, we use the reduced set of functions and an adversarial method which has offered consistent performance across functions and parameters (DDL2). Five runs were respectively executed for the EAs that use the ordinary neural network, and the tricked network.

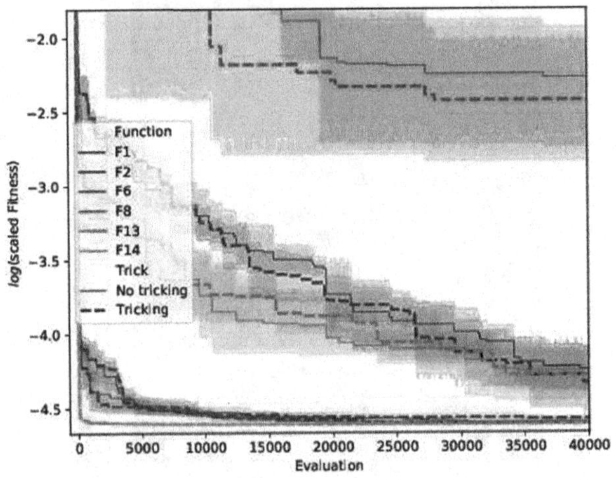

Fig. 3. Best fitness value along generations for EAs that use ordinary and tricked neural network.

Figure 3 shows details of the evolution for the six different functions (in different colors). Each line represents the mean of the best found individual in terms of fitness (y axis) at each point in the evolution (x axis) of each set of five runs. Dashed lines represent runs in which network modification took place. The clear lines accompanying the opaque ones represent the standard deviation of the 5 runs.

As in previous experiments, results are problem-dependent. For functions F1, F6, F8, and F13, performing the network modification produces an improvement in the performance of the algorithm, whereas this was not the case for F2 and F14. What is more, in this case, no pattern about certain characteristics fitting problem particularities can be deduced, since the best results for the unimodal (F1 and F2), and the expanded multimodal functions (F13 and F14) were obtained with different approaches. Taking all into account, however, tricking the network improved the evolutionary process on four of the six sets of runs.

7 Conclusions

In this paper we have proposed the use of adversarial perturbations as a way to guide the search for optimal solutions in an evolutionary algorithm. Our method combines the use of a neural network that acts as a predictor of promising versus poor solutions, with the application of algorithms originally conceived to deceive neural networks. Our results show that it is indeed possible to use the adversarial perturbations to improve the quality of the solutions from the first generation. However, the perturbations stop improving the results after a relatively small number of generations.

Our results are also of interest for research on neural networks. We have shown how evolutionary optimization can serve as test-bed to evaluate different methods for deceiving the networks. In this sense, our work opens another avenue for the investigation of synergies between neural networks and optimization algorithms.

References

1. Baluja, S.: Deep learning for explicitly modeling optimization landscapes. CoRR abs/1703.07394 (2017). http://arxiv.org/abs/1703.07394
2. Dong, Y., et al.: Boosting adversarial attacks with momentum. In: Proceedings of the IEEE Conference on Computer Vision and Pattern Recognition, pp. 9185–9193. IEEE Press (2008)
3. Garciarena, U., Mendiburu, A., Santana, R.: Envisioning the benefits of back-drive in evolutionary algorithms. In: 2020 IEEE Congress on Evolutionary Computation (CEC), pp. 1–8. IEEE (2020)
4. Glorot, X., Bengio, Y.: Understanding the difficulty of training deep feedforward neural networks. In: Proceedings of the Thirteenth International Conference on Artificial Intelligence and Statistics, pp. 249–256 (2010)
5. Goodfellow, I.J., Shlens, J., Szegedy, C.: Explaining and harnessing adversarial examples (2014)
6. Jin, Y.: Surrogate-assisted evolutionary computation: recent advances and future challenges. Swarm Evolut. Comput. **1**(2), 61–70 (2011)
7. Jin, Y., Olhofer, M., Sendhoff, B.: On evolutionary optimization with approximate fitness functions. In: Proceedings of the 2nd Annual Conference on Genetic and Evolutionary Computation, pp. 786–793 (2000)
8. Kurakin, A., Goodfellow, I., Bengio, S.: Adversarial examples in the physical world. CoRR abs/1607.02533 (2016). http://arxiv.org/abs/1607.02533
9. Larrañaga, P., Karshenas, H., Bielza, C., Santana, R.: A review on probabilistic graphical models in evolutionary computation. J. Heuristics **18**(5), 795–819 (2012). https://doi.org/10.1007/s10732-012-9208-4
10. Linden, A., Kindermann, J.: Inversion of multilayer nets. In: Proceedings of the International Joint Conference on Neural Networks, vol. 2, pp. 425–430 (1989)
11. Marti, L., García, J., Berlanga, A., Molina, J.M.: Introducing MONEDA: scalable multiobjective optimization with a neural estimation of distribution algorithm. In: Proceedings of the 10th Annual Conference on Genetic and Evolutionary Computation GECCO-2008, pp. 689–696. ACM, New York (2008). http://doi.acm.org/10.1145/1389095.1389228
12. Moosavi-Dezfooli, S.M., Fawzi, A., Frossard, P.: DeepFool: a simple and accurate method to fool deep neural networks. In: Proceedings of the IEEE Conference on Computer Vision and Pattern Recognition, pp. 2574–2582 (2016)
13. Pal, S.K., Mitra, S.: Multilayer perceptron, fuzzy sets, and classification. IEEE Trans. Neural Netw. **3**(5), 683–697 (1992)

14. Polyak, B.T.: Some methods of speeding up the convergence of iteration methods. USSR Comput. Math. Math. Phys. **4**(5), 1–17 (1964)
15. Probst, M., Rothlauf, F.: Deep Boltzmann machines in estimation of distribution algorithms for combinatorial optimization. CoRR abs/1509.06535 (2015). http://arxiv.org/abs/1509.06535
16. Probst, M., Rothlauf, F., Grahl, J.: Scalability of using restricted Boltzmann machines for combinatorial optimization. Eur. J. Oper. Res. **256**(2), 368–383 (2017)
17. Rony, J., Hafemann, L.G., Oliveira, L.S., Ayed, I.B., Sabourin, R., Granger, E.: Decoupling direction and norm for efficient gradient-based l2 adversarial attacks and defenses. In: Proceedings of the IEEE Conference on Computer Vision and Pattern Recognition, pp. 4322–4330 (2019)
18. Stork, J., Eiben, A.E., Bartz-Beielstein, T.: A new taxonomy of global optimization algorithms. Nat. Comput., 1–24 (2020)
19. Suganthan, P.N., et al.: Problem definitions and evaluation criteria for the CEC 2005 special session on real-parameter optimization. Technical report, Nanyang Technological University, Singapore (2005)
20. Szegedy, C., et al.: Intriguing properties of neural networks. CoRR abs/1512.1312.6199 (2015). http://arxiv.org/abs/1312.6199
21. Tang, H., Shim, V., Tan, K., Chia, J.: Restricted Boltzmann machine based algorithm for multi-objective optimization. In: 2010 IEEE Congress on Evolutionary Computation (CEC), pp. 1–8. IEEE (2010)
22. Wessing, S.: Optproblems: infrastructure to define optimization problems and some test problems for black-box optimization. Python package version 0.9 (2016)
23. Yuan, X., He, P., Zhu, Q., Li, X.: Adversarial examples: attacks and defenses for deep learning. IEEE Trans. Neural Netw. Learn. Syst. **30**(9), 2805–2824 (2019)

Cascaded Classifier for Pareto-Optimal Accuracy-Cost Trade-Off Using Off-the-Shelf ANNs

Cecilia Latotzke(✉), Johnson Loh, and Tobias Gemmeke

RWTH Aachen University, 52062 Aachen, Germany
{latotzke,loh,gemmeke}@ids.rwth-aachen.de

Abstract. Machine-learning classifiers provide high quality of service in classification tasks. Research now targets cost reduction measured in terms of average processing time or energy per solution. Revisiting the concept of cascaded classifiers, we present a first of its kind analysis of optimal pass-on criteria between the classifier stages. Based on this analysis, we derive a methodology to maximize accuracy and efficiency of cascaded classifiers. On the one hand, our methodology allows cost reduction of 1.32× while preserving reference classifier's accuracy. On the other hand, it allows to scale cost over two orders while gracefully degrading accuracy. Thereby, the final classifier stage sets the top accuracy. Hence, the multi-stage realization can be employed to optimize any state-of-the-art classifier.

Keywords: Cascaded classifier · Machine learning · Edge devices · Preliminary classifier · Pareto analysis · Design methodology

1 Introduction

Machine learning techniques are on the rise for classification tasks since they achieve higher accuracies than traditional classification algorithms. Especially, Deep Neural Networks (DNN) have proven highly effective in benchmark competitions such as the ImageNet Challenge [1]. However, this success comes at a price - computational complexity and thereby energy has skyrocketed [5].

Today, automated data analysis is used in sensitive medical or industrial applications. These tasks vary over a wide range such as image classification, voice recognition or medical diagnostic support. As misclassification in these areas has a high price, classification accuracy is expected to be on par with a trained human.

In some application cases, raw data contains sensitive private information, i.e., streaming to and processing in the cloud is not always an option, letting alone the energy cost of transferring raw data as compared to classification results. This together gives rise to classification on edge devices. Featuring a limited energy

C. Latotzke and J. Loh—Contribute equally to the paper.

G. Nicosia et al. (Eds.): LOD 2021, LNCS 13164, pp. 423–435, 2022.
https://doi.org/10.1007/978-3-030-95470-3_32

budget, the computational cost of classification is constrained [18]. Preferably, the classification algorithm needs to be adjustable to balance between the available energy budget and application specific accuracy requirements. Furthermore, operating in real-time drives the need for high throughput. At the same time, live data contains identical or even irrelevant information, i.e., the data-sets are skewed [6,9]. Thus, a promising approach for cost reduction is preprocessing in wake-up or reduced-complexity classifiers that forward only seemingly interesting samples towards the next stage. This was successfully applied in embedded sensor nodes [3] and for speech recognition [2,21]. The goal is to achieve lowest possible energy for data processing on the edge device while maintaining a high quality of service of the classification algorithm [15]. They can be summarized as hierarchical classifiers.

The general concept of hierarchical classifiers was first introduced by Ouali and King [20]. The various design dimensions such as applied algorithm, architecture, or quantization span a large design space for cascaded classifiers. Adding the substantial variety of existing datasets and their complex feature space, benchmarking different solutions becomes a challenge by itself [8].

This work focuses on the optimization of cascaded classifiers to accelerate highly energy efficient classification on edge devices. The main contributions of this manuscript are: 1) a first of its kind in-depth analysis of various existing and novel pass-on criteria between the stages of a cascaded classifier, used to qualify a classifier's output and steer the pass-on-rate, the rate of samples not classified in the current classifier stage and passed to the succeeding classifier stage; 2) a detailed discussion of different design choices of cascaded classifiers as low-cost in-place substitutes of an existing classifier; 3) a design methodology based on the accuracy-cost trade-off for Pareto-optimal cascaded classifier constellations, with cost in terms of average processing time or energy per sample and 4) the composition of accuracy-efficiency trade-off driven cascaded classifiers, of in-house trained classifiers for both datasets, MNIST [17] and CIFAR 10 [14]. Their Pareto-optimal settings provide a graceful trade-off between classification accuracy and cost per classification - matching the specific use case. We provide the template scripts to reproduce the results of this paper in https://git.rwth-aachen.de/ids/public/cascaded-classifier-pareto-analysis.

The paper is organized as follows. Section 2 discusses the state-of-the-art of cascaded classifiers and introduces its design parameters. Based on the analysis of the pass-on criteria, the design methodology for an efficiency-accuracy driven cascaded classifier is presented in Sect. 3 In Sect. 4 the validation of the methodology for CIFAR 10 is shown. Conclusions and remarks are drawn in Sect. 5.

2 Related Work and Background

The concept of hierarchical classification comprises various ways to combine classifiers including trees or cascades [26]. This paper focuses without any loss of generality on the latter subdomain, i. e. classifier structures being activated sequentially. The idea is to pass a sample over consecutive stages of increasing accuracy and cost until a stage's specific pass-on criterion is above a preset threshold. The number of samples in the cascaded classifier is reduced by every decision made in a previous stage (grey arrows in Fig. 1), while undecided samples are passed on to the next stage. Per convention, the indexing starts with $i = 0$ at the last stage - the reference classifier, that sets the maximum accuracy.

An overview of state-of-the-art cascaded classifier architectures is shown in Fig. 1. The approach to develop classifier stages, pass samples and optimize the system varies in the different structures.

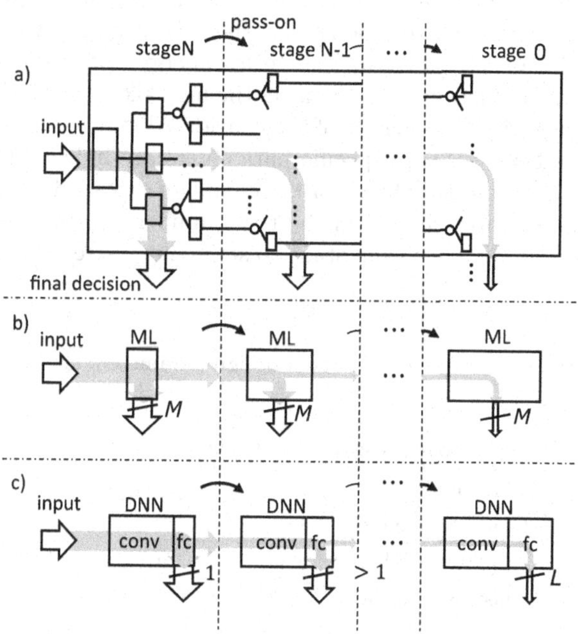

Fig. 1. Concept of state-of-the-art cascaded classifiers with increasing (a) complexity of referenced features [6], (b) quality of multi-class classifier [13,24] and (c) number of output classes [9], with M and L being the maximum number of classes for each model correspondingly.

In Fig. 1a) the stages are using fuzzy trees to determine the classified label [6]. By increasing the number of features utilized by the decision tree, the cost is increased as well as its capability to achieve higher accuracy. The fuzzy classifier computes initially only a subset of all features, since not all samples are equally complex to classify. Additional and more complex features are added in later stages to enable proper classification of less frequent, more difficult to classify samples. Thereby, earlier extracted features are propagated resulting in a classifier with on-demand generation of computationally more expensive features [25]. As a drawback, this multi-stage scheme is rather inflexible, as the architecture has to be revised for any change in the distribution of features contained in the samples.

A more general approach was earlier proposed in [24] (cf. Fig. 1b). The general concept is the concatenation of independent Machine Learning classifiers (ML), whose complexity is progressively increasing. It focuses on ensembles of binary "one-class-expert" classifiers determining the confidence value based on relative affiliation of a sample to a specific class.

Another principle is to modify the number of classes processed in each stage (cf. Fig. 1c), [9]). With increasing stage index the networks are trained to differentiate a larger number of classes in the set. This approach can be especially beneficial in real-life scenarios with highly skewed input data, since many irrelevant, but frequent, samples are processed in the earlier stages. Various examples exist that introduce such "wake-up" stage for energy optimization [2,3,21,22] achieving dramatic reductions by exploiting the high skewness of the sample set.

A summary of design parameters of the considered multi-stage classifier principles is given in Table 1. Stage specific parameters are indicated via '×' while parameters, which are globally fixed for the complete cascaded classifier are indicated via '–'. For example the '×' in the row 'number of labels' stipulates that the number of labels can vary from stage to stage, i.e., increase towards the last stage.

Table 1. Stage specific versus globally fixed design parameters for cascaded classifiers

Design parameters	[9]	[6]	[24]	[13]	*Ours*
Threshold	×	–	×	–	×
Confidence metric	–	–	–	–	×
Number of labels	×	×	–	–	–
Classifier algorithm	–	–	×	–	×

As earlier mentioned, the approaches in Fig. 1a) and Fig. 1c) are beneficial for skewed datasets, while the approach in Fig. 1b) is best suited for unskewed datasets, like MNIST, CIFAR 10 or ImageNet. These datasets are commonly used in benchmarking of machine learning architectures. Hence, the state-of-the-art comparison of cascaded classifiers in Table 2 focuses on these. The key criterion for the comparison is the cost reduction achieved with the cascaded classifier either by upholding the accuracy of the reference classifier or by tolerating 99% preservation of the reference accuracy.

3 Methodology

3.1 Architecture and Quantitative Optimization

This work is based on the general architecture shown in Fig. 1b). The presented methodology can also be used as in-place replacement of any single stage in Fig. 1c) in order to combine the benefits of both cascaded classifier schemes.

Table 2. State of the art cascaded classifiers for unskewed datasets

Paper	[24]	[13]	Ours	Ours
Accuracy	Top-1	Top-5	Top-1	Top-1
Benchmark	MNIST	ImageNet	MNIST	CIFAR 10
Cost function	$\text{Energy}_{\text{norm}}$	GOps/s	MOps	GOps
Classifier	SVM	CNN	CNN	CNN
acc_{ref} [a]	–	87%	99.45%	94.66%
Cost @ acc_{ref}	–	299.07	30.94	3.48
Cost reduction cascaded classifier @ acc_{ref} [b]	1.05×	0.7×	5.71×	1.32×
Cost reduction cascaded classifier @ 99% acc_{ref} [c]	2.85×	1.45×	263.17×	2.55×

[a] reference accuracy: accuracy of stand alone final stage classifier
[b] achieves same accuracy as reference accuracy
[c] achieves 99% of reference accuracy

Our methodology makes use of readily available state-of-the-art classifier architectures reusing the available outputs to derive a confidence metric. With this approach, we are generalizing from the ensemble of one-class-expert classifiers as in [24] to a set of existing, and upcoming, machine learning classifiers freely chosen by their key performance indicators, e.g. cost and accuracy.

Given the complexity of comparing different solutions, cost is assumed to scale proportional to the number of Multiply-and-Accumulate (MAC) operations per inference and according to the quantization. Starting with a 32-bit floating point (fp32) reference implementation, scaling factors for quantized 32-bit fixed point (fx32) and binary (bin) operations have to be identified. To scale to fx32, we use a factor of $\alpha_{\text{fp32}\rightarrow\text{fx32}} = 0.7$ computed as ratio of sums of energy numbers of addition and multiplication given in [10]. Scaling to binary representation according to [23] would result in a factor of $\alpha_{\text{fx32}\rightarrow\text{bin}} > 2000$. Well aware of the crudeness of this assessment, we decided to adopt as upper bound a rather conservative scaling factor of $\alpha_{\text{fp32}\rightarrow\text{bin}} = 10^3$. The above assumption implies that the average cost for memory accesses and data transfers scales accordingly to the MAC operations. Focusing on relative gains between different architectures, this is considered a sufficiently accurate first order approximation of the involved cost in terms of average processing time or energy per sample.

Cost reduction is obtained if the pass-on-probability ρ_i of a sample in a classifier stage i is selected as to meet inequality $C_i + \rho_i \cdot C_{i-1} \leq C_{i-1}$ or $\rho_i \leq 1 - C_i/C_{i-1}$. So, the cost introduced by an additional stage is overcompensated by its savings. At the same time, ρ_i needs to be adjusted individually per classifier stage to achieve the targeted accuracy while minimizing the cost. If all samples would pass through all N stages, total cost would increase to $C_{\text{tot}} = \Sigma_{i=0}^{N} C_i$.

3.2 Vehicle

To elaborate the methodology, we use the basic MNIST digit recognition example. The study features three different Artificial Neural Network (ANN) architectures trained to classify MNIST samples. For this case of balanced datasets, the quality of an ANN is commonly measured in terms of accuracy defined by the number of correctly identified samples divided by the number of all samples. As a baseline, we use LeNet5 [16] with small modifications in order to achieve a competitive accuracy of 99.45% (cf. Table 3). The second ANN is a Multi-Layer Perceptron (MLP) with a single hidden layer (512 neurons) further denoted as FC3. The smallest ANN is a result of binarization (cf. [7]) of the FC3 denoted as FC3_bin. With increasing number of classifiers, the savings diminish [9]. Hence, in this study we limit the maximum number of stages to three. A summary of the key performance indicators of the three ANNs is given in Table 3. The hyperparameters for training the ANNs are found using grid search, since the iterations on small datasets and networks are relatively short. For larger datasets, we would recommend optimization tools (such as Hyperopt [4]) to converge to a good solution faster. This was not performed within the scope of this work. The ANNs in this paper are state-of-the-art architectures for hardware implementations, which achieve state-of-the-art accuracies without extensive data augmentation.

Table 3. Key performance indicators of MNIST classifiers

ANN	Benchmark	Cost C	Accuracy %
FC3_bin	MNIST	410	95.29
FC3	MNIST	410 k	98.43
LeNet5	MNIST	31 M	99.45

3.3 Analyses of Pass-On Criteria

Confidence Metric. An essential step to reuse existing single-stage classifiers in a cascade is the definition of a pass-on criterion, i. e. a way to reject decisions made by a classifier at a specific stage for further analyses in a succeeding classifier stage. The pass-on criterion consists of a specific confidence metric and its corresponding threshold level. Applying the softmax function (cf. Eq. 1) to the raw output vector $\mathbf{x}$ of the last an ANN layer results in the vector of probabilities $\mathbf{x}^{\mathrm{s}}$ indicating the probability of a sample to belong to class i, with L being the total number of labels.

$$x_i^{\mathrm{s}} = \frac{e^{x_i}}{\sum_{j=1}^{L} e^{x_j}} \tag{1}$$

To provide a less computationally complex alternative to the softmax function, we propose a linear normalization of the output vector $\mathbf{x}$ by means of Eq. 2.

This normalization guarantees, that the resulting vector $\mathbf{x}^{\mathrm{l}}$ has the properties of a probability distribution, i.e., $x_i^{\mathrm{l}} \in [0,1]$ and $\sum_{i=1}^{L} x_i^{\mathrm{l}} = 1$.

$$x_i^{\mathrm{l}} = \frac{(x_i - x_{\mathrm{min}})}{\sum_{j=1}^{L}(x_j - x_{\mathrm{min}})} \tag{2}$$

Both normalization schemes enable the utilization of statistical confidence metrics. The most used confidence metric is the absolute maximum value of the normalized output vector $\mathbf{x}^{\mathrm{s}}$ or $\mathbf{x}^{\mathrm{l}}$ here called (ABS). Usually, this value is compared to a given threshold. We extend the selection of confidence metrics, which are applied to the normalized output vector, to the following selection:

- 'Best guess Versus the Second Best guess' (BVSB)
- Variance (VAR)
- Entropy (ENT)
- Kullback-Leibler Divergence (KL_DIV)
- Kurtosis (KURT)

The BVSB computes the absolute difference between the top-two results [12]. Its generalized form [13] is not considered further, as we saw no relevant improvement in our benchmarks. This paper is the first to the authors knowledge, which introduces the KURT and KL_DIV as confidence metrics for cascaded classifier. For all statistical metrics, the elements of the output vector are treated as samples from a probability density function. KURT, defined as the fourth standardized moment, is linearly normalized between the worst case (uniform distribution) and the best-case (delta distribution). Regarding KL_DIV, the reference distribution is defined as a delta distribution, representing full confidence into one label.

Pass-On-Probability. Based on the previously introduced confidence metrics, a pass-on criterion can be set to identify otherwise incorrectly classified samples, that should be passed on to the next classification stage with a more capable classifier. To benchmark the different confidence metrics introduced in the previous chapter, we set up a 2-stage cascaded classifier but focus on the behavior in the first stage. For the first stage, either one of FC3 or FC3_bin is used. In this case, LeNet5 acts as last stage (reference) making any final decision on unclassified samples. Sweeping the threshold for various confidence metrics, the accuracy of the first and last stage will be reached at thresholds $th = 0$ and $th = 1$, respectively. At the same time, the cost scales from the first stage's to the sum of both stages. An example of such sweep is shown in Fig. 2 for the FC3. It visualizes the total error ($e_{\mathrm{tot}} = 1$ - accuracy) of the cascaded classifier vs. the normalized cost C_{norm}, i.e., the total cost divided by the cost of the reference classifier $C_{\mathrm{norm}} = C_{\mathrm{tot}}/C_0$.

Considering confidence metrics entropy ENT or Kullback-Leibler divergence KL_DIV, the error of the cascaded classifier only drops for very large thresholds, i. e., when the majority of samples are passed on to the reference classifier. This

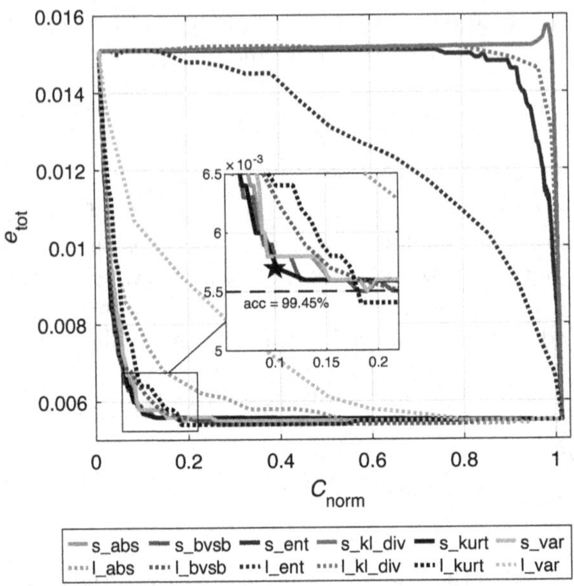

Fig. 2. Pareto analysis of pass-on criteria for softmax 's' and linear 'l' output of 2-stage classifier using FC3 as initial stage followed by reference LeNet5.

is true for any of the two first stage classifiers and appears largely independent whether the confidence metric is based on $\mathbf{x}^l$ or $\mathbf{x}^s$. Hence, these confidence metrics are excluded from further considerations.

Between the remaining four confidence metrics, there is a less apparent difference. The inset in the figure shows a zoom at the knee point of the curves. Out of the remaining confidence metrics, the ones based on BVSB and KURT have the tendency of showing better results, i. e., they achieve a better trade-off in terms of error vs. cost as they are closer to the lower left corner of the figure. For instance the cost reduces by 90% while tolerating an increase of 0.02% in the classification error for KURT on $\mathbf{x}^s$ (black star in Fig. 2). In the MNIST case, this corresponds to another two (out of 10k) misclassified samples. The combination of both classifier stages shows for KURT on $\mathbf{x}^s$ a slight improvement in classification accuracy compared to the stand alone final stage classifier (cf. inset in Fig. 2). This leads to the conclusion that for selected samples the first stage classifier provides correct labeling, whereas in a following stage these samples experience a small probability of being falsely classified.

This experiment shows that the simple confidence metric, maximum value ABS, already provides a good selection of relevant samples, but can be improved by integrating more information available from the output vector. As much as the computation of the presented confidence metrics might not be relevant in terms of total cost, they add to latency especially when the employed transcendental functions are computed iteratively on hardware. In the end, as the difference between the two best confidence metrics (KURT and BVSB) is rather small (being based on linear or softmax output vectors), a final selection will be based on the CIFAR 10 validation vehicle as discussed in Sect. 4.

In conclusion, a threshold applied to an appropriate confidence metric can identify correctly classified samples so that only a minimal number of samples has to be passed on to the following stage.

3.4 Generation of Multi-stage Classifiers

Based on the earlier assessment, a multi-stage classifier has to be created in a trial-and-error fashion iterating through various sequences of ANNs. Thereby, the accuracy and cost should increase going deeper in the hierarchy. In a first step, the earlier introduced LeNet5 is combined with either FC3 or FC3_bin to result in a 2-stage cascade. As before the threshold is swept from 0 to 1 to determine the achievable trade-offs between accuracy and cost as a function of realized pass-on-probability $\rho_i \in [0, 1]$.

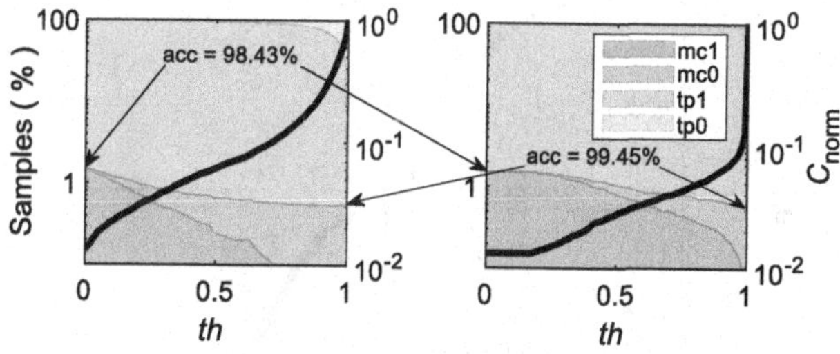

Fig. 3. In-depth analysis of true positives and misclassified samples of the initial classifier FC3 and the reference classifier LeNet5 for the confidence metrics BVSB (left) and KURT (right) with a softmax activation.

Firstly, the resulting classifier behavior is visualized in Fig. 3 for the case of the FC3-LeNet5 combination. As expected, the percentage of misclassified samples (redish coloring) diminishes for higher thresholds, i. e., when passing more samples to the final stage. Also visible is the trend of decreasing number of misclassifications in stage 1 and their related increase in stage 0. However, as stage 0 is a superior classifier the overall error count is reduced. This trend is shown for the two overall best performing confidence metrics BVSB and KURT for $\mathbf{x}^s$ on the left- and right-hand-side, respectively. As much as the reduction in error occurs for lower thresholds in the case of BVSB, it can be attributed to an earlier increase in ρ. The cost level (solid black line) appears almost identical for same accuracy levels. However, for equidistant sampling of the threshold th the pass-on-probability ρ is more evenly distributed in the case of BVSB, which is not the case for KURT. We conclude, that BVSB offers a finer tuning range.

In a second step, all three ANNs are sequentially operated resulting in a 3-stage classifier. Hereby, the classifiers are staged sorted by their increasing accuracy and cost. The thresholding is performed using the softmax activation function and BVSB as a confidence metric. For the earlier cases of 2-stage classifiers basically all values of the threshold resulted in a potential operating point. In the case of multiple stages the sweep results in a large collection of design points. Hence, Pareto-optimization is applied to the results of a full-factorial analysis across a fine granular sweep of both classifier stages. The resulting Pareto-optimal points are dominant combinations of error and cost (cf. Fig. 4). Consequently, these combinations achieve the lowest error for a certain cost. Analogous to the previous experiments, the cost is normalized against the final stage C_0. For reference, the previous 2-stage classifiers FC3_bin-LeNet5 and FC3-LeNet5 are indicated in red and blue, respectively. The operating points of the 3-stage classifier are shown as black dots. As much as the 2-stage variants provide a trade-off between the extreme points, they lack the much wider, useful scaling range offered by the 3-stage implementation. In all cases, the 3-stage classifier achieves a significantly better operating point across the 5 orders of cost - offering graceful degradation in accuracy as a function of available cost.

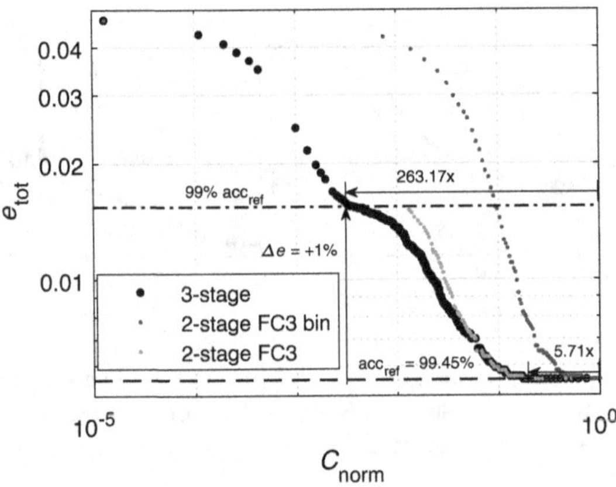

Fig. 4. Pareto plot comparing the 2-stage classifier (red and blue) with the 3-staged classifier in use case MNIST (black dots enlarged for better legibility). (Color figure online)

4 Case Study: CIFAR 10

To validate the methodology developed using the basic MNIST example, it is now applied to the case of CIFAR 10 classification. As before, three ANNs are selected, each capable to perform the classification task stand-alone. Please note, the training of the ANNs for CIFAR 10 (summarized in Table 4[1]) utilized only basic data augmentation methods like padding and shifting from the center. Further improvement in classification accuracy is reached with hyperparameter optimization and heavy data augmentation (cf. [19]). Since DenseNet [11]

[1] Note that the width and depth of the ANNs used for this case study are adjusted for the CIFAR 10 dataset.

achieves the best accuracy (at highest cost), it is selected as reference classifier, i. e. as last stage in the cascade and reference of cost.

Table 4. Key performance indicators of CIFAR 10 classifiers

ANN	Benchmark	Cost C	Accuracy %
LeNet5	CIFAR 10	32.5 M	83.55
VGG7	CIFAR 10	2.6 G	92.78
DenseNet	CIFAR 10	3.5 G	94.66

As for MNIST, an optimal confidence metric has to be selected for CIFAR 10. There are two candidate confidence metrics for MNIST, which hold promising results, but in the CIFAR 10 case, the BVSB shows a clear edge over KURT. Hence, this confidence metric is used in the following experiment. Please note, as for the MINST example, also CIFAR 10 shows a slight improvement in accuracy for intermediate threshold levels. Bear in mind, that all points with $C_{\text{norm}} \geq 1$ are indicating higher cost than a stand-alone reference classifier.

Figure 5 depicts the Pareto plot of the 3-staged classifier. The gap in the Pareto plot indicates that there is no reduction in classification error for this cost interval. With the chosen classifiers, it is possible to reduce the cost over two orders of magnitude. A wider range, as obtained for the 3-stage MNIST classifier, would be possible using an additional ANN of lower cost and reasonable accuracy. While the accuracy of the reference classifier is maintained, the cascaded classifier achieves a cost reduction of 1.32× by using low-cost classification of easy samples. The cost reduction becomes 2.55× maintaining 99% of the reference accuracy.

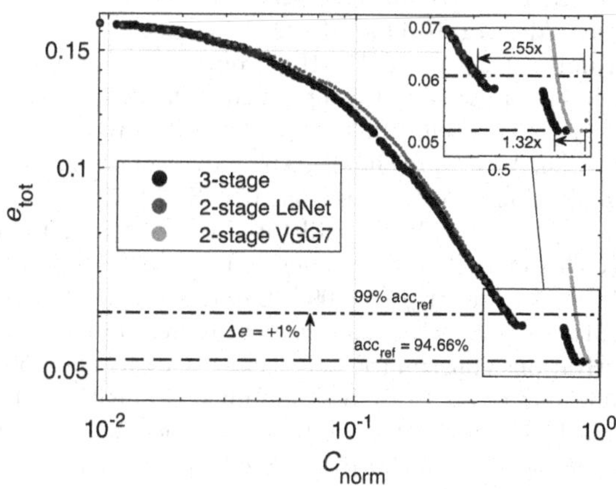

Fig. 5. Pareto plot comparing the 2-staged classifier (red and blue) with the 3-staged classifier (black dots enlarged for better legibility) in use case CIFAR 10. (Color figure online)

5 Conclusion

Ever more powerful machine learning algorithms surpass human performance, yet are prohibitive for embedded devices due to their high computational complexity. Combining classifiers of varying accuracy-to-cost ratios in a cascade, as presented in this paper, provides cost reduction, while preserving top accuracy. Alternatively, graceful degradation is enabled to provide dramatic cost reduction, including throughput and energy, with bounded drop in accuracy. The realization of such cascaded classifier can be reproduced in the presented structured methodology, which makes use of a confidence metric requiring only basic mathematical operations. Pareto-optimization provides optimal settings for the threshold level to adapt classification on embedded devices according to the available energy budget. The presented methodology is derived using MNIST and validated with CIFAR 10. This work achieves with a three stage classifier for MNIST a cost reduction of 5.71× at 99.45% accuracy and remarkable 263.17× at 98.46%. In the case of CIFAR 10, the reduction is 1.32× at 94.66%, and 2.55× at 93.7%.

Acknowledgment. This work was partially funded by the German BMBF project NEUROTEC under grant no. 16ES1134.

References

1. ImageNet. https://www.image-net.org/. Accessed 01 Feb 2021
2. Badami, K., Lauwereins, S., Meert, W., Verhelst, M.: Context-aware hierarchical information-sensing in a 6μW 90nm CMOS voice activity detector. In: ISSCC, vol. 58, pp. 430–432 (2015). https://doi.org/10.1109/ISSCC.2015.7063110
3. Benbasat, A.Y., Paradiso, J.A.: A framework for the automated generation of power-efficient classifiers for embedded sensor nodes. In: SenSys, vol. 5, pp. 219–232 (2007). https://doi.org/10.1145/1322263.1322285
4. Bergstra, J., Yamins, D., Cox, D.D.: Hyperopt: a python library for optimizing the hyperparameters of machine learning algorithms. In: Proceedings of the 12th Python in Science Conference, vol. 13, pp. 13–19 (2013)
5. Canziani, A., Paszke, A., Culurciello, E.: An analysis of deep neural network models for practical applications. CoRR, abs/1605.07678 (2017)
6. Cocaña-Fernández, A., Ranilla, J., Gil-Pita, R., Sánchez, L.: Energy-conscious fuzzy rule-based classifiers for battery operated embedded devices. In: FUZZ-IEEE, pp. 1–6 (2017). https://doi.org/10.1109/FUZZ-IEEE.2017.8015483
7. Courbariaux, M., Bengio, Y.: BinaryNet: training deep neural networks with weights and activations constrained to +1 or -1. CoRR, abs/1602.02830 (2016)
8. Goens, A., Brauckmann, A., Ertel, S., Cummins, C., Leather, H., Castrillon, J.: A case study on machine learning for synthesizing benchmarks. In: MAPL, pp. 38–46 (2019). https://doi.org/10.1145/3315508.3329976
9. Goetschalckx, K., Moons, B., Lauwereins, S., Andraud, M., Verhelst, M.: Optimized hierarchical cascaded processing. JETCAS **8**, 884–894 (2018). https://doi.org/10.1109/JETCAS.2018.2839347
10. Horowitz, M.: 1.1 Computing's energy problem (and what we can do about it). In: ISSCC, pp. 10–14 (2014). https://doi.org/10.1109/ISSCC.2014.6757323

11. Huang, G., Liu, Z., van der Maaten, L., Weinberger, K.Q.: Densely connected convolutional networks. In: CVPR, pp. 2261–2269 (2016). https://doi.org/10.1109/CVPR.2017.243
12. Joshi, A.J., Porikli, F., Papanikolopoulos, N.: Multi-class active learning for image classification. In: CVPR, pp. 2372–2379 (2009). https://doi.org/10.1109/CVPR.2009.5206627
13. Kouris, A., Venieris, S.I., Bouganis, C.: Cascade CNN: pushing the performance limits of quantisation in convolutional neural networks. In: FPL, pp. 155–1557 (2018). https://doi.org/10.1109/FPL.2018.00034
14. Krizhevsky, A., Nair, V., Hinton, G.: CIFAR-10 (Canadian Institute for Advanced Research). http://www.cs.toronto.edu/~kriz/cifar.html
15. Latotzke, C., Gemmeke, T.: Efficiency versus accuracy: a review of design techniques for DNN hardware accelerators. IEEE Access **9**, 9785–9799 (2021). https://doi.org/10.1109/ACCESS.2021.3050670
16. Lecun, Y., Bottou, L., Bengio, Y., Haffner, P.: Gradient-based learning applied to document recognition. Proc. IEEE **86**(11), 2278–2324 (1998). https://doi.org/10.1109/5.726791
17. Lecun, Y., Cortes, C., Burges, C.: The MNIST Database of handwritten digits. http://yann.lecun.com/exdb/mnist/
18. Li, L., Topkara, U., Coskun, B., Memon, N.: CoCoST: a computational cost efficient classifier. In: ICDM, pp. 268–277 (2009). https://doi.org/10.1109/ICDM.2009.46
19. Lin, Z., Memisevic, R., Konda, K.R.: How far can we go without convolution: improving fully-connected networks. CoRR, abs/1511.02580 (2015)
20. Ouali, M., King, R.D.: Cascaded multiple classifiers for secondary structure prediction. Protein Sci., 1162–1176 (2000). https://doi.org/10.1110/ps.9.6.1162
21. Price, M., Glass, J., Chandrakasan, A.P.: A scalable speech recognizer with deep-neural-network acoustic models and voice-activated power gating. In: ISSCC, pp. 244–245 (2017). https://doi.org/10.1109/ISSCC.2017.7870352
22. Rossi, D., et al.: 4.4 A 1.3TOPS/W @ 32GOPS fully integrated 10-core SoC for IoT end-nodes with 1.7μW cognitive wake-up from MRAM-based state-retentive sleep mode. In: ISSCC, vol. 64, pp. 60–62 (2021). https://doi.org/10.1109/ISSCC42613.2021.9365939
23. Stadtmann, T., Latotzke, C., Gemmeke, T.: From quantitative analysis to synthesis of efficient binary neural networks. In: ICMLA, vol. 19, pp. 93–100 (2020). https://doi.org/10.1109/ICMLA51294.2020.00024
24. Venkataramani, S., Raghunathan, A., Liu, J., Shoaib, M.: Scalable-effort classifiers for energy-efficient machine learning. In: DAC, vol. 67, pp. 1–6 (2015). https://doi.org/10.1145/2744769.2744904
25. Xu, Z., Kusner, M., Weinberger, K., Chen, M.: Cost-sensitive tree of classifiers. In: PMLR, vol. 28, no. 1, pp. 133–141 (2013)
26. Xu, Z., Kusner, M.J., Weinberger, K.Q., Chen, M., Chapelle, O.: Classifier cascades and trees for minimizing feature evaluation cost. JMLR **15**(1), 2113–2144 (2014)

Conditional Generative Adversarial Networks for Speed Control in Trajectory Simulation

Sahib Julka[1(✉)], Vishal Sowrirajan[1], Joerg Schloetterer[2], and Michael Granitzer[1]

[1] University of Passau, Passau, Germany
sahib.julka@uni-passau.de
[2] University of Duisburg-Essen, Duisburg, Germany

Abstract. Motion behaviour is driven by several factors - goals, neighbouring agents, social relations, physical and social norms, the environment with its variable characteristics, and further. Most factors are not directly observable and must be modelled from context. Trajectory prediction, is thus a hard problem, and has seen increasing attention from researchers in the recent years. Prediction of motion, in application, must be realistic, diverse and controllable. In spite of increasing focus on multimodal trajectory generation, most methods still lack means for explicitly controlling different modes of the data generation. Further, most endeavours invest heavily in designing special mechanisms to learn the interactions in latent space. We present Conditional Speed GAN (CSG), that allows controlled generation of diverse and socially acceptable trajectories, based on user controlled speed. During prediction, CSG forecasts future speed from latent space and conditions its generation based on it. CSG is comparable to recent GAN methods in terms of the benchmark distance metrics, with the additional advantage of controlled simulation and data augmentation for different contexts. Furthermore, we compare the effect of different aggregation mechanisms and demonstrate that a naive approach of concatenation works comparable to its attention and pooling alternatives. (Open source code available at: https://github.com/ConditionalSpeedGAN/CSG).

Keywords: Conditional generative models · Trajectory simulation

1 Introduction

Modelling social interactions and the ability to forecast motion dynamics is crucial to several applications such as robot planning systems [1], traffic operations [2], and autonomous vehicles [3]. However, doing so remains a challenge due to the subjectivity and variability of interactions in real world scenarios.

Recently we have witnessed a shift in perspective from the more deterministic approaches of agent modelling with handcrafted features [4–8], to the latent learning of variable outcomes via complex data-driven deep neural network

G. Nicosia et al. (Eds.): LOD 2021, LNCS 13164, pp. 436–450, 2022.
https://doi.org/10.1007/978-3-030-95470-3_33

architectures [9–12]. State-of-the-art systems are able to generate variable or multimodal predictions that are socially acceptable (adhere to social norms), spatially aware and similar to the semantics in training data. Most systems can sufficiently generate outcomes according to the original distribution, but lack means for controlling different modes of data generation, or to be able to extrapolate to unseen contexts. Consequently, controlled simulation is a challenge.

We propose that these systems need to be: a) *Spatio-temporal context aware*: aware of space and temporal dynamics of surrounding agents to anticipate possible interactions and avoid collision, b) *Control-aware*: compliant to external and internal constraints, such as kinematic constraints, and simulation control, and c) *Probabilistic*: able to anticipate multiple forecasts for any given situation, beyond those in the training data.

To model the implicit behaviour and predict sudden nuances, it is imperative that these systems understand not only the spatial but also the temporal context. This context should be identifiable, and adaptable. For instance, in urban simulations, it is important to simulate trajectories with different characteristics specific to the location and time, e.g., slow pedestrians in malls vs fast in busy streets. Adaptation and extrapolation to different contexts thus is imperative for simulation in such scenarios.

In this work, we propose a generative neural network framework called CSG (Conditional Speed GAN) that takes into account the aforementioned requirements. We leverage the conditioning properties offered by conditional GANs [13], to induce temporal structure to the latent space in sequence generation system inspired by previous works [9,11,14]. Consequently, CSG can be conditioned for controlled simulation. CSG is trained in a self-supervised setting, on multiple contexts such as speed and agent-type in order to generate trajectories specific to those conditions, without the need for inductive bias in the form of explicit aggregation methods used extensively in previous works [9,15–19]. The main contributions of this work are as follows:

1. A generative system that can be conditioned on agent speed and semantic classes of agents, to simulate multimodal and realistic trajectories based on user defined control.
2. A trajectory forecaster that uses predicted speeds from the latent space to generate conditional future moves that are socially acceptable, without special aggregation mechanisms like pooling or attention.

2 Related Work

There is a plethora of scientific work done previously in the field of trajectory forecasting. Based on structural assumptions [20], the existing literature can broadly be classified as: a) *Ontological*, which are mechanics-based, such as the Cellular Automata model [8], Reciprocal Velocity Obstacles (RVO) method [7], or the Social Forces (SF) model [4] - that use dynamic systems to model the forces that affect human motion. For instance, SF models dynamics with newtonian controls, like, attraction towards goal, and repulsion against other agents; these methods make strong structural assumptions, and often fail to capture the intricate motion dynamics [4,5,21], and b) *Phenomonological*, that are data driven and aim to implicitly learn complex relationships and distributions. These include methods such as GPR (Gaussian Process Regression) [22], Inverse Reinforcement learning [23], and the more recent RNN based methods [24,25], due to their acclaimed success in modelling long sequences, yielding improved prediction over deterministic methods. However, these methods were still restrictive in their inability to model multiple modes of outcome (multimodal).

2.1 Generative Models

Generative methods, with recent advancements, became the natural choice for modelling trajectories, since they offer distribution learning, rather than optimising on a single best outcome. Most related works employ some kind of deep recurrent base with latent variable model, such as the Conditional Variational Autoencoder (CVAE) [20,26] to explicitly encode multimodality or Generative Adversarial Networks (GAN) to implicitly do so [9,11]. A few interesting GAN variants have been developed to tackle some of the aforementioned challenges, such as the Social GAN (SGAN) [9], which can produce socially acceptable trajectories, and encouraged multimodal generation of trajectories by introducing a variety loss. Additionally, with the pooling module, using permutation invariant max-pooling, a form of neighbourhood spatial embedding was introduced, that demonstrated improvement over local grid-based encoding, such as the kind used in Social-LSTM [24]. This was improved with an attention mechanism proposed in SoPhie [11], which was explored by numerous following works [18,19,27–29] and improved in Social Ways [10] and Social BiGAT [14].

In the current state, the generative models can effectively learn distributions to forecast diverse and acceptable trajectories. However, there still exist open questions as to how to decide which mode is best, or if the mean is good enough for changing scenarios. Existing methods do not tackle the problem of mode control, which is an essential characteristic needed for simulation and adaptation to different scenarios. Further, a key challenge is to find an ideal strategy to aggregate information in scenes with variable neighbours, and it remains unanswered whether special mechanisms like pooling or attention are really needed.

2.2 Conditioned Generation

The objective of generative models is to approximate the true data distribution, with which one can generate new samples similar to real data. GANs are useful frameworks for learning from a latent distribution with only a small number of samples, yet they can suffer in regards to output mode control. The mode control of the network requires some additional constraints that force the network to sample from particular segments in the distribution. Such as in the case of conditional GANs [13]. Conditional generative models have previously been explored in the context of trajectory prediction, conditioning on motion planning, weather effects or final goal [30–32] in order to improve prediction. However, most of these works do not consider the interaction of agents in the scene. In [33,34] the authors develop a conditional Variational Autoencoder that predicts the endpoint goal and conditions generation on it, also considering the subjective human aspect of the problem. We use a GAN, and enforce additional constraints on the latent vector so as to predict the next frame speed, and subsequently condition generation on it. The goal of this research is to learn a structured latent space, sampling from which would enable controlled generation. We use speed as a condition for our experiments where we try to disentangle its representation from the latent space, to modularise it and allow explicit control. Further, controlling this vector can allow extrapolation in scenarios with unseen speed context, e.g., simulating fast speeds having trained only on slow and medium speeds.

2.3 Problem Formulation

Trajectory prediction or forecasting is the problem of predicting the path $<(x^t, y^t)|t = t_{obs}+1, \ldots, T>$ that some agent (e.g., pedestrian, cyclist, or vehicle- we omit a subscript to indicate the agent here for better readability) will move along in the future given the trajectory $<(x^t, y^t)|t = 0, \ldots, t_{obs}>$ that the agent moved along in the past.

The objective of this work is to develop a deep generative system that can accurately forecast motions and trajectories for multiple agents simultaneously with user controlled speeds.

Given (x^t, y^t) as the coordinates at time t, L as the agent type and speed S, we seek a function f to generate the next timesteps (x^{t+1}, y^{t+1}) as follows:

$$(x^{t+1}, y^{t+1}) = f(x^t, y^t|S, L), \tag{1}$$

where the generation of future timesteps is conditioned on speed S and agent type L. While the agent type remains constant over time, speed may vary per timestep. In simulation environments, speed S is a user-controlled variable, while in prediction environments, the speed of future timesteps is typically unknown. In order to be able to condition on the speed of the whole timeframe, including future speeds of the yet to be generated trajectories, an estimate $\hat{S}$, learned from the data can be used.

3 Methodology

CSG consists of two main blocks the Generator block (G) and the Discriminator block (D). The generator is comprised of: a) Feature Extraction module, b) Speed Forecasting module, c) Aggregation module and d) Decoder, and the Discriminator is composed of an LSTM encoder module that classifies the conditional inputs as "real" or "fake" (cf. Fig. 1 for a detailed overview).

3.1 Preprocessing

We first calculate the relative positions as the displacement from the previous timeframe $\delta x_i^t = x_i^t - x_i^{t-1}$, $\delta y_i^t = y_i^t - y_i^{t-1}$ with $\delta x_i^0 = \delta y_i^0 = 0$ from the observed

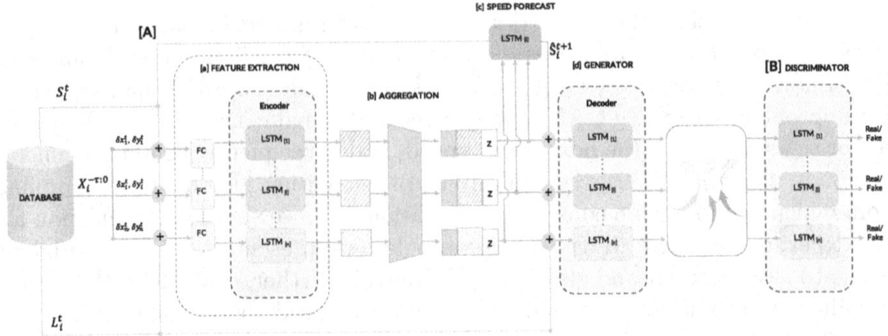

Fig. 1. Overview of the CSG approach: the pipeline comprises of two main blocks: A) Generator Block, comprising of the following sub-modules: (a) Feature Extraction, that encodes the relative positions and speeds of each agent with LSTMs, (b) Aggregation, that jointly reasons multi agent interactions, (c) Speed forecast, that predicts the next timestep speed, (d) Decoder, that conditions on the next timestep speed, agent label and the agent-wise trajectory embedding to forecast next timesteps, and, the B) Discriminator Block, that classifies the generated outputs as "real" or "fake", specific to the conditions.

trajectory for each agent i. We use relative distance for translational invariance, and although the internal computation is based on them, we still use (x_i^t, y_i^t) throughout the paper to ease readability. We calculate the speed labels using Euclidean distance between any two consecutive timeframes from the dataset and scale them in the range (0, 1). For the second condition, i.e., *agent type*, we assign nominal labels and one-hot encode them.

3.2 Feature Extraction

We concatenate the relative positions (x_i^t, y_i^t) of each agent i with their derived speeds S_i^t and agent-labels L_i^t, and embed this vector to a fixed length vector, e_i^t, using a single layer fully connected (FC) network, expressed as:

$$e_i^t = \alpha_e((x_i^t, y_i^t) \oplus S_i^t \oplus L_i^t; W_{\alpha_e}), \tag{2}$$

where, α_e is the embedding function, and W_{α_e} denote the embedding weights.

Encoder: To capture the temporal dependencies of all states of an agent i, we pass the fixed length embeddings as input to the encoder LSTM, with the following recurrence for each agent:

$$h_{ei}^t = LSTM_{enc}(e_i^t, h_{ei}^{t-1}; W_{enc}), \tag{3}$$

where the hidden state is initialised with zeros, e_i^t is the input embedding, and W_{enc} are the shared weights among all agents in a scene.

3.3 Aggregation Methods

To jointly reason across agents in space, and their interaction, we experiment with aggregation mechanisms used widely in previous research works [9,11,14, 35]. In addition, we explore a simple concatenation mechanism. The aggregation vector is computed using either of the three mechanisms per agent, and concatenated to its latent space.

Pooling: Similar to [9], we consider the positions of each agent, relative to all other agents in the scene, and pass it through an embedding layer, followed by a symmetric function.

Let r_i^t be the vector with relative position of an agent i to all other agents in the scene, the social features are calculated as:

$$f_i^t = \alpha_p(r_i^t; W_{\alpha_p}), \tag{4}$$

where W_{α_p} denotes the embedding weight. The social features are concatenated with the hidden states h_{ei}^t and passed through a multi-layer FC network followed by max-pooling to obtain the final pooling vectors as:

$$a_i^t = \gamma_p(h_{ei}^t \oplus f_i^t; W_{\gamma_p}), \tag{5}$$

with W_{γ_p} as the weights of the FC network.

Attention: We implement a soft-attention mechanism on N nearest agents for each agent in the scene. We compute the social features Eq. (4), and pass them to the attention module, with the respective hidden states from the encoder, as:

$$\begin{aligned} f_i^t &= \alpha_a(r_i^t; W_{\alpha_a}), \\ a_i^t &= Attn_{so}(h_{ei}^t \oplus f_i^t; W_{so}), \end{aligned} \tag{6}$$

where $Attn_{so}$ is the soft attention with W_{so} as weights.

Concatenation: For each agent i, we calculate N nearest neighbours and concatenate their final hidden states. The concatenated hidden states are passed through a FC network that learns the nearby agents interaction, as:

$$a_i^t = \gamma_c(h_{ei}^t \oplus [h_{en}^t | \forall n \in N]; W_{\gamma_c}), \tag{7}$$

where h_{ei}^t and h_{en}^t refer to the final encoder hidden states of the current agent and N nearest agents respectively,

Finally, we concatenate and embed the final hidden states of the encoder LSTMs h_{ei}^t along with the respective aggregation function a_i^t to a compressed size vector using a multi-layer FC network and add gaussian noise z to induce stochasticity.

$$h_i^t = \gamma(h_{ei}^t \oplus a_i^t, W_\gamma) \oplus z, \tag{8}$$

where γ denotes the multi-layer FC embedding function with ReLU non-linearity, and embedding weights W_γ.

We treat these vectors as latent spaces to sample from for conditional generation in the following stages.

3.4 Speed Forecasting

In order to forecast the future speeds for each agent in prediction environments, we use a module comprised of LSTMs. We initialise the hidden states of the speed forecaster h_{si}^t with the latent vectors h_i^t. The input is the current timestep speed S_i^t and the future speed estimate $\hat{S}_i^{t+1}$ is calculated by passing the hidden state through a FC network with sigmoid activation in the following way:

$$\begin{aligned} h_{si}^t &= LSTM_{sp}(S_i^t, h_{si}^{t-1}; W_{sp}), \\ \hat{S}_i^{t+1} &= \gamma_{sp}(h_{si}^t; W_{\gamma_{sp}}), \end{aligned} \tag{9}$$

The forecasting module is trained simultaneously with the other components, using ground truth S_i^{t+1} as feedback signal.

3.5 Decoder

As we want the decoder to maintain the characteristics of the past sequence, we initialise its hidden state h_{di}^t with h_i^t, and input the embedded vector of relative positions with the conditions for control during training and simulation as:

$$d_i^t = \alpha_d((x_i^t, y_i^t) \oplus S_i^{t+1} \oplus L_i; W_{\alpha_d}). \tag{10}$$

In prediction environments, we replace S_i^{t+1} with the estimate $\hat{S}_i^{t+1}$ from the forecasting module. The hidden state of the LSTM is fed through a FC network that outputs the predicted relative position of each agent:

$$\begin{aligned} h_{di}^t &= LSTM_{dec}(d_i^t, h_{di}^{t-1}; W_{dec}), \\ (\hat{x}_i^{t+1}, \hat{y}_i^{t+1}) &= \gamma_d(h_{di}^t, W_{\gamma_d}), \end{aligned} \tag{11}$$

where W_{dec} are the LSTM weights and W_{γ_d} are the weights of the FC network.

3.6 Discriminator

We use an LSTM encoder block as the Discriminator. The real input to D can be formulated as:

$$O_i = <(x_i^t, y_i^t), S_i^t, L_i | t = 0, \ldots, T>, \tag{12}$$

including the observed ($t = 0, \ldots, t_{obs}$) and future ground truth ($t = t_{obs} + 1, \ldots, T$) relative positions. The fake input can be formulated as:

$$\begin{aligned} \hat{O}_i =& <(x_i^t, y_i^t), S_i^t, L_i | t = 0, \ldots, t_{obs}> \\ &\oplus <(\hat{x}_i^t, \hat{y}_i^t), S_i^t, L_i | t = t_{obs} + 1, \ldots, T>, \end{aligned} \tag{13}$$

including the observed and predicted relative positions.

The discriminator equation can be framed as:

$$h_{dsi}^t = LSTM_{dsi}(\alpha_{di}(o_i^t; W_{\alpha_{di}}), h_{dsi}^{t-1}; W_{dsi}), \tag{14}$$

where α_{di} is the embedding function with corresponding weights $W_{\alpha_{di}}$, W_{dsi} are the LSTM weights. o_i^t is the input element from the real or fake sequence O_i or $\hat{O}_i$. The real or fake classification scores are calculated by applying a multi-layer FC network with ReLu activations on the final hidden state of the LSTMs, as:

$$\hat{C}_i = \gamma_{di}(h_{dsi}^t; W_{\gamma_{di}}), \tag{15}$$

3.7 Losses

In addition to optimising the GAN minimax game, we apply the L2 loss on the generated trajectories, and L1 loss for the speed forecasting module.

The network is trained by minimising the following losses, taking turns: The Discriminator loss is framed as:

$$\ell_D(\hat{C}_i, C_i) = -C_i \log(\hat{C}_i) - (1 - C_i)\log(1 - \hat{C}_i), \tag{16}$$

The Generator loss together with L2 and L1 loss:

$$\ell_G(\hat{O}_i) + \ell_2((x_i^t, y_i^t), (\hat{x}_i^t, \hat{y}_i^t)) + \ell_1(S_i^t, \hat{S}_i^t), \tag{17}$$

for $t = t_{obs} + 1, \ldots, T$. The Generator loss is the Discriminator's ability to correctly classify data generated by G as "fake", expressed as:

$$\ell_G(\hat{O}_i) = -\log(1 - \hat{C}_i), \tag{18}$$

where $\hat{C}_i$ is the discriminator's classification score.

4 Experiments

4.1 Datasets

For single agent predictions, we perform experiments using publicly available datasets: ETH [36] and UCY [37], that contain complex pedestrian trajectories. These datasets also cover challenging group behaviours such as couples walking together, groups crossing each other and groups forming and dispersing in some scenes, and contain several other non-linear trajectories. In order to test the model on multiple classes of agents, we utilise the Argoverse motion prediction dataset [38]. The labels available in the dataset are *av*, for autonomous vehicles, *agent* for other vehicles and *other* includes other agents present in the scene. We convert the real-world coordinates from the datasets to image coordinates and plot them in real-world maps, so as to qualitatively evaluate the predictions. All plots are best viewed in colour.

4.2 Simulation

Speed Extrapolation: We split the data into three folds according to derived speeds, i.e., slow, medium and fast with 0.33 and 0.66 thresholds. Using CSG with concatenation as the aggregation mechanism, we train on two folds at a time and simulate the agents in the test set of these folds with controlled speeds from the fold left out. We observe controllability in all three segments, indicating

Fig. 2. (a) Pedestrians sampled from the fast fold simulated at slow speeds. (b) Pedestrians from fast fold simulated at medium speeds, and (c) Pedestrians from slow simulated at fast speeds. Ground truth values are marked in blue. The network extrapolates to unseen speed contexts. (Color figure online)

the ability to extrapolate to unseen contexts. In Fig. 2(a), pedestrians from the medium fold are simulated at slow speed. We clearly observe the pedestrians adapt in a meaningful way, traversing less distance compared to the ground truth. In Fig. 2(b) and (c), similarly, we simulate at medium and fast speed unseen in the training set.

Table 1. A comparison of CSG-concatenation vs SGAN in speed extrapolation for Slow, Medium and Fast folds, using Earth mover's distance.

Method	Slow	Medium	Fast
CSG-C	**0.0138**	**0.0058**	**0.0269**
SGAN	0.0439	0.0214	0.0504

Regardless of the properties present in the training set, we observe the network being able to extrapolate the contextual features, indicating distributional changes due to localised causal interventions. We compare the extrapolation performance to SGAN using Earth mover's distance in Table 1. The simulated speeds are compared to the distributions of the expected speeds in the test fold. We observe an increased control in achieving the desired speeds in the unseen folds compared to the baseline.

Multimodal and Socially Aware: To further evaluate the simulations, we assess qualitatively whether they adhere to social constraints and can preserve multi modal dynamics. Figure 3 illustrates the different speed control for agents predicted for 8 timeframes: (a) shows a fast moving pedestrian simulated at medium speeds with K = 5 samples, expressing a diverse generation for the controlled mode. Figure 3(b) illustrates preservation of social dynamics: two pedestrians simulated at different speeds circumvent a possible collision by walking around stationary people. Figure 3(c) depicts group walking behaviour with slow and fast simulations: the pedestrians continue to walk together, and adjust their paths in order to be able to do so. Figure 3(d) depicts another complex collision avoidance scenario: the pedestrians decide to split up and walk around the approaching pedestrian, when simulated at fast speeds.

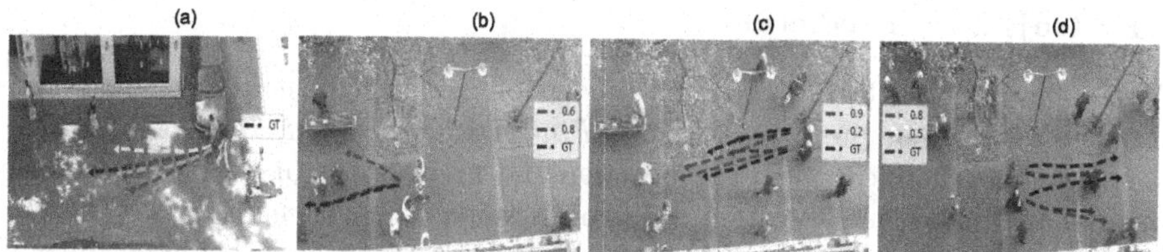

Fig. 3. (a) A fast moving pedestrian simulated at medium speeds, with K = 5, shows a diverse selection of paths (Multimodality). (b) Two pedestrians simulated at different speeds (0.6 and 0.8) walk around stationary people (Collision avoidance). (c) Two pedestrians walking together, simulated at fast and slow speeds, find corresponding paths in order to walk together (Group walking). (d) Two pedestrians walking together adjust their paths in order to circumvent the approaching pedestrian. All ground truth trajectories are marked in blue. (Color figure online)

4.3 Effect of Aggregation Method

We evaluate the performance of our method with different aggregation strategies, one at a time, keeping all other factors constant. CSG, CSG-P, CSG-C and CSG-A refer to our method without aggregation, with pooling, with concatenation and with attention respectively. If two or more pedestrians are closer than an euclidean distance of 0.10m, we consider it as a collision. We observe (cf. Table 2) that the concatenation strategy consistently outperforms all others on collision avoidance, followed by the attention and max-pooling methods, in that order. Regardless of the choice, CSG reduces collisions compared to SGAN, indicating that the speed forecasting module might yield some natural structure in latent space.

Table 2. Average % collisions per predicted frame. A collision is detected if the distance between two pedestrians is less than 0.10 m. Lower is better, and best is in bold. CSG-C reduces overall collisions.

Dataset	SGAN	CSG	CSG-P	CSG-A	CSG-C
ETH	0.2237	0.3167	0.2603	0.2373	**0.1881**
HOTEL	0.2507	0.2143	0.1773	0.2177	**0.0917**
UNIV	0.5237	0.5338	0.6064	0.6425	**0.5025**
ZARA1	0.1103	0.0464	0.0660	0.0680	**0.0328**
ZARA2	0.5592	0.2184	0.2768	0.2258	**0.1988**
AVG	0.3335	0.2659	0.2774	0.2783	**0.2228**

4.4 Trajectory Prediction

We compare the results on trajectory prediction using the benchmark metrics followed extensively by previous works: a) *Final Displacement Error (FDE)*, which computes the euclidean distance between the final points of the ground truth and the predicted final position, and, b) *Average Displacement Error (ADE)*, which averages the distances between the ground truth and predicted output across all timesteps. We generate K samples per prediction step and report the distance metrics for the best out of the K samples.

Single Agent Type (Pedestrian): We evaluate our model on the five sets of ETH and UCY data, with a hold-one-out approach (i.e., training of four sets at a time and evaluating on the set left out) and compare with the following baseline methods:

SGAN [9]: GAN with a pooling module to capture the agent interactions,
SoPhie [11]: GAN with physical and social attention,
S-Ways [10]: GAN with Information loss instead of the L2,
S-BIGAT [14]: Bicycle-GAN augmented with Graph Attention Networks (GAT), and
CGNS [39]: CGAN with variational divergence minimisation.

Table 3 depicts the final metrics for 12 predicted timesteps (4.8 s). Similar to other methods [9,14], we generate K = 20 samples for a given input trajectory, and report errors for the best one. On a quantitative comparison with other GAN models, we observe that our model CSG without an explicit aggregation methods outperforms SGAN, Sophie, and CGNS overall in the ETH datasets, while on the UCY it performs better than most and at par with CGNS.

Table 3. A comparison of GAN based methods on ADE/FDE scores for 12 predicted timesteps (4.8 s) with K = 20. For CSG, we report the metrics with the mean and variance for 20 runs. Lower is better, and best is in bold. For more detailed results please refer to project repository.

Dataset	SGAN [9]	SoPhie [11]	S-Ways [10]	S-BIGAT [14]	CGNS [39]	CSG (Ours)
ETH	0.77/1.49	0.73/1.55	**0.39/0.65**	0.59/1.16	0.66/1.16	0.58 ± 0.01/ 1.07 ± 0.02
UCY	0.51/1.01	0.40/0.83	0.50/0.95	0.40/0.89	**0.38**/0.84	**0.38** ± 0.01/ 0.90 ± 0.01
AVG	0.64/1.25	0.56/1.19	**0.44/0.80**	0.49/1.02	0.52/1.00	0.48/0.98

Multi Agent Type (Argoverse Dataset): With respect to multi agent problem, we compare our model with:

CS-LSTM [40]: Combination of CNN network with LSTM architecture,
SGAN [9]: GAN network with max pooling approach to predict future human trajectories, and
Graph-LSTM [41]: Graph convolution LSTM network using dynamic weighted traffic-graphs that predicts future trajectories and road-agent behavior.

Similar to [42], We utilise the first 2 s as observed input to predict the next 3 s and report the metrics in Table 4. We observe that our model outperforms SGAN by a large margin and performs better than CS-LSTM in terms of FDE but doesn't perform as well when compared with ADE. Although Graph-LSTM performs best overall, CSG offers the added advantage of explicit control.

Table 4. ADE/FDE scores on Argoverse Dataset. We report our score as an average of 20 runs.

Dataset name	CS-LSTM [40]	SGAN [9]	Graph-LSTM [41]	CSG (ours)
Argoverse	1.050/3.085	3.610/5.390	**0.99/1.87**	1.39 ± 0.02/2.95 ± 0.05

5 Conclusion and Future Work

We present a method for generation and controlled simulation of diverse multi agent trajectories in realistic scenarios. We show that our method can be used to explicitly condition generation for greater control and ability to adapt context. Further, we demonstrate with our experiments the efficacy of the model in forecasting mid-range sequences (5 s) with an edge over most existing GAN based variants. It may be that most models are optimised to reduce the overall distance metrics, but not collisions. The models are expected to learn the notion of collision avoidance implicitly. By focussing explicitly on relative velocity predictions, we obtain more domain knowledge driven control over the design of the interaction order. Further, we observe that a simple concatenation of the final hidden state vectors of N neighbours is a good enough strategy for aggregating information across agents in a scene. While this aggregation approach is relatively simple, it is efficient and removes the need to design complex mechanisms. In the future, we aim to extend our method by learning context vectors of variation (e.g., behaviour, speed) automatically and conditioning on static scene information to improve interactions in space.

Acknowledgements. Results obtained have been partially funded by European Union's Horizon 2020 research and innovation programme under grant agreement No. 654208 (EUROPLANET 2020) as well as by the German Federal Ministry of Transport and Digital Infrastructure under the grant KIMONO (No. 45KI01A011).

References

1. Chen, C., Liu, Y., Kreiss, S., Alahi, A.: Crowd-robot interaction: crowd-aware robot navigation with attention-based deep reinforcement learning. In: 2019 International Conference on Robotics and Automation (ICRA), pp. 6015–6022. IEEE (2019)
2. Horni, A., Nagel, K., Axhausen, K.W.: The Multi-agent Transport Simulation MATSim. Ubiquity Press (2016)
3. Rasouli, A., Tsotsos, J.K.: Autonomous vehicles that interact with pedestrians: a survey of theory and practice. IEEE Trans. Intell. Transp. Syst. **21**(3), 900–918 (2019)
4. Helbing, D., Molnar, P.: Social force model for pedestrian dynamics. Phys. Rev. E **51**(5), 4282 (1995)
5. Antonini, G., Bierlaire, M., Weber, M.: Discrete choice models of pedestrian walking behavior. Transp. Res. Part B Methodol. **40**(8), 667–687 (2006)
6. Wang, J.M., Fleet, D.J., Hertzmann, A.: Gaussian process dynamical models for human motion. IEEE Trans. Pattern Anal. Mach. Intell. **30**(2), 283–298 (2007)
7. Van den Berg, J., Lin, M., Manocha, D.: Reciprocal velocity obstacles for real-time multi-agent navigation. In: 2008 IEEE International Conference on Robotics and Automation, pp. 1928–1935. IEEE (2008)
8. Elfring, J., Van De Molengraft, R., Steinbuch, M.: Learning intentions for improved human motion prediction. Robot. Auton. Syst. **62**(4), 591–602 (2014)
9. Gupta, A., Johnson, J., Fei-Fei, L., Savarese, S., Alahi, A.: Social GAN: socially acceptable trajectories with generative adversarial networks. In: CVPR, pp. 2255–2264 (2018)
10. Amirian, J., Hayet, J.-B., Pettré, J.: Social ways: learning multi-modal distributions of pedestrian trajectories with GANs. In: CVPR Workshops, p. 0 (2019)
11. Sadeghian, A., Kosaraju, V., Sadeghian, A., Hirose, N., Rezatofighi, H., Savarese, S.: Sophie: an attentive GAN for predicting paths compliant to social and physical constraints. In: CVPR, pp. 1349–1358 (2019)
12. Salzmann, T., Ivanovic, B., Chakravarty, P., Pavone, M.: Trajectron++: dynamically-feasible trajectory forecasting with heterogeneous data. arXiv preprint arXiv:2001.03093 (2020)
13. Mirza, M., Osindero, S.: Conditional generative adversarial nets. arXiv preprint arXiv:1411.1784 (2014)

14. Kosaraju, V., Sadeghian, A., Martín-Martín, R., Reid, I., Rezatofighi, H., Savarese, S.: Social-BIGAT: multimodal trajectory forecasting using bicycle-GAN and graph attention networks. In: Advances in Neural Information Processing Systems, pp. 137–146 (2019)
15. Varshneya, D., Srinivasaraghavan, G : Human trajectory prediction using spatially aware deep attention models. arXiv preprint arXiv:1705.09436 (2017)
16. Lee, N., Choi, W., Vernaza, P., Choy, C.B., Torr, P.H.S., Chandraker, M.: DESIRE: distant future prediction in dynamic scenes with interacting agents. In: CVPR, pp. 336–345 (2017)
17. Xue, H., Huynh, D.Q., Reynolds, M.: SS-LSTM: a hierarchical LSTM model for pedestrian trajectory prediction. In: 2018 IEEE Winter Conference on Applications of Computer Vision (WACV), pp. 1186–1194. IEEE (2018)
18. Haddad, S., Wu, M., Wei, H., Lam, S.K.: Situation-aware pedestrian trajectory prediction with spatio-temporal attention model. arXiv preprint arXiv:1902.05437 (2019)
19. Fernando, T., Denman, S., Sridharan, S., Fookes, C.: GD-GAN: generative adversarial networks for trajectory prediction and group detection in crowds. In: Jawahar, C.V., Li, H., Mori, G., Schindler, K. (eds.) ACCV 2018. LNCS, vol. 11361, pp. 314–330. Springer, Cham (2019). https://doi.org/10.1007/978-3-030-20887-5_20
20. Ivanovic, B., Leung, K., Schmerling, E., Pavone, M.: Multimodal deep generative models for trajectory prediction: a conditional variational autoencoder approach. IEEE Robot. Autom. Lett. **6**(2), 295–302 (2020)
21. Tay, M.K.C., Laugier, C.: Modelling smooth paths using Gaussian processes. In: Laugier, C., Siegwart, R. (eds.) Field and Service Robotics. Springer Tracts in Advanced Robotics, vol. 42, pp. 381–390. Springer, Heidelberg (2008). https://doi.org/10.1007/978-3-540-75404-6_36
22. Wilson, A.G., Knowles, D.A., Ghahramani, Z.: Gaussian process regression networks (2011)
23. Ng, A.Y., Russell, S.J., et al.: Algorithms for inverse reinforcement learning. In: ICML, vol. 1, p. 2 (2000)
24. Alahi, A., Goel, K., Ramanathan, V., Robicquet, A., Fei-Fei, L., Savarese, S.: Social LSTM: human trajectory prediction in crowded spaces. In: CVPR, pp. 961–971 (2016)
25. Morton, J., Wheeler, T.A., Kochenderfer, M.J.: Analysis of recurrent neural networks for probabilistic modeling of driver behavior. IEEE Trans. Intell. Transp. Syst. **18**(5), 1289–1298 (2016)
26. Schmerling, E., Leung, K., Vollprecht, W., Pavone, M.: Multimodal probabilistic model-based planning for human-robot interaction. In: 2018 IEEE International Conference on Robotics and Automation (ICRA), pp. 3399–3406. IEEE (2018)
27. Sun, J., Jiang, Q., Lu, C.: Recursive social behavior graph for trajectory prediction. In: CVPR, pp. 660–669 (2020)
28. Tao, C., Jiang, Q., Duan, L., Luo, P.: Dynamic and static context-aware LSTM for multi-agent motion prediction. arXiv preprint arXiv:2008.00777 (2020)
29. Li, J., Yang, F., Tomizuka, M., Choi, C.: EvolveGraph: multi-agent trajectory prediction with dynamic relational reasoning. In: Proceedings of the Neural Information Processing Systems (NeurIPS) (2020)
30. Barbi, T., Nishida, T.: Trajectory prediction using conditional generative adversarial network. In: 2017 International Seminar on Artificial Intelligence, Networking and Information Technology (ANIT 2017). Atlantis Press (2017)

31. Pang, Y., Liu, Y.: Conditional generative adversarial networks (CGAN) for aircraft trajectory prediction considering weather effects. In: AIAA Scitech 2020 Forum, pp. 1853 (2020)
32. Dendorfer, P., Osep, A., Leal-Taixé, L.: Goal-GAN: multimodal trajectory prediction based on goal position estimation. In: Proceedings of the Asian Conference on Computer Vision (2020)
33. Mangalam, K., et al.: It is not the journey but the destination: endpoint conditioned trajectory prediction. In: Vedaldi, A., Bischof, H., Brox, T., Frahm, J.-M. (eds.) ECCV 2020. LNCS, vol. 12347, pp. 759–776. Springer, Cham (2020). https://doi.org/10.1007/978-3-030-58536-5_45
34. Mangalam, K., An, Y., Girase, H., Malik, J.: From goals, waypoints & paths to long term human trajectory forecasting. arXiv preprint arXiv:2012.01526 (2020)
35. Vemula, A., Muelling, K., Oh, J.: Social attention: modeling attention in human crowds. In: 2018 IEEE International Conference on Robotics and Automation (ICRA), pp. 1–7. IEEE (2018)
36. Pellegrini, S., Ess, A., Van Gool, L.: Improving data association by joint modeling of pedestrian trajectories and groupings. In: Daniilidis, K., Maragos, P., Paragios, N. (eds.) ECCV 2010. LNCS, vol. 6311, pp. 452–465. Springer, Heidelberg (2010). https://doi.org/10.1007/978-3-642-15549-9_33
37. Leal-Taixé, L., Fenzi, M., Kuznetsova, A., Rosenhahn, B., Savarese, S.: Learning an image-based motion context for multiple people tracking. In: CVPR, pp. 3542–3549 (2014)
38. Chang, M.-F., et al.: Argoverse: 3D tracking and forecasting with rich maps. In: CVPR (2019)
39. Li, J., Ma, H., Tomizuka, M.: Conditional generative neural system for probabilistic trajectory prediction. arXiv preprint arXiv:1905.01631 (2019)
40. Deo, N., Trivedi, M.M.: Convolutional social pooling for vehicle trajectory prediction. In: CVPR Workshops, pp. 1468–1476 (2018)
41. Chandra, R., et al.: Forecasting trajectory and behavior of road-agents using spectral clustering in graph-LSTMs. IEEE Robot. Autom. Lett. **5**(3), 4882–4890 (2020)
42. Chandra, R., et al.: Forecasting trajectory and behavior of road-agents using spectral clustering in graph-LSTMs (2019)

An Integrated Approach to Produce Robust Deep Neural Network Models with High Efficiency

Zhijian Li[1(✉)], Bao Wang[2], and Jack Xin[1]

[1] University of California, Irvine, CA 92697, USA
{zhijil2,jack.xin}@uci.edu
[2] The University of Utah, Salt Lake City, UT 84112, USA

Abstract. Deep Neural Networks (DNNs) need to be both efficient and robust for practical uses. Quantization and structure simplification are promising ways to adapt DNNs to mobile devices, and adversarial training is one of the most successful methods to train robust DNNs. In this work, we aim to realize both advantages by applying a convergent relaxation quantization algorithm, i.e., Binary-Relax (BR), to an adversarially trained robust model, i.e. the ResNets Ensemble via Feynman-Kac Formalism (EnResNet). We discover that high-precision quantization, such as ternary (tnn) or 4-bit, produces sparse DNNs. However, this sparsity is unstructured under adversarial training. To solve the problems that adversarial training jeopardizes DNNs' accuracy on clean images and break the structure of sparsity, we design a trade-off loss function that helps DNNs preserve natural accuracy and improve channel sparsity. With our newly designed trade-off loss function, we achieve both goals with no reduction of resistance under weak attacks and very minor reduction of resistance under strong adversarial attacks. Together with our model and algorithm selections and loss function design, we provide an integrated approach to produce robust DNNs with high efficiency and accuracy. Furthermore, we provide a missing benchmark on robustness of quantized models.

Keywords: Quantization · Channel pruning · Adversarial training

1 Introduction

1.1 Background

Deep Neural Networks (DNNs) have achieved significant success in computer vision and natural language processing. Especially, the residual network (ResNet)[8] has achieved remarkable performance on image classification and has become one of the most important neural network architectures in the current literature.

This work was partly supported by NSF Grants DMS-1854434, DMS-1924548, DMS-1952644, DMS-1924935, and DMS-1952339.

G. Nicosia et al. (Eds.): LOD 2021, LNCS 13164, pp. 451–465, 2022.
https://doi.org/10.1007/978-3-030-95470-3_34

Despite the tremendous success of DNNs, researchers still try to strengthen two properties of DNNs, robustness and efficiency. In particular, for security-critic and on the edge applications. Robustness keeps the model accurate under small adversarial perturbation of input images, and efficiency enables us to fit DNNs into embedded system, such as smartphone. Many adversarial defense algorithms [7,11,14,25] have been proposed to improve the robustness of DNNs. Among them, adversarial training is one of the most effective and powerful methods. On the other hand, quantization [4,16], producing models with low-precision weights, and structured simplification, such as channel pruning [9,26], are promising ways to make models computationally efficient.

2 Related Work

2.1 Binary Quantization

Based on the binary-connect (BC) [24] proposed an improvement of BC called Binary-Relax, which makes the weights converge to the binary weights from the floating point weights gradually. Theoretically, [24] provided the convergence analysis of BC, and [12] presented an ergodic error bound of BC. The space of m-bit quantized weights $\mathcal{Q} \subset \mathbb{R}^n$ is a union of disjoint one-dimensional subspaces of R^n. When $m \geq 2$, it is formulated as:

$$\mathcal{Q} = \mathbb{R}_+ \times \{0, \pm 1, \pm 2, \cdots, \pm 2^{m-1}\}^n = \bigsqcup_{l=1}^{p} \mathcal{A}_l$$

where $\mathbb{R}_+$ is a float scalar. For the special binary case of $m = 1$, we have $\mathcal{Q} = \mathbb{R}_+ \times \{\pm 1\}^n$.

We minimize our objective function in the subspace $\mathcal{Q}$. Hence, the problem of binarizing weights can be formulated in the following two equivalent forms:

- I.

$$\operatorname*{argmin}_{u \in \mathcal{Q}} \mathcal{L}(u)$$

- II.

$$\operatorname*{argmin}_{u \in \mathbb{R}^n} \mathcal{L}(u) + \chi_{\mathcal{Q}}(u) \quad \text{where} \quad \chi_{\mathcal{Q}}(u) = \begin{cases} 0 & u \in \mathcal{Q} \\ \infty & else \end{cases} \tag{1}$$

Based on the alternative form II, [24] relaxed the optimization problem to:

$$\operatorname*{argmin}_{u \in \mathbb{R}^n} \mathcal{L}(u) + \frac{\lambda}{2} dist(u, \mathcal{Q})^2 \tag{2}$$

Observing (2) converges to (1) as $\lambda \to \infty$, [24] proposed a relaxation of BC:

$$\begin{cases} w_{t+1} = w_t - \gamma \nabla \mathcal{L}_t(u_t) \\ u_{k+1} = \operatorname{argmin}_{u \in \mathbb{R}^n} \frac{1}{2} \|w_{t+1} - u\|^2 + \frac{\lambda}{2} dist(u, \mathcal{Q})^2 \end{cases}$$

It can be shown that u_{k+1} above has a closed-form solution

$$u_{t+1} = \frac{\lambda \text{proj}_{\mathcal{Q}}(w_{t+1}) + w_{t+1}}{\lambda + 1}$$

Algorithm 1. Binary-Relax quantization algorithm

Input: mini-batches $\{(\mathbf{x}_1, \mathbf{y}_1), \cdots (\mathbf{x}_m, \mathbf{y}_m)\}$, $\lambda_0 = 1$, growth rate $\rho > 1$, learning rate γ, initial float weight w_0, initial binary weight $u_0 = w_0$, cut-off epoch M
Output: a sequence of binary weights $\{u_t\}$
for $t = 1, \cdots, N$ **do** ▷ N is the number of epochs
  if $t < M$ **then**
    for $k = 1, \cdots, m$ **do**
      $w_t = w_{t-1} - \gamma_t \nabla_k \mathcal{L}(u_{t-1})$
      $u_t = \frac{\lambda_t \cdot \text{Proj}_{\mathcal{Q}}(w_t) + w_t}{\lambda_t + 1}$
      $\lambda_{t+1} = \rho \cdot \lambda_t$
  else
    for $k = 1, \cdots, m$ **do**
      $w_t = w_{t-1} - \gamma_t \nabla \mathcal{L}(u_{t-1})$
      $u_t = \text{Proj}_{\mathcal{Q}}(w_t)$ ▷ This is precisely Binary-Connect
return quantized weights u_N

2.2 Adversarial Attacks

As [18] discovered the limited continuity of DNNs' input-output mapping, the outputs of DNNs can be changed by adding imperceptible adversarial perturbations to input data. The methods that generate perturbed data can be adversarial attacks, and the generated perturbed data are called adversarial examples. In this work, we focus on three benchmark adversarial attacks Fast gradient sign method (FGSM) [6], iterative FGSM (IFGSM), and Carlini and Wagner method (C&W) [2]. In this study, we denote attacks FGSM, IFGSM, and CW to be A_1, A_2, and A_3 respectively.

2.3 Adversarial Training

[5] rigorously established a rigorous benchmark to evaluate the robustness of machine learning models and investigated almost all current popular adversarial defense algorithms. They conclude that adversarially trained models are more robust models with other types of defense methods. [25] also shows that adversarial training is more powerful than other methods such as gradient mask and gradient regularization [11]. While a natural training has objective function:

$$\mathcal{L}(\omega) = \frac{1}{N} \sum_{i=1}^{N} l(f(w, \mathbf{x}_i), y_i) \tag{3}$$

Adversarial training [1,14] generates perturbed input data and train the model to stay stable under adversarial examples. It has the following objective function:

$$\mathcal{L}(w) = \frac{1}{N}\sum_{n=1}^{N} \max_{\tilde{x}_n \in D_n} l(f(w, \tilde{x}_n), y_n) \tag{4}$$

where $D_n = \{x | \|x - x_n\|_\infty < \delta\}$. l and f are the loss function and the DNN respectively. In this work, we denote (3) to be $\mathcal{R}_{nat}$ and (4) to be $\mathcal{R}_{rob}$. A widely used method to practically find $\tilde{x}_n$ is the projected gradient descent (PGD) [14]. [17] investigated the properties of the objective function of the adversarial training, and [22] provided convergence analysis of adversarial training based on the previous results.

Feynman-Kac Formalism Principled Robust DNNs: Neural ordinary differential equations (ODEs) [3] are a class of DNNs that use an ODE to describe the data flow of each input data. Instead of focusing on modeling the data flow of each individual input data, [13,20,21] use a transport equation (TE) to model the flow for the whole input distribution. In particular, from the TE viewpoint, [21] modeled training ResNet as finding the optimal control of the following TE

$$\begin{cases} \frac{\partial u}{\partial t}(\boldsymbol{x}, t) + G(\boldsymbol{x}, \boldsymbol{w}(t)) \cdot \nabla u(\boldsymbol{x}, t) = 0, & \boldsymbol{x} \in \mathbb{R}^d, \\ u(\boldsymbol{x}, 1) = g(\boldsymbol{x}), & \boldsymbol{x} \in \mathbb{R}^d, \\ u(\boldsymbol{x}_i, 0) = y_i, & \boldsymbol{x}_i \in T, \quad \text{with } T \text{ being the training set.} \end{cases} \tag{5}$$

where $G(\boldsymbol{x}, \boldsymbol{w}(t))$ encodes the architecture and weights of the underlying ResNet, $u(\boldsymbol{x}, 0)$ serves as the classifier, $g(\boldsymbol{x})$ is the output activation of ResNet, and y_i is the label of $\boldsymbol{x}_i$.

Based on Eq. (5), [21] interpreted adversarial vulnerability of ResNet as arising from the irregularity of $u(\boldsymbol{x}, 0)$ of the above TE. To enhance $u(\boldsymbol{x}, 0)$'s regularity, they added a diffusion term, $\frac{1}{2}\sigma^2 \Delta u(\boldsymbol{x}, t)$, to the governing equation of (5) which results to the convection-diffusion equation (CDE). By the Feynman-Kac formula, $u(\boldsymbol{x}, 0)$ of the CDE can be approximated by the following two steps:

- Modify ResNet by injecting Gaussian noise to each residual mapping.
- Average the output of n jointly trained modified ResNets, and denote it as En_nResNet.

[21] have noticed that EnResNet, comparing to ResNet with the same size, can significantly improve both natural and robust accuracies of the adversarially trained DNNs. In this work, we leverage the robust advantage of EnResNet to push the robustness limit of the quantized adversarially trained DNNs.

TRADES: It is well-known that adversarial training will significantly reduce the accuracy of models on clean images. For example, ResNet20 can achieve about 92% accuracy on CIFAR10 dataset. However, under the PGD training using $\epsilon = 0.031$ with step-size 0.007 and the number of iterations 10, it only has 76% accuracy on clean images of CIFAR10. [25] designed a trade-off loss function,

TRADES, that balances the natural accuracy and adversarial accuracy. The main idea of TRADES is to minimize the difference of adversarial error and the natural error:

$$\min_f \mathbb{E}\Big(l(f(X), Y) + \max_{X' \in \mathbb{B}(X,\epsilon)} l(f(X), f(X')\lambda)\Big) \tag{6}$$

TRADES is considered to be the state-of-the-art adversarial defense method, it outperforms most defense methods in both robust accuracy and natural accuracy. [5] investigated various defense methods, and TRADES together with PGD training outperforms all other methods they tested. In this work, we will design our own trade-off function that works with PGD training. TRADES will provide important baselines for us.

3 Quantization of EnResNet

We know that the accuracy of a quantized model will be lower than its counterpart with floating point weights because of loss of precision. However, we want to know that whether a quantized model is more vulnerable than its float equivalent under adversarial attacks? In this section, we study this question by comparing the accuracy drops of the natural accuracy and robust accuracy from float weights to binary weights. Meanwhile, we also investigate the performances between two quantization methods BC and BR.

3.1 Experimental Setup

Dataset. We use one of the most popular datasets CIFAR-10 [10] to evaluate the quantized models, as it would be convenient to compare it with the float models used in [21,25].

Baseline. Our baseline model is the regular ResNet. To our best knowledge, there has been work done before that investigated the robustness of models with quantized weights, so we do not have an expected accuracy to beat. Hence, our goals are to compare the robustness of binarized ResNet and binarized EnResNet and to see how close the accuracy of quantized models can be to the float models in [21,25].

Evaluation. We evaluate both natural accuracy and robust accuracy for quantized adversarial trained models. We examine the robustness of models FGSM (A_1), IFGSM (A_2), and C&W (A_3). In our recording, N denotes the natural accuracy (accuracy on clean images) of models. For FGSM, we use $\epsilon = 0.031$ as almost all works share this value for FGSM. For IFGSM, we use $\alpha = 1/255$, $\epsilon = 0.031$, and number of iterations 20. For C&W, we have learning rate 0.0006 and the number of iterations 50.

Algorithm and Projection. We set BC as our baseline algorithm, and we want to examine that whether the advantage of the relaxed algorithm 13 in [24] is preserved under adversarial training. In both BC and BR, we use the wildly used binarizing projection proposed by [16], namely:

$$\mathrm{Proj}_{\mathcal{Q}}(w) = \mathbb{E}[|w|] \cdot sign(w) = \frac{||w||_1}{n} \cdot sign(w)$$

where $sign(\cdot)$ is the component-wise sign function and n is the dimension of weights.

3.2 Result

First, we verify that the Ensemble ResNet consistently outperforms ResNet when binarized. We adversarially train two sets of EnResNet and ResNet with the similar number of parameters. As shown in Table 1, EnResNet has much higher robust accuracy for both float and quantized models.

Second, we investigate the performances of Binary-Connect method (BC) and Binary-Relax method (BR). We verify that BR outperforms BC (Table 2). A quantized model trained via BR provides higher natural accuracy and robust accuracy. As a consequence, we use this relaxed method to quantize DNNs in all subsequent experiments in this paper.

Table 1. Ensemble ResNet vs ResNets. We verify that EnResNet outperforms ResNet with the similar number of parameters with both float and binary weights.

Net(#params)	Model	N	A_1	A_2	A_3
En_1ResNet20 (0.27M)	BR	**69.60%**	**47.17%**	**43.89%**	**58.79%**
ResNet20 (0.27M)	BR	66.81%	43.37%	40.72%	52.14%
En_2ResNet20 (0.54M)	BR	**72.58%**	**49.29%**	**44.72%**	**60.36%**
ResNet34 (0.56M)	BR	70.31%	46.42%	43.26%	54.75%

Table 2. Binary Connect vs Binary Relax. Models quantized by Binary-Relax have higher natural accuracy and adversarial accuracies than those quantized by Binary-Connect

Model	Quant	N	A_1	A_2	A_3
En_1ResNet20	Float	78.31%	56.64%	49.00%	66.84%
	BC	68.84%	46.31%	42.45%	58.52%
	BR	**69.60%**	**47.17%**	**43.89%**	**58.79%**
En_2ResNet20	Float	80.10%	57.48%	49.55%	66.73%
	BC	71.48%	47.83%	43.03%	59.09%
	BR	**72.58%**	**49.29%**	**44.72%**	**60.36%**
En_5ResNet20	Float	80.64%	58.14%	50.32%	66.96%
	BC	75.54%	51.03%	46.01%	60.92%
	BR	**75.40%**	**51.60%**	**46.91%**	**61.52%**

4 Trade-Off Between Robust Accuracy and Natural Accuracy

4.1 Previous Work and Our Methodology

It is known that adversarial training will decrease the accuracy for classifying the clean input data. This phenomenon is verified both theoretically [19] and experimentally [11,21,25] by researchers. [25] proposed a trade-off loss function (6) for robust training to balance the adversarial accuracy and natural accuracy. In practice, it is formulated as following:

$$\mathcal{L} = \mathcal{L}_{nat} + \beta \cdot \frac{1}{N} \sum_{n=1}^{N} l(f(x_n), f(\tilde{x}_n)) \tag{7}$$

Motivated by [25], we study the following trade-off loss function for our quantized models:

$$\mathcal{L} = \alpha \cdot \mathcal{L}_{nat} + \beta \cdot L_{rob} \tag{8}$$

Note that adversarial training is a special case $\alpha = 0$, $\beta = 1$ in (8). Loss function (7), TRADES, improves the robustness of models by pushing the decision boundary away from original data points, clean images in this case. However, intuitively, if the model classifies a original data point wrong, the second term of (7) will still try to extend this decision boundary, which can prevent the first term of the loss function from leading the model to the correct classification. In this section, we will experimentally compare (7) and (8) and theoretically analyze the difference between them.

Table 3. Comparison of loss function. Trade-off loss function (7) outperforms TRADES (8) in most cases.

Model	Loss	N	A_1	A_2	A_3
En_1ResNet20	(7) ($\beta = 1$)	**84.49%**	45.96%	34.81%	51.94%
En_1ResNet20	(8) ($\alpha = 1, \beta = 1$)	83.47%	**54.46%**	**43.86%**	**64.04%**
En_1ResNet20	(7) ($\beta = 4$)	80.05%	51.24%	45.43%	58.85%
En_1ResNet20	(8) ($\alpha = 1, \beta = 4$)	**80.91%**	**55.92%**	**47.17%**	**66.53%**
En_1ResNet20	(7) ($\beta = 8$)	75.82%	51.63%	46.95%	59.31%
En_1ResNet20	(8) ($\alpha = 1, \beta = 8$)	**79.31%**	**56.28%**	**48.02%**	**66.07%**

4.2 Experiment and Result

To compare the performances of two loss functions, we choose our neural network and dataset to be En_1ResNet20 and CIFAR-10 respectively. Based on [25], who studied β of (7) in the range [1, 10], we vary the trade-off parameter β, the weight of adversarial loss, in the set [1, 4, 8] to emphasize the robustness in different levels.

The experiment results are listed in Table 3. We observe that, when the natural loss and the adversarial loss are equally treated, the model trained by (7) has higher natural accuracy while (8) makes its model more robust. As the trade-off parameter β increases, natural accuracy of the model trained (7) drops rapidly, while (8) trades a relatively smaller amount of natural loss for robustness. As a result, when $\beta = 4$ and $\beta = 8$, the model trained by (8) has both higher natural accuracy and higher adversarial accuracy. Hence, we say that (8) has better trade-off efficiency than (7)

4.3 Analysis of Trade-Off Functions

In this subsection, we present a theoretical analysis that why (8) outperforms (7). Let us consider the binary classification case, where have our samples $(\mathbf{x}, y) \in \mathcal{X} \times \{-1, 1\}$. Let $f : \mathcal{X} \to \mathbb{R}$ be a classifier and $\sigma(\cdot)$ be an activation function. Then our prediction for a sample is $\hat{y} = sign(f(\mathbf{x}))$ and the corresponding score is $\sigma(f(\mathbf{x}))$. Above is the theoretical setting provided by [25]. Then, we have the errors $\mathcal{R}_\phi(f)$ and $\mathcal{R}^*_\phi(f)$ corresponding to (7) and (8) respectively:

$$\mathcal{R}_\phi(f) = \mathbb{E}[\phi(\sigma \circ f(\mathbf{x}) \cdot y)] + \beta \cdot \mathbb{E}[\phi(\sigma \circ f(\mathbf{x}) \cdot \sigma \circ f(\mathbf{x}'))]$$

$$\mathcal{R}^*_\phi(f) = \alpha \cdot \mathbb{E}[\phi(\sigma \circ f(\mathbf{x}) \cdot y)] + \beta \cdot \mathbb{E}[\phi(\sigma \circ f(\mathbf{x}') \cdot y)]$$

Table 4. Trade-off loss function for binarized models with different parameters. When $\alpha = 1$ and $\beta = 8$, models can achieve about the same accuracies under FGSM and C&W attacks as solely adversarially trained models, while the natural accuracy is improved.

Model	Loss	N	A_1	A_2	A_3
En_1ResNet20	$\alpha = 0, \beta = 1$	69.60%	**47.81%**	**43.89%**	58.79%
En_1ResNet20	$\alpha = 1, \beta = 4$	**73.40%**	47.41%	41.86%	57.83%
En_1ResNet20	$\alpha = 1, \beta = 8$	71.35%	47.42%	42.46%	**59.01%**
En_2ResNet20	$\alpha = 0, \beta = 1$	71.58%	49.29%	**44.62%**	60.36%
En_2ResNet20	$\alpha = 1, \beta = 4$	**75.92%**	48.97%	43.41%	59.40%
En_2ResNet20	$\alpha = 1, \beta = 8$	74.72%	**49.66%**	43.96%	**60.65%**
En_5ResNet20	$\alpha = 0, \beta = 1$	75.40%	51.60%	**46.91%**	**61.52%**
En_5ResNet20	$\alpha = 1, \beta = 4$	**78.50%**	50.85%	45.02%	60.96%
En_5ResNet20	$\alpha = 1, \beta = 8$	77.35%	**51.62%**	45.63%	61.11%

Now, we consider several common loss functions: the hinge loss ($\phi(\theta) = \max\{1 - \theta, 0\}$), the sigmoid loss ($\phi(\theta) = 1 - \tanh \theta$), and the logistic loss ($\phi(\theta) = \log_2(1 + e^{-\theta})$). Note that we want a loss function to be monotonically deceasing in the interval $[-1, 1]$ as -1 indicates that the prediction is completely wrong, and 1 indicates the prediction is completely correct. Since our classes is 1 and -1, we will choose hyperbolic tangent as our activation function.

Table 5. Sparsity and structure of sparsity of quantized models. We use $\alpha = 1$ and $\beta = 8$ in trade-off loss (8). While models under both natural training and adversarial have large proportion of sparse weights, sparsity of adversarially trained models are much less structured. Trade-off loss function (6) can improve the structure.

Model	Quant	Loss	Weight Sparsity	Channel Sparsity	N	A_1	A_2
ResNet20	tnn	Natural	53.00%	11.16%	90.54%	12.71%	0.00%
En_1ResNet20	tnn	Natural	52.19%	9.57%	90.61%	26.21%	0.71%
ResNet20	tnn	AT	50.71%	2.55%	68.30%	44.80%	42.53%
En_1ResNet20	tnn	AT	50.31%	4.14%	71.30%	48.17%	43.27%
En_1ResNet20	tnn	(8)	55.66%	7.02%	73.05%	48.10%	42.65%
ResNet20	4-bit	Natural	42.79%	9.53%	91.75%	12.38%	0.00%
En_1ResNet20	4-bit	Natural	44.73%	10.52%	91.42%	27.99%	0.62%
ResNet20	4-bit	AT	43.93%	2.55%	71.49%	47.63%	44.08%
En_1ResNet20	4-bit	AT	48.35%	4.94%	73.05%	51.43%	45.10%
En_1ResNet20	4-bit	(8)	55.57%	7.42%	76.61%	51.92%	44.39%
ResNet56	tnn	Natural	60.96%	31.86%	91.91%	15.58%	0.00%
En_1ResNet56	tnn	Natural	60.66%	28.97%	91.46%	38.22%	0.36%
ResNet56	tnn	AT	54.21%	15.37%	74.56%	51.73%	46.62%
En_1ResNet56	tnn	AT	54.70%	16.74%	76.87%	53.16%	47.89%
En_1ResNet56	tnn	(8)	58.89%	21.36%	77.24%	52.96%	46.01%
ResNet56	4-bit	Natural	67.94%	39.16%	93.09%	16.02%	0.00%
En_1ResNet56	4-bit	Natural	71.07%	41.10%	92.39%	39.79%	0.33%
ResNet56	4-bit	AT	55.29%	17.10%	77.67%	52.43%	48.22%
En_1ResNet56	4-bit	AT	55.09%	18.11%	78.25%	55.48%	49.03%
En_1ResNet56	4-bit	(8)	67.31%	33.18%	79.44%	55.41%	47.80%

Proposition 1. *Let ϕ be any loss function that is monotonically decreasing on $[-1, 1]$ (all loss functions mentioned above satisfy this), and $\sigma(\theta) = \tanh\theta$. Define $B = \{\mathbf{x}|f(\mathbf{x})y \geq 0, f(\mathbf{x}')y \geq 0\}$ as in proposition 1. Then:*

$$\mathcal{R}_\phi(f) \geq \mathcal{R}^*_\phi(f) \text{ on } B \text{ and } \mathcal{R}_\phi(f) \leq \mathcal{R}^*_\phi(f) \text{ on } B^C$$

Proof: We first define a set E:

$$E = \{\mathbf{x}|f(\mathbf{x})y < 0, f(\mathbf{x}')y > 0\}$$

By the definition of adversarial examples in (4),

$$\mathbb{E}[\mathbf{1}\{E\}] = 0$$

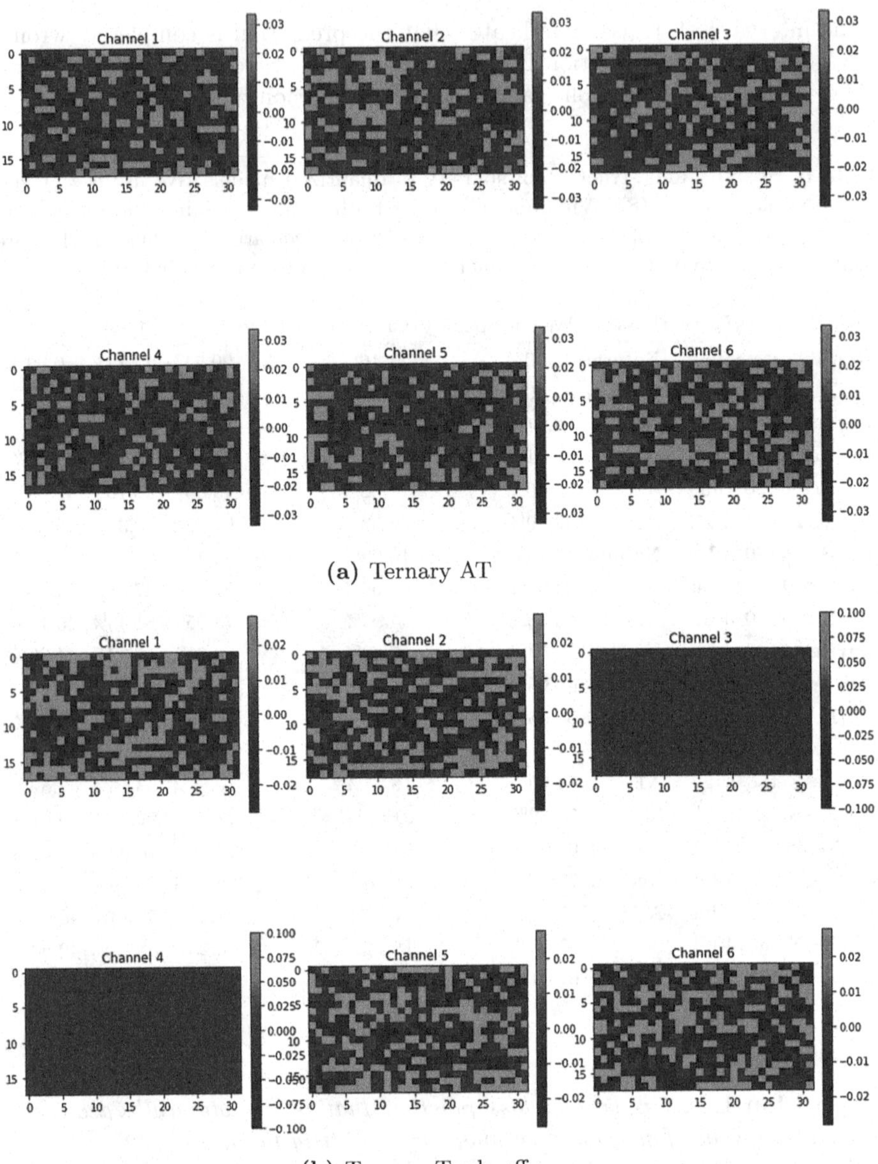

(a) Ternary AT

(b) Ternary Trade-off

Fig. 1. Visualization of 4 ternary channels (reshaped for visualization) of a layer in En_1ResNet56 under natural training. There are less 0 weights in non-sparse channels. Nonzero weights of ternary channels under natural training are more concentrated. As a result, there are more sparse channels under natural training.

where $\mathbf{1}\{E\}$ *is the indicator function of the set. That is, the set that the classifier predicts original data point wrong but the perturbed data point correctly should have measure* 0. *Now, we define the following sets:*

$$D = \{\mathbf{x}|f(\mathbf{x})y \geq 0, f(\mathbf{x}')y < 0\}$$

$$F = \{\mathbf{x}|f(\mathbf{x})y < 0, f(\mathbf{x}')y < 0\}$$

We note that the activation function $\sigma(x)$ *preserves the sign of* x, *and* $|\sigma(x)| \leq 1$.

On the set B, $f(\mathbf{x})$, $f(\mathbf{x}')$, *and* y *have the same sign, so are* $\sigma(f(\mathbf{x}))$, $\sigma(f(\mathbf{x}'))$ *and* y. *Therefore*

$$\phi(\sigma(f(\mathbf{x}')) \cdot \sigma(f(\mathbf{x})) \geq \phi(\sigma(f(\mathbf{x}') \cdot y)$$

as $0 \leq \sigma(f(\mathbf{x}')) \cdot \sigma(f(\mathbf{x})) \leq \sigma(f(\mathbf{x}')) \cdot y$. *This shows*

$$\mathcal{R}_\phi(f) \geq \mathcal{R}_\phi^*(f) \text{ on } B$$

We note that $B^C = E \cup D \cup F$. *Since set* E *has measure zero, we only consider* D *and* F.

On D, *as* f *classifies* $\mathbf{x}$ *correct and* $\mathbf{x}'$ *wrong, we have*

$$\sigma(f(\mathbf{x}')) \cdot y \leq \sigma(f(\mathbf{x}')) \cdot \sigma(f(\mathbf{x})) \leq 0$$

$$\Rightarrow \phi(\sigma(f(\mathbf{x}')) \cdot \sigma(f(\mathbf{x})) \leq \phi(\sigma(f(\mathbf{x}') \cdot y)$$

On F, *as* f *classifies both* $\mathbf{x}$ *and* $\mathbf{x}'$ *wrong, we have*

$$\sigma(f(\mathbf{x}')) \cdot y \leq 0 \leq \sigma(f(\mathbf{x}')) \cdot \sigma(f(\mathbf{x}))$$

$$\Rightarrow \phi(\sigma(f(\mathbf{x}')) \cdot \sigma(f(\mathbf{x})) \leq \phi(\sigma(f(\mathbf{x}') \cdot y)$$

In summary, we have

$$\mathcal{R}_\phi(f) \leq \mathcal{R}_\phi^*(f) \text{ on } B^C$$

□

We partition our space into several sets based on a given classifier f, and we examine the actions of loss functions on those sets. We see that (7) penalize set B heavier than (8), but the classifier classifies both the natural data and the perturbed data correct on B. On the other hand, (7) does not penalize sets E and F, where the classifier makes mistakes, enough, especially on set F. Therefore, (8) as a loss function is more on target. Based on both experimental results and theoretical analysis, we believe (8) is a better choice to balance natural accuracy and robust accuracy.

As our experiments on the balance of accuracies with different parameters in our loss function (8) in Table 4. We find that it is possible to increase the natural accuracy while maintaining the robustness under relatively weak attacks (FGSM & CW), as the case of $\alpha = 1$ and $\beta = 8$ in Table 4. However, the resistance under relatively strong attack (IFGSM) will inevitably decrease when we trade-off.

5 Further Balance of Efficiency and Robustness: Structured Sparse Quantized Neural Network via Ternary/4-Bit Quantization and Trade-Off Loss Function

5.1 Sparse Neural Network Delivered by High-Precision Quantization

When we quantize DNNs with precision higher than binary, such as ternery and 4-bit quantization, zero is in quantization levels. In fact, we find that a large proportion of weights will be quantized to zero. This suggests that a ternary or 4-bit quantized model can be further simplified via channel pruning. However, such simplification requires structure sparsity of DNN architecture. In our study, we use the algorithm 1 as before with the projection replaced by ternary and 4-bit respectively. As shown in Table 5, we find that sparsity of quantized DNNs under regular training are significantly more structured than those under adversarial training. For both ternary and 4-bit quantization, quantized models with adversarial training have very unstructured sparsity, while models with natural training have much more structured sparsity. For example, 50.71% (0.135M out of 0.268M) of weights in convolutional layers are zero in a ternary quantized ResNet20, but there are only 2.55% (16 out of 627) channels are completely zero. If the sparsity is unstructured, it is less useful for model simplification as channel pruning cannot be applied. A fix to this problem is our trade-off loss function, as factor natural loss into adversarial training should improve the structure of sparsity. Our experiment (Table 5) shows that a small factor of natural loss, $\alpha = 1$ and β in (8), can push the sparsity to be more structured. Meanwhile, the deepness of models also has an impact on the structure of sparsity. The deeper the more structured the sparsity is. We see in Table 5 that, under the same settings, the structure of sparsity increases as the model becomes deeper. Figure 1 shows the difference between a unstructured sparsity of a ternary ResNet20 with adversarial training and a much more structured sparsity of ResNet56 with natural training. The trade-off function not only improves the natural accuracy of models with merely minor harm to robustness but also structures the sparsity of high-precision quantization, so further simplification of models can be done through channel pruning.

6 Benckmarking Adversarial Robustness of Quantized Model

Based on previous discussions, integrating the relaxation algorithm, ensemble ResNet, and our trade-off loss function, we can produce very efficient DNN models with high robustness. To our best knowledge, there is no previous work that systematically study the robustness of quantized models. As a result, we do not have any direct baseline to measure our results. Therefore, we benchmark our results by comparing to models with similar size and current state-of-the-art

Table 6. Generalization of quantized models to more datasets

Model	Size	Dataset	Loss	Quant	Ch sparsity	N	A_1	A_2	A_3
En_2ResNet20	0.54M	MNIST	(8)	BR	N/A	99.22%	98.90%	98.90%	99.12%
ResNet44	0.66M	MNIST	TRADES	Float	N/A	99.31%	98.98%	98.91%	99.14%
En_2ResNet20	0.54M	MNIST	(8)	Float	N/A	99.21%	99.02%	98.91%	99.14%
En_2ResNet20	0.54M	FMNIST	(8)	BR	N/A	91.69%	87.85%	87.22%	89.74%
ResNet44	0.66M	FMNIST	TRADES	Float	N/A	91.37%	88.13%	87.98%	90.12%
En_2ResNet20	0.54M	FMNIST	(8)	Float	N/A	92.74%	89.35%	88.68%	91.72%
En_1ResNet56	0.85M	Cifar10	(8)	4-bit	33.18%	79.44%	55.71%	47.81%	65.50%
ResNet56	0.85M	Cifar10	TRADES	Float	N/A	78.92%	55.27%	50.40%	59.48%
En_1ResNet56	0.85M	Cifar10	(8)	Float	N/A	81.63%	56.80%	50.17%	66.56%
En_1ResNet110	1.7M	Cifar100	(8)	4-bit	23.93%	53.08%	30.76%	25.73%	42.54%
ResNet110	1.7M	Cifar100	TRADES	Float	N/A	51.65%	28.23%	25.77%	40.79%
En_1ResNet110	1.7M	Cifar100	(8)	Float	N/A	56.63%	32.24%	26.72%	43.99%
En_2ResNet56	1.7M	SVHN	(8)	4-bit	49.03%	91.21%	70.91%	57.99%	72.44%
ResNet110	1.7M	SVHN	TRADES	Float	N/A	88.33%	64.80%	56.81%	69.79%
En_2ResNet56	1.7M	SVHN	(8)	Float	N/A	93.33%	78.08%	59.11%	75.79%

defense methods. We verify that the performance of the quantized model with our approach on popular datasets, including Cifar 10, Cifar 100 [10], MNIST, Fashion MNIST (FMNIST) [23], and SVHN [15]. In our experiments, we learn SVHN dataset without utilizing its extra training data. Our results are displayed in Table 6. In this table, size refers to the number of parameters. We find that quantization has very little impact on learning small datasets MNIST and FMNIST. En_2ResNet20 and ResNet40 have about the same performance while the previous is binarized. In fact, a binary En_2ResNet20 has about the same performance as its float equivalent, which means we get efficiency for 'free' on these small datasets. We benchmark the robustness of quantized models on large datasets, SVHN, Cifar10, and Ciar100, using models with 4-bit quantization. As in Table 6, 4-bit EnResNets with loss function (8) have better performance than float TRADES-trained ResNets with the same sizes. However, quantized models on these larger datasets are outperformed by their float equivalents. In another word, efficiency of models are not 'free' when learning large datasets, we have to trade-off between performance and efficiency. Although 4-bit quantization requires higher precision and more memories, the highly structured sparsity can compensate the efficiency of models. The 4-bit quantized models of En_1ResNet56 for Cifar10, En_1ResNet110 for Cifar 100, and En_2ResNet56 for SVNH in Table 6 have sizes of 0.58M, 1.43M, and 0.80M respectively if the sparse channels are pruned. Our codes as well as our trained quantized models listed in Table 6 are available at https://github.com/lzj994/Binary-Quantization.

7 Conclusion

In this paper, we study the robustness of quantized models. The experimental results suggest that it is totally possible to achieve both efficiency and robustness as quantized models can also do a good job at resisting adversarial attack. Moreover, we discover that adversarial training will make the sparsity from high-precision quantization unstructured, and a trade-off function can improve the sparsity structure. With our integrated approach to balance efficiency and robustness, we find that keeping a model both efficient and robust is promising and worth paying attention to. We hope our study can serve as a benchmark for future studies on this interesting topic.

References

1. Athalye, A., Carlini, N., Wagner, D.: Obfuscated gradients give a false sense of security: circumventing defenses to adversarial examples. arXiv preprint arXiv:1802.00420 (2018)
2. Carlini, N., Wagner, D.: Towards evaluating the robustness of neural networks. In: 2017 IEEE Symposium on Security and Privacy (SP), pp. 39–57. IEEE (2017)
3. Chen, T., Rubanova, Y., Bettencourt, J., Duvenaud, D.: Neural ordinary differential equations. In: Advances in Neural Information Processing Systems, pp. 6571–6583 (2018)
4. Courbariaux, M., Bengio, Y., David, J.-P.: BinaryConnect: training deep neural networks with binary weights during propagations. In: Advances in Neural Information Processing Systems, pp. 3123–3131 (2015)
5. Dong, Y., et al.: Benchmarking adversarial robustness on image classification. In: IEEE/CVF Conference on Computer Vision and Pattern Recognition (2020)
6. Goodfellow, I.J., Shlens, J., Szegedy, C.: Explaining and harnessing adversarial examples. arXiv preprint arXiv:1412.6572 (2014)
7. Guo, C., Rana, M., Cisse, M., Van Der Maaten, L.: Countering adversarial images using input transformations. arXiv preprint arXiv:1711.00117 (2017)
8. He, K., Zhang, X., Ren, S., Sun, J.: Deep residual learning for image recognition. In: Proceedings of the IEEE Conference on Computer Vision and Pattern Recognition, pp. 770–778 (2016)
9. He, Y., Zhang, X., Sun, J.: Channel pruning for accelerating very deep neural networks. In: Proceedings of the IEEE International Conference on Computer Vision, pp. 1389–1397 (2017)
10. Krizhevsky, A., et al.: Learning multiple layers of features from tiny images (2009)
11. Kurakin, A., Goodfellow, I., Bengio, S.: Adversarial machine learning at scale. arXiv preprint arXiv:1611.01236 (2016)
12. Li, H., De, S., Xu, Z., Studer, C., Samet, H., Goldstein, T.: Training quantized nets: a deeper understanding. In: Advances in Neural Information Processing Systems, pp. 5811–5821 (2017)
13. Li, Z., Shi, Z.: Deep residual learning and PDEs on manifold. arXiv preprint arXiv:1708.05115 (2017)
14. Madry, A., Makelov, A., Schmidt, L., Tsipras, D., Vladu, A.: Towards deep learning models resistant to adversarial attacks. arXiv preprint arXiv:1706.06083 (2017)
15. Netzer, Y., Wang, T., Coates, A., Bissacco, A., Wu, B., Ng, A.Y.: Reading digits in natural images with unsupervised feature learning (2011)

16. Rastegari, M., Ordonez, V., Redmon, J., Farhadi, A.: XNOR-net: imagenet classification using binary convolutional neural networks. In: Leibe, B., Matas, J., Sebe, N., Welling, M. (eds.) ECCV 2016. LNCS, vol. 9908, pp. 525–542. Springer, Cham (2016). https://doi.org/10.1007/978-3-319-46493-0_32
17. Sinha, A., Namkoong, H., Duchi, J.: Certifying some distributional robustness with principled adversarial training. arXiv preprint arXiv:1710.10571 (2017)
18. Szegedy, C., et al.: Intriguing properties of neural networks. arXiv preprint arXiv:1312.6199 (2013)
19. Tsipras, D., Santurkar, S., Engstrom, L., Turner, A., Madry, A.: Robustness may be at odds with accuracy. arXiv preprint arXiv:1805.12152 (2018)
20. Wang, B., Luo, X., Li, Z., Zhu, W., Shi, Z., Osher, S.: Deep neural nets with interpolating function as output activation. In: Advances in Neural Information Processing Systems, pp. 743–753 (2018)
21. Wang, B., Shi, Z., Osher, S.: ResNets ensemble via the Feynman-Kac formalism to improve natural and robust accuracies. In: Advances in Neural Information Processing Systems, pp. 1655–1665 (2019)
22. Wang, Y., Ma, X., Bailey, J., Yi, J., Zhou, B., Gu, Q.: On the convergence and robustness of adversarial training. In: International Conference on Machine Learning, pp. 6586–6595 (2019)
23. Xiao, H., Rasul, K., Vollgraf, R.: Fashion-MNIST: a novel image dataset for benchmarking machine learning algorithms. arXiv preprint arXiv:1708.07747 (2017)
24. Yin, P., Zhang, S., Lyu, J., Osher, S., Qi, Y., Xin, J.: BinaryRelax: a relaxation approach for training deep neural networks with quantized weights. SIAM J. Imaging Sci. **11**(4), 2205–2223 (2018)
25. Zhang, H., Yu, Y., Jiao, J., Xing, E.P., El Ghaoui, L., Jordan, M.I.: Theoretically principled trade-off between robustness and accuracy. In: International Conference on Machine Learning (2019)
26. Zhuang, Z., et al.: Discrimination-aware channel pruning for deep neural networks. In: Advances in Neural Information Processing Systems, pp. 875–886 (2018)

Leverage Score Sampling for Complete Mode Coverage in Generative Adversarial Networks

Joachim Schreurs[1(✉)], Hannes De Meulemeester[1], Michaël Fanuel[2], Bart De Moor[1], and Johan A. K. Suykens[1]

[1] ESAT -STADIUS, KU Leuven, Kasteelpark Arenberg 10, 3001 Leuven, Belgium
{joachim.schreurs,hannes.demeulemeester,bart.demoor, johan.suykens}@kuleuven.be

[2] Univ. Lille, CNRS, Centrale Lille, UMR 9189 - CRIStAL, 59000 Lille, France
michael.fanuel@univ-lille.fr

Abstract. Commonly, machine learning models minimize an empirical expectation. As a result, the trained models typically perform well for the majority of the data but the performance may deteriorate in less dense regions of the dataset. This issue also arises in generative modeling. A generative model may overlook underrepresented modes that are less frequent in the empirical data distribution. This problem is known as complete mode coverage. We propose a sampling procedure based on ridge leverage scores which significantly improves mode coverage when compared to standard methods and can easily be combined with any GAN. Ridge leverage scores are computed by using an explicit feature map, associated with the next-to-last layer of a GAN discriminator or of a pre-trained network, or by using an implicit feature map corresponding to a Gaussian kernel. Multiple evaluations against recent approaches of complete mode coverage show a clear improvement when using the proposed sampling strategy.

Keywords: GANs · Leverage score sampling · Complete mode coverage

1 Introduction

Complete mode coverage is a problem of generative models which has been clearly defined and studied in [26]. In layman's terms, a mode is defined as a local maximum of the data probability density. A closely related problem is mode collapse in GANs [7], which happens when a generative model is only capable of generating samples from a subset of all the modes. Multiple GAN variants have

M. Fanuel—Most of this work was done when MF was at KU Leuven.

Supplementary Information The online version contains supplementary material available at https://doi.org/10.1007/978-3-030-95470-3_35.

G. Nicosia et al. (Eds.): LOD 2021, LNCS 13164, pp. 466–480, 2022.
https://doi.org/10.1007/978-3-030-95470-3_35

been proposed as a solution to this problem, however proposed solutions often assume that every mode has an (almost) equal probability of being sampled, which is often not the case in realistic datasets. Regularly, in critical applications, datasets contain a mixture of different subpopulations where the frequency of each subpopulation can be vastly different. The role of less abundant subpopulations in machine learning data has been discussed recently in [6]. Also, it is often common to presume that an algorithm does not know the abundance of subpopulations. It is however important that a machine learning model performs well on all subpopulations. A standard example is medical data where some rare diseases are less abundant than common diseases. To illustrate the approach presented in this paper, a motivating example containing one majority mode and two minority modes is given in Fig. 1. When sampling a mini-batch from the Probability Density Function (PDF) p, the side modes can be missed. We observe empirically that this is resolved by sampling from the ridge leverage score (RLS) distribution (see Sect. 2), which has been extensively used in randomized linear algebra and kernel methods. Figure 1 shows that the samples from the minority modes have larger RLSs. Thus, when sampling from the RLS distribution, there is a higher probability of including the minority modes.

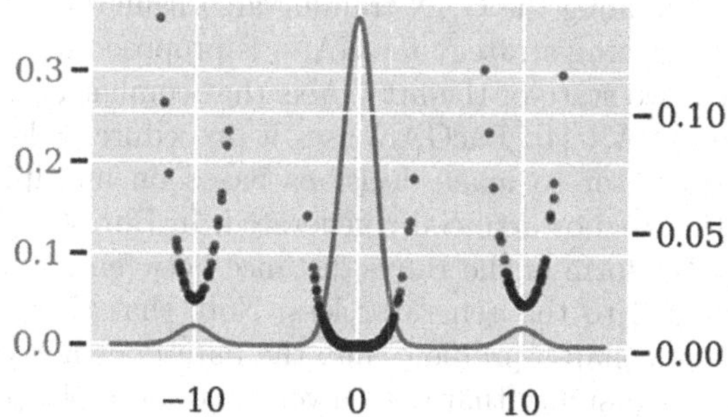

Fig. 1. Probability density function (orange) and RLS of a sample of this PDF (blue). We take the motivating example from [26], which consists of a 1D target PDF p with 1 majority mode and 2 minority modes: $p = 0.9{\cdot}\mathcal{N}(0,1) + 0.05{\cdot}\mathcal{N}(10,1) + 0.05{\cdot}\mathcal{N}(-10,1)$ given in orange. The RLS distribution is calculated using a Gaussian kernel with $\sigma = 3$ and $\gamma = 10^{-3}$. When sampling a mini-batch from the PDF p, the side modes can be missed. This is resolved by sampling from the RLS distribution. (Color figure online)

This paper is motivated by two situations where minority modes can occur: 1) the observed empirical distribution is different from the true distribution (biased data), and the data needs to be rebalanced. 2) The observed empirical distribution approximates the true distribution sufficiently well, but minority modes consist out of infrequent but very important points, e.g. rare diseases in a medical dataset.

Contribution. When training classical GANs, an empirical expectation of a loss $\mathbb{E}_{x\sim p_d}[\mathcal{L}(x)]$ is optimized in the context of a min-max problem. In this work, we propose a sampling procedure that promotes sampling out of minority modes by using ridge leverage scores. The common algorithmic procedure simulates the empirical distribution over the dataset $\mathcal{D} = \{x_1, \ldots, x_n\}$ by uniformly sampling over this set. We intentionally *bias* or *distort* this process by sampling x_i with probability $p(x_i) \propto \ell_i$, where ℓ_i is the i-th ridge leverage score, defined in Sect. 2. Empirical evidence shows that our procedure rebalances the training distribution, as a result, the GAN model generates samples more uniformly over all modes. RLS sampling can easily be applied to any GAN. In particular, using our procedure in combination with a state-of-the-art method for complete mode coverage [26] shows a clear improvement. Finally, RLS sampling is combined with BuresGAN [4] and a state-of-the-art StyleGAN2 with differentiable data augmentations [25], which in both cases improves mode coverage[1].

Related Work. Several works discuss alternative sampling strategies in machine learning. In the context of risk-averse learning, the authors of [2] discuss an adaptive sampling algorithm that performs a stochastic optimization of the Conditional Value-at-Risk (CVaR) of a loss distribution. This strategy promotes models which do not only perform well on average but also on rare data points. In the context of generative models, AdaGAN [20] is a boosting approach to solve the missing mode problem, where at every step a new component is added into a mixture model by running the GAN training algorithm on a re-weighted sample. A supervised weighting strategy for GANs is proposed in [5]. In this paper, we compare against two state-of-the-art GANs that combat mode collapse, PacGAN [10] and BuresGAN [4]. PacGAN uses a procedure called packing. This modifies the discriminator to make decisions based on multiple samples from the same class, either real or artificially generated. In BuresGAN, an additional diversity metric in the form of the Bures distance between real and fake covariance matrices is added to the generator loss. Note that these methods tackle the traditional mode collapse problem, i.e., the data does not include minority modes. In [23,24], it was shown that the convergence speed of stochastic gradient descent can be improved by actively selecting mini-batches using DPPs. In [19], coreset-selection is used to create mini-batches with a 'coverage' similar to that of the large batch - in particular, the small batch tries to 'cover' all the same modes as are covered in the large batch.

Before proceeding further, we discuss two main competitors more in-depth. The authors of [5] propose a solution to reduce selection bias in training data named Importance Weighted Generative Networks. A rescaling of the empirical data distribution is performed during training by employing a weighted Maximum Mean Discrepancy (MMD) loss such that the regions where the observed and the target distributions differ are penalized more. Each sample $i \in \{1, \ldots, n\}$ is scaled by $1/M(x_i)$, where M is the known or estimated Radon-Nykodym derivative between the target and observed distribution. A version of the vanilla GAN with importance weighting is introduced (IwGAN), as well as the weighting combined with MMDGAN (IwMmdGAN). Another approach to complete mode coverage by [26] and dubbed MwuGAN in this paper, iteratively trains a

[1] Code and supplementary at https://github.com/joachimschreurs/RLS_GAN.

mixture of generators. At each iteration, the sampling probability is pointwise normalized so that the probability to sample a missing mode is increased. Hence, this generates a sequence of generative models which constitutes the mixture. More precisely, a weight $w_i > 0$ is given for each $i \in \{1, \dots, n\}$ and initialized such that $w_i = p(x_i)$ for some distribution[2] p. Next, a generative model is trained and the probability density $p_g(x_i)$ of each $i \in \{1, \dots, n\}$ is computed. If $p_g(x_i) < \delta p(x_i)$ for some threshold value $\delta \in (0, 1)$, the weight is updated as follows: $w_i \leftarrow 2w_i$, otherwise the weight is not updated. The probability is then recalculated as follows: $p(x_i) = w_i / \sum_j w_j$ for each $i \in \{1, \dots, n\}$. Another generative model is then trained by using $p(x_i)$ and the procedure is repeated.

Classical Approach. A GAN consists of a discriminator $D : \mathbb{R}^d \to \mathbb{R}$ and a generator $G : \mathbb{R}^\ell \to \mathbb{R}^d$ which are typically defined by neural networks, and parametrized by real vectors. The value $D(x)$ gives the probability that x comes from the empirical distribution, while the generator G maps a point z in the latent space $\mathbb{R}^\ell$ to a point in input space $\mathbb{R}^d$. A typical training scheme for a GAN consists in solving, in an alternating way, the following problems:

$$\begin{aligned} &\max_D \mathbb{E}_{x \sim p_d}[\log(D(x))] + \mathbb{E}_{\tilde{x} \sim p_g}[\log(1 - D(\tilde{x}))], \\ &\min_G -\mathbb{E}_{\tilde{x} \sim p_g}[\log(D(\tilde{x}))], \end{aligned} \tag{1}$$

which include the vanilla GAN objective associated with the cross-entropy loss. In (1), the first expectation is over the empirical data distribution p_d and the second is over the generated data distribution p_g, implicitly given by the mapping by G of the latent prior distribution $\mathcal{N}(0, \mathbb{I}_\ell)$. The data distribution p_d is estimated using the empirical distribution over the training data $\hat{p}_d(x) = \frac{1}{n} \sum_{x_i \in \mathcal{D}} \delta(x - x_i)$ as follows: $\mathbb{E}_{p_d(x)}[\mathcal{L}(x)] \approx \mathbb{E}_{\hat{p}_d(x)}[\mathcal{L}(x)] = \frac{1}{n} \sum_{x_i \in \mathcal{D}} \mathcal{L}(x_i)$, where $\mathcal{L}$ is a general loss function. As noted by [21], positive weights w_i for $1 \leq i \leq n$ can be used to construct a weighted empirical distribution $\hat{p}_d^w(x) = \sum_{x_i \in \mathcal{D}} w_i \delta(x - x_i)$, then one can apply a weighting strategy to use samples distributed according to $\hat{p}(x)$ to estimate quantities with respect to $\hat{p}_d^w(x)$ as follows:

$$\mathbb{E}_{\hat{p}_d^w(x)}[\mathcal{L}(x)] = \mathbb{E}_{\hat{p}_d(x)}\left[\frac{\hat{p}_d^w(x)}{\hat{p}_d(x)} \mathcal{L}(x)\right] = \sum_{x_i \in \mathcal{D}} w_i \mathcal{L}(x_i). \tag{2}$$

A stochastic procedure is applied for minimizing the above expectation over $\hat{p}_d^w$. In this paper, mini-batches are sampled according to the distribution $\hat{p}_d^w$ with w_i given by the normalized RLSs (3) for $1 \leq i \leq n$.

2 Sampling with Ridge Leverage Scores

We propose to use a sampling procedure based on ridge leverage scores (RLSs) [1, 11]. RLSs correspond to the correlation between the singular vectors of a matrix

[2] In [26], this initial distribution is uniform. We discuss in Sect. 3.1 a choice of weights based on RLSs and initialize MwuGAN with the normalized RLSs in (3).

and the canonical basis elements. The higher the score, the more *unique* the point. A sample from a minority mode would thus get a higher RLS. These RLSs are used to bias the sampling, which in turn results in a more uniform sampling over all the modes, regardless of the original weight of the mode in the data distribution. Given a feature map $\varphi(\cdot)$, the corresponding kernel function is $K(x, y) = \varphi(x)^\top \varphi(y)$. Let the regularization parameter be $\gamma > 0$. Then, the γ-RLSs are defined for all $1 \leq i \leq n$ as:

$$\ell_i(\gamma) = \left(K(K + n\gamma\mathbb{I})^{-1}\right)_{ii} = \varphi(x_i)^\top (C + n\gamma\mathbb{I})^{-1}\varphi(x_i), \tag{3}$$

where $C = \sum_{i=1}^n \varphi(x_i)\varphi(x_i)^\top$ and $K_{ij} = \varphi(x_i)^\top \varphi(x_j)$ for $1 \leq i, j \leq n$. They have both a primal and a dual expression that can be leveraged when the size of the feature map or batch-size respectively are too large. When both the batch-size and feature map dimensions are large, one can rely on fast and reliable approximation algorithms with guarantees such as RRLS [15] and BLESS [17]. The role of $\gamma > 0$ is to filter the small eigenvalues of K in the spirit of Tikhonov regularization. RLSs induce the probability distribution: $p_i = \ell_i / \sum_{j=1}^n \ell_j$, for $1 \leq i \leq n$, which is classically used in randomized methods [1]. Figure 2 illustrates the interpretation of RLSs on two artificial datasets used in this paper. The datasets consist of a mixture of Gaussians. In the RING example, the first 4 modes are minority modes (starting on top and going further clockwise). In the GRID example, the first 10 modes are minority modes (starting left). Similar to the first illustration (see Fig. 1), large RLSs are associated with minority modes. More information on the artificial datasets is given in Sect. 3.1.

Fig. 2. RLS distribution using a Gaussian feature map with $\sigma = 0.15$ and regularization $\gamma = 10^{-3}$ on the unbalanced RING (left) and GRID (right) data. The darker the shade, the higher the RLS. Dark modes correspond to minority modes.

RLS sampling has a rich history in kernel methods and randomized linear algebra but has not been used in the context of GANs. One of the key contributions of this paper is to illustrate the use of RLSs in this setting. To do so, we propose the use of different feature maps so that RLS sampling can be used both for low dimensional and high dimensional data such as images. In what follows, the feature map construction is first discussed. Next, two approximation schemes are introduced.

Choice of the Feature Map. Three choices of feature maps are considered in this paper to compute leverage scores:

- Fixed implicit feature map. In low dimensional examples, the feature map can be chosen implicitly such that it corresponds to the Gaussian kernel: $\varphi(x)^\top \varphi(y) = \exp\left(-\|x-y\|^2/\sigma^2\right)$, the bandwidth σ is a hyperparameter.
- Fixed explicit feature map. For image-based data, more advanced similarity metrics are necessary. Therefore, the next-to-last layer of a pre-trained classifier, e.g. the Inception network, is used to extract meaningful features. Note that the classifier does not need to be trained on the exact training dataset, but simply needs to extract useful features.
- Discriminator-based explicit feature map. The feature map can be obtained from the next-to-last layer of the discriminator. Let $D(x) = \sigma(w^\top \varphi_D(x))$, where $w \in \mathbb{R}^m$ contains the last dense layer's weights and σ is the sigmoid function. This feature map is useful in situations where no prior knowledge is available about the dataset.

For a fixed feature map, the RLSs only need to be calculated once before training. The discriminator-based explicit feature map changes throughout the training. Therefore the RLSs are recalculated at every step. Nonetheless, due to approximation schemes which are discussed hereafter, the computational cost stays low. The full algorithm is given in Supplementary Material.

2.1 Approximation Schemes

Current day models are high dimensional, e.g., a DCGAN yields a feature space $\mathbb{R}^m$ of high dimension $m = 10^3$. Moreover, the size of datasets is commonly thousands up to millions of images. Therefore, two approximation schemes are proposed to speed up the computation of RLSs when using explicit feature maps:

- For the discriminator-based explicit feature map, we propose a two-stage sampling procedure in combination with a Gaussian sketch to reduce the dimension of high dimensional feature maps.
- For the fixed explicit feature map, the well-known UMAP is used to reduce the dimensionality [13].

Two-stage Sampling Procedure. For the explicit discriminator-based feature map, φ_D has to be re-calculated at each training step. To speed up the sampling procedure, we propose a sampling procedure in two stages. First, a subset of the data is uniformly sampled, e.g. equal to 20 times the desired batch size. Afterward, the RLSs are calculated only for the uniformly sampled subset, which are then used to sample the final batch used for training. This two-stage sampling procedure is similar to the core-set selection used in smallGAN [19]. A first difference is that core-sets are selected by combining a Gaussian sketch and a greedy selection in [19], while we use a randomized approach. Second, in this reference, cores-sets are used to reduce the batch size to improve scalability. In contrast, RLS sampling is used here to bias the empirical distribution.

Sketching the Discriminator Feature Map. Gaussian sketching is a commonly used method to reduce data dimension and was also used in [19] and [22] to reduce the dimension of large neural nets. Let S be a sketching matrix of size $m \times k$ such that $S = A/\sqrt{k}$ with A a matrix with i.i.d. zero-mean standard normal entries. Consider the following random projection: let a batch be $\{i_1, \ldots, i_b\} \subset \{1, \ldots, n\}$. A random projection of this batch in feature space is then defined as follows:

$$\varphi(x_{i_\ell}) = S^\top \varphi_D(x_{i_\ell}) \in \mathbb{R}^k, \tag{4}$$

for all $\ell \in \{1, \ldots, b\}$. This random projection preserves approximately (squared) pairwise distances in the dataset and is motivated by an isometric embedding result in the spirit of Johnson-Lindenstrauss lemma. Let $0 < \epsilon < 1$ and any integer $b > 0$. Let k be an integer such that $k \geq 4(\epsilon^2/2 - \epsilon^3/3)^{-1} \log b$. Then, for any set $\{\mathsf{x}_1, \ldots, \mathsf{x}_b\}$ in $\mathbb{R}^m$ there is a map $f : \mathbb{R}^m \to \mathbb{R}^k$ such that for any $\ell, \ell' \in \{1, \ldots, b\}$ we have

$$(1-\epsilon)\|\mathsf{x}_\ell - \mathsf{x}_{\ell'}\|_2^2 \leq \|f(\mathsf{x}_\ell) - f(\mathsf{x}_{\ell'})\|_2^2 \leq (1+\epsilon)\|\mathsf{x}_\ell - \mathsf{x}_{\ell'}\|_2^2.$$

The idea of this work is to consider the set of points given by the batch in the discriminator feature space $\mathsf{x}_\ell = \varphi_D(x_{i_\ell})$ for $1 \leq \ell \leq b$. It is proved in [3] that f exists and can be obtained with high probability with a random projection of the form (4). For a more detailed discussion, we refer to [16].

Dimensionality Reduction of the Fixed Explicit Feature Map by UMAP. The Gaussian sketch is a simple and fast method to reduce the dimension of the feature map. This makes it a perfect candidate to reduce the dimension of the proposed discriminator feature map, which has to be recalculated at every iteration. Unfortunately, this speed comes at a price, namely, the gaussian sketch is deemed too simple to reduce the dimension of highly complex models like the Inception network. Therefore, UMAP is proposed [13]. This nonlinear dimensionality reduction technique can extract more meaningful features. UMAP is considerably slower than the Gaussian sketch, therefore the use of UMAP is only advised for a fixed feature map like a pre-trained classifier or the Inception network, where the RLSs are only calculated once before training.

3 Numerical Experiments

The training procedure is evaluated on several synthetic and real datasets, where we artificially introduce minority modes. The GANs are evaluated by analyzing the distribution of the generated samples. Ideally, the models should generate samples from every mode as uniformly as possible. A re-balancing effect should be visible. The proposed methods are compared with vanilla GAN, PacGAN, MwuGAN, BuresGAN, IwGAN, and IwMmdGAN. In particular, MwuGAN outperforms AdAGAN [20] on complete mode coverage problems [26]. BuresGAN [4] promotes a matching between the covariance matrices of real and generated data in a feature space defined thanks to the discriminator. Recall that the discriminator is $D(x) = \sigma(w^\top \varphi_D(x))$, where w is a weight vector and the sigmoid function is denoted by σ. The normalization $\bar{\varphi}_D(x) = \varphi_D(x)/\|\varphi_D(x)\|_2$ is used, after the centering of $\varphi_D(x)$. Then, the covariance matrix is defined as follows: $C(p) = \mathbb{E}_{x\sim p}[\bar{\varphi}_D(x)\bar{\varphi}_D(x)^\top]$. The real data and generated data covariance matrices are denoted by $C_d = C(p_d)$ and $C_g = C(p_g)$, respectively. In BuresGAN, the Bures distance is added to the generator loss: $\min_G -\mathbb{E}_{\tilde{x}\sim p_g}[\log(D(\tilde{x}))] + \lambda\mathcal{B}(C_d, C_g)^2$, with the Bures distance $\mathcal{B}(C_d, C_g)^2 = \mathrm{Tr}(C_d + C_g - 2(C_d C_g)^{\frac{1}{2}})$, depends implicitly on $\varphi_D(x)$ (see e.g., [12]). The loss of the discriminator remains the same.

Overview of Proposed Methods. The RLS sampling procedure can easily be integrated into any GAN architecture or model. In this spirit, we used RLS sampling with a classical vanilla GAN (**RLS GAN**) and BuresGAN (**RLS BuresGAN**). The classical BuresGAN has been shown to outperform competitors in mode collapse problems [4]. We noticed empirically that RLS BuresGAN outperformed RLS GAN on the synthetic data (the comparison is shown in Supplementary Material). Therefore we only continue with RLS BuresGAN in the rest of the experiments. Likewise, RLS sampling is combined with MwuGAN, which is considered state-of-the-art in complete mode coverage. The method, called **RLS MwuGAN**, uses RLSs as initial starting weights to sample as opposed to uniform weights. Besides the initialization, the method remains unchanged. The number of generators in the mixture is always displayed in brackets. Unless specified otherwise, the models are trained for 30k iterations with a batch size of 64, by using the Adam [9] optimizer. Unless specified otherwise, we report the means and standard deviations for 10 runs. The largest mean is depicted in black, a * represents significance using a one-tailed Welch's t-test between the best performing proposed model and best performing competitor at a 0.05 confidence level. Further information about the used architectures, hyperparameters, and timings are given in Supplementary Material.

Fig. 3. Visualization of the generation quality on Ring and Grid. Each column shows 2.5k samples from the trained generator in blue and 2.5k samples from the true distribution in green. The vanilla GAN (first and third) does not cover the minority modes. This is not the case for the RLS BuresGAN Discr. (second and fourth). (Color figure online)

3.1 Synthetic Data

Unbalanced versions of two classical synthetic datasets are generated: an unbalanced ring with 4 minority modes (RING) and an unbalanced grid (GRID) with 10 minority modes (see Fig. 2). RING is a mixture of eight two-dimensional isotropic Gaussians in the plane with means $2.5 \times (\cos((2\pi/8)i), \sin((2\pi/8)i))$ and std 0.05 for $i \in \{1, \ldots, 8\}$. The probability of sampling from the first 4 consecutive Gaussians is only 0.05 times the probability of sampling from the last 4 modes. GRID is a mixture of 25 two-dimensional isotropic normals with standard deviation 0.05 and with means on a square grid with spacing 2. The first rectangular blocks of 2×5 adjacent modes are depleted with a factor 0.05.

Table 1. Experiments on the synthetic datasets RING and GRID. Two RLS BuresGAN are considered: RLSs calculated with the Gaussian kernel (Gauss.) and the next-to-last layer of the discriminator (Discr.). RLS MwuGAN is initialized with RLSs using an implicit feature map associated with the Gaussian kernel.

	Ring with 8 modes		Grid with 25 modes	
	Nb modes (↑)	% in 3σ (↑)	Nb modes (↑)	% in 3σ (↑)
GAN	5.0(1.1)	**0.92**(0.02)*	8.3(3.4)	0.29(0.3)
PacGAN2	5.4(1.4)	**0.92**(0.03)	10.3(2.6)	0.13(0.02)
BuresGAN	5.8(1.4)	0.76(0.27)	16.7(0.9)	**0.82**(0.01)*
RLS GAN Gauss	7.4(0.9)	0.86(0.05)	13.8(7.4)	0.51(0.32)
RLS GAN Discr	7.6(0.8)	0.90(0.02)	20.4(2.6)	0.81(0.03)
IwGAN	**8**(0)	0.85(0.08)	13.4(5.7)	0.29(0.25)
IwMmdGAN	**8**(0)	0.84(0.02)	1.7(5.1)	0.03(0.05)
MwuGAN (15)	7.9(0.3)	0.86(0.02)	15.2(1.7)	0.47(0.11)
RLS MwuGAN (15) (ours)	**8**(0)	0.84(0.06)	22.3(1.9)	0.60(0.1)
RLS BuresGAN Gauss. (ours)	**8**(0)	0.90(0.02)	24.0(1.5)	0.76(0.11)
RLS BuresGAN Discr. (ours)	**8**(0)	0.90(0.02)	**24.4**(0.92)*	0.78(0.06)

Evaluation. The evaluation is done by sampling 10k points from the generator network. High-quality samples are within 3 standard deviations of the nearest mode. A mode is covered if there are at least 50 generated samples within 3 standard deviations of the center of the mode. The knowledge of the full Radon-Nikodym derivative M is given to IwMmdGAN and IwGAN by dividing the true probability of each sample (a Gaussian mixture with equal weights) by the adapted Gaussian mixture containing several minority modes. The results of the experiments are given in Table 1. For RLS sampling, we use a Gaussian kernel with bandwidth $\sigma = 0.15$, and the discriminator network as feature extractor, both with regularization parameter $\gamma = 10^{-3}$. The models are trained using a fully connected architecture (see Supplementary Material). As the models are rather simple, no dimensionality reduction is needed.

Generated samples from models trained with and without RLS sampling are displayed in Fig. 3. One can clearly see that training a GAN with uniform sampling results in missing the first 4 minority modes. This is solved by using RLS sampling and can be interpreted by comparing the two sampling distributions on Fig. 4 (uniform) and Fig. 2 (RLS). The RLSs are larger for samples in minority modes, which results in a more uniform mini-batch over all modes. Note that the RLS sampling procedure, given the feature map, is completely unsupervised and has no knowledge of the desired unbiased distribution. The evaluation metrics in Table 1 confirm our suspicions. Only methods designed for complete mode coverage can recover (almost) all modes for the RING dataset. For the unbalanced GRID, only the proposed method has an acceptable performance. Our method even outperforms multiple generator architectures like MwuGAN and RLS MwuGAN, which are considerably more costly to train. Moreover, IwGAN, with full knowledge of M, is not capable of consistently capturing all modes. This was pointed out by the authors in [5]: the method may still experience high variance if it rarely sees data points from a class it wants to boost.

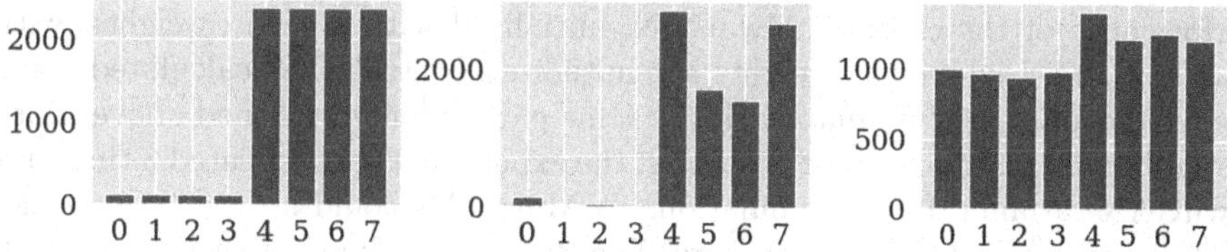

Fig. 4. RING. Number of training samples in each mode for the RING dataset (left) Generated samples in each mode by a vanilla GAN (middle). Generated samples by RLS BuresGAN Discr. (right). A rebalancing effect is visible.

3.2 Unbalanced MNIST

For this experiment, we create two unbalanced datasets out of MNIST. The first modified dataset, named UNBALANCED 012-MNIST, consists of only the digits 0, 1 and 2. The class 2 is depleted so that the probability of sampling 2 is only 0.05 times the probability of sampling from the digit 0 or 1. The second dataset, named UNBALANCED MNIST, consists of all digits. The classes 0, 1, 2, 3, and 4 are all depleted so that the probability of sampling out of the minority classes is only 0.05 times the probability of sampling from the majority digits. For these experiments, we use a DCGAN architecture. The following metrics are used for performance evaluation: the number of generated digits in each mode, which measures mode coverage, and the KL divergence [14] between the classified labels of the generated samples and a balanced label distribution, which measures sample quality. The mode of each generated image is identified by using a MNIST classifier which is trained up to 98.43% accuracy (see Supplementary Material). The metrics are calculated based on 10k generated images for all the models. For the RLS computation, we use both the discriminator with a Gaussian sketch and

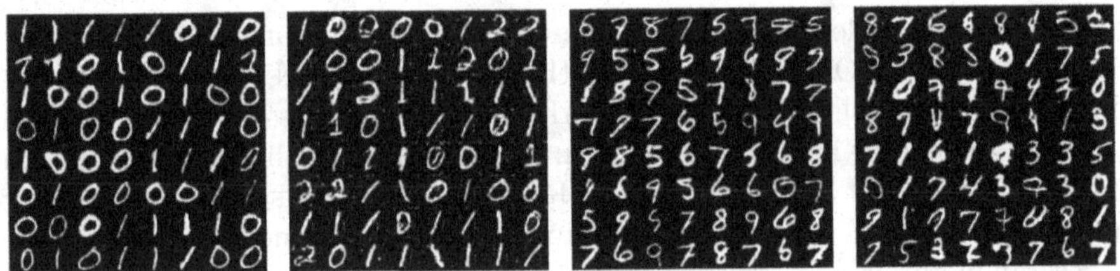

Fig. 5. Generated images from UNBALANCED 012-MNIST by a vanilla GAN (first), by RLS BuresGAN Discr. (second) and generated images from UNBALANCED MNIST by a vanilla GAN (third), by RLS BuresGAN Class. (fourth). The minority digits are generated more frequently in the proposed methods that include RLS sampling.

the next-to-last layer of the pre-trained classifier with a UMAP dimensionality reduction as a feature map. For the UNBALANCED 012-MNIST, both feature maps are reduced to $k = 25$ and the regularization parameter is $\gamma = 10^{-4}$. In the UNBALANCED MNIST, we take $k = 10$ and $\gamma = 10^{-4}$. An ablation study over different k and γ is given in Supplementary Material. We also compare the performance of the classical MwuGAN, initialized with uniform weights, with RLS MwuGAN where the weights are initialized by the RLSs calculated using the fixed explicit feature map with the same parameters mentioned above. Both methods contain a mixture of 15 GANs, the experiments are repeated 3 times for MwuGAN variants. In our simulations, IwMmdGAN could not be trained successfully with a DCGAN architecture. The Radon-Nikodym derivative M, which is used by IwMmdGAN and IwGAN, is defined as follows: $M_i = 1$ for digits 0 and 1 and $M_i = 0.05$ for digits 2, analogous for the UNBALANCEDMNIST dataset. Only the proposed models trained with RLS sampling are capable of covering all modes consistently. The diversity of images generated by RLS BuresGAN can be visualized in Fig. 5 where digits from minority modes appear more frequently. A quantitative analysis of mode coverage and sample quality is reported in Tables 2 and 3. In the UNBALANCED 012-MNIST dataset, there is a clear advantage in using RLS sampling with BuresGAN since mode coverage and the KL divergence are improved compared to the other methods. The second best method is RLS MwuGAN which outperforms RLS with uniform starting weights in KL. For the more difficult UNBALANCED MNIST dataset, using the fixed explicit feature map to calculate the RLSs clearly outperforms other methods.

3.3 Unbalanced CIFAR10

We conclude this section with an experiment on colored images, namely the CIFAR10 dataset. This highly diverse dataset contains 32×32 color images from 10 different classes. We consider two unbalanced variations. The first modified dataset, named UNBALANCED 06-CIFAR10, consists of only the classes 0 and 6 or images of airplanes and frogs respectively. The class 0 is depleted with a factor 0.05. The second dataset, named UNBALANCED 016-CIFAR10, consists of the classes 0,1 and 6. Compared to the previous dataset, we add images from

Table 2. Experiments on the UNBALANCED 012-MNIST dataset. Two RLS BuresGAN are considered: RLSs calculated with an explicit feature map obtained from a pre-trained classifier (Class.) and the next-to-last layer of the discriminator (Discr.). RLS MwuGAN is initialized with RLSs using the explicit feature maps obtained from a pre-trained classifier. Minority modes are highlighted in black in the first row.

	Mode 1	Mode 2	Mode 3	KL
GAN	4381(172)	5412(179)	129(36)	0.31(0.01)
PacGAN2	4492(237)	5328(242)	123(29)	0.32(0.01)
BuresGAN	4586(287)	5190(292)	142(19)	0.30(0.01)
IwGAN	4368(295)	5414(287)	147(32)	0.32(0.01)
IwMmdGAN	34(12)	0(0)	69(12)	0.56(0.10)
MwuGAN (15)	4886(473)	4865(466)	176(14)	0.31(0.01)
RLS MwuGAN (15) (ours)	3982(218)	4666(164)	870(65)	0.14(0.01)
RLS BuresGAN Class. (ours)	3414(161)	4862(134)	1461(183)	**0.08**(0.01)*
RLS BuresGAN Discr. (ours)	5748(172)	2416(268)	**1566**(293)*	0.16(0.02)

the class automobile. Now, the class 6 consisting of frogs is depleted with a factor 0.05. We show the improvement of RLS sampling in a StyleGAN2 with differentiable data augmentation (StyleGAN2 + Aug.) [25][3]. By clever use of various types of differentiable augmentations on both real and fake samples, the GAN can match the top performance on CIFAR10 with only 20% training data and is considered state-of-the-art. The StyleGAN2 models are trained for 156k iterations with a mini-batch size of 32 using 'color, translation, and cutout' augmentations, which is suggested by the authors when only part of the CIFAR10 dataset is used. All the other parameters remained the same, only the sampling strategy is changed to RLS sampling in RLS StyleGAN2 + Aug. For the RLS computation, we use the discriminator feature map with Gaussian sketching and a fixed explicit feature map given by the next-to-last layer of the Inception network where UMAP is used to reduce the dimension. For both the RLSs, the dimension is reduced to $k = 25$ and the regularization parameter is $\gamma = 10^{-4}$. The performance is assessed using 10k generated samples at the end of training by the Inception Score (IS) and the Fréchet inception distance (FID) between the generated fake dataset and the balanced dataset. Mode coverage is evaluated by the number of generated samples in each class. The class of a generated sample is evaluated by a trained CIFAR10 classifier using a resnet56 type architecture [8] which is trained up to 93.77% accuracy[4]. The results of the experiments are given in Table 4, examples of generated images are given in Fig. 6. Including RLS sampling clearly improves the performance in unbalanced datasets, this is especially the case for the fixed feature map given by the Inception network. The minority mode is oversampled by approximately a factor

[3] Code taken from https://github.com/mit-han-lab/data-efficient-gans.

[4] Classifier is available at https://github.com/gahaalt/ResNets-in-tensorflow2.

Table 3. Experiments on the UNBALANCED MNIST dataset. Two RLS BuresGAN variants are considered: RLSs calculated with an explicit feature map obtained from a pre-trained classifier (Class.) and the next-to-last layer of the discriminator (Discr.). RLS MwuGAN is initialized with RLSs using the explicit feature maps obtained from a pre-trained classifier. Only the number of samples in the minority modes are visualized. The number of samples in the remaining modes are given in Supplementary Material.

	Mode 1	Mode 2	Mode 3	Mode 4	Mode 5	KL
GAN	123(22)	137(103)	81(31)	161(97)	161(23)	0.48(0.02)
PacGAN2	109(29)	142(70)	89(23)	147(100)	152(40)	0.48(0.02)
BuresGAN	126(31)	157(97)	108(30)	153(62)	147(26)	0.46(0.02)
IwGAN	117(33)	139(33)	97(31)	212(69)	154(34)	0.46(0.02)
IwMmdGAN	1(1)	0(0)	23(16)	**1140**(367)*	3(3)	1.92(0.1)
MwuGAN (15)	144(29)	113(28)	146(13)	172(18)	167(28)	0.46(0.02)
RLS MwuGAN (15) (ours)	336(47)	283(32)	191(23)	381(38)	276(33)	0.30(0.02)
RLS BuresGAN Class. (ours)	**875**(112)*	**663**(122)*	**360**(198)*	831(59)	**615**(82)*	**0.09**(0.01)*
RLS BuresGAN Discr. (ours)	235(62)	183(141)	264(44)	255(109)	219(54)	0.37(0.02)

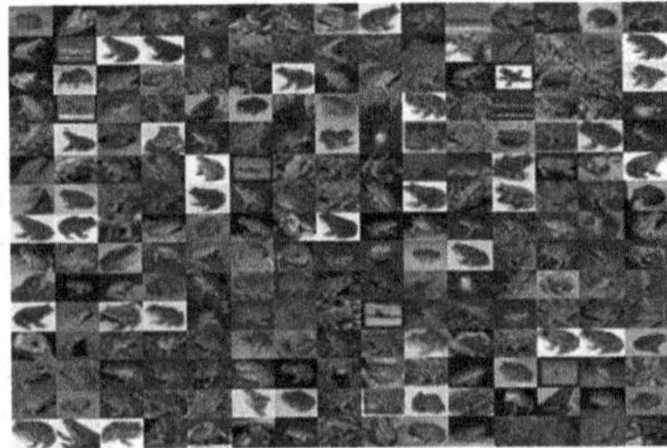
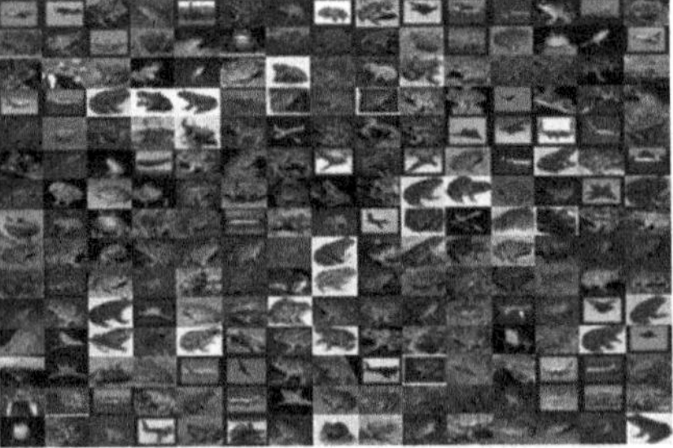

Fig. 6. Generated images from 06-CIFAR10 by a StyleGAN2 + Aug. (left) and by RLS StyleGAN2 + Aug. (right). Including RLS sampling promotes sampling from the minority class. Generated samples classified as planes are marked by a red border.

10 or even 100 in the case of the unbalanced 06-CIFAR10 and 016-CIFAR10 datasets respectively. Both the IS and FID also improve significantly. Note that the maximum achievable performance for IS and FID is lower when only a subset of classes is included, as pointed out by [18].

Table 4. Experiments on the UNBALANCED 06-CIFAR10 and UNBALANCED 016-CIFAR10 dataset. Including RLS sampling in the StyleGAN2 + Aug. clearly improves the performance. Minority modes are highlighted in black in the second row.

	06-CIFAR10				016-CIFAR10				
	Mode 1	Mode 2	IS (↑)	FID (↓)	Mode 1	Mode 2	**Mode 3**	IS (↑)	FID (↓)
StyleGAN2 + Aug	261	9500	4.8	67.5	4526	5206	18	4.3	48.8
RLS StyleGAN2 + Aug. Disc. (ours)	994	8659	5.7	46.4	4449	5132	139	4.6	44.4
RLS StyleGAN2 + Aug. Class. (ours)	**2438**	7212	**6.2**	**31.3**	4156	4393	**1155**	**5.7**	**27.2**

4 Conclusion

We introduced the use of RLS sampling for training GANs. This 'diverse' sampling procedure was motivated by a notion of complete mode coverage in the presence of minority modes. RLS sampling is easy to integrate into any GAN model. Three feature maps have been discussed. An implicit feature map performs well for low-dimensional data. A fixed explicit feature map, such as a pre-trained classifier, achieves good results in high-dimensional cases. Lastly, the discriminator can be used as a feature map when no prior knowledge exists about the data. Two approximation methods for the explicit feature maps are also discussed: dimensionality reduction of explicit feature maps and a two-stage sampling procedure to efficiently speed up online RLS computation. We demonstrated empirically that the use of RLS sampling in GANs successfully combats the missing mode problem.

Acknowledgments. EU: ERC Advanced Grants(787960, 885682). This paper reflects only the authors' views and the Union is not liable for any use that may be made of the contained information. Research Council KUL: projects C14/18/068, C16/15/059, C3/19/053, C24/18/022, C3/20/117, Industrial Research Fund: 13-0260, IOF/16/004; Flemish Government: FWO: projects: GOA4917N, EOS Project no G0F6718N (SeLMA), SBO project S005319N, Infrastructure project I013218N, TBM Project T001919N; PhD Grants (SB/1SA1319N, SB/1S93918, SB/1S1319N), EWI: the Flanders AI Research Program. VLAIO: Baekeland PhD (HBC.20192204) and Innovation mandate (HBC.2019.2209), CoT project 2018.018. Foundation 'Kom op tegen Kanker', CM (Christelijke Mutualiteit). Ford KU Leuven Research Alliance Project KUL0076.

References

1. Alaoui, A., Mahoney, M.W.: Fast randomized kernel ridge regression with statistical guarantees. In: Neural Information Processing Systems, pp. 775–783 (2015)
2. Curi, S., Levy, K., Jegelka, S., Krause, A., et al.: Adaptive sampling for stochastic risk-averse learning. Neural Inf. Process. Syst. (2020)
3. Dasgupta, S., Gupta, A.: An elementary proof of a theorem of Johnson and Lindenstrauss. Random Struct. Algorithms **22**(1), 60–65 (2003)
4. De Meulemeester, H., Schreurs, J., Fanuel, M., De Moor, B., Suykens, J.A.K.: The bures metric for generative adversarial networks. In: Oliver, N., Pérez-Cruz, F., Kramer, S., Read, J., Lozano, J.A. (eds.) ECML PKDD 2021. LNCS (LNAI), vol. 12976, pp. 52–66. Springer, Cham (2021). https://doi.org/10.1007/978-3-030-86520-7_4
5. Diesendruck, M., Elenberg, E.R., Sen, R., Cole, G.W., Shakkottai, S., Williamson, S.A.: Importance weighted generative networks. In: Brefeld, U., Fromont, E., Hotho, A., Knobbe, A., Maathuis, M., Robardet, C. (eds.) ECML PKDD 2019. LNCS (LNAI), vol. 11907, pp. 249–265. Springer, Cham (2020). https://doi.org/10.1007/978-3-030-46147-8_15
6. Feldman, V.: Does learning require memorization? a short tale about a long tail. In: 52nd Annual ACM SIGACT Symposium on Theory of Computing, pp. 954–959. STOC 2020 (2020)

7. Goodfellow, I., et al.: Generative adversarial nets. In: Neural Information Processing Systems, pp. 2672–2680 (2014)
8. He, K., Zhang, X., Ren, S., Sun, J.: Deep residual learning for image recognition. In: IEEE Conference on Computer Vision and Pattern Recognition, pp. 770–778 (2016)
9. Kingma, D.P., Ba, J.: Adam: a method for stochastic optimization. In: International Conference on Learning Representations (ICLR) 2015 (2014)
10. Lin, Z., Khetan, A., Fanti, G., Oh, S.: Pacgan: the power of two samples in generative adversarial networks. In: Neural Information Processing Systems vol. 31, pp. 1498–1507 (2018)
11. Ma, P., Mahoney, M., Yu, B.: A statistical perspective on algorithmic leveraging. In: 31st International Conference on Machine Learning. Proceedings of Machine Learning Research, vol. 32, pp. 91–99 (2014)
12. Massart, E., Absil, P.A.: Quotient geometry with simple geodesics for the manifold of fixed-rank positive-semidefinite matrices. SIAM J. Matrix Anal. Appl. **41**(1), 171–198 (2020)
13. McInnes, L., Healy, J., Melville, J.: UMAP: uniform manifold approximation and projection for dimension reduction. preprint arXiv:1802.03426 (2018)
14. Metz, L., Poole, B., Pfau, D., Sohl-Dickstein, J.: Unrolled generative adversarial networks. In: International Conference on Learning Representations (ICLR) (2017)
15. Musco, C., Musco, C.: Recursive sampling for the Nystrom method. In: Neural Information Processing Systems, pp. 3833–3845 (2017)
16. Oymak, S., Recht, B., Soltanolkotabi, M.: Isometric sketching of any set via the restricted isometry property. Inf. Infer. J. IMA **7**(4), 707–726 (2018)
17. Rudi, A., Calandriello, D., Carratino, L., Rosasco, L.: On fast leverage score sampling and optimal learning. In: Neural Information Processing Systems, pp. 5672–5682 (2018)
18. Sajjadi, M.S.M., Bachem, O., Lucic, M., Bousquet, O., Gelly, S.: Assessing generative models via precision and recall. In: 32nd International Conference on Neural Information Processing Systems, pp. 5234–5243 (2018)
19. Sinha, S., Zhang, H., Goyal, A., Bengio, Y., Larochelle, H., Odena, A.: Small-GAN: speeding up GAN training using core-sets. In: 37th International Conference on Machine Learning, vol. 119, pp. 9005–9015 (2020)
20. Tolstikhin, I.O., Gelly, S., Bousquet, O., Simon-Gabriel, C.J., Schölkopf, B.: Adagan: boosting generative models. In: Neural Information Processing Systems, pp. 5424–5433 (2017)
21. Tripp, A., Daxberger, E., Hernández-Lobato, J.M.: Sample-efficient optimization in the latent space of deep generative models via weighted retraining. Neural Inf. Process. Syst. **33** (2020)
22. Yang, Z., et al.: Deep fried convnets. In: IEEE International Conference on Computer Vision, pp. 1476–1483 (2015)
23. Zhang, C., Kjellstrom, H., Mandt, S.: Determinantal point processes for mini-batch diversification. Uncertainty Artif. Intell. (2017)
24. Zhang, C., Öztireli, C., Mandt, S., Salvi, G.: Active mini-batch sampling using repulsive point processes. In: AAAI Conference on Artificial Intelligence, vol. 33, pp. 5741–5748 (2019)
25. Zhao, S., Liu, Z., Lin, J., Zhu, J.Y., Han, S.: Differentiable augmentation for data-efficient GAN training. Neural Inf. Process. Syst. (2020)
26. Zhong, P., Mo, Y., Xiao, C., Chen, P., Zheng, C.: Rethinking generative mode coverage: a pointwise guaranteed approach. Neural Inf. Process. Syst. **32**, 2088–2099 (2019)

Public Transport Arrival Time Prediction Based on GTFS Data

Eva Chondrodima[1(✉)], Harris Georgiou[1], Nikos Pelekis[2], and Yannis Theodoridis[1]

[1] Department of Informatics, University of Piraeus, Piraeus, Greece
{evachon,hgeorgiou,ytheod}@unipi.gr

[2] Department of Statistics and Insurance Science, University of Piraeus, Piraeus, Greece
{npelekis@unipi.gr}

Abstract. Public transport (PT) systems are essential to human mobility. PT investments continue to grow, in order to improve PT services. Accurate PT arrival time prediction (PT-ATP) is vital for PT systems delivering an attractive service, since the waiting experience for urban residents is an urgent problem to be solved. However, accurate PT-ATP is a challenging task due to the fact that urban traffic conditions are complex and changeable. Nowadays thousands of PT agencies publish their public transportation route and timetable information with the General Transit Feed Specification (GTFS) as the standard open format. Such data provide new opportunities for using the data-driven approaches to provide effective bus information system. This paper proposes a new framework to address the PT-ATP problem by using GTFS data. Also, an overview of various ML models for PT-ATP purposes is presented, along with the insightful findings through the comparison procedure based on real GTFS datasets. The results showed that the neural network -based method outperforms its rivals in terms of prediction accuracy.

Keywords: Estimated time of arrival · GTFS · GTFS-RT · GTFS validation · Machine learning methods · Mobility data mining · Neural networks · Public transport

1 Introduction

Public transport (PT) offers significant social and environmental benefits. More specifically, high quality PT services lead to: (a) a considerable improvement on the quality of citizens' life, e.g. areas with access to public transportation help social inclusion, and (b) environmental benefits related to minimizing the CO2 emissions of private vehicles. Public transportation goal is to provide efficient, reliable, and high quality services, in order to attract more passengers. The planning of high quality PT systems is a difficult task. PT networks are highly complex systems, due to the large number of passengers that are transported each day, due to the number of employees and because they are affected by technical and organizational complications.

G. Nicosia et al. (Eds.): LOD 2021, LNCS 13164, pp. 481–495, 2022.
https://doi.org/10.1007/978-3-030-95470-3_36

The wide adoption of GPS tracking systems in PT provides new opportunities for using data-driven approaches to fit the demand of passengers and provide effective bus information system. Arrival time prediction (ATP) of PT vehicles is an important part of intelligent transportation systems. Accurate prediction can assist passengers in planning their travels and may improve travel efficiency. Also, efficient PT-ATP is necessary in order to eliminate passengers' long waiting time for the arrival of a new vehicle and the existence of delays during a trip. However, accurate PT-ATP is a challenging task due to a variety of factors, including stochastic variables such as traffic conditions, weather conditions, etc.

In order to deal with the PT-ATP problem it is necessary to collect moving PT vehicles information, manage big mobility data, and address spatio-temporal prediction problems. The collection and data management related to PT vehicles is accomplished through PT agencies. However, the inherent difficulty of managing PT data poses challenges in terms of storing data in Big Data platforms, as well as further analysis to extract useful and usable knowledge. Indeed, valuable knowledge is hidden in big mobility data, which can be fully exploited through Machine Learning (ML) techniques, such as Neural Networks (NN).

The key to the implementation and validation of transport models are the real-world data. Their availability and quality can significantly affect the reliability of the resulting estimates [3]. Data availability and data quality are of equal importance. Nowadays thousands of PT providers employ a common format for publishing their public transportation schedules and associated geographic information, called General Transit Feed Specification (GTFS). GTFS data are composed of two types of feeds: a) the GTFS feed (also known as static GTFS), which contains static timetabling information, and b) the GTFS-real time (GTFS-RT) feed, which contains real time information about the transit network.

However, processing PT raw data, even standard GTFS, is challenging. Particularly, GTFS data (static and real-time) often contain missing information and errors, such as missing timetable information (e.g. times of operations), invalid stops coordinates, invalid vehicle coordinates, e.t.c. Thus, GTFS data should pass through a set of validation steps. In order to validate GTFS and GTFS-RT feeds, open-sources have been introduced [11]. However, GTFS validator tools cannot guarantee that the validated data are appropriate for using ML methods for ATP purposes. To address this problem and to make GTFS and GTFS-RT feeds appropriate for learning ATP purposes, we introduce a tool for Cleansing and Reconstructing GTFS data, called CR-GTFS tool.

In the previous years, a number of prediction algorithms have been applied to moving objects [6–8,20]. Furthermore, various studies have been conducted that use ML techniques in predicting the transit travel time by using GPS traces from transport vehicles [9,16,28], or the so-called Live Automatic Vehicle Locations (AVL) data [12,19,21]. There are also some works that employ AVL with GTFS feed. In [15] the purpose was to train an NN to predict the travel times of buses based on open data collected in real-time, while the model evaluation was conducted on data derived from Sao Paulo City bus fleet location, real-time traffic data, and traffic forecast from Google Maps. By mining Live AVL data that buses provided by the Toronto Transit Commission along with schedules retrieved from GTFS, and weather data, Alam et.al [1] found that their proposed recurrent NN-based architecture predicts the occurrence of arrival time irregularities accurately. In the literature there are limited works regarding the PT-ATP problem by using GTFS and GTFS-RT feeds. Particularly, Sun et al. [24] combined clustering analysis with Kalman filters to predict arrival times at various bus stops on Nashville, TN, USA, by calculating the delay versus a scheduled time, based on GTFS and GTFS-RT as well as historical bus timing data.

The main contribution of this work is to propose a new framework to clean-reconstruct GTFS and GTFS-RT feeds and simultaneously address the PT-ATP problem by using GTFS data. Also, we examine various ML models for PT-ATP purposes and provide insightful findings through the comparison procedure.

The rest of this paper is organized as follows: Sect. 2 presents the employed GTFS data along with the proposed GTFS data preprocessing method for ATP purposes. Section 3 formulates the problem definition and briefly summarizes typical solutions for PT-ATP. Section 4 presents the experimental setup, the results of our approach, and compares the performance of different solutions, followed by conclusions in Sect. 5.

2 Preprocessing Static and Real-Time GTFS Data

In this section, we present some preliminary terms for the GTFS and GTFS-RT feeds. Also, we describe the employed GTFS data, the GTFS errors and the proposed GTFS data preprocessing method for ATP purposes.

2.1 PT Provider and GTFS Data

The American Public Transportation Association named Metro Transit is the primary PT operator in the Minneapolis-Saint Paul of the U.S. state of Minnesota and the largest operator in the state. Metro Transit provides an integrated network of buses, light rail, and commuter trains, and has adopted GTFS format to share information with the public. Particularly, the website at svc.metrotransit.org is well maintained with frequent updates of GTFS and GTFS-RT feeds. More specifically, GTFS static data are downloaded as a zip file[1] and are updated weekly, but are subject to change at any time and daily checks are recommended. As far as the GTFS-RT feeds are concerned, they are updated every 15 s and include three feeds: the TripUpdate feed[2], the VehiclePosition feed[3] and the ServiceAlerts feed[4].

[1] https://svc.metrotransit.org/mtgtfs/gtfs.zip.
[2] https://svc.metrotransit.org/mtgtfs/tripupdates.pb.
[3] https://svc.metrotransit.org/mtgtfs/vehiclepositions.pb.
[4] https://svc.metrotransit.org/mtgtfs/alerts.pb.

In general, a GTFS feed is a collection of at least six comma-separated values (CSV) files (agency, routes, trips, stops, stop times, calendar) and seven optional ones. Metro Transit provides 11 files: agency, routes, trips, stops, stop times, calendar, shapes, calendar dates, feed info, linked datasets, vehicles. In a PT network, stops represent the available stations at which the PT vehicles can stop to pick up or drop off passengers. A sequence of stops constitutes a route. Multiple routes may use the same stop. Each route has a schedule that is followed by a PT vehicle and each route is composed of many trips, which follow the same route, but occur at a specific time throughout a day. Moreover, shapes describe the path that a vehicle travels along a route alignment, are associated with trips, and consist of a sequence of points through which the vehicle passes in order. Stops on a trip should lie within a small distance of the shape for that trip.

As far as the GTFS-RT feed is concerned, it allows PT agencies to provide real-time updates about their fleet through three different types of live feed-trip updates: Trip Update (provide information about predicted arrival/departure times for stops along the operating trips), Vehicle Position (provide information about the locations of the vehicles, e.g. GPS coordinates), and Service Alerts (provide human-readable descriptions regarding disruptions on the network). More specifically, the Metro Transit Trip Update feed includes information about vehicle's timestamp, trip id, route id, direction id, start time, start date, vehicle id, vehicle label, stop sequence, stop id, arrival time, departure time. Note that the arrival/departure times at stops are the predicted ones; at least in Metro Transit, the actual arrival/departure times are not included in the feed. Also, the Metro Transit Vehicle Position includes information about vehicle's timestamp, trip id, route id, direction id, start time, start date, vehicle id, vehicle label, position latitude, position longitude, bearing, odometer, speed. Finally, since the Metro Transit Service Alerts feed includes human-readable descriptions, which are not easily manageable automatically this feed is not used in this work.

2.2 GTFS Data Errors and Proposed Solutions

GTFS data often contain errors, such as misrepresentations of the actual network, stops could be encoded imprecisely and have incorrect coordinates. Several GTFS errors for the static counterpart need to be resolved in order to be useful for data analytic purposes, where the most common errors along with the provided solutions, are presented below:

- Duplicate trip information: Each trip should be unique within a route (i.e. same trips, at the same times, should not occur), otherwise is eliminated.
- Incorrect or duplicate stop timestamps: Scheduled arrival and departure times should increase for each stop along the trip and should not be the same at three or more consecutive stops, otherwise the respective trip is eliminated.
- Incorrect stops: Stops coordinates and sequence should match the road network coordinates and direction, respectively, generated from the available shapes. The incorrect stops are eliminated and if a large number of stops within a trip is incorrect, then the whole trip is eliminated.

- Incorrect shapes: Shapes coordinates should match the actual road network coordinates and direction generated from OpenStreetMap [18], otherwise the respective shape is eliminated.

Moreover, GTFS-RT feed contains errors, such as mismatches with the scheduled data, that need to be resolved in order to be useful for data analytic purposes, where the most common errors along with the provided solutions, are presented below:

- Route/Trip/Stop ids mismatching: The vehicle route/trip/stop ids in GTFS-RT should be included in GTFS static feed, otherwise the respective points are eliminated.
- Unrealistic alighting times: Multiple consecutive timestamps for one vehicle position may occur when arrival or departure events occur. These timestamps must occur in short times otherwise the ids are invalid and are eliminated.
- Multiple vehicle positions for one timestamp: For each timestamp only one vehicle position should be recorded, otherwise the most reasonable value is kept.
- Incorrect vehicle position data: Vehicle positions should match the available GTFS shapes (i.e. vehicle positions should be within a buffer surrounding the GTFS shapes) of the trip that the vehicle operates and the road distances between consecutive positions should result in reasonable values, otherwise the points are invalid and are eliminated.
- Invalid timestamps: Timestamps should be strictly sorted for a specific vehicle operating on a specific trip and the time differences between consecutive positions should result in reasonable values, otherwise the timestamps are invalid and are eliminated.
- Invalid vehicle speed: The vehicle speed is calculated by using the road distance between two vehicle positions and the time horizon between the corresponding timestamps. The vehicle speed should follow the minimum and maximum speed limits defined by Metro Transit, otherwise the validity of the associated vehicle positions and timestamps should be investigated.
- Invalid trip start times: The GTFS-RT trip start timestamp should match the scheduled arrival time of the first stop of the trip provided by the "stop times" file, otherwise the trip is invalid and is eliminated.
- Invalid trip start/end times and start/end stop positions: The vehicles may report timestamped positions long time before the scheduled trip start time and/or on a different position of the first stop location of the trip. Also, the vehicles may report timestamped positions many kilometres from the last reported location before going out of service. These are considered extreme errors and are resolved by a) deleting the vehicle's positions with distance from the stop location higher than 50 m, b) deleting the timestamps that differ from the scheduled trip start time more than a specific amount of time which is equal to the maximum delay time of the trips of the same road id at the first stop, and c) deleting the timestamps that differ from the scheduled trip end time more than a specific amount of time which is equal to the maximum delay time of the trips of the same road id at the last stop.

2.3 Cleansing and Reconstructing GTFS Data (CR-GTFS) Tool

GTFS static and real-time data main purpose is to share PT information and are not appropriate for being used by ML methods for data analytic purposes directly. To address this problem and to use GTFS and GTFS-RT feeds effectively for learning ATP purposes, we propose the CR-GTFS tool, which downloads, saves, cleans and reconstructs the GTFS data (both static and real-time).

In this work, GTFS and GTFS-RT feeds are retrieved from the Metro Transit. Particularly, CR-GTFS tool downloads the available schedule data every 2 h and the real-time feeds every 5 s to ensure that all data are collected, and stores them in a PostgreSQL database, which is spatially enable by using PostGIS. Then the process to automatically identify the problematic and missing information, and to reconstruct GTFS and GTFS-RT feeds follows and is described subsequently.

As far as the GTFS static feed is concerned, the respective files are merged according to the following flow: agency is merged with routes (using "agency id" as the key), then merged with trips (using "route id" as the key), then merged with stop times (using "trip id" as the key), then merged with stops (using "stops id" as the key), then merged with shapes (using "shape id" as the key), then merged with calendar and calendar dates (using "service id" as the key). Subsequently, the GTFS data errors described in Sect. 2.2 are resolved and the final GTFS static dataset is a complete dataset that includes all the available timetable information.

As far as the GTFS-RT data are concerned, the TripUpdate feed is merged with the VehiclePosition feed by using a number of different combination of keys, e.g. a) "trip id" and "vehicle id", or b) "route id", "direction id" and "vehicle id" (in the case of missing/faulty values of "trip id"), or c) "route id", "direction id" and vehicle timestamp occur within the time frame of a scheduled trip with "route id" (in the case of missing/faulty values of "vehicle id"). By merging the VehiclePosition with the TripUpdate and by using various combination of merging keys, the proposed tool fills the missing information concerning the common features: vehicle's timestamp, trip id, route id, direction id, start time, start date, vehicle id and vehicle label. Subsequently, the abovementioned reconstructed GTFS static dataset is merged with the GTFS-RT data by using as keys the "road id", the "direction id" and the "trip id". Then, the GTFS-RT data errors described in Sect. 2.2 are resolved and the resulted dataset is a complete set of information about the transit system.

Due to the fact that the GTFS-RT feed is designed to provide only updates on operating vehicles, each reported vehicle is in service and is operating on a trip and is assigned a unique trip id. The trip ends when a) the assigned vehicle on the specific trip does not operate anymore (i.e. the vehicle does not appear in the GTFS-RT feeds), and/or b) the assigned vehicle on the specific trip reports a different route for consecutive timestamps, or a different direction. For each completed trip, CR-GTFS tool saves the related information to the database and process it. The resulted dataset is a complete dataset that includes all the available real-time information for each vehicle operating on a trip. However, this dataset does not include the information whether a vehicle actually arrived or departed the stop.

For ATP purposes, we need to know the vehicle's arrival time at a stop. This can be addressed by matching the stops from the schedule data to the GTFS-RT stop times based on the timestamps of the vehicles passing those stops, for each GTFS-RT completed trip. However, this is a challenging task due to a number of reasons such as the fact that sometimes no location is reported near a stop as the vehicle passes the stop quickly. In order to solve this problem, we need to determine which stops were passed and estimate their times of arrival and departure. Thus, the CR-GTFS, for each trip, calculates the distance between the timetable stops locations and the available GTFS-RT vehicle's positions. If the distance between a specific stop location and a vehicle's position falls within the range of 30 m, then the recorded vehicle timestamp matches the stop. Subsequently, the timestamp for each remaining stop (that has not a matched timestamp) can be estimated by using the road network distance and the speed of the vehicle from the previous and next positions of the specific stop's location.

For the experimental purposes of this study we created a one month GTFS data (March 2021), by using the proposed CR-GTFS tool. Note that both bus and rail services were included.

3 Methodology

3.1 Problem Formulation

Using the processed dataset as described above, the problem can be formulated as follows:

- **Given**: An input vector $V = \{v_{t-k}, \dots, v_{t-1}, \hat{v_t}\}$, where t is the current bus stop and v_{t-k} contains sequential information about passing through stop $t-k$,
- **Predict**: The arrival time or $dT_{t,t+1}$ towards the next bus stop in sequence.

For each of the previous k bus stops, the sequential information gathered is stop identifier "stop id", actual distance $dS_{t-j,t-j+1}$ travelled from previous stop (GPS-based), actual time $dT_{t-j,t-j+1}$ for this transition and estimated (mean) speed, i.e., $dS_{t-j,t-j+1}/dT_{t-j,t-j+1}$.

The reason for including redundant information with speed is that some regression models become simplified and easier to train, as the pair distance-and-time introduces non-linearities in the input compared to distance-and-speed. Furthermore, speed may also be included in the next-step sub-vector $\hat{v_t}$ where it is typically not available ($dT_{t,t+1}$ not realized yet), if instead some globally available estimation of it can be attained from historic data, i.e., the mean time it usually takes to travel between these two bus stops in that specific direction. For more accurate comparison of the examined models, not such additional estimations were made for next-speed elements and, hence, the next-step sub-vector $\hat{v_t}$ contains only "stop id" and actual distance $dS_{t,t+1}$ that are available at any given t.

Although any N-step look-ahead setup can be used in this core regression task, the choice of one-step look-ahead was based on two reasons. First, the purpose here is to compare the methods in the pure short-term sense, i.e., limit the effects of noise and stationarity shifts caused by exogenous localized factors like unstable road traffic. Second, it is straight-forward to extend the one-step look-ahead approach to any N-step option by iterating the same process multiple times and employing a sliding window that incorporates every new prediction; for robust regressors, the expected N-step look-ahead error of this process is typically bounded by N times the one-step error of the model. Therefore, the one-step look-ahead results are a very good indication of how these models will behave if used iteratively and, most importantly, what is their performance if used continuously in an online fashion, e.g. with streaming data as they become available.

3.2 Machine Learning Methods Compared

As a performance baseline, two variants of the Linear Regression [17] were employed in this study: (a) the standard method based on the least squares error minimization, (b) the same method but enhanced with the M5 algorithm [26] for attribute selection-elimination. Since the regression task at hand is clearly non-linear, every other regression method with realistically usable application should perform better than this baseline.

A simple model that is often used as density-based estimators is the k nearest neighbour (k-nn) [25] approach, more specifically the Instance Based learners (IBk). In regression problems of high or unknown intrinsic dimensionality, or when the underlying target distribution is suspected to be skewed or clustered, the IBk algorithm is often used as a very good indicator of the expected performance of other, properly trained and robust models. However, like the k-nn, it does not include per-se any attribute-selective process and the distance metric may be negatively affected by a few heavily skewed or correlated dimensions. Moreover, the model itself always requires the complete training dataset or a very extensive representation of it, in order to make each decision during the evaluation phase. This is why these models are often called 'lazy' learners. Nevertheless, as density-based, models, they do not depend on a 'global' functional approximation of the entire manifold, but rather good approximation of arbitrary local neighborhoods of the data. In this study, IBk was included as such a representative of 'lazy' learners based on the k-nn approach, using the Manhattan distance metric and inverse distance as weighting factors within each local neighborhood of k samples.

Focusing on the data space partitioning and implicit attribute selection characteristics of decision tree algorithms, the 'Reduced Error Prunning Tree' or REPtree algorithm [13] was used as a representative of this category. It is a fast decision tree model that can be used for classification or regressions tasks, similarly to the classic Classification and Regression Tree (CART) algorithms [25]. During training, it builds several trees based on a loss minimization criterion, typically the information gain or reduction of variance with each new node split. Due the sorting of numeric features only once, the overall speed of the training process is improved. Subsequently, the tree is pruned for improving its generalization based on an error function, typically mean squared error (MSE) or minimum absolute error (MAE). Finally, it selects the best-performing tree from all the candidates as the representative model.

As a representative of the ensemble methods [5,14,27], Additive Regression was also employed in this study. It is a realization of the boosting approach for regression tasks, i.e., training separate subsequent models upon the residuals (errors) of the previous iterations. The final result is an aggregation of all the trained models, which are typically some weak classifier/regressor, e.g. decision trees. It also includes a shrinkage factor for the learning rate, in order to accomplish smoother trained manifold and avoid over-fitting. In this study, REPtree was used as the base regression model for the ensembles.

Finally, a neural network model was employed as a classic 'universal approximator'. in particular, a multi-layer perceptron (NN-MLP) [25] architecture was used with topologies of one, two and three hidden neurons with softmax activation functions. The main advantages of NN-MLP over many other types of regression models is that the (one or more) hidden layer provides a data space partitioning feature similar to the decision trees and at the same time incorrorate a non-linear aggregation scheme for producing the final output. Instead of the classic back-propagation training, the Broyden-Fletcher-Goldfarb-Shanno (BFGS) [2,4,10,22] optimization algorithm was employed instead, due to its enhanced stability and faster convergence. The BFGS algorithm, used in a wide range of (mostly) unconstrained optimization tasks, is based on directional preconditioning of the descent gradient and it is one order faster than the classic Newton methods. It uses localized curvature information via gradual improvements upon the Hessian matrix of the loss function without the need for matrix inversions or analytical gradient definitions.

4 Experimental Study

The proposed method was formulated and experimentally validated over a real-world PT dataset, which was created by using the proposed CR-GTFS tool as presented in Sect. 2.

4.1 Experimental Protocol - Parameter Selection

After the full pre-processing of the raw data, the training and testing datasets were prepared according to the specific problem formulation for the regression task, according to the input-output schema described in Sect. 3.1. In practice, collected sequential data (stop id, elapsed time, distance, speed) from the previous four bus stops, as well as the available data towards the next stop (stop id, distance), are used as input vector; the arrival time to the next stop is the output, i.e., the conditioned variable in the regression.

For proper evaluation, a 5-fold cross-validation process [25] was employed for all trained models, using exactly the same randomization (seed) in order to avoid any partitioning side-effects. Hence, all performances were evaluated on 80% training and 20% testing splits of the initial dataset, in five iterations, and the final numbers are the mean values over these splits. Due to the significant differences in complexity, the training cycles of the various models ranged from 1–2 seconds (Linear Regression) to more than six hours (NN-MLP).

Regarding model parameterization, several aspects of each model type were taken into account and optimized with intermediate experiments before the final performance assessment, with the exception of Linear Regression which is essentially non-parametric. For IBk, the distance metric (Manhattan) and the weighting factor (inverse distance), as well as the size of the neighborhood between $k = \{5, \ldots, 10\}$, were selected as optimal for this task. Similarly, for REPtree the node splitting criterion was selected to 1e-3 and at least four instances per leaf for the pruning, but with no prior constraint for the expansion depth, in order to accommodate the large dataset size without imposed approximation deficiencies. The REPtree was also used as the base weak learner in the Additive Regression model, used without shrinkage factor and with 10 iterations.

Finally, for NN-MLP the main focus of the optimization was in its topology, i.e., the number and size of the hidden layers employed. As expected, in regression tasks any feature space partitioning beyond a single hidden layer does not provide improvements in the final accuracy, single any subsequent aggregation steps may actually increase, instead of decreasing, the approximation error. In other words, and in contrast to the multiple-layer NNs used in deep learning approaches like with auto-encoders used in classification tasks, a properly designed first hidden layer is more than adequate to address an arbitrary regression tasks. It should be noted that the choice of the size of the single (or first of multiple) hidden layers in NN-MLP regressors can be examined in combination with a preliminary clustering step, in order to make a rough estimation of the level of non-uniformity of the input data space which the model can exploit. In this study, up to three hidden layers were tested in NN-MLP topologies, but the best candidates were those with a single hidden layer of size within a range of $n_L = \{10, \ldots, 50\}$. The lower bound was based on clustering estimations via k-means and EM algorithms [25] that yielded a total of 8–10 clusters, 3–4 of which were mapping 8–10% of the data, i.e., can be considered 'outliers' clusters. The upper bound is mostly constrained by the training time required, as well as by the fact that the increase in accuracy (MAE) versus hidden layer size increases only logarithmically (very slow), as it is explained and illustrated below in Sect. 4.2.

4.2 Results and Discussion

Figure 1 illustrates the distribution of the target for arrival time in the dataset, i.e., true $dT_{t,t+1}$, which, as expected, follows a highly skewed Gaussian or a Generalized Extreme Value (GEV) profile [23], with heavy positive tail. This is due to the fact that the dataset contains a few, very large time differences in specific bus routes, i.e., with very sparse bus stops. In order to test the robustness of the models and their training, it was decided not to remove any such extreme values but instead use it as-is, simulating a real-world requirement of having to produce ATP for any given input vector, including extremes.

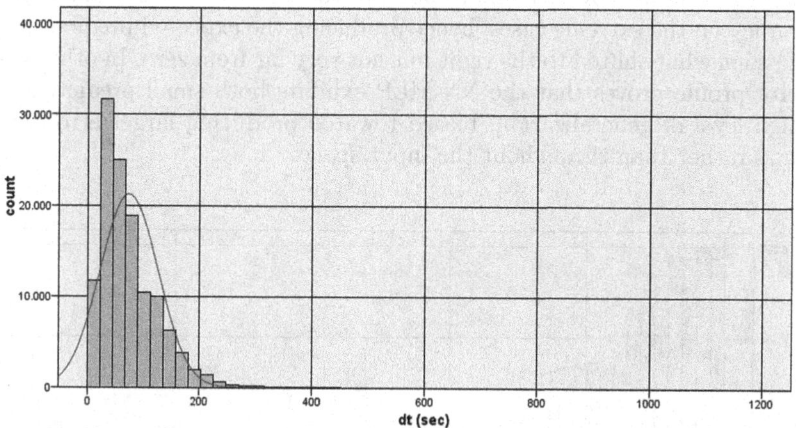

Fig. 1. Distribution of regression target for arrival time in the dataset, i.e., true $dT_{t,t+1}$.

Table 1. Results for all the implemented methods

Method	MAE (sec)	RMSE (sec)	R
Linear regression (std)	29.493	41.021	0.6058
Linear regression (M5)	26.949	38.487	0.6654
IBk (k-nn)	21.119	31.744	0.7891
REPtree	21.503	32.092	0.7829
Additive regression	21.427	32.013	0.7842
NN-MLP (hiddenL=1)	**21.050**	**31.243**	**0.7956**
NN-MLP (hiddenL=2)	24.619	35.731	0.7216

The experimental protocol employed was the same for all models, as described previously in Sect. 4.1. Table 1 presents the results for all the implemented methods, each with its best-performing configuration. For IBk this is with $k = 9$; for NN-MLP this is with a single hidden layer of size $n_L = 50$ (at least $n_L \geq 35$). Bold indicates the overall-best performance given the Mean Absolute Error (MAE), Root Mean Squared Error (RMSE) and Pearson's pairwise correlation coefficient (R) between true and predicted values.

Using the best topology for NN-MLP (single-layer, $n_L = 50$), Fig. 2 presents the distribution of errors (MAE) against the regression variable (arrival time).

Again, it is obvious that it follows a skewed Gaussian or a Generalized Extreme Value (GEV) profile [23], with moderate positive tail, as expected. However, in contrast to Fig. 1, the distribution is much more 'packed' towards zero and the positive tail is suppressed. This essentially means that the resulting NN-MLP 'prefers' to generalize over the main body of the input space, evidently lacking in accuracy on the extreme cases, hence producing the expected prediction error (MAE) somewhat shifted to the right but not very far from zero. In other words, the error profile proves that the NN-MLP exhibits both small prediction error and high level of generalization, biased towards producing larger errors in the extremes rather than throughout the input space.

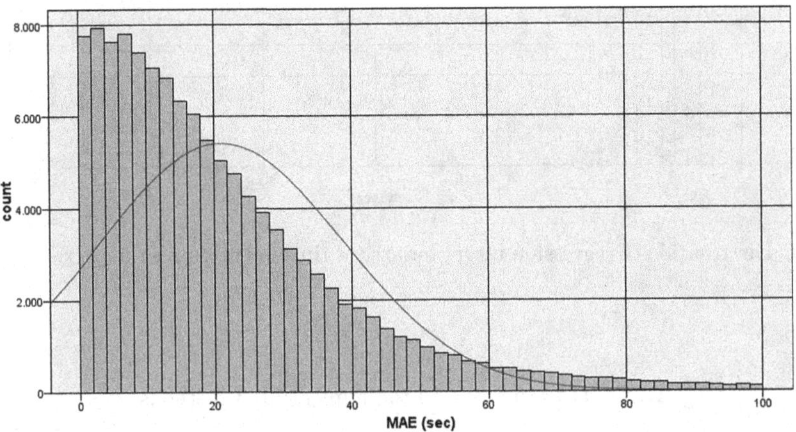

Fig. 2. Distribution of regression error (MAE) for one-step look-ahead ATP using NN-MLP.

Regarding the single-layer NN-MLP, which is the overall-best regressor in this task, it is worth noting that as a model it exhibits a very high level of information 'packing' in its trained parameters: the specific topology of $n_L = 50$ translates to a total of 1,251 weight parameters, including one weight per input attribute ($n_{inp} = 23$) plus one bias coefficient per neuron, i.e., $|W| = n_L(n_{inp}+1)+(n_L+1)$. In contrast, the second-best model which is IBk requires the complete dataset used for k-nn lookups with $k = 9$, a process that is much slower and two orders of magnitude more space-demanding. Similarly, single and ensemble REPtree (Additive Regression) comes close in terms of accuracy, but again the space complexity (tree sizes) is at least 4–5 times larger than the NN-MLP model. Furthermore, the ensemble option (multiple trees) are usually required in order to cope with the inherent noise sensitivity (instability) of single decision trees and the improvement of generalization.

Regarding the trade-off of the size of the single hidden layer in NN-MLP versus the performance improvement, Fig. 3 illustrates some reference points and the corresponding trend in terms of logarithmic fit. The exact formula of the fit is: $MAE \approx f(n_L) = \alpha \cdot \ln(n_L - 8) + \beta$, where $\alpha = -0.640232$ and $\beta = +23.427965$. It is obvious that the trend fit is good, very close to the actual reference points, and the performance gain beyond $n_L \geq 35$ becomes negligible. Nevertheless, even with $n_L = 50$ the NN-MLP topology translates to a model several times smaller in size than the next best alternative.

In summary, the single-layer NN-MLP model outperforms all the other techniques. Besides Linear Regression and two-layer NN-MLP, the performances of the rest of the models differ only marginally; however, due to the very large number of training samples (122,320 in total), the relative ranking of the tested models can be considered as statistically significant and valid for performance comparison. Finally, it should be noted that an increase in the number of samples and the problem dimensionality results in higher computational times in NN-MLP models. However, the performance of the employed NN-MLP model indicates that the algorithm can handle GTFS datasets with large amounts of samples and take advantage of the high prediction accuracy, in contrast to the non NN-based methods, which provide less accurate predictions.

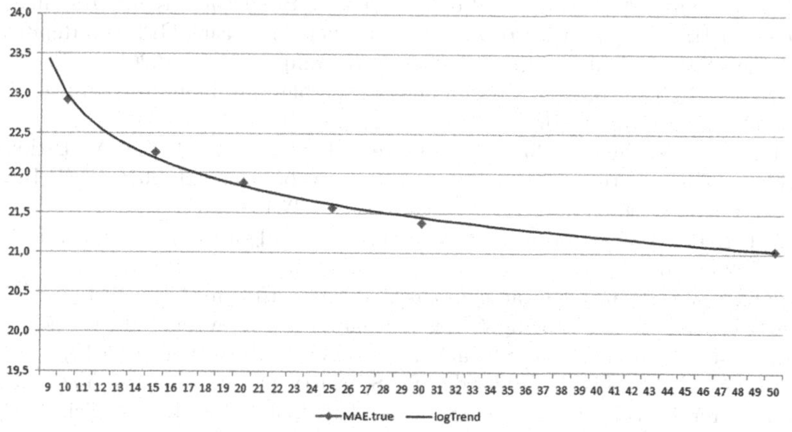

Fig. 3. NN-MLP topology (hidden layer size) versus one-step look-ahead ATP error (MAE); blue dots are real test points (training results) and red line is the estimated trend (logarithmic fit).

5 Conclusion

Due to recent advances in position broadcasting technology and the adoption of a common transit feed format by thousands of PT agencies, PT movement information has become increasingly available. An effective estimation of PT-ATP is substantial for improving the quality and the reliability of the PT services. Taking advantage of the GTFS data, this work proposes a new framework to address the PT-ATP problem. Also, various ML models are tested in solving the

PT-ATP problem. As a result, insightful findings through the comparison procedure are provided. The results showed that the NN-based method outperforms its rivals in terms of prediction accuracy.

Future work includes the investigation of weather information impact on the PT-ATP problem. Also, we plan to experiment with further ML algorithms, such as recurrent NN architectures, taking into account the training computational times. Finally, we plan to focus on a larger prediction time horizon, as well as extending the prediction's window length by including more stops.

Acknowledgements. This paper is one of the deliverables of the project with MIS 5050503, of the Call entitled "Support for researchers with emphasis on young researchers - cycle B" (Code: EDBM103) which is part of the Operational Program "Human Resources Development, Education and Lifelong Learning", which is co-financed by Greece and the European Union (European Social Fund).

References

1. Alam, O., Kush, A., Emami, A., Pouladzadeh, P.: Predicting irregularities in arrival times for transit buses with recurrent neural networks using GPS coordinates and weather data. J. Ambient Intell. Humanized Comput. 1–14 (2020)
2. Broyden, G.: The convergence of a class of double-rank minimization algorithms. J. Inst. Math. Appl. **6**, 76–90 (1970)
3. Caiati, V., Bedogni, L., Bononi, L., Ferrero, F., Fiore, M., Vesco, A.: Estimating urban mobility with open data: a case study in bologna. In: 2016 IEEE International Smart Cities Conference (ISC2), pp. 1–8 (2016)
4. Fletcher, R.: A new approach to variable metric algorithms. Comput. J. **13**(3), 317–322 (1970)
5. Friedman, J.: Stochastic Gradient Boosting. Stanford University, Tech. rep. (1999)
6. Georgiou, H., et al.: Moving Objects Analytics: Survey on Future Location & Trajectory Prediction Methods. arXiv e-prints arXiv:1807.04639 (2018)
7. Georgiou, H., Pelekis, N., Sideridis, S., Scarlatti, D., Theodoridis, Y.: Semantic-aware aircraft trajectory prediction using flight plans. Int. J. Data Sci. Analytics **9**(2), 215–228 (2019). https://doi.org/10.1007/s41060-019-00182-4
8. Georgiou, H., et al.: Future location and trajectory prediction. In: Big Data Analytics for Time-Critical Mobility Forecasting, pp. 215–254. Springer, Cham (2020). https://doi.org/10.1007/978-3-030-45164-6_8
9. Ghanim, M., Shaaban, K., Miqdad, M.: An artificial intelligence approach to estimate travel time along public transportation bus lines. In: The International Conference on Civil Infrastructure and Construction, pp. 588–595 (2020)
10. Goldfarb, D.: A family of variable metric updates derived by variational means. Math. Comput. **24**(109), 23–26 (1970)
11. Google: Testing gtfs feeds, June 2021. https://developers.google.com/transit/gtfs/guides/tools
12. Hua, X., Wang, W., Wang, Y., Ren, M.: Bus arrival time prediction using mixed multi-route arrival time data at previous stop. Transport **33**(2), 543–554 (2018)
13. Kalmegh, S.: Analysis of weka data mining algorithm reptree, simple cart and randomtree for classification of Indian news. Int. J. Innov. Sci. Eng. Technol. **2**(2), 438–446 (2015)

14. Kuncheva, L.: Combining Pattern Classifiers: Methods and Algorithms. Wiley-Interscience, USA (2004)
15. Larsen, G.H., Yoshioka, L.R., Marte, C.L.: Bus travel times prediction based on real-time traffic data forecast using artificial neural networks. In: 2020 International Conference on Electrical, Communication, and Computer Engineering (ICECCE), pp. 1–6 (2020)
16. Liu, H., Xu, H., Yan, Y., Cai, Z., Sun, T., Li, W.: Bus arrival time prediction based on LSTM and spatial-temporal feature vector. IEEE Access **8**, 11917–11929 (2020)
17. Montgomery, D., Runger, G.: Applied Statistics and Probability for Engineers (7th/Ed.). John Wiley & Sons (2018)
18. OpenStreetMap contributors: Planet dump retrieved from (2017). https://planet.osm.org,https://www.openstreetmap.org
19. Petersen, N.C., Rodrigues, F., Pereira, F.C.: Multi-output bus travel time prediction with convolutional LSTM neural network. Expert Syst. Appl. **120**, 426–435 (2019)
20. Petrou, P., Tampakis, P., Georgiou, H., Pelekis, N., Theodoridis, Y.: Online long-term trajectory prediction based on mined route patterns. In: International Workshop on Multiple-Aspect Analysis of Semantic Trajectories, pp. 34–49. Springer, Cham (2019)
21. Ranjitkar, P., Tey, L.S., Chakravorty, E., Hurley, K.L.: Bus arrival time modeling based on Auckland data. Transp. Res. Rec. **2673**(6), 1–9 (2019)
22. Shanno, D.: Conditioning of quasi-newton methods for function minimization. Math. Comput. **24**(111), 647–656 (1970)
23. Spiegel, M., Schiller, J., Srinivasan, R.: Probability and Statistics, (3rd/Ed.). McGraw-Hill (2009)
24. Sun, F., Pan, Y., White, J., Dubey, A.: Real-time and predictive analytics for smart public transportation decision support system. In: 2016 IEEE International Conference on Smart Computing (SMARTCOMP), pp. 1–8 (2016)
25. Theodoridis, S., Koutroumbas, K.: Pattern Recognition, 4th edn. Academic Press, Cambridge, November 2008
26. Witten, I., Frank, E.: Data Mining: Practical Machine Learning Tools and Techniques with Java Implementations. Morgan Kaufmann Publishers, USA (2000)
27. Yuksel, S., Wilson, J., Gader, P.: Twenty years of mixture of experts. IEEE Trans. Neural Netw. **23**(8), 1177–1192 (2012)
28. Čelan, M., Lep, M.: Bus arrival time prediction based on network model. Procedia Comput. Sci. **113**, 138–145 (2017). the 8th International Conference on Emerging Ubiquitous Systems and Pervasive Networks (EUSPN 2017) / The 7th International Conference on Current and Future Trends of Information and Communication Technologies in Healthcare (ICTH-2017) / Affiliated Workshops

The Optimized Social Distance Lab

A Methodology for Automated Building Layout Redesign for Social Distancing

Des Fagan(✉) and Ruth Conroy Dalton

Lancaster University, Bailrigg, Lancaster, UK
{d.fagan,r.dalton1}@lancaster.ac.uk

Abstract. The research considers buildings as a test case for the development and implementation of multi-objective optimized social distance layout redesign. This research aims to develop and test a unique methodology using software Wallacei and the NSGA-II algorithm to automate the redesign of an interior layout to automatically provide compliant social distancing using fitness functions of social distance, net useable space and total number of users. The process is evaluated in a live lab scenario, with results demonstrating that the methodology provides an agile, accurate, efficient and visually clear outcome for automating a compliant layout for social distancing.

Keywords: Social distancing · Architecture · Optimization · Signage · Wayfinding

1 Introduction

COVID-19 has had an unprecedented impact on the day-to-day use of buildings [1]. These effects are likely to have an enduring medium and long- term impact on the arrangement of building layouts to comply with social distancing, posing immediate and ongoing risks to both the personal health of users through non-compliance and to the financial viability of building operation due to increased circulation and distancing requirements [2]. The cost in person-hours to the global economy represented by the millions of concurrent and disparate exercises in building layout replanning during the pandemic has been truly significant [3]. To ameliorate against further substantial cost to the economy through both abortive space planning and non-compliant layouts [4], we propose a unique automated methodology for building operators to redesign their layouts to comply with social distancing. This will reduce timescales for reopening and adaptation in the event of revised government advice, local lockdown, or further variant outbreaks [5]; benefitting user health through verification of distances, whilst improving the efficiency of building operation through optimization of capacity. Our key research question is: can social distancing guidance be effectively automated for building layout plans?

G. Nicosia et al. (Eds.): LOD 2021, LNCS 13164, pp. 496–501, 2022.
https://doi.org/10.1007/978-3-030-95470-3_37

2 Related Work

Our approach was to build a multi-criteria optimization definition using parametric software Grasshopper and Wallacei [6] to generate a redesigned floor layout with minimal human design input. A review of existing research reveals that no existing study provides practical development, testing and evaluation of optimized floor layout design in relation to social distancing; an expected consequence of the short time since the start of the pandemic. Of the papers that explore spatial layouts in the context of COVID-19, most refer to speculative or theoretical guidance as opposed to means-tested outcomes: Fischetti et al. [7] considers a mathematics-based approach to social distancing, exploring the effect of aerosol spread on spatial layouts. Banon et al. [8] investigate shape grammar optimization by mathematical formula. Yet no existing research evaluates the complexity of practical application considering real-world influences including multiple paths, wayfinding and unpredictable user behavior.

Of the significant research completed on multi-criteria optimization of design for spatial layout studies pre-pandemic, Guo et al. [9] explore a multi-agent evolutionary optimization process to define office and housing layouts. The introduction of pedestrian flow for multi-objective optimization presented by Huang et al. [10] provides insight into the potential of agent-based modelling on wayfinding cognition. Recent research by Dubey et al. [11] proposes a new system for the automated positioning of signage based on a multi-criteria optimization approach; referencing theories of Space Syntax and behavioral and cognitive science. Yet, as a consequence of the rapid onset of the pandemic, none have investigated automated optimization of layouts in the setting of social distancing restrictions. In the context of this gap of knowledge, the work proposes a new methodology to bridge between theory and practical evaluation in the new context of the pandemic.

3 Research Methodology

To evaluate the methodology, the project tested a 'live' site, automating the design of the interior layout and wayfinding signage of the ground floor of a public building complex owned by Lancaster City Council (LCC) - the Storey Building in Lancaster City center. The *Social Distance Lab* opened to key stakeholders for three weeks in May 2020, providing opportunity for local business owners to explore a building altered to comply with social distance restrictions, with the dual purpose of collecting evaluation data from users active in the space.

To generate the redesigned and optimized building layout incorporating a) user routes b) user destinations (e.g. seating / toilets) and c) signage locations, a simplified AutoCAD 2D building plan of the building was used as input. The workflow method is summarized in Fig. 1. Three fitness functions were defined: (i) Social distance (in meters) (ii) Net useable space (m2) (iii) Total number of users. Using the Wallacei plugin, fitness functions were tested using the NSGA-II algorithm. The analysis tools contained within Wallacei, including the Parallel Coordinate Plot (PCP), were used to identify preferred outputs on the basis of fitness.

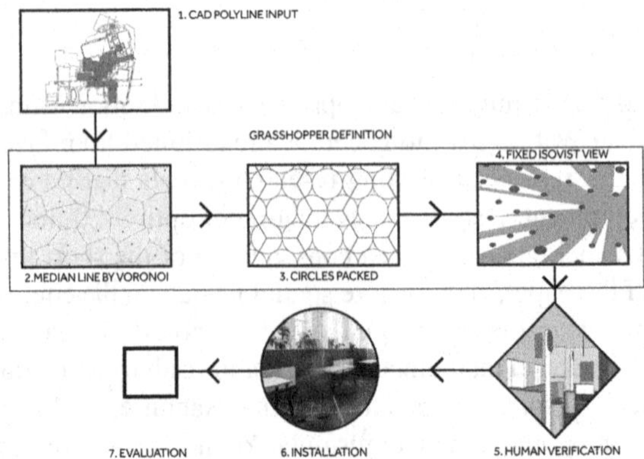

Fig. 1. Generative social distance framework overview

3.1 User Route Generation

A Voronoi offset was applied to the 2D DWG building plan, generating a median line to establish a user route centered between adjacent fixed structures. An *exclusion zone* representing the social distance offset *(fitness function 1)* either side of the median line was tracked onto plans, and areas highlighted at risk of non-compliance were identified using attractor points checking collision on an analysis surface. This provided an early visual risk analysis through color codification of existing non-compliant spaces and routes. Using an Isovist definition [12] a visibility graph analysis was generated to indicate the visibility of walled surfaces using a restricted field of view Isovist in the direction of the path of movement. (Fig. 2). This subsequently defined the physical location of wayfinding signage.

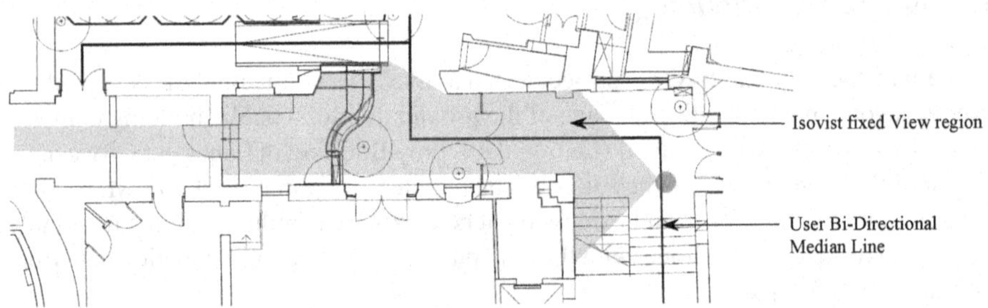

Fig. 2. Visibility graph analysis indicating fixed view projected onto surfaces

3.2 Layout Optimization

The seating region was defined by subtracting the established *exclusion zone* from the net building outline *(fitness function 2)*. To provide optimal seating capacity within this region, circles offset from a point (representing each user) were packed to fit wholly within each region *(fitness function 3)*. Signage typologies and layouts were developed concurrently in collaboration with the client, LCC. Directed by both the optimized lay-out outcomes from Wallacei and associated isovist visibility graphs, the design team verified the final signage design and location with clients for fabrication and installation.

Fig. 3. Installation of the signage in the Storey building, Lancaster, UK.

4 Results

In order to quantify the differences between human designed and automated layouts, five store designers were asked to draw a plan with identical parameters of input *prior* to viewing the automated outcome [Fig. 4]. Human-designed plans included, on average, 32 seated locations compared with 40 of Plan B, a 25% increase in total capacity using the automated methodology. The percentage of useable space defined as the seating region is improved by 12% in the generated layout. On verifying accuracy, the human-designed

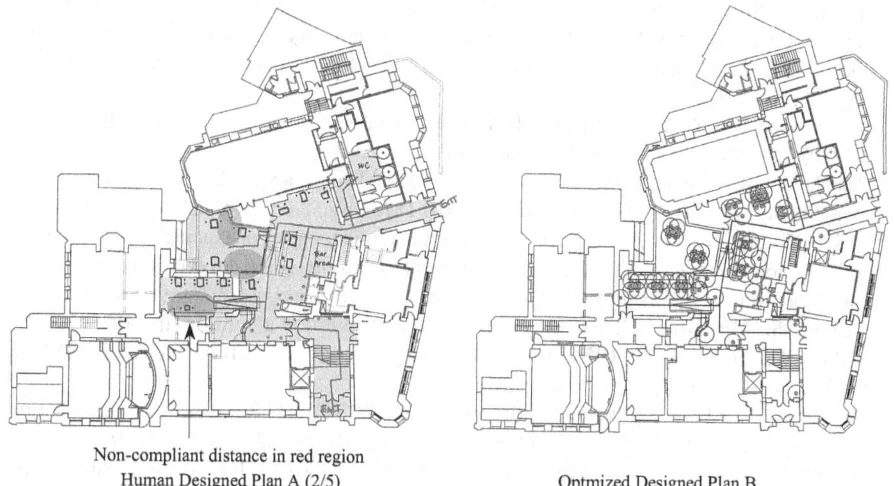

Fig. 4. Comparison of human-designed plan (A) and automated Grasshopper definition (B)

plans (e.g. *Plan A,* Fig. 4*)* include an average of three locations that infringed upon the 2 m social distance.

Table 1. Result of plan comparisons and survey of designer and key stakeholders

	Human Designed	Generative Designed	Diff
Useable Net Seating Area (m2)	210	224	14
Length of Path (m)	52	48	-4
Number of Seats	32	40	8
Instances of sub 2m compliance	3	0	-3

5 Conclusion

The research has provided a methodology that successfully automates social distancing guidance using optimization software in the context of the case study building. The method provides automated socially complaint plan designs, delivering improved capacity and net useable area in comparison with human designed layouts. Subsequent user evaluation in the live lab proves the method presents visually clear and effective social distancing measures. As the definition retains variable fitness functions, crucially social distance, the layout may be redesigned instantly to comply with any value of distance, providing an agile and responsive means to comply with changing social distance advice, providing an essential resource for resilience against future viral variants.

References

1. Megahed, N.A., Ghoneim, E.M.: Antivirus-built environment: Lessons learned from Covid-19 pandemic. Sustain. Cities Soc. **61**, 102350 (2020). https://doi.org/10.1016/j.scs.2020.102350
2. Parker, L.D.: The COVID-19 office in transition: cost, efficiency and the social responsibility business case. Account. Audit. Account. J. **33**, 1943–1967 (2020). https://doi.org/10.1108/AAAJ-06-2020-4609
3. Greenstone, M., Nigam, V.: [PREPRINT] Does social distancing matter? SSRN Electron. J. (2020). https://doi.org/10.2139/ssrn.3561244
4. McNeish, J.E.: Retail signage during the COVID-19 pandemic. Interdiscip. J. Signage Wayfinding. **4**, 67–89 (2020). https://doi.org/10.15763/issn.2470-9670.2020.v4.i2.a64
5. Wilder-Smith, A., Freedman, D.O.: Isolation, quarantine, social distancing and community containment: pivotal role for old-style public health measures in the novel coronavirus (2019-nCoV) outbreak. J. Travel Med. **27**, 1–4 (2020). https://doi.org/10.1093/jtm/taaa020
6. Saremi, S., Mirjalili, S., Lewis, A.: Grasshopper optimisation algorithm: theory and application. Adv. Eng. Softw. **105**, 30–47 (2017). https://doi.org/10.1016/j.advengsoft.2017.01.004.2
7. Fischetti, M., Ab, V., Fischetti, M., Stoustrup, J.: [PREPRINT] Mathematical optimization for social distancing. 1–20 (2020). https://doi.org/10.13140/RG.2.2.35799.91049
8. Bañón, L., Bañón, C.: Improving room carrying capacity within built environments in the context of COVID-19. Symmetry (Basel). **12**, 1–13 (2020). https://doi.org/10.3390/sym12101683

9. Guo, Z., Li, B.: Evolutionary approach for spatial architecture layout design enhanced by an agent-based topology finding system. Front. Archit. Res. **6**, 53–62 (2017). https://doi.org/10.1016/j.foar.2016.11.003
10. Huang, H., et al.: Automatic optimization of wayfinding design. IEEE Trans. Vis. Comput. Graph. **24**, 2516–2530 (2018). https://doi.org/10.1109/TVCG.2017.2761820
11. Dubey, R.K., Thrash, T., Kapadia, M., Hoelscher, C., Schinazi, V.R.: Information theoretic model to simulate agent-signage interaction for wayfinding. Cogn. Comput. **13**(1), 189–206 (2019). https://doi.org/10.1007/s12559-019-09689-1
12. Bielik, M., Schneider, S., König, R.: Parametric Urban Patterns. eCAADe **1**, 701–708 (2012)

Distilling Financial Models by Symbolic Regression

Gabriele La Malfa[1(✉)], Emanuele La Malfa[2(✉)], Roman Belavkin[3(✉)], Panos M. Pardalos[4(✉)], and Giuseppe Nicosia[5(✉)]

[1] University of Cambridge, Cambridge, UK
g.lamalfa@jbs.cam.ac.uk
[2] University of Oxford, Oxford, UK
emanuele.lamalfa@trinity.ox.ac.uk
[3] Middlesex University London, London, UK
R.Belavkin@mdx.ac.uk
[4] University of Florida, Gainesville, FL, USA
pardalos@ufl.edu
[5] University of Catania, Catania, Italy
giuseppe.nicosia@unict.it, gn263@cam.ac.uk

Abstract. Symbolic Regression has been widely used during the last decades for inferring complex models. The foundation of its success is due to the ability to recognize data correlations, defining non-trivial and interpretable models. In this paper, we apply Symbolic Regression to explore possible uses and obstacles for describing stochastic financial processes. Symbolic Regression (SR) with Genetic Programming (GP) is used to extract financial formulas, inspired by the theory of financial stochastic processes and Itô Lemma. For this purpose, we introduce in the model two operators: the derivative and the integral. The experiments are conducted on five market indices that are reliable at defining the evolution of the processes in time: Tokyo Stock Price Index (TOPIX), Standard & Poors 500 Index (SPX), Dow Jones (DJI), FTSE 100 (FTSE) and Nasdaq Composite (NAS). To avoid both trivial and not interpretable results, an error-complexity optimization is accomplished. We perform computational experiments to obtain and investigate simple and accurate financial models. The Pareto Front is used to select between multiple candidates removing the over specified ones. We also test Eureqa as a benchmark to extract invariant equations. The results we obtain highlight the limitations and some pursuable paths in the study of financial processes with SR and GP techniques.

Keywords: Symbolic regression · Stochastic processes · Finance

Supplementary Information The online version contains supplementary material available at https://doi.org/10.1007/978-3-030-95470-3_38.

G. Nicosia et al. (Eds.): LOD 2021, LNCS 13164, pp. 502–517, 2022.
https://doi.org/10.1007/978-3-030-95470-3_38

1 Introduction

In Genetic Programming (GP), the automatic research of laws governing nature has taken advantage of the advent of machine learning [1,2]. Several disciplines have been influenced by these advancements [3], with co-evolution [4,5] that played a fundamental role for physics [6], medicine [7,8], robotics [9,10] and chemistry [14]. Other methodologies exploit partial differential equations [15] or dynamical system equations [16]. In Symbolic Regression (SR), which has often been combined with GP[18], optimal candidates must fit the data while being relatively interpretable. In addition, while fitting solutions are possibly infinite in numbers and easy to obtain, natural laws are on the other hand rare and difficult to extract [17]. SR evolves mathematical expressions that fit well the data. The evolution of the initial random population follows GP rules [19], using a fitness function to measure the errors and show the "convenient" paths of evolution. Scientific literature has marked a step forward with algorithms capable of searching and selecting conservation laws and enhancing generalization [20,21]. Such methods propose to select valuable candidates predicting the derivative relationships among the system variables over time [4,17]. In machine learning SR has been combined with convolutional neural networks [23] as well as recurrent architectures [24,25] for data analysis [6]: function approximation with deep learning has been also investigated by many other authors, e.g., [26,27]. In general, applying machine learning to SR is a major research subject in GP [28,29]. The paper is structured as follows: Sect. 2 presents the theoretical basis of financial stochastic processes; Sect. 3 presents SR with GP method; Sect. 4 explains the cross-disciplinary relation between financial model theory and SR with GP; Sect. 5 and Sect. 6 are respectively dedicated to the experiments and to the conclusions.

1.1 Related Works

SR has been applied recently to finance and time series forecasting [22]: in this sense, the study of the financial markets presents inherent non-trivial challenges, as in example financial forecasting (indices, stocks, commodities, currencies). Many prominent works in literature propose statistical-based forecasting approaches and signal modeling. Many of them fall into the ARMA (Auto Regressive Moving Average) [30] and GARCH (Generalized Autoregressive Conditional Heteroskedasticity) [31] model families [32,33]. One of the biggest challenge of the field is relative to the lack of information of the market and the modified perceptions of the agents [34,35], while distilling financial formulae from time series - which is the objective of our study - is still an open problem, as outlined by the limited results in recent literature. A seminal approach to symbolic stock market prediction has been made through multi-gene symbolic regression genetic programming [22], with a few successive works that leveraged the idea to forecast economic growth with tools proper of evolutionary computation [11]. More recently, strongly typed GP has been adapted to generate forecasting trading rules [12], while the advent of deep learning has inspired similar works with reinforcement learning [13].

2 Financial Stochastic Processes - Itô Formula for Brownian Motion

Financial process theory was developed with the application in continuous time of stochastic calculus [36–38]. Itô Integral and Itô Formula/Lemma [39] have been fundamental for setting the bases of the financial stochastic processes [40, 41]. In this section, we discuss the ideas about the evolution in time of a financial process, with a focus on Itô Lemma. The goal is to review some basic elements that are a fundamental support of the experimental setup of the paper. The notation used is taken from Shreve [40]. Defining a function $f(t,x)$ that depends on both time t and a generic process x. The partial derivatives $f_t(t,x)$, $f_x(t,x)$ and $f_{xx}(t,x)$ are defined and continuous. Defining $W(t)$ as a Brownian Motion (BM). The Itô Formula for a BM, with $T > 0$ is:

$$\begin{aligned} f(T,W(T)) = f(0,W(0)) + \int_0^T f_t(t,W(t))dt \\ + \int_0^T f_x(t,W(t))dW(t) + \frac{1}{2}\int_0^T f_{xx}(t,W(t))dt \end{aligned} \tag{1}$$

Without a detailed demonstration of Eq.(1), the idea is to model a "step" difference of the function $f()$ w.r.t. time t and a generic process x (that it will be turned to be a BM). More formally, for a given interval $(j, j+1]$ with $j > 0$:

$$\begin{aligned} f(t_{j+1},x_{j+1}) - f(t_j,x_j) =& f_t(t_j,x_j)(t_{j+1}-t_j) + f_x(t_j,x_j)(x_{j+1}-x_j) \\ &+ \frac{1}{2}f_{xx}(t_j,x_j)(x_{j+1}-x_j)^2 + f_{tx}(t_j,x_j) \\ &+ (t_{j+1}-t_j)(x_{j+1}-x_j) + \frac{1}{2}f_{tt}(t_j,x_j)(t_{j+1}-t_j)^2 + ... \end{aligned} \tag{2}$$

that is the Taylor expansion of the function f given a partition $(j, j+1]$.

Replacing the process x with a BM $W(t)$ in the partition:

$$\begin{aligned} f(T,W(T)) - f(0,W(0)) =& \sum_{j=0}^{n-1}[f(t_{j+1},W(t_{j+1})) - f(t_j,W(t_j))] \\ =& \sum_{j=0}^{n-1} f_t(t_j,W(t_j))(t_{j+1}-t_j) + \sum_{j=0}^{n-1} f_x(t_j,W(t_j)) \cdot \\ & \cdot (W(t_{j+1}) - W(t_j)) + \frac{1}{2}\sum_{j=0}^{n-1} f_{xx}(t_j,W(t_j)) \cdot \\ & \cdot (W(t_{j+1}) - W(t_j))^2 + \sum_{j=0}^{n-1} f_{tx}(t_j,W(t_j))(t_{j+1}-t_j) \cdot \\ & \cdot (W(t_{j+1}) - W(t_j)) \end{aligned} \tag{3}$$

Considering n as the number of subintervals and $||\Pi||$ as the partitions of $[0, T]$. As $n \to \infty$, then $||\Pi|| \to 0$. Looking at the terms of Eq.(3):

- $\lim_{||\Pi|| \to \infty} \sum_{j=0}^{n-1} f_t(t_j, W(t_j))(t_{j+1} - t_j) = \int_0^T f_t(t, W(t))dt$
- $\lim_{||\Pi|| \to \infty} \sum_{j=0}^{n-1} f_x(t_j, W(t_j))(W(t_{j+1}) - W(t_j)) = \int_0^T f_x(t, W(t))dW(t)$, that is an Itô Integral
- $\lim_{||\Pi|| \to \infty} \frac{1}{2} \sum_{j=0}^{n-1} f_{xx}(t_j, W(t_j))(W(t_{j+1}) - W(t_j))^2 = \frac{1}{2} \int_0^T f_{xx}(t, W(t))dt$
- The fourth term converges to 0 as $\lim_{||\Pi|| \to \infty}$
- The fifth term converges to 0 as $\lim_{||\Pi|| \to \infty}$
- The other higher-order terms converge to 0 as $\lim_{||\Pi|| \to \infty}$

Eq.(3) becomes:

$$\begin{aligned} f(T, W(T)) - f(0, W(0)) = & \lim_{||\Pi|| \to \infty} \sum_{j=0}^{n-1} f_t(t_j, W(t_j))(t_{j+1} - t_j) \\ & + \lim_{||\Pi|| \to \infty} \sum_{j=0}^{n-1} f_x(t_j, W(t_j))(W(t_{j+1}) - W(t_j)) \\ & + \frac{1}{2} \lim_{||\Pi|| \to \infty} \sum_{j=0}^{n-1} f_{xx}(t_j, W(t_j))(W(t_{j+1}) - W(t_j))^2 \end{aligned} \tag{4}$$

That is:

$$\begin{aligned} f(T, W(T)) - f(0, W(0)) = \int_0^T f_t(t, W(t))dt + \int_0^T f_x(t, W(t))dW(t) \\ + \frac{1}{2} \int_0^T f_{xx}(t, W(t))dt \end{aligned} \tag{5}$$

Equation (5) provides an explanation on how the process evolves over time, considering the application to a BM. Itô Formula, can be applied to Itô processes which are more general processes than BMs. This analysis will be used as a pathway to SR, since it states the level of the relationships between the variation of the variables. The model based on a BM has limitations in the description of financial phenomena [42]. However, for the purpose of the paper it represents a first approach to SR with GP.

3 Symbolic Regression by Genetic Programming

Symbolic Regression is a learning technique that formalizes mathematical models (formulas) for the relations between variables [43]. SR with GP derives generations of models through an evolutionary approach based on optimization.

GP randomly generates a first population of formulas. Each formula suggests a possible interdependence between variables. The selection from the first to next generation is based on a fitness function that is an appropriate measure of how well each formula fits the data. The evolutionary approach aims at optimizing fitness. From one generation to the next, the individuals are subject to standard evolutionary changes: selection, mutation and crossover. The algorithm is free to explore the possible combinations of variables, functions and constants. The first-stage representation of the model is called primitive. A classic representation of the primitives is the syntax tree: the terminal nodes are the combination of variables and constants, while the internal nodes are the operators. The symbolic solutions are called candidates. The optimization is oriented towards complexity. Too simple candidates are often a poor fitting, while too complex solutions are hardly interpretable. The research of dominant non-trivial solutions transforms the problem into a multi-objective optimization.

The fitness function is an indicator of how well the algorithms fits the data. In SR the fitness function is a guide to optimal solutions. SR with GP drives the performance improvement over generations. This is called the convergence, and it aims at exploring the space of solutions that achieved better results in the past. At the same time, the approach leaves a high degree of exploratory freedom. The fitness function proposed in this paper considers the error between a candidate function f and the response variable y. The fitness is calculated as follows:

$$-\frac{1}{2}\sum_{i=1}^{n} log(1 + abs(y - f(x, z))) \tag{6}$$

where x and z are the sample variable of the function f. The fitness function in Eq.(6) is called Mean Log Absolute Error (MLAE). Candidates' evolution exhibits both error and complexity optimization, discarding the solutions that perform worse according to the dominance of Pareto.

4 Methodology

SR with GP is the main method of the paper. Eureqa [17] is used as second method to compare the differences. The characteristics of the first approach are described below, while those of Eureqa can be found in [17].

GP with SR evolves candidate functions to optimal solutions. The standard practice is a random initialization to avoid any bias. The population is composed by individuals of symbolic formulae, where each variable corresponds to an entry of the dataset (in Table 1 we define each variable name). SR with GP evolves the individuals throughout the generations and the process is guided by a multi-objective optimization of error and complexity to select the best candidates. The operators used to create and combine the symbolic formulae are as follows: *add*, *sub*, *mul*, *div*, *sin*, *cos*, *log*, D, D^2, *int*. First and second order gradients D, D^2 and the integral, *int* are inspired by Eq.(5).

We use 'half and half' initialization method to allow both "grow" and "full" growths. We choose the Mean Log Absolute Error (MLAE) as the fitness metric, as reported in (3.1). We report both the length of the solution and the complexity calculations as by Eureqa [5]. Error and complexity are minimized at the same time. The multi-objective optimization helps to prevent the "bloat", which is the research for more complex functions, without effective enhancing of the fitness.

We perform the evolution of the model with the following methods: crossover, subtree mutation, hoist mutation and point mutation. The best candidates are chosen at the end of each generation. Non-Dominated Sorting Genetic Algorithm (NSGA-II) [44] is used to create the Pareto Front. It selects the next generation according to non-dominated sorting and crowding distance. Eureqa [4] introduces the difference between the numerical partial derivatives respectively calculated from the data and from the candidate law equation for the computation of the fitness. Given a time-series described by two variables x and y, the partial derivatives of the system variables are represented as follows:

$$\frac{\Delta x}{\Delta y} \approx \frac{dx}{dt} \Big/ \frac{dy}{dt} \tag{7}$$

Defining the candidate equation $f(x, y)$, the ratio between the partial derivatives can be expressed as:

$$\frac{\delta x}{\delta y} = \frac{\delta f}{\delta y} \Big/ \frac{\delta f}{\delta x} \tag{8}$$

The difference between Eq.(7) and Eq.(8) defines the fitness function of the model. A solution is an invariant law if derived w.r.t. the variables of the system to obtain Eq.(8). Equation(8) is then compared with the numerical derivatives of Eq.(7) and the functions with the best fitness values are selected.

4.1 Symbolic Regression and Financial Processes Variation

The formulae reported in Sect. 2 describe the evolution of a financial process variation over time. Performing SR with GP for catching the variation in financial processes, as in Eq.(5), means exploring the function f at the foundations of the process. The selection of a proper set of variables to obtain significant results is crucial. In this sense, Itô Formula applies to the Stochastic Differential Equation (SDE) of a diffusion process, that is:

$$dX(t) = \mu(t)d(t) + \sigma(t)dW(t) \tag{9}$$

where $\mu(t)$ and $\sigma(t)$ are two adapted processes. Informally, $\mu(t)$ is the pace of the Stochastic Differential Equation (SDE) while $\sigma(t)$ is the drift (also interpreted as the volatility of the process). The algorithm should be able to recognize the patterns between the variables' derivatives. Inspired by the theory of Sect. 2, SR with GP is equipped with the operators of derivative and integral.

5 Experimental Results

The experiments are performed on five financial indices, respectively: Tokyo Stock Price Index ($TOPIX$), Standard & Poors 500 Index (SPX), Dow Jones (DJI), FTSE 100 ($FTSE$), Nasdaq Composite (NAS). The data cover the period from 2010 to 2019 (10-years data). The aim is to extract a non-trivial solution that fits well the time-series. To avoid over-fitting, we choose very classic explanatory variables: time (t), the standard deviation of the index (Sd), the daily rolling average of the index (tdi) and the index values. The explained variable is the index returns (y) which reflect the differences between the index phase of a single time step. In Eq.(4) and Eq.(5) of Sect. 2, the left portion represents index returns. Index returns depend on time, and with some approximation, the stochastic variation can be modelled with a BM or an Itô Process. With respect to Eq.(5) and Eq.(9), the standard deviation and the daily rolling average of the index represent respectively the volatility and the pace of the process. The second term of the right-hand side of Eq.(5) is interpreted as the variation of the process, and it depends on the BM $W(t)$, while the third term of the equation is the pace of the process. The variables of the experiments are reported in Table 1.

Table 1. Datasets

Variables	
Name	Description
y	Topix, SPX, DJI, FTSE, NAS index returns
t	Time
Sd	Standard deviation
tdi	Index window average
Top/SPX/DJI/FTSE/NASDAQ	Indices values

5.1 Experimental Protocol

We apply SR with GP and Eureqa to discover explanatory non-trivial models for the time-series indices. The first method uses SR with GP. The second one is detailed in [17] and uses Eureqa.

First Method: Genetic Programming. We conduct the research in the direction of explaining y w.r.t the other variables as follows:

$$y = f(t, Sd, tdi, idx \in \{Top, SPX, DJI, FTSE, NAS\}) \tag{10}$$

Eq.(10) expresses y as a function of: time t, standard deviation Sd and rolling average tdi of the respective index idx. All the simulations use the classic operator of SR to elaborate the formulae. Furthermore, we introduce the numeric first and second derivatives $D()$ and $D^2()$. Table 2 reports the hyperparameters and the operators used for each simulation. A first simulation with $population = 1000$ and $generations = 3000$ is conducted with the method of SR with GP. The experiments are extended to $generations = 5000$. We set up another round of simulations with $population = 1000$ for both $generations = 3000$ and 5000 without the use of trigonometric functions. We are interested in a model that is as unrelated as possible to repeated patterns over time. The last simulation introduces numerical integral *int*() as a new evolutionary operator. Also, in this case, we conduct the experiments with *population* = 1000 and *generations* = 3000 and 5000. The best candidates are found according to the multi-objective optimization, where error (Mean Log Absolute Error) and complexity are minimized. Specifically, the complexity of a model is calculated according to its length. For simplicity, we will call this complexity "node depth". We report both the node depth and the complexity according to Eureqa in the results table. For this second complexity, specific weights are assigned to each operation, constant and variable that compose a solution. This is called the building blocks complexity. Table 3 reports the weights of the building blocks complexity for the two methods. Since integral and derivative are not available in Eureqa, we assigned them a value of complexity strictly greater than any other operator (i.e., 5).

Table 2. Symbolic Regression with Genetic Programming: hyperparameters and operators

Hyperparameters	
Population	1000
Generations	3000/5000
Fitness	MLAE (Mean Log Absolute Error) and Formula Length
Initialization Method	half half
Selection Method	NSGA-II Error/Complexity
Prob. of Crossover	0.5
Prob. of Subtree Mutation	0.1
Prob of Hoist Mutation	0.1
Prob. of Point Mutation	0.1
Init. Tree Depth	range [2:20]
Operators	
Sim. 1a, 2a	*add*, *sub*, *mul*, *div*, *sin*, *cos*, *exp* *log*, D, D^2
Sim. 1b, 2b	*add*, *sub*, *mul*, *div*, *log*, *exp* D, D^2
Sim. 1c, 2c	*add*, *sub*, *mul*, *div*, *sin*, *cos*, *exp* *log*, D, D^2, *int*

Table 3. Building blocks complexity: Eureqa and SR with GP

Building Blocks/Weigths - Eureqa	
Constant	1
Input Variable	1
Addition/Subtraction/Multiplication	1
Division	2
Sine/Cosine	3
Exponential/Logarithm	4

Building Block/Weigths - SR with GP	
Constant	1
Input Variable	1
Addition/Subtraction/Multiplication	1
Division	2
Sine/Cosine	3
Exponential/Logarithm	4
Derivative/Integral	5

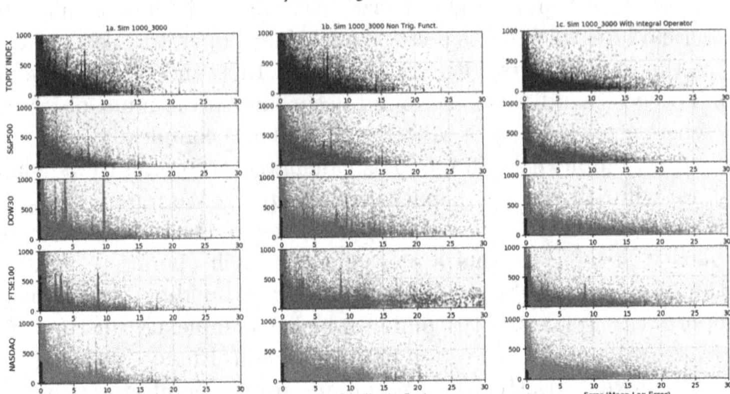

Fig. 1. Method: GP with SR. Pareto Fronts and evolutionary steps in simulations 1a 1b and 1c (columns). The Pareto Fronts are the results of the multi-objective optimization where Error (x-axis) and Complexity (y-axis) are optimized. The five indices are represented on the subplot rows with different colours. Each single point is a solution. The highlighted blue points are the Pareto solutions distilled from 3000 generations. (Color figure online)

Simulation 1: population 1000 and 3000 generations. We report three representative solutions of the experiments with 3000 generations in Tables 1a,2a and 3a, for each of the indices under consideration. The logic is to present to the reader the top-1 solutions that minimize the error maintaining acceptable levels of interpretability. The complexities of the selected models remain in the ranges $0 - 100$ when this is possible. The same logic is also applied to the next simulations. Figure 1 represents The Pareto Front Solutions and the evolutionary steps for each index and the experiments $1a$, $1b$ and $1c$. The experiment $1a$ reaches error levels between 0.43% and 0.41% for Topix index, while the error is between 0.33% and 0.31% for SPX, DJI and FTSE. The solutions of NAS index present an error between 0.39% and 0.37%. The trigonometric functions are mostly used to explain FTSE index, while for the other indices we do not register their presence. The first and second derivative operators are widely used and among all the variables, the time t seems to be the least explanatory among all the indices. The experiment $1b$ without trigonometric functions, does not suffer their absence: for the same level of complexity, the results reach almost the same lower bound errors. The introduction of integral as a new operator, conditions the results of the Pareto Fronts and integrals appear in almost all solutions with a complexity higher than 100. The time t variable assumes a new role with the integral operator. The third solution of FTSE index is an example. We denote in the empirical results the important role of the derivatives w. r. t. time and to the Brownian Motion.

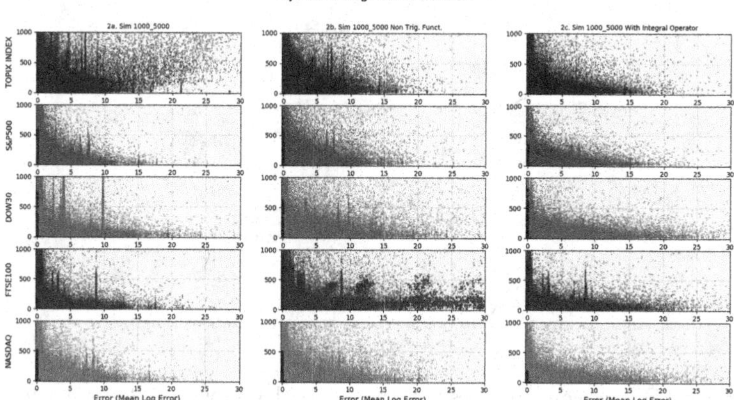

Fig. 2. Method: GP with SR. Pareto Fronts and evolutionary steps in simulations 1a 1b and 1c (columns). The Pareto Fronts are the results of the multi-objective optimization where Error (x-axis) and Complexity (y-axis) are optimized. The five indices are represented on the subplot rows with different colours. Each single point is a solution. The highlighted blue points are the Pareto solutions distilled from 5000 generations. (Color figure online)

Simulation 2: population 1000 and 5000 generations. We report the top-1 representative solutions of the experiments with 5000 generations in Tables 2a, 2b, 2c for each of the indices under consideration. The representative logic of the solutions is the same of Simulation 1 (Fig. 1). Figure 2 represents The Pareto Front Solutions and the evolutionary steps for each index and the experiments 2*a*, 2*b* and 2*c*. The experiment 2*a* reaches error levels between 0.43% and 0.39% for Topix index, while the error is between 0.34% and 0.29% for SPX, DJI and FTSE. The solutions of NAS index present an error between 0.40% and 0.36%. The trigonometric functions do not have a crucial role to reach higher error levels to model the indices. The first and second derivative operators are widely used and among all the variables. The introduction of integral as a new operator, conditions the results of the Pareto Fronts and integrals appear in almost all solutions with a complexity higher than 100.

Convergence of the Simulations. To conclude the examination of GP with SR experiments, we present the Pareto Fronts convergences to the origin of the axes for each of the five indices under consideration. One of the cornerstones of GP is that by increasing the number of generations, it also increases the convergence towards optimal Pareto solutions. Figure 3 shows for simulations 1a, 1b, 1c, and 2.a, 2b, 2c the convergence of the solutions. The convergence to the

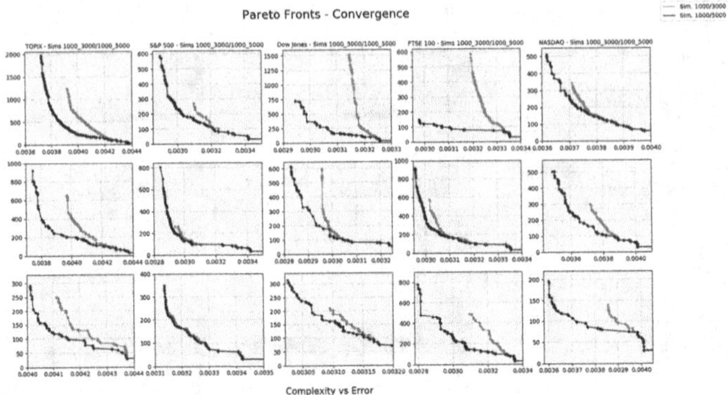

Fig. 3. Pareto Fronts convergences for the method GP with SR: for each financial index, we plot the convergences of simulation 1 and simulation 2 to the origin of the axes. The cyan Pareto Fronts are the solutions of the experiments with 3000 generations while the blue Pareto Fronts are from the experiments with 5000 generations. (Color figure online)

origin of the axes is evident in almost all experiments and indexes. The error tends to decrease with the increase of the complexity of the solutions, however the presence of solutions distinctly belonging to the two frontiers is recorded with the same levels of complexity and lower error for those related to the simulations with 5000 generations. This result leads to the conclusion that there is a real improvement of the optimization and that it is not reduced to the only error/complexity trade-off.

Second Method: Eureqa. The second method proposes SR using Eureqa [17] on Eq.(10). The error calculated by Eureqa is the Mean Log Absolute Error (MLAE). The results obtained with Eureqa are reported for each index in Table 7 and the experiments reached around 1.6e7 generations and around 100 h of time computation. Eureqa finds a considerably low number of solutions for the same research done with the first method. Furthermore, the error level is around the double of the one reached with the first method for all the indices. We observe the tendency of the algorithm to make low use of the index values and of the standard deviation Sd. Conversely, the rolling average tdi and time variables are essential parts of all the solutions. In the conclusive part of Sect. 5, in Table 4 we show the error performances of SR with GP and Eureqa.

Table 4. Error comparison MLAE: SR with GP (NSGA-II) vs Eureqa

Error Comparison: SR with GP and Eureqa		
Index	SR with GP	Eureqa
Top	0.38%	0.85%
SPX	0.295%	0.64%
DJI	0.295%	0.61%
FTSE	0.295%	0.675%
NAS	0.36%	0.75%

Table 5. Pareto solutions Eq.(10): Symbolic formula - Error - Complexity (Length compl. - Eureqa compl.)

1a. Pareto Fronts Solutions: Pop 1000; Gen 3000		
TOPIX		
Symbolic Formula	Error (MLAE)	Complexity
$D(Top - D(Top))/(Top + (D(Top)/ - 0.217))$	0.004318	63 — 21
SPX		
$(D(SPX) - D^2(SPX))/((0.468/D^2(Sd)) + SPX)$	0.003334	62 — 23
DJI		
$D(log((D((DJI + (tdi \cdot log(D((Sd - D(tdi))))))) - DJI)))$	0.003246	76 — 26
FTSE		
$D(log((D(FTSE) - (FTSE + (sin(D^2(Sd)) \cdot D^2((Sd + FTSE)))))))$	0.003327	76 — 27
NASDAQ		
$(D(NAS) - D^2(NAS))/((((NAS \cdot (0.452 \cdot Sd)) + NAS) \cdot D^2(t)) + NAS)$	0.003900	89 — 27
1b. Pareto Fronts Solutions: Pop 1000; Gen 3000 - Without Trigonometric Functions		
TOPIX		
Symbolic Formula	Error (MLAE)	Complexity
$D(Top - D(Top))/(D^2(D(((Sd - t) \cdot Top))) + Top)$	0.004286	75 — 27
SPX		
$(D(SPX) + tdi)/SPX$	0.004783	27 — 10
DJI		
$D(DJI - D(DJI))/(DJI - D(DJI))$	0.003250	50 — 18
FTSE		
$D(log((D(FTSE) - (Sd + ((t/tdi) + FTSE)))))$	0.003350	57 — 19
NASDAQ		
$(D(NAS) - D^2(NAS))/(NAS - (log(-0.347)/D^2((0.082 \cdot t))))$	0.003892	80 — 33
1c. Pareto Fronts Solutions: Pop 1000; Gen 3000 - With Integral Operator		
TOPIX		
Symbolic Formula	Error (MLAE)	Complexity
$(tdi + D(Top))/Top$	0.006174	27 — 10
SPX		
$D((SPX - D(SPX)))/SPX$	0.003420	34 — 10
DJI		
Symbolic Formula	Error (MLAE)	Complexity
$sin((D((DJI - D(DJI)))/(DJI - D(DJI))))$	0.003249	55 — 23
FTSE		
$D((sin(t) + log(FTSE)))$	0.004825	28 — 10
NASDAQ		
$D(log((D(NAS) - (NAS + log(D^2((Sd + t)))))))$	0.003998	60 — 25

Table 6. Pareto solutions Eq.(10): Symbolic formula - Error - Complexity (Length compl. - Eureqa compl.)

2a. Pareto Fronts Solutions: Pop 1000; Gen 5000		
TOPIX		
Symbolic Formula	Error (MLAE)	Complexity
$D(Top) + D^2((tdi - Top)/Top)$	0.004370	43 — 13
SPX		
$D(SPX) - D^2(SPX)/SPX$	0.003421	34 — 10
DJI		
$D(log((D((DJI + tdi)) - (DJI + Sd))))$	0.003256	48 — 16
FTSE		
$D(log((D(FTSE) - (FTSE + (0.147/tdi)))))$	0.003351	51 — 17
NASDAQ		
$((D(NAS) + cos((NAS \cdot tdi))) - D^2(NAS))/NAS$	0.004006	57 — 22
2b. Pareto Fronts Solutions: Pop 1000; Gen 5000 - Without Trigonometric Functions		
TOPIX		
Symbolic Formula	Error (MLAE)	Complexity
$D((Top - D(Top))/(D^2(D((Sd \cdot Top))) + Top))$	0.004289	66 — 25
SPX		
$D(log(((SPX + tdi) - D(SPX))))$	0.003421	39 — 14
DJI		
$D((DJI - D(DJI)))/(DJI - D(DJI))$	0.003250	50 — 18
FTSE		
$D(log((D(FTSE) - ((D^2(FTSE) \cdot D^2(Sd)) + FTSE))))$	0.003336	62 — 26
NASDAQ		
$D(((Sd + NAS) - D(NAS)))/NAS$	0.004010	43 — 12
2c. Pareto Fronts Solutions: Pop 1000; Gen 5000 - With Integral Operator		
TOPIX		
Symbolic Formula	Error (MLAE)	Complexity
$D((Top + D((tdi - Top))))/Top$	0.004369	43 — 12
SPX		
$D((SPX - D(SPX)))/(SPX - (-0.454/D^2(int(t, Sd))))$	0.003301	27 — 103
DJI		
$sin(((D(DJI) - D^2(DJI))/DJI))$	0.003255	39 — 16
FTSE		
$D(log((FTSE - D((FTSE + D(Sd))))))$	0.003353	46 — 19
NASDAQ		
$D(log((D(NAS) - (NAS + (D^2(NAS) \cdot D(log(D(NAS))))))))$	0.003947	74 — 35

Table 7. Pareto solutions Eq.(10): Symbolic formula - Error - Complexity (Length compl. - Eureqa compl.)

Eureqa Solutions		
TOPIX		
Symbolic Formula	Error (MLAE)	Complexity
$0.0005278 + 1.54369e - 5/(t - 0.161362)$	0.00873661	66 — 7
SPX		
$0.001980 \cdot tdi - 0.000351 - 0.000644 \cdot t$	0.006433	71 — 12
DJI		
$2.906536e - 6 \cdot Sd \cdot tdi - 0.001418$	0.006168	48 — 12
FTSE		
$0.000693 \cdot t \cdot tdi^2$	0.006828	32 — 8
NASDAQ		
$0.000466 \cdot tdi + 4.672516e - 6/(t - 0.189731)$	0.007576	70 — 11

6 Concluding Remarks

We present the application of Symbolic Regression (SR) with Genetic Programming (GP) to extract models for the description of five financial indices' variations. The approach is inspired by the theory of stochastic processes for the empirical section. We introduce the derivative and the integral operators for SR with GP. We automatically retrieve Pareto Front solutions from five financial time-series that represent American, European and Asian markets. The experimental part consists of two simulations: the first one with 3000 generations and the second one with 5000 generations. We show the Pareto Fronts convergence for all the simulations and all the indices: the results validate the optimization process of the three GP variants. The omission of trigonometric operators does not prevent to reach the same error levels for all the indices w.r.t. the experiments that use them. The variant with the integral makes use of this operator in the results, especially for higher complexities. For the research of natural and theoretical formulas, both methods seem to be promising, however the results suggest that there are margins to improve the model in the direction of a diffusion oriented process. The next steps of the research might be directed to new approximations of the random walks used in the experiments, with the introduction of GP stochastic elements.

References

1. Koza, J.R.: 1992 Genetic Programming: On the Programming of Computers by Means of Natural Selection. MIT Press, Cambridge (1992)
2. Wolf, J.B.: 2007 evolutionary computation: a unified approach. In: A De Jong, K., (ed.) The Quarterly Review of Biology. A Bradford Book, vol. 82, p. 46. MIT Press, Cambridge (2006). ISBN: 0-262-04194-4 82

3. Villaverde, A.F., Banga, J.R.: Reverse engineering and identification in systems biology: strategies, perspectives and challenges. J. R. Soc. Interface **11**, 20130505 (2014)
4. Schmidt, M., Lipson, H.: Coevolution of fitness predictors. IEEE Trans. Evol. Comput. **12**, 736–749 (2008)
5. Schmidt, M., Lipson, H.: Co-evolution of fitness maximizers and fitness predictors. GECCO Late Breaking Paper (2005)
6. Udrescu, S.M., Tegmark, M.: AI Feynman: a physics-inspired method for symbolic regression. Sci. Adv. **6**(16), eaay2631 (2020)
7. Saxena, A., Lipson, H., Valero-Cuevas, F.J.: Functional inference of complex anatomical tendinous networks at a macroscopic scale via sparse experimentation. PLoS Comput. Biol. (2012)
8. Pandey, S., Purohit, G.N., Munshi, U.M.: In: Munshi, U.M., Verma, N. (eds.) Data Science Landscape. SBD, vol. 38, pp. 321–326. Springer, Singapore (2018). https://doi.org/10.1007/978-981-10-7515-5_24
9. Tan, K.C., Wang, L.F., Lee, T.H., Vadakkepat, P.: Evolvable hardware in evolutionary robotics. Autonom. Robot. **16**, 5–21 (2004). https://doi.org/10.1023/B:AURO.0000008669.57012.88
10. Trudeau, A., Clark, C.M.: Multi-robot path planning via genetic programming. ARMS 2019 Workshop (AAMAS), arXiv:1912.09503v1 (2019)
11. Claveria, O., Enric, M., Salvador, T.: Evolutionary computation for macroeconomic forecasting. Comput. Econ. **53**(2), 833–849 (2019)
12. Michell, K., Kristjanpoller, W.: Generating trading rules on US Stock Market using strongly typed genetic programming. Soft. Comput. **24**(5), 3257–3274 (2019). https://doi.org/10.1007/s00500-019-04085-1
13. Taghian M., Asadi A., Safabakhsh R.: Learning financial asset-specific trading rules via deep reinforcement learning. arXiv preprint arXiv:2010.14194 (2020)
14. Butler, K.T., Davies, D.W., Cartwright, H.: Machine learning for molecular and materials science. Nature **559**, 547–555 (2018)
15. Rudy, S.H., Brunton, S.L., Proctor, J.L., Kutz, J.N.: Data-driven discovery of partial differential equations. Sci. Adv. **3**(4), e1602614 (2017)
16. Brunton, S.L., Proctor, J.L., Kutz, J.N.: Discovering governing equations from data by sparse identification of nonlinear dynamical systems. PNAS **113**(15), 3932–3937 (2016)
17. Schmidt, M., Lipson, H.: Distilling free-form natural laws from experimental data. Science **324**(5923), 81–85 (2009)
18. Koza, J.R., Keane, A., Rice, J.P.: Performance improvement of machine learning via automatic discovery of facilitating functions as applied to a problem of symbolic system identification. In: IEEE International Conference on Neural Networks, San Francisco, IEEE, pp. 191–198 (1993)
19. Forrest, S.: Genetic algorithms: principles of natural selection applied to computation. Science **261**(5123), 872–878 (1993)
20. Chen, Q., Xue, B., Zhang, M.: Improving generalization of genetic programming for symbolic regression with angle-driven geometric semantic operators. IEEE Trans. Evolution. Comput. **23**(3) (2019)
21. Chen, Q., Zhang, M., Xue, B.: Structural risk minimization-driven genetic programming for enhancing generalization in symbolic regression. IEEE Trans. Evolution. Comput. **23**(4) (2019)
22. Sheta, F., Ahmed, S.E.M., Farid, H.: Evolving stock market prediction models using multi-gene symbolic regression genetic programming. Artif. Intell. Mach. Learn. **15**(1), 11–20 (2015)

23. Wick, C.: Deep learning. Informatik-Spektrum **40**(1), 103–107 (2016). https://doi.org/10.1007/s00287-016-1013-2
24. Hochreiter, S., Schmidhuber, J.: Long short-term memory. Neural Comput. **9**(8), 1735–1780 (1997)
25. Graves A.: Supervised Sequence Labelling with Recurrent Neural Networks, vol. 385. Springer, Heidelberg (2012). https://doi.org/10.1007/978-3-642-24797-2
26. Lin, H.W., Tegmark, M., Rolnick, D.: Why does deep and cheap learning work so well? J. Stat. Phys. **168**(6), 1223–1247 (2017). https://doi.org/10.1007/s10955-017-1836-5
27. Wu, T., Tegmark, M.: Toward an AI physicist for unsupervised learning. Phys. Rev. E **100**(3) arXiv:1810.10525v4 (2018)
28. McRee, R.K.: Symbolic regression using nearest neighbor indexing. In: GECCO 2010: Proceedings of the 12th Annual Conference Companion on Genetic and Evolutionary Computation, pp. 1983–1990 (2010)
29. Stijven, S., Minnebo, W., Vladislavleva, K.: Separating the wheat from the chaff: on feature selection and feature importance in regression random forests and symbolic regression. In: GECCO 2011: Proceedings of the 13th Annual Conference Companion on Genetic and Evolutionary Computation (2011)
30. Cavicchioli, M.: Higher Order Moments of Markov Switching Varma Models. Cambridge University Press, Cambridge (2016)
31. Charles, A., Darné, O.: The accuracy of asymmetric GARCH model estimation. Int. Econ. **157** (2019)
32. Hyndman, R.J., Athanasopoulos, G.: Forecasting: Principles and Practice. oTexts, Monash University, Australia (2015)
33. Tsay, R.S.: Multivariate Time Series Analysis: With R and Financial Applications. Wiley, Hoboken (2013)
34. Black, F.: Noise. Wiley, Hoboken (1986)
35. Fama, E.F.: Random walks in stock market prices. Financ. Anal. J. **21**(5), 55–59 (1965)
36. Merton, R.C.: Lifetime portfolio selection under uncertainty: the continuous-time case. Rev. Econ. Stat. **51**, 247 (1969)
37. Merton, R.C.: Continuous-Time Finance. Basil Blackwell, Oxford (1990)
38. Black, F., Scholes, M.: The pricing of options and corporate liabilities. J. Polit. Econ. **81**(3), 637–654 (1973)
39. Itô, K.: 1944 Stochastic Integral. Proc. Imperial Acad. **20**(8), 519–524 (1944)
40. Shreve, S.: Stochastic Calculus for Finance II: Continuous-Time Models. Springer, New York (2004)
41. Stochastic Differential Equations. Springer, Heidelberg (2003). https://doi.org/10.1007/978-3-642-14394-6_9
42. Hull, J.C.: Options, Futures, and Other Derivatives. 10th edn., Pearson, London (2018)
43. Vladislavleva, E.J., Smits, G.F., Hertog, D.: Order of nonlinearity as a complexity measure for models generated by symbolic regression via pareto genetic programming. IEEE Trans. Evolution. Comput. **13**(2) (2009)
44. Deb, K., Pratap, A., Agarwal, S., Meyarivan, T.: A fast and elitist multiobjective genetic algorithm: NSGA-II. IEEE Trans. Evolution. Comput. **6**(2) (2002)

Analyzing Communication Broadcasting in the Digital Space

Giovanni Giuffrida[1], Francesco Mazzeo Rinaldi[1,2], and Andrea Russo[3(✉)]

[1] Department of Political and Social Science, University of Catania, Catania, Italy
{giovanni.giuffrida,fmazzeo}@unict.it

[2] School of Architecture and The Built Environment, KTH, Royal Institute of Technology, Stockholm, Sweden

[3] Department of Physics and Astronomy, University of Catania, Catania, Italy
Andrea.russo@phd.unict.it

Abstract. This paper aims to understand complex social events that arise when communicating general concepts in the digital space. Today, we get informed through many different channels, at different times of the day, in different contexts, and on many different devices. In addition to that, more complexity is added by the *bidirectional* nature of the communication itself. People today react very quickly to specific topics through various means such as rating, sharing, commenting, tagging, icons, tweeting, etc. Such activities generate additional metadata to the information itself which become part of the original message. When planning proper communication we should consider all this. In such a complicated environment, the likelihood of a message's real meaning being received in a distorted or confused way is very high.

However, as we have seen recently during the Covid-19 pandemic, at times, there is the need to communicate something, somewhat complicated in nature, while we need to make sure citizens fully understand the actual terms and meaning of the communication. This was the case faced by many governments worldwide when informing their population on the rules of conduct during the various lockdown periods.

We analyzed trends and structure of social network data generated as a reaction to those official communications in Italy. Our goal is to derive a model to estimate whether the communication intended by the government was properly understood by the large population. We discovered some regularities in social media generated data related to "poorly" communicated issues.

We believe it is possible to derive a model to measure how well the recipients grasp a specific topic. And this can be used to trigger real-time alerts when the need for clarification arises.

Keywords: Information entropy · Social media data · Big Data analysis · Information structure · Covid-19 pandemic

G. Nicosia et al. (Eds.): LOD 2021, LNCS 13164, pp. 518–530, 2022.
https://doi.org/10.1007/978-3-030-95470-3_39

1 Introduction

Spreading the right information to the right audience has always been the underlying goal of proper communication. However, it is hard to estimate whether the intended addressees of the communication have understood the information correctly, in particular when the communication is irregular, disseminated among different channels and/or complex in nature.

Today, public discourses around current events, especially those that occur online, are increasingly influenced by those issues that news media companies choose to concentrate on [1]. Not only do news media have the power to create public awareness around social issues [2,3], but they can also influence how public perceives the importance of those issues [4,5].

However, today's news site are mostly designed for a *one-to-many mono-directional* style of communication. Even though, on some sites, readers can comment, share, and rate articles, most of the public reactions about specific issues do not take place on the newspaper itself. As a matter of facts, readers who want to express opinions on particular articles tend to transfer that discussion on social media (Twitter, Facebook, etc.) connecting to the originating source through the usage of "hashtags."

Nowadays, social media are by large the most relevant online communication mean designed for networking [6] and interactive communication. When used effectively, their applications can promote dialogue [7], facilitate information transfer and understanding [8], engage stakeholders [9], and improve communication and collaboration in online environments [10]. In contrast to online news sites, social media are designed for a *many-to-many* and *bi-directional* communication among participants.

In many cases, especially when important issues are communicated to a large public audience through online news, it becomes important to make sure that people fully understand the relevant intended facts. In this paper, we combine the dynamics of online news communication with social media to understand whether or not the public *understands* the intended message. We study the *structure* of Twitter data generated as reaction on certain topics to estimate whether the message was properly understood. Basically, depending on how people react on Twitter we estimate whether the public message was understood or needs additional explanation.

In some cases, the originator of the public message may intentionally create social media hype on certain topics as part of a well designed communication strategy. This is particularly true in political scenarios when having people (just) talking about something becomes important, even though they do not fully understand the real meaning of the original message. In this case, the communication to be diffused is properly designed to trigger social media activity.

Here we are not interested in this type of scenario. We focus on those public messages that need to be properly and quickly grasped by the readers. In such cases, it is important for the source to make sure the audience understands what was intended by the communication. Thus, for the source it becomes important to be alerted, as quickly as possible, if this is not the case. And, if necessary, the source may produce additional communication to clarify (e.g., through the usage of Frequently Asked Questions, FAQ.)

From a methodological standpoint, users on social networks often use *hashtags* embedded in their discussion. Thus, we use hashtag related data combined with an unusual usage of the question mark symbols (denoting questions asked) in the tweets to estimate the need for additional explanation about a particular issue [11]. Such analysis are performed within a specific time frame as the readers' interest decay degree over time is also important.

We use an *information entropy based model* on Twitter data to estimate the readers' need for more clarification. As defined by Shannon in 1948, the entropy-rate is the hypothetical average level of "surprise" or "unexpectedness" contained in a message. We measure the entropy of a specific newspaper topic by observing the related data trends on Twitter. A *high entropy topic* is something that, potentially, needs further clarification. Basically, if people are interested in something, and did not yet understand all its terms, they pose lots of questions about it in their tweets. Conversely, "standard" news, such as sport events or flash news, exhibits an interest natural temporal decay that leads to a low entropy information value: they are easy to understand and do not trigger high-entropy reaction on Twitter. Therefore, we use "entropy" to measure the *unbalance* between intended information (by publishers) and perceived information (by readers).

We applied our model to a recent official, sometimes complicated, communication by the Italian government about rules of conduct during the Covid-19 lockdown restrictions. In general, those official communications have been followed by an intense social network activity discussing terms and conditions of those rules. People have been very creative in *interpreting* those terms. The government often had to turn to FAQ or additional official clarification to make sure Italians grasped the right meaning.

2 Data and Methodological Approach

2.1 Information Entropy

In information theory, the *information value* contained in a message is directly related to how "surprising" or "unexpected" the message is for the reader [12,13].

Suppose we have a biased coin with probability p of landing on heads and probability $1-p$ of landing on tails. For what value of p do we have the maximum "surprise" or "uncertainty" on the outcome of a coin toss? If $p = 1$, the outcome of a coin toss is expected to be always heads, so there is no surprise or uncertainty. Similarly for $p = 0$, when we always expect the coin to land on tails. If instead $p = 0.5$, then we have the maximum surprise or uncertainty.

After discussion with John Von Neumann, Shannon decided to use the term "entropy" in place of the word "uncertainty." Claude Shannon mathematically formalized this value of "surprise" or "uncertainty" in 1948 as part of his communication theory.

Formally, the entropy of our biased coin is given by:

$$H(coin) = -(p * log(p) + (1 - p) * log(1 - p)) \tag{1}$$

where the base of the logarithm can be chosen arbitrarily. If the base is 2, the entropy is measured in *bits*. If instead the base is e, the entropy is measured in *nats*. Finally, if the base is 10, the entropy is measured in *dits*.

When $p = 0.5$, $H(coin)$ is maximal, and it is equal to 1 bit. When instead $p = 0$ or $p = 1$, $H(coin)$ is minimal, and it is equal to 0 bits.

The concept of entropy can be generalized from the simplest case of a coin to the more complex case of a discrete probability distribution. A discrete probability distribution over n possible outcomes $x_1, ..., x_n$ is given by n probability values $p(x_1), ..., p(x_n)$, where $0 <= p(x_i) <= 1$, and the sum of all $p(x_i)$ is equal to 1. Note that a coin is a probability distribution over two possible outcomes.

Formally, the entropy of a discrete probability distribution is defined by:

$$H(P) = -(p_1 * log(p_1) + ... + p_n * log(p_n)) \tag{2}$$

where, as before, we can chose arbitrarily the base of the logarithm.

2.2 Entropy and Twitter Trends

Our study models the concept of information entropy based on the tweets distribution over time generated as a reaction to certain official communications that appeared on various media. In general, immediately after certain news becomes public, either on TV or in newspapers, social media activity related to that news has a sudden hype. In our work we concentrate specifically on Twitter activity due to its open approach to share data with researchers. We use *hashtags* to identify the relevant topics on Twitter.[1]

We believe that if a topic is well-communicated, people do not (necessarily) need to comment about it on Twitter or, at least, no more than usual. Conversely, as mentioned above, if the message was unclear, people will look for better understanding by tweeting comments, expressing opinions, and asking questions.

Therefore, depending on the *information value* contained in the official communication, we observe different trends on Twitter. A clearly communicated concept does not lead to a "surprised" or "unexpected" reaction: its *information value* is low, which corresponds to a low entropy level–based on Shannon's entropy concepts. Conversely, a poorly communicated concept leads to high entropy (i.e., more surprising effect) on the reader. A poorly understood message triggers a sudden need on the readers for better understanding. And this triggers different trends on Twitter.

For the sake of our analysis, we distinguish two types of event:

1. Low entropy events (LEE)
2. High entropy events (HEE)

[1] We collected tweets related by using Tweepy and the Twitter archive API, both services need permission from Twitter. However, downloaded topics need further cleaning and normalization before being processed.

LEEs are events that are quickly *grasped* by the public (i.e., properly communicated.) They tend not to generate an intense activity on Twitter on the associated hashtags. We have analyzed many different topics-spawn events on social networks like everyday events (*sports*, *weather*, etc.), periodic events (*mondaysmotivation*, etc.), or simple news. We arbitrarily labelled such events as LEE. In general, some Twitter activity follows those events and usually it decays almost entirely in about 1 day. Information value on such types of events exhibits a low entropy value.

Conversely, HEEs are in general more complicated events. Information dissemination for these events is inherently more difficult compared to LEE. Twitter's reaction on this type of communication follows an entirely different pattern. We selected and studied some issues which have created a significant social reaction in Italy. Following the official communication by the government on public media, Italians started to debate among friends and family, posting on social media, or researching on Google. Either because those events are more difficult in nature and/or they are poorly communicated, people reacted to those by creating lots of social media activity and posing many questions. This denotes an obvious interests for those issues by the public and, we believe, a lack of proper understanding. In any case, it signals a need for better understanding by additional communication.

These events' natural decay shows a complicated and longer curve that grows during the day and slows down at nighttime. But they tend to remain active for at least 60 h. After about 3 days, people tend to stop discussing and commenting on those. During this time, an intense activity takes place on Internet. Entropy measured on such trends is in general significantly higher than LEEs.[2]

HEEs remain rare compared to LEEs. They are the goal of our study as they indicate an *unbalance* between the senders' intended information and receivers' assimilated information. We believe that having an automated way to quickly identify those can surely help governments, institutions, and private companies to create better and more informative communication.

In Fig. 1 we show the different trends for LEE and HEE (Thousands of tweets) starting right after the official communication was released to the public. The HEE curve spawns for a period greater than 24 h so it includes some overnight effect.

2.3 Text Analysis

As discussed above, HEE events exhibit a high entropy value due to their intense activity on Twitter, and this hype of Twitter activity denotes clear interest for those particular issues. In Table 1 we report the entropy value for various events we classified as HEE or LEE. Here you can notice that the two events we classified as HEE exhibit the highest entropy values compared to the others.

[2] For our trend analysis we collected data from https://getdaytrends.com/italy/. GetDayTrends offers both recent and historical data trends in major countries or even cities worldwide.

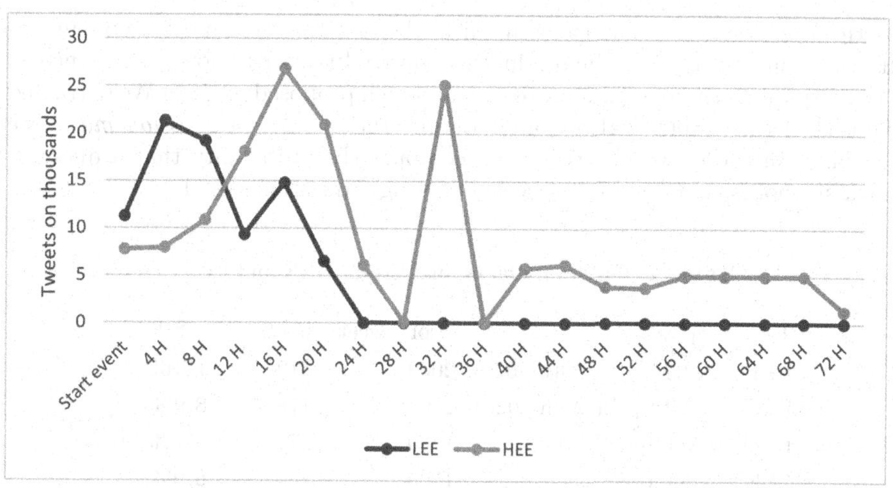

Fig. 1. Low entropy events (LEE) vs High entropy events (HEE) trends. Median for 72 h and relative number of tweets.

Table 1. Information Entropy and average tweets/hours.

Type	Topic	N. of Tweets	Time-span (H)	Tweets/hour	Entropy
HEE	DPCM (Apr 26th 2020)	894.500	78	11.468	**5.6113**
HEE	DPCM (Oct 25th 2020)	618.150	64	9.658	**4.1972**
LEE	SuperLega	45.400	27	1.681	**4.0541**
LEE	G.dellibro	11.600	14	555	**3.8232**
LEE	JuveInter	48.500	16	3.031	**3.6126**
LEE	MilanNapoli	16.800	10	1.680	**3.5672**
LEE	DPCM (Oct 13th 2020)	26.700	12	2.225	**3.1569**
LEE	Dupasquier	<10.000	7	1.428	**3.0201**
LEE	GazaU.Attack	623.800	8	44.557	**2.9740**
LEE	SuperLeague	616.700	48	12.838	**2.5417**

In our study we are mostly interested in topics that are (1) interesting for the people and (2) have not been clearly understood; thus they need more explanation.

However, certain topics may trigger a HEEs type of reaction but not necessarily be poorly understood by the public. Certain controversial issues may be clearly communicated. Thus, they do not need additional explanation, and still generate an intense social reaction, thus we would classify those as HEE. Authors in [18] have studied the dynamics of information diffusion across social networks, showing like a topic with controversial subjects, seem to lead to longer discussions representing personal opinions.

However, we need a way to distinguish when a HEE event needs to be further discussed and clarified by the originating source. In order to do so, we turned to a quite simple technique (similar to the approach proposed in [20].) We extracted the text of the tweets analysed and counted the number of *question marks* we encounter in those tweets. We simply compute the probability that a question mark symbol is contained in a tweet, denoting a question posed.

Table 2. Question marks analysis on HEE and LEE.

Type	Topics	N of Tweets	N of "?"	"?" Prob
HEE	DPCM (Apr 26th 2020)	2000	312	**15.6%**
HEE	DPCM (Oct 25th 2020)	2000	178	**8.9%**
LEE	JuveInter	1000	71	**7.1%**
LEE	G.dellibro	1000	69	**6.9%**
LEE	SuperLega	1000	68	**6.8%**
LEE	MilanNapoli	1000	61	**6.1%**
LEE	DPCM (Oct 13th 2020)	1000	60	**6%**
LEE	SuperLeague	1000	55	**5.5%**
LEE	MothersDay	1000	31	**3.1%**
LEE	GazaU.Attack	1000	14	**1.4%**
LEE	Dupasquier	1000	14	**1.4%**

In Table 2 we show the results of this analysis for the trends above discussed.[3]

Now, by looking at both tables, it is interesting to notice that the two events classified as HEE show the highest values for both Entropy (see Table 1) and ?-probability (see Table 2). And this supports our theory that by simply observing those two metrics we may easily spot complex events that need more explanation. Basically, according to our model, if *both* the entropy value and the ?-probability are high the topic was not properly understood and it needs additional explanation.

However, it is interesting to notice that the big event "SuperLeague" spans for a total of 48 h which could make us think of a complex event; however its entropy is very low and its ?-probability is also on the lower side. Similarly, the event "GazaU.Attack" generated an extremely high number of tweets in a short period of time (just 8 h) with both a very low value of entropy and ?-probability. According to our model, people have grasped that concept pretty well and they do not need additional explanation, even though the reaction on Twitter was pretty intense.

[3] We used Tweepy to extract text from Twitter. For each hashtag we extracted a certain number of tweets. Such number is limited by the Twitter API itself, so we computed the probability on a smaller number of tweets compared to those used for trend analysis.

3 Case Study Discussion: Information on the Restrictions Procedure During Covid-19 Pandemic in Italy

On April 26th, 2020, after seven weeks of severe Covid-19 related lockdown, Italians were anxiously waiting for government news about legal procedures to finally meet parents, relatives, friends, and significant ones.

Already many days before the government's official announcement, this topic was largely speculated and discussed in newspapers and social media. That was the first one during the pandemic, and many others took place in the following months during the Covid- 19 epidemic.

Typically, these official announcements, which described the emergency decrees (formally called DPCM: Decree of the President of the Council of Ministers) took place on national TV around 9 PM in Italy after the evening news. In each of those, the Italian prime minister instructed Italians on various social restrictions to which they had to obey. Those DPCMs were presented around the same time of the day, and this simplified our study as we did not need to normalize for day/night impact on Twitter data generation.

We used web data on these events as a real case scenario for our study. We collected API data from Twitter related to those decrees, and we started at the same time the prime minister was presenting the DPCMs on TV. And we kept on collecting those for many hours/days after the event. We were interested in analysing the information dissemination quality about the post-pandemic lockdown release procedures. That is, how well Italians understood the prime minister's official announcements details.

We noticed that Italian people had poorly understood some terms and concepts and, as a matter of fact, there was an intense social media activity on those terms following the official announcement by the Italian prime minister. Most comments about those topics were ironic and doubtful, caused by sudden and unexpected and, with some regards, unclear information. People were just confused about some term's definitions and, consequently, they looked around for a better explanation on important information affecting their expected behavior. Unsurprisingly, right after the TV official announcement, there was a sudden intense social media activity on this topic.

We collected data on three DPCM events to measure how well Italians understood the terms of the decree. We matched the Twitter trend data against our LEE/HEE model discussed above. We observed a LEE trend in one case, suggesting that Italian people easily understood information about that decree. In two cases, we noticed that the DPCM related Twitter data followed an HEE trend, denoting confusion of Italian people in understanding some terms of the decree itself. We measured entropy levels for some terms used in the official announcements.

During the first DPCM on April 26th 2020, we started collecting Twitter data about people discussing it (see Fig. 2.) Italians did not understand the decree well. In particular, the term generating great confusion was "#Congiunti" ("joint people" in English.) This archaic word, present in the that DPCM in force since

May 4th 2020, had an extremely important role to understand whom Italians were allowed to visit during the restriction period. This has triggered a myriad of reactions. Doubts and perplexities have been raised by politicians and jurists, and associations that protect LGBT rights because of fears about the possible discrimination that this term may introduce. Italians were primarily confused about this word, and its meaning was essential to understand the restrictions properly. Also, because many wonder how to demonstrate a "stable affection". In Italy, "Congiunti" does not have a legal definition, so it was interpretable in many ways. Can an unmarried couple be considered "Congiunti?" What about family members not living together? What about second/third cousins? What about son/daughter living in separate cities. What about a couple of the same sex (formally cannot be married in Italy.) And the list goes on. The government promised clarification in the following days.

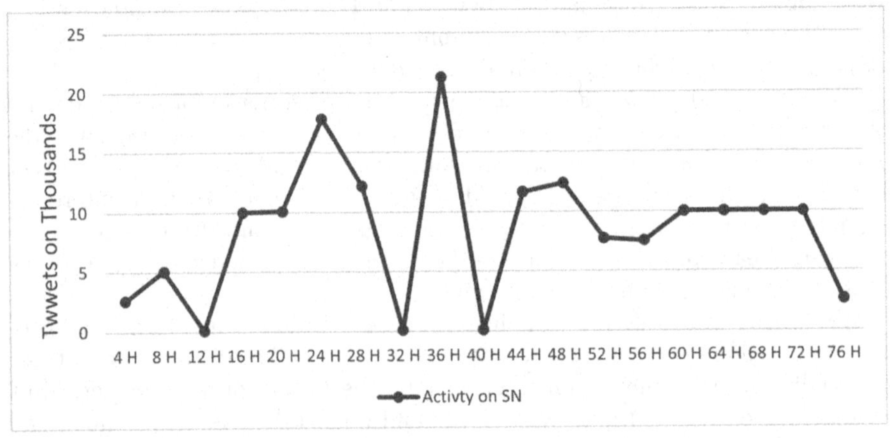

Fig. 2. Twitter trend data following the DPCM on 26th April 2020

As an additional verification, we measured Google trends data on the word "Congiunti" in Italy immediately following the TV's official announcement. This is shown in Fig. 3, which clearly states an interest in that word in Italy. Five days after the official announcement, the Italian government published some FAQs to clarify the meaning of the term "Congiunti."

As already mentioned, the "Congiunti" case of April 2020 was not the only one we observed and measured. At the end of October 2020, during the second wave of the Covid-19 pandemic, the Italian government issued another DPCM (October 25th 2020) with new rules and restrictions for fighting the pandemic's second wave. We again used web data on this event against our model. This new DPCM was defining a three-tier system based on a color code according to the intensity of the epidemic danger for the different areas in Italy. Each of the 20

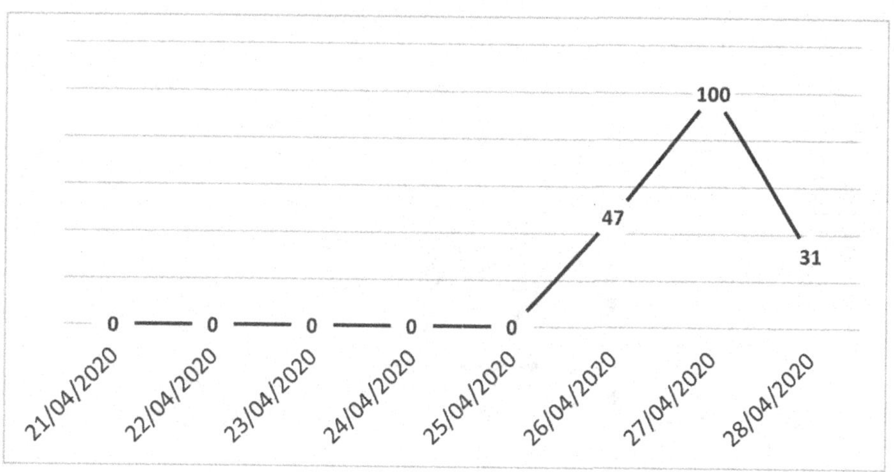

Fig. 3. Word "Congiunti" Google search trends in Italy

Italian regions was assigned a specific color: Red (high-risk), Orange (medium risk), and Yellow (low risk). Each color defines a specific set of restrictions. Here again, we started collecting Twitter trend data while the prime minister presented the new procedures on TV. Even in this case, Italians were confused, and the primary source of confusion was how to interpret the different restrictions for the different colors. Once again, the trend shows a high entropy depicting confusion about proper understanding the terms of the communication. Even in this case, the Italian government and the press, in general, made those terms clearer by further announcements in the following days.

For a side-by-side comparison and to reinforce our model, we also collected data about a third DPCM, presented earlier by the Prime Minister on October 13th, 2020. Once again it was describing a new set of restrictions in force during the second wave of the pandemic that started right after the Summer. This time the communication was very clear, Italians correctly understood its contents, and we observed a LEE type of trends for its related Twitter data.

We show the trends of the three DPCMs discussed in Fig. 4. Here we can easily spot the different curves, which again describe two different scenarios. The LEE trend (Oct 13th) indicates straightforward and easy-to-understand communication from the Prime Minister. In contrast, the HEE type denotes a less clear communication, which triggered a sudden need for further explanation on Twitter.

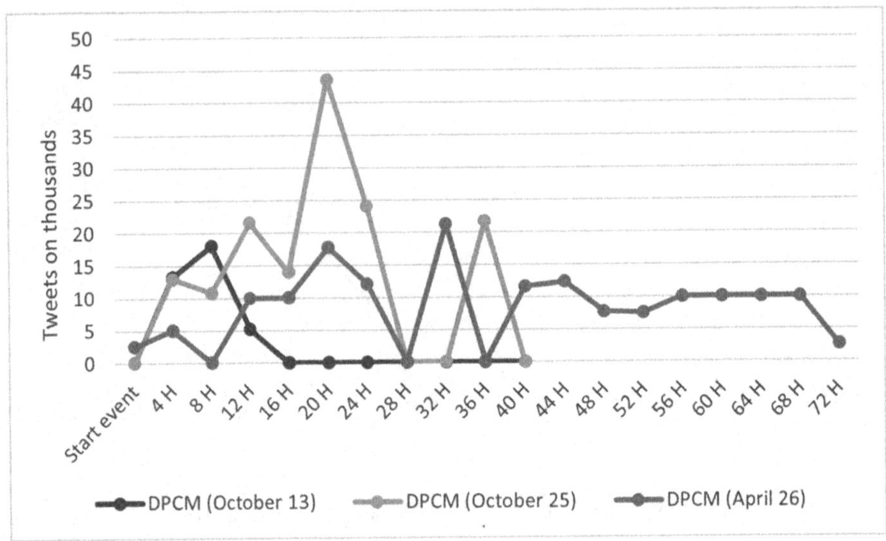

Fig. 4. Twitter data trend comparison for the three DPCMs discussed.

Thus, by real-time measuring information entropy on social media, we could promptly comprehend how easy to understand a particular communication was. And we believe this may have various practical applications in real life.

4 Conclusions

Thanks to the rapid development of technology, online communication channels are becoming increasingly popular platforms for communicating information of various kinds [14]. Millions of people get informed on online news and interact with each other on social media: Blogs, social networks, and online newspapers are now social relationship tools [15]. The usage of Social media to communicate prompt information during specific situations is now a common practice. In particular, the one-to-many nature of Twitter created an opening for governments to disseminate relevant, and often critical, messages [16].

When news break—as in our case study—people turn quickly to Internet to get informed. As we have observed, the proper management of information communicated through digital platforms is essential to prevent misunderstandings and confusion about the message. The more controversial a topic or message we intend to communicate is, the more articulated and rich the reaction on social media is. This becomes more relevant when, as in the case study presented, politicians and decision-makers communicate in times of crisis.

Crises are socially constructed phenomena, and Governments tried to shape the attitudes and behaviors of citizens following their policies. Maintaining the public's attention and preserving government credibility proved a hard challenge for many government leaders during the COVID-19 pandemic [21]. In times of crisis, political leaders have to communicate clearly about what is going on, what is at stake, what governments are doing in response, what people can do to keep themselves and others safe [22].

The model we presented in this paper can help in this context and, in general, in all those situations where it is essential to obtain real-time feedback on the quality of communication.

4.1 Future Work

In our study, the data collection process has been very labor intensive. It is quite challenging to coordinate hashtags between newspapers and Twitter, we mostly did that manually. Moreover, tools to collect trends and extract text from Twitter have various limitations which impose certain time-consuming manual workarounds. We intend to improve our model by automating as much as possible the data gathering and synchronization process. This will allow us to test our model on a much larger data sets.

We also intend to improve the performances of our model by collecting data from other sources such as different social media and online newspapers. We will also integrate with Natural Language Processing tools for a more accurate content analysis of people's messages and comments.

References

1. Marcus, M.: Mediatization and the technologization of discourse: exploring official discourse on the Internet and information and communications technology within the Evangelical Lutheran Church of Finland. New Media Soc. **20**(2), 515–531 (2018)
2. Lei, G., McCombs, M.: Toward the third level of agenda setting theory: a network agenda setting model. Annual convention of the Association for Education in Journalism & Mass Communication. St. Louis, Missouri (2011)
3. Leticia, B.: Political news in the news feed: learning politics from social media. Mass Commun. Soc. **19**(1), 24–48 (2016)
4. Russell Neuman, W., et al.: The dynamics of public attention: agenda-setting theory meets big data. J. Commun. **64**(2), 193–214 (2014)
5. Feezell Jessica, T.: Agenda setting through social media: the importance of incidental news exposure and social filtering in the digital era. Polit. Res. Q. **71**(2), 482–494 (2018)
6. Kaplan, A.M., Haenlein, M.: Users of the world, unite! The challenges and opportunities of Social Media. Busi. Horiz. **53**(1), 59–68 (2010)
7. Cristina, C., Huang, L.: Social media in an alternative marketing communication model. J. Market. Development Compet. **6**(1), 117–134 (2012)
8. Lei, H.: Social contagion effects in experiential information exchange on bulletin board systems. J. Mark. Manag. **26**(3–4), 197–212 (2010)
9. Robert, N., Dale, A.: Meeting the climate change challenge (MC3): the role of the internet in climate change research dissemination and knowledge mobilization. Environ. Commun. **9**(2), 208–227 (2015)

10. Glen, M., Salomone, S.: Using social media to facilitate knowledge transfer in complex engineering environments: a primer for educators. Eur. J. Eng. Educ. **38**(1), 70–84 (2013)
11. Mazzeo Rinaldi, F., Russo, A., Giuffrida, G.: Information balance between newspapers and social networks. iN: CARMA 2020: 3rd International Conference on Advanced Research Methods and Analytics (2020)
12. Shannon Claude Elwood: A mathematical theory of communication. ACM SIGMOBILE Mobile Comput. Commun. Rev. **5**(1), 3–55 (2001)
13. Shannon, C.E.: A mathematical theory of communication. Board of Trustees of the University of Illinois (1949)
14. Lei, H., et al.: The communication role of social media in social marketing: a study of the community sustainability knowledge dissemination on LinkedIn and Twitter. J. Market. Analyt. **7**(2), 64–75 (2019)
15. Bennato, D.: Il computer come macroscopio. Big data e approccio computazione per comprendere i (2015)
16. Olteanu, A., Vieweg, S., Castillo, C.: What to expect when the unexpected happens: Social media communications across crises. In: Proceedings of the 18th ACM Conference on Computer Supported Cooperative Work & Social Computing (2015)
17. Olga, B., et al.: Entropy-based approach for the detection of changes in Arabic newspapers' content. Entropy **22**(4), 441 (2020)
18. Minkyoung, K., Newth, D., Christen, P.: Modeling dynamics of diffusion across heterogeneous social networks: news diffusion in social media. Entropy **15**(10), 4215–4242 (2013)
19. Mingsheng, T., Mao, X.: Information entropy-based metrics for measuring emergences in artificial societies. Entropy **16**(8), 4583–4602 (2014)
20. Mazzeo, V., Rapisarda, A., Giuffrida, G.: Detection of fake news on CoViD-19 on web search engines. arXiv preprint arXiv:2103.11804 (2021)
21. Boin A., McConnell A., Hart P.: Crafting crisis narratives. In: Governing the Pandemic. Palgrave Pivot, Cham (2021). https://doi.org/10.1007/978-3-030-72680-5-4
22. Kim, D., Kreps, G.: An analysis of government communication in the United States during the COVID-19 pandemic: recommendations for effective government health risk communication. World Med. Health Policy **12**(4), 398–412 (2020)

Multivariate LSTM for Stock Market Volatility Prediction

Osama Assaf[1(✉)], Giuseppe Di Fatta[1], and Giuseppe Nicosia[2]

[1] Department of Computer Science, University of Reading, Reading, UK
o.assaf@pgr.reading.ac.uk

[2] Department of Biomedical and Biotechnological Sciences – School of Medicine, University of Catania, Catania, Italy

Abstract. Volatility is a measure of fluctuation in financial asset returns, practical measurement of risk, and a key variable for calculating options prices. Accurate prediction of volatility is crucial to maintaining profitable investments and trading strategies. Statistical models such as GARCH are used today to predict volatility and time series, though new methods are actively being researched to improve the prediction accuracy to cope with the rapidly increasing trading volumes and stock market influencing factors. The aim of this paper is to investigate a new method to improve market volatility forecasting accuracy by innovatively introducing a new setup of the Recurrent Neural Network (RNN) algorithm. In particular, the proposed model is a stacked Long Short-Term Memory (LSTM) with multivariate input composed of multiple asset daily prices of different lag time-steps. The proposed model is used to predict volatility under different market conditions and is compared to the predictions obtained with GARCH as well as to the actual volatility of the same forecasting period. The results show that the prediction of the future realized volatility using a single feature LSTM has comparable accuracy to GARCH. They also indicate that a stacked LSTM can significantly improve the volatility prediction accuracy when configured with multivariate input of more than one asset and a lagging period of more than a day. A stacked multivariate LSTM setup enables the prediction model to capture complex patterns in the time series data of assets prices and provides a superior alternative to statistical models in volatility modelling and prediction. The proposed multivariate LSTM architecture clearly shows faster and more accurate modelling of daily volatility and therefore can be used for intra-day modelling specifically for high frequency trading environments.

Keywords: Deep Learning · Recurrent Neural Network · Long Short-Term Memory · Financial markets · Volatility · Day trading

1 Introduction

Risk management is key for any successful investment and trading strategy. There are different types of risks that can adversely impact an investment or a

G. Nicosia et al. (Eds.): LOD 2021, LNCS 13164, pp. 531–544, 2022.
https://doi.org/10.1007/978-3-030-95470-3_40

trading portfolio which require the adoption of different methodologies to mitigate and hedge against. Figure 1 illustrates typical probabilities that can lead to returns losses.

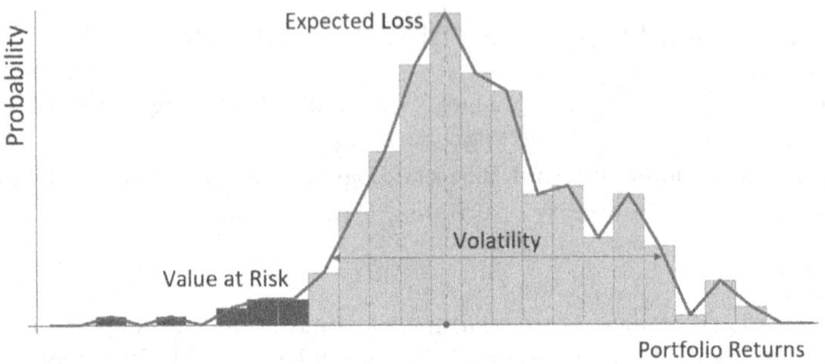

Fig. 1. Portfolio loss probability. JP Morgan daily returns, 2017

Volatility is a measure of fluctuation in a security or a market index price, a fundamental measure of risk, and a practical indicator for managing uncertainty in the financial markets. There are different types of volatility, such as the volatility of a security compared to the benchmark index (Beta), the volatility of a security calculated from historical prices (Historical Volatility), and the volatility forecast over the life of an option (Implied Volatility) [1], which is used in the Black-Scholes formula for options pricing [2]. Forecasting volatility has been researched by practitioners and academics for decades. Better prediction accuracy of market volatility leads to improved management of risk and pricing models, enabling profit-maximizing trading and investment strategies. Nowadays, financial applications typically use statistical models such as GARCH [3] to forecast volatility and price movements in the stock market. Trading volumes and market influencing factors have increased significantly in the recent years due to faster Internet technologies and mobile cellular networks, alternative methods to model volatility are desirable to improve processing speed and accuracy over vast volumes of complex data. Artificial Neural Networks (ANN) and Deep Learning have recently been given particular attention due to algorithmic improvements, widespread availability of very large data sets and the advancements in the computing hardware (e.g., TensorFlow cores) required to process large amounts of data with this specific type of algorithms. This research examines the use of a special type of Recurrent Neural Networks (RNN) to forecast volatility. RNN are feed-forward artificial neural networks that make use of loops allowing information to persist: they provide a memory mechanism to keep track of the information observed over time, hence, making them very good in predicting sequences in time series data [4]. In particular, this works adopts the Long Short-Term Memory (LSTM) networks [5], which are a very effective type of RNN.

This paper verifies two hypotheses: (1) market volatility prediction using LSTM is equivalent to statistical models such as GARCH; (2) stacked LSTM with multivariate input composed of multiple stock prices improves prediction accuracy of future realized volatility. Several tests were performed using a tool built in Python and TensorFlow libraries to verify the efficiency of the model. Results observed confirms both hypotheses and show very strong evidence that stacked multivariate LSTM are a strong alternative to statistical methods used to predict volatility.

The rest of this paper is organized as follows. In Sect. 2 the concept of volatility in the stock market analysis and prediction is introduced and different definitions are provided. In Sect. 3 some related work is discussed. Section 4 provides a brief description of the LSTM model adopted in this work. Section 5 describes the methodology used for the experimental analysis of the proposed approach and the results. Finally, Sect. 6 provides some conclusive remarks.

2 Volatility

Volatility is a key measure for trading and risk management. There are different methods used to calculate or derive volatility. *Historical Volatility*[1] uses the time series of historical prices of the securities within a portfolio to calculate returns. Return for each asset is calculated independently. The average return R_{avg} is defined as:

$$R_{avg} = \frac{\sum_{i=1}^{n} R_i}{n} \tag{1}$$

where and n is the sample size of the time-series and R_i is the security return. The volatility σ_{Sec} of a single asset and the *Annualized Volatility* σ_{An} are given by the following formulas.

$$\sigma_{Sec} = \sqrt{\frac{\sum_{i=1}^{n} (R_i - R_{avg})^2}{n-1}} \tag{2}$$

$$\sigma_{An} = \sigma_{Sec}\sqrt{260} \tag{3}$$

For a portfolio with multiple assets, the correlation coefficient and weights are required to measure volatility. For example, the volatility σ_p of a two-asset portfolio is calculate by

$$\sigma_p = \sqrt{W_1^2\sigma_1^2 + W_2^2\sigma_2^2 + 2W_1W_2\sigma_1\sigma_2\rho_{1,2}} \tag{4}$$

where $\mathbf{W}$ is the proportion of each asset in relation to the whole portfolio, σ_1, σ_2 and $\rho_{1,2}$ are, respectively, the volatility of the portfolio asset 1, the volatility of the asset 2 and the correlation coefficient of assets 1 and 2.

[1] Parametric Value at Risk (VaR) is calculated using the same method of *Historical Volatility*.

If the portfolio includes more than two assets, the volatility can be represented by the general formula below.

$$\sigma_p^2 = [W_1 \cdots W_n] \begin{bmatrix} \sigma_{11} & \cdots & \sigma_{1n} \\ \vdots & \ddots & \vdots \\ \sigma_{n1} & \cdots & \sigma_{nn} \end{bmatrix} \begin{bmatrix} W_1 \\ \vdots \\ W_n \end{bmatrix} \tag{5}$$

Implied Volatility is derived from the Black-Scholes formula [3] below by substituting the call or option value using the actual market price of the derivative. This formula is essential to options traders as it can be used to evaluate options and to derive *Implied Volatility*:

$$d_1 = \frac{\ln \frac{S_0}{K} + \left(r + \frac{\sigma^2}{2} T\right)}{\sigma \sqrt{T}} \tag{6}$$

$$d_2 = d_1 - \sigma\sqrt{T}$$

$$c = S_0 N(d1) - K \exp -rTN(d_2)$$

$$p = K \exp -rTN(-d_2) - S_0 N(-d_1)$$

where S_0 is the security price, K the strike price, r the interest free rate, T the time to expiration in years, σ the volatility, $N()$ the normal distribution, c the call option price, and p is the put option price.

In this paper, Eqs. 1, 2 and 3 were used to calculate the volatility of one stock portfolio. Future work will include Eqs. 4 and 5 for multiple stocks portfolio and Eq. 6 to estimate option prices by predicting implied volatility of underlying stocks.

3 Related Work

Different methodologies are used today to model market movement and volatility. In the recent years, and due to the surge of high volumes of data and the fear effect caused by the financial crisis of 2008, demand to review and improve existing pricing models and risk management has never been so high. In many studies, ANN have been compared against statistical models used for volatility forecasting like GARCH and it has been shown that they can be more effective and accurate. Due to their high non-linear, large scale, continuous, time-based and dynamic nature [6], ANN are among the models often chosen for time series prediction and have been shown to be very promising in this particular application domain. The authors in [7] reviewed and highlighted the potential of ANN computation methods in modelling time-varying data, showing that ANN are embedded in more traditional time-series theory and have a potential to provide powerful alternatives to traditional models, especially with respect to non-linearity. The work in [8] confirmed time-series capabilities of ANN and highlighted the importance of the sampling window size; from the experiments reported, it was concluded that optimal performance is clearly obtained at the correct embedding dimension and variation either side of this window size diminishes performance. In [9] the authors conducted an experiment to predict the Japanese Nikkei 255 index using the classical Back Propagation Neural Network (BPNN) against one supporting Genetic Algorithm (GA) and Simulated Annealing (SA) and concluded that classical BPNN would gain more accuracy and speed when combined with one of the global search techniques. Other studies also showed promising stock price prediction results for Deep Learning [10]. In this study [11] the authors used a Wavelet De-noising-based Back Propagation (WDBP) neural network. In this model, the original data are first decomposed into multiple layers by the wavelet transform.

The work in [12] used LSTM with 23 years of SP500 daily index prices and 240 days sequences to predict the market movement. Results showed that LSTM provided better prediction capabilities compared to random forest and study concluded that LST can be used to construct profitable trading strategies. Cao and Li [13], combined a multi-layer LSTM with Empirical Mode Decomposition (EMD) to improve time series prediction. The resulting model provided better forecasting capabilities compared to Support Vector Machine (SVM). The work in [14], proposed an LSTM model to predict SP500 with over-fitting prevention capabilities. The results suggest excellent forecasting accuracy. In [15], the authors introduced Deep LSTM (DLSTM), an architecture of LSTM to be used for modelling time series effectively. This model was tested against ARIMA and the results confirmed the superiority of the model. The paper also examines different lagging configuration; however, no evidence suggests an optimal setup. The work in [16] and [17] used stacked LSTM architecture to model time series data, while the earlier confirms great improvements in prediction performance compared to single LSTM model, the later confirms the same and suggests more accuracy than traditional statistical methods in modelling time series data. The approach in [18] examined combining traditional ANN with LSTM and bidirectional LSTM (BLSTM) to improve time series forecasting accuracy. The results suggest that BLSTM works better for three-week ahead volatility forecast while LSTM works better with shorter time horizons. The work in [19] combined LSTM with traditional ANN to improve gold volatility forecasting. The experiments conducted with different lagging setup to allow the model to learn from temporal structures. The hybrid model suggests that better accuracy can be achieved when compared to classic GARCH and LSTM.

4 Long Short-Term Memory Networks

Figure 2 illustrates the structure of an LSTM network, where X indicates the input, Y the output, and C the cell state.

The cell state in LSTM is controlled by gates that performs multiplication and summation operations on top the sigmoid output. These operations control what information to keep and what to purge, hence, making quite effective in predicting sequenced data. Figure 3 illustrates LSTM cell structure.

In particular, Fig. 4 illustrates the stacked multivariate LSTM architecture that is adopted in this work. The setup is similar to the models used in [16] and [17] which is optimised for time series modelling.

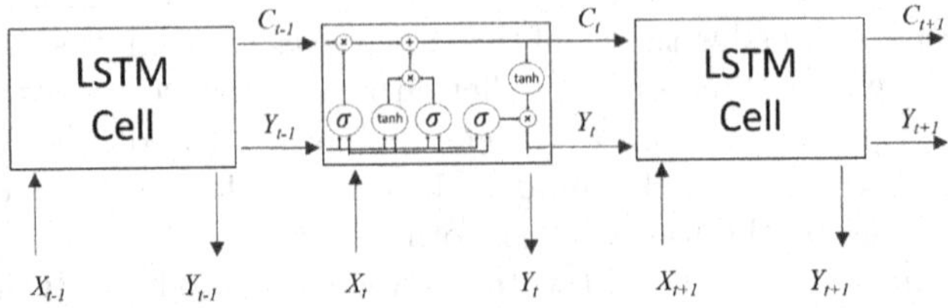

Fig. 2. Long Short-Term Memory RNN architecture.

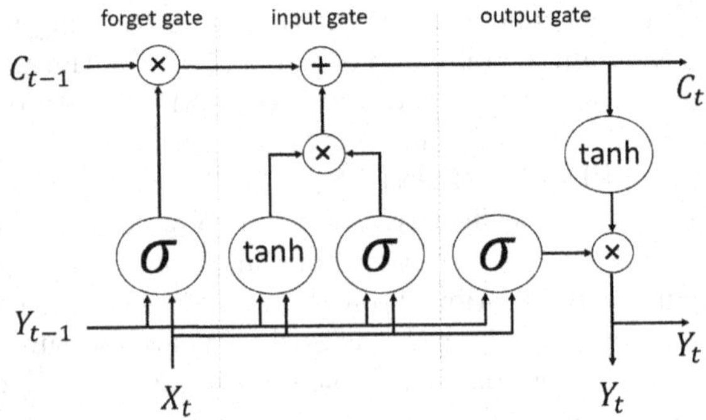

Fig. 3. Long Short-Term Memory RNN cell architecture.

5 Methodology and Experimental Analysis

The stacked LSTM architecture discussed in 4 has been implemented and used to test different combinations of LSTM networks using a TensorFlow GPU. The application architecture and the work-flow are illustrated in Fig. 5. 'Test GARCH' option is added to switch between benchmark and normal testing. The test server and all the computational resources used to perform the experiments are listed in Tables 1 and 2.

Two sets of experiments were carried out. The first set (Test 1) is aimed at testing the hypothesis that a stacked LSTM provides equivalent volatility prediction accuracy to statistical methods such as GARCH. The second (Test 2) is aimed at testing the hypothesis that multivariate input stacked LSTM composed of multiple assets prices can further improve the volatility prediction accuracy. To validate the model efficiency, each set of experiments predicted volatility in two market conditions, (1) declining prices (Bear) for which the year 2009 was used, and (2) rising prices (Bull) for which the year 2017 was

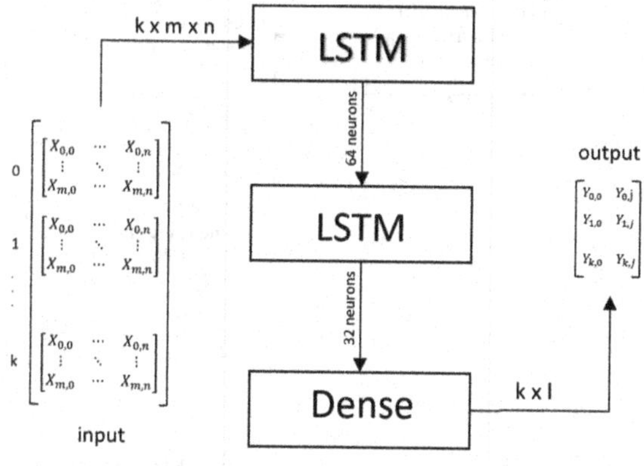

Fig. 4. Stacked LSTM model.

Table 1. Test environment - server specifications.

CPU	Intel Xeon CPU E5-2640 @3.0GHz
Memory	32 GB
Environment	UBUNTU 18.04
Language	Python 3.7
ML Framework	TensorFlow 2.0

used. The daily stock price datasets used in the experiments were obtained from Yahoo Finance and are described in Tables 3 and 4.

The adopted LSTM model with multivariate input used in all the experiments is implemented in Python 3.7 and follows the stacked model illustrated in Fig. 4, where k represents the sample size, m is the number of feature assets used to train the model, and n is the lag parameter which represents the previous number of days used for every prediction. The GARCH statistical model used for benchmark testing is using ARCH libraries within Python 3.7 and executed by the same testing application. The historical stock prices of the test asset are used as input for the model to predict the volatility.

Table 2. Test environment - GPU specifications.

GPU	GeForce RTX 2060 SUPER
Memory	8 GB
CUDA Cores	2176
GPU Clock	1650 MHz

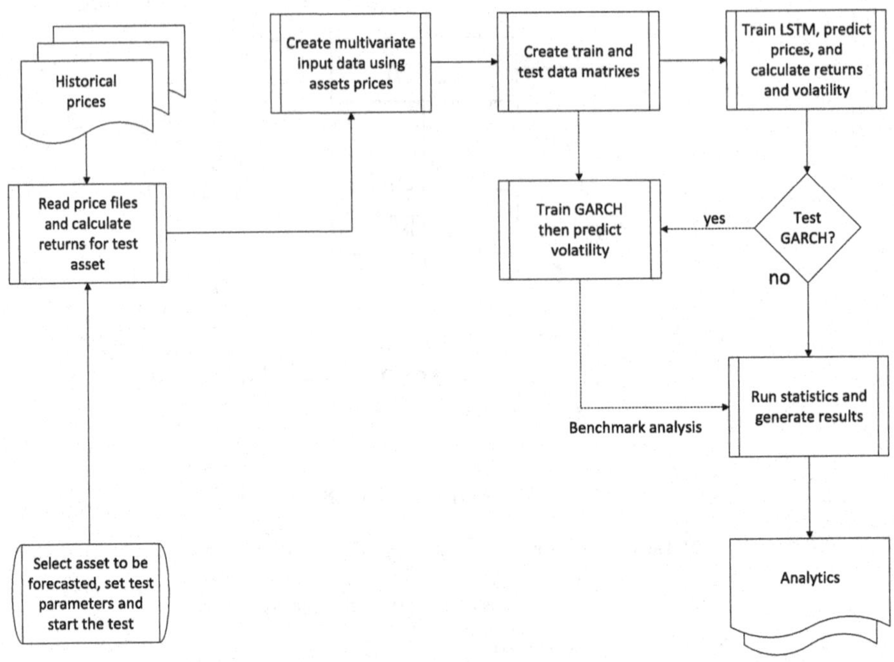

Fig. 5. The general data workflow.

The hold-out performance estimation method was adopted in all the experiments. Historical stock prices were used to train the LSTM network and GARCH. The models then were used to predict future realized volatility for multiple consequent days, unlike previous approaches that make a single prediction for the subsequent day only. The average volatility over the prediction period was compared against the actual one for the same time frame.

5.1 Test 1 - Benchmark Testing

GARCH(1,1) and a stacked LSTM on a single asset (no auxiliary assets) were used to predict future realized volatility for Bank of America stock (BAC) using

Table 3. Test data definition - time series.

Market conditions	Prediction	Training data
Bear	1,5,10,20 future days	03/01/2000–03/03/2009
Bull	1,5,10,20 future days	03/01/2000–01/10/2017

Table 4. Test data definition - stocks.

Symbol	Company	Sector
BAC	Bank of America	Financial services
JPM	JP Morgan Chase	Financial services
WFC	Wells Fargo	Financial services
C	City bank	Financial services

a lag of 1 previous day time-step, an optimal setup for GARCH. The test parameters used for LSTM are illustrated in Table 5.

Table 5. Test 1 - LSTM model parameters.

Experiment	Epochs	Neurons	Batch size	Prev. Days	Auxiliary assets
Benchmark test	10	64	500	1	0

The results in Tables 6 and 7 provide the Mean Squared Error (MSE) of the estimated volatility, respectively, for LSTM and GARCH against the actual market volatility for bear and bull markets. It is observed that LSTM is faster (about 88% faster for bull market) and also offers better prediction for single day volatility forecasting. However, GARCH prediction accuracy is superior for more than one day forecasting.

Table 6. Forecast accuracy of BAC daily volatility in a bear market.

Prediction days	LSTM		GARCH	
	MSE	Time (s)	MSE	Time (s)
1	3.29	26	6.01	92
5	9.18	26	5.42	92
10	7.73	27	5.83	94
20	10.89	27	7.70	98

Plotting bear and bull 1-day actual volatility against predicted for both models illustrates the findings. Figures 6(a) and 6(b) illustrates that the error variance follows an auto-regressive moving average process, which is one of the characteristics of the GARCH model that explains the better accuracy for forecasting of more than one day.

On the other hand, LSTM illustrated in Figs. 7(a) and 7(b) clearly indicates superior modelling of daily volatility and therefore can be used for intra-day modelling for volatility, specifically in high frequency trading environments.

Table 7. Forecast accuracy of BAC daily volatility in a bull market.

Prediction days	LSTM		GARCH	
	MSE	Time (s)	MSE	Time (s)
1	0.69	27	0.43	220
5	0.93	27	0.46	213
10	0.94	27	0.64	216
20	0.93	27	0.71	227

5.2 Test 2 - Multivariate Input LSTM Testing

This experiment is aimed at testing the multivariate LSTM model. Table 8 illustrates the testing parameters.

Table 8. Test 2 - LSTM model parameters.

Experiment	Epochs	Neurons	Batch Size	Prev Days	Auxiliary Assets
Multiple Assets	10	64	500, 2000	5	0 to 3

Table 9. LSTM forecast accuracy of BAC daily volatility in a bear market.

No. Features	Future prediction (Days)	MSE	Time (s)
3	1	2.55	67
3	5	6.17	65
3	10	13.21	66
3	20	18.01	71
2	1	2.56	69
2	5	8.11	65
2	10	13.31	66
2	20	15.96	69
1	1	3.98	76
1	5	5.79	66
1	10	12.62	67
1	20	45.38	68

The test is repeated multiple times to predict future realised volatility using a lag of 5 days and different combinations of positively correlated stock prices for multivariate input. Tables 9 and 10 illustrates the test results collected.

Table 10. LSTM forecast accuracy of BAC daily volatility in a bull market.

No. Features	Future prediction (Day)	MSE	Time (s)
3	1	0.47	76
3	5	0.64	76
3	10	0.70	78
3	20	0.89	77
2	1	0.22	72
2	5	0.53	70
2	10	0.73	73
2	20	1.14	72
1	1	0.36	71
1	5	0.55	68
1	10	0.76	69
1	20	1.01	69

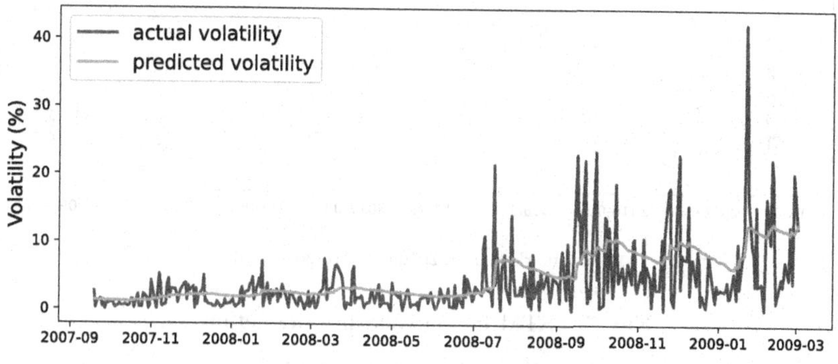

(a) bear market: from 2007-09-19 to 2009-03-02

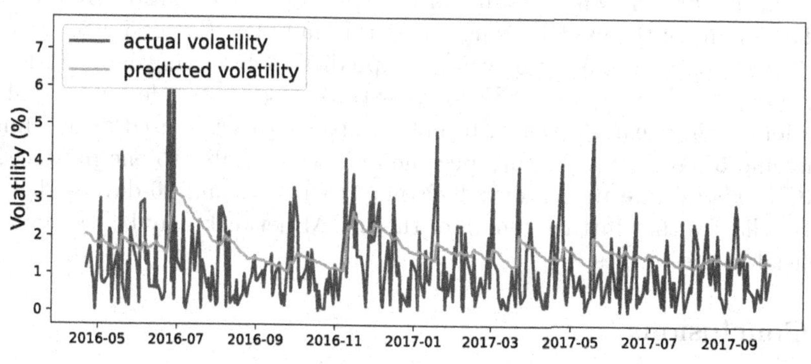

(b) bull market: from 2016-04-22 to 2017-10-02

Fig. 6. GARCH model for BAC volatility prediction

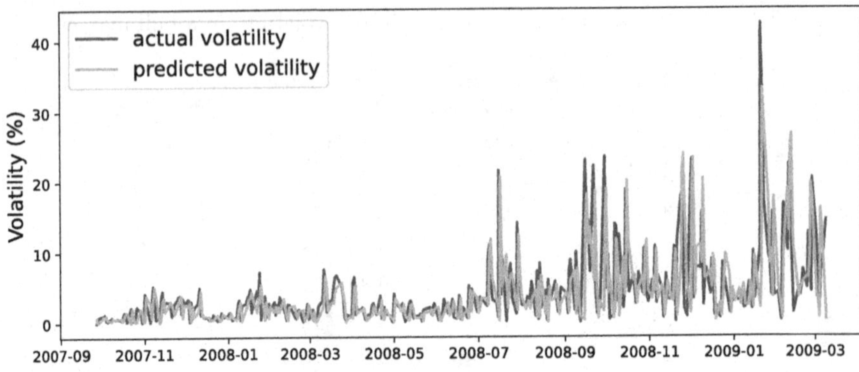

(a) bear market: from 2007-09-19 to 2009-03-02

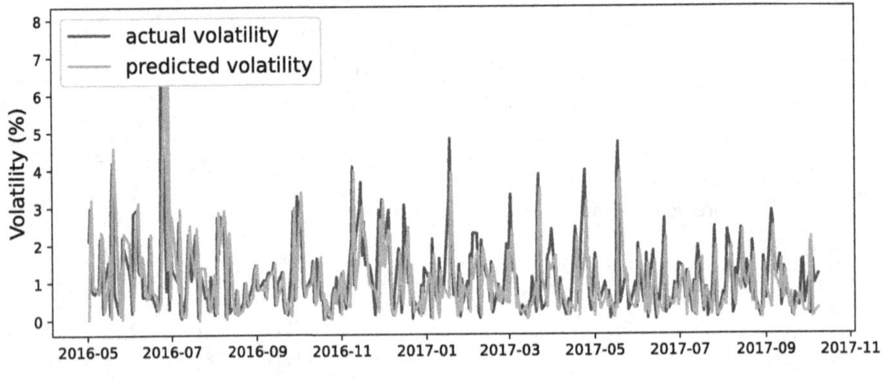

(b) bull market: from 2016-04-22 to 2017-10-02

Fig. 7. LSTM model volatility prediction

It is observed that LSTM prediction accuracy for 1 day and 5 days has improved significantly when features with high positive correlation are used during the training of the model. Using previous data points (lag of 5 days) has also contributed to the accuracy of volatility prediction by giving more insight of pattern changes. The models MSE drops with the use of more features. MSE for 1-day forecasting nearly dropped to half compared to setup used in benchmark testing and 5-day forecasting has become more comparable to one produced by GARCH. There were no significant changes to 10-day and 20-day predictions, which indicates that further tuning of the LSTM network may be required and beneficial in these cases.

6 Conclusions

The first objectives of this work is to answer the question on the possibility of using LSTM as an alternative method to statistical models such as GARCH to predict market volatility and build more profitable trading portfolios and options pricing models. The second objective is to propose and investigate a multivariate LSTM architecture to improve volatility forecasting accuracy.

The benchmark testing has confirmed the first hypothesis and clearly indicates that LSTM 1-Day forecasting of volatility is superior to GARCH. This also shows strong evidence that LSTM is a suitable method to intra-day volatility modelling due to its strength in detecting short-term market influencing factors.

The second set of tests indicates that a multivariate LSTM can significantly improve the forecasting of the volatility in both bear and bull markets. Both 1-day and 5-day predictions were improved when more assets prices were used in the multivariate input, which confirms the second hypothesis. It was also observed that a lag window is required to improve prediction accuracy of LSTM, especially during bear market condition, this helps training the model to adapt to drastic changes in the market and accommodate missing data points if incurred.

Further work is required to further improve prediction accuracy of LSTM and to lengthen the prediction horizon. This will involve changes to the multivariate input such as configuring different lagging periods and introducing new classes of time series input data such as news and weather. Future research will also be dedicated to use the proposed LSTM model to predicts implied volatility, which is a key parameter in options pricing.

References

1. Alexander, C.: Market Risk Analysis, Volume IV, Value at Risk Models. Wiley, Hoboken (2009)
2. Alexander, C.: Market Models: A Guide to Financial Data Analysis. Wiley, Hoboken (2001)
3. Black, F., Scholes, M.: The pricing of options and corporate liabilities. J. Polit. Econ. **81**(3), 637–654 (1973)
4. Rumelhart, D.E., Hinton, G.E., Williams, R.J.: Learning internal representations by error propagation. Cogn. Sci. **1**, 318–362 (1986)
5. Hochreiter, S., Schmidhuber, J.: Long short term memory. Neural Comput. **9**(8), 1735–1780 (1997)
6. Kai, F., Wenhua, X.: Training neural network with genetic algorithms for forecasting the stock price index. In: IEEE International Conference on Intelligent Processing Systems (Cat. No.97TH8335), vol. 1, pp. 401–403 (1997)
7. Dorffner, G.: Neural computation and applications in time series and signal processing. J. Signal Process. Syst. (1996)
8. Frank, R.J., Davey, N., Hunt, S.P.: Time series prediction and neural networks. J. Intell. Robot. Syst. **31**, 91–103 (2001)
9. Qiu, M., Song, Y., Akagi, F.: Application of artificial neural network for the prediction of stock market returns: the case of the Japanese stock market. Chaos, Solitons Fractals **85**, 1–7 (2016)
10. Chong, E., Han, C., Park, F.C.: Deep learning networks for stock market analysis and prediction: methodology, data representations, and case studies. Expert Syst. Appl. **83**, 187–205 (2017)
11. Wang, J.Z., Wang, J.J., Zhang, Z.G., Guo, S.P.: Forecasting stock indices with back propagation neural network. Expert Syst. Appl. **38**(11), 14346–14355 (2011)

12. Fischer, T., Krauss, C.: Deep learning with long short-term memory networks for financial market predictions, Eur. J. Oper. Res. **270**(2), 654–669 (2018)
13. Cao, J., Li, Z., Li, J.: Financial time series forecasting model based on CEEMDAN and LSTM. Phys. A Stat. Mech. Appl. **519**, 127–139 (2019)
14. Baek, Y., Kim, H.Y.: ModAugNet: a new forecasting framework for stock market index value with an overfitting prevention LSTM module and a prediction LSTM module. Expert Syst. Appl. **113**, 457–480 (2018)
15. Sagheer, A., Kotb, M.: Time series forecasting of petroleum production using deep LSTM recurrent networks. Neurocomputing **323**, 203–213 (2019)
16. Krstanovic, S., Paulheim, H.: Stacked LSTM snapshot ensembles for time series forecasting. In: Valenzuela, O., Rojas, F., Pomares, H., Rojas, I. (eds.) ITISE 2018. CS, pp. 87–98. Springer, Cham (2019). https://doi.org/10.1007/978-3-030-26036-1_7
17. Sagheer, A., Kotb, M.: Unsupervised pre-training of a deep LSTM-based stacked autoencoder for multivariate time series forecasting problems. Expert Systems with Applications. Sci. Rep. **9**, 1–16 (2019)
18. Hu, Y., Ni, J., Wen, L.: A hybrid deep learning approach by integrating LSTM-ANN networks with GARCH model for copper price volatility prediction. Statist. Mech. Appl. Phys. A (2020)
19. Vidal, A., Kristjanpoller, W.: Gold volatility prediction using a CNN-LSTM approach. Expert Syst. Appl. (2020)

Author Index

Printed in the United States
by Baker & Taylor Publisher Services